Advances in Experimental Medicine and Biology

Volume 919

More information about this series at http://www.springer.com/series/5584

Hamid Mirzaei • Martin Carrasco
Editors

Modern Proteomics – Sample Preparation, Analysis and Practical Applications

Springer

Editors
Hamid Mirzaei
UT Southwestern Medical Center
Dallas, TX, USA

Martin Carrasco
Biotech Division
Neurophagy Therapeutics, INC
Odessa, TX, USA

ISSN 0065-2598 ISSN 2214-8019 (electronic)
Advances in Experimental Medicine and Biology
ISBN 978-3-319-41446-1 ISBN 978-3-319-41448-5 (eBook)
DOI 10.1007/978-3-319-41448-5

Library of Congress Control Number: 2016960751

Printed on acid-free paper

This Springer imprint is published by Springer Nature
The registered company is Springer International Publishing AG
The registered company address is: Gewerbestrasse 11, 6330 Cham, Switzerland

Preface

With significant advances made in recent years, there has been an increasing demand for proteomics technology to help pave the way for hypothesis-driven sciences or produce data for data-driven hypothesis generation. Clinical proteomics is another area of research that has received a significant amount of attention in recent years with the great promise of developing biomarkers for the early detection of fatal diseases such as cancer. Other than scientists who are involved either in method development or complex proteomics applications such as biomarker discovery, there is an increasing number of scientists without training experience in the field who wish to use proteomics in their research. Due to the significant cost associated with mass spectrometer acquisition, maintenance, and operation, many educational institutes and companies have formed core facilities to provide access to proteomics technology. Faculties specializing in proteomics have also been experiencing an increased demand for collaboration in recent years as the potential of proteomics technology has become more apparent due to an increasing number of high-impact publications. A major bottleneck in collaborations between the proteomics community and biologists is the lack of efficient communication. Biologists often don't have a clear understanding of what it takes to carry out a proteomics experiment successfully and reproducibly; this lack of understanding often leads to unrealistic expectations or poor experimental design and execution. Proteomics sample preparation is highly variable and experiment dependent and as such not comparable to DNA/RNA sample preparation methods. There are many different ways to perform mass spectrometry and process data. Often it is hard to identify the best method when biologists do not have a clear understanding of how proteomics works.

We decided to write this book to help all scientists interested in using proteomics in their research and those who want to become experts in the field. This book shall be a resource for experimental design starting from sample source, sample preparation, and mass spectrometry to data analysis and interpretation. With this purpose in mind, we contacted scientists who will be considered leaders in their field and asked for chapter contributions. Since authors of various chapters do not communicate with each other, there is some redundancy between chapters. We believe this redundancy is necessary as it reflects different experiences and viewpoints. It is also helpful for scientists who are not familiar with proteomics to learn the different methods and tools used in various steps in the proteomics pipeline to bolster their experimental design and execution capabilities. We are hoping that this book will serve as a tool for understanding how to design a practical, successful experiment with desirable results. Furthermore, we believe this work can also be used as a manual for the execution of the various steps in a proteomics experiment. It is not practical to include every known proteomics protocol in one book, as there is no way of verifying every protocol for reproducibility. Protocols presented in this book were provided by leaders in the field and represent a good starting point for method development. For more complex proteomics experiments, we recommend that those who are not experts in the field work with an experienced proteomics team.

Every field benefits from a centralized source of information; proteomics is no different. By taking the first step to create what could become a primary reference for proteomics, we are providing a resource for scientists in their own research and encouraging other leaders in the field to unite and support our cause. In turn, this resource could be updated periodically as new technology and techniques arise, thus assisting future scientists in their endeavors.

Contents

Part I

Sample Preparation Strategies for Proteomics

Proteomes, Their Compositions and Their Sources

1

Anna Kwasnik, Claire Tonry, Angela Mc Ardle,
Aisha Qasim Butt, Rosanna Inzitari,
and Stephen R. Pennington

Abstract

Biological samples of human and animal origin are utilized in research for many purposes and in a variety of scientific fields, including mass spectrometry-based proteomics. Various types of samples, including organs, tissues, cells, body fluids such as blood, plasma, cerebrospinal fluid, saliva and semen, can be collected from humans or animals and processed for proteomics analysis. Depending on the physiological state and sample origin, collected samples are used in research and diagnostics for different purposes. In mass spectrometry-based proteomics, body fluids and tissues are commonly used in discovery experiments to search for specific protein markers that can distinguish physiological from pathophysiological states, which in turn offer new diagnosis strategies and help developing new drugs to prevent disease more efficiently. Cell lines in combination with technologies such as Stable Isotope Labeling by Amino Acids in Cell Culture (SILAC) have broader application and are used frequently to investigate the mechanism of a disease or to investigate for the mechanism of a drug function. All of these are important components for defining the mechanisms of disease, discovering new pharmaceutical treatments and finally testing side effects of newly discovered drugs.

Keywords

Sample origin • Cell culture • Biological fluids

1.1 Cell

1.1.1 Cell Culture

The human body is composed of an average of 37.2 trillion types of cells that differ in morphology, size and function. When the same or

A. Kwasnik • C. Tonry • A.M. Ardle • A.Q. Butt
R. Inzitari • S.R. Pennington (✉)
School of Medicine and Medical Sciences, UCD Conway Institute of Biomolecular and Biomedical Research, University College Dublin, Dublin 4, Ireland
e-mail: stephen.pennington@ucd.ie

H. Mirzaei and M. Carrasco (eds.), *Modern Proteomics – Sample Preparation, Analysis and Practical Applications*, Advances in Experimental Medicine and Biology 919, DOI 10.1007/978-3-319-41448-5_1

different types of cells are interconnected with each other to carry out specific functions within an organism they are referred to as tissues. Organs are formed from the combination of at least two tissues in the human body, which in turn, are interconnected to form specialised body systems. The interplay between such systems contributes to the maintenance of homeostasis of the entire body [1].

Cell culture can be divided into three basic types, the culture of primary and secondary cells and of cell lines. Cells directly isolated from mammalian tissues that have not been sub-cultured are called primary cells. Once a primary cell culture has been sub-cultured (propagated *in vitro*), it is termed as a secondary cell culture. Such cells have a limited life span in culture. Primary or secondary cells that have been immortalised to expand their life span are called cell lines.

Many steps have to be carefully considered before initiating a primary cell culture. Thoughtful planning of how samples will be obtained, how tissue will be isolated, what method(s) will be used to isolate cells from tissue and finally what method of cell culture will be used after cell isolation, have to be undertaken before setting up cell cultures. Additionally, any work performed on animal or human samples has to be carried out according to proper legislation on experimentation with animals [2] or medical ethical rules in the case of human samples [3]. While whole organs can be isolated from animals, the most common cell sources from humans are biopsies from specific organs or tissues usually obtained during diagnostic examination or surgery. Work with and/or sample collection from animals for research requires ethical approval from the appropriate research ethics committee, while obtaining human samples requires consent from the local hospital ethics committee, from the doctor or surgeon responsible for the patient and from the patient or the patient's relatives. Biopsies taken during surgery are collected using sterile containers in appropriate physiological solution or culture medium. It is important to note that human biopsies carry a risk of infection such as hepatitis, HIV or tuberculosis, so human samples have to be handled with care, according to biohazard rules [4]. In the laboratory, biopsies are dissected and/or disaggregated in sterile conditions by either mechanical techniques or enzymatic methods to establish primary cell lines. Depending on their origin (tissues or body fluids) animal cell cultures can grow either as an adherent monolayer or in suspension. Adherent cells may grow as a monolayer attached to a cell culture vessel (plate or flask surface) and this attachment is often essential for cell growth and proliferation. Most tissue derived cell lines are anchorage-dependent. In contrast, hematopoietic cells (cells derived from blood, spleen, or bone marrow) are anchorage-independent and can grow and proliferate without being attached to a substratum. Interestingly, some transformed and malignant tumour cell lines can grow in anchorage-independent conditions. Over the years various techniques have been established to initiate specific types of primary cell culture. The original method that was used in cell culture initiation was a method called the 'spillage technique'. The name of the method reflects how tissue was processed to isolate cells. Slices of tissue section were placed in medium and shaken to allow cells to migrate into the medium. The cell suspension was further used to establish new primary cell culture [5]. Nowadays the most common methods used to establish a primary cell culture from tissues utilise enzymatic digestion to separate cells from tissues [6]. Cells that are isolated from mammalian tissues and are grown until sub-culture are known as primary cell cultures. Cells isolated directly from tissues are usually heterogeneous but closely represent tissue specific properties and the protein expression profile of parental cells. Primary cell cultures after several sub-cultures into fresh media will either die out or transform to become a continuous cell line. This ability is commonly observed among rodent cells but not in human cells, as cells derived from humans have a limited lifespan and can only be cultured for a limited time period before becoming 'senescent'. To extend the lifespan of cell cultures, the cells are commonly immortalised by viral transfection so that the cells continue to divide for more 'passages'. Bioresources of immortalised human and animal cell lines are commercially available. One of the biggest resources of cell lines is provided the ATCC Cell Biology Collection (http://www.lgcstandards-atcc.org/en.aspx). Transformed cell cultures show similar phenotypic and molecular properties to neoplastic cells, including changes in morphology or chromosomal variations. Moreover, transformed cells have an ability to form tumours when injected to animals/hosts with weak immune systems.

Conducting research on cell lines has many advantages, with the major one being the consistency and reproducibility of results that can be achieved from experiments performed on cell lines when compared to tissues, organs or body fluids. For this reason, cell culture is commonly used in the field of mass spectrometry, especially during method development, where consistency and reproducibility between experiments is necessary. Easy accessibility, rapid growth rate and ease of manipulation both genetically and biochemically (through chemical and pharmacological treatment) make cell lines an attractive model in research. Thus, cell lines are commonly used in mass spectrometry discovery experiments to investigate differences between normal and aberrant (for example cancerous) phenotypes [7–9] or to investigate different stages of the disease [10, 11]. Cell lines are also commonly applied in mass spectrometry research to investigate the signal transduction of molecular pathways of specific types of cells [12, 13] and post-translational modifications such as phosphorylation [14, 15] or ubiquitinylation [16, 17]. Cell lines are also used to test the effect of various chemical compounds (for example inhibitors or activators) or pharmacological drugs on different cellular systems. This approach is commonly used in the mass spectrometry field to determine pathways that are affected by treatment or to investigate side effect(s) of treatment [18]. Another advantage of using cell culture for mass spectrometry based research is the analysis of proteomes of specific cellular organelles. This is referred to as subcellular proteomics and is based on the identification of proteins specifically expressed in cellular organelles such as the nucleus, ribosome or mitochondria [19, 20].

The simplicity of maintaining cell lines under various media conditions has led to the invention of a mass spectrometry technique called stable isotope labelling with amino acids in cell culture (SILAC) [21]. In this method, culture media is supplemented with light (normal) and heavy labelled amino acids that are incorporated into newly synthesised proteins. The heavy amino acid contain a ^{2}H instead of H, a ^{13}C instead of ^{12}C or a ^{15}N instead of ^{14}N. The incorporation of heavy labelled amino acids into the proteins increases the molecular weight/size of the protein compared to the light (normal) proteins. This rule also applies to the peptides generated after enzymatic digestion of proteins, and leads to a known mass shift compared to the respective unlabelled peptide. This SILAC technique allows direct experiment comparison of different proteomes in a single tube experiment with minimal introduction of sample preparation errors. Indeed SILAC is broadly used in the mass spectrometry field [22, 23].

As one may expect, cell culture is not without limitations. The most common problems with cell lines include; infection with microorganisms, the cross-contamination of cell lines with other cell types and genomic and phenotypic instability [24–26]. Contamination with microorganisms is a serious problem worldwide among laboratories, as the presence of microbes in culture media can inhibit cell proliferation and growth and in the most extreme infections lead to cellular death. The most common animal culture contaminations are caused by bacteria, yeast, fungi, mold and mycoplasma [27]. Microbial contamination is most often caused by poor cell culture technique and by the use of contaminated media, reagents or equipment. Microbes can also be present in incubators, refrigerators, laminar flow hoods or on the skin of researchers working under laminar flow hoods. The infection can also be introduced when cell cultures are received from external sources such as other laboratories, or cells that have been isolated from infected animals or humans. Several features of microbial contamination can be visually observed, such as a change in pH that usually leads to a change in colour of the medium or makes the medium appear cloudy. Also, careful inspection of the cultured cells under the microscope can indicate some changes in infected cultures such as cell death. The presence of rods, cocci or thin filamentous mycelia that may form clumps or spores may indicate bacterial or yeast/fungi contamination, respectively. While bacterial or yeast/fungus contamination can be detected easily, the cells contaminated with mycoplasma may grow undetected for several passages as there is no obvious evidence of mycoplasma infection such as a pH change or the presence of cellular death [28]. To overcome problems with microbial contamination, cell culture media are commonly supplemented with antibiotics such as penicillin, streptomycin and amphotericin b. Although these reduce the risk of microbial contamination, the routine use of antibiotics may affect the phenotype of the cells. Moreover, microbes may become resistant to the antibiotics and grow despite the presence of antibiotics. Most of the cell culture laboratories routinely apply sensitive tests to detect mycoplasma infections, which are based on the detection of rRNA or DNA of mycoplasma or the visualisation of mycoplasma-specific polyclonal antibodies [29]. A common problem that is not easy to detect, and was subsequently ignored for many years in science research, is cross-contamination of the cells with different cell lines. Cross-contamination is more difficult to recognise compared to microbial infections such as bacterial or fungal, as there is no physical indication of cross-contamination such as the change of medium colour or cell death. Often the most common sources of cross-contamination are poor cell culture techniques and human mistakes such as simple errors made during sample labelling. The first reports of cross-contamination came from the research conducted by Nelson-Ress who reported that many of the cell lines used in research have been switched or cross-contaminated with *HeLa* cells [30]. Since then the problem of cross-contamination has become widely recognised by scientists and much research has been undertaken to address this issue [31–33]. Several methods, such as karyotyping [34], isozyme analysis [35], HLA (human leukocyte antigen) typing [36] and DNA fingerprinting [37] have been applied for the identification of cross-contamination. One problem that has been noticed during long term culturing of cell lines is that rapidly growing cell lines (such as tumour cell lines) are likely to undergo genomic fluctuations and this often leads to phenotypic and/or genotypic instability as well as gene or protein expression changes between different passages of the same cell line or sub-lines that are derived from the same parental population of cells [38]. These genotypic and phenotypic changes can have an effect on gene expression and are caused by many environmental factors, such as culturing conditions, including different types of media, serum, trypsin, CO_2 levels or the temperature used to culture cells between different or even within the same laboratory. Genetic changes can occur spontaneously when a small population of cells divide at a higher rate than the other cells. This will result in the natural selection of the smaller population of the cells within a short period of time. While the culturing conditions can be kept constant, there is not much that can be done about natural drifting except that stocks of the low cell passages must be prepared and stored in liquid nitrogen and these stocks should be used at regular time intervals, for example every ten passages. Research has shown that maintaining cell lines under identical culture conditions results in a more stable genotype and phenotype over a long period of time [39, 40].

Overall, cell-cell cross contamination and contamination with microorganisms as well as genomic and phenotypic instability are common problems in cell culture. All of these factors affect experimental results, which in turn forbid reliable comparison of the results within and between laboratories due to the lack good reproducibility. Regular quality controls of cross-contamination and microbial contamination in the cell culture laboratory can help to overcome these problems. Maintaining stable culturing conditions and renewing cell cultures at low passages over short periods of time is a good way to retain good genomic and phenotypic stability of cultured cells. All of these disadvantages of cell culture may cause serious problems when conducting research but can be easily avoided when good aseptic culturing technique is used and cells are carefully monitored.

1.1.2 Tissue Culture

Tissue culture is the growth of animal tissue outside of the organism, in a culture medium. For tissue culture, cells are grown in a medium supplemented with nutrients and energy sources that are essential for cell survival. To prevent contamination or infection, tissue culture medium is also supplemented with antibiotics and/or fungicides [41]. Tissue culture provides an *in vitro* model of animal tissue that can be easily manipulated for research investigations pertaining to disease progression. Furthermore, tissue culturing allows the analysis of single cell populations (e.g. fibroblasts or macrophages) as well as mixed cell populations, similar to what would be found in the *in vivo* environment [42, 43]. Cultured tissue cells can be frozen down and stored over long periods of time for future study. Freezing cultures prevents genetically induced changes and the loss of cultures due to senescence or accidental contamination [44]. Tissue culture is classified as a primary cell culture when cells are extracted directly from human tissues and grown in culture medium, however, once the cells are sub-cultured and immortalised, they are then classified as cell lines [45]. As primary cultured cells retain the unique characteristics of the original tissue from which they were extracted, they are of significant use in investigations designed towards the understanding of disease origin and malignant progression at a cellular level [46].

The proteome refers to all proteins that are expressed by a cell, tissue or organism under defined conditions, at a certain time. The proteome is highly influenced by both environmental stimuli and disease processes, which is why proteomic-based investigations are key to understanding the biological mechanisms which underline disease [47]. To this end, tissue culture models are useful for the investigation of the role of certain genes, the effects of drug treatment and the effects of viral infection at the protein level [48]. Once tissue cells have been treated to induce a desired phenotype, cells are lysed to extract protein for comprehensive analysis of protein expression and protein-protein interactions [48]. The major downside of performing proteomic investigations on cultured cells is that they cannot provide accurate insight into disease progression *in vivo*. Successful *in vitro* investigations of the pathobiology of disease are therefore enhanced if cells are grown in an environment that mimics the 3D architecture of human tissue [49]. Growing tissue cells *in vitro* in 3D heterogeneous co-culture systems, which allow for interactions between disease (e.g. tumour) cells, stromal cells, endothelial cells, fibroblasts, immune cells and the extracellular matrix (ECM), is thought to overcome the limitations of the standard 2D monolayer culture systems [49–51]. Various 3D model systems have been extensively utilized in the field of cancer research [52, 53]. The 'multicellular tumour spheroid model' refers to the culturing of cells under non adherent conditions, 'tumour spheres' mimic cancer cell expansion in serum-free conditions with supplemented growth factors, while 'tissue-derived tumour spheres and organotypic multicellular spheroids' are derived from mechanical dissociation and cutting of tumour tissue [51].

Aside from culturing of tissue cells, proteomic experiments can also be performed on fresh tissue specimens. This method is slightly more challenging, as harvesting and processing tissue specimens must be performed as quickly as possible to avoid any protein degradation [54]. Tissue samples can, however, be snap frozen to preserve their proteome integrity if, for example, samples have to be retrospectively analysed [55]. Another way to preserve tissue samples is to fix them using formalin and embed them in paraffin wax. Such samples are referred to as formalin fixed paraffin embedded (FFPE) tissues samples. This is a universal method of preparation to preserve and stabilize tissue samples for histological evaluation. Protein extraction from FFPE material has proven difficult in the past, due to the molecular cross-linking that occurs during formalin fixing. However, numerous protocols have been optimized for efficient protein extraction from FFPE material for subsequent proteomic analysis via both antibody and mass spectrometry based techniques [56, 57]. Techniques for harvesting cells directly from tissue samples have evolved in the last number of years. Laser capture microdissection (LCM) is a popular technique in which cells from specified regions of interest within a tissue section can be obtained, using a microscope to guide a laser beam that attaches cells from the tissue to an adhesive film [58]. This technique is particularly useful when, for example, comparing tumour tissue to surrounding benign or stromal tissue from the same patient.

Tissue samples for either culturing or direct harvesting are generally obtained during routine surgery (human) or following euthanasia of animal models. In this way they are useful for proteomic profiling of the disease state and/or the surrounding area, for molecular investigations of the disease process.

1.1.3 Organ

Organs encompass a variety of different tissue and cell types. Therefore they provide a much more heterogeneous sample source for proteomic investigation. As with tissue samples, organ cells can be extracted and grown under cell culturing conditions, or extracted and digested directly from organ tissue (generally obtained post-mortem), as described for tissue culture. However, it is difficult to routinely obtain organ samples from humans. Because proteins can be routinely measured in easily accessible biological fluids, thus they are more attractive candidates as biomarkers; many biomarker discovery and validation experiments for organ-associated pathologies are therefore conducted on bio-specimens, which are secreted from the organ of interest. For example, blood, bile, stool and urine are attractive sources for the identification of protein biomarkers related to the heart, liver, intestine and kidney/pancreas, respectively [59–63]. However, these efforts have failed to result in clinically applicable disease biomarkers. As such, there is still a reliance on informative animal models to accelerate the progress in clinical proteomics [64].

The use of animal models overcomes limitations regarding the organ and tissue sampling from humans, which is particularly restricting in the study of neurological disorders [64]. For disease-focused investigations, animal models provide a much more controlled system which allows for proteomic profiling at set times/disease points with less influence from the external environment [65, 66]. Animals such as mice, rats, pigs, dogs, zebra-fish and fruit flies are considered useful models for human disease, based on the overall conservation of their proteome with the human proteome of interest [67, 68]. Similar to cell culture experiments, a disease or disease-like state can be induced in animals, which would be housed under identical conditions as control healthy animals. The disease phenotype can be induced genetically, therapeutically or with environmental stimuli [69–71]. When animals are eventually euthanized, the differences in protein expression observed in their organ/tissue material compared to that of the control healthy animals, can be more confidently associated with disease, which is less true for human samples due to the inherent heterogeneity of humans [66, 72]. Indeed, the comprehensive quantitative proteomic analysis of whole animals is now achievable by implementing an *in vivo* SILAC technique (see chap. 13). This is achieved by feeding animals with a $^{13}C_6$-lysine diet for *in vivo* labelling of proteins, which, when extracted from animal organ/tissue material, can be analysed using mass spectrometry based techniques to make comparisons between healthy and diseased tissues [73].

1.1.4 Exosomes

In addition to profiling the proteins expressed within the cell, it is widely accepted that proteins that are secreted by various cells are also a valuable source of pathobiological information [74]. The global study of proteins that are secreted by cells is defined as secretomics [75]. Secreted proteins can be found in both biological fluids and conditioned media from cell cultures [76]. The secretome is largely represented by membranous vesicles, of which there are many types including exosomes, exosome-like vesicles, microparticles, microvesicles, membrane-bound particles, apoptotic bodies and apoptotic microparticles [77]. Of these, exosomes are considered particularly attractive for proteomic research. Exosomes are small membrane vesicles derived from the luminal membranes of multivesicular bodies. They are actively released by fusion of microvesicular bodies with the cell membrane [78]. They are differentiated from other membranous vesicles by their characteristic size (approximately 30–100 nm) and expression of CD81 protein [79, 80]. Exosomes are likely to be enriched in low abundant and membrane proteins that are difficult to detect in standard cell or biological fluid material. They contain a conserved set of common proteins which are essential for the biogenesis, structure and trafficking of the biovesicles. Moreover, they contain proteins which would be specific to the cell/biological fluid from which they are isolated and are therefore considered a valuable sample source for disease-specific biomarker discovery [81, 82].

Due to the growing popularity of exosomes in proteomic research, there are numerous optimized protocols available for exosome isolation from biofluids (conditioned media, serum/plasma, urine etc.). Generally, exosome isolation can be achieved following a series of ultracentrifugation steps. However, there are also a number of commercial kits available for exosome isolation and purification which are applicable for proteomic profiling of exosome material [83].

1.2 Biological Fluid

1.2.1 Serum and Plasma

Blood is a bodily fluid that circulates through arteries and veins, supplying the tissue with oxygen and taking away carbon dioxide to be excreted. It is also responsible for providing nutrients to tissues, hormones to cells, and is an important part of the immune system. Blood constitutes up to 8 % of total body weight in humans and it contains components such as serum, plasma, red blood cells (RBCs), white blood cells (WBCs) and clotting factors. Serum is the liquid fraction of whole blood that is collected by centrifugation after the blood is allowed to clot and it does not contain RBCs, WBCs or clotting factors. Plasma is the pale yellow liquid component of blood that holds the blood cells in suspension, thus acting as an extracellular matrix for blood cells. Plasma is collected by centrifugation of whole blood collected in tubes that are treated with anticoagulant. Both serum and plasma contain similar components such as glucose, electrolytes, antibodies, antigens, hormones, proteins, enzymes, nutrients and certain other molecules whereas clotting factors are only present in plasma [84].

Both serum and plasma represent ideal biological samples, as they are readily accessible body fluids and contain many proteins that are synthesized, secreted, shed or lost from the cells and tissues throughout the body. Fluctuations in the expression levels of these proteins in serum and plasma can reflect a pathophysiological condition [85] and thus they are routinely used for blood testing in hospitals and clinics [86–88]. Serum is preferentially used for the determination of an individual's blood group and for various diagnostic blood tests such as determining the levels of hCG, cholesterol, proteins, sugar, etc. in blood. However, plasma is primarily used for transfusion in patients suffering from haemophilia and other blood-clotting disorders, immunodeficiency, shock or burns [84]. Serum is favourably used for diagnostic testing in medicine due to the presence of more antigens as compared to plasma. Moreover, anticoagulants in plasma may interfere with the chemical reactions that are employed in diagnostic tests to measure levels of the blood constituents. Furthermore, anticoagulants in plasma may draw water out of cells, thus diluting the sample and changing the test results. Whilst plasma may not be the preferential body fluid for diagnostic tests, it presents various benefits for patients suffering from blood-clotting disorders requiring transfusion, as plasma can be frozen and stored for up to a year and is easy to transport. Moreover, plasma is replaced in the human body after every 2–3 days, thus it can be donated more frequently while whole blood cannot be donated very frequently. Therefore, while the anti-coagulants present in plasma makes it undesirable for certain diagnostic tests; serum cannot be used for transfusions, due to the absence of blood clotting factors. Thus, both serum and plasma have different advantages and disadvantages and are fit to serve different applications in medicine.

Serum and plasma have been used for multiple proteomics based biomarker discovery studies [89–93] as they represent readily accessible and clinically relevant samples. However, there appears to be a lack of understanding of the issues critical for the processing of plasma and serum samples for analysis. Often, the most basic yet crucial aspects of serum and plasma sample collection are neglected, such as uniformity in collection of samples using a standard operating procedure (SOP), sample processing, and storage conditions. It is only by doing this that one is able to assure reproducibility of samples and to allow some rational comparison of data from various laboratories [94]. Until that is accomplished, any kind of data analysis is questionable. The next problem with the use of serum and plasma samples for proteomic analysis is the analytical challenge that these samples present due to the presence of the wide dynamic qualitative and quantitative range of proteins that spans over 12 orders of magnitude [95]. In fact, 96 % of total serum or plasma protein content represents a small number of highly abundant proteins such as albumin, immunoglobulins, alpha-1-antitrypsin, haptoglobulin, etc. that can mask potential biomarkers. Thus, prior to proteomic analysis, it is essential to deplete these highly abundant proteins with the use of columns or matrices such as Multiple immuno-Affinity Removal System (MARS, Agilent Technologies) [96–98], ProteomeLab IgY system (Beckman Coulter) [99], hexapeptide combinatorial library beads [100], ImmunoAffinity Subtraction Chromatography resin (IASC) [101] and others. It is clear that during affinity depletions of high abundant proteins, these columns also remove other components of serum and plasma by 'non-specific' binding. Since most proteomic biomarker discovery studies don't have a specific target protein, it is not possible to know whether a biomarker of interest is lost upon the removal of high abundant proteins.

1.2.2 Cerebrospinal Fluid

Cerebrospinal fluid (CSF) is a transparent body fluid (mean volume 150 ml) contained within the brain ventricles (25 ml) and the central and spinal subarachnoid spaces (125 ml). It is produced predominantly in the choroid plexus and plays a protective role in the central nervous system (CNS) [102–104]. Historically it was believed that the main role of CSF was to provide mechanical protection to the CNS, acting as a shock absorber. However it is now well understood that in addition to this function CSF has an essential role in maintaining homeostasis within the interstitial fluid of the brain parenchyma as well as regulating neuronal functioning [103, 104].

Due to the proximity to the brain and spinal cord, CSF is a common matrix for monitoring and assessing neurodegenerative disorders. Molecular changes that occur in the CNS such as changes in protein expression levels serve as objective markers of CNS-associated disease. Indeed over the past decade, considerable effort has been extended to the discovery of putative protein biomarkers of neurodegenerative disease in human CSF. Proteomic analysis of CSF is typically performed using high resolution liquid chromatography mass spectrometry (LC-MS/MS) [104]. Many mass spectrometry based studies have identified CSF biomarkers with potential diagnostic utility in neurodegenerative disease including Alzheimer's, multiple sclerosis and Parkinson's [105, 106].

Despite the advantages of CSF there are some difficulties associated with using this body fluid. Firstly the collection of CSF requires an invasive procedure referred to as a lumbar puncture or a spinal tap. A lumbar puncture must be performed by a physician, it is an uncomfortable procedure and can be associated with postdural puncture headaches [107]. Moreover, traumatic punctures can introduce red blood cells into the CSF and artificially increase the white blood cell count and protein expression levels and thereby skew a diagnosis [108]. Secondly CSF is a complex matrix, 80 % of the CSF proteome originates from plasma yielding a highly dynamic range of protein concentrations (spanning 10 orders of magnitude) [109, 110]. As with serum and plasma, the presence of highly abundant proteins precludes the identification of potentially interesting analytes present in lower concentrations. To facilitate a greater depth of analysis it is necessary to remove the highly abundant proteins from the sample before analysis and many methods for protein depletion have been established [109]. Alternatively, fractionation methods can be employed for improved coverage and deeper proteome analysis [104, 111]. While depletion and fractionation methods enhance our ability to identify lower abundant proteins, they are neither time or cost effective techniques and these step wise procedures can add variation during sample preparation leading to dubious findings [102].

1.2.3 Urine

Human urine has been used for decades by physicians to diagnose various disorders. Urine is produced by the kidney during the elimination of waste produced by the human body, which accumulates in the blood. The kidney also fulfils other roles such as maintaining whole body homeostasis and producing hormones including renin and erythropoietin [112, 113]. The human kidney is composed of one million units called nephrons, which can be divided into two functional parts: the glomerulus and the renal tubule. The glomerulus is responsible for the first filtration of the plasma to generate the "primitive" urine. The renal tubule is dedicated to reabsorb most of the primitive urine to generate the "final" urine that exits the kidney through the ureter into the bladder. In 24 h, about 900 L of plasma flows through the kidneys. 150–180 L is filtered as 'primitive' urine but more than 99 % of this urine is reabsorbed. The remaining unabsorbed plasma generates the "final" urine. Analysis of the urinary proteome may therefore contain information not only from the kidney and the urinary tract but also from other organs of plasma obtained by glomerular of plasma, making it a good source of biomarkers for urogenital and systemic diseases [114, 115].

Under normal conditions, urinary proteins are stored in different compartments that can be isolated by sequential centrifugation. The separate populations of proteins are identified as soluble proteins, urinary sediment proteins and urinary exosomes. Soluble proteins are derived by glomerular filtration of plasma proteins while some are also excreted by epithelial cells. The urinary sediment proteins are mainly sloughed epithelial cells and casts. The urinary exosomes are derived from the epithelial lining and the urinary tract but can also be derived from many other cell types, which can be identified in plasma and may be filtered in urine. Urine has several advantages compared to that of other body fluids: they can be obtained in large quantities using a non-invasive procedure, urinary peptides and lower molecular weight proteins are generally soluble and can be analysed in a mass spectrometer without any digestion. Moreover, the urinary protein composition is relatively stable, probably due to the presence of endogenous proteases in the bladder, while urine is being stored there. Stability studies have shown that the urinary proteome does not change significantly when urine is stored at 4 °C for several days or while stored at room temperature for up to 6 h [116, 117]. In addition, urine can be stored for several years at −20 °C without significant alterations to its proteome. Studies of urinary exosomes, however, indicate that this proteome may be less stable [118].

On the other hand, urine varies widely in protein and peptide concentrations, depending on differences in the daily intake of fluid, however this can be normalized through the consideration of creatinine excretion [119]. In addition, the definition of disease-specific biomarkers in urine is complicated, due to the significant changes in the proteome throughout the day that can be connected with the time of collection, diet, exercise, circadian rhythms and circulatory levels of various hormones [120, 121]. These variations seem to affect only a limited fraction of the urinary proteome while a large portion shows high reproducibility [122]. The Human Kidney and Urine Proteome Project (www.hkupp.org/), under the directive of the World Human Proteome Organisation (www.hupo.org/), is currently establishing standardized procedures to avoid this variability.

Currently, the common preparation method for biomarker identification in urine involves centrifugation of the urine sample and collection of the soluble fraction or the urinary exosomes, followed by 1 or 2 separation steps before mass spectrometry analysis [123]. However, the pellet fraction is also of biological interest as it contains information from proximal tissue or organs and also from organisms that colonize or infect the urogenital tract. Filter-aided sample preparation (FASP) has been used in shotgun proteomics for the lysis of cells presents in urinary pellets, after the solubilisation of proteins derived from cell pellets [124].

1.2.4 Saliva

Saliva is a clear liquid that originates mainly from three major glands (parotid, submandibular, and sublingual) with a small fluid contribution from several minor glands and from the gingival crevicular fluid (GCF) [125]. Most salivary proteins are synthesized in the acinar cells of the salivary glands and follow a well establish secretory pathway. For the majority of salivary proteins this common secretory pathway includes transit in the Golgi apparatus and storage in secretory granules, release from the cell into the duct system and secretion into the mouth [126].

During the different steps of the secretory pathway, proteins are subjected to a number of changes such as removal of the signal peptide as well as various post-translational modifications (PTMs) including proteolytic cleavage, glycosylation, phosphorylation, and sulfation. Further modifications of the proteins and peptides occur during transition into the ducts before secretion and additional modifications occur in the oral cavity after secretion from the cells as a result of the action of a number of proteolytic enzymes of different origin [127].

Saliva is composed mostly of water containing electrolytes, immunoglobulins, proteins and enzymes and plays an important part in the health of the oral cavity [128]. The basic role of saliva is protection of the oral mucous membrane of the oral cavity and digestive tract through the following functions: maintaining lubrication, buffering action and clearance, maintenance of tooth and mucosal integrity, and also facilitating the repair of the mucosal layer. Saliva also contains components that show antibacterial and antiviral activity as well as playing an important role in taste and the first phase of food digestion [129].

In healthy subjects the production of saliva is up to 1 to 2 L a day. Saliva secretion follows circadian rhythms and production is usually highest in the late afternoon while it is lowest during the night [130, 131].

The production of low amounts of saliva is related to a number of different pathologies and is indicated by the general term of dry mouth (xerostomia). Certain medication can also affect saliva production (low or over production) [132].

Human saliva contains proteins of clinical relevance and about 30 % of blood proteins are also present in saliva, making this biological sample an important tool for clinical application. Moreover, the simplistic nature of sampling, the non-invasiveness, ease of collection and the possible multiple collections by untrained professionals are some of the advantages of saliva sampling. On the other hand, due to the dynamics of the salivary proteome, sample preparation needs to be coupled to a well-controlled study design in order to allow saliva to enter clinical practice as an alternative to blood-based methods.

Human saliva reflects the health and well-being of the body, and most of the biomolecules that are usually detected in urine and blood can also be found in salivary secretions, however, the concentration of proteins range in saliva is 10–1000x lower than in blood [133, 134]. The low concentration of highly interesting proteins and the high concentration of some classes of proteins (i.e. mucins), along with the technology used for their characterization and analysis, significantly influence the preparation method. For the purpose of precision and accuracy of a measurement, specimen collection, handling and processing are of vital importance. For example the use of a cocktail of protease inhibitors after collection and during the processing of saliva samples must be standardised.

Proteolytic activity plays a fundamental role in the secretion pathway, which allows fully mature proteins to be secreted and be functional in the oral cavity. However, different proteases, both endogenous (derived either from the salivary glands or from the exfoliating cells) and exogenous (oral flora) contribute to the overall proteolytic activity in saliva samples post collection. The action of these proteases may result in misleading information about the saliva proteome. The use of protease inhibitors can help to avoid incorrect identification of a pre-secretory event due to the post secretory proteolytic activity [135], however, their use, especially when the inhibitors are peptides, can increases the complexity of the sample and interfere with proteomic analysis.

It is also generally recommended to use low-protein binding tubes made of plastic to avoid the adsorption of analytes to the tubes or the release of polymers from the plastic that can interfere with the subsequent analysis. It is important to use a standardized saliva collection and processing protocol from both diseased patients and healthy controls. For example, it is recommended to discard the initial 2 min of parotid secretion due to its large inter-individual variability. Moreover, for proteomic analysis, it is really important to keep samples on ice during collection and processing because protein degradation in whole saliva is very rapid at room temperature and this may occur during saliva collection and handling [136]. One way to minimize misleading or artificial degradation of proteins is to minimize the processing time between sample collection and final storage. Saliva samples in research projects are often stored for long time periods before they can be analysed. The recommended storage temperature is below −20 °C until analysis. Some researchers freeze samples in liquid nitrogen to avoid problems of slow freezing of biological samples and protein dishomogeneity. The recommended temperature for saliva sample storage is −80 °C as unusual post-translational modifications have been reported for samples stored for 3 days at −20 °C, demonstrating that protease activity is still present at −20 °C. This activity was not observed when sample were stored at −80 °C [135].

To subject saliva samples to proteomic analysis, samples are typically collected on ice as whole saliva or as selective saliva from specific salivary gland. The samples are then centrifuged to remove insoluble material and the supernatant is stored at below −20 °C until analysis [137–141]. The centrifugation step needs to be evaluated in terms of length and speed applied, because the extent of centrifugation has been shown to cause co-precipitation of specific classes of proteins such as PRP, cystatins and statherin [142]. Some researchers report the use of centrifugation in conjunction with protein precipitation (e.g. 10 % TCA/acetone/20 mM DTT) to prevent loss of proteins [143, 144], while another study reports the use of acid to eliminate mucins and acidic insoluble protein to generate a sample that can be directly analysed by mass spectrometry [135]. Centrifugation may sometimes be avoided if the samples are collected from single glands using canniculation, or for ductal secretion collections using a Carlson–Crittenden cup [138, 145] over the orifice of the Stenson's duct [146, 147]. The main goal of a well-established and rigorous process for processing the salivary sample is to minimize artificial changes after sample collection that could lead to a 'false salivary proteome'. Low abundant salivary proteins have been extensively studied by applying sample preparation methods involving separation and enrichment strategies, and the same strategies have been applied for the characterization of post translational modification (PTM), with a special focus on phosphorylation [148, 149] and glycosylation [150]. Enrichment strategies typically involve the use of a solid phase matrix with affinity for the PTM being studied (i.e. TiO2 for phosphorylation). There are a number of commercially available kits for biomarker discovery on saliva samples and identified biomarkers are strictly related to the type of sample acquired, such as whole salivary samples or samples selected from a single gland [151]. A number of commercially available kits that are applicable to research or diagnostic purposes for the study of saliva include DNA Genotek (www.dnagenotek.com); Salimetrics oral swabs (http://www.salimetrics.com); Oasis Diagnostics® VerOFy® I/II; DNA SALTM (http://www.4saliva.com); OraSure Technologies OraSure HIV specimen collection device (http://www.orasure.com); CoZart® drugs of abuse collection devices (http://www.concateno.com); and the Greiner Bio-One Saliva Collection System (http://www.gbo.com) [152].

Standard proteomic analysis can be performed using either a 'bottom-up' or 'top-down' approach. The top-down approach is used for analysis of intact proteins without protease digestion and can lead to the unbiased detection of isoforms and variants from sequence polymorphisims, splice variants and post-translational modifications as compared to a digested peptide mixture against a specific protein database. Application of this top-down proteomics approach to saliva samples allows the identification of single nucleotide polymorphisms and new sites of phosphorylation on cystatin SN and PRP3 [153]. Moreover, small proteins and peptides are abundant in saliva, and the relatively small size of these components has enabled top-down analytical approaches to profile their abundances and identify PTMs including phosphorylation, Gln to pyro-Glu conversion and glycosylation [154].

The bottom-up proteomic approach minimizes sample complexity, increases sensitivity and is the traditional approach for PTM characterization following protease digestion or PTM release. Top-down and bottom-up proteomics approaches require sample preparation to separate or fractionate components before detection by a mass spectrometer. These separation procedures can include SDS-PAGE, liquid chromatography, isoelectric focusing, affinity chromatography for depletion or enrichment, and release of PTMs.

For the detection of low abundant disease specific biomolecules in human saliva, which are mainly derived from blood or GCF, an enrichment strategy needs to be implemented to enrich for low abundant proteins by the removal of high abundant proteins. Enrichment strategies include pre-fractionation methods, such as sequential extraction of proteins with varying buffer conditions [155], sub-cellular fractionation [156] and selective removal of high abundant proteins via affinity methods [157].

1.2.5 Semen

Human semen is a greyish coloured body fluid that is composed of a variety of components produced by male gonads during a process called ejaculation. The main component of semen is the spermatozoa, which are ejaculated in the presence of enzymes and nutrients (seminal fluid) that help spermatozoa to survive and enable fertilization. Seminal fluid is produced by multiple male accessory glands such as the prostate, seminal vesicles, the epididymis and Cowper's gland. Seminal fluid contains acid phosphatase, inositol, citric acid, calcium, magnesium, zinc, fructose, ascorbic acid, prostaglandins, L-carnitine and neutral alpha-glucosidase [158]. Moreover, seminal fluid contains high amounts of proteins and amino acids that range from 35 to 55 g/L and is therefore a good and easily accessible source for protein identification. However, similar to other body fluids, semen contains a number of highly abundant proteins that mask the low abundant proteins and this makes proteomic analysis of seminal fluid difficult.

Semen samples have applications in research areas such as reproduction [159, 160] and prostate cancer and are used for many purposes in the diagnosis of male fertility, for example, for the assessment of spermatozoa morphology, motility and concentration [161, 162].

For diagnostic or research purposes, semen is collected by ejaculation into a non-toxic and clean plastic or glass container. The collection, transport and processing of semen samples should be kept at an ambient temperature of 20–37 °C [163]. An essential step in semen sample preparation for mass spectrometry analysis is the purification of seminal fluid from sperm cells and any other semen containing cells. This step is usually achieved by density gradient centrifugation by using PureSperm or Percoll. An alternative method; through swim-up has also been described [164]. The non-invasive collection of seminal fluid and the specificity of seminal fluid to male glands make it a potentially good source for discovery of new biomarkers in prostate cancer and research towards infertility. Indeed, the application of seminal fluid in both the prostate cancer research [165] and reproduction [166, 167] has increased over the last few years.

1.2.6 Circulating Tumour Cells

Body fluids, in addition to aqueous solution, also contain solid cells. For example, 45 % of the blood is composed of the mixture of red blood cells (erythrocytes), white blood cells (lymphocytes) and pellets (thrombocytes). The complete count of blood cells is routinely used in diagnosis to screen for a wide range of conditions and diseases. Any variations from normal cell morphology, composition of the cells or differences in expression of cell surface markers may indicate various disease conditions, thus an evaluation of blood cells has a practical application in diagnosis and disease treatment. Blood cells are also routinely used in research to investigate the molecular mechanism of various disease states or to develop new disease treatments. Blood for cell-based research is usually collected into tubes with anticoagulants such as heparin, EDTA or acid citrate dextrose (ACD) to assure that the coagulation cascade is blocked and cells stay in a suspension rather than as a clotted blood sample. An initial and important step in research based on blood cells, is isolation of the cells from blood plasma. Several methods to isolate specific subsets of the blood cells, erythrocytes [168, 169], thrombocytes [170] and lymphocytes [171] have been described.

Over the last few years, the application of circulating tumour cells (CTC), that are present in the blood of patients with metastatic cancer, have become very popular models to investigate certain aspects of metastatic disease. CTCs are used to determine the prognosis of metastatic progression or relapse, to monitor anti-cancer treatments, to understand the mechanism of metastatic disease and finally to use this knowledge to develop new strategies in disease treatment [172, 173]. Several methods to isolate CTCs from blood have been developed and optimised including density gradient centrifugation [174], size-dependent selection [175], positive selection of cells based on expression of the membrane antigen EpCAM [175] or negative selection of cells based on the depletion of cells with the CD45 antigen [176]. Although CTCs are an excellent model to investigate the metastatic state of disease, the biggest disadvantage of CTCs is their very low abundancy in the blood, with a yield of 1 cell per 10^6–10^7 leukocytes. Another limitation of research conducted on CTC cells is cell heterogeneity, which make it difficult to isolate the whole CTC population from the blood. Both the low yield of cells and the heterogeneity of the cells make research with CTCs challenging and limited, thus only few advances in the field have been made. To improve isolation of CTCs from blood, combined isolation techniques have been used. Further *ex-vivo* culture of CTCs increases the amount of available material [177].

References

1. Sherwood L (2015) Human physiology: from cells to systems. Cengage Learning, Andover
2. McGrath J et al (2010) Guidelines for reporting experiments involving animals: the ARRIVE guidelines. Br J Pharmacol 160(7):1573–1576
3. Sciences C.f.I.O.o.M (2002) International ethical guidelines for biomedical research involving human subjects. Bull Med Ethics (182):17
4. Miller MJ et al (2012) Guidelines for safe work practices in human and animal medical diagnostic laboratories. MMWR Surveill Summ 6(61):1–102
5. McCallum HM, Lowther GW (1996) Long-term culture of primary breast cancer in defined medium. Breast Cancer Res Treat 39(3):247–259
6. Rittie L, Fisher GJ (2005) Isolation and culture of skin fibroblasts. Methods Mol Med 117:83–98
7. Patwardhan AJ et al (2005) Comparison of normal and breast cancer cell lines using proteome, genome, and interactome data. J Proteome Res 4 (6):1952–1960
8. He J et al (2014) Fingerprinting breast cancer vs. normal mammary cells by mass spectrometric analysis of volatiles. Sci Rep 4:5196
9. Rubporn A et al (2009) Comparative proteomic analysis of lung cancer cell line and lung fibroblast cell line. Cancer Genomics-Proteomics 6(4):229–237
10. Masayo Y et al (2009) The proteomic profile of pancreatic cancer cell lines corresponding to carcinogenesis and metastasis. J Proteome Bioinforma 2:1–18
11. Wu W et al (2002) Identification and validation of metastasis-associated proteins in head and neck cancer cell lines by two-dimensional electrophoresis and mass spectrometry. Clin Exp Metastasis 19 (4):319–326
12. Lewis TS et al (2000) Identification of novel MAP kinase pathway signaling targets by functional proteomics and mass spectrometry. Mol Cell 6 (6):1343–1354
13. Choudhary C, Mann M (2010) Decoding signalling networks by mass spectrometry-based proteomics. Nat Rev Mol Cell Biol 11(6):427–439
14. Salomon AR et al (2003) Profiling of tyrosine phosphorylation pathways in human cells using mass spectrometry. Proc Natl Acad Sci 100(2):443–448
15. Zhang Y et al (2005) Time-resolved mass spectrometry of tyrosine phosphorylation sites in the epidermal growth factor receptor signaling network reveals dynamic modules. Mol Cell Proteomics 4 (9):1240–1250
16. Meierhofer D et al (2008) Quantitative analysis of global ubiquitination in HeLa cells by mass spectrometry. J Proteome Res 7(10):4566–4576
17. Xu P, Peng J (2006) Dissecting the ubiquitin pathway by mass spectrometry. Biochim Biophys Acta (BBA)-Protein Proteomics 1764(12):1940–1947
18. Bose R et al (2006) Phosphoproteomic analysis of Her2/neu signaling and inhibition. Proc Natl Acad Sci U S A 103(26):9773–9778
19. Dreger M (2003) Proteome analysis at the level of subcellular structures. Eur J Biochem 270 (4):589–599
20. Drissi R, Dubois ML, Boisvert FM (2013) Proteomics methods for subcellular proteome analysis. FEBS J 280(22):5626–5634
21. Mann M (2014) Fifteen years of Stable Isotope Labeling by Amino Acids in Cell Culture (SILAC). Methods Mol Biol 1188:1–7
22. Zanivan S et al (2013) SILAC-based proteomics of human primary endothelial cell morphogenesis unveils tumor angiogenic markers. Mol Cell Proteomics 12(12):3599–3611
23. Geiger T et al (2013) Initial quantitative proteomic map of 28 mouse tissues using the SILAC mouse. Mol Cell Proteomics 12(6):1709–1722
24. Markovic O, Markovic N (1998) Cell cross-contamination in cell cultures: the silent and neglected danger. In Vitro Cell Dev Biol Anim 34 (1):1–8
25. Burdall SE et al (2003) Breast cancer cell lines: friend or foe? Breast Cancer Res 5(2):89–89
26. Masters JR (2000) Human cancer cell lines: fact and fantasy. Nat Rev Mol Cell Biol 1(3):233–236
27. Langdon SP (2004) Cell culture contamination: an overview. Methods Mol Med 88:309–317
28. Drexler HG, Uphoff CC (2002) Mycoplasma contamination of cell cultures: incidence, sources, effects, detection, elimination, prevention. Cytotechnology 39(2):75–90
29. Uphoff CC, Gignac SM, Drexler HG (1992) Mycoplasma contamination in human leukemia cell lines: I. Comparison of various detection methods. J Immunol Methods 149(1):43–53
30. Nelson-Rees WA, Flandermeyer RR, Hawthorne PK (1975) Distinctive banded marker chromosomes of human tumor cell lines. Int J Cancer 16(1):74–82
31. Drexler HG et al (2003) False leukemia–lymphoma cell lines: an update on over 500 cell lines. Leukemia 17(2):416–426

32. Yoshino K et al (2006) Essential role for gene profiling analysis in the authentication of human cell lines. Hum Cell 19(1):43–48
33. MacLeod RA et al (1999) Widespread intraspecies cross-contamination of human tumor cell lines arising at source. Int J Cancer 83(4):555–563
34. van Bokhoven A et al (2001) TSU-Pr1 and JCA-1 cells are derivatives of T24 bladder carcinoma cells and are not of prostatic origin. Cancer Res 61 (17):6340–6344
35. Nims RW et al (1998) Sensitivity of isoenzyme analysis for the detection of interspecies cell line cross-contamination. In Vitro Cell Dev Biol Anim 34(1):35–39
36. Masters J et al (1988) Bladder cancer cell line cross-contamination: identification using a locus-specific minisatellite probe. Br J Cancer 57(3):284
37. van Helden PD et al (1988) Cross-contamination of human esophageal squamous carcinoma cell lines detected by DNA fingerprint analysis. Cancer Res 48(20):5660–5662
38. Nowell PC (1976) The clonal evolution of tumor cell populations. Science 194(4260):23–28
39. Wistuba II et al (1999) Comparison of features of human lung cancer cell lines and their corresponding tumors. Clin Cancer Res 5(5):991–1000
40. Wistuba II et al (1998) Comparison of features of human breast cancer cell lines and their corresponding tumors. Clin Cancer Res 4 (12):2931–2938
41. Phelan K, May KM (2015) Basic techniques in mammalian cell tissue culture. Curr Protoc Cell Biol 1.1. 1–1.1. 22
42. Seluanov A, Vaidya A, Gorbunova V (2010) Establishing primary adult fibroblast cultures from rodents. J Vis Exp: JoVE(44)
43. Legouis D et al (2015) Ex vivo analysis of renal proximal tubular cells. BMC Cell Biol 16(1):1–11
44. Phelan MC Basic techniques for mammalian cell tissue culture. Curr Protoc Cell Biol
45. Martin BM (1994) Tissue culture techniques: an introduction. Springer Science & Business Media
46. Laschi M et al (2015) Establishment of four new human primary cell cultures from Chemo-Naïve Italian Osteosarcoma patients. J cell physiol 230:2718
47. Hood LE et al (2012) New and improved proteomics technologies for understanding complex biological systems: addressing a grand challenge in the life sciences. Proteomics 12(18):2773–2783
48. Ahmad Y, Lamond AI (2014) A perspective on proteomics in cell biology. Trends Cell Biol 24 (4):257–264
49. Kim JB (2005) Three-dimensional tissue culture models in cancer biology. In: Seminars in cancer biology. Elsevier
50. Olechnowicz SW, Edwards CM (2014) Contributions of the host microenvironment to cancer-induced bone disease. Cancer Res 74 (6):1625–1631
51. Weiswald L-B, Bellet D, Dangles-Marie V (2015) Spherical cancer models in tumor biology. Neoplasia 17(1):1–15
52. Ellem SJ, De-Juan-Pardo EM, Risbridger GP (2014) In vitro modeling of the prostate cancer microenvironment. Adv Drug Deliv Rev 79:214–221
53. Fang X et al (2013) Novel 3D co-culture model for epithelial-stromal cells interaction in prostate cancer. PLoS One 8(9):e75187
54. Lexander H et al (2006) Evaluation of two sample preparation methods for prostate proteome analysis. Proteomics 6(13):3918–3925
55. Micke P et al (2006) Biobanking of fresh frozen tissue: RNA is stable in nonfixed surgical specimens. Lab Investig 86(2):202–211
56. Guo H et al (2012) An efficient procedure for protein extraction from formalin-fixed, paraffin-embedded tissues for reverse phase protein arrays. Proteome Sci 10(1):56
57. Scicchitano MS et al (2009) Protein extraction of formalin-fixed, paraffin-embedded tissue enables robust proteomic profiles by mass spectrometry. J Histochem Cytochem 57(9):849–860
58. Kerk NM et al (2003) Laser capture microdissection of cells from plant tissues. Plant Physiol 132(1):27–35
59. Dos Remedios C et al (2003) Genomics, proteomics and bioinformatics of human heart failure. J Muscle Res Cell Motil 24(4–6):251–261
60. Folli F et al (2010) Proteomics reveals novel oxidative and glycolytic mechanisms in type 1 diabetic patients' skin which are normalized by kidney-pancreas transplantation. Plos one 5(3):e9923
61. Kienzl K et al (2009) Proteomic profiling of acute cardiac allograft rejection. Transplantation 88 (4):553–560
62. Mas VR et al (2009) Proteomic analysis of HCV cirrhosis and HCV-induced HCC: identifying biomarkers for monitoring HCV-cirrhotic patients awaiting liver transplantation. Transplantation 87 (1):143
63. Vidal BC, Bonventre JV, I-Hong Hsu S (2005) Towards the application of proteomics in renal disease diagnosis. Clin Sci (Lond) 109(5):421–430
64. Bendixen E (2014) Animal models for translational proteomics. PROTEOMICS Clin Appl 8 (9–10):637–639
65. Terp MG, Ditzel HJ (2014) Application of proteomics in the study of rodent models of cancer. PROTEOMICS Clin Appl 8(9–10):640–652
66. Bousette N, Gramolini AO, Kislinger T (2008) Proteomics-based investigations of animal models of disease. PROTEOMICS Clin Appl 2(5):638–653
67. Conn PM (2013) Animal models for the study of human disease. Academic, Amsterdam
68. Kooij V et al (2014) Sizing up models of heart failure: proteomics from flies to humans. PROTEOMICS Clin Appl 8(9–10):653–664

69. Götz J, Ittner LM (2008) Animal models of Alzheimer's disease and frontotemporal dementia. Nat Rev Neurosci 9(7):532–544
70. Edinger M et al (1999) Noninvasive assessment of tumor cell proliferation in animal models. Neoplasia 1(4):303–310
71. Raeburn D, Underwood SL, Villamil ME (1992) Techniques for drug delivery to the airways, and the assessment of lung function in animal models. J Pharmacol Toxicol Methods 27(3):143–159
72. Sowell RA, Owen JB, Butterfield DA (2009) Proteomics in animal models of Alzheimer's and Parkinson's diseases. Ageing Res Rev 8(1):1–17
73. Flintoft L (2008) Animal models: proteomics goes live in the mouse. Nat Rev Genet 9(9):655–655
74. Stastna M, Van Eyk JE (2012) Secreted proteins as a fundamental source for biomarker discovery. Proteomics 12(4–5):722–735
75. Hathout Y (2007) Approaches to the study of the cell secretome
76. Pavlou MP, Diamandis EP (2010) The cancer cell secretome: a good source for discovering biomarkers? J Proteome 73(10):1896–1906
77. Théry C, Ostrowski M, Segura E (2009) Membrane vesicles as conveyors of immune responses. Nat Rev Immunol 9(8):581–593
78. Bijnsdorp IV et al (2013) Exosomal ITGA3 interferes with non-cancerous prostate cell functions and is increased in urine exosomes of metastatic prostate cancer patients. J Extracell Vesicles 2
79. Jeppesen DK et al (2014) Quantitative proteomics of fractionated membrane and lumen exosome proteins from isogenic metastatic and nonmetastatic bladder cancer cells reveal differential expression of EMT factors. Proteomics 14(6):699–712
80. Hosseini-Beheshti E et al (2012) Exosomes as biomarker enriched microvesicles: characterization of exosomal proteins derived from a panel of prostate cell lines with distinct AR phenotypes. Mol Cell Proteomics: MCP. M111. 014845
81. Duijvesz D et al (2011) Exosomes as biomarker treasure chests for prostate cancer. Eur Urol 59 (5):823–831
82. Raimondo F et al (2011) Advances in membranous vesicle and exosome proteomics improving biological understanding and biomarker discovery. Proteomics 11(4):709–720
83. Kang G-Y et al (2014) Exosomal proteins in the aqueous humor as novel biomarkers in patients with neovascular age-related macular degeneration. J Proteome Res 13(2):581–595
84. Anthea M, H J, McLaughlin CW, Johnson S, Warner MQ, LaHart D, Wright JD (1993) Human biology and health. Prentice Hall, Englewood Cliffs
85. Anderson NL, Anderson NG (2002) The human plasma proteome: history, character, and diagnostic prospects. Mol Cell Proteomics 1(11):845–867
86. Rodak BS, WB (2002) Hematology: clinical principles and applications, 2nd edn. Philadelphia
87. Palkuti HS (1998) Specimen control and quality control. In: Corriveau DMAF, Fritsma GA (eds) Hemostasis and thrombosis in the clinical laboratory. Lippincott, Philadelphia, pp 67–91
88. Kratz A, Ferraro M, Sluss PM, Lewandrowski KB (2004) Laboratory reference values. N Engl J Med 351:1548–1563
89. Keshishian H et al (2015) Multiplexed, quantitative workflow for sensitive biomarker discovery in plasma yields novel candidates for early myocardial injury. Mol Cell Proteomics 14:2375
90. Morrissey B et al (2013) Development of a label-free LC-MS/MS strategy to approach the identification of candidate protein biomarkers of disease recurrence in prostate cancer patients in a clinical trial of combined hormone and radiation therapy. Proteomics Clin Appl 7(5–6):316–326
91. Mullan RH et al (2007) Early changes in serum type II collagen biomarkers predict radiographic progression at one year in inflammatory arthritis patients after biologic therapy. Arthritis Rheum 56 (9):2919–2928
92. Sekigawa I et al (2008) Protein biomarker analysis by mass spectrometry in patients with rheumatoid arthritis receiving anti-tumor necrosis factor-alpha antibody therapy. Clin Exp Rheumatol 26 (2):261–267
93. Zhao J et al (2015) Identification of potential plasma biomarkers for esophageal squamous cell carcinoma by a proteomic method. Int J Clin Exp Pathol 8 (2):1535–1544
94. Lundblad R (2003) Considerations for the use of blood plasma and serum for proteomic analysis. Internet J Genomics and Proteomics 1(2)
95. Millioni R et al (2011) High abundance proteins depletion vs low abundance proteins enrichment: comparison of methods to reduce the plasma proteome complexity. PLoS One 6(5):e19603
96. Cyr DD et al (2011) Characterization of serum proteins associated with IL28B genotype among patients with chronic hepatitis C. PLoS One 6(7): e21854
97. Haslene-Hox H et al (2011) A new method for isolation of interstitial fluid from human solid tumors applied to proteomic analysis of ovarian carcinoma tissue. PLoS One 6(4), e19217
98. Smith MP et al (2011) A systematic analysis of the effects of increasing degrees of serum immunodepletion in terms of depth of coverage and other key aspects in top-down and bottom-up proteomic analyses. Proteomics 11(11):2222–2235
99. Levreri I et al (2005) Separation of human serum proteins using the Beckman-Coulter PF2D system: analysis of ion exchange-based first dimension chromatography. Clin Chem Lab Med 43(12):1327–1333
100. Sennels L et al (2007) Proteomic analysis of human blood serum using peptide library beads. J Proteome Res 6(10):4055–4062

101. Pieper R et al (2003) Multi-component immunoaffinity subtraction chromatography: an innovative step towards a comprehensive survey of the human plasma proteome. Proteomics 3 (4):422–432
102. Huang JT, McKenna T, Hughes C, Leweke FM, Schwarz E, Bahn S (2007) CSF biomarker discovery using label-free nano-LC-MS based proteomic profiling: technical aspects. J Sep Sci 30:214–225
103. Sakka L, Coll G, Chazal J (2011) Anatomy and physiology of cerebrospinal fluid. Eur Ann Otorhinolaryngol Head Neck Dis 123:309–316
104. Percy AJ, Yang J, Chambers AG, Simon R, Hardie DB, Borchers CH (2014) Multiplexed MRM with internal standards for cerebrospinal fluid candidate protein biomarker quantitation. J Proteome Res 13:3733
105. Choi YS, Choe LH, Lee KH (2010) Recent cerebrospinal fluid biomarker studies of Alzheimer's disease. Expert Rev Proteomics 7(6):919–929
106. Stoop MP et al (2008) Multiple sclerosis-related proteins identified in cerebrospinal fluid by advanced mass spectrometry. Proteomics 8(8):1576–1585
107. Claveau D, Dankoff J (2013) Is lumbar puncture still needed in suspected subarachnoid hemorrhage after a negative head computed tomographic scan? CJEM 15:1–3
108. Seehusen DA, Reeves MM, Fomin DA (2003) Cerebrospinal fluid analysis. Am Fam Physician 68 (6):1103–1108
109. Shores KS et al (2008) Use of peptide analogue diversity library beads for increased depth of proteomic analysis: application to cerebrospinal fluid. J Proteome Res 7(5):1922–1931
110. Thouvenot E et al (2008) Enhanced detection of CNS cell secretome in plasma protein-depleted cerebrospinal fluid. J Proteome Res 7(10):4409–4421
111. Lehnert S, Jesse S, Rist W, Steinacker P, Soininen H, Herukka SK, Tumani H, Lenter M, Oeckl P, Ferger B, Hengerer B, Otto M (2012) iTRAQ and multiple reaction monitoring as proteomic tools for biomarker search in cerebrospinal fluid of patients with Parkinson's disease dementia. Exp Neurol 234 (2):499–505
112. Iorio L, Avagliano F (1999) Observations on the Liber medicine orinalibus by Hermogenes. Am J Nephrol 19(2):185–188
113. Moe OW, Berry CA, Rector FC (2000) The kidney. W. B. Saunders, Philadelphia
114. Ling XB et al (2010) Urine peptidomic and targeted plasma protein analyses in the diagnosis and monitoring of systemic juvenile idiopathic arthritis. Clin Proteomics 6(4):175–193
115. Wu T et al (2013) Urinary angiostatin-a novel putative marker of renal pathology chronicity in lupus nephritis. Mol Cell Proteomics 12(5):1170–1179
116. Schaub S et al (2004) Urine protein profiling with surface-enhanced laser-desorption/ionization time-of-flight mass spectrometry. Kidney Int 65 (1):323–332
117. Theodorescu D et al (2006) Discovery and validation of new protein biomarkers for urothelial cancer: a prospective analysis. lancet oncol 7(3):230–240
118. Zhou H et al (2006) Collection, storage, preservation, and normalization of human urinary exosomes for biomarker discovery. Kidney Int 69 (8):1471–1476
119. Vestergaard P, Leverett R (1958) Constancy of urinary creatinine excretion. J Lab Clin Med 51 (2):211–218
120. Mischak H (2005) Capillary electrophoresis coupled to mass spectrometry for clinical diagnostic purposes. Electrophoresis 26:2708–2716
121. Pisitkun T, Johnstone R, Knepper MA (2006) Discovery of urinary biomarkers. Mol Cell Proteomics 5 (10):1760–1771
122. Weissinger EM et al (2004) Proteomic patterns established with capillary electrophoresis and mass spectrometry for diagnostic purposes. Kidney Int 65 (6):2426–2434
123. Hortin GL et al (2006) Proteomics: a new diagnostic frontier. Clin Chem 52(7):1218–1222
124. Yu Y, Pieper R (2015) Urinary pellet sample preparation for shotgun proteomic analysis of microbial infection and host–pathogen interactions. Proteomic Profiling: Methods and Protocols 65–74
125. Riva A et al (2000) A high resolution sem study of human minor salivary glands. Eur J Morphol 38 (4):219–226
126. Castle D, Castle A (1998) Intracellular transport and secretion of salivary proteins. Crit Rev Oral Biol Med 9(1):4–22
127. Messana I et al (2008) Trafficking and postsecretory events responsible for the formation of secreted human salivary peptides: a proteomics approach. Mol Cell Proteomics 7(5):911–926
128. Humphrey SP, Williamson RT (2001) A review of saliva: normal composition, flow, and function. J Prosthet Dent 85(2):162–169
129. Edgar WM (1992) Saliva: its secretion, composition and functions. Br Dent J 172(8):305–312
130. Hansen AM, Garde AH, Persson R (2008) Measurement of salivary cortisol–effects of replacing polyester with cotton and switching antibody. Scand J Clin Lab Invest 68(8):826–829
131. Hansen AM, Garde AH, Persson R (2008) Sources of biological and methodological variation in salivary cortisol and their impact on measurement among healthy adults: a review. Scand J Clin Lab Invest 68(6):448–458
132. Tzioufas AG, Kapsogeorgou EK (2015) Biomarkers. Saliva proteomics is a promising tool to study

Sjogren syndrome. Nat Rev Rheumatol 11 (4):202–203
133. Schafer CA et al (2014) Saliva diagnostics: utilizing oral fluids to determine health status. Monogr Oral Sci 24:88–98
134. Heflin L, Walsh S, Bagajewicz M (2009) Design of medical diagnostics products: a case-study of a saliva diagnostics kit. Comput Chem Eng 33 (5):1067–1076
135. Messana I et al (2008) Facts and artifacts in proteomics of body fluids. What proteomics of saliva is telling us? J Sep Sci 31(11):1948–1963
136. Esser D et al (2008) Sample stability and protein composition of saliva: implications for its use as a diagnostic fluid. Biomark Insights 3:25–27
137. Yan W et al (2009) Systematic comparison of the human saliva and plasma proteomes. Proteomics Clin Appl 3(1):116–134
138. Navazesh M, Kumar SK (2008) Measuring salivary flow: challenges and opportunities. J Am Dent Assoc 139 Suppl:35s–40s
139. Atkinson KR et al (2008) Rapid saliva processing techniques for near real-time analysis of salivary steroids and protein. J Clin Lab Anal 22(6):395–402
140. Michishige F et al (2006) Effect of saliva collection method on the concentration of protein components in saliva. J Med Investig 53(1–2):140–146
141. Vitorino R et al (2004) Identification of human whole saliva protein components using proteomics. Proteomics 4(4):1109–1115
142. Saunte C (1983) Quantification of salivation, nasal secretion and tearing in man. Cephalalgia 3 (3):159–173
143. Jessie K et al (2010) Proteomic analysis of whole human saliva detects enhanced expression of interleukin-1 receptor antagonist, thioredoxin and lipocalin-1 in cigarette smokers compared to non-smokers. Int J Mol Sci 11(11):4488–4505
144. Soares S et al (2011) Reactivity of human salivary proteins families toward food polyphenols. J Agric Food Chem 59(10):5535–5547
145. Carlson A, Crittenden A (1909) The relation of ptyalin concentration to the diet and to the rate of salivary secretion. Exp Biol Med 7(2):52–54
146. Heft MW, Baum BJ (1984) Unstimulated and stimulated parotid salivary flow rate in individuals of different ages. J Dent Res 63(10):1182–1185
147. Lashley K (1916) Reflex secretion of the human parotid gland. J Exp Psychol 1(6):461
148. Nita-Lazar A, Saito-Benz H, White FM (2008) Quantitative phosphoproteomics by mass spectrometry: past, present, and future. Proteomics 8 (21):4433–4443
149. Thingholm TE, Jensen ON, Larsen MR (2009) Analytical strategies for phosphoproteomics. Proteomics 9(6):1451–1468
150. Zielinska DF et al (2010) Precision mapping of an in vivo N-glycoproteome reveals rigid topological and sequence constraints. Cell 141(5):897–907
151. Al-Tarawneh SK et al (2011) Defining salivary biomarkers using mass spectrometry-based proteomics: a systematic review. OMICS J Integr Biol 15 (6):353–361
152. Pfaffe T et al (2011) Diagnostic potential of saliva: current state and future applications. Clin Chem 57 (5):675–687
153. Whitelegge JP et al (2007) Protein-sequence polymorphisms and post-translational modifications in proteins from human saliva using top-down Fourier-transform ion cyclotron resonance mass spectrometry. Int J Mass Spectrom 268(2):190–197
154. Vitorino R et al (2011) Finding new posttranslational modifications in salivary proline-rich proteins. PROTEOMICS Clin Appl 5(3–4):197–197
155. Molloy MP et al (1999) Extraction of Escherichia coli proteins with organic solvents prior to two-dimensional electrophoresis. Electrophoresis 20 (4–5):701–704
156. Pasquali C, Fialka I, Huber LA (1999) Subcellular fractionation, electromigration analysis and mapping of organelles. J Chromatogr B Biomed Sci Appl 722 (1):89–102
157. Krief G et al (2011) Improved visualization of low abundance oral fluid proteins after triple depletion of alpha amylase, albumin and IgG. Oral Dis 17 (1):45–52
158. Owen DH, Katz DF (2005) A review of the physical and chemical properties of human semen and the formulation of a semen simulant. J Androl 26 (4):459–469
159. Bartoov B et al (1999) Quantitative ultramorphological analysis of human sperm: fifteen years of experience in the diagnosis and management of male factor infertility. Arch Androl 43(1):13–25
160. Pizzol D et al (2014) Genetic and molecular diagnostics of male infertility in the clinical practice. Front Biosci (Landmark Ed) 19:291–303
161. Liu DY, Baker HW (1992) Tests of human sperm function and fertilization in vitro. Fertil Steril 58 (3):465–483
162. Liu DY, Baker HW (2002) Evaluation and assessment of semen for IVF/ICSI. Asian J Androl 4 (4):281–285
163. World Health Organization (1999) WHO laboratory manual for the examination of human semen and sperm-cervical mucus interaction. Cambridge University Press, Cambridge
164. Amaral A et al (2014) The combined human sperm proteome: cellular pathways and implications for basic and clinical science. Hum Reprod Update 20 (1):40–62
165. Drake RR et al (2010) In-depth proteomic analyses of direct expressed prostatic secretions. J Proteome Res 9(5):2109–2116
166. Milardi D et al (2013) Proteomics of human seminal plasma: identification of biomarker candidates for fertility and infertility and the evolution of technology. Mol Reprod Dev 80(5):350–357

167. Milardi D et al (2012) Proteomic approach in the identification of fertility pattern in seminal plasma of fertile men. Fertil Steril 97(1):67–73.e1
168. Thompson CB et al (1984) A method for the separation of erythrocytes on the basis of size using counterflow centrifugation. Am J Hematol 17(2):177–183
169. Van der Vegt SGL et al (1985) Counterflow centrifugation of red cell populations: a cell age related separation technique. Br J Haematol 61(3):393–403
170. Dhurat R, Sukesh M (2014) Principles and methods of preparation of platelet-rich plasma: a review and author's perspective. J Cutan Aesthet Surg 7(4):189–197
171. Godoy-Ramirez K et al (2004) Optimum culture conditions for specific and nonspecific activation of whole blood and PBMC for intracellular cytokine assessment by flow cytometry. J Immunol Methods 292(1–2):1–15
172. de Bono JS et al (2008) Circulating tumor cells predict survival benefit from treatment in metastatic castration-resistant prostate cancer. Clin Cancer Res 14(19):6302–6309
173. Schmidt U et al (2004) Quantification of disseminated tumor cells in the bloodstream of patients with hormone-refractory prostate carcinoma undergoing cytotoxic chemotherapy. Int J Oncol 24(6):1393–1399
174. Gertler R et al (2003) Detection of circulating tumor cells in blood using an optimized density gradient centrifugation. Recent Results Cancer Res 162:149–155
175. Farace F et al (2011) A direct comparison of cell search and ISET for circulating tumour-cell detection in patients with metastatic carcinomas. Br J Cancer 105(6):847–853
176. Liu Z et al (2011) Negative enrichment by immunomagnetic nanobeads for unbiased characterization of circulating tumor cells from peripheral blood of cancer patients. J Transl Med 9:70
177. Yu M et al (2014) Cancer therapy. Ex vivo culture of circulating breast tumor cells for individualized testing of drug susceptibility. Science 345(6193):216–220

Protein Fractionation and Enrichment Prior to Proteomics Sample Preparation

2

Andrew J. Alpert

Abstract

Proteins may be considered as polypeptides large enough to have a well-defined tertiary, or three-dimensional structure. In aqueous media, this structure is typically one in which polar and charged amino acid residues are on the surface while hydrophobic residues tend to be sequestered in the core and reasonably inaccessible to the aqueous environment. Proteins that are not normally found free in aqueous media, such as membrane proteins and apolipoproteins, can have tertiary structures that deviate from this model. In general, the biological activity of proteins requires the preservation of their tertiary structure, and this sets more limits upon the chromatography than is true of peptides. In proteomics, the concern is with which proteins are present and in what quantity rather than maintaining biological activity. Such applications are freer to use mobile and stationary phases that denature protein structure. However, considerations of solubility and recovery may still set more limits on the chromatography than is the case with peptides.

Keywords

Protein fractionation • Protein chromatography • Ion-exchange chromatography (IEX) • Hydrophobic Interaction Chromatography (HIC) • Size-Exclusion Chromatography (SEC) • Reversed-Phase Chromatography (RPC) • Hydrophilic Interaction Chromatography (HILIC) • Affinity chromatography • Multi-dimensional chromatography for top-down proteomics

2.1 Overall Requirements

Proteins may be considered as polypeptides large enough to have a well-defined tertiary, or three-dimensional structure. In aqueous media, this

A.J. Alpert (✉)
PolyLC Inc., Columbia, MD, USA
e-mail: aalpert@polylc.com

H. Mirzaei and M. Carrasco (eds.), *Modern Proteomics – Sample Preparation, Analysis and Practical Applications*, Advances in Experimental Medicine and Biology 919, DOI 10.1007/978-3-319-41448-5_2

structure is typically one in which polar and charged amino acid residues are on the surface while hydrophobic residues tend to be sequestered in the core and reasonably inaccessible to the aqueous environment. Proteins that are not normally found free in aqueous media, such as membrane proteins and apolipoproteins, can have tertiary structures that deviate from this model. In general, the biological activity of proteins requires the preservation of their tertiary structure, and this sets more limits upon the chromatography than is true of peptides. In proteomics, the concern is with which proteins are present and in what quantity rather than maintaining biological activity. Such applications are freer to use mobile and stationary phases that denature protein structure. However, considerations of solubility and recovery may still set more limits on the chromatography than is the case with peptides.

Proteomics can involve the analysis of minor variants of individual proteins as well as the identification and quantitation of the proteins in a complex mixture. Accordingly, this chapter presents examples of quality control and clinical applications as well as separations for basic research.

2.2 Modes of Chromatography

2.2.1 Ion-Exchange Chromatography (IEX)

Proteins have charged residues on the surface of their structures and so are attracted electrostatically to a stationary phase of the opposite charge. Figure 2.1 shows the separation of variants of ovalbumin that have differing numbers of phosphate groups. At the isoelectric point (pI) of a protein, the amount of positive (+) and negative (−) charge is in balance. At a pH higher than the pH corresponding to the pI, a protein has a net (−) charge and is retained by an anion-exchange column. At a lower pH, it has a net (+) charge and is retained by a cation-exchange column. However, chromatography is a surface interaction. This distinguishes it from electrophoresis, which involves field effects. It is

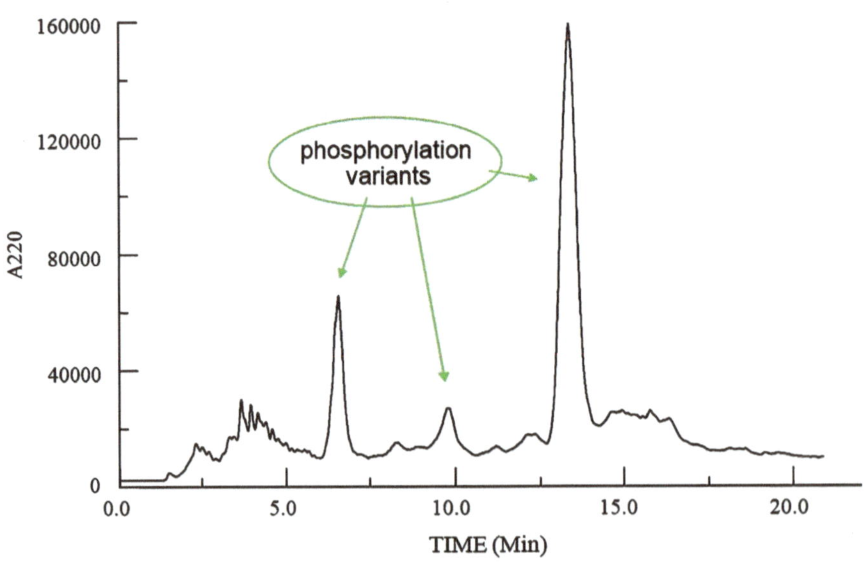

Fig. 2.1 Anion-exchange of phosphorylation variants of ovalbumin. Sample: Ovalbumin (Sigma Grade VI [99 %]). *Column*: PolyWAX LP (100 × 4.6 mm; 5-μm, 1000-Å). *Gradient*: 10 mM K-PO_4, pH 7.0, with 60–300 mM NaCl in 20’

possible for a protein to have non-uniform distribution of the charged residues on its surface. If there is a cluster of residues of the same charge, then the protein can bind through that cluster to an ion-exchange column of the opposite charge from the residues. The result is retention of many proteins at pH ranges one unit or more beyond their pI value, despite their having the same net charge as the column. Accordingly, some proteins are retained by both anion- and cation-exchange columns. In addition, since the protein can be highly oriented in its binding to the surface, some residues will be closer to the surface than others and can have a greater effect on the interaction. Accordingly, chromatography can distinguish between variants of a protein that differ in the position of a residue that has been derivatized, oxidized, or otherwise modified, as shown in Fig. 2.2. Such positional variants would not be separated by electrophoresis.

In general, the conditions used in IEX are mild and do not denature proteins. Many membrane and structural proteins are not readily soluble in the aqueous media that are normally used for IEX. In such cases, organic solvents or solubilizing agents such as hexafluoro-2-propanol or trifluoroethanol can be included in the mobile phases. See Figs. 2.9 and 2.11 for examples of the use of such solubilizing agents.

When counting numbers of proteins, approximately 50 % of mammalian proteins have pI values above 7 and 50 % below 7, with a minimum at pH 7 itself [1]. When counting protein abundance rather than protein numbers, though, one finds that acidic proteins, with pI values below 7, are about 4x more abundant in serum and cell lysates than are basic proteins. For this reason, anion-exchange is generally more useful for fractionation of complex mixtures of mammalian proteins when compared to cation-exchange. Cation-exchange is more useful for certain specific applications, such as quality control (QC) analysis of monoclonal antibodies and the clinical analysis of hemoglobin variants.

Elution in ion-exchange usually involves a gradient of increasing salt concentration. If an absorbance detector is used at a wavelength below 230 nm, then salts should be used that are transparent in this range. Such salts would include NaCl or KCl with phosphate, MES (2-(N-morpholino)ethanesulfonic acid), or HEPES (4-(2-hydroxyethyl)-1-piperazineethanesulfonic acid) used as buffers. If the application requires a mobile phase that is volatile, then one

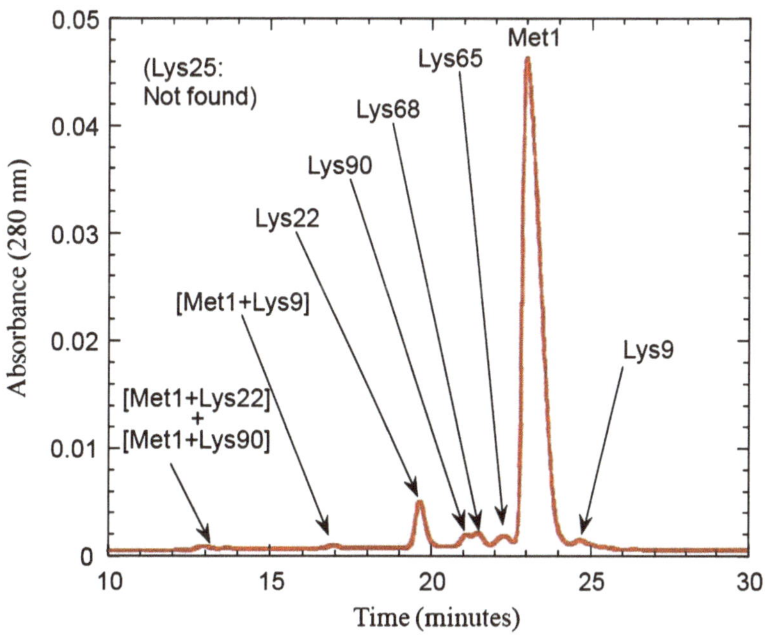

Fig. 2.2 Separation by cation-exchange of PEGylation positional variants of tumor necrosis factor soluble receptor Type I. The main product was PEGylated at the N-terminus (Met1). Side products were PEGylated at the lysine residues indicated instead of or in addition to the N-terminus. *Column*: PolyCAT A (200 × 4.6 mm; 5-μm, 1000-Å). *Gradient*: 20 mM sodium acetate, pH 5.0, with 20 % ACN, and a gradient of 1–160 mM KCl in 30'. (Adapted from J.E. Seely et al., *BioPharm Intl. 18* (March 2005) 30)

can use ammonium acetate as both the buffering and the gradient salt. However, most protein applications require concentrations of salt for elution that are so high as to be incompatible with applications such as mass spectrometry that require volatile mobile phases. Alternatively, a pH gradient can be used to change the net charge of a protein to one closer to that of the stationary phase, decreasing the amount of salt required for elution. With cation-exchange, this involves a gradient of increasing pH (usually to one above pH 6.5, above which histidine residues lose their (+) charge). Conversely, the pH change with anion-exchange usually involves a decreasing pH gradient.

The charge density of the ion-exchange material varies with pH. The titration curve of a simple amine or carboxylic acid in solution features a sharp inflection point. That is not true of polyelectrolytes, in which the charge on one functional group affects the ease of charging neighboring groups. This is true whether the polyelectrolyte is in solution or immobilized on the surface of a stationary phase. As a result, titration curves of ion-exchange stationary phase materials in suspension feature a continuum of charge density varying over a wide pH range [2]. An ion-exchange material is considered to be "weak" or "strong" based just on the pH range where it starts to lose charge, not on the degree of attraction of a charged analyte to the material. Weak anion-exchange (WAX) materials are fully charged below pH 5 but are only about 5 % charged at pH 9, with a continuum of variation of charge density in between. Such materials feature primary, secondary or tertiary amines as functional groups or a mixture of the three. A strong anion-exchange (SAX) material has quaternary amine functional groups and retains most of its charge density as high as pH 12. A weak cation-exchange (WCX) material has carboxyl- functional groups which can be uncharged at pH < 4. A strong cation-exchange (SCX) material retains most (−) charge density down to pH 2. At a pH where a weak and a strong ion-exchange material both have their full charge density, there is no significant difference in performance between them, assuming all other variables are the same.

For protein applications, it is important to use ion-exchange materials that have been manufactured for the purpose. The least expensive ion-exchange materials generally feature charged groups attached to a polymeric resin. While there are resins that are hydrophilic enough for the purpose, many are not, such as those with a polystyrene-divinylbenzene base. Such materials may perform well with small analytes, but are so hydrophobic that many proteins will not elute from columns of such materials. In general, an ion-exchange material for proteins must have a thick, hydrophilic coating that hides the base material from proteins in solution. Another important property is that porous materials, such as those based on silica, must have pores wide enough for protein diffusion in and out to be facile. This requires pores at least 300 Å wide, and many proteins afford sharper, more symmetrical peaks with pores in the range 1000–1500 Å. Such materials have lower surface area than do 300-Å pore materials, but the degree of retention is usually not a problem in ion-exchange of proteins.

An exception to these general trends is the recent use of weak cation-exchange (WCX) materials with a gradient to a pH low enough to uncharge the carboxyl- groups in the coating. This can be performed with a gradient from dilute ammonium formate, pH ~ 5, to unbuffered formic acid (typically in the range 0.5–2 %). This will be discussed in more detail in Sect. 2.3.5.

If one is not sure whether to use an anion- or cation-exchange column, one solution would be to use a mixed-bed column that contains both materials. In principle, a mixed-bed column will retain all proteins. Such columns have proved to be useful for fractionation of complex mixtures of proteins. Figure 2.3 shows the uniform distribution of proteins such columns can afford. Strong retention of a protein on a mixed-bed IEX column is facilitated by either of two structural characteristics:

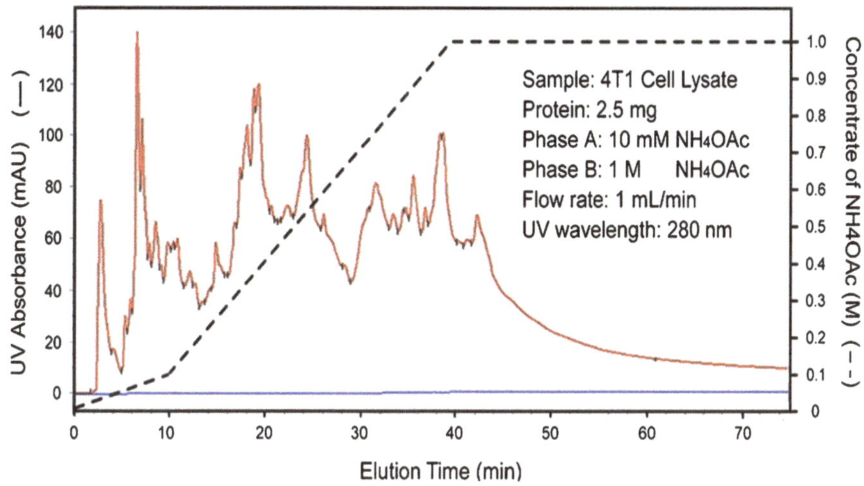

Fig. 2.3 Fractionation of proteins from a lysate of 4 T1 cells (mouse mammary tumor) by a mixed-bed ion-exchange column. *Column*: PolyCATWAX (200 × 4.6-mm; 5-μm, 1000-Å). *Gradient*: 10–1000-mM ammonium acetate, as shown (Adapted from Ref. [3])

(a) An extreme pI value in either direction
(b) A high percentage of charged residues of either sign [3]

The higher the percentage of charged residues, the more likely that the protein surface will contain at least one patch with several residues of the same sign through which strong binding can occur, as discussed above. A protein with a patch of this sort will be strongly retained even if the pI value is near neutrality.

High concentrations of organic solvents in the mobile phases will generally denature water-soluble proteins. However, more modest concentrations of solvents can sometimes improve selectivity, depending on the protein involved. Figure 2.4 shows an example of this with a set of closely-related glycoproteins.

2.2.2 Hydrophobic Interaction Chromatography (HIC)

Chromatography in the HIC mode starts with a high concentration of a salt whose ions are surrounded with a strongly retained sphere of hydration, such as a sulfate, phosphate or citrate. This leaves less of the water free to hydrate the polar residues of proteins. At this point the solvation of proteins is marginal, since a solute must surround itself with molecules of the solvent in order to remain in solution. When now exposed to a modestly hydrophobic surface, the protein will adsorb to the surface, thereby partitioning out of the aqueous phase. A gradient is then run of decreasing salt concentration. Proteins are resolvated – or, rather, rehydrated – and elute in order of their increase in hydrophobic character of the surface of their tertiary structure. This elution order appears to be the same as that in RPC. The difference is that the stationary phase is appreciably less hydrophobic than is true with an RPC material, and the mobile phase lacks any denaturing components. Consequently, in general HIC is a nondenaturing method. An exception would be with protein complexes in which the subunits interact through electrostatic interactions rather than hydrophobic interactions; this attraction can be disrupted by the high salt concentrations used in HIC.

HIC compares well with IEX in its high capacity and selectivity. The basis of selectivity is complementary to that of IEX, operating via hydrophobic interaction with the hydrophobic residues. Consequently, the two modes can fruitfully be used in sequence for isolating a protein

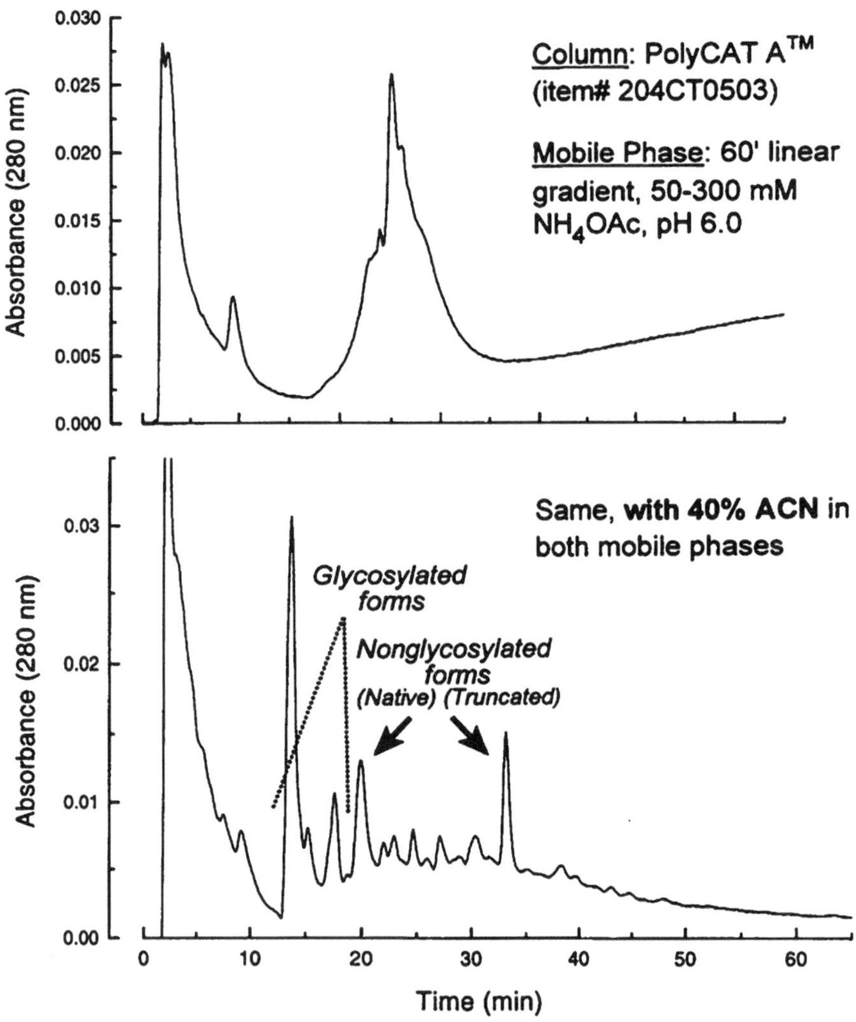

Fig. 2.4 Separation by cation-exchange of glycosylation variants of recombinant α–bungarotoxin expressed in *P. pastoris*. *Column*: PolyCAT A (200 × 4.6-mm; 5-μm, 300-Å). *Gradient*: 60' linear, 50–300 mM ammonium acetate, pH 6.0. *Top*: No ACN. *Bottom*: 40 % ACN in both mobile phases (Data courtesy of Robert Rogowski and Edward Hawrot, Brown University)

from a complex mixture or simply dividing it into fractions with fewer components per fraction. This is discussed later in the Proteomics section. HIC can be extremely sensitive to minor variations in polarity, as is the case in the example in Fig. 2.5. This is helpful in quality control analysis of proteins.

HIC is also often used as the "capture" step for the initial collection of a recombinant protein in solution in a fermentation vat. A suitable salt such as ammonium sulfate is added in an amount sufficient to promote binding to a HIC material and the liquid is pumped into a HIC cartridge. It can then be eluted in a volume much lower than that of the original sample.

2.2.3 Size-Exclusion Chromatography (SEC)

SEC of proteins is performed with hydrophilic stationary phases with well-defined pore

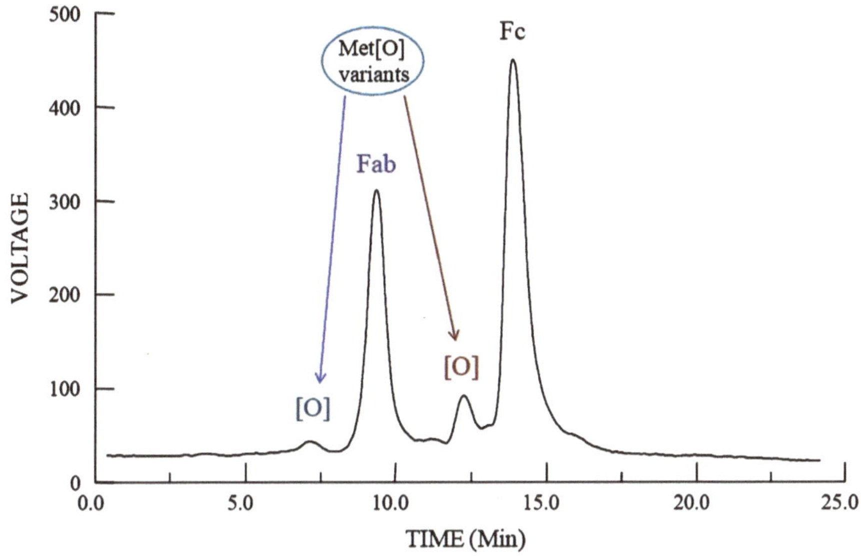

Fig. 2.5 Separation by HIC of Fab and Fc antibody fragments and their oxidation products. The minor peaks indicated correspond to the major peaks eluting after them but with a single methionine residue oxidized to the sulfoxide form. *Column*: PolyPROPYL A (100 × 4.6-mm; 3-μm, 1500-Å). *Gradient*: Decreasing ammonium sulfate concentration in 20 mM K-PO_4, pH 7.0

diameters. This is a nondenaturing mode and can be performed with moderate concentrations (100–200 mM) of volatile salts such as ammonium acetate. Being an isocratic mode, it is easy to implement. The main limitation is that it is a low-resolution mode. A general rule is that for two proteins to be resolved to baseline in SEC, they must differ in molecular weight by at least a factor of two, a characteristic that does not predispose this mode to separations based on fine differences. Given this limitation, the histogram in Fig. 2.6 suggests that the entire human proteome would yield only 5 or 6 baseline-resolved peaks in SEC. That is something of an underestimate; a good SEC column can produce about eight baseline-resolved peaks within the fractionation range, including the Total Exclusion Volume (V_o) peak and the Total Inclusion Volume (V_t) peak.

In recent years some applications have started using SEC as a filter to separate very large molecules from very small ones. These include the following:

(a) SEC-MS of intact proteins: A column is chosen with a pore diameter narrow enough to insure that the protein of interest elutes in the V_o peak, which is then directed to the mass spectrometer. Small molecules such as nonvolatile salts elute later and are directed to waste. The mobile phase must be volatile. Solvents used to date include 50 mM formic acid, in which case proteins elute in denatured forms (Fig. 2.7), or 200 mM ammonium acetate, in which case they elute with their tertiary structures intact (for "native" mass spectrometry) (Fig. 2.8).

(b) Top-down proteomics: In mass spectra of intact proteins, the signal-to-noise ratio decreases as the protein molecular weight increases [4]. Consequently, small proteins interfere with the detection of large proteins in the same sample. Identification of large proteins is facilitated by preliminary separation of the proteins < 40 KDa from proteins > 40 KDa using an SEC spin cartridge.

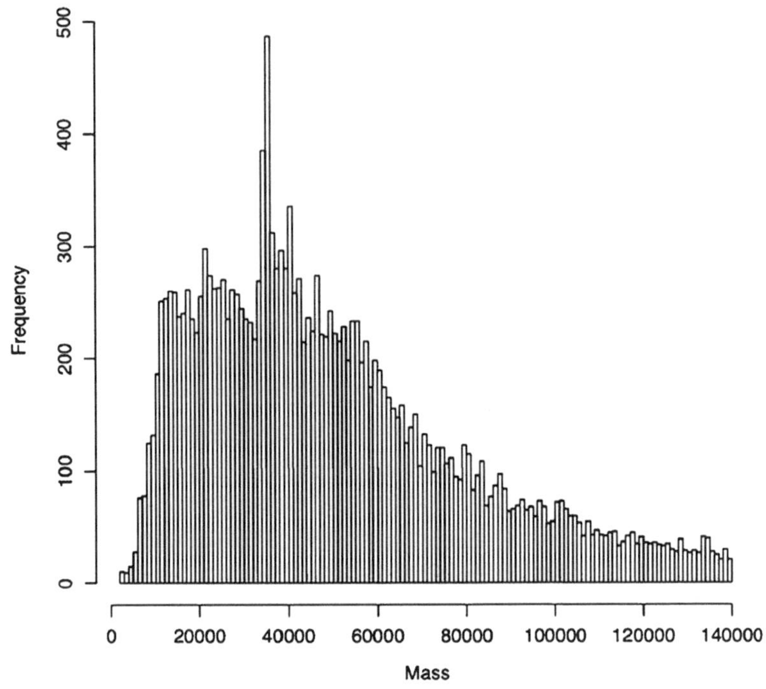

Fig. 2.6 Frequency of occurrence of proteins with specific masses in the human proteome (Adapted from Ref. [4])

2.2.4 Reversed-Phase Chromatography (RPC)

RPC is the most widely used mode of HPLC. In general, it is not well-suited to protein applications. Proteins tend to denature when exposed to the hydrophobic surface. Subsequent exposure to organic solvents, sometimes featuring extremes of pH and chaotropes such as TFA, causes even more thorough loss of tertiary structure. Small proteins can tolerate these conditions. The denaturation of large proteins could expose more than a hundred hydrophobic residues for simultaneous interaction with the stationary phase. The result may be elution in peaks 15 min wide or no elution at all.

A large percentage of RPC applications with proteins involve columns with C-4 or C-8 functional groups [5]. Elution from such materials is more facile than from more hydrophobic materials. The greater retention capacity of more hydrophobic materials is not needed in any case, since proteins contain more than enough hydrophobic residues to guarantee retention. The pore diameter should be at least 300 Å, and a diameter of 500–1000 Å should be considered. Such materials are uncommon, but Polymer Laboratories offers PLRP materials with pore diameters in this range. Recently, columns of PLRP material have been shown to afford higher protein recovery and sharper peaks than does a silica-based C-4 column [6].

2.2.5 Hydrophilic Interaction Chromatography (HILIC)

The use of HILIC for intact proteins has been limited to date. The main obstacle is the tendency of many proteins to precipitate from the predominantly organic solvents used for binding in the HILIC mode. To date, applications of pure HILIC have involved membrane proteins [7, 8] and apolipoproteins [9] that do not normally freely occur in aqueous media and which therefore are compatible with the organic solvents used in HILIC, as in the example in Fig. 2.9.

A combination that has been more widely used is an IEX column eluted with a predominantly organic mobile phase. Under these conditions

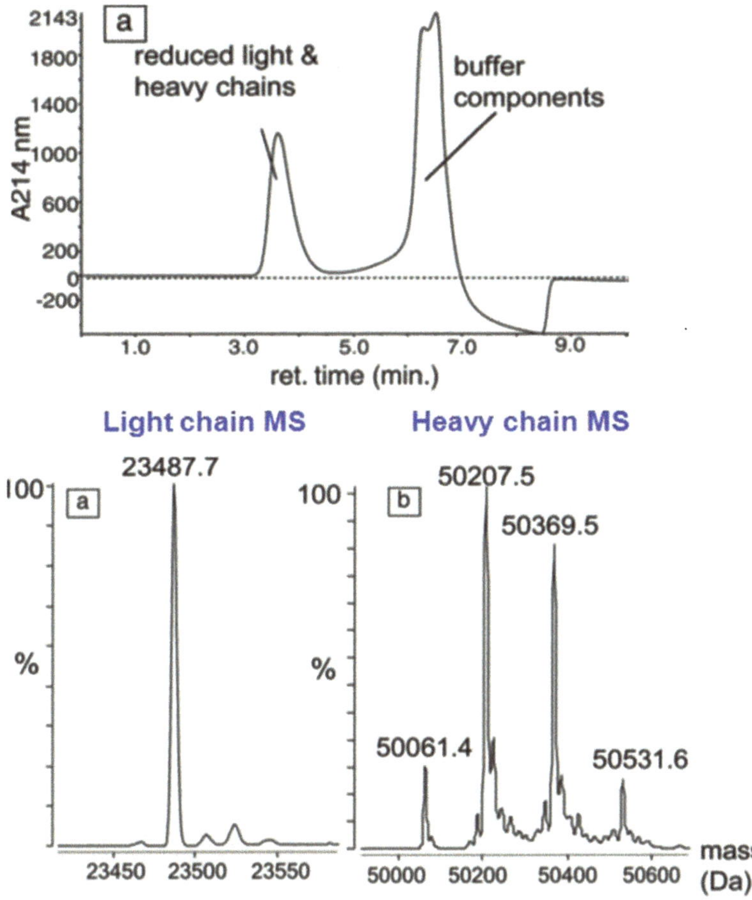

Fig. 2.7 SEC of antibody chains under denaturing conditions. (**a**) Protein components elute in the Vo peak at 3.5' which is directed to the mass spectrometer, while the rest of the eluate is directed to waste; (**b**) Resulting mass spectra. *Column*: PolyHYDROXYETHYL A, 250 × 2.1-mm; 5-μm, 300-Å. Mobile phase: 0.1 % formic acid (Adapted from L.J. Brady et al., *J. Am. Soc. Mass Spectrom. 19* (2008) 502)

hydrophilic interaction is superimposed upon the electrostatic effects. Accordingly, the column will be sensitive to variations in polarity as well as in charge. For example, in the absence of hydrophilic interaction, a column will be sensitive to the acetylation of lysine residues, which reduces the number of (+) charges. In the presence of hydrophilic interaction, it will also be sensitive to the methylation of lysine residues, which affects polarity but not charge. This combination has been used for separation of histone variants with numerous possible combinations of lysine acetylation, methylation, and other post-translational modifications (Fig. 2.10). This combination can also be used for high-resolution separation of variants of other types proteins that do not normally occur in aqueous solution, as in the example in Fig. 2.11 of an emulsion of pulmonary surfactant proteins. Treatment of this sort removes lipids and detergents from protein samples.

2.2.6 Affinity Chromatography

This mode involves a stationary phase with some immobilized compound that has an unusually

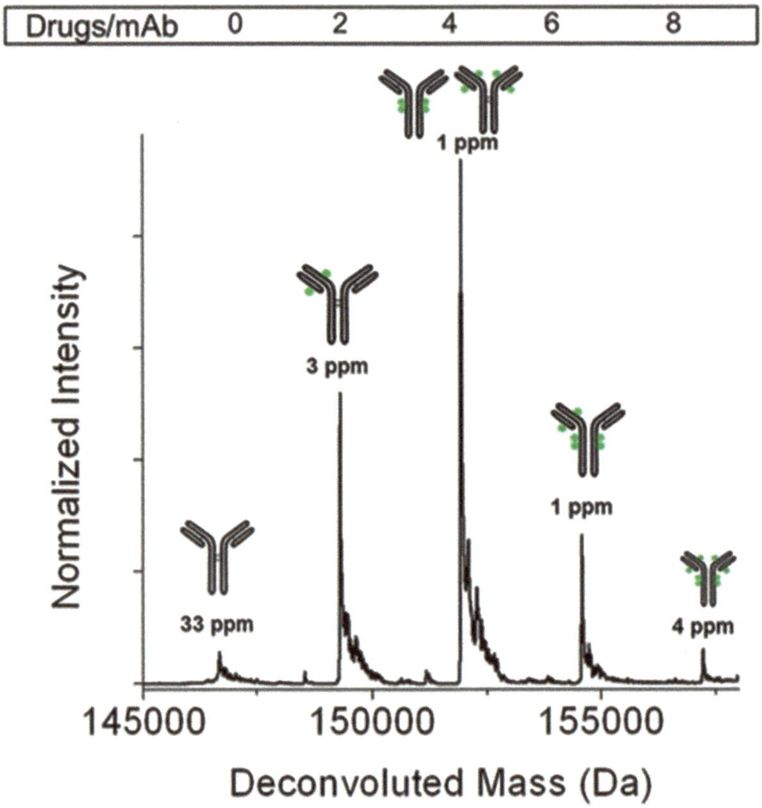

Fig. 2.8 SEC of antibody-drug conjugates (ADC's) under nondenaturing conditions. The conjugates are attached to free thiol group resulting from reduction of disulfide bridges and so conjugate content of the antibody is in multiples of two. *Column*: PolyHYDROXYETHYL A, 150 × 0.3-mm capillary; 5-μm, 300-Å. Mobile phase: 200 mM ammonium acetate (From: S.M. Hengel et al., *Anal. Chem. 86* (2014) 3420)

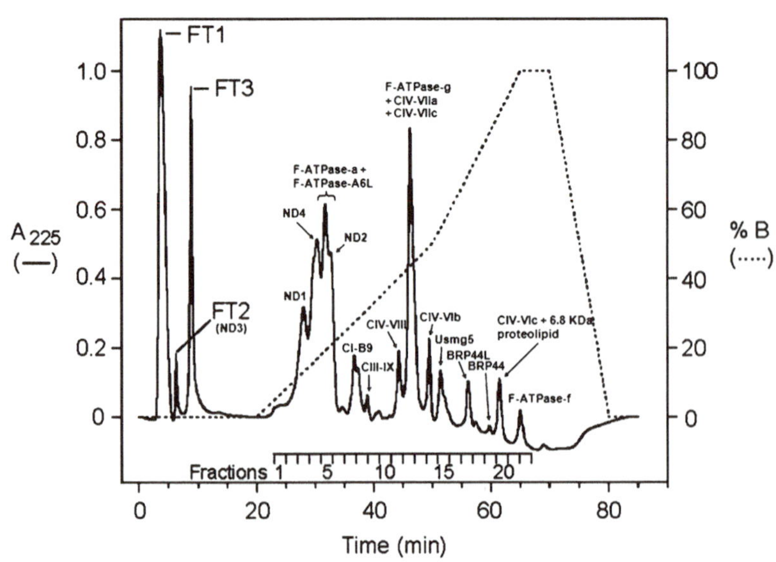

Fig. 2.9 HILIC of intact mitochondrial membrane proteins. Proteins were identified by direct analysis by MS. Sample: Extract of bovine heart mitochondria. *Column*: PolyHYDROXYETHYL A (100x2.1-mm; 5-μm, 300-Å). *Gradient*: (**a**) 20 mM ammonium formate, pH 3.7, containing 0.5 % hexafluoro-2-propanol [a solubilizing agent], with 63 % 2-propanol + 22.5 % ACN; (**b**) Same but with no ACN and with 30 % 2-propanol (Adapted from Ref. [8])

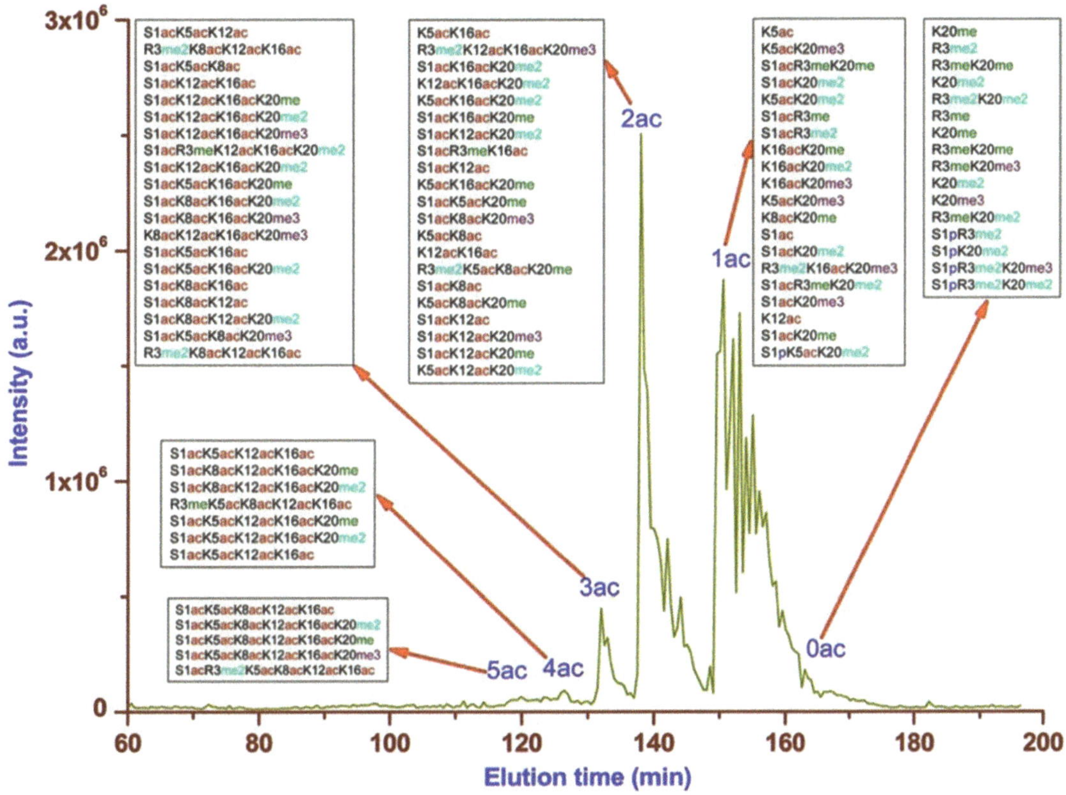

Fig. 2.10 Separation of Histone H4 isoforms by WCX-HILIC. *Column*: PolyCAT A capillary, 500 × 0.1 mm; 5-μm, 1000-Å. *Gradient*: 1–8 % formic acid in 70 % ACN (From Ref. [10])

strong and selective interaction with a specific protein or class of proteins. The selectivity can be quite high, as with an immobilized antibody or lectin, or more general, as with the interaction of an immobilized boronic acid group with glycoproteins that contain carbohydrate residues with cis-diol groups. One can obtain a high enrichment factor with an affinity column. This is helpful when the protein of interest is a minor component in a large volume. An alternative situation is one in which an affinity column is used to deplete a sample of the proteins of highest abundance in order to facilitate the identification of the remaining proteins. The Agilent MARS column has been widely used for this.

There are two main drawbacks to affinity chromatography:

1. An affinity column may not exist for a specific separation or purification of interest;
2. The interaction with the affinity ligand is so strong that it tends to dominate the chromatography.

Consequently, affinity columns are poor at separating the retained proteins from each other. Instead, they are used in a version of solid phase extraction, which separates a mixture into components that are retained and components that are not retained. The retained components can then be separated from each other with a more general mode of chromatography, such as IEX or RPC. An example of this is presented in Fig. 2.12.

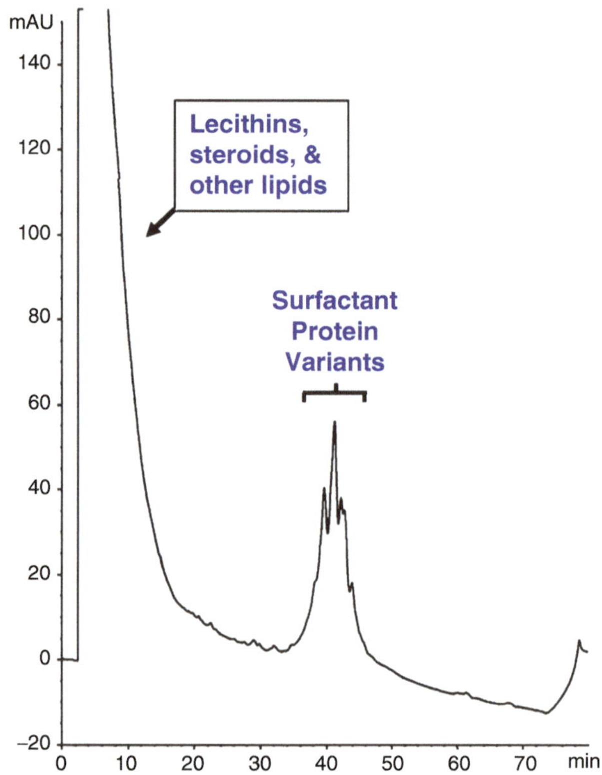

Fig. 2.11 SCX-HILIC of pulmonary surfactant protein (SP). Sample: Emulsion of 500 parts lipids (lecithins, steroids, etc.): 1 part bovine SP. The lipids eluted in the void volume. Some of the SP isn't soluble in water but was soluble in this mobile phase and eluted within the salt gradient. The retained peaks presumably correspond to the different SP proteins present *in vivo*. *Column*: PolySULFOETHYL A, 200 × 4.6-mm; 5-μm, 1000-Å. Mobile phase: (**a**) 0.1 % methylphosphonic acid + 5 mM $NaClO_4$, pH 3.0, with 70 % ACN; (**b**) Same but with 100 mM $NaClO_4$. *Gradient*: 5' hold, then 0–100 % B in 60'

2.3 Examples of Applications

2.3.1 Multi-Dimensional Chromatography for Top-Down Proteomics

When an affinity method does not exist for the isolation of a protein, then the alternative is to perform sample simplification: Distribution of the components of a mixture into subsets by collection of fractions from a column used in a general-purpose mode of chromatography. The protein of interest would then represent a greater percentage of the protein in the fraction in which it resides. Extracts of biological fluids generally contain so many proteins that no single method suffices for purification of an individual component. In such cases, then, each fraction from the first run is subdivided further on the basis of properties complementary to those that governed retention in the first run. Proteins that eluted together from an IEX column because they had the same charge will probably differ in polarity and can be separated by a HIC column, for example. Common sequences are IEX followed by HIC or the opposite sequence (HIC-IEX).

An example is shown in Fig. 2.13. There is some sense in using IEX as the first dimension since sample processing is minimized; one can simply add salt to the collected fractions and load them onto a HIC column directly with no need for desalting (although it may be beneficial to concentrate the sample). Fractions collected from the HIC column would probably have to be desalted prior to subsequent analysis by some other mode.

The more dimensions of fractionation used and the more fractions collected for each one, the fewer proteins there will be in each fraction. That facilitates their identification, especially the ones of low abundance. The drawback to this approach is that the number of fractions multiplies rapidly as one adds additional

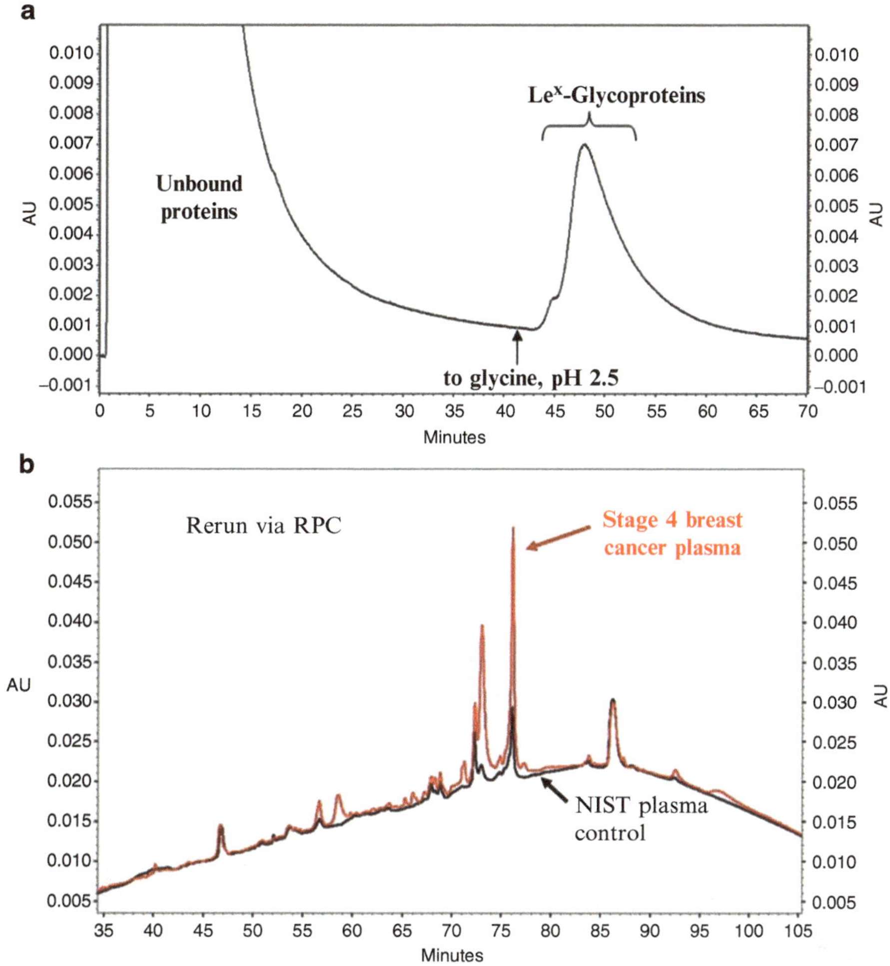

Fig. 2.12 *Top*: Affinity isolation of plasma glycoproteins containing the Lewis x antigen by a column with immobilized anti-Le^x antibody.
Bottom: Separation of the retained glycoproteins on a nonporous C-18 RPC column with an ACN gradient (Adapted from: W. Cho, K. Jung, and F.E. Regnier, *Anal. Chem. 80* (2008) 5286)

fractionation steps. There is a tradeoff, then, between how many proteins one wishes to isolate or identify and how much time and work will be involved. In the example in Fig. 2.13, for example, the addition of an extra dimension of separation increased the number of fractions to be processed 35 times while increasing the number of nonredundant proteins identified from 47 to 201, an increase by a factor of 4.3. Most of the additional identifications were of proteins of low abundance.

Even if one is performing bottom-up proteomics, preliminary fractionation of the intact proteins can generate a significant increase in identifications. The fractions collected from the separation in Fig. 2.3 were individually digested with trypsin and then further fractionated with an SCX-RPC sequence. This resulted in the identification of 3135 proteins. Omitting the protein fractionation step reduced the protein identifications to 1292. In this case, then, the extra step increased the number of fractions

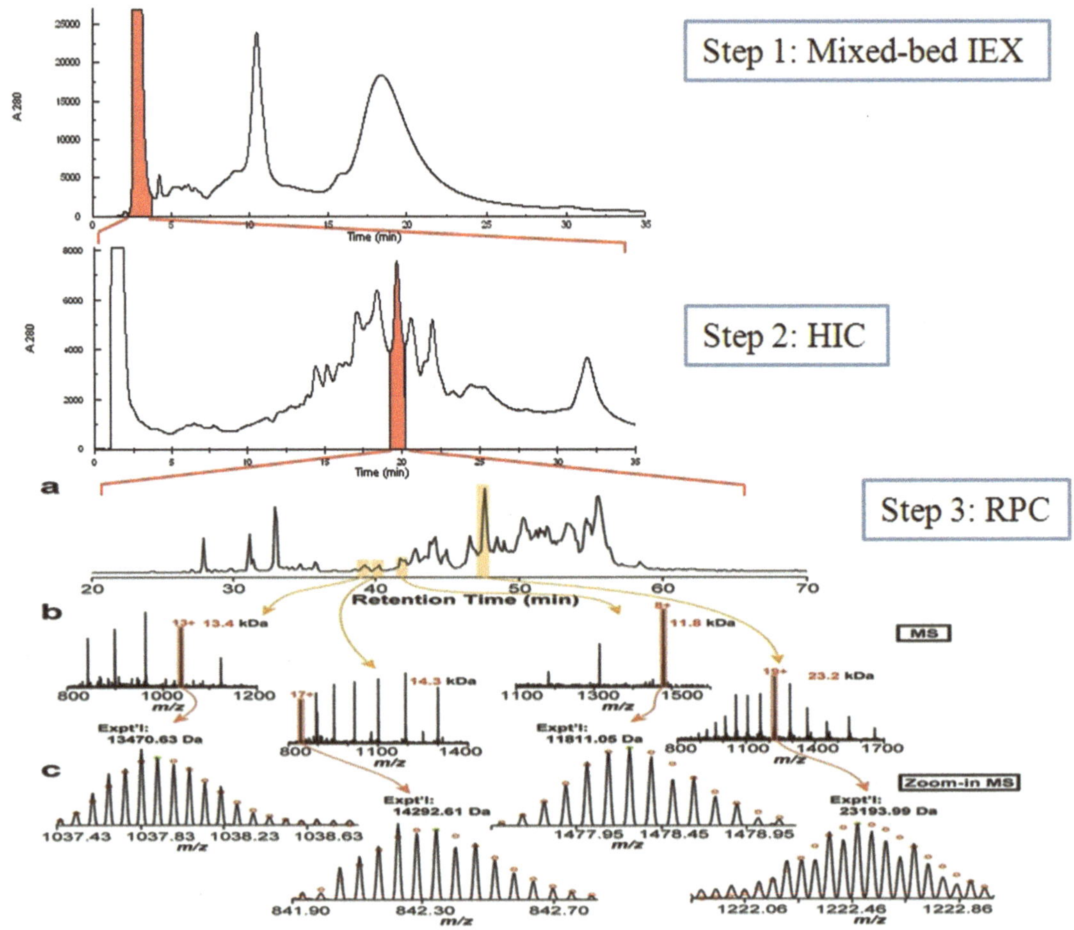

Fig. 2.13 An IEX-HIC-RPC sequence for 3-D fractionation of intact proteins for top-down proteomics. Sample: HEK 293 cell lysate. Step 1: Mixed-Bed IEX, with 35 1-min fractions collected. Fraction #3 [*colored*] was selected for further processing. Step 2: HIC. Again, 35 fractions were collected, and fraction #20 [*colored*] was selected for further processing. Step. 3: RPC. Proteins were eluted into a Q Exactive Orbitrap mass spectrometer. Representative mass spectra are shown for the indicated peaks, along with zoom-in spectra with unit mass isotopic resolution. Starting with all 35 HIC fractions from IEX fraction #3, 201 nonredundant proteins were identified (Adapted from: S.G. Valeja *et al.*, *Anal. Chem. 87* (2015) 5363)

12x while increasing the protein identifications 2.4x.

2.3.2 Location and Isolation of a Pure Protein from a Mixture

Isolation of an individual protein from a complex mixture usually requires several sequential steps. If no affinity method is available, then a succession of general-purpose chromatography steps is required. In general, three successive purification steps suffice for purification of proteins from complex mixtures, provided a suitable bioassay or location method is available. Success requires the ability to determine where in the eluate is the protein of interest. If the protein has unique absorption or fluorescence characteristics – for example, the absorption of light at 415 nm by proteins with a heme ring – then its location is clear. Otherwise, its presence must be ascertained by a bioassay, fraction by fraction. Figure 2.14 demonstrates one method to assess the location of a protein that binds a specific

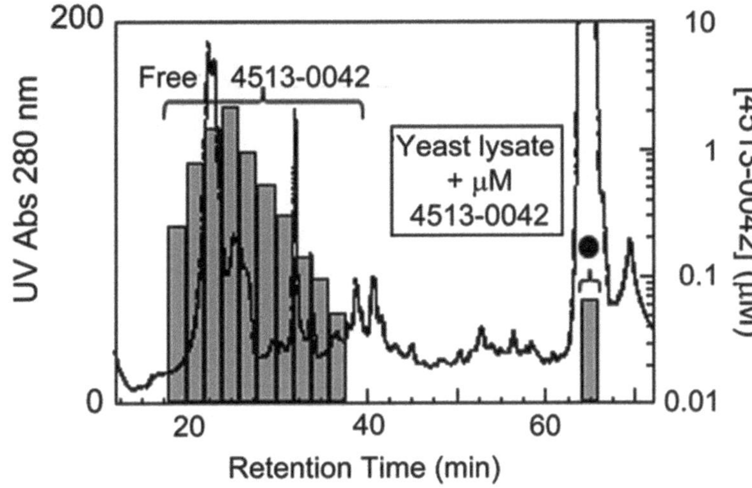

Fig. 2.14 Identification of a target protein in a lysate that binds a drug. An antifungal compound, 4513–0042, reportedly disrupts ergosterol synthesis. It was incubated with whole yeast extract and the proteins were then fractionated on a mixed-bed IEX column. Fractions were collected and analyzed via LC-MS for the presence of 4513–0042. The shaded bars on the left represent the elution position of free 4513–0042. The single shaded bar at 65' indicates the elution position of a probable complex of 4513–0042 and Erg6p, a yeast protein in the ergosterol pathway (Adapted from: J.N.Y. Chan et al., *Mol. Cell. Prot. 11* (2012) M111.016642)

drug. The drug is added to a mixture of proteins, which are then separated. The collected fractions are individually analyzed for the drug via mass spectrometry to locate the elution position of any complexes formed by the selective binding of the drug by specific proteins in the mixture.

An alternative approach is evident in Fig. 2.15. Here, the elution position of interleukin-6 (IL-6) from a mixed-bed IEX column was ascertained. The corresponding fraction in serum was collected and digested with trypsin. Peptides unique to IL-6 were measured via Selected Reaction Monitoring (SRM) mass spectrometry. The Limit of Detection (LOD) was 50 ng IL-6/ml serum. This concentration is too high for measurement of normal levels of IL-6 in serum but may be low enough for its measurement in cases of disease.

2.3.3 QC Analysis

In contrast to the situation described above, QC applications frequently involve a protein at a high state of purity. The objective is usually to separate closely-related variants of the protein from each other. Examples of different types of applications follow:

1. Assessing the degree of deamidation of a protein.

   Deamidation of susceptible asparagine (Asn) residues is the most significant nonenzymatic reaction affecting the shelf life of biologically active proteins. The consequences for biological activity of deamidation of any particular Asn residue range from trivial to crucial, depending on the protein and the location of the residue. Deamidation of Asn is promoted by the following factors:

   (a) Elevated temperature and pH
   (b) The presence of a sterically unhindered residue on the C-terminal side of the Asn in question.

   The kinetics of deamidation are fastest with an Asn-Gly sequence. Other sequences that are frequently involved in deamidation are Asn-Ala, Asn-Asp, and Asn-Ser. Nonenzymatic deamidation proceeds via loss of

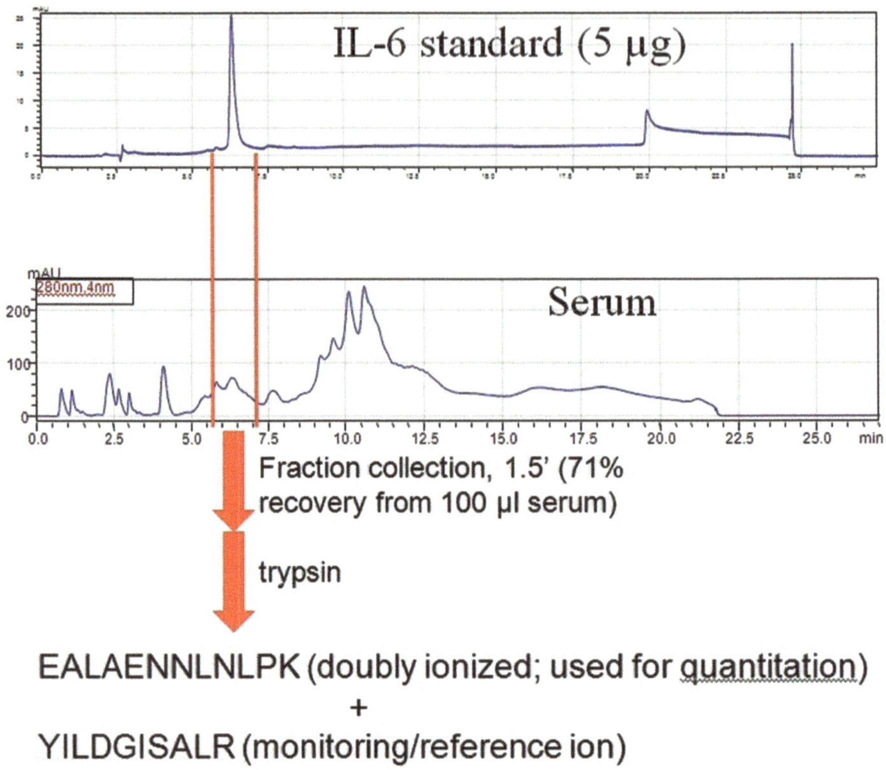

Fig. 2.15 Measurement of interleukin-6 (IL-6) in whole serum. *Top*: Elution of an IL-6 standard from a mixed-bed IEX column. *Bottom*: Fractionation of whole serum on the same column under the same conditions. The 1.5' fraction corresponding to the elution window of IL-6 was collected, digested with trypsin, and then analyzed via LC-MS for measurement of peptides unique to IL-6 (MRM) (Adapted from: L. Bian, M. Kukula, J. Barrera, and K.A. Schug, ASMS 2015 conference, poster Th 597)

ammonia from the Asn side-chain with subsequent formation of a succinimide ring. This ring can hydrolyze unsymmetrically to form either an n-aspartyl (n-Asp) residue or an isoaspartyl (isoAsp) residue, in a ratio between 1:2 and 1:3. Conversion of a neutral Asn residue to an Asp residue adds one additional (−) charge to the protein. Accordingly, IEX is a good way to separate the native protein from various deamidation variants.

A susceptible Asp residue can also form a succinimide ring, this time proceeding via dehydration rather than deamidation. Again, a sterically unhindered residue on the C-terminal side of the Asp residue tends to promote the reaction. In contrast with deamidation, though, dehydration of susceptible Asp residues is promoted by neutral or acidic pH. The products of hydrolysis of the ring are the same as with deamidation; n-Asp and isoAsp variants, one of which is identical to the starting protein. It is possible to separate even these closely-related variants. The pKa of an n-Asp residue is around 3.9, while the pKa of an isoAsp residue is about 3.1. Consequently, at pH 4.0, an n-Asp residue has lost about half of its (−) charge while an isoAsp residue retains most of its charge. This can cause a protein variant with an isoAsp residue to elute earlier from a cation-exchange column than the same protein with an n-Asp variant.

2. Assessing the position of derivatization. Some reactions are not limited to the target residue. An example is shown in Fig. 2.2. Here, a PEG (polyethylene glycol) chain was attached

covalently to the N-terminus of the protein. There were also significant side reactions with most of the lysine residues. A cation-exchange column was able to separate these positional variants because a lysine residue is a good binding site in cation-exchange. The column is sensitive to anything that affects that binding. Some lysine residues are more important than others to the overall binding, which accounts for the sensitivity to the position of the PEG's attachment.

3. Analysis of monoclonal antibodies. This is usually performed by cation-exchange. The heavy chains have a lysine or arginine as the C-terminal residue. A basic residue in a terminal position is readily available for interaction with a stationary phase, and so those residues play a significant role in retention. Loss of the basic residue from the end of one heavy chain causes the antibody to elute significantly sooner, and loss of the basic residues from the ends of both heavy chains leads to even earlier elution. Consequently, cation-exchange of monoclonal antibodies characteristically results in a pattern of three major peaks. The minor peaks eluting earlier than each of the major ones are generally deamidation variants. Some antibody producers treat their antibodies with carboxypeptidase B to cleave off the terminal basic residues. This does not affect the biological activity; the motive is solely to simplify the pattern in Quality Control analysis. Another concern is the degree of aggregation of antibodies. This is generally measured by Size Exclusion Chromatography.

Recently there has been considerable interest in the diagnostic and therapeutic potential of antibodies with covalently-attached drugs or toxins. These are called antibody-drug conjugates, or ADC's. Antibody molecules vary in the number and position of conjugate molecules attached. The product of the synthesis must be analyzed to ascertain the composition of the product in this regard. In Fig. 2.8, SEC with "native" MS analysis is used to determine the number of conjugates per molecule based on mass differences. Figure 2.16 shows the physical separation of ADC's via HIC.

2.3.4 Example of a Clinical Analysis: Hemoglobins

This may be the most widespread application in the world involving the analysis of a protein by HPLC. The analyses play a role in the control of two significant problems in public health:

(a) Glycated hemoglobin: Hemoglobin A1c (Hb A1c) has a residue of glucose covalently attached to the N-terminus of the

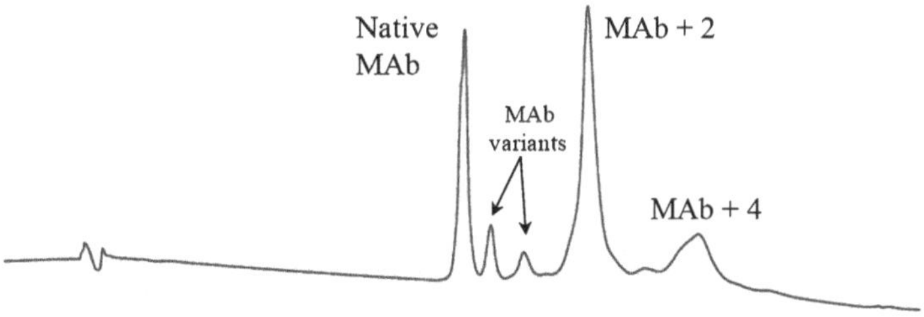

Fig. 2.16 Analysis via HIC of an antibody-drug conjugate (ADC). The two minor peaks eluting after the native antibody peak are variants that contain a single conjugate in different positions. The subsequent major peaks contain 2 or 4 molecules of the conjugate. *Column*: PolyPROPYL A, 100 × 4.6 mm (3-μm, 1500-Å). A decreasing gradient of ammonium sulfate was used

beta-chain. Its concentration is proportionate to the average glucose level in the blood during approximately a 1-month period. Such information is useful for diagnosis of diabetes and monitoring its treatment. About 4–5 % of the hemoglobin of a normal individual is in the form of Hb A1c. In a case of uncontrolled diabetes, the level can be as high as 15–16 %. There are a number of different assays for Hb A1c. The most common one that involves chromatography is to pass a sample through a column with an immobilized ligand of phenylboronic acid. Boronic acids form a covalent but transient 5-member ring with compounds containing *cis*-diol groups, including glucose residues attached to proteins. The resulting chromatogram features just two peaks: A major early peak consisting of the hemoglobins that lack sugar adducts and a minor peak, eluted with a step to lower pH, that causes the glycated hemoglobins to elute. The area under the two peaks is then integrated to determine the percentage of glycated hemoglobin. This method does not distinguish between glycation of hemoglobin at the N-terminus of the beta chain and the less frequent glycation at a lysine residue.

(b) Analysis of hemoglobin variants: Certain parts of the world have a significant occurrence of genetic mutants of hemoglobin in the local gene pool. The occurrence of such mutations tends to coincide with a high incidence of malaria; it is speculated that the carriers of the mutations are more resistant to the effects of the disease. However, people who are homozygous for these mutations suffer effects that shorten their lives significantly. This is a significant public health problem in such countries. In subsaharan Africa the major variants, Hb S and Hb C, cause sickle cell anemia. A similar syndrome occurs in the India-Pakistan area (Hb D) and in southeast Asia (Hb E). In the Mediterranean basin and a belt across the Middle East through Iran, the major hemoglobinopathy is beta-thalassemia, which is diagnosed via an elevation in the percentage of hemoglobin A_2 (Hb A_2).

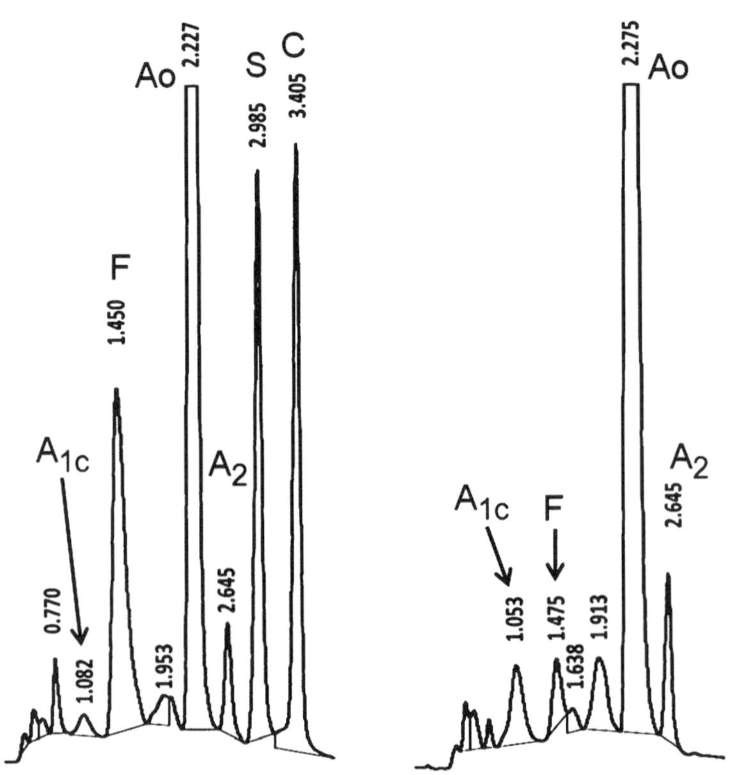

Fig. 2.17 Analysis of hemoglobins via cation-exchange. *Left*: A composite standard, including the S and C variants associated with sickle cell anemia. *Right*: A clinical sample from an individual with an elevated level of hemoglobin A_2. All of these variants, including hemoglobins A1c and F, are completely separated in less than 3.5'. *Column*: PolyCAT A, 35 × 4.6-mm; 3-μm, 1500-Å. *Gradient*: An increasing NaCl gradient in a Bis-tris buffer that contains 2 mM NaCN

Erythrocytes are isolated by centrifugation of a blood sample and then lysed. The resulting solution, a hemolyzate, can be analyzed directly. An even simpler method involves blotting a drop of blood on filter paper, punching out the blot, and solubilizing and analyzing the hemoglobins. There are various tests for the variants of interest, but the most widely employed is HPLC separation via cation-exchange. Hemoglobin has an absorption maximum at 415 nm, which makes it convenient to analyze with a minimum of sample processing. Figure 2.17 shows some examples.

2.3.5 Alternatives to RPC for Direct LC-MS

Examples of SEC-MS were presented in the section on SEC, and the section on HILIC has an example of HILIC-MS of some membrane proteins.

One of the more widely used alternatives is WCX-HILIC. This is the use of a weak cation-exchange column with a gradient to a pH low enough to uncharge the carboxyl- groups. While this can be performed in strictly aqueous media, the most popular combination starts with a concentration of acetonitrile in the range 60–70 %. This superimposes a significant degree of hydrophilic interaction on the electrostatic effects, promoting the retention of proteins with a net charge of either sign. Along with the decreasing pH gradient, then, there is a gradient of decreasing ACN concentration, which tunes down the hydrophilic interaction. The eluting proteins are readily analyzed directly via mass spectrometry. This combination is widely used for analysis of histones, both "top-down" [10] and "middle-down" [11, 12]. Figure 2.10 [*above*] shows an example of this.

2.4 Summary

The ability of bottom-up proteomics to identify more than 30–40 peptides was made possible by increasing the degree of separation prior to the mass spectrometer. At present the separation methods are a major bottleneck in the development of top-down mass spectrometry of proteins. Given the examples described above, there is reason to be optimistic that appropriate methods will be forthcoming and that progress will then depend on advances in the mass spectrometry instrumentation.

References

1. Wang H, Qian W-J, Chin MH, Petyuk VA, Barry RC, Liu T, Gritsenko MA, Mottaz MA, Moore RJ, Camp DG II, Khan AH, Smith DJ, Smith RD (2006) J Proteome Res 5:361
2. Alpert AJ, Regnier FE (1979) J Chromatogr 185:375
3. Zhang L, Yao L, Zhang Y, Xue T, Dai G, Chen K, Hu X, Xu LX (2012) J Chromatogr B 905:96
4. Compton PD, Zamdborg L, Thomas PM, Kelleher NL (2011) Anal Chem 83:6868
5. Zhang J, Roth MJ, Chang AN, Plymire DA, Corbett JR, Greenberg BM, Patrie SM (2013) Anal Chem 85:10377
6. Vellaichamy A, Tran JC, Catherman AD, Lee JE, Kellie JF, Sweet SMM, Zamdborg J, Thomas PM, Ahlf DR, Durbin KR, Valaskovic GA, Kelleher NL (2010) Anal Chem 82:1234
7. Jenö P, Scherer PE, Manningkrieg U, Horst M (1993) Anal Biochem 21:292
8. Carroll J, Fearnley IM, Walker JE (2006) Proc Natl Acad Sci U S A 103:16170
9. Tetaz T, Detzner S, Friedlein A, Molitor B, Mary J-L (2011) J Chromatogr A 1218:5892
10. Tian Z, Tolić N, Zhao R, Moore RJ, Hengel SM, Robinson EW, Stenoien DL, Wu S, Smith RD, Paša-Tolić L (2012) Genome Biol 13:R86
11. Young NL, DiMaggio PA, Plazas-Mayorca MD, Baliban RC, Floudas CA, Garcia BA (2009) Mol Cell Prot 8:2266
12. Sidoli S, Lin S, Karch KR, Garcia BA (2015) Anal Chem 87:3129

Sample Preparation for Mass Spectrometry-Based Proteomics; from Proteomes to Peptides

3

John C. Rogers and Ryan D. Bomgarden

Abstract

Mass spectrometry (MS) has become the predominant technology to analyze proteins due to it ability to identify and characterize proteins and their modifications with high sensitivity and selectivity (Aebersold and Mann, Nature 422(6928):198–207, 2003; Han et al., Curr Opin Chem Biol 12(5):483–490, 2008). While mass spectrometry instruments have improved rapidly over the past couple of decades, mass spectrometry results have remained largely dependent on sample preparation and quality. Sample ionization and mass measurements are susceptible to a wide variety of interferences, including buffers, salts, polymers, and detergents. These contaminants also impair MS system performance, often requiring time consuming maintenance or costly repairs to restore function. The goal of this chapter is to describe the rationale, considerations, and general techniques used to prepare samples for proteomic mass spectrometry analysis.

Keywords

Protein chromatography • Protein extraction • Lysis • Protein depletion or enrichment • Digestion • In-gel digestion • In-solution digestion • Filter-assisted sample preparation (FASP) • Digestion comparison

3.1 Overview

Due to the complexity of proteomic samples and the wide variety of sample preparation techniques, a proteomics researcher must first determine the right experimental strategy. A successful proteomics experiment requires the integration of good sample preparation, instrumentation, and software (Fig. 3.1). Therefore, it is important to understand the goals and expectations of the project and to choose and optimize the best sample preparation method accordingly. For example, the sample preparation requirements for protein identification from

J.C. Rogers (✉) • R.D. Bomgarden
Thermo Fisher Scientific, Rockford, IL, USA
e-mail: john.rogers@thermofisher.com

H. Mirzaei and M. Carrasco (eds.), *Modern Proteomics – Sample Preparation, Analysis and Practical Applications*, Advances in Experimental Medicine and Biology 919, DOI 10.1007/978-3-319-41448-5_3

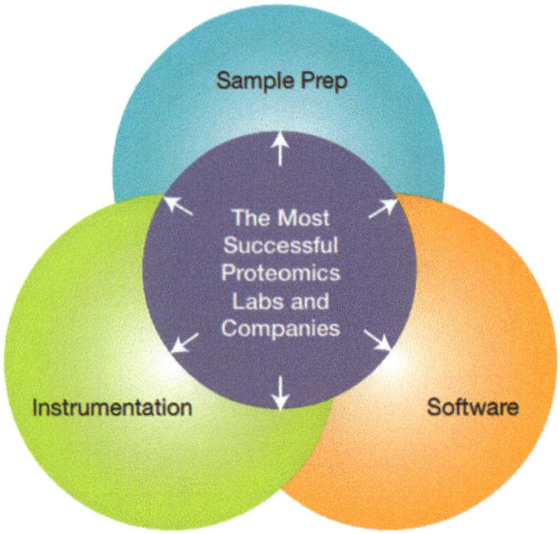

Fig. 3.1 The key to proteomics success. Successful proteomics laboratories and companies recognize the importance of sophisticated sample preparation, instrumentation, and software technologies and skills. Workflows designed to maximize the overlap between these complementary technologies are an effective means of improving proteomics research

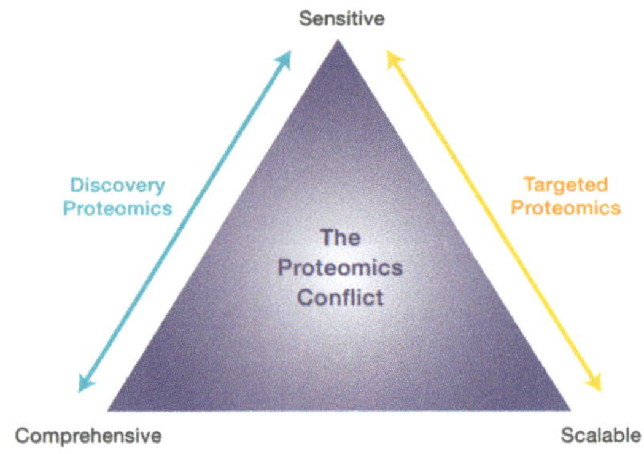

Fig. 3.2 The proteomics conflict. It is impossible to optimize sensitivity, throughput and comprehensiveness simultaneously. Discovery proteomics strategies optimize sensitivity and comprehensiveness with few samples. Targeted proteomics strategies optimize sensitivity and scalability by limiting the number of monitored features. Note that comprehensive analysis with reasonable throughput is enabled by sample multiplexing with mass tag reagents

a gel slice are very different from the requirements to identify protein interaction networks, measure changes in the mitochondrial proteome, understand protein phosphorylation and signaling in cancer, or identify protein biomarkers of cancer metastasis in plasma [3–6]. Unlike genomic or transcriptomic research, there is no "standard" universal sample preparation method for proteomics.

Additionally, proteomics experiments must balance the competing needs for sensitive and complete proteome coverage with the scalability of analyses (Fig. 3.2). Proteomic strategies to improve proteome coverage require multidimensional fractionation; however, this fractionation increases the sample analysis time and sacrifices throughput [7, 8]. Alternatively, MS acquisition strategies that improve the sensitivity, reproducibility, and throughput of protein quantification, such as selected reaction monitoring (SRM) or parallel reaction monitoring (PRM), limit the number of features that can be monitored [9, 10]. For this reason, proteomics research is generally divided into three categories: protein identification and characterization, proteome profiling, and targeted protein analysis.

Protein identification and characterization is commonly performed to identify protein isoforms, splice variants, post-translational modifications, and interacting proteins [11]. These studies are typically performed after protein separation using SDS polyacrylamide gel electrophoresis (SDS-PAGE) and may also involve a protein enrichment step, such as immunoprecipitation. In contrast, proteomic profiling is typically performed on whole protein or sub-proteome extracts digested in solution. This comprehensive approach requires more instrument analysis time per sample to maximize the number of protein identifications at the expense of the number of samples that can be analyzed. Isobaric mass tags (e.g. iTRAQ and TMT) can help to address this sample throughput limitation by allowing multiple samples to be combined into a single LC-MS analysis [12–14]. Targeted protein analysis limits the number of features that are monitored to a pre-selected list of target peptides and their transitions. These methods optimize sample preparation, chromatography, instrument tuning, and fragmentation to achieve the highest sensitivity and throughput for hundreds of samples. Ultimately, a sample preparation strategy should be chosen which generates the most biologically relevant or useful data possible for a given experiment.

Protein analysis using tandem mass spectrometry (MS/MS, or MS^n) can be performed on intact proteins ("top-down" proteomics) or protein digests ("bottom-up" proteomics). Top-down proteomics is a growing field, as it permits nearly complete protein sequence coverage and enables simultaneous characterization of protein isoforms and modifications [15, 16]. However, top-down analysis is currently limited to proteins less than ~50,000 Da and requires high resolution MS instrumentation (>100,000 resolving power) to accurately identify proteins and protein isoforms. Recently, "middle-down" strategies have also been developed to reduce the sizes of intact proteins through partial digestion or using proteases that cleave at rare sites or at specific positions within a protein (e.g. antibodies, [17, 18]). Sample preparation for intact proteins typically involves multi-dimensional protein fractionation to reduce sample complexity and protein desalting to remove residual salts or other impurities that may form adducts during ionization.

Bottom-up proteomic strategies represent the vast majority of MS proteomic analyses. These methods use proteases to digest proteins at specific amino acids into peptides with a predictable terminus. Unlike proteins, peptides are more easily separated by reverse phase HPLC and ionize well by electrospray or matrix-assisted laser desorption ionization (MALDI). Importantly, peptides fragment during MS/MS to yield amino acid sequence information. Similar to proteins, multi-dimensional fractionation of peptides can be used to reduce sample complexity [19] but removal of salts, detergents and other impurities can be more difficult at the peptide level than the protein level. As peptide fractionation, liquid chromatography (LC), and MS analysis are addressed in other chapters, this chapter will primarily focus on bottom-up protein sample preparation strategies prior to LC-MS/MS analysis.

The quality and consistency of sample preparation influences the time and cost of MS analysis and the reliability of the results. For MS-based proteomics to reach its full potential as a routinely used detection technology in research and clinical settings, variability associated with the sample preparation steps that precede MS analysis must be addressed. Despite extensive literature describing various MS sample preparation methods explained below and elsewhere, there is little standardization among methods. This results in confusion for those new to MS sample preparation techniques and high variability in MS analysis results, even among expert MS laboratories.

3.2 Protein Extraction

Tissue or cell lysis is the first step in protein extraction and solubilization. Numerous techniques have been developed to obtain the highest protein yield for different organisms, sample types, subcellular fractions, or specific proteins. Due to the diversity of tissue and cell types, both physical disruption and reagent-based methods are often required to extract cellular proteins. Physical lysis equipment, such as homogenizers, bead beaters, and sonicators, are commonly used to disrupt tissues or cells in order to extract cellular contents and shear DNA. In contrast, reagent-based methods use denaturants or detergents to lyse cells and solubilize proteins. Cell lysis also liberates proteases and other catabolic enzymes so broad-spectrum protease and phosphatase inhibitor cocktails are typically included during sample preparation to prevent nonspecific proteolysis and loss of protein phosphorylation, respectively.

Through the use of different buffers, detergents and salts, cell lysis protocols can be optimized for the best protein extraction for a particular sample or protein fraction. Strong denaturants (e.g. urea or guanidine) and ionic detergents (e.g. sodium dodecyl sulfate (SDS) or deoxycholate (SDC)) solubilize membrane proteins and denature proteins. Non-ionic or zwitterionic detergents (e.g. Triton X-100, NP-40, digitonin, or CHAPS) have a lower critical micelle concentration and require lower detergent concentrations to solubilize proteins [20, 21]. These detergents generally solubilize membrane proteins and protein complexes with less denaturation and disruption of protein-protein interactions [21].

Unfortunately, many detergents used to solubilize proteins cause significant problems during downstream mass spectrometry analysis if they are not completely removed. In addition to cell lysis buffers, detergents used to clean laboratory glassware may also contaminate samples and LC solvents. Detergents present in the sample can:

1. Contaminate and foul autosampler needles, valves, connectors, and lines
2. Affect liquid chromatography by reducing column capacity and performance
3. Affect crystallization prior to matrix assisted laser desorption ionization (MALDI) sample analysis;
4. Suppress electrospray ionization (ESI) prior to introduction into the mass spectrometer
5. Deposit in the mass spectrometer, interfering with the spectra and reducing sensitivity of the instrument.

Flexible tubing or poor quality plastic consumables can also leach phthalates and other contaminants that can interfere with downstream LC-MS analysis [22]. Both phthalates and detergents ionize very well and overwhelm peptide signals. Polydisperse detergents, such as Triton X-100, Tween or NP-40, contain a distribution of variable length polyethylene glycol (PEG) chains that often elute throughout the LC gradient as a family of peaks separated by 44 Da mass units and overwhelm the LC-MS results. Fortunately, these leachables and detergents can often be removed by gel electrophoresis, protein precipitation, or filter-assisted sample preparation (FASP) techniques described later in this chapter.

While all detergents can affect downstream LC-MS analysis, N-octyl-beta-glucoside and octylthioglucoside are considered more compatible with mass spectrometry because they are dialyzable and monodisperse (i.e. homogeneous) [23]. In addition, a variety of mass spectrometry-compatible detergents are commercially available. Invitrosol (Thermo Scientific) contains several monodisperse detergents that elute in regions of the HPLC gradient that do not interfere with peptides or their chromatography. Cleavable detergents, such as ProteaseMax (Promega), Rapigest (Waters), PPS Silent Surfactant (Expedeon), or Progenta (Protea), degrade with heat or at low pH into products that do not interfere with LC-MS. As digestion requires incubation at 37 °C and LC-MS loading buffers contain formic acid or trifluoroacetic acid, sample preparation workflows do not require any significant modification to use these MS-compatible detergents [24].

3.3 Protein Depletion or Enrichment

Depending on the protein source and the copy number per cell, there can be a tremendous difference in the concentration between the lowest and most abundant proteins. For mammalian tissues and cell lines, protein expression can range over 6–9 orders of magnitude. For serum and plasma samples, the dynamic range can be greater than 12 orders of magnitude with serum albumin representing over 50 % of the protein content [25]. In order to get an adequate depth of protein coverage in serum, to identify relevant biomarkers, abundant protein depletion is required. Although affinity chromatography using Cibacron blue dye can be used to remove albumin, immunoaffinity using antibodies is typically required to remove other abundant proteins such as immunoglobulins, transferrin, fibrinogen, and apo-lipoproteins [26]. One advantage of using antibodies for immunodepletion is that one sample preparation technique can be used to remove the top 2–20 most abundant proteins depending on the product used. Another is that the depletion resins can be regenerated for multiple uses; though this can affect protein depletion reproducibility over time.

Protein enrichment techniques are commonly overlooked during protein sample preparation but may be necessary in order to identify and quantify biologically relevant proteins which are typically in lower abundance. One method of protein enrichment is subcellular fractionation, which separates proteins by location in a particular cellular compartment or organelle. Subcellular fractionation using sucrose density gradient centrifugation can separate vesicles and organelles including the nucleus, mitochondria, or chloroplasts from cytosolic and vesicle proteins [27, 28]. Differential extraction is another subcellular fractionation technique which uses detergents to selectively solubilize nuclear, chromatin-bound, membrane, cytosolic, and cytoskeletal proteins [29]. Another method of protein enrichment is through protein modifications. Cell surface proteins which are glycosylated can be enriched by chemical labeling of oxidized glycans, metabolic incorporation of azide-containing sugars [30–32], or lectin affinity [33]. Phosphoproteins can be enriched with immobilized metal affinity chromatography [34]. Activity-based chemical probes are another method for enrichment of enzyme subclasses such as kinases, hydrolases, and oxidases [35, 36]. Finally, affinity capture using immunoprecipitation is the method of choice for enrichment of specific protein targets or protein complexes as this technique provides the highest selectivity and sensitivity for the lowest abundant proteins [37].

3.4 Protein Preparation

Unfortunately, many protein extraction, fractionation, enrichment and depletion methods introduce salts, buffers, detergents, and other contaminants which are not MS compatible. Because of the relative difference in molecular weight, it is simplest and preferable to remove these small molecule contaminants before protein digestion. There are a variety of options to remove these small molecules, including gel electrophoresis, chromatography, dialysis, buffer exchange, size exclusion, and protein precipitation [38, 39]. Gel electrophoresis is an inexpensive, straightforward method for the removal of salts, detergents, and other small molecules prior to in-gel digestion. However, keratins from skin and dust are common contaminants which can be introduced when pouring and handling gels so it is imperative to always wear gloves and to use MS grade reagents to minimize this contamination.

Reverse phase C4 or C8 cartridges can remove salts from proteins but concentrate non-ionic detergents and may have poor recovery of hydrophilic proteins. Strong cation exchange resins can remove anionic detergents, like deoxycholate or sodium dodecyl sulfate (SDS), but typically require salts for protein elution which then have to be removed before LC-MS analysis. Dialysis membranes and cassettes are available with a variety of molecular weight cut-offs (MWCO) and can effectively exchange buffer components to remove contaminants; but dialysis is relatively slow, requires multiple buffer changes, and may be difficult with small volumes. Spin columns or stirred-cell pressure devices with MWCO membranes can rapidly exchange buffers to remove small molecule contaminants and concentrate samples. These MWCO devices allow sequential buffer exchange steps to be performed and can be used for complete MS sample preparation in the filter-assisted sample preparation (FASP) methods. Size exclusion resins retain small molecules in porous beads while excluding proteins enabling rapid and efficient buffer exchange with minimal sample loss, especially in a spin column format. Notably, of all of the desalting methods available, precipitation with organic solvents such as acetone or methanol/chloroform with or without organic acids (e.g. TCA or TFA) is the most common method for desalting proteins prior to MS sample preparation as it the least expensive, simplest and most scalable option.

3.5 Protein Digestion

Trypsin is the most commonly used protease for MS sample preparation because of its high activity, selectivity and relatively low cost. Trypsin cleaves proteins to generate peptides with a lysine or arginine residue at the carboxy terminus [40]. These basic amino acids at the end of every tryptic peptide improve peptide ionization and MS/MS fragmentation for peptide identification. Although trypsin is the most popular enzyme used for protein digestion, some protein sequences are not efficiently cleaved by trypsin or do not contain basic amino acids spaced close or far enough apart to generate peptides which can be used for protein identification. Trypsin digestion is less efficient at lysine and arginine residues followed by proline, repeated basic residues (e.g. KK, RK), or in the presence of post translational modifications (e.g. methylation, acetylation), resulting in missed cleavages [41]. Some tryptic peptides may be too small to retain on reversed phase LC columns or are not unique for a particular protein. Others may be too large and hydrophobic to identify by LC-MS. For example, 56 % of the tryptic peptides in yeast are ≤6 amino acids long, while 97 % of peptides identified by LC-MS are 7–35 amino acids [42]. These short or extremely long unidentified peptides result in incomplete protein sequence coverage, resulting in missing specific peptide sequences or sites of posttranslational modifications.

For more comprehensive proteome coverage, alternative proteases are often used to generate different peptide sequences that may not be identified from tryptic digests. Partial digestion with specific or non-selective proteases, like elastase or proteinase K, have been used to increase protein sequence coverage; but these proteases also increase the complexity and variability of digestion, making it more difficult to reproducibly identify the same peptides and proteins in replicate samples [43, 44]. Proteases with distinct cleavage specificities, such as ArgC, AspN, chymotrypsin, GluC, LysC, or LysN, produce complementary sequence information which can be combined to improve sequence coverage. This multi-enzyme approach has been used successfully by multiple laboratories to increase the number of protein identifications >10–15 % and improve the average sequence coverage by 60–160 % [42, 45–47]. Different proteases have also been shown to provide a unique repertoire of phosphopeptides which are not observed in tryptic digests [48]. Therefore, a multiple enzyme strategy is recommended for comprehensive analysis of single proteins or complex proteomes.

Multiple studies have demonstrated that chaotropes, solvents and detergents increase the efficiency of protein digestion [49, 50]. These reagents assist in the solubilization and unfolding of proteins, especially integral and transmembrane proteins or hydrophobic stretches of protein sequence. Efficient digestion is important to maximize the number of peptides and proteins identified in a sample, and complete digestion permits the reproducible quantitation of peptides. Organic solvent additives, such as 5–20 % acetonitrile (ACN), trifluoroethanol, and methanol have been shown to improve digestion efficiency and only require vacuum centrifugation or dilution to be compatible with LC-MS analysis. Urea and guanidine chaotropes also improve protein solubilization and digestion efficiency. These salts are easily removed from proteins by desalting on dialysis, or from peptides by using reverse phase C18 tips, cartridges, or trap columns. However, urea can modify lysine residues, resulting in carbamylation artifacts [51] and some proteases are not active in guanidine. Finally, some detergents which are used for protein extraction have also been shown to aid protein digestion. Depending on the detergent, these reagents can be removed after digestion by phase transfer, detergent removal resins, or hydrolysis with low pH [24, 50, 52]. Interestingly, it is reported that a combination of 1 M guanidine and 20 % ACN with any MS compatible detergent greatly improves the digestion efficiency and specificity over any one of these additives alone [24]. While the effects of solvents, chaotropes, and detergents have been well studied for trypsin digestion, and to a lesser extent for LysC digestion, the effects of these additives on other proteases are not well understood.

3.6 Peptide Preparation: In-Gel Digestion

Once the proteins in a complex sample are solubilized, there are three general approaches to prepare protein digests: in-gel digestion, in-solution digestion, and filter-assisted sample preparation (Fig. 3.3). All three of these methods remove contaminating detergents and other small molecules, reduce and alkylate proteins, digest proteins to peptides, and prepare peptides for mass spectrometry analysis. Sodium dodecyl sulfate-polyacrylamide gel electrophoresis (SDS-PAGE) is the most common technique for protein analysis [39, 53]. Gel electrophoresis is a simple, inexpensive and a relatively high resolution protein separation method that can be employed in either one dimension (1D) to resolve proteins by molecular weight or two dimensions (2D) to resolve proteins by isoelectric point and molecular weight [54]. Although 2D PAGE is not compatible with salts and ionic detergents, 1D SDS-PAGE can easily remove these and other substances which may interfere with LC-MS analysis. In fact, many academic proteomic core labs prefer or require samples to be provided in gels or gel slices because this method is so effective for sample clean up. Depending on the depth of analysis, a single band can be excised or a complex sample can then be excised as a set of gel slices in a method often referred to as GeLC-MS [55]. Another advantage of gel-based fractionation methods is that they can reduce sample complexity and separate highly abundant proteins from lower abundant proteins. Since all of the peptides from the respective protein(s) are contained in a single gel band, spot or fraction, protein sequence coverage and posttranslational modification mapping is also improved.

After gel electrophoresis, separated proteins are detected and visualized with a variety of gel stains, including Coomassie Blue, Colloidal Coomassie, and glutaraldehyde-free silver stain. Gel bands containing protein(s) of interest are then excised, destained, reduced, and alkylated to improve digestion and peptide extraction [39]. Disulfide bonds prevent complete protein unfolding and limit proteolytic digestion. Peptides that remain linked by disulfides are also difficult to identify due to the complexity of the peptide fragment ion spectra. Protein disulfides are typically reduced with either dithiothreitol (DTT) or tris 2-carboxyethylphosphine (TCEP) in the presence of other denaturants (i.e. heat, SDS, urea, guanidine, etc.). Reduced cysteines are then alkylated with iodoacetamide, iodoacetic acid, chloroacetamide, 4-vinyl pyridine, or N-ethyl maleimide (NEM) to prevent oxidation [56–58]. Haloacetyl-containing alkylating agents are light sensitive and must be made fresh. Alkylation reactions should be performed at pH 8.0 to avoid alkylation at other amino acids, and excess reagent should be quenched with DTT to prevent side reactions and over-alkylation of proteins. After reduction and alkylation, gel bands are digested with a protease; and the peptides are extracted using standard techniques [39]. While in-gel digestion is more prone to incomplete or less reproducible digestion and lower recovery of peptides relative to in-solution option (50–70 % recovery), gel electrophoresis remains an important sample preparation technique prior to MS analysis (Fig. 3.3, and Supplement Method 1).

3.7 Peptide Preparation: In-Solution Digestion

In-solution digestion is a popular alternative to in-gel digestion, because it requires fewer steps and can be scaled for the analysis of samples containing less than 10 μg or greater than 1 mg of protein. For this method, proteins are first denatured with detergents and heat or with urea or guanidine chaotropes. Disulfide bonds between cysteine residues are reduced and alkylated and then sample contaminants are typically removed by precipitation prior to digestion and cleanup. As stated above, urea has been used for many years but is not recommended because it must be made fresh as the formation of isocyanic acid over time increases the likelihood of protein carbamylation [51]. Protein solubilization and denaturation with

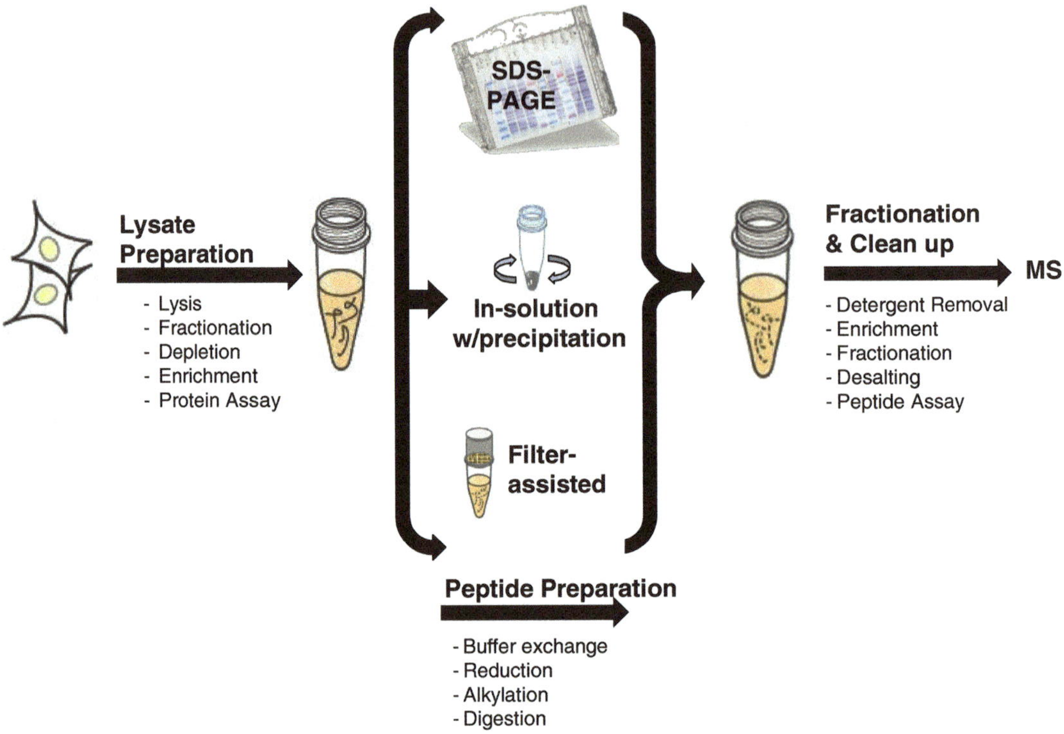

Fig. 3.3 General protein sample preparation workflow. There are many options for the extraction of proteins from tissue and cell lysates, protein fractionation and enrichment, and digestion to peptides for MS analysis

SDS or SDC is more effective than urea, and these detergents permit heating during the reduction of disulfides improving protein denaturation before digestion.

Once disulfides have been reduced and alkylated, contaminating salts, reducing and alkylating reagents, detergents, and small molecule metabolites present in the sample matrix should be removed from the sample before digestion. Depending on the sample source and extraction technique, small molecule contaminants may include excess protein labeling reagents, lipids, nucleotides, and phosphoryl- or amine-containing metabolites (e.g. phosphocholine, aminoglycans, etc.) that could interfere with downstream peptide enrichment or chemical tagging [59]. These contaminants can be removed by buffer exchange using gel filtration resins, dialysis, gel electrophoresis, filtration with a molecular weight cutoff filter, or most commonly by precipitation with an acid or an organic solvent [59–69]. Polydisperse detergents must be removed prior to digestion in order to prevent downstream contamination of LC-MS equipment. Most detergents can be removed by protein precipitation with four volumes of cold (−20 °C) acetone. Precipitation with dilute deoxycholate and trichloroacetic acid, methanol, a 4:1:3 ratio of methanol:chloroform: water, followed by an additional three volumes of methanol, or partitioning with ethyl acetate are alternative methods of detergent removal [65, 67–69]. As an alternative workflow, digestion can be performed in 0.1 % SDS or SDC, and these detergents may be removed from the peptides after digestion using a detergent removal spin column or by acidification to precipitate SDC [52, 70, 71].

Detergents, chaotropes, and organic solvent additives improve trypsin digestion efficiency and dramatically increase peptide and protein identifications in complex protein mixtures [49, 52, 71]. For tryptic digestion, the protein is dissolved in a buffered solution at pH 8.0 (e.g. 50–100 mM ammonium bicarbonate), and digestion is performed for 4–16 h at 37 °C with agitation. Low concentrations of acetonitrile, urea, SDS, SDC, or MS-compatible detergents may be included to solubilize the precipitated protein pellets and partially denature the protein to improve digestion efficiency. Endoproteinase LysC is an enzyme which cleaves after lysines similar to trypsin. Unlike trypsin, LysC can cleave at lysine residues followed by proline and is active under denaturing conditions (e.g. 8 M urea). LysC digestion is often performed for 1–4 h before tryptic digestion for more complete and reproducible digestion [72]. After digestion, peptides may be desalted off-line using reverse phase solid phase extraction cartridges, tips, or on-line using a trap column before MS analysis, as described in another chapter of this book.

3.8 Peptide Preparation: Filter-Assisted Sample Preparation (FASP)

Molecular weight cutoff (MWCO) filters have been used for decades to concentrate and exchange buffers for protein samples. Protocols for protein sample preparation with MWCO filters prior to MS were introduced in 2005 by Manza et al., and improved upon in 2009 and over subsequent years by the laboratory of Matthias Mann [63, 73, 74]. Filter-assisted sample preparation (FASP) utilizes SDS, heat, and urea to solubilize and denature proteins before transfer to a MWCO spin column which is used for protein collection, concentration, and digestion. An advantage of FASP is that detergents, salts, and small molecules can be easily removed through multiple rounds of washing. Concentrated proteins are then alkylated, washed and digested on the membrane before elution and desalting. FASP is compatible with a wide variety of samples and has been applied to 0.2–200 μg protein samples in a wide variety of applications, including brain tissue samples, formalin fixed paraffin embedded slices, *C. elegans*, phosphoproteomic, and glycoproteomic samples [73, 75–77]. Recently some proposed enhancements to the FASP protocol have been reported including: 1) simultaneous reduction and alkylation to eliminate several centrifugation steps and improve alkylation specificity; 2) prior passivation of the MWCO membrane with Tween-20 for higher peptide recovery, and; 3) the replacement of urea with deoxycholate for improved tryptic digestion [78].

3.9 Peptide Preparation Comparison

As described previously, many proteomic sample preparation methods have been described in the literature (Figs. 3.3 and 3.4), and these methods are modified further by members of the same lab or by other laboratories. This makes it extremely difficult for new MS users to identify the best protocol and generate consistent results. Each of these protocols described here has advantages and disadvantages. GeLC-MS simplifies protein fractionation and maintains peptides from the proteins from a gel band in a single fraction, but it is limited by scale, protein digestion efficiency, and peptide recovery. In-solution digestion with urea can carbamylate lysine residues, requires desalting to remove urea after digestion, and can suffer from poor protein extraction recovery without detergents. FASP is compatible with a wide variety of samples but requires many centrifugation steps, resulting in low sample processing throughput. Finally, digestion in the presence of detergent and subsequent removal of the detergent with a resin, precipitation, or phase transfer extraction may not be scalable or reproducible. Since sample preparation is the most problematic area of MS-based proteome analysis, it is important to have robust, reproducible methods that can be easily adopted by novice and expert MS labs alike.

We have compared the sample preparation results from FASP and three solution-based

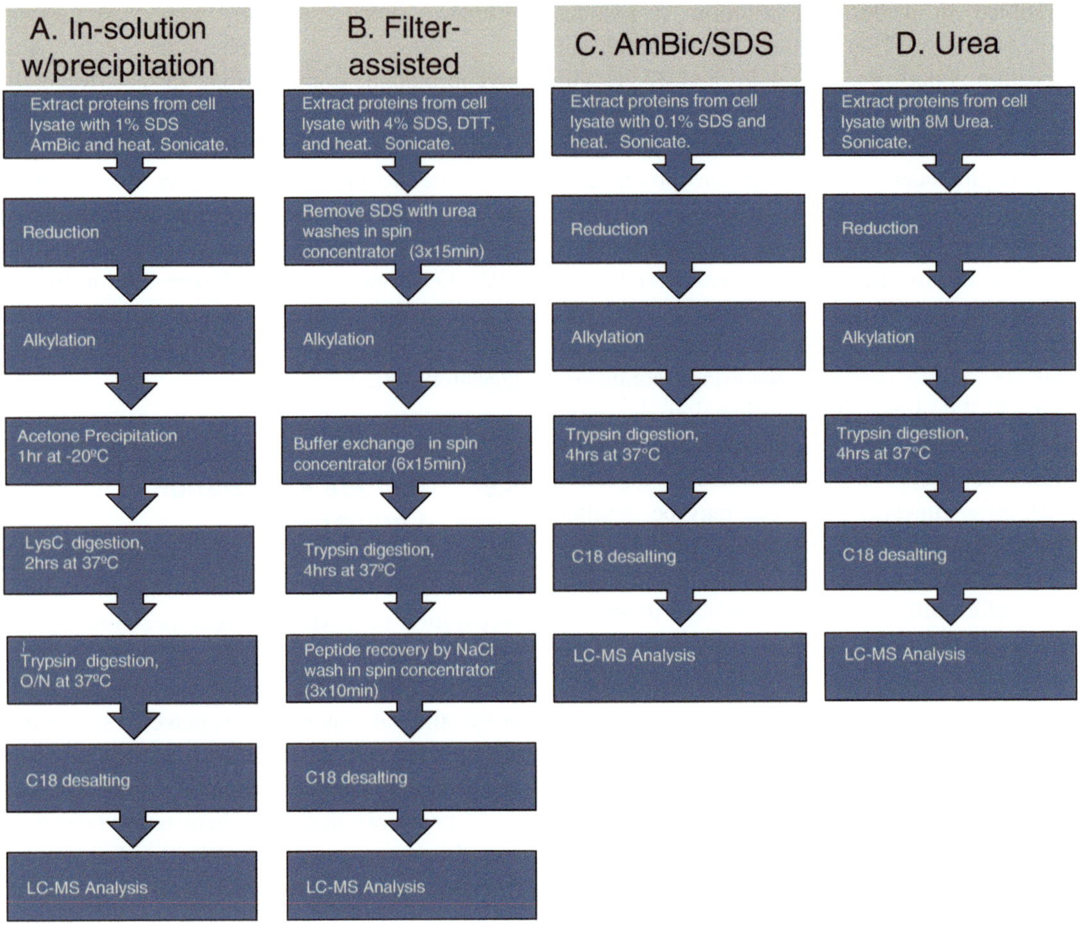

Fig. 3.4 Comparison of standard sample preparation workflows. A summary of the optimized Pierce sample preparation protocol is compared to three other popular standard proteomic sample prep methods that were evaluated

sample preparation methods (Fig. 3.4, [79]). We first used a step-wise approach to optimize a lysis protocol for high protein recovery from mammalian cell lysates. Protein solubilization with 0.1–4 % SDS yielded 5–40 % more protein than solubilization with 8 M urea [79]. Next, the completeness of disulfide reduction, the selectivity of alkylation at cysteine residues, and the digestion efficiency was assessed with single or double digestion (LysC-trypsin) routines. During this analysis, we discovered that improved chromatography resins and columns combined with fast, high resolution instruments often reveal longer, more highly charged peptides with missed cleavages that are not detected on lower resolution or slower mass spectrometers. By optimizing protocols to minimize non-selective alkylation or incompletely digested peptides, we could significantly improve the reproducibility and the number of peptide and protein identifications (Tables 3.1 and 3.2).

Reproducibility of digestion was assessed by the number of identified peptides and proteins identified, by the sequence coverage of a digestion indicator internal standard (Table 3.1), and by the targeted quantitative analysis of peptides from a digestion indicator internal standard. To address this, we spiked a non-mammalian protein in each lysate, processed triplicate samples according to the optimized protocol, and then quantified five peptides by targeted product ion monitoring on a Thermo Scientific Velos ion trap. The coefficients of variation (CV) were

Table 3.1 Reproducibility of LC-MS/MS results from three biological replicates

	Sample 1	Sample 2	Sample 3
Number of Proteins	3382	3228	3376
Number of Unique Peptides	16,333	15,939	17,048
Missed Cleavages (%)	7.8	8.8	8.6
Disulfide Bond Reduction (%)	100	100	100
Cysteine Alkylation (%)	100	100	100
Over Akylation (%)	0.1	0.3	0.9
Digestion Indicator Protein Sequence Coverage (%)	62.50	62.93	65.09

HeLa cell lysate (200 μg) in 200 μL lysis buffer was spiked with 2 μg Digestion Indicator processed by the Pierce Mass Spec Sample Prep Kit for Culture Cells and then analyzed by LC-MS/MS on a Q Exactive mass spectrometer

Table 3.2 Comparison of peptide and protein identification results between sample preparation methods

	Pierce	FASP	AmBic-SDS	Urea
Number of Proteins	3964 ± 22	3894 ± 13	3716 ± 79	3756 ± 91
Number of Unique Peptides	19,902 ± 190	18,738 ± 128	17,401 ± 587	19,398 ± 689
Missed Cleavages (%)	7.3 ± 0.1	13.9 ± 1.2	17.5 ± 1.3	9.8 ± 1.0
Disulfide Bond Reduction (%)	100	100	100	100
Methionine Oxidation (%)	3.0 ± 0.1	11.3 ± 1.5	2.6 ± 0.1	5.3 ± 0.5
Cysteine Alkylation (%)	99.8 ± 0.4	99.8 ± 0.3	100.0 ± 0.0	100.0 ± 0.0
Over Akylation (%)	0.7 ± 0.2	0.1 ± 0.1	0.8 % ± 0.6	2.4 ± 0.4

Hela lysate samples (100 μg) were prepared according to each protocol and 500 ng was analyzed in triplicates by LC-FT MS/IT MS2 CID on an Orbitrap Elite mass spectrometer

4–15 % with a mean CV of 7 % [79]. This quantitative analysis further demonstrated the high reproducibility of sample processing using the optimized protocol.

To assess the scalability of this sample preparation protocol, 10 μg to 5 mg of HeLa cell lysate was processed according to the protocol. Analysis of equivalent volumes of peptide samples by LC-MS/MS resulted in identical chromatograms, demonstrating the scalability of this protocol over a 500x dynamic range of sample amounts (Fig. 3.5). This sample preparation protocol was also used for brain tissue and resulted in reproducible, high quality peptide sample preparations, demonstrating the versatility of this method for different cell and tissue sample types (Fig. 3.6).

We found that the acetone precipitation protocol with optimized reduction, alkylation, and digestion reproducibly yielded high quality peptide samples for LC-MS/MS analysis (Table 3.1). This method yields more protein lysate from cultured cells, is highly reproducible, is scalable, is simpler and faster than FASP, has no risk of carbamylation by urea, and results in higher protein identification rates than other popular "standard" sample preparation methods (Fig. 3.3 and Table 3.2).

3.10 Methods

3.10.1 Protein Extraction

Duplicate or triplicate HeLa S3 cell pellets, each containing 2 × 10^6 cells, were re-suspended in: (a) 0.2 mL of 0.1 M Tris–HCl, 4 % SDS, 0.1 M DTT, pH 7.6 (FASP method); (b) 0.05 M ammonium bicarbonate, 0.1 % SDS, pH 8.0 (AmBic/SDS method); (c) 0.1 M Tris–HCl, 8 M urea, pH 8.5 (urea method), or (d) Lysis Buffer from the Thermo Scientific Pierce Mass Spec Sample Prep Kit for Cultured Cells. Samples were incubated at 95 °C for 5 min except the urea sample, which was incubated at RT for 30 min. Each cell suspension was sonicated on ice for 20 s. The cell debris was removed by

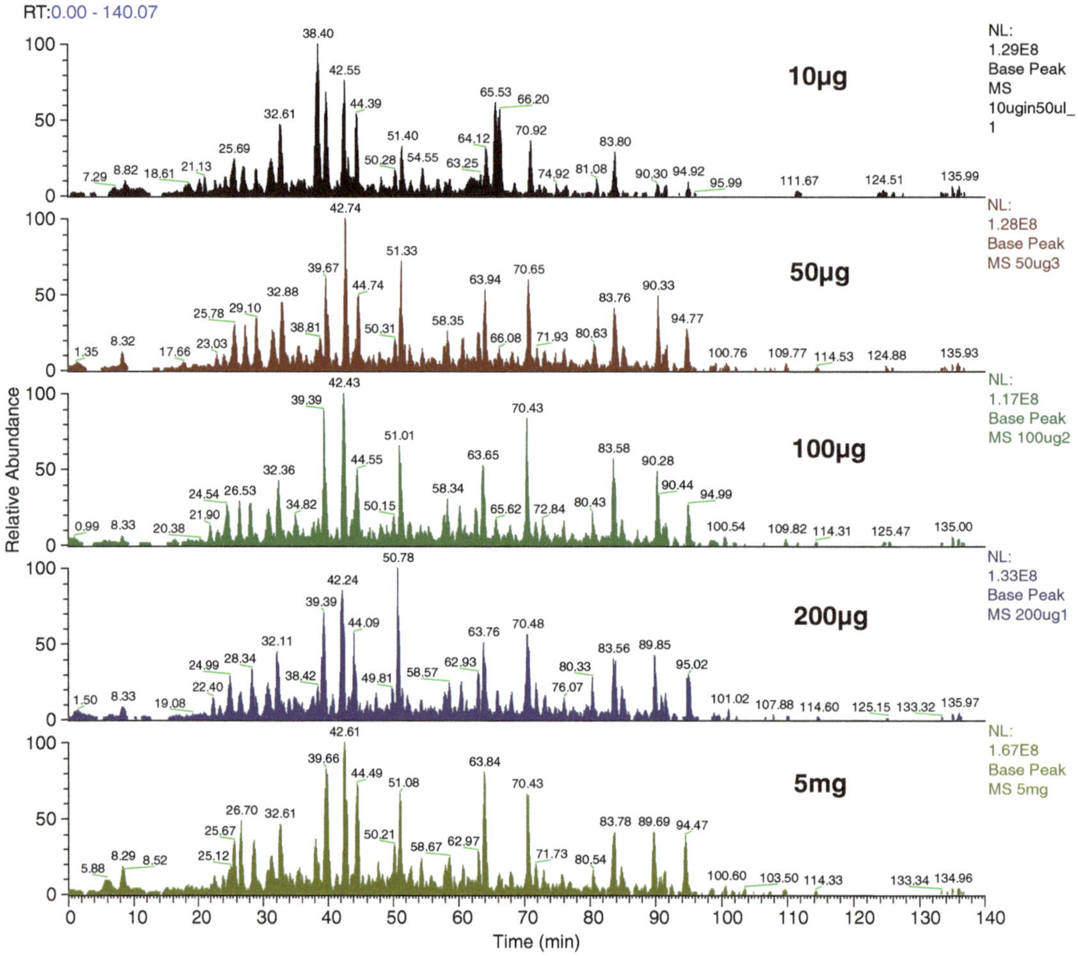

Fig. 3.5 Scalability of new MS sample prep kit protocol. *HeLa* lysate samples (10 μg–5 mg) were prepared according to protocol. Samples (500 ng) subjected to LC-MS/MS analysis on a Thermo Scientific Velos Pro ion trap mass spectrometer

centrifugation at 16,000 × g for 10 min and the supernatant was assayed for protein concentration using Thermo Scientific Pierce BCA Protein Assay or Thermo Scientific Pierce BCA Protein Assay Kit-Reducing Agent Compatible Assay.

3.10.2 Sample Preparation

HeLa cell lysate (100 μg) with digestion indicator (1 %, w/w) was reduced with 10 mM DTT for 45 min at 50 °C and alkylated with 50 mM iodoacetamide for 20 min in dark at RT. Excess iodoacetamide and other contaminants were removed by acetone precipitation at -20 °C for 1 h. The protein was re-suspended in digestion buffer and digested with Lys-C (1:100, enzyme: substrate) for 2 h at 37 °C followed by digestion with trypsin (1:50, enzyme:substrate) overnight at 37 °C. Peptide samples were also prepared according to standard urea, FASP1, and AmBic/SDS workflow.

3.10.3 LC-MS and Data Analysis

A Thermo Scientific EASY-nLC 1000 HPLC system and Thermo Scientific EASYSpray Source with Thermo Scientific EasySpray Column (25 cm × 75 μm i.d., PepMap C18) was

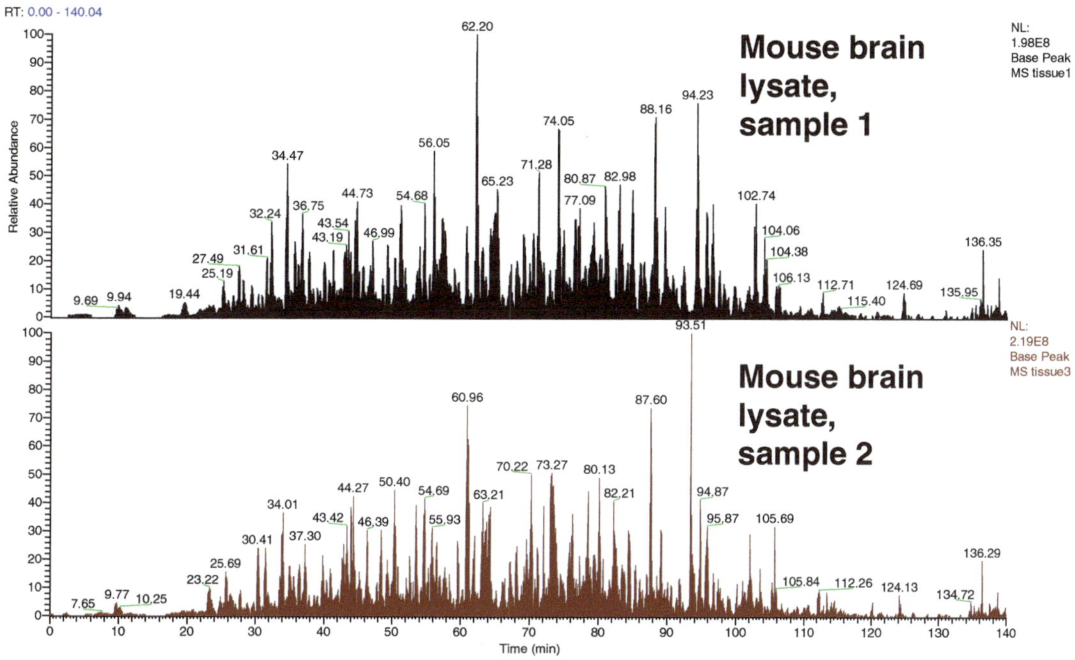

Fig. 3.6 Evaluation of sample preparation workflow with tissue samples. Mouse brain tissue (0.25 g) was homogenized with a tissue tearer and the proteins were extracted using the Thermo Scientific Pierce Mass Spec Sample Prep Kit for Cultured Cells. Tissue lysate (100 μg) was subjected to sample preparation workflow and sample (500 ng) was analyzed by LC-MS/MS on a Thermo Scientific Velos Pro ion trap mass spectrometer

used to separate peptides (500 ng) with a 30 % acetonitrile gradient in 0.1 % formic acid over 100–140 min at a flow rate of 300 nL/min. The samples were analyzed using a Thermo Scientific Velos Pro, a Q Exactive hybrid quadrupole-Orbitrap or an Thermo Scientific Orbitrap Elite mass spectrometers. For data analysis, Thermo Scientific Proteome Discoverer software version 1.4 was used to search MS/MS spectra against the uniprot human database using SEQUEST* search engine with a 1 % false discovery rate. Static modifications included carbamidomethyl (C) and dynamic modifications included oxidation (M). The data set was screened by Preview software (Protein Metrics) for assessment of sample preparation quality. To assess the digestion efficiency, the Digestion Indicator protein sequence was included in the protein database. Five digestion indicator peptides were quantified manually with extracted ion-chromatograms of the raw LC-MS/MS data or automatically with Thermo Scientific Pinpoint 1.2 software.

3.11 Conclusions

A variety of sample preparation methods have been described, along with a brief comparison of several in-solution and filter-assisted sample preparation methods. While each of these methods has advantages and disadvantages, all of these methods are capable of providing contaminant-free peptide samples compatible with mass spectrometric analysis. Unfortunately, none of these sample preparation methods is sufficiently simplified, standardized, or automated to enable rapid adoption and widespread use by novice or non-MS users.

In order to identify thousands of proteins from a complex lysate, it is essential to have robust sample preparation methods for protein extraction, reduction, alkylation, digestion, and cleanup. It is also essential to optimize LC and MS instrument performance, and to regularly (daily or weekly) assess instrument performance with a standard, well understood positive control samples. A variety of such standards are commercially available, including mixtures of isotopically labeled heavy peptides to assess chromatography, standard digests of common proteins or protein mixtures (e.g. bovine serum albumin and cytochrome C), as well as standard digests of complex proteomes from bacteria, yeast, or human cell lines from several MS reagent vendors. Regular use of standards is critical to ensuring that the instrumentation is working properly before precious samples are analyzed.

Ideally, it would be best to have a simpler, universal sample preparation method, as it would permit standardization of methods and would improve the reproducibility of results across laboratories and over time. For example, decades ago ion exchange-based DNA preparation kits rapidly supplanted the use of ultracentrifugation for plasmid DNA sample preparation. That simplification enabled broader adoption, higher throughput, and standardization of nucleic acid preparation methods. In contrast to DNA extraction from bacteria, the variety of protein sources, the diversity of proteins themselves, and protein biology in general are perhaps too complex to permit similar improvements that simplify, standardize, and automate protein sample preparation. Nevertheless, continued improvements in sample preparation robustness and ease of use are necessary for proteomics methods to be more widely adopted and to successfully advance protein MS beyond academic research or specialized MS labs and into individualized, bench top point of use or large clinical applications.

Supplementary Protocols

1. In-gel Digestion

Materials

- 25 mM ammonium bicarbonate: Dissolve 80 mg ammonium bicarbonate in 40 mL ultrapure water
- Destain solution: 25 mM ammonium bicarbonate/50 % acetonitrile (ACN). Mix 80 mg of ammonium bicarbonate with 20 mL of acetonitrile (ACN) and 20 mL of ultrapure water.
  Note: if destaining glutaraldehyde-free silver stained gels, prepare separate 100 mM sodium thiosulfate and 30 mM potassium ferricyanide solutions, then make destaining solution by mixing them in a 1:1 (v:v) ratio. Protect ferricyanide solution from light.
- DTT stock solution: 10 mM DTT in 25 mM ammonium bicarbonate
- Iodoacetamide (IAM) stock solution: 20 mM in 25 mM ammonium bicarbonate (always prepare fresh, protect from light)
- 10 ng/μl Trypsin, sequencing-grade (use 25 mM ice cold ammonium bicarbonate to dilute stock trypsin solution, immediately before adding to gel pieces)

Equipment

- Gloves! (to minimize keratin contamination)
- Clean glass plate (large enough to place entire gel on and room for a working area, 8" × 8")
- Gel-cutting devices: clean steel razor blades or surgical scalpel
- Low protein binding micro-centrifuge tubes (0.65 mL or 1.5 mL)
- Gel-loading pipette tips
- Autosampler vials with perforated caps
- SpeedVac Concentrator

Sample Processing

1. Place the gel on a clean glass plate. Cover the gel with just enough ultrapure water to prevent dehydration during the slicing process.
2. Cut the gel lane using (new, if possible) scalpel or razor blade.
3. Cut each of the excised bands into 1–2 mm cubes and transfer these cubes to a 0.65 mL low protein binding microcentrifuge tube.
4. Add ~100 μL (or enough to cover gel slices) of 25 mM ammonium bicarbonate/50 % ACN and vortex for 10 min.

5. Using gel loading pipet tip, extract the supernatant and discard. The procedure should be repeated until the stain is completely removed. Two additional washes should be sufficient for moderately intense bands.
6. Add 100 μL of 5 mM DTT and incubate for 30 min at 50 °C. Spin. Discard all the liquid afterwards.
7. Allow samples to cool to room temperature.
8. Add 100 μL of 20 mM iodoacetamide and incubate the gel pieces in the dark for 45 min at room temperature. Spin. Discard the liquid afterward.
9. Wash the gel pieces with 100 μL of 25 mM ammonium bicarbonate, vortex 10 min, spin. Discard the liquid afterwards.
10. Wash the gel pieces with ~100 μL (or enough to cover) of 25 mM ammonium bicarbonate in 50 % ACN, vortex 10 min, spin. Discard the liquid.
11. Dehydrate the gel pieces in 100 % ACN for 10 min, spin and discard the liquid afterwards.
12. Dry the sample in a speed-vac for 10 min. The gel pieces are now ready for tryptic digestion.
13. Just before use, dilute or reconstitute trypsin with 50 mM ice cold ammonium bicarbonate to give final concentration of the 10 ng/μL.
14. Add trypsin solution to just cover the gel pieces.
15. Verify that the gel pieces are covered with trypsin solution.
16. Add 25 mM ammonium bicarbonate as needed to cover the gel pieces.
17. Spin briefly and incubate at 37 °C for 4 h – overnight.
18. Stop digestion by adding 20 μL of 5 % formic acid.
19. Vortex 15–20 min, spin, and transfer the digest solution (aqueous extraction) into a clean autosampler vial appropriate for LC/MS-MS.
20. To the gel pieces, add 30 μL (enough to cover) of 50 % ACN/1 % formic acid, vortex 15–20 min., spin, and transfer solution to the tube used above. Repeat this step once.
21. Concentrate peptide extracts using a speed-vac concentrator to a volume that is slightly larger than will be used for injection during LC-MS/MS analysis.
22. Store the vial with the extracted peptides at −20 °C if the samples will not be run the same day.

2. In-Solution Sample Preparation With Acetone Precipitation

Materials

- 100ABCS: 100 mM NH4HCO3 with 0.1 % sodium dodecyl sulfate, pH 8.0, 5 mL
- 50ABC: 50 mM NH4HCO3, pH 8.0, 5 mL
- 500 mM DTT in 50ABC, 0.5 mL
- 500 mM Iodoacetamide (IAM) in 50ABC, 0.5 mL (protect solution from light)
- 0.1 % acetic acid in water, 250 μL
- Lys-C Protease, MS Grade, 20 μg
- MS-Grade Trypsin Protease, MS Grade, 20 μg
- Pre-chilled 90 % acetone: Prepare 90 % acetone in ultrapure water (e.g., mix 45 mL of 100 % acetone with 5 mL of ultrapure water) and store at −20 °C.
- Pre-chilled 100 % acetone: Store 100 % acetone at −20 °C.
- Trifluoroacetic acid (TFA)
- Phosphate-buffered saline (PBS)

Equipment

- Low protein binding microcentrifuge tubes
- Microtip probe sonicator or nuclease (e.g., Thermo Scientific™ Pierce™ Universal Nuclease for Cell Lysis, Product No. 88700)
- Heating block
- SpeedVac Concentrator

Procedure

Cell Lysis

1. Culture cells to harvest at least 100 μg of protein. For best results, culture a minimum of 1×10^6 cells.

Note: Rinse cell pellets 2–3 times with 1X PBS to remove cell culture media. Pellet cells using low-speed centrifugation (i.e., < 1000 × g) to prevent premature cell lysis.

2. Lyse the cells by adding five cell-pellet volumes of 100ABCS (i.e. 100 μL of 100ABCS for a 20 μL cell pellet). Pipette sample up and down to break up the cell clumps and gently vortex sample to mix.
3. Incubate the lysate at 95 °C for 5 min.
4. Cool the lysate on ice for 5 min.
5. Sonicate lysate on ice using a microtip probe sonicator to reduce the sample viscosity by shearing DNA.
6. Centrifuge lysate at 14,000 × g for 10 min at 4 °C.
7. Carefully separate the supernatant and transfer into a new tube.
8. Determine the protein concentration of the supernatant using established methods such as the BCA Protein Assay Kit

Reduction, Alkylation and Acetone Precipitation

Note: This procedure is optimized for 100 μg of cell lysate protein at 1 mg/mL concentration; however, the procedure may be used for 10–200 μg of cell lysate protein with an appropriate amount of reagents (DTT, IAM, Lys-C and trypsin). When using 10 μg of cell lysate, a protein concentration of 0.2–1 mg/mL may be used.

1. Add 100 μg of lysate protein to a polypropylene microcentrifuge tube and adjust the sample volume to 100 μL using 100ABCS to a final concentration of 1 mg/mL.
2. Add 2.1 μL of DTT solution to the sample (final DTT concentration is ~10 mM). Mix and incubate at 50 °C for 45 min. Discard any unused DTT solution.
3. Cool the sample to room temperature for 10 min.
4. Add 11.5 μL of IAM solution to the sample (final IAM concentration is ~50 mM). Mix and incubate at room temperature for 20 min protected from light. Discard any unused IAM solution.
5. After alkylation with IAM, immediately add 460 μL (4 volumes) of pre-chilled (−20 °C) 100 % acetone to sample. Vortex tube and incubate at −20 °C for 1 h to overnight to precipitate proteins.
6. Centrifuge at 14,000 × g for 10 min at 4 °C. Carefully remove acetone without dislodging the protein pellet.
7. Add 50 μL of pre-chilled (−20 °C) 90 % acetone, vortex to mix and centrifuge at 14,000 × g for 5 min at 4 °C.
8. Carefully remove acetone without dislodging the protein pellet. Allow the pellet to dry for 2–3 min and immediately proceed to Protein Digestion.

   Note: Do not dry the acetone-precipitated protein pellet for more than 2–3 min; excess drying will make the pellet difficult to re-suspend in the Digestion Buffer.

Enzymatic Protein Digestion

9. Add 100 μL of 50ABC to the acetone-precipitated protein pellet and resuspend by gently pipetting up and down to break the pellet.

   Note: An acetone-precipitated protein pellet may not completely dissolve; however, after proteolysis at 37 °C, all the protein will be solubilized.
10. Immediately before use, add 40 μL of ultrapure water to the bottom of the vial containing lyophilized Lys-C and incubate at room temperature for 5 min. Gently pipette up and down to dissolve. Store any remaining 0.5 μg/μL Lys-C solution in single-use volumes at −80 °C.
11. Add 2 μL of Lys-C (1 μg, enzyme-to-substrate ratio = 1:100) to the sample. Mix and incubate at 37 °C for 2 h.
12. Immediately before use, add 40 μL of 0.1 % acetic acid to the bottom of the vial containing trypsin and incubate at room temperature for 5 min. Gently pipette up and down to dissolve. Store

any remaining 0.5 μg/μL trypsin solution in single-use volumes at −80 °C for long-term storage.
13. Add 4 μL of trypsin (2 μg, enzyme-to-substrate ratio = 1:50) to the sample. Mix and incubate overnight at 37 °C.
14. Freeze samples at −80 °C to stop digestion. (Optional: stop digestion by acidifying with TFA)
15. Speed vac sample to 1–5 μL.
16. Resuspend the sample in an appropriate buffer (e.g., 0.1 % TFA) for LC-MS analysis.
    Note: Proteolytic digests prepared using this protocol are directly compatible with LC-MS analysis. Clean-up of samples with C18 spin tips or columns is optional.

3. Filter-assisted Sample Preparation (FASP)

Materials

- UABC: 8 M urea in 100 mM NH_4HCO_3 (ABC) pH 8.0. Prepare fresh, 1 mL per sample.
- IAM solution: 55 mM iodoacetamide in UABC. Prepare 100 μL per sample.
- DTT solution: 50 mM DTT in UABC. Prepare 100 μL per sample
- Trypsin: MS grade Modified Trypsin, 0.5 μg/μL in 50 mM NH_4HCO_3 in water
- 50ABC: 50 mM NH_4HCO_3 in water. Prepare 0.5 mL per sample
- 25ABC: 25 mM NH_4HCO_3 in water. Prepare 0.25 mL per sample
  Note: UABC and IAM solutions must be freshly prepared and used within a day. IAM is light sensitive, so protect from light

Equipment

- Low protein binding tubes
- 10 or 30 kDa cut off filter (Vivacon 500, cat # VN01H02)
- Bench-top centrifuge
- Temperature-controlled incubator or heat block at 50 °C
- Thermo-mixer at 37 °C
- SpeedVac Concentrator

Procedure

1. Combine up to 30 μL of a protein extract (0.2–400 μg) with 200 μL of UABC in the filter unit and centrifuge at 14,000 × g for 15 min.
2. Add 200 μL of UABC to the filter unit and centrifuge at 14,000 × g for 15 min.
3. Discard the flow-through from the collection tube.
4. Add 100 μL DTT solution and mix at 600 rpm in a thermo-mixer for 1 min and incubate at 50 °C without mixing for 45 min.
5. Centrifuge the filter units at 14,000 × g for 10 min.
6. Add 100 μL IAM solution, cover with foil, mix by gentle vortexing for 1 min, and incubate in dark at room temperature without mixing for 30 min.
7. Centrifuge the filter units at 14,000 × g for 10 min.
8. Add 100 μL of UABC to the filter unit and centrifuge at 14,000 × g for 15 min. Repeat this step one more time.
9. Add 100 μL of 50ABC to the filter unit and centrifuge at 14,000 × g for 10 min. Repeat this step one more time.
10. Transfer the filter units to new collection tubes.
11. Add 100 μL of 50ABC with trypsin (enzyme to protein ratio 1:50) and mix at 600 rpm in thermo-mixer at 37 °C for 4–18 h.
12. Centrifuge the filter units at 14,000 × g for 10 min.
13. Add 50 μL of 25ABC and centrifuge the filter units at 14,000 × g for 10 min.
14. Add 50 μL of 10ABC and centrifuge the filter units at 14,000 × g for 10 min.
15. Concentrate down to ~5 μL and add 0.1 % FA to a final volume of ~20–25 μL.

References

1. Aebersold R, Mann M (2003) Mass spectrometry-based proteomics. Nature 422(6928):198–207
2. Han X, Aslanian A, Yates JR 3rd (2008) Mass spectrometry for proteomics. Curr Opin Chem Biol 12 (5):483–490
3. Washam CL, Byrum SD, Leitzel K, Ali SM, Tackett AJ, Gaddy D et al (2013) Identification of PTHrP (12–48) as a plasma biomarker associated with breast cancer bone metastasis. Cancer Epidemiol Biomark Prev Publ Am Assoc Cancer Res Cosponsored Am Soc Prevent Oncol 22(5):972–983
4. Huttlin EL, Ting L, Bruckner RJ, Gebreab F, Gygi MP, Szpyt J et al (2015) The BioPlex network: a systematic exploration of the human interactome. Cell 162(2):425–440
5. Meisinger C, Sickmann A, Pfanner N (2008) The mitochondrial proteome: from inventory to function. Cell 134(1):22–24
6. Bryson BD, White FM (2012) Signaling for death: tyrosine phosphorylation in the response to glucose deprivation. Mol Syst Biol 8:591
7. Wolters DA, Washburn MP, Yates JR 3rd (2001) An automated multidimensional protein identification technology for shotgun proteomics. Anal Chem 73 (23):5683–5690
8. Wang Y, Yang F, Gritsenko MA, Wang Y, Clauss T, Liu T et al (2011) Reversed-phase chromatography with multiple fraction concatenation strategy for proteome profiling of human MCF10A cells. Proteomics 11(10):2019–2026
9. Picotti P, Aebersold R (2012) Selected reaction monitoring-based proteomics: workflows, potential, pitfalls and future directions. Nat Methods 9 (6):555–566
10. Peterson AC, Russell JD, Bailey DJ, Westphall MS, Coon JJ (2012) Parallel reaction monitoring for high resolution and high mass accuracy quantitative, targeted proteomics. Mol Cell Proteomics MCP 11 (11):1475–1488
11. Stastna M, Van Eyk JE (2012) Analysis of protein isoforms: can we do it better? Proteomics 12 (19–20):2937–2948
12. Savitski MM, Reinhard FB, Franken H, Werner T, Savitski MF, Eberhard D et al (2014) Tracking cancer drugs in living cells by thermal profiling of the proteome. Science (New York, NY) 346(6205):1255784
13. Weekes MP, Tomasec P, Huttlin EL, Fielding CA, Nusinow D, Stanton RJ et al (2014) Quantitative temporal viromics: an approach to investigate host-pathogen interaction. Cell 157(6):1460–1472
14. Klein T, Fung SY, Renner F, Blank MA, Dufour A, Kang S et al (2015) The paracaspase MALT1 cleaves HOIL1 reducing linear ubiquitination by LUBAC to dampen lymphocyte NF-kappaB signalling. Nat Commun 6:8777
15. Catherman AD, Skinner OS, Kelleher NL (2014) Top down proteomics: facts and perspectives. Biochem Biophys Res Commun 445(4):683–693
16. McLafferty FW, Breuker K, Jin M, Han X, Infusini G, Jiang H et al (2007) Top-down MS, a powerful complement to the high capabilities of proteolysis proteomics. FEBS J 274(24):6256–6268
17. Fornelli L, Ayoub D, Aizikov K, Beck A, Tsybin YO (2014) Middle-down analysis of monoclonal antibodies with electron transfer dissociation orbitrap fourier transform mass spectrometry. Anal Chem 86 (6):3005–3012
18. Wu C, Tran JC, Zamdborg L, Durbin KR, Li M, Ahlf DR et al (2012) A protease for 'middle-down' proteomics. Nat Methods 9(8):822–824
19. Wu CC, MacCoss MJ (2002) Shotgun proteomics: tools for the analysis of complex biological systems. Curr Opin Mol Ther 4(3):242–250
20. Seddon AM, Curnow P, Booth PJ (2004) Membrane proteins, lipids and detergents: not just a soap opera. Biochim Biophys Acta 1666(1–2):105–117
21. Feist P, Hummon AB (2015) Proteomic challenges: sample preparation techniques for microgram-quantity protein analysis from biological samples. Int J Mol Sci 16(2):3537–3563
22. Keller BO, Sui J, Young AB, Whittal RM (2008) Interferences and contaminants encountered in modern mass spectrometry. Anal Chim Acta 627(1):71–81
23. Loo RR, Dales N, Andrews PC (1996) The effect of detergents on proteins analyzed by electrospray ionization. Methods Mol Biol (Clifton, NJ) 61:141–160
24. Waas M, Bhattacharya S, Chuppa S, Wu X, Jensen DR, Omasits U et al (2014) Combine and conquer: surfactants, solvents, and chaotropes for robust mass spectrometry based analyses of membrane proteins. Anal Chem 86(3):1551–1559
25. Anderson NL, Anderson NG (2002) The human plasma proteome: history, character, and diagnostic prospects. Mol Cell Proteomics MCP 1(11):845–867
26. Polaskova V, Kapur A, Khan A, Molloy MP, Baker MS (2010) High-abundance protein depletion: comparison of methods for human plasma biomarker discovery. Electrophoresis 31(3):471–482
27. Huber LA, Pfaller K, Vietor I (2003) Organelle proteomics: implications for subcellular fractionation in proteomics. Circ Res 92(9):962–968
28. Dunkley TP, Watson R, Griffin JL, Dupree P, Lilley KS (2004) Localization of organelle proteins by isotope tagging (LOPIT). Mol Cell Proteomics MCP 3 (11):1128–1134
29. Ramsby ML, Makowski GS, Khairallah EA (1994) Differential detergent fractionation of isolated hepatocytes: biochemical, immunochemical and two-dimensional gel electrophoresis characterization of cytoskeletal and noncytoskeletal compartments. Electrophoresis 15(2):265–277
30. Gu B, Zhang J, Wang W, Mo L, Zhou Y, Chen L et al (2010) Global expression of cell surface proteins in embryonic stem cells. PLoS One 5(12):e15795
31. Weekes MP, Antrobus R, Lill JR, Duncan LM, Hor S, Lehner PJ (2010) Comparative analysis of techniques to purify plasma membrane proteins. J Biomol Tech JBT 21(3):108–115

32. Yang L, Nyalwidhe JO, Guo S, Drake RR, Semmes OJ (2011) Targeted identification of metastasis-associated cell-surface sialoglycoproteins in prostate cancer. Mol Cell Proteomics MCP 10(6): M110.007294
33. Deeb SJ, Cox J, Schmidt-Supprian M, Mann M (2014) N-linked glycosylation enrichment for in-depth cell surface proteomics of diffuse large B-cell lymphoma subtypes. Mol Cell Proteomics MCP 13(1):240–251
34. Nilsson CL, Dillon R, Devakumar A, Shi SD, Greig M, Rogers JC et al (2010) Quantitative phosphoproteomic analysis of the STAT3/IL-6/HIF1alpha signaling network: an initial study in GSC11 glioblastoma stem cells. J Proteome Res 9 (1):430–443
35. Patricelli MP, Szardenings AK, Liyanage M, Nomanbhoy TK, Wu M, Weissig H et al (2007) Functional interrogation of the kinome using nucleotide acyl phosphates. Biochemistry 46(2):350–358
36. Lemeer S, Zorgiebel C, Ruprecht B, Kohl K, Kuster B (2013) Comparing immobilized kinase inhibitors and covalent ATP probes for proteomic profiling of kinase expression and drug selectivity. J Proteome Res 12 (4):1723–1731
37. ten Have S, Boulon S, Ahmad Y, Lamond AI (2011) Mass spectrometry-based immuno-precipitation proteomics – the user's guide. Proteomics 11 (6):1153–1159
38. Evans DR, Romero JK, Westoby M (2009) Concentration of proteins and removal of solutes. Methods Enzymol 463:97–120
39. Gundry RL, White MY, Murray CI, Kane LA, Fu Q, Stanley BA et al (2009) Preparation of proteins and peptides for mass spectrometry analysis in a bottom-up proteomics workflow. In: Frederick MA et al (eds) Current protocols in molecular biology. Chapter 10: Unit10.25
40. Olsen JV, Ong SE, Mann M (2004) Trypsin cleaves exclusively C-terminal to arginine and lysine residues. Mol Cell Proteomics MCP 3(6):608–614
41. Benore-Parsons M, Seidah NG, Wennogle LP (1989) Substrate phosphorylation can inhibit proteolysis by trypsin-like enzymes. Arch Biochem Biophys 272 (2):274–280
42. Swaney DL, Wenger CD, Coon JJ (2010) Value of using multiple proteases for large-scale mass spectrometry-based proteomics. J Proteome Res 9 (3):1323–1329
43. Wu CC, MacCoss MJ, Howell KE, Yates JR 3rd (2003) A method for the comprehensive proteomic analysis of membrane proteins. Nat Biotechnol 21 (5):532–538
44. Niessen S, McLeod I, Yates JR 3rd (2006) Direct enzymatic digestion of protein complexes for MS analysis. CSH Protoc 2006(7)
45. Bian Y, Ye M, Song C, Cheng K, Wang C, Wei X et al (2012) Improve the coverage for the analysis of phosphoproteome of HeLa cells by a tandem digestion approach. J Proteome Res 11(5):2828–2837
46. Biringer RG, Amato H, Harrington MG, Fonteh AN, Riggins JN, Huhmer AF (2006) Enhanced sequence coverage of proteins in human cerebrospinal fluid using multiple enzymatic digestion and linear ion trap LC-MS/MS. Brief Funct Genomic Proteomic 5 (2):144–153
47. Choudhary G, Wu SL, Shieh P, Hancock WS (2003) Multiple enzymatic digestion for enhanced sequence coverage of proteins in complex proteomic mixtures using capillary LC with ion trap MS/MS. J Proteome Res 2(1):59–67
48. Giansanti P, Aye TT, van den Toorn H, Peng M, van Breukelen B, Heck AJ (2015) An augmented multiple-protease-based human phosphopeptide atlas. Cell Rep 11(11):1834–1843
49. Leon IR, Schwammle V, Jensen ON, Sprenger RR (2013) Quantitative assessment of in-solution digestion efficiency identifies optimal protocols for unbiased protein analysis. Mol Cell Proteomics: MCP 12 (10):2992–3005
50. Chen EI, Cociorva D, Norris JL, Yates JR 3rd (2007) Optimization of mass spectrometry-compatible surfactants for shotgun proteomics. J Proteome Res 6(7):2529–2538
51. Kollipara L, Zahedi RP (2013) Protein carbamylation: in vivo modification or in vitro artefact? Proteomics 13(6):941–944
52. Proc JL, Kuzyk MA, Hardie DB, Yang J, Smith DS, Jackson AM et al (2010) A quantitative study of the effects of chaotropic agents, surfactants, and solvents on the digestion efficiency of human plasma proteins by trypsin. J Proteome Res 9(10):5422–5437
53. Laemmli UK (1970) Cleavage of structural proteins during the assembly of the head of bacteriophage T4. Nature 227(5259):680–685
54. Rabilloud T, Chevallet M, Luche S, Lelong C (2010) Two-dimensional gel electrophoresis in proteomics: past, present and future. J Proteome 73 (11):2064–2077
55. Schirle M, Heurtier MA, Kuster B (2003) Profiling core proteomes of human cell lines by one-dimensional PAGE and liquid chromatography-tandem mass spectrometry. Mol Cell Proteomics MCP 2(12):1297–1305
56. Sechi S, Chait BT (1998) Modification of cysteine residues by alkylation. A tool in peptide mapping and protein identification. Anal Chem 70 (24):5150–5158
57. Nielsen ML, Vermeulen M, Bonaldi T, Cox J, Moroder L, Mann M (2008) Iodoacetamide-induced artifact mimics ubiquitination in mass spectrometry. Nat Methods 5(6):459–460
58. Jiang X, Shamshurin D, Spicer V, Krokhin OV (2013) The effect of various S-alkylating agents on the chromatographic behavior of cysteine-containing peptides in reversed-phase chromatography. J Chromatogr B Anal Technol Biomed Life Sci 915–916:57–63
59. Ruhaak LR, Zauner G, Huhn C, Bruggink C, Deelder AM, Wuhrer M (2010) Glycan labeling strategies and

their use in identification and quantification. Anal Bioanal Chem 397(8):3457–3481
60. Arnold U, Ulbrich-Hofmann R (1999) Quantitative protein precipitation from guanidine hydrochloride-containing solutions by sodium deoxycholate/trichloroacetic acid. Anal Biochem 271(2):197–199
61. Bensadoun A, Weinstein D (1976) Assay of proteins in the presence of interfering materials. Anal Biochem 70(1):241–250
62. Buxton TB, Crockett JK, Moore WL 3rd, Moore WL Jr, Rissing JP (1979) Protein precipitation by acetone for the analysis of polyethylene glycol in intestinal perfusion fluid. Gastroenterology 76(4):820–824
63. Manza LL, Stamer SL, Ham AJ, Codreanu SG, Liebler DC (2005) Sample preparation and digestion for proteomic analyses using spin filters. Proteomics 5 (7):1742–1745
64. Peterson GL (1977) A simplification of the protein assay method of Lowry et al. which is more generally applicable. Anal Biochem 83(2):346–356
65. Wessel D, Flugge UI (1984) A method for the quantitative recovery of protein in dilute solution in the presence of detergents and lipids. Anal Biochem 138 (1):141–143
66. Barritault D, Expert-Bezancon A, Guerin MF, Hayes D (1976) The use of acetone precipitation in the isolation of ribosomal proteins. Eur J Biochem/FEBS 63(1):131–135
67. Crowell AM, Wall MJ, Doucette AA (2013) Maximizing recovery of water-soluble proteins through acetone precipitation. Anal Chim Acta 796:48–54
68. Yeung YG, Nieves E, Angeletti RH, Stanley ER (2008) Removal of detergents from protein digests for mass spectrometry analysis. Anal Biochem 382 (2):135–137
69. Yeung YG, Stanley ER (2010) Rapid detergent removal from peptide samples with ethyl acetate for mass spectrometry analysis. Current protocols in protein science/editorial board, John EC et al. Chapter 16: Unit 16.2
70. Antharavally BS, Mallia KA, Rosenblatt MM, Salunkhe AM, Rogers JC, Haney P et al (2011) Efficient removal of detergents from proteins and peptides in a spin column format. Anal Biochem 416(1):39–44
71. Bereman MS, Egertson JD, MacCoss MJ (2011) Comparison between procedures using SDS for shotgun proteomic analyses of complex samples. Proteomics 11(14):2931–2935
72. Glatter T, Ludwig C, Ahrne E, Aebersold R, Heck AJ, Schmidt A (2012) Large-scale quantitative assessment of different in-solution protein digestion protocols reveals superior cleavage efficiency of tandem Lys-C/trypsin proteolysis over trypsin digestion. J Proteome Res 11(11):5145–5156
73. Wisniewski JR, Zougman A, Mann M (2009) Combination of FASP and StageTip-based fractionation allows in-depth analysis of the hippocampal membrane proteome. J Proteome Res 8(12):5674–5678
74. Wisniewski JR, Zougman A, Nagaraj N, Mann M (2009) Universal sample preparation method for proteome analysis. Nat Methods 6(5):359–362
75. Wisniewski JR, Nagaraj N, Zougman A, Gnad F, Mann M (2010) Brain phosphoproteome obtained by a FASP-based method reveals plasma membrane protein topology. J Proteome Res 9(6):3280–3289
76. Zielinska DF, Gnad F, Jedrusik-Bode M, Wisniewski JR, Mann M (2009) Caenorhabditis elegans has a phosphoproteome atypical for metazoans that is enriched in developmental and sex determination proteins. J Proteome Res 8(8):4039–4049
77. Zielinska DF, Gnad F, Wisniewski JR, Mann M (2010) Precision mapping of an in vivo N-glycoproteome reveals rigid topological and sequence constraints. Cell 141(5):897–907
78. Erde J, Loo RR, Loo JA (2014) Enhanced FASP (eFASP) to increase proteome coverage and sample recovery for quantitative proteomic experiments. J Proteome Res 13(4):1885–1895
79. Antharavally B, Jiang X, Cunningham R, Bomgarden R, Zhang Y, Viner R et al (2013) Versatile Mass Spectrometry Sample Preparation Procedure for Complex Protein Samples [cited 2015 November 8]. Available from: https://www.thermofisher.com/us/en/home/life-science/protein-biology/protein-biology-learning-center/protein-biology-resource-library/protein-biology-application-notes/mass-spectrometry-sample-preparation-procedure-protein-samples.html

Plant Structure and Specificity – Challenges and Sample Preparation Considerations for Proteomics

4

Sophie Alvarez and Michael J. Naldrett

Abstract

Plants are considered as a simple structured organism when compared to humans and other vertebrates. The number of organs and tissue types is very limited. Instead the origin of the complexity comes from the high number and variety of plant species that exist, with >300,000 compared to 5000 in mammals. Proteomics, defined as the large-scale study of the proteins present in a tissue, cell or cellular compartment at a defined time point, was introduced in 1994. However, the first publications reported in the plant proteomics field only appeared at the beginning of the twenty-first century. Since these early years, the increase of proteomic studies in plants has only followed a linear trend. The main reason for this stems from the challenges specific to studying plants, those of protein extraction from cells with variously strengthened cellulosic cell walls, and a high abundance of interfering compounds, such as phenolic compounds and pigments located in plastids throughout the plant. Indeed, the heterogeneity between different organs and tissue types, between species and different developmental stages, requires the use of optimized plant protein extraction methods as described in this section. The second bottleneck of plant proteomics, which will not be discussed or reviewed here, is the lack of genomic information. Without sequence databases of the >300,000 species, proteomic studies of plants, especially of those that are not considered economically relevant, are impossible to accomplish.

Keywords

Plant proteomics • Plant cell lysis • Plant secretome • Plant organs • Plant meristem and suspension culture cells • Green algae and plastids • Plant protein extraction

Plant proteomics has been and is still an interesting approach to studying the content and abundance of proteins (protein expression) in response to changes

S. Alvarez (✉) • M.J. Naldrett
Center for Biotechnology, University of Nebraska–Lincoln, Beadle Center, 1901 Vine St, Lincoln, NE 68588, USA
e-mail: salvarez@unl.edu

H. Mirzaei and M. Carrasco (eds.), *Modern Proteomics – Sample Preparation, Analysis and Practical Applications*, Advances in Experimental Medicine and Biology 919, DOI 10.1007/978-3-319-41448-5_4

of the environment, such as drought, cold or a challenge by pathogens. Understanding how plants grow and interact with the environment is a pre-requisite for crop improvement and increased productivity. The study of plant growth is of course not seen to be as important as studying the more immediate ailments of our human condition. Though, as we move further into the twenty-first century and the world's population now exceeds seven billion, the question of how to produce enough food and feedstock to sustain this growth is an ever more significant challenge, which could endanger the human species if this issue is not given the effort that it deserves. In addition, many plants have been used for thousands of years for their healing properties in traditional medicine. Even more recently, since the 1980s, plants have been bioengineered for drug production by pharmaceutical companies as they often offer faster and cheaper production alternatives. The most pertinent recent example is the production of ZMapp in tobacco (*Nicotiana benthamiana*) for the treatment of patients suffering from the Ebola virus disease [82].

Proteomic studies have focused on many different plant species, with one of the first and most studied being *Arabidopsis thaliana*, a model system in the plant field. Its popularity amongst plant biologists comes from its small genome of 135 Mbp, which was fully sequenced in 2000 (The Arabidopsis Genome Initiative [1]), the known 27,029 protein coding genes [98], and its rapid life cycle under controlled conditions (6 weeks from germination to mature seed). Although the study of *Arabidopsis* affords a look into most of the basic biological processes in plants, many of these plant mechanisms and their molecular regulation are not always transposable into other plants, especially to crops, because of genomic and physiological differences. Among the crops, the most studied are maize, wheat, and soybean, because of their economic relevance in food production, and more recently for their biofuel possibilities.

Below, we describe some specific structures found in plants and highlight the challenges for proteomic studies. The specific requirements and procedures for protein sample preparation of these different types of plant material are also described to help scientists overcome the unique challenges of working with plant samples.

4.1 Plant Cell Wall and Secretome

The cell wall is a structure found in eukaryotic plant and algal cells. It is composed of two layers called the primary and the secondary cell wall, which surround the plasma membrane (Fig. 4.1a). The composition of the cell wall, which varies between plant families and tissues, consists of a very complex network of cellulose microfibrils embedded in a polysaccharide matrix composed of pectin, hemicellulose and glycoproteins for the primary cell wall and a rigid skeleton of cellulose, hemicellulose and lignin for the secondary cell wall. The cell wall is not surrounded by any additional physical barrier, but the space between two cell walls from adjacent cells, called apoplast, allows small molecules and proteins to circulate between cells, through pores in the cell wall made of proteins, called plasmodesmata (Fig. 4.1a). While the cell wall has the essential roles of maintaining the turgor pressure of plant cells by providing rigidity and as a physical barrier to avoid pathogen invasion, the cell wall proteins not only contribute to the structural role of the cell wall, but are also involved in cell-cell and cell-pathogen communication, contributing to plant development and growth and also to the plant's response to environmental changes and its adaptation. However, the heterogeneity of the cell wall and the complex network of the polysaccharides, make protein extraction from the cell wall challenging. The use of mechanical tissue disruption to release proteins from plant cells is a prerequisite to the study of the total protein content of the cell. It is not, however, always suitable for studying cell wall proteins (CWP). Here, different isolation and elution methods are necessary, depending on the type of protein that is to be studied. Three different types of cell wall proteins are defined, depending on the strength of their interaction with the cell wall. Proteins with no or little interaction with cell wall components (CWP1) and proteins

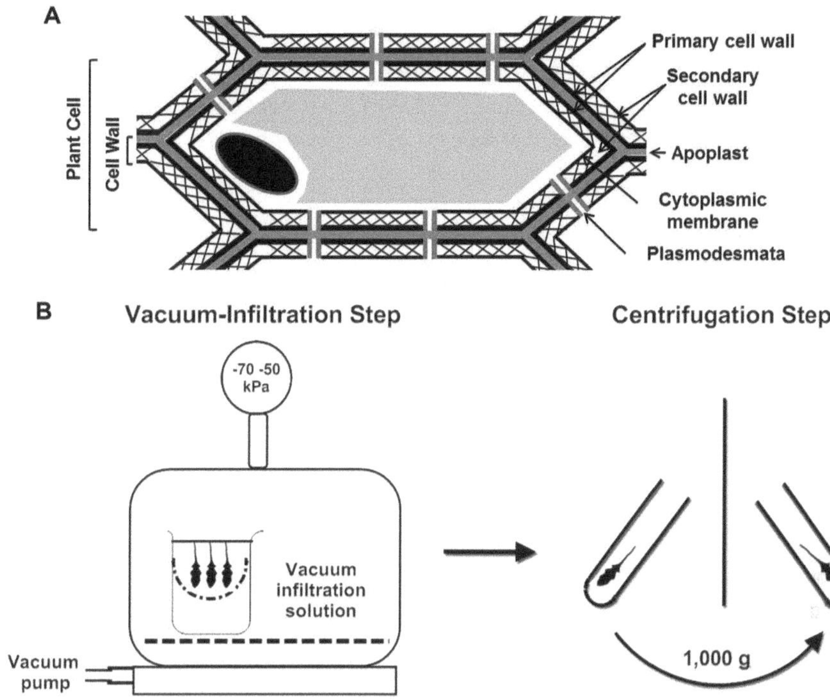

Fig. 4.1 (**a**) Plant cell wall diagram; (**b**) Vacuum infiltration and centrifugation workflow for cell wall protein isolation

weakly interacting with the matrix by van der Waals' interactions, hydrogen bonds, and hydrophobic or ionic interactions (CWP2) are extractable with salts, using non-disruptive methods to avoid damaging the plasma membrane. These proteins, found in the apoplast, constitute the apoplastic proteome or plant secretome (i.e. secreted proteins). These have an essential function in plant-pathogen interactions and adaptation to stress. The third category of cell wall proteins are those proteins strongly bound to the cell wall (CWP3) which require disruptive methods for extraction. These methods have been optimized to reduce the contamination from cell wall polysaccharides, phenolic compounds and the intracellular proteins.

Suspension culture cells (SCCs) are typically the method of choice to study the secretome and cell wall proteins. To isolate the apoplastic cell wall proteins (CWP1 and 2), non-disruptive methods are essential to preserve intact plasma membranes and avoid contamination from intracellular components. Two approaches can be used depending on the sample type: recovery of the SCCs' medium and washing of cells with liquid medium and low salt solutions [14, 20]; or using vacuum infiltration-centrifugation which consists of infusing a solution of salts into the intercellular space by use of reduced pressure (−75 to −50 kPa) for a short time (5–15 min) and harvesting the apoplastic fluid containing the proteins by gentle centrifugation (Fig. 4.1B). To dissolve the loosely bound CWP1, a low ionic strength (KCl) solution is used, while salts such as $CaCl_2$, LiCl and NaCl are used to isolate the weakly bound CWP2 (Table 4.1). The first method, applying only to suspension cells, cannot be applied to actual plant tissues which are more relevant for understanding responses to environmental interactions. Here the second method is more suitable. Although these methods are non-disruptive, the salt concentration has to be very carefully optimized to the type of sample under investigation in order to minimize cytosolic contamination. Excessive salt concentrations can cause the membrane to rupture and release not

Table 4.1 Salt solutions to isolate CWP1, CWP2 or CWP3 according to the sample type and species

Salt solutions	Cell wall group	Sample type and species	References
10 mM sodium phosphate, pH 6.0	**CWP1**	Root of onion (*Allium sepa* L.)	[28]
10 mM phosphate buffer, pH 6.0, containing 0.2 M KCl	**CWP1**	Leaf blades of maize (*Zea mays*)	[31]
0.01 M Mes buffer pH 5.5, containing 0.2 M KCl	**CWP1**	Leaf blades of tall fescue (*Festuca arundinacea*) Root of maize	[64] [123]
50 mM of $LaCl_3$	**CWP1**	Pea (*Pisum sativum* L.) internodes	[70]
0.3 M mannitol	**CWP1**	*Arabidopsis* rosettes	[15]
0.6 M NaCl	**CWP1 and CWP2**	Potato (*Solanum tuberosum* L.) tubers	[74]
1 M NaCl	**CWP1 and CWP2**	Maize endosperm	[24, 94]
50 mM phosphate buffer, 200 mM NaCl, pH 7.5	**CWP1 and CWP2**	Tobacco (*Nicotiana tabacum*) leaves	[30]
25 mM Tris–HCl pH 7.4, 50 mM EDTA (Ethylenediamine-tetraacetic acid), 150 mM $MgCl_2$	**CWP1 and CWP2**	*Arabidopsis* seedlings	[19]
0.1 M potassium phosphate (pH 7.8) or 66 mM potassium phosphate (pH 7.6) containing 10 mM $MgCl_2$ and 14 mM 2-ME	**CWP1 and CWP2**	Leaves from *Arabidopsis*, wheat (*Triticum aestivum*) and rice (*Oryza sativa*)	[41]
20 mM ascorbic acid and 20 mM $CaCl_2$ (pH 3.0)	**CWP1 and CWP2**	Winter rye (*Secare cereale*) leaves	[43]
1 M NaCl followed by 0.4 M $CaCl_2$	**CWP1 and CWP2**	Protoplasts from flax (*Linum usitatissimum*) hypocotyls	[85]
200 mM $CaCl_2$, 50 mM Na-acetate, pH 5.5, followed by 3 M LiCl, 50 mM Na-acetate, pH 5.5	**CWP3**	*Medicago sativa* stem	[108, 109]
200 mM $CaCl_2$	**CWP3**	*Arabidopsis* suspension cells	[12, 26]
5 mM acetate buffer, pH 4.6, 0.2 M $CaCl_2$ and 10 μL protease inhibitor cocktail followed by 5 mM acetate buffer, pH 4.6, 2 M LiCl and 10 μL protease inhibitor cocktail, followed by 62.5 mM Tris, 4 % SDS, 50 mM DTT, pH 6.8 (HCl)	**CWP3**	*Arabidopsis* hypocotyls	[34]

only intracellular proteins but also phenolic compounds, which can lead to downstream issues with proteomic experiments as we will describe in the next section. Therefore, cytosolic contamination must be checked for, when using the non-disruptive and disruptive methods described in Table 4.1, by measuring the activity of glucose-6-phosphate dehydrogenase (G6PDH) [60].

An enzymatic approach can be used to study CWP1 and CWP2 from suspension cells or plant tissue. A mixture of enzymes, cellulase and pectinase, is used to digest the cell wall carbohydrates, supplemented with an osmoticum medium containing salts and sorbitol [58] or glucose [85] to allow plasmolysis, which will lead to the obtention of protoplasts, *e.g.* plant cells which have had their cell wall removed. These protoplasts can be placed onto cell wall regeneration medium, where newly synthesized and secreted cell wall proteins are released using washes with low ionic salt solutions (Table 4.1).

For proteins embedded in the cell wall (CWP3), the cell wall can be described as recalcitrant to study. Disruptive methods must be used to study CWP3. The disruption is used to isolate the cell wall fraction first, followed by stringent washes with aqueous and organic solutions to remove contaminating proteins and small molecules from the intracellular compartment, prior to protein extraction using the same salts mentioned before, $CaCl_2$ and LiCl (Table 4.1). Another approach described previously for use with *Arabidopsis* suspension cells [26] consists of adding a sedimentation step in glycerol followed by centrifugation to remove the intracellular compartment which is less dense than the cell wall. This protocol was more recently adopted to extract cell wall proteins from sugarcane suspension cells [17]. In order to decrease cytosolic contamination, additional modifications have been considered. The modified protocol introduced by Feiz et al. [34], which was adopted for the study of cell wall proteins from alfalfa stem [102] and *Arabidopsis* suspension cells [49] consists of combining sequential steps of sedimentation using different concentrations of sucrose, followed by centrifugation, with an extensive wash with 5 mM acetate buffer, pH 4.6, to remove the sucrose, followed by two extraction steps with two low ionic salts of 0.2 M $CaCl_2$ and 2 M LiCl, including protease inhibitors, followed by a final extraction with detergent (4 % SDS).

One should keep in mind that cell wall protein extraction is not the only challenging aspect, the downstream procedure for protein identification or characterization can be far from straightforward depending on the origin and nature of the proteins extracted, such as the arabinogalactoproteins, or the amino acid Gly- or Pro-rich proteins found in the cell wall matrix.

4.2 Plant Organs and Organ Systems

Plants, unlike most mammals, have a limited number of organs, split between two categories, *vegetative* (i.e. root, stem and leaf) and *reproductive* (i.e. flower, fruit and seed). The "typical" plant body only consists of two organ systems:

- The root system (below the ground) which functions to anchor the plant to the soil, to absorb water and minerals and to store some of the products of photosynthesis
- The shoot system (above the ground) which includes the stem and conducts minerals and water from the root to the leaves with leaves being the critical location of photosynthesis for energy production and synthesis of organic and other compounds. At the reproductive stage, the shoot also includes flowers which when fertilized develop into fruit carrying the seeds for dispersion.

The shoot system and its specific components, because of their important role in energy production and reproduction, but also because of their ready availability and ease of harvesting, have resulted in the most proteomic publications. However, the shoot system also represents the most challenging structure because of the presence of high levels of phenolic compounds and derived pigments. The phenolic compounds are secondary small molecules with at least one hydroxy-substituted aromatic ring and a large diversity of structures, the properties of which are related to specific functions and location. Some phenolic compounds can be widespread throughout the plant while others are specific to certain plants, plant organs and developmental stages. Examples showing their extreme complexity in structure and distribution range from phenolic acids such as caffeic acid (critical precursor in the phenylpropanoid biosynthesis pathway) to flavonoids such as anthocyanins (pigments found in flowers and fruit involved in color and flavor) and lignin (polymerized aromatic alcohols constituent of the secondary cell wall). These compounds present a major challenge for protein extraction. Another abundant pigment found in plants and concentrated in the leaves is chlorophyll. It has a critical role in photosynthesis, functioning as the antenna, assisted by other components, that absorbs photons from light and converts them into energetic electrons further captured by coenzymes involved in Calvin Cycle reactions. In addition to chlorophyll, another abundant component of the photosynthesis process responsible for the fixation of CO_2 is the enzyme Rubisco (Ribulose-1,5-bisphosphate carboxylase/oxygenase), also known as the most abundant protein on earth. The abundance of this protein is a major issue in proteomics, since its presence makes the study of low level proteins very challenging. Various Rubisco depletion strategies have been developed. Amongst them, small molecules such as phytate or polyethylenimine (PEI), which interact with specific proteins, have been used to precipitate Rubisco from protein samples. In the case of phytate, the interaction, done in the presence of Ca^{2+} at a defined pH of 6.8, removed 85 % of Rubisco from soybean leaves [56]. PEI was successfully used in combination with fractionation to increase the protein resolution, an approach called PARC (PEI-assisted Rubisco cleanup) [118]. However these interactions are not specific to Rubisco and can also precipitate non-targeted proteins. Commercial kits using immunoaffinity removal of Rubisco are also available (IgY Rubisco columns from Sigma; Rubisco depletion kit from Agrisera). However, the antibodies in these kits do not work as well for every plant species. A different approach, using differential PEG (polyethylene glycol) precipitation, was used to isolate Rubisco from a specific fraction [113]. PEG present in the fractions now containing only low levels of Rubisco could then be cleaned up using one of the protein extraction methods used for plants as described below. Clearly, this type of fractionation can also lead to losses of other groups of proteins that coprecipitate with Rubisco. There clearly is no perfect solution for increasing the dynamic range of proteins identified from tissues containing Rubisco without suffering from protein losses one way or another. An alternative strategy to increase the resolution and the dynamic range is to increase the fractionation of protein samples or consider subcellular fractionation, both of which incur time penalties. Similar challenges are faced in the study of seeds and their germination process. Seeds, essential for propagation and multiplication of plants, are mainly composed of an endosperm, which is the storage compartment consisting of starch and storage proteins (the source of nutrients during germination) and the embryo, which will germinate into a seedling when dormancy is abolished. When the seed is studied as a whole, it can be a challenge to identify low abundance proteins amongst the abundant storage proteins. In rice, a PEG-assisted fractionation method has been used to remove the abundant storage proteins [3]. More recently, a different approach was also developed to remove the storage proteins found in soybean (i.e. glycinin and beta-conglycinin) using a precipitation step with 10 mM Ca^{2+} [57].

The root system is a simple structure that has been largely ignored in plant proteomics, having lost out to the shoot system. Only over the past 6 years has interest in roots increased, because of their direct involvement in the perception of stresses related to water and nutrient availability in soils. Roots do not represent any greater difficulty in terms of protein extraction, though yields are somewhat lower than from leaves. Rather, issues arise with the isolation of clean, soil or media-free, roots. The time between harvest and the subsequent soil removal steps can delay the freezing of samples, which in turn can be responsible for biological variation in the downstream studies. One alternative is to grow plants and harvest roots using sterile medium on plates, or to use hydroponic cultures. In both cases, this makes harvesting easier and faster.

In addition, some specific organs and tissues found in certain species are considered recalcitrant to protein extraction. In particular, woody plant material or tissues containing large amounts of pigments, phenolic compounds and carbohydrates (i.e. flowers and fruits). These often bring with them significant difficulties and require specific sample preparation to remove interfering compounds. The presence of the cell wall also presents another issue. There are two main methods for total protein extraction described in the literature that aim to solve these challenges.

These are the TCA (trichloroacetic acid)/acetone precipitation method and the phenol extraction method. Both have been used successfully for various plant organ and tissue samples of a large variety of species. Although these protocols were first optimized to be compatible with two-dimensional polyacrylamide gel electrophoresis (2D-PAGE) as this was the main proteomics separation tool used 15 years ago, nowadays these protein extraction protocols are still being used for LC-based separations after adjustment of the solubilization buffer to make it mass spectrometry compatible.

The TCA/acetone precipitation method was used as early as the 1980s for protein separation using 1D and 2D-PAGE [29, 111]. Here, removal of pigments and phenolics, but also lipids and nucleic acids that could interfere with gel resolution was essential. The process consists of denaturing and precipitating proteins. Addition of 2-mercaptoethanol (2-ME) into the buffer inhibits the formation of new disulfide bonds. Most of the interfering compounds will stay soluble in the acetone, which is removed, leaving the pelleted proteins which can be solubilized using any type of 2D-PAGE or MS-compatible solubilization buffer, depending on later steps. The protocol adapted from Damerval et al. [29] with modifications is described in Table 4.2. Tissue disruption is done before the addition of precipitation buffer to improve total protein release through better tissue homogenization and cell wall breakage.

The phenol extraction method is commonly used by molecular biologists for nucleic acid extraction because it efficiently removes protein from samples. This feature meant it could be adapted for protein extraction from plant tissues [48], taking advantage of its ability to remove the interfering compounds, but also for its ability to remove DNA, to give improved 2D-PAGE separations compared to the TCA/acetone extraction. In this adapted phenol extraction method, the first step consists of separating proteins from salts, nucleic acids and carbohydrates with a Tris-phenol buffered solution mixed with a sucrose buffer. This inverts the phases putting the phenol phase at the top. The proteins, denatured and dissolved in the phenol phase, are then precipitated with a solution of methanolic ammonium acetate. The pellet is then successively washed with methanolic ammonium acetate, followed by acetone and methanol washes to remove pigments and lipids. The pellet can then be re-suspended in any buffer desired. The protocol used in our lab modified from [48] is described in Table 4.2.

It is important to note that:

1. The phenol extraction method is more favorable for solubilizing membrane-associated proteins, not like the TCA/acetone method which favors water soluble proteins.
2. For both the TCA/acetone and phenol extraction protocols, the use of protease inhibitors is not mandatory, because these methods use denaturing conditions which have been reported to inhibit protease activity. Also throughout, the samples are kept at 4 °C or −20 °C as a precaution. However, because some proteases can still be active even in harsh conditions, it is recommended to add a cocktail of protease inhibitors, either in the acetone wash for the TCA/acetone procedure, or in the aqueous extraction buffer for the phenol extraction method.
3. Although manual grinding for tissue disruption leads in our hands to better protein yields, this step can be automated for higher throughput if laboratories are equipped with a bead beating tool which uses grinding balls and a shaking homogenizer to disrupt the tissues or cells. The use of these tools also provides better reproducibility between samples; however, attention has to be placed on keeping samples frozen during the homogenization steps to avoid protease activity.
4. Protein solubilization from the pellet is more efficient after phenol extraction than using the TCA/acetone method, resulting in reduced protein loss.

Table 4.2 TCA/acetone precipitation and phenol extraction procedures

TCA/acetone precipitation	Phenol extraction
Buffers:	***Buffers:***
Precipitation buffer: 10 % trichloroacetic acid (TCA) in acetone solution. Store at −20 °C.	**Extraction buffer** (stock solution): 0.1 M Tris–HCl pH 8.8, 10 mM EDTA, 0.4 % 2-ME, 0.9 M sucrose – 2-ME must be omitted from the stock solution. Store at 4 °C.
Wash buffer: 100 % acetone. Store at −20 °C.	**Phenol buffer:** Phenol buffered with Tris–HCl, pH 8.8 (commercial product). Store at 4 °C.
	Wash buffer I: 0.1 M ammonium acetate in methanol, **II**: 80 % acetone and **III**: 70 % methanol. Store at −20 °C.
Procedure:	***Procedure:***
1. Aliquot the volume needed of COLD **precipitation buffer** and COLD **wash buffer**, and add 0.07 % 2-mercaptoethanol (2-ME) in both before use. Store at −20 °C.	1. Aliquot the volume of **extraction buffer** needed based on the number of samples and add 2-ME at 0.4 % and the cOmplete protease cocktail inhibitor (Roche) at 1x. Keep the buffer on ice.
2. Transfer the frozen samples (150–200 mg) to the liquid nitrogen in a mortar and grind into fine powder using the pestle. Note: Liquid nitrogen should be used to cool the equipment needed (mortar, pestle and spatula) just prior to use.	2. Transfer the frozen samples (100–150 mg) to the liquid nitrogen in the mortar and grind into a fine powder using the pestle. Note: Liquid nitrogen is used to cool down the equipment needed (mortar, pestle and spatula) before being used.
3. Using the chilled spatula transfer the ground sample into 2 mL centrifuge tubes and place them in liquid nitrogen until all the samples are ready.	3. Using the chilled spatula transfer the sample into 2 mL centrifuge tubes and place them in liquid nitrogen until all the samples are ready.
4. Remove the tubes from the liquid nitrogen and add 1.5 mL of COLD **precipitation buffer** to each.	4. Remove the tubes from the liquid nitrogen one by one and add 600 μL of extraction buffer and 600 μL of **phenol buffer** to each. Note: Use phenol and extraction buffer in the hood. If phenol droplets get on your gloves, change them as soon as possible.
5. Precipitate proteins by placing the tubes at −20 °C for at least 2 h or overnight.	5. Vortex the tubes immediately after adding the buffers and place on ice until all the samples are ready for the next step.
6. Centrifuge the tubes at 4 °C for 15 min at 16,000 × g.	6. Vortex the tubes on high for 30 min using a vortexer equipped with a tube adapter, placed in a cold room at 4 °C.
7. Remove the supernatant and wash the pellet by adding 1.5 mL COLD **wash buffer** and transferring them to −20 °C for at least 2 h.	7. Centrifuge the tubes for 15 min at 16,000 × g at 4 °C.
8. Repeat steps 6 and 7 one more time or until the pellet gets discolored.	8. Remove the phenol phase (top phase) and transfer it to a new 2 mL tube. Place the tubes on ice until all the samples are ready for the next step.
9. Dry the pellet under vacuum or air dry in the hood.	9. Back-extract the aqueous phase by adding 400 μL of phenol buffer.
10. Resuspend the pellet using the appropriate buffer (see Sect. 4.5. for solubilization buffer recommendations).	10. Repeat steps 5–8.
	11. Remove the phenol phase and combine it with the first extraction.
	12. Precipitate the phenol phase by adding 5 volumes of COLD **wash buffer I**. Split the sample into several 2 mL tubes if necessary before adding the **wash buffer I**.
	13. Vortex the tubes and incubate them at −20 °C for at least 2 h or overnight.
	14. Centrifuge the tubes at 4 °C for 15 min at 16,000 × g.

(continued)

Table 4.2 (continued)

TCA/acetone precipitation	Phenol extraction
	15. Remove the supernatant and wash the pellet by adding 1.5 mL of COLD **wash buffer I** and transferring to −20 °C for at least 20 min.
	16. Repeat steps 12 and 13 one more time with COLD **wash buffer I**, then one time with COLD **wash buffer II** and one time with COLD **wash buffer III**.
	17. Dry the pellet under vacuum or air.
	18. Resuspend the pellet using the appropriate buffer (see Sect. 4.5. for solubilization buffer recommendations).

4.3 Plant Meristem and Suspension Culture Cells (SCCs)

Suspension culture cells (SCCs) are widely used as a model system in plant biology to investigate the molecular basis of different mechanisms and their regulation, because they bypass the complexity of regulation from an entire plant. SCCs can be grown to any quantity required and the homogeneity of the population offers greater reproducibility. For proteomic studies, they also present an easily harvestable material which often does not contain the interfering compounds found in whole plants. SCCs are produced from plant tissue culture, which is widely used for gene transformation and regeneration of modified plants, or for the propagation of identical plants without the need for crossing and seed production. The cells can be prepared from different types of explants (hypocotyls, leaves and roots) by dedifferentiation techniques which yield unorganized cell masses called callus. These calluses can then be transplanted into a new medium for plant regeneration via somatic embryogenesis, or can be transferred into flasks containing liquid culture medium for growth of cell biomass to maintain suspension cells. Cell growth goes through an exponential phase before it slows down to a plateau, when the nutrients in the medium are no longer sufficient. Regularly transferring an aliquot of the cells into new medium allows suspension cells to be maintained for years.

Undifferentiated cells are also present *in planta* throughout their entire life. These rapidly dividing cells located at the tips of the root and shoot, which are called apical meristems are essential for producing new cells and tissue for plant growth. In contrast to suspension cells, meristems in plants are very limited in amount and can be difficult to isolate. However, they do not contain phenolics and other interfering compounds, so can be easier to study than other types of plant samples. Meristematic cells, like stem cells can also be used to seed new SCCs without the need for going through dedifferentiation first, providing an even faster way to obtain cell suspension material.

Protein extraction from suspension cells does not require the tedious extractions described previously for organ and organ systems; however, disruption of the cells is still required to break the cell wall structure to release the proteins. Homogenization of cells in the presence of a protein solubilization buffer can be used. This method does not require precipitation of the proteins first, instead, the proteins are directly solubilized and the cell debris is removed by centrifugation. A wider diversity of buffers can be used to extract soluble proteins, but buffers suitable for extracting membrane or membrane-bound proteins can also be used, depending on the requirements of the proteomics analytical approach selected (e.g. 2D-PAGE or LC-MS-based methods, with or without offline pre-fractionation, or multi-dimensional protein identification technology (MudPIT)).

Direct solubilization of proteins from cells without prior cleanup was inspired by the O'Farrell lysis buffer which contained 9 M urea, 2 % Nonidet P-40 (NP-40), 2 % 2-ME and 2 % carrier ampholytes (any desired pH interval) [73]. This buffer consists of a mixture of solubilizing components: a chaotrope to disrupt intra- and interprotein interactions and unfold structure by breaking hydrogen bonds; a detergent to disrupt hydrophobic bonds and improve protein solubilization; and a reducing reagent to break and prevent reformation of disulfide cross-links. Modifications to the lysis buffer recipe according to sample type and approach were made using mixtures of urea and thiourea with non-ionic (NP-40 or Triton X-100) or zwitterionic (3-[(3-Cholamidopropyl) dimethylammonio]-1-propanesulfonate; CHAPS) detergents containing reducing reagents (Dithiothreitol; DTT). Others replaced urea with the anionic detergent SDS (sodium dodecyl sulfate) [40]. The use of detergents, such as SDS and Triton X-100, gives increasingly better membrane protein solubilization when compared to the TCA/acetone and phenol extraction methods previously described. However, this does oversimplify the situation somewhat. Detergents perform differently depending on sample type and whether denaturing chaotropes such as urea and thiourea are present. Therefore, if membrane protein solubilization efficiency is the goal, optimization using the biological system of interest is required [62]. Some examples of lysis buffer compositions found in the literature are given in Table 4.3 to provide a useful starting point for single step protein solubilization during homogenization. Because of the efficiency of lysis buffers at directly solubilizing proteins from tissues, this method has not only been applied to suspension cells or meristems, but also to other organs. Similarly, the TCA/acetone and phenol precipitation extractions have their place, when the removal of interfering compounds is important from suspension cells and meristems, which have been used to elicit the synthesis of secondary small molecules into the liquid medium.

It is important to note that:

1. The use of protease inhibitors is critical. Examples are: ethylenediamine-tetraacetic acid (EDTA) or ethylene glycol tetraacetic acid (EGTA) – chelators; pepstatin A and phenylmethylsulfonyl fluoride (PMSF) – site blocking reagents; or a combination of both chelators and blocking site reagents; or the use of commercial mixtures such as the cOmplete protease inhibitor cocktail from Roche in the buffers to avoid protein degradation during solubilization.
2. In most cases this homogenization is followed by an acetone precipitation or desalting step to remove detergents and/or the high concentrations of salts and chaotropes which can interfere with downstream steps such as isoelectrofocusing (IEF) or trypsin digestion.
3. The presence of detergents in the final stages of sample analysis causes severe suppression of ionization in the mass spectrometer and major contamination of the separation systems that are usually online to such equipment.

Table 4.3 Examples of lysis buffer compositions used for total protein solubilization during tissue disruption and homogenization for different sample types and applications

Buffer composition	Sample type	Platform	Reference
0.5 M EDTA, pH 8.0, 1 M Tris–HCl, pH 6.8, 10 % SDS, 100 % 2-ME, 100 % glycerol and bromophenol blue powder	*Arabidopsis* cells	SDS-PAGE	[99]
8 M urea, 2 M thiourea, 2 mM disodium EDTA salt, 4 % CHAPS, 65 mM DTT, 2 % ampholytes (pH 3–10), and 1 % TBP (tributylphosphine)	Ginseng (*Panax ginseng*) cells	2D-PAGE	[97]
9.5 M urea, 2 % NP-40, 2 % ampholine (pH 3.5–10), 5 % 2-ME and 0.05 % polyvinylpyrrolidone (PVP-40)	Rice suspension cells	2D-PAGE	[55]
50 mM Tris–HCl, 15 mM EDTA, 1 mM NaF, 0.5 mM Na_3VO_3, 15 mM β-glycerophosphate, 1 mM PMSF, 1 mM DTT, 2 μg/mL pepstatin, 2 μg/mL aprotinin, 2 μg/mL leupeptin, pH 7.5	Rice suspension cells	DIGE and iTRAQ	[61]

4.4 Green Algae and Plastids

Plants are not the only organisms with a cell wall capable of photosynthesis and therefore posing similar challenges in the study of their protein content. Algae are eukaryotic organisms more recently classified in the kingdom of protists. They are found as unicellular or multicellular organisms living in salt or fresh water environments. Most of them are photosynthetic and able to fix CO_2 from the atmosphere or from organic sources using light and water to produce energy and biomass while releasing oxygen into the atmosphere. The cell wall found in algae presents a huge variety in its composition that has evolved throughout the taxa, with the later divergence in the algal taxa presenting a very similar cell wall composition to that of land plants. Their cell wall acts as a physical protection and defense against microbial attack and also, just like in plants, has a great implication in cell-cell and cell-substrate sensing and communication. Furthermore, in contrast to plants, the algal cell wall is involved in sexual reproduction. The study of the cell wall composition found in the different algal taxa has been recently reviewed [32].

In addition to the cell wall, algal cells also contain plastids that are capable of photosynthesis. Depending on the pigment accumulated, three out of the seven phyla of algae are easily identifiable: the green algae or Chlorophyta consists of plastids containing chlorophyll a and b, similar to plant chloroplasts, and also some carotenoids; the red algae or Rhodophyta contains chlorophyll a and accumulates phycobilins in their plastids; the brown algae or Phaeophyta has chlorophyll a and c plus fucoxanthin. In addition to algae, because cyanobacteria are also photosynthetic and live in water, these have been named the blue-green algae in reference to the pigments accumulating in their cells. However, cyanobacteria are prokaryotic organisms with no plastids, which instead use the pigment phycocyanin to capture light for photosynthesis. It is this group of organisms that is thought to be the ancestral precursor of the chloroplast found in plant cells. Because their cell wall is similar to gram negative bacteria, this organism will not be addressed in this section.

The green algae have been shown greater interest over the past decade because of their potential to solve the global food and fuel crisis. Indeed, some families of algae produce high levels of lipid and can be grown fast without the use of valuable farmland. However, the cost for production of energy-rich oil from algae is very expensive and more efforts still have to be made to increase production and lower the cost in order to compete with the cost of crude oil. Understanding how to increase lipid production and biomass is critical and proteomics tools are among those that are used to tease out the pathways involved. Because of the presence of chlorophyll and lipids the TCA/acetone precipitation and phenol extraction methods have been shown to be the most successful in proteomic studies [105].

Plastids like chloroplasts have their own DNA and protein biosynthesis machinery. Understanding the origin of chloroplast proteins and how they are involved in chloroplast development, signaling and interaction networks is of great interest. Plastids and chloroplasts can be isolated from algal cells or plant tissue, respectively, by using sucrose [45] or Percoll density gradients [51, 52]. Much has already been done to identify the chloroplast proteomes in plants, and the specific subproteomes of the thylakoid membrane and lumen [78, 79], and the chloroplast envelope [16].

4.5 Recommendations for Selection of the Optimal Protein Extraction Method According to Sample Type

The "best" method of protein extraction for each sample type and species (i.e. the method leading to the highest number of proteins in a reproducible manner) will be one that has been fully optimized for the particular system in hand

Table 4.4 Summary table of references using one or more of the three main methods used for plant protein extraction according to sample type and species

Sample type	TCA/acetone precipitation	Phenol extraction	Tissue homogenization in buffer
Seedlings	*Arabidopsis* [27]; Rice (*Oryza sativa*) [100]		*Arabidopsis* [11, 42]
Shoots	*Arabidopsis* [9]		Maize (*Zea mays*) [93]
Leaves	Banana (*Musa* spp.) [18]; Cucumber (*Cucumis sativus* L.) [27]; Maize [10, 27, 103]; Rice [27, 100, 115]; Sugar beet (*Beta vulgaris* L.) [39, 116]; Wheat (*Triticum aestivum* L.) [27, 80]	Banana [18]; *Brachypodium distachyon* [63]; Grape (*Vitis vinifera*) [66]; Olive (*Olea europaea* L.) [106]; Potato (*Solanum tuberosum* L.) [18]	Maize [93]; Rice [53, 54, 91]; Soybean (*Glycine max*) [95]; Wheat [67]
Stem/ hypocotyl	Rice [100];		Soybean [95]
Roots	*Arabidopsis* [9, 50, 59]; Barley (*Hordeum vulgare* L.) [110]; Cucumber (*Cucumis sativus*) [33]; Rice [71, 100, 114]; Wheat [80]; Sugar beet [116]	*Arabidopsis* [6]; Avocado (*Persea Americana*) [2]; *Brassica juncea* [4]; Grape [66]; Rice [25]; Soybean [104]; Wheat [8]	Rice [11, 121]; Soybean [95]
Flowers	*Cannabis sativa* [83]		
Fruit		Apple (*Malus x domestica*) [92, 120]; Avocado (*Persea americana*), Banana and Tomato (*Solanum lycopersicum*) [87];	Tomato [89]
Seeds (including endosperm and embryos)	European beech (*Fagus sylvatica* L.) [76]; Maize [69]; Norway maple (*Acer platanoides* L.) [77]; Rice [100];	Cacao (*Theobroma cacao* L.) [72]; Soybean [23, 38]; Maize [94]	*Arabidopsis* [35, 36]; *Camellia sinensis* [21]; Rice [27, 53, 117]; Maize [22]
Meristems	Banana [18]; *Medicago truncatula* [44]	Apple [18]; Banana [18, 86]; Cacao [81]	
Suspension Cells	Sugar beet [75]; Ginseng (*Panax ginseng*) [97]; Grape [90]	Sugar beet [75]; Ginseng [97]	Rice [55, 61]; Ginseng [97]
Others	Horse gram [13]; Maize xylem sap [5]; Rice bran, chaff and callus [100]. Tomato flower bud (Zhao et al. 2013); Cactus and houseleek [75]	Cactus and houseleek [75]; Cotton (*Gossypium barbadense*) ovules [46]; Horse gram [13]; Poplar wood [101]; Wheat callus [122]	Horse gram [13];

before committing to any further proteomics experiments. However, given no prior knowledge, it is most often the case that scientists will adopt one of the previously described methods (TCA/acetone precipitation, phenol extraction and direct extraction using lysis buffer) and stick with it. Table 4.4 summarizes the references describing the use of one or more of these methods to help with the selection of the method most likely to offer initial success for each sample type and species. Only a few studies have systematically compared methods for specific sample types, such as recalcitrant plant tissue of banana leaves [18], various fruits (banana, avocado and tomato) [87, 119], horse gram [13] and grapewine leaves, pine needles and cork oak ectomycorrhizal roots [88]. They concluded that while both TCA/acetone and phenol extractions are more suitable for recalcitrant plant tissue and give similar protein yields, the phenol extraction method gives higher quality 2D-PAGE results, showing that the cleaning steps lead to cleaner samples with less interfering contaminants, aided by the use of phenol which acts as a dissociating reagent decreasing the interactions between proteins and other compounds [18]. These studies also noted that the phenol extraction method gave a more glycoprotein-rich sample with reduced levels of Rubisco [18, 87]. Other studies in samples of sugar beet cells, cactus and houseleek [75], and ginseng cells [97] showed that phenol extraction has a higher cleaning capacity, which also led to better protein yields. This result is consistent with the observation noted in Table 4.4 where the largest contribution of samples extracted by the phenol extraction method comes from the group of plant tissues considered recalcitrant.

We have to keep in mind that even if the phenol extraction method is known to give cleaner protein samples, its time-consuming use may not nowadays be required with the evolution of downstream experiments having moved away from 2D-PAGE to gel-free methods. Here, offline HPLC fractionation is often combined with LC-MS/MS. However, there is still a distinct lack of literature on the comparison of extraction methods using gel-free shotgun proteomics approaches in the plant field.

Only a few plant studies have tried to find alternative extraction protocols for gel-based approaches that match the ones previously described here. Their goal has been to combine the benefits of each. In 2006, a study performed on leaf tissue of various species (*Arabidopsis*, rice, wheat, maize and cucumber) showed that using a lysis buffer (composed of 7 M urea, 2 M thiourea, 4 % CHAPS, 18 mM Tris–HCl, pH 8.0) for the solubilization of protein pellets obtained from TCA/acetone precipitation improved the number of spots resolved by 2D-PAGE when compared to the use of lysis buffer without pre-TCA/acetone precipitation [27]. The addition of the TCA/acetone precipitation removes most of the non-protein compounds allowing better solubilization of the pelleted material. Another study from 2006 compared the use of TCA/acetone precipitation and the lysis buffer method, to the use of SDS buffer (2 % SDS, 60 mM DTT, 20 % glycerol and 40 mM Tris–HCl, pH 8.5) on the recalcitrant fruit tissue of apple and banana [96]. After boiling the samples in SDS-containing buffer, the proteins were precipitated using TCA/acetone and washed to remove the SDS. Although the use of SDS improves membrane protein solubilization, even small amounts of SDS left after precipitation can impair many downstream analytical approaches through the presence of the negative charge. The use of SDS has been combined with phenol extraction and with the benefits of TCA/acetone precipitation from various recalcitrant plant tissues [107]. This combination allows removal of many of the interfering phenolic compounds and pigments present in recalcitrant plant tissues – the presence of SDS helps with the solubilization of proteins before phenol extraction, but becomes very time consuming. A variation of the same protocol without the TCA/acetone extraction was more recently tested and optimized on soybean roots using the extraction buffer: 0.1 M Tris–HCl, pH 8.0, 2 % SDS, 5 % 2-ME, 30 % sucrose, 1 mM PMSF [84]. The comparison of both protocols, with or without TCA/acetone precipitation before SDS/phenol extraction did not show significant differences either in the 2D-PAGE profiles, which were very well resolved and streakless, or in protein yield and reproducibility. However for recalcitrant plant tissues, this protocol can be recommended and has been more recently optimized and described [112]. In summary, although some efforts have been made to find a faster, more reproducible and higher yield protein extraction method, the methods established in the 1980s are still very popular and have not been completely superseded.

In contrast to lysis buffers used for direct protein extraction during tissue homogenization, so-called solubilization buffers used to dissolve the protein pellets after the TCA/acetone or phenol precipitation steps have evolved with the move from 2D-PAGE to mass spectrometry-based approaches. At the point where IPG strips were introduced for 2D-PAGE the solubilization buffers used after TCA/acetone extraction were modified from the original O'Farrell lysis buffer for compatibility. These modifications involved combining chaotropes such as urea with thiourea, adding detergents such as Triton X-100, CHAPS

Table 4.5 Examples of buffer composition for protein solubilization according to the proteomics platform used

Buffer composition	Sample type	Platform	References
7 M urea, 2 M thiourea, 4 % CHAPS, 18 mM Tris-HCI (pH 8.0), 14 mM Trizma® base, two EDTA-free protease inhibitor cocktail tablets, 0.2 % Triton X-100 (R), and 50 mM DTT, to a final volume of 100 mL	*Arabidopsis* seedlings, leaves of rice, maize, wheat, cucumber	2D-PAGE	[27]
5 M urea, 2 M thiourea, 2 % CHAPS, 2 % Sulfobetaine 3–10, 20 mM DTT, 5 mM TCEP, 0.75 % carrier ampholytes	Maize endosperm	2D-PAGE	[69]
0.5 M bicine buffer, pH 8.5 containing 0.1 % SDS	Grapevine leaf	iTRAQ/LC-MS/MS	[65]
1 M urea, 0.5 M bicine, 0.09 % SDS	*Arabidopsis* roots	iTRAQ/LC-MS/MS	[7]
8 M urea in 500 mM triethylammonium bicarbonate (TEAB)	Wheat roots	iTRAQ/LC-MS/MS	[8]
8 M urea, 25 mM TEAB, 0.2 % Triton X-100, 0.1 % SDS, pH 8.5	Cotton (*Gossypium barbadense*) ovules	iTRAQ/LC-MS/MS	[47]

or sulfobetaine 3–10, and reducing reagents such as DTT and tributylphosphine [68]. Some examples of buffer compositions used are found in Table 4.5. These buffers were also found useful for solubilization of protein pellets after phenol extraction prior to 2D-PAGE, but are incompatible with gel-free LC-MS/MS approaches. The high concentration of chaotropes such as urea, without prior removal or dilution, can deactivate proteases such as trypsin. Native proteases can also be affected and deactivated by high urea, a useful feature, though offset by the fact that urea, if not prepared correctly, can highly carbamylate amino groups and sulfhydryls making peptide identification difficult. The use of detergents like CHAPS, Triton X-100 and SDS are also incompatible with liquid chromatography and mass spectrometry and must be removed prior to analysis. An acetone precipitation is often used at the protein level to remove the interfering reagents from the sample, however protein losses are observed and membrane proteins, for example, do not redissolve as well in buffers that do not contain detergents.

A new workflow was introduced in 2008 to remove chaotropes and detergents efficiently before mass spectrometry, known as GeLC-MS/MS [37]. In this method, pellet solubilization is done using the standard Laemmli buffer (0.1 % 2-ME, 0.0005 % Bromophenol blue,10 % glycerol, 2 % SDS, 63 mM Tris–HCl, pH 6.8), followed by SDS-PAGE to fractionate and separate the proteins from the detrimental reagents. The gel lane is then cut up into many consecutive pieces, the proteins digested and the released peptides analyzed by LC-MS/MS. This workflow is widely used for shotgun proteomics experiments as well as for label-free quantitative proteomics by spectrum counting. However, it is not compatible with quantitative proteomics platforms using isobaric tags (i.e. iTRAQ and TMT) for multiplex labeling. The labeling step, here required after protein digestion and before subsequent fractionation, has additional buffer compatibility requirements (absence of primary amines). The iTRAQ kit provides a dissolution buffer consisting of 0.5 M triethylammonium bicarbonate (TEAB, pH 8.5) where SDS is added to 1 %. However, this buffer has very limited efficiency at completely solubilizing protein pellets after phenol extraction, acetone precipitation or TCA/acetone extraction. Additionally, since the SILAC approach is not suitable for plants because of limited isotope amino acid incorporation into plant cells and tissues, iTRAQ and TMT labeling is more widely used. Solubilizing these protein pellets in a manner compatible with the labeling reagents still remains a challenge. Some examples of compatible buffer compositions for use with these quantitative labeling platforms are given in Table 4.5. The pellets do not fully redissolve, but they have been shown to give good protein recovery. Beyond the challenges of cell walls and plant protein extraction, once the protein is digested the techniques that can be applied at the peptide level are the same as for any other organism.

References

1. The Arabidopsis Genome Initiative (2000) Analysis of the genome sequence of the flowering plant Arabidopsis thaliana. Nature 408:796–815
2. Acosta-Muniz CH, Escobar-Tovar L, Valdes-Rodriguez S, Fernandez-Pavia S, Arias-Saucedo LJ, de la Cruz Espindola Barquera M, Gomez Lim MA (2012) Identification of avocado (Persea americana) root proteins induced by infection with the oomycete Phytophthora cinnamomi using a proteomic approach. Physiol Plant 144:59–72
3. Ahsan N, Lee SH, Lee DG, Lee H, Lee SW, Bahk JD, Lee BH (2007) Physiological and protein profiles alternation of germinating rice seedlings exposed to acute cadmium toxicity. C R Biol 330:735–746
4. Alvarez S, Berla BM, Sheffield J, Cahoon RE, Jez JM, Hicks LM (2009) Comprehensive analysis of the Brassica juncea root proteome in response to cadmium exposure by complementary proteomic approaches. Proteomics 9:2419–2431
5. Alvarez S, Goodger JQ, Marsh EL, Chen S, Asirvatham VS, Schachtman DP (2006) Characterization of the maize xylem sap proteome. J Proteome Res 5:963–972
6. Alvarez S, Hicks LM, Pandey S (2011) ABA-dependent and -independent G-protein signaling in Arabidopsis roots revealed through an iTRAQ proteomics approach. J Proteome Res 10:3107–3122
7. Alvarez S, Roy Choudhury S, Hicks LM, Pandey S (2013) Quantitative proteomics-based analysis supports a significant role of GTG proteins in regulation of ABA response in Arabidopsis roots. J Proteome Res 12:1487–1501
8. Alvarez S, Roy Choudhury S, Pandey S (2014) Comparative quantitative proteomics analysis of the ABA response of roots of drought-sensitive and drought-tolerant wheat varieties identifies proteomic signatures of drought adaptability. J Proteome Res 13:1688–1701
9. Alvarez S, Zhu M, Chen S (2009) Proteomics of Arabidopsis redox proteins in response to methyl jasmonate. J Proteome 73:30–40
10. Amiour N, Imbaud S, Clement G, Agier N, Zivy M, Valot B, Balliau T, Armengaud P, Quillere I, Canas R, Tercet-Laforgue T, Hirel B (2012) The use of metabolomics integrated with transcriptomic and proteomic studies for identifying key steps involved in the control of nitrogen metabolism in crops such as maize. J Exp Bot 63:5017–5033
11. Ban Y, Kobayashi Y, Hara T, Hamada T, Hashimoto T, Takeda S, Hattori T (2013) Alpha-tubulin is rapidly phosphorylated in response to hyperosmotic stress in rice and Arabidopsis. Plant Cell Physiol 54:848–858
12. Bayer EM, Bottrill AR, Walshaw J, Vigouroux M, Naldrett MJ, Thomas CL, Maule AJ (2006) Arabidopsis cell wall proteome defined using multidimensional protein identification technology. Proteomics 6:301–311
13. Bhardwaj J, Yadav SK (2013) A common protein extraction protocol for proteomic analysis: horse gram a case study. Am J Agric Biol Sci 8:293–301
14. Borderies G, Jamet E, Lafitte C, Rossignol M, Jauneau A, Boudart G, Monsarrat B, Esquerre-Tugaye MT, Boudet A, Pont-Lezica R (2003) Proteomics of loosely bound cell wall proteins of Arabidopsis thaliana cell suspension cultures: a critical analysis. Electrophoresis 24:3421–3432
15. Boudart G, Jamet E, Rossignol M, Lafitte C, Borderies G, Jauneau A, Esquerre-Tugaye MT, Pont-Lezica R (2005) Cell wall proteins in apoplastic fluids of Arabidopsis thaliana rosettes: identification by mass spectrometry and bioinformatics. Proteomics 5:212–221
16. Brautigam A, Hoffmann-Benning S, Weber AP (2008) Comparative proteomics of chloroplast envelopes from C3 and C4 plants reveals specific adaptations of the plastid envelope to C4 photosynthesis and candidate proteins required for maintaining C4 metabolite fluxes. Plant Physiol 148:568–579
17. Calderan-Rodrigues MJ, Jamet E, Bonassi MB, Guidetti-Gonzalez S, Begossi AC, Setem LV, Franceschini LM, Fonseca JG, Labate CA (2014) Cell wall proteomics of sugarcane cell suspension cultures. Proteomics 14:738–749
18. Carpentier SC, Witters E, Laukens K, Deckers P, Swennen R, Panis B (2005) Preparation of protein extracts from recalcitrant plant tissues: an evaluation of different methods for two-dimensional gel electrophoresis analysis. Proteomics 5:2497–2507
19. Casasoli M, Spadoni S, Lilley KS, Cervone F, De Lorenzo G, Mattei B (2008) Identification by 2-D DIGE of apoplastic proteins regulated by oligogalacturonides in Arabidopsis thaliana. Proteomics 8:1042–1054
20. Charmont S, Jamet E, Pont-Lezica R, Canut H (2005) Proteomic analysis of secreted proteins from Arabidopsis thaliana seedlings: improved recovery following removal of phenolic compounds. Phytochemistry 66:453–461
21. Chen Q, Yang L, Ahmad P, Wan X, Hu X (2011) Proteomic profiling and redox status alteration of recalcitrant tea (Camellia sinensis) seed in response to desiccation. Planta 233:583–592
22. Chen ZY, Brown RL, Rajasekaran K, Damann KE, Cleveland TE (2006) Identification of a maize kernel pathogenesis-related protein and evidence for its

involvement in resistance to Aspergillus flavus infection and Aflatoxin production. Phytopathology 96:87–95
23. Cheng L, Gao X, Li S, Shi M, Javeed H, Jing X, Yang G, He G (2010) Proteomic analysis of soybean [Glycine max (L.) Meer.] seeds during imbibition at chilling temperature. Mol Breed 26:1–17
24. Cheng WH, Taliercio EW, Chourey PS (1996) The miniature1 seed locus of maize encodes a cell wall invertase required for normal development of Endosperm and maternal cells in the Pedicel. Plant Cell 8:971–983
25. Chitteti BR, Peng Z (2007) Proteome and phosphoproteome differential expression under salinity stress in rice (Oryza sativa) roots. J Proteome Res 6:1718–1727
26. Chivasa S, Ndimba BK, Simon WJ, Robertson D, Yu XL, Knox JP, Bolwell P, Slabas AR (2002) Proteomic analysis of the Arabidopsis thaliana cell wall. Electrophoresis 23:1754–1765
27. Cho K, Torres NL, Subramanyam S, Deepak SA, Sardesai N, Han O, Williams CE, Ishii H, Iwahashi H, Rakwal R (2006) Protein extraction/solubilization protocol for monocot and dicot plant gel-based proteomics. J Plant Biol 49:413–420
28. Cordoba-Pedregosa M, Gonzalez-Reyes JA, Canadillas M, Navas P, Cordoba F (1996) Role of apoplastic and cell-wall peroxidases on the stimulation of root elongation by Ascorbate. Plant Physiol 112:1119–1125
29. Damerval C, De Vienne D, Zivy M, Thiellement H (1986) Technical improvements in two-dimensional electrophoresis increase the level of genetic variation detected in wheat-seedling proteins. Electrophoresis 7:52–54
30. Dani V, Simon WJ, Duranti M, Croy RR (2005) Changes in the tobacco leaf apoplast proteome in response to salt stress. Proteomics 5:737–745
31. de Souza IR, MacAdam JW (2001) Gibberellic acid and dwarfism effects on the growth dynamics of B73 maize (Zea mays L.) leaf blades: a transient increase in apoplastic peroxidase activity precedes cessation of cell elongation. J Exp Bot 52:1673–1682
32. Domozych DS, Ciancia M, Fangel JU, Mikkelsen MD, Ulvskov P, Willats WG (2012) The cell walls of green algae: a journey through evolution and diversity. Front Plant Sci 3:82
33. Du CX, Fan HF, Guo SR, Tezuka T, Li J (2010) Proteomic analysis of cucumber seedling roots subjected to salt stress. Phytochemistry 71:1450–1459
34. Feiz L, Irshad M, Pont-Lezica R, Canut H, Jamet E (2006) Evaluation of cell wall preparations for proteomics: a new procedure for purifying cell walls from Arabidopsis hypocotyls. Plant Methods 2:10
35. Gallardo K, Job C, Groot SP, Puype M, Demol H, Vandekerckhove J, Job D (2001) Proteomic analysis of arabidopsis seed germination and priming. Plant Physiol 126:835–848
36. Gallardo K, Job C, Groot SP, Puype M, Demol H, Vandekerckhove J, Job D (2002) Proteomics of Arabidopsis seed germination. A comparative study of wild-type and gibberellin-deficient seeds. Plant Physiol 129:823–837
37. Gao BB, Stuart L, Feener EP (2008) Label-free quantitative analysis of one-dimensional PAGE LC/MS/MS proteome: application on angiotensin II-stimulated smooth muscle cells secretome. Mol Cell Proteomics MCP 7:2399–2409
38. Hajduch M, Ganapathy A, Stein JW, Thelen JJ (2005) A systematic proteomic study of seed filling in soybean. Establishment of high-resolution two-dimensional reference maps, expression profiles, and an interactive proteome database. Plant Physiol 137:1397–1419
39. Hajheidari M, Abdollahian-Noghabi M, Askari H, Heidari M, Sadeghian SY, Ober ES, Salekdeh GH (2005) Proteome analysis of sugar beet leaves under drought stress. Proteomics 5:950–960
40. Harder A, Wildgruber R, Nawrocki A, Fey SJ, Larsen PM, Gorg A (1999) Comparison of yeast cell protein solubilization procedures for two-dimensional electrophoresis. Electrophoresis 20:826–829
41. Haslam RP, Downie AL, Raventon M, Gallardo K, Job D, Pallett KE, John P, Parry MAJ, Coleman JOD (2003) The assessment of enriched apoplastic extracts using proteomic approaches. Ann Appl Biol 143:81–91
42. Hebeler R, Oeljeklaus S, Reidegeld KA, Eisenacher M, Stephan C, Sitek B, Stuhler K, Meyer HE, Sturre MJ, Dijkwel PP, Warscheid B (2008) Study of early leaf senescence in Arabidopsis thaliana by quantitative proteomics using reciprocal 14 N/15N labeling and difference gel electrophoresis. Mol Cell Proteomics MCP 7:108–120
43. Hiilovaara-Teijo M, Hannukkala A, Griffith M, Yu XM, Pihakaski-Maunsbach K (1999) Snow-mold-induced apoplastic proteins in winter rye leaves lack antifreeze activity. Plant Physiol 121:665–674
44. Holmes P, Farquharson R, Hall PJ, Rolfe BG (2006) Proteomic analysis of root meristems and the effects of acetohydroxyacid synthase-inhibiting herbicides in the root of Medicago truncatula. J Proteome Res 5:2309–2316
45. Hopkins JF, Spencer DF, Laboissiere S, Neilson JA, Eveleigh RJ, Durnford DG, Gray MW, Archibald JM (2012) Proteomics reveals plastid- and periplastid-targeted proteins in the chlorarachniophyte alga Bigelowiella natans. Genome Biol Evol 4:1391–1406
46. Hu G, Koh J, Yoo M-J, Grupp K, Chen S, Wendel JF (2013) Proteomic profiling of developing cotton fibers from wild and domesticate Gossypium barbadense. New Phytol 200:570–582
47. Hu G, Koh J, Yoo MJ, Pathak D, Chen S, Wendel JF (2014) Proteomics profiling of fiber development and domestication in upland cotton (Gossypium hirsutum L.). Planta 240:1237

48. Hurkman WJ, Tanaka CK (1986) Solubilization of plant membrane proteins for analysis by two-dimensional gel electrophoresis. Plant Physiol 81:802–806
49. Irshad M, Canut H, Borderies G, Pont-Lezica R, Jamet E (2008) A new picture of cell wall protein dynamics in elongating cells of Arabidopsis thaliana: confirmed actors and newcomers. BMC Plant Biol 8:94
50. Jiang Y, Yang B, Harris NS, Deyholos MK (2007) Comparative proteomic analysis of NaCl stress-responsive proteins in Arabidopsis roots. J Exp Bot 58:3591–3607
51. Kamal AH, Cho K, Kim DE, Uozumi N, Chung KY, Lee SY, Choi JS, Cho SW, Shin CS, Woo SH (2012) Changes in physiology and protein abundance in salt-stressed wheat chloroplasts. Mol Biol Rep 39:9059–9074
52. Kamal AH, Cho K, Komatsu S, Uozumi N, Choi JS, Woo SH (2012) Towards an understanding of wheat chloroplasts: a methodical investigation of thylakoid proteome. Mol Biol Rep 39:5069–5083
53. Komatsu S, Kajiwara H, Hirano H (1993) A rice protein library: a data-file of rice proteins separated by two-dimensional electrophoresis. TAG Theor Appl Genet Theoretische und angewandte Genetik 86:935–942
54. Komatsu S, Muhammad A, Rakwal R (1999) Separation and characterization of proteins from green and etiolated shoots of rice (Oryza sativa L.): towards a rice proteome. Electrophoresis 20:630–636
55. Komatsu S, Rakwal R, Li Z (1999) Separation and characterization of proteins in rice (Oryza sativa) suspension cultured cells. Plant Cell Tissue Organ Cult 55:183–192
56. Krishnan HB, Natarajan SS (2009) A rapid method for depletion of Rubisco from soybean (Glycine max) leaf for proteomic analysis of lower abundance proteins. Phytochemistry 70:1958–1964
57. Krishnan HB, Oehrle NW, Natarajan SS (2009) A rapid and simple procedure for the depletion of abundant storage proteins from legume seeds to advance proteome analysis: a case study using Glycine max. Proteomics 9:3174–3188
58. Kwon HK, Yokoyama R, Nishitani K (2005) A proteomic approach to apoplastic proteins involved in cell wall regeneration in protoplasts of Arabidopsis suspension-cultured cells. Plant Cell Physiol 46:843–857
59. Lan P, Li W, Wen TN, Shiau JY, Wu YC, Lin W, Schmidt W (2011) iTRAQ protein profile analysis of Arabidopsis roots reveals new aspects critical for iron homeostasis. Plant Physiol 155:821–834
60. Li ZC, McClure JW, Hagerman AE (1989) Soluble and bound apoplastic activity for peroxidase, beta-d-glucosidase, malate dehydrogenase, and nonspecific Arylesterase, in barley (Hordeum vulgare L.) and oat (Avena sativa L.) primary leaves. Plant Physiol 90:185–190
61. Liu D, Ford KL, Roessner U, Natera S, Cassin AM, Patterson JH, Bacic A (2013) Rice suspension cultured cells are evaluated as a model system to study salt responsive networks in plants using a combined proteomic and metabolomic profiling approach. Proteomics 13:2046–2062
62. Luche S, Santoni V, Rabilloud T (2003) Evaluation of nonionic and zwitterionic detergents as membrane protein solubilizers in two-dimensional electrophoresis. Proteomics 3:249–253
63. Lv DW, Subburaj S, Cao M, Yan X, Li X, Appels R, Sun DF, Ma W, Yan YM (2014) Proteome and phosphoproteome characterization reveals new response and defense mechanisms of Brachypodium distachyon leaves under salt stress. Mol Cell Proteomics MCP 13:632–652
64. Macadam JW, Sharp RE, Nelson CJ (1992) Peroxidase activity in the leaf elongation zone of tall fescue: II. Spatial distribution of apoplastic peroxidase activity in genotypes differing in length of the elongation zone. Plant Physiol 99:879–885
65. Marsh E, Alvarez S, Hicks LM, Barbazuk WB, Qiu W, Kovacs L, Schachtman D (2010) Changes in protein abundance during powdery mildew infection of leaf tissues of Cabernet Sauvignon grapevine (Vitis vinifera L.). Proteomics 10:2057–2064
66. Marsoni M, Vanini C, Campa M, Cucchi U, Espen L, Bracale M (2005) Protein extraction from grape tissues by two-dimensional electrophoresis. Vitis 44:181–186
67. Maytalman D, Mert Z, Baykal AT, Inan C, Gunel A, Hasancebi S (2013) Proteomic analysis of early responsive resistance proteins of wheat (Triticum aestivum) to yellow rust (Puccinia striformis f. sp. tritici) using ProteomeLab PF2D. Plant OMICS J 6:24–35
68. Mechin V, Consoli L, Le Guilloux M, Damerval C (2003) An efficient solubilization buffer for plant proteins focused in immobilized pH gradients. Proteomics 3:1299–1302
69. Mechin V, Thevenot C, Le Guilloux M, Prioul JL, Damerval C (2007) Developmental analysis of maize endosperm proteome suggests a pivotal role for pyruvate orthophosphate dikinase. Plant Physiol 143:1203–1219
70. Morrow DL, Jones RL (1986) Localization and partial characterization of the extracellular proteins centrifuged from pea internodes. Physiol Plant 67:397–407
71. Nam MH, Huh SM, Kim KM, Park WJ, Seo JB, Cho K, Kim DY, Kim BG, Yoon IS (2012) Comparative proteomic analysis of early salt stress-responsive proteins in roots of SnRK2 transgenic rice. Proteome Sci 10:25
72. Noah AM, Niemenak N, Sunderhaus S, Haase C, Omokolo DN, Winkelmann T, Braun HP (2013) Comparative proteomic analysis of early somatic and zygotic embryogenesis in Theobroma cacao L. J Proteome 78:123–133

73. O'Farrell PH (1975) High resolution two-dimensional electrophoresis of proteins. J Biol Chem 250:4007–4021
74. Olivieri F, Godoy AV, Escande A, Casalongue CA (1998) Analysis of intercellular washing fluids of potato tubers and detection of increased proteolytic activity upon fungal infection. Physiol Plant 10:232–238
75. Pavokovic D, Kriznik B, Krsnik-Rasol M (2012) Evaluation of protein extraction methods for proteomic analysis of non-model recalcitrant plant tissues. Croat Chem Acta 85:177–183
76. Pawlowski TA (2007) Proteomics of European beech (Fagus sylvatica L.) seed dormancy breaking: influence of abscisic and gibberellic acids. Proteomics 7:2246–2257
77. Pawlowski TA (2009) Proteome analysis of Norway maple (Acer platanoides L.) seeds dormancy breaking and germination: influence of abscisic and gibberellic acids. BMC Plant Biol 9:48
78. Peltier JB, Cai Y, Sun Q, Zabrouskov V, Giacomelli L, Rudella A, Ytterberg AJ, Rutschow H, van Wijk KJ (2006) The oligomeric stromal proteome of Arabidopsis thaliana chloroplasts. Mol Cell Proteomics MCP 5:114–133
79. Peltier JB, Friso G, Kalume DE, Roepstorff P, Nilsson F, Adamska I, van Wijk KJ (2000) Proteomics of the chloroplast: systematic identification and targeting analysis of lumenal and peripheral thylakoid proteins. Plant Cell 12:319–341
80. Peng Z, Wang M, Li F, Lv H, Li C, Xia G (2009) A proteomic study of the response to salinity and drought stress in an introgression strain of bread wheat. Mol Cell Proteomics MCP 8:2676–2686
81. Pirovani CP, Carvalho HA, Machado RC, Gomes DS, Alvim FC, Pomella AW, Gramacho KP, Cascardo JC, Pereira GA, Micheli F (2008) Protein extraction for proteome analysis from cacao leaves and meristems, organs infected by Moniliophthora perniciosa, the causal agent of the witches' broom disease. Electrophoresis 29:2391–2401
82. Qiu X, Wong G, Audet J, Bello A, Fernando L, Alimonti JB, Fausther-Bovendo H, Wei H, Aviles J, Hiatt E, Johnson A, Morton J, Swope K, Bohorov O, Bohorova N, Goodman C, Kim D, Pauly MH, Velasco J, Pettitt J, Olinger GG, Whaley K, Xu B, Strong JE, Zeitlin L, Kobinger GP (2014) Reversion of advanced Ebola virus disease in nonhuman primates with ZMapp. Nature 514:47
83. Raharjo TJ, Widjaja I, Roytrakul S, Verpoorte R (2004) Comparative proteomics of Cannabis sativa plant tissues. J Biomol Tech JBT 15:97–106
84. Rodrigues EP, Torres AR, da Silva Batista JS, Huergo L, Hungria M (2012) A simple, economical and reproducible protein extraction protocol for proteomics studies of soybean roots. Genet Mol Biol 35:348–352
85. Roger D, David A, David H (1996) Immobilization of flax protoplasts in agarose and alginate beads. Correlation between ionically bound cell-wall proteins and morphogenetic response. Plant Physiol 112:1191–1199
86. Samyn B, Sergeant K, Carpentier S, Debyser G, Panis B, Swennen R, Van Beeumen J (2007) Functional proteome analysis of the banana plant (Musa spp.) using de novo sequence analysis of derivatized peptides. J Proteome Res 6:70–80
87. Saravanan RS, Rose JK (2004) A critical evaluation of sample extraction techniques for enhanced proteomic analysis of recalcitrant plant tissues. Proteomics 4:2522–2532
88. Sebastiana M, Figueiredo A, Monteiro F, Martins J, Franco C, Coelho AV, Vaz F, Simoes T, Penque D, Pais MS, Ferreira S (2013) A possible approach for gel-based proteomic studies in recalcitrant woody plants. SpringerPlus 2:210
89. Shah P, Powell AL, Orlando R, Bergmann C, Gutierrez-Sanchez G (2012) Proteomic analysis of ripening tomato fruit infected by Botrytis cinerea. J Proteome Res 11:2178–2192
90. Sharathchandra RG, Stander C, Jacobson D, Ndimba B, Vivier MA (2011) Proteomic analysis of grape berry cell cultures reveals that developmentally regulated ripening related processes can be studied using cultured cells. PLoS One 6: e14708
91. Shen S, Sharma A, Komatsu S (2003) Characterization of proteins responsive to gibberellin in the leaf-sheath of rice (Oryza sativa L.) seedling using proteome analysis. Biol Pharm Bull 26:129–136
92. Shi Y, Jiang L, Zhang L, Kang R, Yu Z (2014) Dynamic changes in proteins during apple (Malus x domestica) fruit ripening and storage. Hortic Res 1:6
93. Shoresh M, Harman GE (2008) The molecular basis of shoot responses of maize seedlings to Trichoderma harzianum T22 inoculation of the root: a proteomic approach. Plant Physiol 147:2147–2163
94. Silva-Sanchez C, Chen S, Zhu N, Li QB, Chourey PS (2013) Proteomic comparison of basal endosperm in maize miniature1 mutant and its wild-type Mn1. Front Plant Sci 4:211
95. Sobhanian H, Razavizadeh R, Nanjo Y, Ehsanpour AA, Jazii FR, Motamed N, Komatsu S (2010) Proteome analysis of soybean leaves, hypocotyls and roots under salt stress. Proteome Sci 8:19
96. Song J, Braun G, Bevis E, Doncaster K (2006) A simple protocol for protein extraction of recalcitrant fruit tissues suitable for 2-DE and MS analysis. Electrophoresis 27:3144–3151
97. Sun J, Fu J, Zhou R (2014) Proteomic analysis of differentially expressed proteins induced by salicylic acid in suspension-cultured ginseng cells. Saudi J Biol Sci 21:185–190
98. Swarbreck D, Wilks C, Lamesch P, Berardini TZ, Garcia-Hernandez M, Foerster H, Li D, Meyer T, Muller R, Ploetz L, Radenbaugh A, Singh S, Swing V, Tissier C, Zhang P, Huala E (2008) The Arabidopsis Information Resource (TAIR): gene

structure and function annotation. Nucleic Acids Res 36:D1009–D1014
99. Tsugama D, Liu S, Takano T (2011) A rapid chemical method for lysing Arabidopsis celss for protein analysis. Plant Methods 7:22
100. Tsugita A, Kawakami T, Uchiyama Y, Kamo M, Miyatake N, Nozu Y (1994) Separation and characterization of rice proteins. Electrophoresis 15:708–720
101. Vander Mijnsbrugge K, Meyermans H, Van Montagu M, Bauw G, Boerjan W (2000) Wood formation in poplar: identification, characterization, and seasonal variation of xylem proteins. Planta 210:589–598
102. Verdonk JC, Hatfield RD, Sullivan ML (2012) Proteomic analysis of cell walls of two developmental stages of alfalfa stems. Front Plant Sci 3:279
103. Vincent D, Lapierre C, Pollet B, Cornic G, Negroni L, Zivy M (2005) Water deficits affect caffeate O-methyltransferase, lignification, and related enzymes in maize leaves. A proteomic investigation. Plant Physiol 137:949–960
104. Wan J, Torres M, Ganapathy A, Thelen J, DaGue BB, Mooney B, Xu D, Stacey G (2005) Proteomic analysis of soybean root hairs after infection by Bradyrhizobium japonicum. Mol Plant-Microbe Interact MPMI 18:458–467
105. Wang H, Alvarez S, Hicks LM (2012) Comprehensive comparison of iTRAQ and label-free LC-based quantitative proteomics approaches using two Chlamydomonas reinhardtii strains of interest for biofuels engineering. J Proteome Res 11:487–501
106. Wang W, Scali M, Vignani R, Spadafora A, Sensi E, Mazzuca S, Cresti M (2003) Protein extraction for two-dimensional electrophoresis from olive leaf, a plant tissue containing high levels of interfering compounds. Electrophoresis 24:2369–2375
107. Wang W, Vignani R, Scali M, Cresti M (2006) A universal and rapid protocol for protein extraction from recalcitrant plant tissues for proteomic analysis. Electrophoresis 27:2782–2786
108. Watson BS, Lei Z, Dixon RA, Sumner LW (2004) Proteomics of Medicago sativa cell walls. Phytochemistry 65:1709–1720
109. Watson BS, Sumner LW (2007) Isolation of cell wall proteins from Medicago sativa stems. Methods Mol Biol 355:79–92
110. Witzel K, Weidner A, Surabhi GK, Borner A, Mock HP (2009) Salt stress-induced alterations in the root proteome of barley genotypes with contrasting response towards salinity. J Exp Bot 60:3545–3557
111. Wu FS, Wang MY (1984) Extraction of proteins for sodium dodecyl sulfate-polyacrylamide gel electrophoresis from protease-rich plant tissues. Anal Biochem 139:100–103
112. Wu X, Xiong E, Wang W, Scali M, Cresti M (2014) Universal sample preparation method integrating trichloroacetic acid/acetone precipitation with phenol extraction for crop proteomic analysis. Nat Protoc 9:362–374
113. Xi J, Wang X, Li S, Zhou X, Yue L, Fan J, Hao D (2006) Polyethylene glycol fractionation improved detection of low-abundant proteins by two-dimensional electrophoresis analysis of plant proteome. Phytochemistry 67:2341–2348
114. Yan S, Tang Z, Su W, Sun W (2005) Proteomic analysis of salt stress-responsive proteins in rice root. Proteomics 5:235–244
115. Yan SP, Zhang QY, Tang ZC, Su WA, Sun WN (2006) Comparative proteomic analysis provides new insights into chilling stress responses in rice. Mol Cell Proteomics MCP 5:484–496
116. Yang L, Zhang Y, Zhu N, Koh J, Ma C, Pan Y, Yu B, Chen S, Li H (2013) Proteomic analysis of salt tolerance in sugar beet monosomic addition line M14. J Proteome Res 12:4931–4950
117. Zhang H, Lian C, Shen Z (2009) Proteomic identification of small, copper-responsive proteins in germinating embryos of Oryza sativa. Ann Bot 103:923–930
118. Zhang Y, Gao P, Xing Z, Jin S, Chen Z, Liu L, Constantino N, Wang X, Shi W, Yuan JS, Dai SY (2013) Application of an improved proteomics method for abundant protein cleanup: molecular and genomic mechanisms study in plant defense. Mol Cell Proteomics MCP 12:3431–3442
119. Zhao X, Ren J, Cui N, Fan H, Yu G, Li T (2013) Preparation of protein extraction from flower buds of Solanum lycopersicum for two-dimensional gel electrophoresis. Br Biotechnol J 3:183–190
120. Zheng Q, Song J, Campbell-Palmer L, Thompson K, Li L, Walker B, Cui Y, Li X (2013) A proteomic investigation of apple fruit during ripening and in response to ethylene treatment. J Proteome 93:276–294
121. Zhong B, Karibe H, Komatsu S, Ichimura H, Nagamura Y, Sasaki T, Hirano H (1997) Screening of rice genes from a cDNA catalog based on the sequence data-file of proteins separated by two-dimensional electrophoresis. Breed Sci 47:245–251
122. Zhou X, Wang K, Lv D, Wu C, Li J, Zhao P, Lin Z, Du L, Yan Y, Ye X (2013) Global analysis of differentially expressed genes and proteins in the wheat callus infected by Agrobacterium tumefaciens. PLoS One 8:e79390
123. Zhu J, Chen S, Alvarez S, Asirvatham VS, Schachtman DP, Wu Y, Sharp RE (2006) Cell wall proteome in the maize primary root elongation zone. I. Extraction and identification of water-soluble and lightly ionically bound proteins. Plant Physiol 140:311–325

Improving Proteome Coverage by Reducing Sample Complexity via Chromatography

5

Uma Kota and Mark L. Stolowitz

Abstract

High performance liquid chromatography (HPLC) is currently one of the most powerful analytical tools that has revolutionized the field of proteomics. Formerly known as high pressure liquid chromatography, this technique was introduced in the early 1960s to improve the efficiency of liquid chromatography separations using small stationary phase particles packed in columns. Since its introduction, continued advancements in column technology, development of different stationary phase materials and improved instrumentation has allowed the full potential of this technique to be realized. The various modes of HPLC in combination with mass spectrometry has evolved into the principal analytical technique in proteomics. It is now common practice to combine different types of HPLC in a multidimensional workflow to identify and quantify peptides and proteins with high sensitivity and resolution from limited amounts of samples. More recently, the introduction of Ultra High Performance Liquid Chromatography (UHPLC) has further raised the level of performance of this technique with significant increases in resolution, speed and sensitivity. The number of applications of HPLC and UHPLC in proteomics has been rapidly expanding and will continue to be a pivotal analytical technique. The aim of the following sections is to familiarize the beginner with the various HPLC methods routinely used in proteomics and provide sufficient practical knowledge regarding each of them to develop a separation and analytical protocol.

5.1 High Performance Liquid Chromatography

High performance liquid chromatography (HPLC) is currently one of the most powerful analytical tools that has revolutionized the field of proteomics. Formerly known as high pressure

U. Kota • M.L. Stolowitz (✉)
Canary Center at Stanford for Cancer Early Detection, Palo Alto, CA 94304, USA
e-mail: mstolowitz@stanford.edu

H. Mirzaei and M. Carrasco (eds.), *Modern Proteomics – Sample Preparation, Analysis and Practical Applications*, Advances in Experimental Medicine and Biology 919, DOI 10.1007/978-3-319-41448-5_5

liquid chromatography, this technique was introduced in the early 1960s to improve the efficiency of liquid chromatography separations using small stationary phase particles packed in columns. Since its introduction, continued advancements in column technology, development of different stationary phase materials and improved instrumentation has allowed the full potential of this technique to be realized. The various modes of HPLC in combination with mass spectrometry has evolved into the principal analytical technique in proteomics. It is now common practice to combine different types of HPLC in a multidimensional workflow to identify and quantify peptides and proteins with high sensitivity and resolution from limited amounts of samples. More recently, the introduction of Ultra High Performance Liquid Chromatography (UHPLC) has further raised the level of performance of this technique with significant increases in resolution, speed and sensitivity. The number of applications of HPLC and UHPLC in proteomics has been rapidly expanding and will continue to be a pivotal analytical technique. The aim of the following sections is to familiarize the beginner with the various HPLC methods routinely used in proteomics and provide sufficient practical knowledge regarding each of them to develop a separation and analytical protocol.

5.2 Reversed-Phase Chromatography

Reversed-Phase chromatography (RPC) is routinely used for the high-resolution separation of proteins, peptides and nucleic acids. Most common applications include desalting, concentrating samples, peptide mapping, purification procedures and determining purity. RPC is one of the most widely applied separation techniques on an analytical scale due to several reasons. It is a robust technique that can be applied to a wide range of molecules including charged and polar molecules. This separation technique allows precise control of variables such as organic solvent type and concentration, pH, temperature. In addition to being highly reproducible, RPC columns are also known to be stable and efficient over long periods of time, thus making it a reliable and cost-effective fractionation method. The resolving power of this RPC is reflected by its frequent use in multidimensional separations [38, 97, 98, 128, 162, 166]. The combined desalting and purification aspects of RP-HPLC makes it suitable as the final step of a multidimensional fractionation protocol, specifically prior to analysis by mass spectrometry of purified solutes [31].

Theory of Reversed- Phase Chromatography The separation of biomolecules by RPC is based on the reversible hydrophobic interaction between the sample in the mobile phase and the stationary phase. The distribution of the sample between the two phases depends on the binding properties of the stationary phase, hydrophobicity of the sample molecule, composition of the mobile phase. RPC is an adsorptive process which relies on partitioning mechanism to effect separation. Separation relies on an equilibrium between the sample molecules in the eluent and the surface of the stationary phase. The stationary phase is more hydrophobic than the mobile phase when an aqueous/organic solvent mobile phase is used. Initial conditions are primarily aqueous, favoring a high degree of organized water structure surrounding the sample molecule and favoring the adsorption of the sample molecule from the mobile phase onto the stationary phase. A small percentage of organic modifier, typically 3–5 % acetonitrile is present in order to achieve a "wetted" surface. As sample binds to the stationary phase, the hydrophobic area exposed to the mobile phase is minimized, thus the degree of organized water structure is diminished [168]. Bound samples are desorbed from the stationary phase by adjusting the polarity of the mobile phase over time by increasing the final concentration of the organic solvent in the final mobile phase, such that the bound molecules tend to dissociate from the stationary phase back into the mobile phase, in the order of increasing hydrophobicity.

RPC almost always uses gradient elution instead of isocratic elution. Peptides and proteins have a mix of accessible hydrophilic and hydrophobic amino acid side chains, hence interaction with the stationary phase has the nature of a multi-point attachment. Furthermore, although these biomolecules adsorb strongly to the surface of a RP matrix under aqueous conditions, they desorb from the matrix within a very narrow window of organic modifier concentration. Any given biological sample typically contains a broad mixture of biomolecules with a diverse range of adsorption affinities and hence the only practical method for reverse phase separation of such samples is by gradient elution. Separation in RPC is due to the different binding properties of the solute present in the sample as a result of the differences in their hydrophobic properties. The degree of solute binding to the stationary phase can be controlled by manipulating the hydrophobic properties of the initial mobile phase. This allows a high degree of flexibility in separation conditions allowing one to resolve solutes that vary only slightly in their hydrophobicity. Because of its excellent resolving power and great flexibility, RPC is an indispensable technique for high performance separation of complex biological samples and purification of desired solutes. Furthermore, since binding under the initial phase is absolute, the starting concentration of the desired solute in the sample is not critical allowing diluted samples to be applied to the column.

5.2.1 Column Characteristics

Stationary Phases and Bonding Chemistries
The RPC system used for analysis of peptides and proteins usually consists of an n-alkyl-silica-based sorbent from which the solutes are eluted with gradients of increasing concentrations of organic solvent such as acetonitrile containing an ionic modifier such as trifluoroacetic acid (TFA) [2]. The chromatographic packing material used in RPC are commonly based on microparticulate porous silica that is chemically modified by a reactive silane containing n-alkyl hydrophobic ligand. The most commonly used ligands are hydrocarbons such as *n*-butyl (C4), *n*-octyl (C8) and *n*-octadecyl (C18). The process of chemical immobilization of the ligands on silica results in approximately only half the silica surface being modified [1]. This is due to steric hindrance from the large and bulky C8 and C18 ligands that often prevents complete derivatization of all the silanol groups. This partial derivatization can lead to undesirable mixed mode ion exchange effects due to the residual polar silanol groups. Therefore, the residual silanols are subjected to further silanization with reactive trimethylsilane reagents to yield a so-called end-capped packing material. Reproducible chemical derivatization of the silanol surface as well as capping is critical for efficient reverse phase chromatography with batch-to-batch reproducibility.

The most important stationary phase properties that have a profound influence on retention and selectivity in RPC are the type of native silica, the silanol content and the carbon loading.

Spherical silica gel is the most commonly used packing material for RPC. While the porous silica beads are chemically stable at low pH and in organic solvents used for RPC, it is chemically unstable in aqueous solution at high pH and not recommended for prolonged exposure above pH 7.5, particularly at elevated temperature. Alternatively, synthetic organic polymer-based columns have become increasingly popular as reversed phase media. The commonly used polymers are polystyrene, methylacrylate, polyethylene and polypropylene. Polystyrene-based columns have particularly been used in large scale preparative chromatography because of their excellent chemical stability, particularly under strong acidic and basic conditions [157]. Polymer-based RPC columns have several intrinsic features that give them key advantages over silica based columns. Apart from their chemical stability, polymer-based columns have uniform particles and high physical stability. The polymeric reverse-phase sorbents allow the mobile phase to perfuse through the sorbent matrix thus allowing the transport of the solutes into the interior of the sorbent particle much more rapidly than by simple diffusion as in the case of silica particles. Consequently separations can be achieved at higher flow rates with shorter re-equilibration times.

The carbon load is dependent on the choice of n-alkyl ligands and its density. The type of ligand has a significant influence on the retention of peptides and proteins. In general proteins and large peptides are best separated on short RPC columns that have less hydrophobic *n*-butyl ligands bonded to wide pore silica gels (e.g. 300 Å). This allows greater protein recovery and conformational integrity. The more hydrophobic longer alkyl chain ligands (C18 ligands) are generally useful for the separation of peptides smaller than ~2000–3000 Da range and are commonly used for separation of peptides resulting from protease digestion of proteins.

Other ligands such as phenyl, including phenyl-hexyl, diphenyl and cyanopropyl ligands have also been used for RPC but afford different retention characteristics and can provide different selectivities [21, 103, 184]. Polar modified reverse phase columns (polar embedded or polar end capped groups) enhance interaction between the peptides and the particles and result in different selectivity. In general, polar-endcapped phases display similar hydrophobic retention characteristics as conventional C18 columns, but express higher hydrogen bonding and silanol activity. While polar-embedded phases display the opposite behavior, with greatly reduced hydrophobic properties compared to both conventional C18 and polar-end capped phases as well as reduced silanol activity [89].

Surface Area and Pore Size While retention is primarily controlled by the bonded phase chemistry and mobile phase chemistry, the surface area of the packing material also plays an important role. The surface area available is dependent on the pore size of the particle used for packing. The pore size of a column is selected so that the sample molecules have easy access to the pores. Smaller-pore columns are desired because of their higher surface area, as long as the analytes are sufficiently small to easily enter the pores. The surface area of a particle is inversely proportional to the pore diameter, so a 3-μm particle size, 100 Å pore column will have approximately three times the surface area as a 3-μm, 300 Å pore column. Particles with pores 100–150 Å are used for peptides and small molecules while particles with ≥300 Å pore size are used for separation of proteins.

Particle Size The separation efficiency of a column depends on the particle size, column length or flow rate. The particle size is defined as the mean diameter of the silica spheres used as the support material. Large particle sizes (10–15 μm) are generally used for large scale preparative and process applications due to their increased capacity and low pressure requirements at high flow rates. Small scale preparative and analytical scale chromatography routinely use 3 μm and 5 μm size particles. Until recently, the practical particle size limit was around 3 μm since smaller particle sizes often limits the use of conventional liquid chromatography (LC) systems with a standard pressure rating of 5800 psi (400 bar). However with the continuous demand for high throughput analysis and higher resolution the use of sub-2 μm stationary-phase support particles was made possible because of the advent of LC systems capable of handling very high back pressures (>1400 bar or 20,000 psi). This technique termed as ultra-high pressure liquid chromatography (UHPLC) allows the use of smaller particle size columns with higher efficiency and wider range of usable flow rates resulting in better resolution, higher sensitivity with significantly faster overall analysis time.

The substantial gain in column efficiency and sample throughput acheived by using sub-2 μm can be explained using the Van Deemter equation that describes the relationship between the column efficiency (measured in terms of plate height) versus the flow rate (linear velocity, μ) [160].

$$H = A + B/\mu + C\mu$$

The term H refers to the plate height and is defined as the distance a compound must travel in a column needed to separate two similar analytes. Plate height (H) is derived by dividing the column length (L) by the calculated number of theoretical plates (N). It is desirable to have the smallest plate height in order to obtain the maximum number of plates. Hence a column with a higher N will provide narrower peak at a given RT than a given column with a lower N number. The term A represents "eddy diffusion or multi-path effect" that the analytes experience as they travel through a packed bed. The A term is directly proportional to particle size (d_p) and is smaller in well-packed columns. The B term represents "longitudinal diffusion" of the solute band in the mobile phase and is proportional to the Diffusion coefficient (D_m) of the solute. Longitudinal diffusion has a negative effect on resolution at lower flow rates and is not significant at higher flower rates. The C term represents "resistance to mass transfer coefficient" of the analyte between the mobile phase and stationary phase. Since it takes time for analytes in the mobile phase to move into the stationary phase and vice versa, at faster flow rates, there is less time for equilibrium to be reached between the phases and the mass transfer effect on peak broadening is directly related to mobile phase velocity.

As seen in the Van Deemter plot (Plate height, H vs Linear Velocity, μ) which is a composite curve from three relatively independent parameters, the minimum point on the curve marks the minimum plate height (H_{min}) and the optimum velocity (V_{opt}), which is the flow rate at maximum column efficiency.

Particles with small d_p have shorter diffusion path lengths, thus allowing the solute to travel in and out of the particle faster. Therefore the analyte spends less time inside the particle where peak diffusion can occur. Hence the contributions from the A and B/μ terms are minimal even at higher flow rates when using columns packed with particles with small d_p (sub-3 μm). At higher flow rates, the van Deemter curve is dominated by contribution from the C term which is proportional to d_p^2. Given the inherent higher efficiency of smaller-particles columns, they have come to dominate modern HPLC and are particularly useful in fast LC and in high-speed applications [41].

An alternative to sub-2 μm particle size particles has been the development of the Fused Core particles. These particles consists of a solid 1.7 μm core with a 0.5 μm porous silica shell surrounding it (d_p = 2.7 μm). These superficially porous particle columns offer significant advantages over conventional porous columns. Because the diffusion occurs in the porous outer shell and not the solid core, the fused core particle columns allows higher flow rates without sacrificing column efficiency [35]. Another advantage of the fused core particles is that their relatively large particle size greatly reduces the backpressure, thus providing a practical alternative to UHPLC. These columns provide similar efficiency, resolution and throughput as the sub-2 μm particle size columns but at conventional HPLC pressure limits [141].

Column Dimensions There are several critical parameters in HPLC that contribute to the resolution and recovery of proteins and peptides. These include column dimensions, flow rate, column temperature, mobile phase composition and gradient used for elution. Column dimensions affect the efficiency, sensitivity and speed of analysis. Further, the choice of column dimensions will depend on the chromatographic application; analytical, semi-preparative, preparative, complexity of the sample, etc. Column dimensions consist of particle size, column length and internal diameter. The effect of particle size on efficiency has been discussed in detail above. While column efficiency in inversely proportional to d_p, it is directly proportional to column length. Increasing the column length not only increases efficiency but also improves resolution. However, longer columns also lead to higher back pressure and longer analysis times. The resolution of larger molecules such as proteins and polypeptides is not significantly impacted by the column length as their interaction with the column packing happens in a single adsorption/desorption step near the top of the column and very little interaction takes place as these molecules elute down the column without affecting resolution. Column lengths play a more important role when resolving smaller peptides such as those generated from enzymatic digests where resolution can be improved by increasing the column length. The column internal diameter affects the sample capacity which is a function of sample volume. Consequently for two columns of equal diameter but differing in length, the longer column has higher sample capacity and higher resolution [142].

In general, short columns of 50–150 mm in length with 2.0–4.6 mm I.D., packed with 3- or 5-μm particles are recommended for the separation of large peptides and proteins. Longer columns, 150–250 mm and I.D. of 2.1–4.6 mm, packed with 1.8–3 μm particles or 2.7 μm fused core particles are recommended for the separation of small peptides and enzymatic digests. Nano (0.075–0.1 mm), capillary (0.2–0.4 mm) or microbore (1–2 mm I.D.) columns are employed when sample is limited and/or higher sensitivity is required. The column dimensions also determine the flow rate used for separation, which in turn affects resolution. Typical analytical scale columns utilize flow rates ranging between 0.5 and 2.0 ml/min. With microbore columns flow rates of 50–250 μl/min are used, whereas capillary and nanobore columns typically utilize flow rates of 1–20 μl/min and 20–300 nl/min, respectively. The separation of large biomolecules is insensitive to flow rate. However, flow rate is an important factor for the separation of small peptides and protein digests in order to achieve good resolution. Additionally, the choice of column dimensions and flow rates is also determined by its compatability with the type of HPLC/UHPLC.

Resolution can also be manipulated by controlling the operating temperature. Although reverse phase separation of proteins and peptides are normally performed at ambient temperature, the retention and/resolution of analytes in RP-HPLC is influenced by temperature by changes in solvent viscosity. In general, an increase in temperature reduces retention in RPC and can have some effects on selectivity. Temperature variations have also been shown to affect the secondary structure of peptides and hence affect selectivity during RP-HPLC [31].

Mobile Phase The most important characteristic of RP-HPLC is the ability to manipulate the solute retention and resolution by changing the composition of mobile phase used during the separation process. Since the peptides and proteins bind to the RPC column under aqueous conditions and elute as the hydrophobicity of the mobile phase increases, high resolution of complex mixtures is often achieved by applying a gradient of increasing organic solvent concentration. The most commonly used organic solvents in the order of their eluotropic strength are acetonitrile, methanol and isopropanol, all of them being readily miscible with water. Acetonitrile is the most popular choice for most peptide and protein fractionation protocols due its lower viscosity (lower back pressure), good "wetting" properties even at low concentrations of organic solvent (% B) in the mobile phase and is highly volatile allowing easy sample preparation for downstream mass spectrometry analysis. Additionally, acetonitrile exhibits high optical transparency in the detection wavelength of proteins and peptides making it suitable for UV detection [1].

Most preparative and analytical, high resolution separations of proteins and peptides are carried out using gradient elution. Method development starts with carrying out the separation with an initial mobile phase which is highly aqueous (3–5 % B) and rapidly increasing the % of organic solvent over a short period of time. The retention and elution of the analytes of interest can be further optimized by adjusting the concentration of organic solvent in the mobile phase and/or modifying the gradient length. High resolution analyses typically use longer gradients in order allow as many components in the analyte mixture as possible to bind to the RP column and then elute them differentially to obtain a comprehensive profile. For preparative applications, the gradient conditions are optimized to allow the separation of the analyte of interest from contaminants. Desalting of samples is a low resolution application and typically done using a step gradient. The hydrophilic contaminants and salt are eluted under low organic conditions and the more hydrophobic components are eluted at a higher concentration of the organic solvent.

Besides organic modifier, altering pH can also improve control over the selectivity and in some cases improve ionization and solubility. RP-HPLC is generally carried out with trifluroacetic acid (TFA). This anionic counter ion interacts with the with protonated groups on the proteins/peptides and suppresses their influence on the overall hydrophobicity and enhance binding to the stationary phase. Thus the use of an ion pairing agent can alter the retention behavior and subsequent selectivity.

Column Sources A partial list of the popular silica-based columns from different vendors can be found in Table 5.1. There are also several

Table 5.1 RPC column manufacturers and products

Manufacturer/vendor	Product name
Agilent	Zorbax StableBond, Eclipse Plus, Bonus, Extend-C18, **Poroshell 120 and 300**
Waters	Atlantis, Symmetry, SunFire, X-Bridge, ACQUITY, XTerra, **CORTECS**
Thermo/Dionex	Acclaim, Hypersil, Hypersil GOLD, Syncronis, **Accucore, Accucore XL**
Sigma-Aldrich (Supelco)	Discovery, Supelcosil, **Ascentis Express, Ascentis Express Pepitde-ES 160 Å**,
GL Sciences	Intersil, InterSustain
The Nest Group Inc.	GRACE/Vydac®, HAISIL, PROTO ™, TARGA
EMD Millipore	Chromolith,CapRod, LiChrospher,
Macherey-Nagel	Nucleosil, **Nucleoshell**
AkzoNobel	Kromasil
MacMod	ProntoSIL, **ACE Ultracore**
Advanced Material Technology	**Halo, Halo Peptide-ES 160 Å**,
Phenomenex	Gemini, Luna, Synergi, **Kinetex**, **Aeris Peptide**

commercially available software and automated systems for HPLC method development and computer aided optimization, some of which are listed below and also reviewed in some key references [114, 151].

- DryLab (http://molnar-institute.com/drylab/)
- ChromSwordAuto (http://www.chromsword.com/en/products/method_development/chromswordauto/automated_hplc_method_development/)
- Fusion LC Method Development (www.smatrix.com/fusion_lc_method_dev.html)
- ACD/AutoChrom (http://www.acdlabs.com/products/com_iden/meth_dev/autochrom/)

Because many factors influence separation (column efficiency, type of stationary phase, flow rate, pH etc.) and challenging samples often require the simultaneous adjustment of several variables, many researchers use computer-facilitated method development. The software uses a small number of experimental runs to simulate the chromatographic separation when any of the several conditions are changed. The experimental data are used to "calibrate" the software for a given sample, after which simulated runs can be carried out by entering new conditions into the computer [151]. Alternatively, the software may summarize the results of a large number of such simulations in the form of convenient resolution maps, allowing the user to analyze the results and identify the most promising chromatographic conditions, which can then be subjected to fine optimization. Computer facilitated method development has been made available for over two decades and have been predominantly used in the pharmaceutical industry. Their application to proteomics workflow has been limited and yet to be fully utilized.

5.3 Ion Exchange Chromatography

Theory The range of concentrations in conjunction with the large number of proteins in any given proteome requires the use of multi-dimensional separation strategies in order to obtain comprehensive profiles. Ion exchange chromatography (IEX) is commonly used as the first fractionation step in chromatographic multi-dimensional separation, involving proteins. Separation in IEX is based on Coulombic interactions between ionic components of proteins/peptides and the charged stationary phase [38]. The stationary phases for IEX are characterized by the nature and strength of the acidic or basic moieties covalently attached to their surfaces and the types of ions they attract. Anion exchangers contain positively charged groups and retain negatively charged analytes, whereas cation exchangers retain positively charged analytes on their negatively charged surface. The binding and elution is based on competition between the charged groups on the proteins/peptides and the charged counter ions in the mobile phase for binding the oppositely charged groups on the stationary phase. Elution of bound proteins/peptides is commonly done by increasing the ionic strength of the mobile phase. The salt concentration in the mobile phase can be controlled such that the anion or cation population in the solution competitively displaces the analytes bound to the stationary phase. Alternatively, a change in the pH of the mobile phase alters the ionic properties of the functional groups on both the stationary phase and analytes. Thus separation of bioanalytes can be performed either by gradient or isocratic elution, allowing more variability in the design of the IEX experiments [155]. Strong ion exchangers bear functional groups that remain ionized over a wide range of pH (includes sulfonic acid and quaternary ammonium moieties) and are used to separate weakly basic and acidic analytes. The bound analytes are eluted by displacement with salts that have a higher affinity to the stationary phase exchange sites, i.e. by salt elution. Weak ion exchangers bear functional groups that are titratable over a narrow pH range (includes carboxylic acid and secondary amine properties) and hence used to retain and separate highly charged analytes. Further details regarding the mechanism of ion exchange chromatography is well reviewed in the following references [144, 155].

Stationary Phases The stationary phases used in IEX consists of a support material synthetic resins, polysaccharides or silica with charged functional groups covalently attached to them. Colloidal Cellulose-based ion exchangers were the first to be used for the separation of proteins [63], but their irregular particle shape led to poor flow properties. Ion exchangers based on dextran (Sephadex), agarose (Sepharose) and cross-linked cellulose (Sephacel) have high porosity and are better suited for the separation of high molecular weight biomolecules.

The agarose and dextran bead based ion – exchangers were first introduced by Pharmacia (now General Electric Health Care BioSciences) and ever since has led to big advances in protein separation [81]. In Sephadex ion exchangers, the charged group is attached to the glucose unit of the dextran. These ion exchangers are derived from either Sephadex G-25 or Sephadex G-50, with the former being more tightly cross-linked and rigid, while the latter is more porous and has better capacity for molecules with molecular weights greater than 300,000. The dextran beads are stable in water, salt solutions, organic solvents, alkaline and weakly acidic solutions. However, very low pH (<2) could hydrolyze the glycosidic linkage, especially at higher temperatures. Sepharose ion exchangers are based on cross-linked agarose gel filtration media Sepharose CL-6B and the functional groups are attached to the gel by ether linkages to the monosaccharide units. They have an exclusion limit for proteins with molecular weight of approximately 4×10^6. Sephacel ion exchangers are based on high-purity micro-crystalline cellulose and the functional groups are attached during their synthesis by ether linkage to glucose units of the polysaccharide chains. While Sephacel is also macroporous with an exclusion limit of 1×10^6, agarose and dextran beads have better flow properties [81]. These soft ion exchange chromatography media are available as dry granular powder or as pre-swollen as well as prepacked columns for HPLC.

Organic polymer-based support material such as styrene/divinyl benzene copolymers, polymethylacrylate and polyvinyl resins. The surface of these non- porous, synthetic polymers is modified with a hydrophilic coating and bonded with a uniform, ion-exchange layer in order to prevent low recovery due to their hydrophobicity. Similar to RP-HPLC columns, a plethora of IEX columns are commercially available, varying in particle size, pore size and other characteristics, depending on the type of application [167].

The charged functional group bound to the matrix determines the useful pH range and the type of ion exchanger. The total number of charged moieties and their availability determines the capacity of the ion exchanger. The different functional groups used in IEX are listed in the table below (Table 5.2). Most widely used anion and cation ion exchangers have immobilized diethylaminoethyl (DEAE) and carboxymethyl (CM) groups, respectively.

Table 5.2 Functional groups and pK_a values of ion exchangers

Ion-exchanger type	Functional group name	Abbreviation	Structure	pK_a*
Anion, weak	Diethylaminoethyl	DEAE	$-O-CH_2-CH_2-NH^+(CH_2-CH_3)_2$	6–9
Anion, weak	Dimethylaminoethyl	DMAE	$-O-CH_2-CH_2-NH^+(CH_3)_2$	~10
Anion, strong	Trimethylaminoethyl	TMAE	$-O-CH_2-CH_2-N^+(CH_3)_3$	–
Anion, strong	Trimethylaminohydroxypropyl	QA	$O-CH_2-CHOH-CH_2-N^+(CH_3)_3$	
Anion, strong	Diethyl-(2-hydroxypropyl) aminoethyl	QAE	$O-CH_2-CH_2-N^+(CH_2-CH_3)_2(CH_2-CHOH-CH_3)$	
Cation, weak	Carboxymethyl	CM	$-O-CH_2-COO^-$	3.5–4.5
Cation, strong	Sulfoethyl	SE	$-O-CH_2-CH_2-SO_3^-$	2
Cation, strong	Sulphopropyl	SP	$-O-CH_2-CHOH-CH_2-O-CH_2-CH_2-CH_2-CH_2SO_3^-$	2–2.5
Cation, strong	Methyl sulphonate	S	$-O-CH_2-CHOH-CH_2-O-CH_2-CH_2-CHOH-CH2SO_3^-$	2

*Karlsson [83]; pKa values are a function of ionic strength. Values reported here are at 0.1 M NaCl

DEAE is a weak base with a net positive charge, while CM is a weak acid containing a negative charge. Sulfonate {Sulfopropyl (SP) and methyl sulfate [S]} and quaternary amino groups are the commonly used strong ion exchangers.

The choice of ion exchanger for separation depends on the isoelectric point of the biomolecule and its stability at various pH values. In practice, proteins are stable and functionally active within a narrow pH range and so the choice of ion exchanger is often determined by the pH stability of the desired protein(s). If the protein(s) are stable at pH values below its pI, then a cation ion exchanger is used and similarly an anion exchanger is used if the protein is stable at pH values above its pI. While both strong and weak ion exchangers have been used for proteomics applications, strong cation exchangers have a considerable advantage for protein and peptide separations as they retain negative charge over the whole range from acidic to neutral pH [22]. However, the high degree of tertiary structure of proteins makes them less tolerant to drastic separation conditions. By contrast, peptides tolerate a much wider range of conditions as their native state is dominated by secondary structures, stabilized mainly by hydrogen bonding. At low pH conditions (< pH 3), under which SCX is performed, peptides are positively charged due to protonation of the N-termini of lysine, arginine and histidine side chains and neutralization of the carboxyl side chains of aspartate and glutamate. The overall positive charge on peptides allows them to bind to the negatively charged strong cationic stationary phase and subsequently eluted using a linear gradient of increasing ionic strength.

Mobile Phase The buffering components used in IEX play an important role in the binding and elution of the analytes. Ion exchange resins are usually have counter ions bound to their functional groups. These are normally Cl^- ions cation exchangers and Na^+ for anion exchangers. The counter ions are held by electrostatic interaction and have specific selectivity for each type of media. The lower the specificity of the counter ion the more readily it can be exchanged with another group with a similar charge. Thus the buffering ions must have the same charge as the functional groups as the ion exchangers, since the opposite charges will be participate in the ion-exchange process. The common cationic buffers used for anion exchangers include Tris, alkylamines, ammonium, triethanoamine etc. Similarly, anionic buffers recommended for cation exchangers include phosphate, acetate, formate, HEPES etc.

Protein/peptide movement down the column can be slowed down by hydrophobic attraction and hydrogen bonding with the ion exchanger. This may lead to irreversible binding or denaturation of proteins and hence poor recovery. The inclusion of acetonitrile (10–25 %) in the ion-exchange mobile phase helps reduce hydrophobic attraction and improves retention of charged peptides.

Experimental Technique Ion exchange columns can be purchased ready to use or can be prepared in the lab by packing a column with loose ion exchange matrix according to manufacturer's instructions. Prior to packing, the matrix must be equilibrated with the working buffer and after packing the column must be washed further with several volumes of the equilibration buffer. Before use, the column must be charged with counter ion by flushing the column with one to two column volumes of the high-salt buffer used for elution. Once charged, the column must be washed thoroughly with the binding buffer to ensure equilibration in the low-salt buffer prior to sample loading. The sample must be prepared in the low salt buffer and should be filtered before applying onto the column to reduce the risk of blocking the column. Many proteins tend to aggregate in solution close to the protein's p*I* and this aggregation is increased at low ionic strength. Hence, a starting buffer of 20–50 mM is recommended [144]. The flow rate used is determined by the particle size and column dimension, however it is typical to use lower flow rates during chromatography compared to the column washing and equilibration steps.

Following sample loading, the column must be washed with 5–10 column-bed volumes of the equilibrating buffer to remove any unbound proteins and other contaminants. All the steps during the chromatography (loading, washing and elution) must be monitored by measuring the optical density (A_{280}/A_{215}) of the flow through. When the washes contain little or no proteins/peptides, elution can be initiated. Target proteins/peptides can be eluted either by increasing the ionic strength or by changing the pH. Proteins and peptides are commonly eluted by increasing the ionic strength of the mobile phase. Elution can also be done by changing the buffer pH (raising the pH to elute from cation exchangers and lowering the pH for anion exchangers), but is not frequently used since many proteins tend to be unstable or precipitate out at certain pH values. Moreover, it is more difficult to produce a continuous pH gradient at constant ionic strength on standard ion-exchange columns, since mixing of buffers of different pH results in simultaneous change in ionic strength [20].

Elution by changing the ionic strength can be performed in a linear or step wise fashion. A linear gradient is achieved by gradually increasing the ionic strength (usually sodium chloride) at a constant rate using a gradient mixer. The gradient mixers allow the formation of a controlled and reproducible salt gradient that is essential for run-to-run consistency. In stepwise elution, the salt solution of the next higher concentration in the step in introduced onto the column and maintained for at least two bed volumes or until the proteins of interest have eluted. This is followed by the next higher concentration and the process is repeated until all proteins are eluted. Since most proteins elute between 0.1 and 0.4 M sodium chloride, steps at 25–50 mM increments until 0.4 M are recommended, during method development [20]. The resolution achieved depends on the type of gradient used. Step gradient being a simpler and more rapid process, usually results in simultaneous elution of multiple protein peaks due to the large increase in ionic strength. Gradient elution is useful when separating proteins/peptides with very close pIs.

The concentrations of salt used for the gradient can be determined by trial and error. After initial purification runs have been analyzed, the gradient can be altered and fine-tuned to optimize the separation of the desired proteins. Following elution, the column is regenerated using a high-salt buffer. Sodium Chloride (1 M) is often used to clean the column between purification runs. The salt removes tightly bound contaminants on the stationary phase and simultaneous charges the column with the counter ion.

Applications The high complexity of most proteomic samples, both in the number of proteins and their concentration range, often exceeds the separation capacity and detection power of most liquid chromatography and mass spectrometry platforms. Hence, it is imperative to use a multidimensional chromatography approach in order to obtain a detailed proteome map. The most common "shot-gun' proteomic approach involves the generation of tryptic peptides in solution and the sample is carried through the first dimension of separation based on charge (SCX, hydrophilic interaction chromatography (HILIC) or iso-electric focusing (IEF)) [17, 25, 30] followed by reverse phased chromatography that is based on hydrophobicity. SCXC is an ideal pre-fractionation method for peptides because at low pH values, the vast majority of tryptic peptides bind to the strong cation exchange column and elution with a salt or pH gradient allows the peptides to be resolved exclusively by their relative charge state [14]. A typical multidimensional chromatography workflow combines SCX with RP, where between 10 and 60 fractions from the SCX are collected, either on-line or off-line. Subsequently, each fraction is injected onto the RP either offline or directly eluted into ESI source with nonpolar buffer. Post translational modifications can also influence the charge properties of peptides and hence their retention and separation. In fact, SCX chromatography has been used for the enrichment of phosphopeptides and N-terminal acetylated peptides from complex mixtures [42, 113]. The counterpart to SCXC, strong anion exchange (SAX), a separation based primarily on negative charge has also been used in multidimensional chromatography workflows both at the protein [187] and peptide level, particularly for characterizing protein phosphorylation [36, 40, 60].

Among the different separation strategies available for peptide fractionation, SCX chromatography is a relatively simple and well established method. Compared to most other enrichment techniques, SCX chromatography is relatively simple, robust, and reproducible that can be performed on small amounts of sample [113].

5.4 Size Exclusion Chromatography

Most proteomic approaches employ one or more methods to reduce sample complexity. These might include fractionation and/or enrichment techniques that are performed either at the protein and/or peptide level. Chromatography-based protein fractionation/enrichment techniques resolve proteins based on their physico-chemical properties. Size-exclusion chromatography (SEC) separates proteins based on their molecular sizes in solution [46]. When an aqueous or aqueous/organic mobile phase is used for separation, this technique is referred to as gel-filtration chromatography. SEC can be applied in two distinct ways, a group separation process and a fractionation process. Group separation refers to where the sample is separated into two major groups such as sample desalting, buffer exchange, removal of low molecular weight contaminants or removal of reagents to stop a reaction. SEC can be applied as a fractionation process where the sample is fractionated according to their molecular size.

Theory SEC is a liquid chromatography technique where the stationary phase consists of spherical porous particles with carefully controlled pore size, through which the biomolecules diffuse based on their molecular size using an aqueous buffer as the mobile phase. The pore size of the stationary phase particles determines the molecular size range within which the separation occurs. Solute molecules larger than the available pore size are excluded from the particles and migrate through the column exclusively in the mobile phase. As the molecular size decreases with respect to the average pore size of the packing material, molecules penetrate the pores at varying degrees with the smallest molecules diffusing furthest into the pore structure and eluting last. Thus, very large molecules elute first, in the void volume of the column followed by smaller molecules, sequentially in the order of decreasing molecular size, with the smallest molecules eluting in the elution volume of the column.

Since separation is based on size, SEC is widely used as an analytical technique to determine the molecular weight distribution of proteins in their native state. The size of the proteins can be determined, provided that the SEC column has been previously calibrated with appropriate molecular weight standards. In the context of proteomics, SEC is mostly used as a fractionation technique. Fractionation at the protein level has the advantage that it allows for maintaining important information such a post-translational modification, polymorphisms, functional groups, cellular location, complexes/aggregates and protein interactions [12].

Stationary Phase The separation of bio-molecules by size-exclusion chromatography was first demonstrated by Lindqvist and Strogårds [96], where they used starch to separate peptides from amino acids. Subsequently, Porath and Flondin [132, 133] developed a cross-linked dextran gel and demonstrated the separation of proteins based on their molecular weight. This gel was made commercially available as Sephadex (GE Healthcare Life Sciences Inc.) and was the standard media for size-based separation for many years. Polymeric resins based on agar/agarose [64, 130], polyacrylamide [65, 90, 154], polyvinylethylcarbitol [90], polyvinylpyrrolidone [90] and derivatized porous silica [44, 137, 139] have been developed and used for SEC. The soft polymeric resins were found to compress under pressure and higher flow rates. This limited the speed and resolution of the chromatographic process. Alternatively, the use of derivatized porous silica for SEC was explored [86]. The high mechanical strength, non-swelling nature and inertness to a fairly wide range of conditions (temperature, solvents) proved valuable for high pressure or high performance SEC (HPSEC) applications. Despite the several significant advantages over organic gels, silica-based media suffers from strong ionic interactions between the proteins and the surface silanol groups. This has been addressed by both surface modifications and the use of mobile phase additives. Diol-modified silica phases is typically used for SEC applications involving proteins and peptides [112, 135, 139]. More recently, porous hybrid materials having a mixed composition of silica and organosiloxanes [171], initially developed for reverse-phase chromatography have now been improved and expanded to SEC. One such example is the bridged ethyl hybrid (BEH) particles with surface modified diols have improved chemical stability and reduced silanol activity over silica columns [123].

The other silica-based media used for SEC are the Zorbax Porous Silica Microspheres (Agilent Technologies Inc.) that consists of extremely uniform colloidal silica beads that are agglutinated to form spherical spheres. The patented polymerization process enables the control of both the particle size and pore size so as to produce column packing that will provide separation over a specific molecular range. Additionally, Zorbax PSM packed columns have excellent bed stability, high-efficiency performance and moderate back pressures [115]. Alternatively, columns packed with superficially porous silica microspheres, called "Poroshell" particles, also from Agilent Technologies Inc. have also been used for SEC [87, 100]. These particles have been described in detail before (RP-HPLC, section-X). Owing to their large surface area, these particles have been shown to have higher sample loading capacities. They allow fast gradient elution separation of proteins and peptides, with good peak shapes well within the operating pressure limits of most modern HPLC systems. Most commercially available SEC columns for protein and peptide separations are silica based, ranging in 3–5 μm particle size and 100–450 Å pore size. The larger pore sizes are best suited for the analysis of monoclonal antibodies, their aggregates, very large proteins and protein complexes.

Method Development SEC is often used for the fractionation of one or more proteins of known molecular weight and is typically the first step in a multidimensional chromatography strategy for sample fractionation. Although a relatively low resolution technique, SEC has the advantages of high reproducibility, stability and relatively short analysis time [149]. It is a robust technique that can be performed in the presence of detergents, denaturing agents, at low or high ionic strength and varying temperatures.

It is important to establish if the aim of the experiment is group separation or high resolution fraction prior to selecting a column. Efficient column packing is essential, particularly for high resolution fractionation. Hence, the use of prepacked columns is recommended to ensure reproducible and high resolution fractionation. Columns must be selected based on the highest flow rate that maintains the resolution and minimum separation time. Gel filtration columns packed with sub-2 μm particles have been shown to have higher efficiency and improved resolution at higher flow rates with short run times [39]. Resolution in SEC is mostly influenced by sample volume and column dimensions. Sample volumes should ideally be between 5 and 10 % of the total column volume as higher volumes beyond this range results in decreased resolution and peak distortion (i.e., tailing) [67]. This technique is independent of sample mass and hence sample concentration. However, the solubility or the viscosity of the sample may limit the concentration of sample used for separation. High viscosity causes irregular flow patterns and inconsistent separation, leading to broad peaks and high back pressure. The viscosity of the sample should be the same as that of the eluent. Column length has a significant effect on resolution in SEC. Since samples are eluted isocratically during gel filtration, increasing the column length provides a means of improving resolution [131, 138]. However, increasing column length also leads to a proportional increase in run time and peak width.

Once the right matrix and column are selected, other chromatographic variables variables that may be manipulated and optimized are the buffer system (type, ionic strength), pH, and solubility additives (e.g. detergents, organic solvents) [3]. Nonbinding interactions between the sample molecules and the stationary phase are dominated by electrostatic and hydrophobic interactions. SEC often employs high salt concentration and/or ionic strength buffers to reduce electrostatic interactions between the stationary phase and proteins/peptides as well as protein-protein interactions [10]. However, at very high concentrations (600–1000 mM), the peaks begin to broaden and to be retained due to hydrophobic effects, especially for peptides and strongly hydrophobic proteins [72, 73, 102]. If detergents are used to stabilize the sample, they should be present both in the sample buffer and mobile phase buffer. Samples are eluted isocratically, hence there is no need to use different buffers during the separation. The pH of the mobile phase has a significant effect on the peak shape and elution time through electrostatic, hydrophobic and solubility effects [138]. The pH-dependent ionic interactions with the stationary phase can be predicted based on the relationship between the pH of the mobile phase and the pI of the sample. Ion-exchange and ion-exclusion effects can occur at pH values below and above the pI of the protein sample, respectively [56]. The ionic strength, composition and pH of the mobile phase can be manipulated to improve resolution as long as it does not affect the stability of the sample or cause conformational changes of proteins.

Applications There are only a few reports in the literature about the application of SEC in proteome research. SEC has been used as a first dimension in multidimensional separation for proteome research both at the peptide [92, 116, 124, 125] and protein level [74, 82, 91, 150]. SEC has also been applied as a fractionation technique to study post translational modifications such as phosphorylation [159] and glycosylation [8]. SEC was used to fractionate complex yeast tryptic digests into pools of peptides based on their size. The large post-tryptic digestion peptides were subjected to a secondary digestion followed by LC − MS/MS analysis that lead to a significant increase in identified proteins and a 32–50 % relative increase in average sequence coverage compared to a single trypsin digestion alone. This secondary digestion strategy was applied to analyze the phosphoproteomes of fission yeast and of a human cell line. SEC has also been applied to enrich N-linked glycopeptides relative to the non-glycosylated peptide from human serum digest. The gylcosylation sites were identified by treating the enriched glycosylation fraction with PNGaseF followed by LC/MS/MS analysis.

5.5 Hydrophilic Interaction Liquid Chromatography (HILIC)

Although RP-HPLC is the most widely applied separation technique, this technique is not suitable for the analysis of highly polar, hydrophilic and ionizable molecules as they are poorly retained on the hydrophobic stationary phase. Polar compounds can be separated by normal phase chromatography (NPC), in which the sample components partition between a polar stationary phase and a less polar mobile phase. The compounds are eluted in the order of decreasing hydrophobicity by increasing the polarity of the mobile phase. The mobile phase in NPC typically utilizes 100 % organic solvent or a blend of miscible organic solvents. Hydrophilic interaction chromatography (HILIC), the term first coined by Alpert [5], is a variation of NPC that still utilizes a polar stationary phase but the mobile phase consists of organic solvents that are water miscible. Since the coining of the term in 1990, HILIC has become an increasingly popular technique for the analysis of polar and hydrophilic compounds, particularly because this method has been shown to provide improved sensitivity compared to RPC when used in combination with electrospray ionization-based mass spectrometry [118, 119]. HILIC has also gained popularity in proteomics and is often used as an orthogonal separation method in conjunction with RP-HPLC in multidimensional separation of peptides, especially for the targeted analysis of post-translational modifications [16, 18].

Theory The HILIC mode of separation had been applied to separate amino acids [105], sugars [94, 120, 161], organic amines [15], basic drugs [47, 75] and nitrogenous bases [34] even before the actual term was coined. The publication by Alpert was the first paper to demonstrate the application of HILIC to the separation of peptides in addition to other polar compounds as well as discuss the separation mechanism in detail. As mentioned earlier, HILIC uses polar stationary phases such as underivatized bare silica or uncharged modified silica (diol, amino, cyano) and high levels of organic solvent. The retention mechanism works on the basis that water adsorbs onto the stationary phase to form an immobilized layer and the analyte partitions between this and the bulk mobile phase. This distinguishes HILIC from NPC where the solutes adsorb directly onto the stationary phase [5]. Additionally, polar analytes can also undergo ion exchange with the charged groups on the silica surface depending on the nature of the stationary phase. For example, underivatized silanol groups on bare silica are themselves both acidic and hydrophilic in nature. The pKa of the surface silanol group is 7.1 ± 0.5 (Hair 1970). These residual silanol groups are partially ionized and can interact with basic analytes through hydrogen bonding and electrostatic interactions. Hence depending on the surface charge on the stationary phase, the retention mechanism can be a combination of partitioning of solutes in aqueous two-phase system and specific interactions with the surface charged groups. Other factors governing retention are hydrogen bonding, which depends on the acidity or basicity of the peptides and the dipole-dipole interactions, which depends on the dipole moments and polarizability of the analytes [28, 37]. As mentioned above, HILIC uses aqueous-organic solvent mobile phases, typically 40–97 % acetonitrile in water or other volatile buffers, thus making it a very mass spectrometry friendly technique [9]. Since partitioning is an important component of the HILIC retention mechanism, the presence of a significant amount of water in the mobile phase is crucial for maintaining an immobilized aqueous layer on the surface of the stationary phase [118]. Unlike RPLC, gradient elution in HILIC begins with low polarity organic solvent and the polar analytes are eluted by increasing the polar aqueous content in the mobile phase. Thus the elution order in HILIC is more or less the inverse of that in RPC [5], which means this separation technique is well suited particularly for those peptides that are poorly retained on RP columns. The reader is directed to a comprehensive review by Hermström and Irgum [62] that provides an excellent background to HILIC and details about its separation mechanism.

Stationary Phases The rising popularity of HILIC as a distinct chromatographic mode for separation of protein and peptide mixtures has coincided with the development of a diverse range of stationary phase materials with different retention and selective properties. Separations are typically performed using packing materials having particle sizes in the ranging from sub-2 μm to 10 μm and average pore size of approximately 120 Å, thus making HILIC a high resolution technique amenable for both HPLC and UHPLC systems. The separation efficiencies of the different commercially available HILIC columns have been studied and comprehensively reviewed in the following articles [70, 84].

The most common stationary phases used for HILIC are silica-based and are available as fully porous, superficially porous, ethylene bridged hybrid (BEH) and monolithic columns [122]. The silica-based phases can be classified into two large groups: unmodified bare silica phases and polar chemically-bonded phases. The first HILIC applications were developed on unmodified bare underivatized silica phases and remain a popular choice for the separation of carbohydrates. The free silanol groups are the key chemical feature of hydrated silica surfaces and their acidity is controlled by the purity of silica itself [69]. Chemically bonded phases are supplied by many manufacturers and include weak and strong cation exchangers [95], diol [137, 156], amino [6, 94, 152, 158], amide [152, 179], polysulfoethyl aspartamide and polyhydroxyethyl aspartamide, pentafluorophenylpropyl and amino-cyano-phases [95, 118]. The chemical derivatization of the surface with polar functional groups is done much the same way as C18 or C8 phases are prepared for RPC. The polar stationary phases can be further classified into neutral, charged and zwitterionic phases based on the charge state of the functional groups. The chemical stability of silica-based phases is limited under extreme pH values and most separations are performed in the pH range between 2.0 and 8.0.

Neutral stationary phases contain polar functional groups that are in neutral form in the range of pH 3–8, usually used for the mobile phase in HILIC. The retention mechanism is mainly based on hydrophilic interactions. Many HILIC stationary phases belong to this category, which comprises a large variety of functional groups, including diol (YMC-Triart Diol, LiChrospher 100 Diol, Intersil Diol), cross-linked diol (Luna HILIC), amide (TSKgel Amide-80, GlycoSep N), aspartamide (PolyHYDROXYETHYL A), cyanopropyl (LiChrospher 100 CN, Altima Cyano HP, Spherisorb CN) and cyclodextrin (Nucleodex β-OH, Cyclobond I 2000) groups. They have found application for the separation of oligosaccharides, peptides, proteins, and oligonucleotides.

Aminopropyl silica phases (Luna NH_2, Hypersil APS-2 (amino), Zorbax NH_2, LiChrospher 100 NH2, TSKgel NH_2-100) are positively charged and among the oldest amine-based phases. The negatively charged derivatized silica mainly consist of stationary–phases having a special poly(peptide) coating. Examples include Poly(aspartic acid)-silica (PolyCAT A), poly(2-sulfoethyl aspartamide) (Polysulfo-ethyl A) and poly(2-hydroxyethyl aspartamide) (Polyhydroxy-ethyl A) all manufactured by PolyLC Inc. (Columbia,MD). Poly(aspartic acid) silica was originally developed as a weak cation exchange material and used for the separation of proteins [4]. The stationary phase consists of silica material with a bonded coating of hydrophilic aspartic acid polymer. While the β-carboxy group of aspartic acid is responsible for the cation exchange capacity, it can also act as an acceptor/donor group for hydrogen bonds between solutes and the stationary phase [95]. It is this feature that makes this poly(peptide) stationary phase well suited for HILIC separations of peptides and proteins. Poly(2-hydroxyethylaspartamide)-silica is made by incorporating ethanolamine into a coating of poly(succinimide) bonded to silica [95]. The material is neutrally charged and retention mechanism is mainly through hydrophilic interaction, thus allowing sharper peaks and better selectivity. Although the poly(2-hydroxyethyl aspartamide) stationary phases was used for the separation of a wide range of biomolecules including peptides [19, 126, 179], this HILIC phase seems to have lost some of its momentum compared to more recent dedicated HILIC phases, due to their lower efficiency [158], limited longtime stability [188], or column bleeding, as recently reported for a poly(succinimide)-based phase [111]. Poly(2-sulfoethyl aspartamide) was originally developed as a strong cation ion exchanger of peptides but has also been used for HILIC separations [7, 95]. It is synthesized by aminolysis of taurine with poly(succinimide) covalently bonded to silica and exhibits mixed-mode effect i.e. hydrophilic interactions and electrostatic effects [95]. Like the poly (2-hydroxyethyl aspartamide), this stationary phase also exhibited column bleeding resulting in several interfering peaks during a two dimensional proteomics study [111].

Zwitterionic derivatized HILIC phases was introduced by Irgum and coworkers [76, 77, 78]. These phases consist of a layer of highly polar switterionic sulfoalkylbetaine groups (Fig. 5.2a) grafted onto wide bore silica (ZIC-HILIC) or a polymeric support (ZIC-pHILIC). More recently, another new stationary phase with phosphorylcholine functional groups bonded to silica (ZIC-cHILIC) (Fig. 5.2b) has been introduced. ZIC-cHILIC is an inverted zwitterionic stationary phase and hence shows different selectivity compared to ZIC-HILIC. The net charge on either of these phases in neutral as the oppositely charge groups are in equal molar ratio, but they still exhibit weak ionic interactions that allow separations to be optimized using low ionic strength buffers. The charge state of the zwitterionic phases is pH independent. However, pH can affect the charge state of peptides, affecting their hydrophilicity and, thereby their retention [37].

5.5.1 Mobile Phase Selection

The typical mobile phase for the HILIC separation of peptides is a water-miscible polar organic solvent such as acetonitrile, methanol and isopropanol at concentrations of up to 85 % [95]. Alcohols can be used as alternative solvents, but a higher concentration is needed in order to achieve the same degree of retention of the analyte relative to an aprotic solvent-water combination [28]. The eluotropic strength of the most commonly used mobile phase solvents are listed below according to their decreasing elution strength: water > methanol > ethanol > isopropanol > acetonitrile. Acetonitrile is highly recommended due to its low viscosity and has the

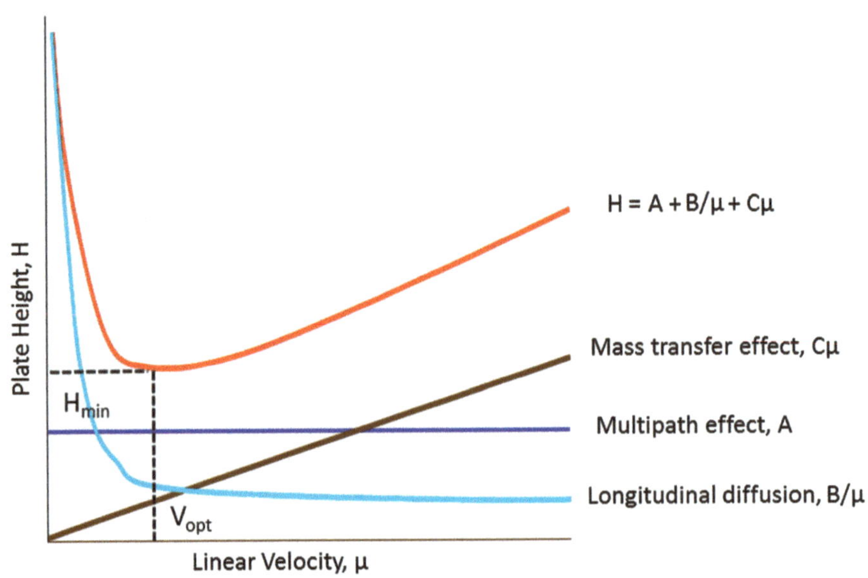

Fig. 5.1 Van Deemter curve showing the relationship between plate height (H) vs. linear velocity (μ). The Van Deemter curve is a composite plot of the A, B/μ and Cμ terms where the each term is plotted to show their individual contributions. H_{min} = minimum plate height, V_{opt} = Optimum velocity (Figure adopted from various web sources)

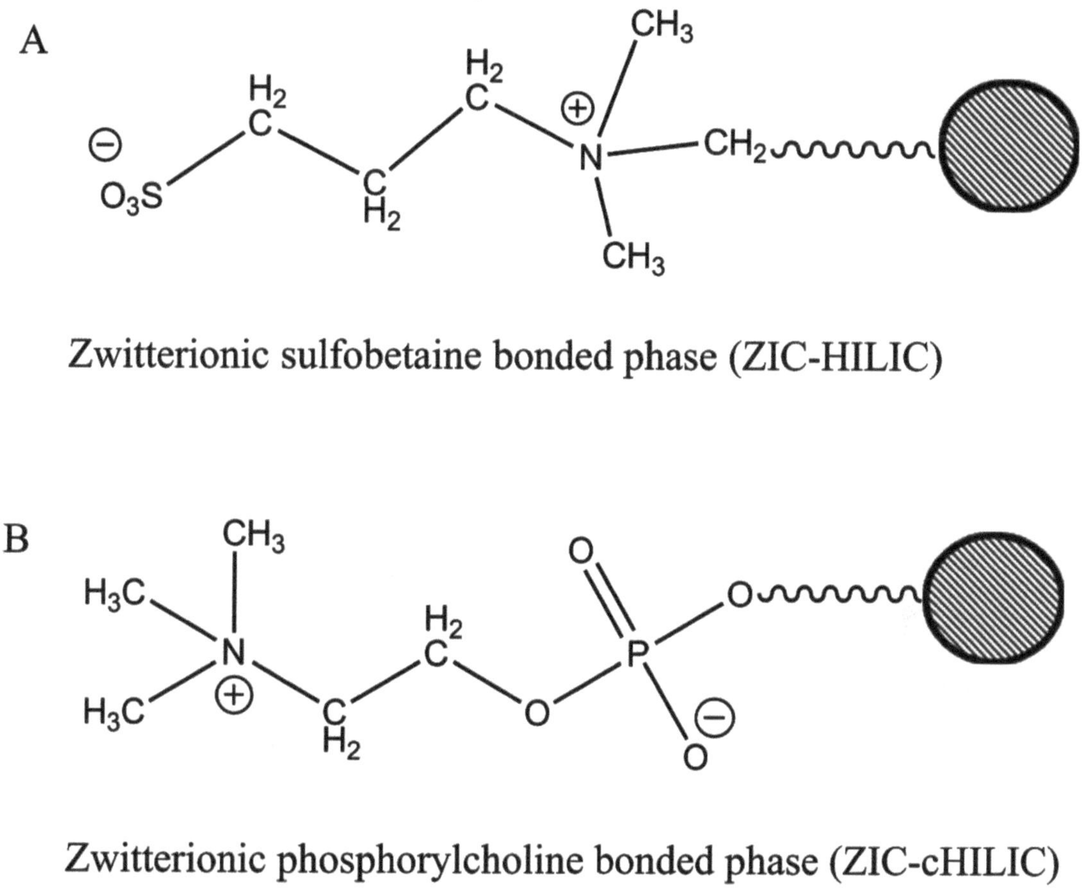

Fig. 5.2 Chemical structure of the zwitterionic bonded phases with (**a**) sulfobetaine functional group (ZIC-HILIC) and (**b**) phopshorylcholine group (ZIC-cHILIC)

lowest absorbance at shorter wavelengths used to measure peptides. It is recommended to try a range of acetonitrile concentrations starting with atleast 60 % to ensure sufficient hydrophilic interaction. Other solvents such as tetrahydrofuran and acetone can also be used. The different organic solvents can also be used in various concentrations to alter retention and selectivity [57].

Separation can be performed either in isocratically or using a gradient. Gradient elution is performed either by increasing the amount of water, i.e. decreasing the organic solvent concentration in the mobile phase or by an increasing salt gradient. In addition to organic solvent, mobile phase pH and buffer/salt concentrations are also critical to HILIC method development. Mobile phase pH affects ionization state of both polar analytes and the stationary phase, consequently having a significant effect on retention and selectivity [57]. Most silica-based HILIC separations are carried out in the pH range of 3–8. Solvent pH is adjusted by the addition of using buffer salts such as ammonium acetate and ammonium formate for acidic pHs and ammonium hydroxide and carbonate for high pHs. The selectivity of the separations can be altered by changing the mobile phase pH, which not only changes the ionization of the functional groups on the stationary phase (e.g. amino) but also affects the relative ionization of the anaytes and thus their retention [107, 108]. Bare silica and silica-based neutral stationary phases are also affected by the mobile phase pH. Normal silanol groups are slightly acidic and can become deprotonated at higher pH values. This could lead to increased electrostatic attraction of basic compounds with the negatively charged silanol groups and results in stronger retention.

While buffers are important to prevent pH fluctuations of the mobile phase, appropriate buffer concentration is important to minimize peak broadening. The most common buffer salts used at ammonium acetate or formate (typically 5–15 mM), because of their high solubility in organic solvents, low UV absorbance and are mass spectrometer friendly. Other buffer salts may be used, however it is important that they are readily soluble in organic solvents and have excellent UV transparency. Ammonium bicarbonate, triethylamine phosphate (TEAP), sodium-methylphosphonate (Na-$MePO_4$), sodium percholate have also been applied in HILIC separations, however these buffers are not volatile and cannot be used with mass spectrometry as a detector. The impact of buffer concentration on retention and selectivity is dependent on the nature of interaction between the analyte and the stationary phase. For non-ionizable compounds, retention is solely dependent on portioning between the immobilized aqueous layer and the hydrophobic mobile phase. Thus, high buffer/salt concentration increases the retention time of these analytes. For ionizable analytes, electrostatic interaction (attractive or repulsive) is an important component of the retention mechanism. In this case, high salt concentration is necessary to disrupt the electrostatic attractions between the analyte and stationary phase. A detailed investigation by McCalley on the HILIC separation of basic compounds using bare silica columns has shown that high salt concentration improves peak shape of charged analytes and also diminishes column overloading effects [107]. In general, it is important to identify the type of electrostatic interactions between the charged analytes and the stationary phases so as to optimize the buffer/salt type and concentration in order to achieve the desired retention and selectivity [57].

Method Development It is important to select the target analytes and the objective of the method prior to starting the method development as this will determine the type of stationary phase and buffers to be used. Commercially available columns are shipped in alcohol or other organic solvents and must be conditioned prior to sample injection. The column must be first washed thoroughly with HPLC grade water (95 % or higher, at least 10–15 column volumes) to remove alcohol. Failure to remove the organic solvent could lead to precipitation of salts that are not soluble in organic solvents and damage the column. The initial washing is followed by rinsing the column with a ~ 10 column volumes of wash buffer. The composition of the wash buffer is typically defined in the manufacturer's instructions. The pH of the wash buffer is not adjusted and used as is. This is followed by flushing the column again with water to remove the salt buffer prior to conditioning the column with the starting mobile phase.

To obtain optimum binding to the stationary phase, samples must be dissolved in organic solvents such as acetonitrile, methanol, ethanol or isopropanol, with acetonitrile being the first choice. Samples are typically dissolved in the buffers having the same organic solvent content as the starting mobile phase. If sample solubility is an issue, then a mixture of the different organic solvents or a small percentage of water or buffer may be used to improve solubility. Water is a strong eluent, hence the amount of water used to dilute and/or dissolve the sample must be carefully adjusted as too much water can lead to peak broadening or splitting. Furthermore, large injection volumes containing high percentage of water causes peak deterioration and loss of sensitivity [57]. Salts (KCl, NaCl etc) may be present due to sample preparation and are insoluble in high organic solvents. Hence samples must be filtered prior to injection to prevent the precipitate from clogging the column.

Several factors influence the retention of peptides on the stationary phase. These include the hydrophilicity of the analyte, solvent pH and buffer concentration and column temperature. The importance of pH and buffer concentration has been discussed in detail above. Most HILIC separations use gradient elution and depending on the elution profile, resolution can be further optimized by changing the slope of the gradient or by increasing the salt concentration in the mobile phase used for elution [95]. Shallow gradients yield enhanced resolution but also increase analysis time. If higher salt concentration is needed for elution, then a two-step gradient elution could be advantageous compared to a one-step linear gradient [95]. Like all other various forms of LC, determining the optimal gradient involves several experiments. Temperature can also be used to optimize resolution and selectivity of the HILIC separation. Increasing temperature leads to decreased retention, if hydrophilic retention is the primary retention mechanism, but deviation from this behavior can occur if other retention mechanisms are involved. In HILIC, column temperature has relatively less of an effect on the retention mechanism compared to the organic solvent content of the mobile phase, its pH and buffer strength [57]. Temperature can be used to optimize the method and achieve high resolution separations.

Applications Because of its selectivity, HILIC is becoming increasingly popular as an "orthogonal" separation technique to RPC and applied in two dimensional separation of complex proteome samples [17, 52, 53, 71, 165]. Evidently, HILIC and RPC mobile phase buffers are not directly compatible and hence the 2D-LC setup is often performed in an off-line mode [16, 18]. HILIC has been shown to have separation power superior to both SEC and SCXC [52, 53]. Although SCXC is the most common first dimensional separation, this method is shown to lead to incomplete recovery of hydrophobic peptides [7]. Even with the addition of organic solvents (e.g. 25 % acetonitrile) in the mobile phase, the recovery of peptides by SCXC has been found to be lower than expected [52, 53]. On the other hand, SEC has low peak capacity, which limits its utility as a first dimensional separation technique in a 2D-LC approach. The HILIC retention mechanism includes both partitioning and electrostatic interactions; hence the separation partially resembles the peptide retention in SCX mode. However, SCX separation power can be limited by the fact that the most prevalent peptides (net charge +2 and +3) are clustered together in a narrow elution window, which is not observed in HILIC separations [16, 18, 52, 53]. The electrostatic interactions ensure that HILIC separation is not merely the reverse of RPC and the hydrophilic interactions allows similarly charged peptides to elute over a wider time window [17].

HILIC is becomingly an increasingly popular technique for the enrichment of post-translational modifications both at the peptide and protein level. The most common applications have been the targeted analysis of phosphorylated, glycosylated and N-terminal acetylated peptides. The attachment of a phospho-moiety to a peptide increases its hydrophilicity and lowers its *pI*. Phosphopeptides can be enriched by SCX at low pH owing to the fact that acidic residues such as aspartic acid and glutamic acid are neutral while the phosphor-serine/tyrosine/threonine residues are negative. Tryptic phosphopeptides elute earlier than the unmodified tryptic peptides. However, multiply phosphorylated peptides are poorly retained by SCX and either poorly retained or even lost. McNulty and Annan [110] were the first to explore HILIC as a first dimension of separation and enrichment for phosphopeptides. When using HILIC for phoshopeptide enrichment, retention is based on overall hydrophilicity of the peptides. Therefore in contrast to SCX, phosphorylated peptides are strongly retained under typical HILIC conditions, allowing the separation of peptides with differing numbers of phosphorylation to be separated using a step-wise gradient [16, 18]. The HILIC separation is usually combined with another enrichment technique such as IMAC or TiO_2 for a more comprehensive analysis of the phosphoproteome [45, 48, 110, 183].

Acetylated N-terminal tryptic peptides behave similarly to phosphorylated peptides during SCX fractionation, i.e., they tend to cluster in the first few fractions. The N-terminal charge is neutralized by acetylation, lowering the net charge of the peptide compared to the unmodified version. Boersema et al. have evaluated the use of ZIC-HILIC for the enrichment of N-acetylated peptides [17]. The neutralized acetylated peptides have reduced hydrophilicity. The polarity is further reduced at pH 3, at which ZIC-HILIC separation was performed and the N-acetylated peptides eluted in the first fractions. Their work also showed that at higher pH conditions (pH 6.8 and 8) ZIC-HILIC has higher separation power, while at pH 3, this separation technique is most orthogonal with RPC.

Following phosphorylation, glycosylation is the second most studied PTM. Although glycoproteins/glycopeptides have been enriched by affinity techniques such as lectin-mediated capture, HILIC is an upcoming promising additional enrichment technique. The glycan group (s) contributes to the overall hydrophilicity of the modified peptide and this physicochemical property is used to separate these peptides from the non-glycosylated peptides. ZIC-HILIC and/or amide-bonded phases are commonly used for glycan and glycopeptide enrichment [13, 29, 58, 88, 106, 164, 169]. There are also several reports of applications of HILIC-SPE for desalting and/or purification of glycans and glycopeptides [104, 140, 145–147]. The advantage of performing HILIC-SPE is the possibility to elute with water, thus providing a salt free and acid-free sample that is ideal for subsequent mass spectrometry or other detection methods [180].

5.6 Mixed Mode Chromatography

Most proteomics workflows utilize two- dimensional liquid chromatography (2D-LC) or multidimensional chromatography to fractionate complex biological samples. This often requires elaborate experimental set up using two or more columns, collecting/processing large number of fractions and analyzing each of them. This process is time consuming and limits the number of samples that be characterized with high resolution and high sensitivity. More recently, mixed mode chromatography (MMC) has received significant attention as an alternative chromatography technique that can enhance selectivity beyond that of single mode separation, performed separately. MMC utilizes more than one type of interaction between the stationary phase and the solutes in the mobile phase. Although most chromatography interactions are considered in terms of single modes, such as ionic or hydrophobic interactions, proteins and peptides are polyions that exhibit both hydrophilic and hydrophobic properties. Their actual chromatographic separation involves multiple modes of interaction and often unintended 'secondary interactions' with the stationary phase lead to peak tailing. Such effects were observed during the development of reversed-phase chromatography where incompletely capped silanol groups exhibited ion exchange activity, causing peak tailing and retention shift for basic compounds [54]. However, it was later realized that this "mixed mode" interaction could be a new technique to improve the resolution and selectivity of the separation by using suitable approaches such as mixing of two types of stationary phase in a single column or using biphasic columns [152, 166, 170].

Theory In contrast to single mode chromatography, MMC uses a stationary phase that is intentionally functionalized with ligands that is capable of multiple modes of interaction with the biomolecules. These multiple modes can include hydrophobic, ion exchange, affinity, electrostatic as well as hydrogen bonding, π-π and thiophilic interactions [32]. There are several MMC separation modes and are named based on at least two types of interactions between the stationary phase and the solutes, that occur either simultaneously or separately. Like all other chromatographic techniques, the retention mechanism in MMC is influenced by the type of ligand, the base matrix, linker and the linker chemistry. Depending on the chemical nature of the ligands (e.g. polar, non-polar, hydrophobic, hydrophilic, acidic and basic groups) and nature of the matrices, different types of mixed-mode stationary phases provide different retention mechanisms. Retention mechanism is also influenced by the size and structure of the solute molecules as well as the mobile phase. The exact mechanism of interactions between the analyte and the multimodal ligands has not been well studied, as most publications are focused on applications of MMC. Theoretical explanations for the MMC retention mechanism, based on the different molecular interactions are classified into three categories detailed explanations of which can be found in [174].

In the case of protein and peptide separations, interaction between the multiple ligands on the MMC matrix and the multiple types of amino acid residues and their charge states at the contact region affects the retention of these biomolecules. The most common mixed mode separations for proteins and peptides are based on ion-exchange and hydrophobic interactions. One such example is the WAX-RP mixed mode separation, where the separation mechanism in an IEX/RP mixed mode is predicted to be a complex interplay of hydrophobic, ion exchange and ion exclusion [129]. The interactions between the different types of chromatography are not independent of each other and the relative contribution of each mechanism depends on the hydrophobicity of the analyte, its charge and also the mobile-phase conditions such as pH, ionic strength and degree of organic modifier. Hence in such a system, increasing the ionic strength will disrupt the ionic bonds but the increasing salt strength will also favor stronger hydrophobic adsorption of the solute. When compared to conventional one dimensional SCX separation, mixed mode IEX/RPC has been shown to have increased fractionation efficiency resulting in a more homogenous distribution of peptides across all fractions. Furthermore, the doubly/triply charged peptides were found to elute over a wide elution window unlike in SCX where the majority of the tryptic peptides elute within a narrow elution window [129].

Hydrophobic interaction chromatography (HIC) is also a type of mixed mode chromatographic process in which the protein of interest in the mixture binds to a dual mode (i.e., one mode for binding and another mode for elution), ionizable ligand. This chromatographic process was pioneered by Porath et al. [134] and Hjertén [66] and is considered a gentle separation technique compared to RPC and also complimentary to other chromatographic modes such as IEX, SEC and affinity chromatography [173]. Hydrophobic ligands such as short alkyl chains and phenyl groups are attached to the stationary phase and separation is based on the reversible interaction between the hydrophobic amino acid side chains on the surface of a protein and the hydrophobic ligand. Proteins bind to the column in high ionic strength buffer and elution is usually performed by decreasing the salt concentration, stepwise or using a gradient. When compared to RPC, the density of the ligands on the stationary phase is much lower and HIC uses milder binding and eluting conditions that allow to maintain the biological activity of the target proteins [136, 173].

The biggest advantage of MMC is that selectivity can be optimized by adjusting the mobile phase ionic strength, pH and/or organic solvent. Additionally, MMC does not require ion pairing agents in the mobile phase for separating highly hydrophilic charged analytes and hence is MS compatible. The adjustable selectivity allows easy separation of analytes of varying charges and hydrophobicity in a single analysis.

Stationary Phases The stationary phase for MMC can be generated either by physical or chemical methods. The simplest form of MMC can be achieved by connecting two different types of columns in series, known as "tandem column" [43]. However, the two mobile phases used for chromatography must be compatible and work synergistically. Tandem columns also lead to high back pressure, especially if high flow rates are used for rapid separation [174]. A second approach is to pack two or more types of stationary phases into a column, termed as "biphasic column" [43], however packing two stationary phases homogeneously can be challenging. Rossi and Horvath [43] compared the performance of both tandem and hybrid columns using commercially available WCX, WAX and SCX and strong- anion exchange (SAX) stationary phases and found their separation efficiency, including resolution to be very similar.

Biphasic columns were first used by Yates et al. for the fractionation of tryptic peptides in their multidimensional protein identification technology (MudPIT) [97, 98, 166, 170]. For the MudPIT approach, peptides were loaded onto a biphasic microcapillary column, packed with 1:1 ratio of SCX and RP (C18) stationary phase. This approach led to the unbiased, comprehensive analysis of the *S. cerevisiae* proteome largely because it was able to detect and identify a wide variety of protein classes including proteins with extremes in p*I*, molecular weight, abundance, and hydrophobicity [170]. The biphasic system was further improved to a three phase MudPIT columns which included an additional reversed-phase column prior to the SCX and RPC column and was used for the online desalting of the sample, prior to fractionation [109].

A third physical approach is to pack two or more types of stationary phases into a column to generate a "mixed bed column" or" hybrid column". The first 'hybrid column' was prepared by Walshe et al. [163] by mixing together SCX and RP (C18) stationary phases and was found to exhibit chromatographic properties of both modes. Motoyama et al. [117] prepared a mixed-bed resin of a blend of anion and cation exchange (ACE) and showed improved recovery of peptides and phosphopeptides compared to SCXC alone.

The biggest advantage of using these physical methods of MMC is that the analytes can be directly transferred from one chromatographic mode to another, thereby reducing dead volume of the system, number of connections and simplifies the overall experimental procedure [174].

In the chemical approach to MMC, the column consists of a single stationary phase that is derivatized with two or more functional groups (or ligands). Most MMC ligands have been designed for the purpose of protein purification, specifically for immunoglobulin purification. Hydroxyapatite is one of the oldest mixed mode chromatographic media that has been used regularly for the purification of antibodies due to its high selectivity and ease of use as it can be performed under neutral conditions [85]. The hydroxyapatite crystals generate a mixed-mode resin, where the separation is achieved by both cation exchange and metal-affinity mechanism. The phosphate groups of the media interact electrostatically with the amines/positively charged amino groups on proteins while the also the calcium ions on the surface of the hydroxyapatite crystals bind to either the carboxyl or phosphoryl groups on proteins. Elution is achieved by either using a phosphate or NaCl gradient [11, 49].

The more recent mixed mode ligands are summarized in Fig. 5.3 and a more comprehensive list can be reviewed here [181]. Some of the first MMC ligands were developed by Yon and coworkers for protein chromatography [175, 176, 178]. These mixed-mode ligands with a net negative charge adsorbed proteins based on the net effect of hydrophobic interactions and electrostatic repulsion for protein purification. Hydrocarbyl ligands are also frequently used in protein chromatography. Examples of this family include the two commercially available adsorbents, hexylamine (HEA) and phenylpropylamine (PPA) Hypercel (Pall LifeScience, NY, USA) [23]. Ligands based on alkyl amines with ω-amino groups [148] as well as the negatively charged counterparts of these ligands, i.e., carboxylic [26, 177] or sulfonic [24, 26, 55] acids have also reported. Secondary interactions such as hydrogen-bonding have also been used as one of the interactions in MMC. The introduction of a hydrogen bonding group in the proximity of ionic groups has been shown to be beneficial for protein binding under high salt conditions [79, 80]. Based on these findings two commercial adsorbents Capto™MMC and Capto™adhere (GE Healthcare, NJ, USA) were developed. Capto™MMC is a weak cation exchanger with a phenyl group for hydrophobic interactions and amide group for hydrogen bonding. Capto™adhere is a strong cation exchanger again with a phenyl group for hydrophobic interaction and a hydroxyl group for hydrogen bonding.

IEXC functional groups such as quaternary ammonium, amino, carboxyl and sulfonic groups can also be adapted to act as mixed mode ligands. Girot and coworkers [24, 55], have used 2-mercapto-5-benzimidazole sulfonic acid, a ligand of MBI Hypercel (Pall Life Sciences,

MMC ligands

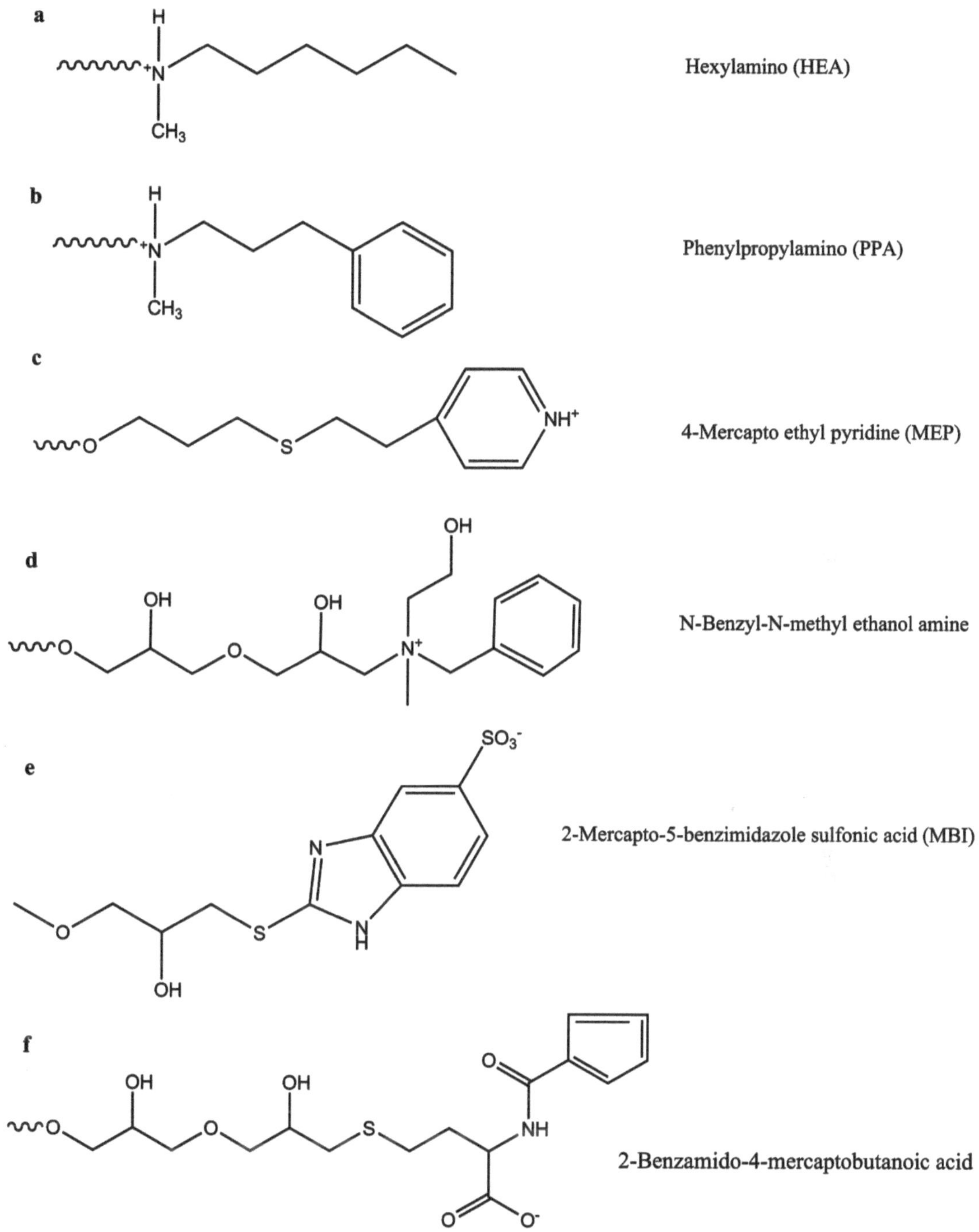

Fig. 5.3 Ligands for mixed mode chromatography (MMC) from selected commercially available mixed mode media. (**a**) Hexylamino (HEA) is a positively charged ligand for HEA Hypercel (Pall Life Sciences, NY, USA). (**b**) Phenylpropylamino (PPA) is a positively charged ligand for PPA Hypercel (Pall Life Sciences, NY, USA). (**c**) 4-Mercapto ethyl pyridine (MEP) is a positively charged ligand for MEP Hypercel (Pall Life Sciences, NY, USA). (**d**) N-benzyl-N-methyl ethanol amine is a positively charged ligand for Capto™adhere (GE Healthcare, NJ, USA). It is a mulitmodal strong cation exchanger. (**e**) 2-mercapto-5-benzimidazole sulfonic acid is a negatively charged ligand for MBI stationary phase (Pall Life Sciences, NY, USA). (**f**) 2-Benzamido-4-mercaptobutamoic acid is a negatively charged ligand for Capto™MMC (GE Healthcare, NJ, USA). It is a mulitmodal weak cation exchanger. (* Positive/Negative charge is reported at physiological pH)

NY, USA) as a multimodal ligand for the purification of antibodies. Heterocyclic mixed mode ligands have also been applied for protein purification [27, 33, 59, 172, 182]. MEP Hypercel (Pall Life Sciences, NY, USA) contains 4-mercapto-ethyl-pyridine (MEP), an ionizable ligand which is uncharged at physiological pH. Protein adsorption is achieved by hydrophobic interactions and eluted by reducing the pH of the mobile phase to 4 or lower, where the ligand is positively charges. This dual mode mechanism forms the principle of hydrophobic charge induction chromatography (HCIC) [27, 50, 51, 172]. A number of new mixed mode chromatographic stationary phases have been commercialized by SIELC (Wheeling, IL) and are available under the trade name Primesep. The HPLC column choices include combinations of RP with anion, cation and zwitter ion functional groups [93].

In summary, the ligand for mixed-mode chromatography should have at least one hydrophobic moiety and one ionic moiety. The hydrophobic moiety must be carefully chosen so as to achieve a sufficiently high capacity and afford reasonable recovery. The pKa of the ionic moiety is essential for the performance of the ligand and should be estimated in ligand screening and design [181]. Secondary interactions such as hydrogen bonding can also contribute to protein-ligand binding and can be introduced either as hydrogen donors to anion-exchange ligands or hydrogen acceptors to cation-exchange ligands.

Method Development As mentioned earlier, most proteomic approaches use either IEX/RPLC [54, 121, 129] and HILIC/IEX [61, 101] combinations of separation modes. While octadecylsilanes (C18) still remains the preferred RP ligand, the choice of the ionic ligand will depend on the class of peptides to be enriched and/or fractionated. Gilar et al. [54] used silica-based pentafluorophenyl (PFP) MMC column to selectively enrich for negatively charged peptides, such as phosphopeptides and sialylated glycopeptides. Stationary phase containing octadecylsilanes and dialkylamines has been used as RPC/AEX mixed mode combination for peptide separation [68]. HILIC/SCX mixed mode approach, first introduced by Hodges and group [185, 186] has proven to be very versatile for peptide separations versus RPC, specifically for separating highly charged species [101]. HILIC/SCX was carried out on a poly (2-sulphoethyl aspartamide)-silica (polysulphoethyl A) (PolyLC, Columbia, MD, USA) strong SCX column. Peptide separation was carried out in the presence of a high organic modifier concentration (60–80 % ACN) to promote hydrophilic interactions between the solute and the hydrophilic/charged SCX stationary phase, with peptides then eluted with a linear salt gradient. Peptides are generally eluted in groups in order of increasing net positive charge; within these groups, peptides are resolved in order of increasing hydrophilicity (decreasing hydrophobicity) [101].

As in all other forms of chromatography, solvent selection, pH, salt concentration and temperature influence sensitivity and resolution. In general adsorption of proteins in MMC occurs under low –to-moderate ionic strength, neutral pH and elution is achieved by electrostatic repulsion when the pH value is lowered below the p*I* of the target and pKa of the ligand. Columns are regenerated with chelating reagents, acid/base wash, high salt concentrations [143].

Summary Examples of peptide fractionation by MMC cited in the previous sections clearly demonstrates the several distinct advantages this form of chromatography has over the single-mode chromatography. As compared with traditional 2D approaches, MMC has shown improved selectivity, resolution and higher sample loading capacity [61, 117, 121]. This approach offers increased separation and degrees of freedom in adjusting separation selectivity compared to any one type of chromatography. Given the limited number of publications that have reported the use of MMC as part of a routine proteomic sample preparation workflow suggests that this method has not yet been fully exploited. Currently, most of the MMC applications are focused on small molecules and proteins, predominantly immunoglobulin purification. This could partly be due to the lack of a deep understanding of the mixed mode retention mechanism, which would otherwise be useful for synthesizing new ligands and stationary phases that could help accelerate development of its applications.

References

1. Aguilar M-I (2004) HPLC of peptides and proteins. HPLC of peptides and proteins, vol 251 M-I Aguilar. Springer, New York, p 3–8
2. Aguilar MI, Hearn MT (1996) High-resolution reversed-phase high-performance liquid chromatography of peptides and proteins. Methods Enzymol 270:3–26
3. Allen DP (1999) 18 – Application of size exclusion-high-performance liquid chromatography for biopharmaceutical protein and peptide therapeutics. Column handbook for size exclusion chromatography. Cs Wu. Academic Press, San Diego, 531–537
4. Alpert AJ (1983) Cation-exchange high-performance liquid chromatography of proteins on poly (aspartic acid)—silica. J Chromatogr A 266 (0):23–37[1]
5. Alpert AJ (1990) Hydrophilic-interaction chromatography for the separation of peptides, nucleic acids and other polar compounds. J Chromatogr A 499(0):177–196[2]
6. Alpert AJ (2007) Electrostatic repulsion hydrophilic interaction chromatography for isocratic separation of charged solutes and selective isolation of phosphopeptides. Anal Chem 80(1):62–76[3]
7. Alpert AJ, Andrews PC (1988) Cation-exchange chromatography of peptides on poly(2-sulfoethyl aspartamide)-silica. J Chromatogr 443:85–96[4]

[1] A simple cation-exchange material for high-performance liquid chromatography of proteins was developed. Poly(succinimide) reacted rapidly with aminopropylsilica and the product was hydrolysed to poly(aspartic acid)—silica. Reaction conditions were optimized to yield a material with an ion-exchange capacity of 430 mg hemoglobin/g material. High-performance liquid chromatographic columns of the material featured excellent performance in terms of capacity, selectivity, recovery of enzyme activity, peak shape and durability. Protein standards and clinical hemoglobin samples were well resolved in minutes. Poly(succinimide)—silica was readily derivatized to give products other than poly (aspartic acid)—silica, and several such materials were prepared. Such materials could be useful for affinity chromatography or enzyme immobilization.

[2] When a hydrophilic chromatography column is eluted with a hydrophobic (mostly organic) mobile phase, retention increases with hydrophilicity of solutes. The term hydrophilic-interaction chromatography is proposed for this variant of normal-phase chromatography. This mode of chromatography is of general utility. Mixtures of proteins, peptides, amino acids, oligonucleotides, and carbohydrates are all resolved, with selectivity complementary to those of other modes. Typically, the order of elution is the opposite of that obtained with reversed-phase chromatography. A hydrophilic, neutral packing was developed for use in high-performance hydrophilic-interaction chromatography. Hydrophilic-interaction chromatography is particularly promising for such troublesome solutes as histones, membrane proteins, and phosphorylated amino acids and peptides. Hydrophilic-interaction chromatography fractionations resemble those obtained through partitioning mechanisms. The chromatography of DNA, in particular, resembles the partitioning observed with aqueous two-phase systems based on polyethylene glycol and dextran solutions.

[3] If an ion-exchange column is eluted with a predominantly organic mobile phase, then solutes can be retained through hydrophilic interaction even if they have the same charge as the stationary phase. This combination is termed electrostatic repulsion-hydrophilic interaction chromatography (ERLIC). With mixtures of solutes that differ greatly in charge, repulsion effects can be exploited to selectively antagonize the retention of the solutes that normally would be the best retained. This permits the isocratic resolution of mixtures that normally require gradients, including peptides, amino acids, and nucleotides. ERLIC affords convenient separations of highly charged peptides that cannot readily be resolved by other means. In addition, phosphopeptides can be isolated selectively from a tryptic digest.

[4] A strong cation-exchange material, poly(2-sulfoethyl aspartamide)-silica (PolySULFOETHYL Aspartamide) was developed for purification and analysis of peptides by high-performance liquid chromatography. All peptides examined were retained at pH 3, even when the amino terminus was the only basic group. Peptides were eluted in order of increasing number of basic residues with a salt gradient. Capacity was high, as was selectivity and column efficiency. This new column material displays modest mixed-mode effects, allowing the resolution of peptides having identical charges at a given pH. The selectivity can be manipulated by the addition of organic solvent to the mobile phases; this increases the retention of some peptides and decreases the retention of others. The retention in any given case may reflect a combination of steric factors and non-electrostatic interactions. Selectivity was complementary to that of reversed-phase chromatography (RPC) materials. Excellent purifications were obtained by sequential use of PolySULFOETHYL Aspartamide and RPC columns for purification of peptides from crude tissue extracts. The new cation exchanger is quite promising as a supplement to RPC for general peptide chromatography.

8. Alvarez-Manilla G, Atwood et al (2006) Tools for glycoproteomic analysis: size exclusion chromatography facilitates identification of tryptic glycopeptides with N-linked glycosylation sites. J Proteome Res 5(3):701–708[5]
9. Appelblad P, Jonsson P et al (2006) A practical guide to HILIC: a tutorial and application book. Merck SeQuant AB, Umeå
10. Arakawa T, Ejima D et al (2010) The critical role of mobile phase composition in size exclusion chromatography of protein pharmaceuticals. J Pharm Sci 99 (4):1674–1692[6]
11. Ayyar BV, Arora S et al (2012) Affinity chromatography as a tool for antibody purification. Methods 56 (2):116–129[7]
12. Barbour J, Wiese S et al (2008) Mass spectrometry. Proteomics sample preparation. Wiley-VCH Verlag GmbH & Co. KGaA, p 41–128[8]
13. Bereman MS, Williams TI et al (2009) Development of a nanoLC LTQ orbitrap mass spectrometric method for profiling glycans derived from plasma from healthy, benign tumor control, and epithelial ovarian cancer patients. Anal Chem 81 (3):1130–1136[9]

[5] Proteomic techniques, such as HPLC coupled to tandem mass spectrometry (LC-MS/MS), have proved useful for the identification of specific glycosylation sites on glycoproteins (glycoproteomics). Glycosylation sites on glycopeptides produced by trypsinization of complex glycoprotein mixtures, however, are particularly difficult to identify both because a repertoire of glycans may be expressed at a particular glycosylation site, and because glycopeptides are usually present in relatively low abundance (2–5 %) in peptide mixtures compared to nonglycosylated peptides. Previously reported methods to facilitate glycopeptide identification require either several pre-enrichment steps, involve complex derivatization procedures, or are restricted to a subset of all the glycan structures that are present in a glycoprotein mixture. Because the N-linked glycans expressed on tryptic glycopeptides contribute substantially to their mass, we demonstrate that size exclusion chromatography (SEC) provided a significant enrichment of N-linked glycopeptides relative to nonglycosylated peptides. The glycosylated peptides were then identified by LC-MS/MS after treatment with PNGase-F by the monoisotopic mass increase of 0.984 Da caused by the deglycosylation of the peptide. Analyses performed on human serum showed that this SEC glycopeptide isolation procedure results in at least a 3-fold increase in the total number of glycopeptides identified by LC-MS/MS, demonstrating that this simple, nonselective, rapid method is an effective tool to facilitate the identification of peptides with N-linked glycosylation sites. Keywords: glycoproteomics; LC/MS/MS; glycopeptides; N-linked glycosylation sites; size excusion chromatography.

[6] Size exclusion chromatography (SEC) is the most widely used method for aggregation analysis of pharmaceutical proteins. However SEC analysis has a number of limitations, and one of the most important ones is protein adsorption to the resin. This problem is particularly severe when using new columns, and often column preconditioning protocols are required. This review focuses on the role that addition of various cosolvents to the mobile phase plays in suppressing that protein adsorption. Cosolvents such as salt, amino acids, and organic solvents are often used for this purpose. Because the protein interaction with the resin surface is highly heterogeneous, different cosolvents affect the protein adsorption differently. We will summarize the various effects of cosolvents on protein adsorption and retention and describe the mechanism of the cosolvent effects. © 2009 Wiley-Liss, Inc. and the American Pharmacists Association J Pharm Sci 99: 1674–1692, 2010.

[7] The global antibody market has grown exponentially due to increasing applications in research, diagnostics and therapy. Antibodies are present in complex matrices (e.g. serum, milk, egg yolk, fermentation broth or plant-derived extracts). This has led to the need for development of novel platforms for purification of large quantities of antibody with defined clinical and performance requirements. However, the choice of method is strictly limited by the manufacturing cost and the quality of the end product required. Affinity chromatography is one of the most extensively used methods for antibody purification, due to its high selectivity and rapidity. Its effectiveness is largely based on the binding characteristics of the required antibody and the ligand used for antibody capture. The approaches used for antibody purification are critically examined with the aim of providing the reader with the principles and practical insights required to understand the intricacies of the procedures. Affinity support matrices and ligands for affinity chromatography are discussed, including their relevant underlying principles of use, their potential value and their performance in purifying different types of antibodies, along with a list of commercially available alternatives. Furthermore, the principal factors influencing purification procedures at various stages are highlighted. Practical considerations for development and/or optimizations of efficient antibody-purification protocols are suggested.

[8] This chapter contains sections titled: * A Practical Guideline to Electrospray Ionization Mass Spectrometry for Proteomics Application * References * Sample Preparation for the Application of MALDI Mass Spectrometry in Proteome Analysis * References * Sample Preparation for Label-Free Proteomic Analyses of Body Fluids by Fourier Transform Ion Cyclotron Mass Spectrometry * References * Sample Preparation for Differential Proteome Analysis: Labeling Technologies for Mass Spectrometry * References * Determining Membrane Protein Localization Within Subcellular Compartments Using Stable Isotope Tagging * References.

[9] We report the development of split-less nano-flow liquid chromatography mass spectrometric analysis of glycans chemically cleaved from glycoproteins in plasma. Porous

14. Betancourt LH, De Bock PJ et al (2013) SCX charge state selective separation of tryptic peptides combined with 2D-RP-HPLC allows for detailed proteome mapping. J Proteomics 91(0):164–171[10]
15. Bidlingmeyer BA, Del Rios JK, 3 et al (1982) Separation of organic amine compounds on silica gel with reversed-phase eluents. Anal Chem 54:442–447
16. Boersema P, Mohammed S et al (2008) Hydrophilic interaction liquid chromatography (HILIC) in proteomics. Anal Bioanal Chem 391(1):151–159
17. Boersema PJ, Divecha N et al (2007) Evaluation and optimization of ZIC-HILIC-RP as an alternative MudPIT strategy. J Proteome Res 6(3):937–946[11]
18. Boersema PJ, Mohammed S et al (2008) Hydrophilic interaction liquid chromatography (HILIC) in proteomics. Anal Bioanal Chem 391(1):151–159[12]

graphitized carbon operating under reverse-phase conditions and an amide-based stationary phase operating under hydrophilic interaction conditions are quantitatively compared for glycan separation. Both stationary phases demonstrated similar column efficiencies and excellent retention time reproducibility without an internal standard to correct for retention time shift. The 95 % confidence intervals of the mean retention times were +/−4 s across 5 days of analysis for both stationary phases; however, the amide stationary phase was observed to be more robust. The high mass measurement accuracy of less than 2 ppm and fragmentation spectra provided highly confident identifications along with structural information. In addition, data are compared among samples derived from 10 healthy controls, 10 controls with a differential diagnosis of benign gynecologic tumors, and 10 diseased epithelial ovarian cancer patients (EOC). Two fucosylated glycans were found to be up-regulated in healthy controls and provided an accurate diagnostic value with an area under the receiver operator characteristic curve of 0.87. However, these same glycans provided a significantly less diagnostic value when used to differentiate EOC from benign tumor control samples with an area under the curve of 0.73.

[10] Multidimensional peptide fractionation is widely used in proteomics to reduce the complexity of peptide mixtures prior to mass spectrometric analysis. Here, we describe the sequential use of strong cation exchange and reversed phase liquid chromatography in both basic and acidic pH buffers for separating tryptic peptides from complex mixtures of proteins. Strong cation exchange exclusively separates peptide by their charge state into neutral, singly and multi-charged species. To further reduce complexity, each peptide group was separated by reversed phase liquid chromatography at basic pH and the resultant fractions were analyzed by LC–MS/MS. This workflow was applied to a soluble protein lysate from mouse embryonic fibroblast cells, and more than 5000 proteins from 29,843 peptides were identified. The high selectivity displayed during the SCX step (93 % to 100 %) and the overlaps between proteins identified from the SCX-separated peptide groups, are interesting assets of the procedure. Biological significance The present work shows how complex mixture of peptides can be selectively separated by SCX based essentially on the net charge of peptides. The proposed workflow results in three well-defined subset of peptides of specific amino acid composition, which are representative of the constituent proteins. The very high selectivity obtained (93 % to 99 %) on the peptide side, underscores for the first time the possibility of SCX chromatography to aid in validating identified peptides.

[11] In proteomics, a digested cell lysate is often too complex for direct comprehensive mass spectrometric analysis. To reduce complexity, several peptide separation techniques have been introduced including very successful two-dimensional liquid chromatography (2D-LC) approaches. Here, we assess the potential of zwitterionic Hydrophilic Interaction Liquid Chromatography (ZIC-HILIC) as a first dimension for the analysis of complex peptide mixtures. We show that ZIC-HILIC separation is dramatically dependent on buffer pH in the range from 3 to 8, due to deprotonation of acidic amino acids. ZIC-HILIC exhibits a mixed-mode effect consisting of electrostatic and polar interactions. We developed a 2D-LC system that hyphenates ZIC-HILIC off-line with reversed-phase (RP). The two dimensions are fairly orthogonal, and the system performs very well in the analysis of minute amounts of complex peptide mixtures. Applying this method to the analysis of 10 mug of a cellular nuclear lysate, we were able to confidently identify over 1000 proteins. Compared to strong cation exchange chromatography (SCX), ZIC-HILIC shows better chromatographic resolution and absence of clustering of prevalent +2 and +3 charged peptides. At pH 3, ZIC-HILIC separation allows best orthogonality with RP and resembles conventional SCX separation. A significant enrichment of N-acetylated peptides in the first fractions is observed at these conditions. ZIC-HILIC separation at high pH (6.8 and 8), however, enables better chromatography, resulting in more comprehensive data acquisition. With this extended flexibility, we conclude that ZIC-HILIC is a very good alternative for the more conventional SCX in multidimensional peptide separation strategies.

[12] In proteomics, nanoflow multidimensional chromatography is now the gold standard for the separation of complex mixtures of peptides as generated by in-solution digestion of whole-cell lysates. Ideally, the different stationary phases used in multidimensional chromatography should provide orthogonal separation characteristics. For this reason, the combination of strong cation exchange chromatography (SCX) and reversed-phase (RP) chromatography is the most widely used combination for the separation of peptides. Here, we review the potential of hydrophilic interaction liquid chromatography (HILIC) as a separation tool in the

19. Boutin JA, Ernould AP et al (1992) Use of hydrophilic interaction chromatography for the study of tyrosine protein kinase specificity. J Chromatogr B Biomed Sci Appl 583(2):137–143[13]
20. Boyer R (2005) Principles and reactions of protein extraction, purification, and characterization: Ahmed, Hafiz. Biochem Mol Biol Educ 33 (2):145–146
21. Boyes BE, Walker DG (1995) Selectivity optimization of reversed-phase high-performance liquid chromatographic peptide and protein separations by varying bonded-phase functionality. J Chromatogr A 691(1–2):337–347[14]
22. Boysen RI, Hearn MTW (2001) HPLC of peptides and proteins. Current protocols in protein science, Wiley[15]
23. Brenac Brochier V, Schapman A et al (2008) Fast purification process optimization using mixed-mode chromatography sorbents in pre-packed mini-columns. J Chromatogr A 1177(2):226–233[16]
24. Brenac V, Ravault V et al (2005) Capture of a monoclonal antibody and prediction of separation conditions using a synthetic multimodal ligand attached on chips and beads. J Chromatogr B 818 (1):61–66[17]

multidimensional separation of peptides in proteomics applications. Recent work has revealed that HILIC may provide an excellent alternative to SCX, possessing several advantages in the area of separation power and targeted analysis of protein post-translational modifications. [figure: see text]

[13] A new HPLC method has been developed to assay tyrosine protein kinase activity. Using hydrophilic interaction chromatography, it is possible to resolve the four components of the incubation medium: substrate peptide, [32P]phosphorylated peptide, unreacted [γ-32P]ATP, and 32P-labelled inorganic phosphate. ATP interacts so strongly with the stationary phase material that it can be removed selectively from the incubation medium with solid-phase extraction cartridges packed with the same type of material. The three remaining components of interest can then be resolved by reversed-phase or hydrophilic interaction HPLC. This procedure permits the evaluation of almost every type of peptide as a substrate of tyrosine protein kinase.

[14] Several chemical bonded-phase modified silicas were prepared using sterically protected monofunctional silane reagents which varied widely in structure and polarity. Since some of these bonded-phase packing materials are highly polar (hydrophilic), resistance to acid-catalyzed bonded-phase loss by hydrolysis was examined, and observed to remain high even for the highly polar Diol bonded-phase functionality. Modification of the surface of 300 Å pore size, fully hydroxylated and base-deactivated silica microspheres with these sterically protected silanes yielded HPLC column packing materials for examination of separation selectivities in reversed-phase separations of peptide and protein mixtures. Distinct separation selectivities were apparent for each bonded-phase functionality. Selectivity differences ranged from limited band spacing changes for steric-protected C18 and C8 bonded-phases, to reversal of elution order for the more polar C3 and CN bonded phases. The use of column-based selectivity differences between sequential reversed-phase separation steps is used for the two-step HPLC isolation of a recombinant human amyloid precursor polypeptide fragment from a crude bacterial extract.

[15] High-performance liquid chromatography (HPLC) is an essential tool for the purification and characterization of biomacromolecules. This unit presents a thorough discussion of the eight types of HPLC currently used, highlighting equipment and start-up procedures, recommendations for running each type of experiment, and theoretical considerations for the separation of peptides and proteins. This is an excellent primer for HPLC users.

[16] Pre-packed MediaScoutÂ® MiniChrom columns of 2.5, 5 and 10 mL were investigated for screening three mixed-mode chromatography sorbents (HEA, PPA and MEP HyperCel™). Packing performance was of good quality and the three sorbents displayed higher capacity than traditional HIC sorbents in physiological-like conditions. Each sorbent offered a unique selectivity. Bovine Î²-lactoglobulin was partially purified after loading milk whey directly on HEA HyperCel sorbent. The combination of small pre-packed columns and SELDI-MS appeared to be a valuable strategy for high-throughput screening of chromatography sorbents and for enabling rapid process development and optimization.

[17] A synthetic ligand called 2-mercapto-5-benzimidazole-sulfonic acid has been successfully used for the specific chromatographic capture of antibodies from a cell culture supernatant. Adsorption occurred at physiological ionic strength and pH range between 5.0 and 6.0, with some binding capacity variations within this pH range: antibody uptake increased when the pH decreased. With very dilute feedstocks, as was the case with the cell culture supernatant under investigation, it was found that the pH had to be slightly lowered to get a good antibody sorption capacity. To optimize separation conditions, a preliminary study was made using ProteinChipÂ® Arrays that displayed the same chemical functionalities as the resin. Arrays were analyzed using SELDIâ€“MS. By this mean, it was possible to cross-over simultaneously different pH conditions at the adsorption and the desorption steps. Best conditions were implemented for preparative separation using regular lab-scale columns. At pH 5.2, antibody adsorption was not complete, while at pH 5.0 the antibody was entirely captured. pH 9 was selected at elution, rather than pH 8.0 or 10.0, and resulted in a complete desorption of antibodies from the column. Benefits of the prediction of separation conditions of antibodies on MBI beads using

25. Brunner E, Ahrens CH et al (2007) A high-quality catalog of the Drosophila melanogaster proteome. Nat Biotech 25(5):576–583[18]
26. Burton SC, Haggarty NW et al (1997) One step purification of chymosin by mixed mode chromatography. Biotechnol Bioeng 56(1):45–55[19]
27. Burton SC, Harding DRK (1998) Hydrophobic charge induction chromatography: salt independent protein adsorption and facile elution with aqueous buffers. J Chromatogr A 814(1â€"2): 71–81[20]
28. Buszewski B, Noga S (2012) Hydrophilic interaction liquid chromatography (HILIC)–a powerful separation technique. Anal Bioanal Chem 402(1):231–247[21]
29. Calvano CD, Zambonin CG et al (2008) Assessment of lectin and HILIC based enrichment protocols for characterization of serum glycoproteins by mass spectrometry. J Proteome 71(3):304–317[22]

SELDIâ€"MS were a significant reduction in analysis time and in sample volume. This was possible because the separation of IgG on the chip surface did mimic very well the separation on beads.

[18] Understanding how proteins and their complex interaction networks convert the genomic information into a dynamic living organism is a fundamental challenge in biological sciences. As an important step towards understanding the systems biology of a complex eukaryote, we cataloged 63 % of the predicted Drosophila melanogaster proteome by detecting 9124 proteins from 498,000 redundant and 72,281 distinct peptide identifications. This unprecedented high proteome coverage for a complex eukaryote was achieved by combining sample diversity, multidimensional biochemical fractionation and analysis-driven experimentation feedback loops, whereby data collection is guided by statistical analysis of prior data. We show that high-quality proteomics data provide crucial information to amend genome annotation and to confirm many predicted gene models. We also present experimentally identified proteotypic peptides matching [sim]50 % of D. melanogaster gene models. This library of proteotypic peptides should enable fast, targeted and quantitative proteomic studies to elucidate the systems biology of this model organism.

[19] Mixed mode Sepharose and Perloza bead cellulose matrices were prepared using various chemistries. These matrices contained hydrophobic (aliphatic and/or aromatic) and ionic (carboxylate or alkylamine) groups. Hydrophobic amine ligands were attached to epichlorohydrin activated Sepharose (mixed mode amine matrices). Hexylamine, aminophenylpropanediol and phenylethylamine were the preferred ligands, on the basis of cost and performance. Other mixed mode matrices were produced by incomplete attachment (0–80 %) of the same amine ligands to carboxylate matrices. The best results were obtained using unmodified or partially ligand-modified aminocaproic acid Sepharose and Perloza. High ligand densities were used, resulting in high capacity. Furthermore, chymosin was adsorbed at high and low ionic strengths, which reduced sample preparation requirements. Chymosin, essentially homogeneous by electrophoresis, was recovered by a small pH change. The methods described were simple, efficient, inexpensive and provided very good resolution of chymosin from a crude recombinant source. The carboxylate matrices had the best combination of capacity and regeneration properties. The performance of Sepharose and Perloza carboxylate matrices was similar, but higher capacities were found for the latter. Because it is cheaper and can be used at higher flow rates, Perloza should be better suited to large scale application. High capacity chymosin adsorption was found with carboxymethyl ion exchange matrices, but low ionic strength was essential for adsorption and the purity was inferior to that of the mixed mode matrices. © 1997 John Wiley & Sons, Inc. Biotechnol Bioeng56: 45–55, 1997.

[20] A new form of protein chromatography, hydrophobic charge induction, is described. Matrices prepared by attachment of weak acid and base ligands were uncharged at adsorption pH. At low ligand densities, protein adsorption was typically promoted with lyotropic salts. At higher ligand densities, chymosin, chymotrypsinogen and lysozyme were adsorbed independently of ionic strength. A pH change released the electrostatic potential of the matrix and weakened hydrophobic interactions, inducing elution. Matrix hydrophobicity and titration range could be matched to protein requirements by ligand choice and density. Both adsorption and elution could be carried out within the pH 5™9 range.

[21] Hydrophilic interaction liquid chromatography (HILIC) provides an alternative approach to effectively separate small polar compounds on polar stationary phases. The purpose of this work was to review the options for the characterization of HILIC stationary phases and their applications for separations of polar compounds in complex matrices. The characteristics of the hydrophilic stationary phase may affect and in some cases limit the choices of mobile phase composition, ion strength or buffer pH value available, since mechanisms other than hydrophilic partitioning could potentially occur. Enhancing our understanding of retention behavior in HILIC increases the scope of possible applications of liquid chromatography. One interesting option may also be to use HILIC in orthogonal and/or two-dimensional separations. Bioapplications of HILIC systems are also presented.

[22] Protein glycosylation is a common post-translational modification that is involved in many biological processes, including cell adhesion, protein–protein and receptor-ligand interactions. The glycoproteome constitutes a source for identification of disease biomarkers since altered protein glycosylation profiles are associated with certain human ailments. Glycoprotein analysis by mass spectrometry of biological samples, such as blood serum, is hampered by sample complexity and the low concentration of the potentially informative

30. Cargile BJ, Bundy JL et al (2004) Gel based isoelectric focusing of peptides and the utility of isoelectric point in protein identification. J Proteome Res 3 (1):112–119[23]
31. Chen Y, Mant CT et al (2003) Temperature selectivity effects in reversed-phase liquid chromatography due to conformation differences between helical and non-helical peptides. J Chromatogr A 1010 (1):45–61[24]
32. Chung WK, Freed AS et al (2010) Evaluation of protein adsorption and preferred binding regions in multimodal chromatography using NMR. Proc Natl Acad Sci 107(39):16811–16816[25]
33. Coffinier Y, Vijayalakshmi MA (2004) Mercaptoheterocyclic ligands grafted on a poly(ethylene vinyl alcohol) membrane for the purification of immunoglobulin G in a salt independent thiophilic chromatography. J Chromatogr B 808(1):51–56[26]

glycopeptides and -proteins. We assessed the utility of lectin-based and HILIC-based affinity enrichment techniques, alone or in combination, for preparation of glycoproteins and glycopeptides for subsequent analysis by MALDI and ESI mass spectrometry. The methods were successfully applied to human serum samples and a total of 86 N-glycosylation sites in 45 proteins were identified using a mixture of three immobilized lectins for consecutive glycoprotein enrichment and glycopeptide enrichment. The combination of lectin affinity enrichment of glycoproteins and subsequent HILIC enrichment of tryptic glycopeptides identified 81 N-glycosylation sites in 44 proteins. A total of 63 glycosylation sites in 38 proteins were identified by both methods, demonstrating distinct differences and complementarity. Serial application of custom-made microcolumns of mixed, immobilized lectins proved efficient for recovery and analysis of glycopeptides from serum samples of breast cancer patients and healthy individuals to assess glycosylation site frequencies.

[23] Here we present the theoretical and experimental evaluation of peptide isoelectric point as a method to aid in the identification of peptides from complex mixtures. Predicted pI values were found to match closely the experimentally obtained data, resulting in the development of a unique filter that lowers the effective false positive rate for peptide identification. Due to the reduction of the false positive rate, the cross-correlation parameters Xcorr and deltaCn from the SEQUEST program can be lowered resulting in 25 % more peptide identifications. This approach was successfully applied to analysis of the soluble fraction of the E. coli proteome, where 417 proteins were identified from 1022 peptides using just 20 microg of material.

[24] In order to characterize the effect of temperature on the retention behaviour and selectivity of separation of polypeptides and proteins in reversed-phase high-performance liquid chromatography (RP-HPLC), the chromatographic properties of four series of peptides, with different peptide conformations, have been studied as a function of temperature (5–80 °C). The secondary structure of model peptides was based on either the amphipathic α-helical peptide sequence Ac-EAEKAAKEXd/lEKAAKEAEK-amide, (position X being in the centre of the hydrophobic face of the α-helix), or the random coil peptide sequence Ac-Xd/lLGAKGAGVG-amide, where position X is substituted by the 19 l- or d-amino acids and glycine. We have shown that the helical peptide analogues exhibited a greater effect of varying temperature on elution behaviour compared to the random coil peptide analogues, due to the unfolding of α-helical structure with the increase of temperature during RP-HPLC. In addition, temperature generally produced different effects on the separations of peptides with different l- or d-amino acid substitutions within the groups of helical or non-helical peptides. The results demonstrate that variations in temperature can be used to effect significant changes in selectivity among the peptide analogues despite their very high degree of sequence homology. Our results also suggest that a temperature-based approach to RP-HPLC can be used to distinguish varying amino acid substitutions at the same site of the peptide sequence. We believe that the peptide mixtures presented here provide a good model for studying temperature effects on selectivity due to conformational differences of peptides, both for the rational development of peptide separation optimization protocols and a probe to distinguish between peptide conformations.

[25] NMR titration experiments with labeled human ubiquitin were employed in concert with chromatographic data obtained with a library of ubiquitin mutants to study the nature of protein adsorption in multimodal (MM) chromatography. The elution order of the mutants on the MM resin was significantly different from that obtained by ion-exchange chromatography. Further, the chromatographic results with the protein library indicated that mutations in a defined region induced greater changes in protein affinity to the solid support. Chemical shift mapping and determination of dissociation constants from NMR titration experiments with the MM ligand and isotopically enriched ubiquitin were used to determine and rank the relative binding affinities of interaction sites on the protein surface. The results with NMR confirmed that the protein possessed a distinct preferred binding region for the MM ligand in agreement with the chromatographic results. Finally, coarse-grained ligand docking simulations were employed to study the modes of interaction between the MM ligand and ubiquitin. The use of NMR titration experiments in concert with chromatographic data obtained with protein libraries represents a previously undescribed approach for elucidating the structural basis of protein binding affinity in MM chromatographic systems.

[26] In this study, we attempted a limited combinatorial approach for designing affinity ligands based on mercaptoheterocyclic components. The template, divinyl

34. Cox GB, Stout RW (1987) Study of the retention mechanisms for basic compounds on silica under pseudo-reversed-phase conditions. J Chromatogr 384:315–336
35. Cunliffe JM, Maloney TD (2007) Fused-core particle technology as an alternative to sub-2-μm particles to achieve high separation efficiency with low backpressure. J Sep Sci 30(18):3104–3109
36. Dai J, Jin WH et al (2006) Protein phosphorylation and expression profiling by Yin-Yang multidimensional liquid chromatography (Yin-Yang MDLC) mass spectrometry. J Proteome Res 6(1):250–262[27]
37. Di Palma S, Boersema PJ et al (2011) Zwitterionic Hydrophilic Interaction Liquid Chromatography (ZIC-HILIC and ZIC-cHILIC) Provide high resolution separation and increase sensitivity in proteome analysis. Anal Chem 83(9):3440–3447[28]
38. Di Palma S, Hennrich ML et al (2012) Recent advances in peptide separation by multidimensional liquid chromatography for proteome analysis. J Proteome 75(13):3791–3813[29]

sulfone structure (DVS), which was grafted on poly(ethylene vinyl alcohol) (PEVA) hollow fiber membrane, has served for the tethering of different heterocyclic compounds as pyridine, imidazole, purine and pyrimidine rings. Their ability to adsorb specifically IgG in a salt independent manner out of pure IgG solution, mixture of IgG/albumin and human plasma was demonstrated. Mercapto methyl imidazole (MMI) has shown the best adsorption of IgG in terms of binding capacity. No subclass discrimination was observed on all tested ligands except for mercapto methyl pyrimidine where the major IgG subclass adsorbed was IgG3. MMI gave an IgG binding capacity of 100 Î¼g/cm2 of hollow fiber membrane surface area.

[27] A system which consisted of multidimensional liquid chromatography (Yin-yang MDLC) coupled with mass spectrometry was used for the identification of peptides and phosphopeptides. The multidimensional liquid chromatography combines the strong-cation exchange (SCX), strong-anion exchange (SAX), and reverse-phase methods for the separation. Protein digests were first loaded on an SCX column. The flow-through peptides from SCX were collected and further loaded on an SAX column. Both columns were eluted by offline pH steps, and the collected fractions were identified by reverse-phase liquid chromatography tandem mass spectrometry. Comprehensive peptide identification was achieved by the Yin-yang MDLC-MS/MS for a 1 mg mouse liver. In total, 14?105 unique peptides were identified with high confidence, including 13?256 unmodified peptides and 849 phosphopeptides with 809 phosphorylated sites. The SCX and SAX in the Yin-Yang system displayed complementary features of binding and separation for peptides. When coupled with reverse-phase liquid chromatography mass spectrometry, the SAX-based method can detect more extremely acidic (pI < 4.0) and phosphorylated peptides, while the SCX-based method detects more relatively basic peptides (pI > 4.0). In total, 134 groups of phosphorylated peptide isoforms were obtained, with common peptide sequences but different phosphorylated states. This unbiased profiling of protein expression and phosphorylation provides a powerful approach to probe protein dynamics, without using any prefractionation and chemical derivation. Keywords: Protein phosphorylation; Protein expression; Strong-cation exchange; Strong-anion exchange; Yin-Yang multidimensional liquid chromatography; pH elution; Mass spectrometry.

[28] The complexity of peptide mixtures that are analyzed in proteomics necessitates fractionation by multidimensional separation approaches prior to mass spectrometric analysis. In this work, we introduce and evaluate hydrophilic interaction liquid chromatography (HILIC) based strategies for the separation of complex peptide mixtures. The two zwitterionic HILIC materials (ZIC-HILIC and ZIC-cHILIC) chosen for this work differ in the spatial orientation of the positive and negative charged groups. Online experiments revealed a pH-independent resolving power for the ZIC-cHILIC resin while ZIC-HILIC showed a decrease in resolving power at an acidic pH. Subsequently, we extensively evaluated the performances of ZIC-HILIC and ZIC-cHILIC as first dimension in an off-line two-dimensional liquid chromatography (2D-LC) strategy in combination with reversed phase (RP), with respect to peptide separation efficiency and how the retention time correlates with a number of peptide physicochemical properties. Both resins allowed the identification of more than 20?000 unique peptides corresponding to over 3500 proteins in each experimental condition from a remarkably low (1.5μg) amount of starting material of HeLa lysate digestion. The resulting data allows the drawing of a comprehensive picture regarding ZIC- and ZIC-cHILIC peptide separation characteristics. Furthermore, the extent of protein identifications observed from such a level of material demonstrates that HILIC can rival or surpass traditional multidimensional strategies employed in proteomics.

[29] Shotgun proteomics dominates the field of proteomics. The foundations of the strategy consist of multiple rounds of peptide separation where chromatography provides the bedrock. Initially, the scene was relatively simple with the majority of strategies based on some types of ion exchange and reversed phase chromatography. The thirst to achieve comprehensivity, when it comes to proteome coverage and the global characterization of post translational modifications, has led to the introduction of several new separations. In this review, we attempt to provide a historical perspective to separations in proteomics as well as indicate the principles of their operation and rationales for their implementation. Furthermore, we provide a guide on what are the possibilities for combining different separations in order to increase peak capacity and proteome coverage. We aim to show how separations enrich

39. Diederich P, Hansen SK et al (2011) A sub-two minutes method for monoclonal antibody-aggregate quantification using parallel interlaced size exclusion high performance liquid chromatography. J Chromatogr A 1218(50):9010–9018[30]
40. Dong M, Wu M et al (2010) Coupling strong anion-exchange monolithic capillary with MALDI-TOF MS for sensitive detection of phosphopeptides in protein digest. Anal Chem 82(7):2907–2915
41. Dong MW (2006) HPLC columns and trends. Modern HPLC for practicing scientists, Wiley 47–75[31]
42. Edelmann MJ (2011) Strong cation exchange chromatography in analysis of posttranslational modifications: innovations and perspectives. J Biomed Biotechnol 2011:7
43. El Rassi Z, Horváth C (1986) Tandem columns and mixed-bed columns in high-performance liquid chromatography of proteins. J Chromatogr A 359 (0):255–264[32]
44. Engelhardt H, Mathes D (1981) High-performance liquid chromatography of proteins using chemically-modified silica supports. Chromatographia 14 (6):325–332
45. Engholm-Keller K, Hansen TA et al (2011) Multidimensional strategy for sensitive phosphoproteomics incorporating protein prefractionation combined with SIMAC, HILIC, and TiO2 chromatography applied to proximal EGF signaling. J Proteome Res 10(12):5383–5397[33]

the world of proteomics and how further developments may impact the field.

[30] In process development and during commercial production of monoclonal antibodies (mAb) the monitoring of aggregate levels is obligatory. The standard assay for mAb aggregate quantification is based on size exclusion chromatography (SEC) performed on a HPLC system. Advantages hereof are high precision and simplicity, however, standard SEC methodology is very time consuming. With an average throughput of usually two samples per hour, it neither fits to high throughput process development (HTPD), nor is it applicable for purification process monitoring. We present a comparison of three different SEC columns for mAb-aggregate quantification addressing throughput, resolution, and reproducibility. A short column (150 mm) with sub-two micron particles was shown to generate high resolution (~1.5) and precision (coefficient of variation (cv) < 1) with an assay time below 6 min. This column type was then used to combine interlaced sample injections with parallelization of two columns aiming for an absolute minimal assay time. By doing so, both lag times before and after the peaks of interest were successfully eliminated resulting in an assay time below 2 min. It was demonstrated that determined aggregate levels and precision of the throughput optimized SEC assay were equal to those of a single injection based assay. Hence, the presented methodology of parallel interlaced SEC (PI-SEC) represents a valuable tool addressing HTPD and process monitoring.

[31] This chapter contains sections titled: * Scope * General Column Description and Characteristics * Column Types * Column Packing Characteristics * Modern HPLC Column Trends * Guard Columns * Specialty Columns * Column Selection Guides * Summary * References * Internet Resources.

[32] By using a cation- and an anion-exchange column in series, mixtures of acidic and basic proteins were separated in a single chromatographic run with increasing salt gradient at pH 7.0. The serial order of the columns was found to affect the chromatographic results, and the effect was attributed to alteration of the salt gradient profile upon traversing the first ion-exchange column. Single columns, packed with a binary mixture of a cation and an anion exchanger gave similar chromatographic results as the tandem columns and thus offered an alternative approach to the separation of both acidic and basic proteins in a single chromatographic run. A ternary mixed phase was obtained by adding a mildly hydrophobic stationary phase to the mixture of the two ion exchangers. This column could be used with increasing salt gradient as a cation exchanger for the separation of basic proteins, or as an anion exchanger for the separation of acidic proteins. Furthermore, it could be used as a "bipolar" electrostatic-interaction column with increasing salt gradient and as a hydrophobic-interaction column with decreasing salt gradient for the separation of both types of proteins in a single chromatographic run. The constituent stationary phases used in the mixed-bed columns were prepared from the same silica support, i.e., they had the same particle and pore dimensions, density, and pore volume. Besides their obvious advantages in analytical applications, appropriate mixed stationary phases, all having retentive properties for the components to be separated, are expected to be useful also in preparative chromatography to "tailor" column selectivity for a given separation problem without loss of separating capacity.

[33] Comprehensive enrichment and fractionation is essential to obtain a broad coverage of the phosphoproteome. This inevitably leads to sample loss, and thus, phosphoproteomics studies are usually only performed on highly abundant samples. Here, we present a comprehensive phosphoproteomics strategy applied to 400 μg of protein from EGF-stimulated HeLa cells. The proteins are separated into membrane and cytoplasmic fractions using sodium carbonate combined with ultracentrifugation. The phosphopeptides were separated into monophosphorylated and multiphosphorylated pools using sequential elution from IMAC (SIMAC) followed by hydrophilic interaction liquid chromatography of the mono- and nonphosphorylated peptides and subsequent titanium dioxide chromatography of the HILIC fractions. This strategy facilitated the identification of >4700 unique phosphopeptides, while 636 phosphosites were changing following short-term EGF stimulation, many of which were not previously known to be involved in

46. Fekete S, Beck A et al (2014) Theory and practice of size exclusion chromatography for the analysis of protein aggregates. J Pharm Biomed Anal(0)[34]
47. Flanagan RJ, Jane I (1985) High-performance liquid chromatographic analysis of basic drugs on silica columns using non-aqueous ionic eluents. I. Factors influencing retention, peak shape and detector response. J Chromatogr 323(2):173–189[35]
48. Fukuda I, Hirabayashi-Ishioka Y et al (2013) Optimization of enrichment conditions on TiO2 chromatography using glycerol as an additive reagent for effective phosphoproteomic analysis. J Proteome Res 12(12):5587–5597[36]
49. Gagnon P, Beam K (2009) Antibody aggregate removal by hydroxyapatite chromatography. Curr Pharm Biotechnol 10(4):440–446[37]

EGFR signaling. We further compared three different data processing programs and found large differences in their peptide identification rates due to different implementations of recalibration and filtering. Manually validating a subset of low-scoring peptides exclusively identified using the MaxQuant software revealed a large percentage of false positive identifications. This indicates that, despite having highly accurate precursor mass determination, peptides with low fragment ion scores should not automatically be reported in phosphoproteomics studies.

[34] Size exclusion chromatography (SEC) is a historical technique widely employed for the detailed characterization of therapeutic proteins and can be considered as a reference and powerful technique for the qualitative and quantitative evaluation of aggregates. The main advantage of this approach is the mild mobile phase conditions that permit the characterization of proteins with minimal impact on the conformational structure and local environment. Despite the fact that the chromatographic behavior and peak shape are hardly predictable in SEC, some generic rules can be applied for SEC method development, which are described in this review. During recent years, some improvements were introduced to conventional SEC that will also be discussed. Of these new SEC characteristics, we discuss (i) the commercialization of shorter and narrower columns packed with reduced particle sizes allowing an improvement in the resolution and throughput; (ii) the possibility of combining SEC with various detectors, including refractive index (RI), ultraviolet (UV), multi-angle laser light scattering (MALLS) and viscometer (IV), for extensive characterization of protein samples and (iii) the possibility of hyphenating SEC with mass spectrometry (MS) detectors using an adapted mobile phase containing a small proportion of organic modifiers and ion-pairing reagents.

[35] The use of silica columns together with non-aqueous ionic eluents provides a stable yet flexible system for the high-performance liquid chromatographic analysis of basic drugs. At constant ionic strength, eluent pH influences retention via ionisation of surface silanols and protonation of basic analytes, pKa values indicating the pH of maximum retention. At constant pH, retention is proportional to the reciprocal of the eluent ionic strength for fully protonated analytes and quaternary ammonium compounds. The addition of water up to 10 % (v/v) has little effect on retention if the protonation of the analytes is unaffected. Thus, it is likely that retention is mediated primarily via cation exchange with surface silanols. However, additional factors must play a part with compounds such as morphine which give tailing peaks at acidic or neutral eluent pHs.

[36] Metal oxide affinity chromatography (MOAC) represented by titanium dioxide (TiO2) chromatography has been used for phosphopeptide enrichment from cell lysate digests prior to mass spectrometry. For in-depth phosphoproteomic analysis, it is important for MOAC to achieve high phosphopeptide enrichment efficiency by optimizing purification conditions. However, there are some differences in phosphopeptide selectivity and specificity enriched by various TiO2 materials and procedures. Here, we report that binding/wash buffers containing polyhydric alcohols, such as glycerol, markedly improve phosphopeptide selectivity from complex peptide mixtures. In addition, the elution conditions combined with secondary amines, such as bis-Tris propane, made it possible to recover phosphopeptides with highly hydrophobic properties and/or longer peptide lengths. To assess the practical applicability of our improved method, we confirmed using PC3 prostate cancer cells. By combining the hydrophilic interaction chromatography (HILIC) with the optimized TiO2 enrichment method prior to LC-MS/MS analysis, over 8300 phosphorylation sites and 2600 phosphoproteins were identified. Additionally, some dephosphorylations of those were identified by treatment with dasatinib for a kinase inhibitor. These results indicate that our method is applicable to understanding the profiling of kinase inhibitors such as anticancer compounds, which will be useful for drug discovery and development.

[37] Hydroxyapatite (HA) has proven in recent years to be one of the most versatile and powerful methods for removing aggregates from antibody preparations. It is effective with IgA, IgG and IgM, and it reduces aggregate levels from above 60 % to less than 0.1 %. Three basic elution strategies have evolved, one that removes aggregates from a modest proportion of clones, another from the majority, and one that appears to be universally effective. Each has distinct development and process ramifications. This review defines what HA is, how it interacts with various classes of biomolecules, how those interactions are controlled by different elution strategies, and how to determine which approach may be most effective for a particular antibody. Consideration is also given to HA's specific strengths and limitations from an industrial perspective.

50. Ghose S, Hubbard B et al (2005) Protein interactions in hydrophobic charge induction chromatography (HCIC). Biotechnol Prog 21(2):498–508
51. Ghose S, Hubbard B et al (2006) Evaluation and comparison of alternatives to Protein A chromatography: mimetic and hydrophobic charge induction chromatographic stationary phases. J Chromatogr A 1122(1â€"2):144–152[38]
52. Gilar M, Olivova P et al (2005) Orthogonality of separation in two-dimensional liquid chromatography. Anal Chem 77(19):6426–6434[39]
53. Gilar M, Olivova P et al (2005) Two-dimensional separation of peptides using RP-RP-HPLC system with different pH in first and second separation dimensions. J Sep Sci 28(14):1694–1703[40]
54. Gilar M, Yu YQ et al (2008) Mixed-mode chromatography for fractionation of peptides, phosphopeptides, and sialylated glycopeptides. J Chromatogr A 1191(1–2):162–170[41]

[38] In this paper Protein A mimetic and hydrophobic charge induction chromatographic (HCIC) stationary phases are characterized in terms of their protein adsorption characteristics and their selectivity is compared with Protein A chromatography using a set of Chinese hamster ovary-derived monoclonal antibodies and Fc-fusion proteins. Linear retention experiments were employed to compare the selectivities of these resins for both non-IgG model proteins as well as antibodies and the fusion proteins. While none of the non-IgG model proteins were observed to bind to the Protein A resin, most of them did in fact bind to the alternative resins. In addition, while the elution pH was similar for the model proteins and antibodies on the HCIC resin, the mimetic resins did exhibit higher binding for the antibodies under these linear pH gradient conditions. A mixed mode preparative isotherm model previously developed for HCIC was shown to accurately describe the adsorption behavior of the mimetic materials as well. Host cell protein clearance profiles were also investigated under preparative conditions using complex biological feeds and the results indicated that while some selectivity was observed for both the HCIC and the mimetic materials, the purification factors were in general significantly less than those obtained with Protein A. It is important to note, however, that the selectivity of the mimetic and HCIC materials was also observed to be antibody specific indicating that further optimization may well result in increased selectivities for these materials.

[39] Two-dimensional liquid chromatography is often used to reduce the proteomic sample complexity prior to tandem mass spectrometry analysis. The 2D-LC performance depends on the peak capacity in both chromatographic dimensions, and separation orthogonality. The peak capacity and selectivity of many LC modes for peptides is not well known, and mathematical characterization for orthogonality is underdeveloped. Consequently, it is difficult to estimate the performance of 2D-LC for peptide separation. The goal of this paper was to investigate a selectivity of common LC modes and to identify the 2D-LC systems with a useful orthogonality. A geometric approach for orthogonality description was developed and applied for estimation of a practical peak 2D-LC capacity. Selected LC modes including various RP, SCX, SEC, and HILIC were combined in 2D-LC setups. SCX-RP, HILIC-RP, and RP-RP 2D systems were found to provide suitable orthogonality. The RP-RP system (employing significantly different pH in both RP separation dimensions) had the highest practical peak capacity of 2D-LC systems investigated.

[40] Two-dimensional high performance liquid chromatography is a useful tool for proteome analysis, providing a greater peak capacity than single-dimensional LC. The most popular 2D-HPLC approach used today for proteomic research combines strong cation exchange and reversed-phase HPLC. We have evaluated an alternative mode for 2D-HPLC of peptides, employing reversed-phase columns in both separation dimensions. The orthogonality of 2D separation was investigated for selected types of RP stationary phases, ion-pairing agents and mobile phase pH. The pH appears to have the most significant impact on the RP-LC separation selectivity; the greatest orthogonality was achieved for the system with C18 columns using pH 10 in the first and pH 2.6 in the second LC dimension. Separation was performed in off-line mode with partial fraction evaporation. The achievable peak capacity in RP-RP-HPLC and overall performance compares favorably to SCX-RP-HPLC and holds promise for proteomic analysis.

[41] A mixed-mode chromatographic (MMC) sorbent was prepared by functionalizing the silica sorbent with a pentafluorophenyl (PFP) ligand. The resulting stationary phase provided a reversed-phase (RP) retention mode along with a relatively mild strong cation-exchange (SCX) retention interaction. While the mechanism of interaction is not entirely clear, it is believed that the silanols in the vicinity of the perfluorinated ligand act as strongly acidic sites. The 2.1 mm × 150 mm column packed with such sorbent was applied to the separation of peptides. Linear RP gradients in combination with salt steps were used for pseudo two-dimensional (2D) separation and fractionation of tryptic peptides. An alternative approach of using linear cation-exchange gradients combined with RP step gradients was also investigated. Besides the attractive forces, the ionic repulsion contributed to the retention mechanism. The analytes with strong negatively charged sites (phosphorylated peptides, sialylated glycopeptides) eluted in significantly different patterns than generic tryptic peptides. This retention mechanism was employed for the isolation of phosphopeptides or sialylated glycopeptides from non-functionalized peptide mixtures. The mixed-mode column was utilized in conjunction with a phosphopeptide enrichment solid phase extraction (SPE) device packed with metal oxide affinity chromatography (MOAC)

55. Girot P, Averty E et al (2004) 2-Mercapto-5-benzimidazolesulfonic acid: an effective multimodal ligand for the separation of antibodies. J Chromatogr B 808(1):25–33[42]
56. Golovchenko NP, Kataeva Ia Fau – Akimenko VK et al (1992) Analysis of pH-dependent protein interactions with gel filtration medium[43]
57. Guo Y, Wang X (2013) Hilic method development. Hydrophilic interaction chromatography, Wiley, 87–110[44]
58. Hägglund P, Bunkenborg J et al (2004) A new strategy for identification of N-Glycosylated proteins and unambiguous assignment of their glycosylation sites using HILIC enrichment and partial deglycosylation. J Proteome Res 3(3):556–566[45]

sorbent. The combination of MOAC and mixed-mode chromatography (MMC) provided for an enhanced extraction selectivity of phosphopeptides and sialylated glycopeptides peptides from complex samples, such as yeast and human serum tryptic digests.

[42] The report describes the use of 2-mercapto-5-benzimidazolesulfonic acid (MBISA) as a ligand for the separation of antibodies by chromatography. The ligand shows a relatively specific adsorption property for antibodies from very crude biologicals at pH 5.0–5.5. At this pH range most of other proteins do not interact with the resin especially when the ionic strength is similar to physiological conditions. Several characterization studies are described such as antibody adsorption in different conditions of ionic strength, pH and temperature. These properties are advantageously used to selectively capture antibodies from very crude feed stocks without dilution or addition of lyotropic salts. Demonstration was made that the adsorption mechanism is neither based on ion exchange nor on hydrophobic associations, but rather as an assembly of a variety of properties of the ligand itself. Binding capacity in the described conditions ranges between 25 and 30 mg/mL of resin. The sorbent does not co-adsorb albumin (Alb) and seems compatible with a large variety of feedstocks. Quantitative antibody desorption occurs when the pH is raised above 8.5. The final purity of the antibody depends on the nature of the feedstock, and can reach levels of purity as high as 98 %. Even with very crude biological liquids such as ascites fluids, cell culture supernatants and Chon fraction II + III from human plasma fractionation where the number of protein impurities is particularly large, immunoglobumins G (IgG) were separated at high purity level in a single step.

[43] A prepacked Superose 12 HR 10/30 column was used to study the effects of elution ionic strength and pH on the chromatographic behaviour of a strong hydrophobic Clostridium thermocellum endoglucanase (1) and two weak hydrophobic proteins, Clostridium thermocellum endoglucanase C and egg white lysozyme. Ion-exclusion or ion-exchange interactions between weakly hydrophobic proteins and the gel matrix were observed at low ionic strength, depending on whether the pH of the elution buffer was higher or lower than the pI values of the proteins. These interactions were due to the presence of negatively charged groups on the surface of Superose and could be eliminated at any pH by adding electrolyte at a concentration determined by its chemical identity. The optimum results were observed with sodium sulphate at a concentration of 100 mM. The chromatographic behaviour of strong hydrophobic endoglucanase (1) on a Superose column as a function of pH was much more complex because of two interplaying effects, electrostatic and hydrophobic. Ideal size-exclusion chromatography could be achieved only in a narrow range of the conditions: first, the mobile phase must contain a weak salting-out electrolyte such as NaCl, and second, the mobile phase pH must be high enough that hydrophobic interactions between the solute and support are balanced by their electrostatic repulsion. At pH greater than pI, the retardation of endoglucanase (1) gradually increased with decreasing pH as a result of lowering of repulsive electrostatic interactions whether or not the buffer ionic strength was high. At pH less than pI a drastic increase in the capacity factor k' was observed owing to the additivity of hydrophobic and ion-exchange effects. (ABSTRACT TRUNCATED AT 250 WORDS) FAU – Golovchenko, N P.

[44] This chapter focuses on method development employing hydrophilic interaction chromatography (HILIC) as the chromatographic technique. Various aspects of method development are discussed including method objectives, sample considerations, systematic method development, column and mobile phase selection, and other operating parameters (e.g., column temperature, sample solvent, and charged aerosol detector (CAD) or mass spectrometric (MS) detectors). The chapter provides general guidance on HILIC method development based on a solid understanding of HILIC basics and the authors' experience with bioanalytical and pharmaceutical methods.

[45] Characterization of glycoproteins using mass spectrometry ranges from determination of carbohydrate-protein linkages to the full characterization of all glycan structures attached to each glycosylation site. In a novel approach to identify N-glycosylation sites in complex biological samples, we performed an enrichment of glycosylated peptides through hydrophilic interaction liquid chromatography (HILIC) followed by partial deglycosylation using a combination of endo-?-N-acetylglucosaminidases (EC 3.2.1.96). After hydrolysis with these enzymes, a single N-acetylglucosamine (GlcNAc) residue remains linked to the asparagine residue. The removal of the major part of the glycan simplifies the MS/MS fragment ion spectra of glycopeptides, while the remaining GlcNAc residue enables unambiguous assignment of the glycosylation site together with the amino acid sequence. We first tested our approach on a mixture of known glycoproteins, and

59. Hamilton GE, Luechau F et al (2000) Development of a mixed mode adsorption process for the direct product sequestration of an extracellular protease from microbial batch cultures. J Biotechnol 79 (2):103–115[46]
60. Han G, Ye M et al (2008) Large-scale phosphoproteome analysis of human liver tissue by enrichment and fractionation of phosphopeptides with strong anion exchange chromatography. Proteomics 8(7):1346–1361[47]
61. Hartmann E, Chen Y et al (2003) Comparison of reversed-phase liquid chromatography and hydrophilic interaction/cation-exchange chromatography for the separation of amphipathic alpha-helical peptides with L- and D-amino acid substitutions in the hydrophilic face. J Chromatogr A 1009 (1–2):61–71[48]
62. Hemström P, Irgum K (2006) Hydrophilic interaction chromatography. J Sep Sci 29(12):1784–1821[49]

---

subsequently the method was applied to samples of human plasma obtained by lectin chromatography followed by 1D gel-electrophoresis for determination of 62 glycosylation sites in 37 glycoproteins. Keywords: proteomics; post-translational modifications ? mass spectrometry ? HILIC ? endoglycosidase ? lectin affinity chromatography ? glycosylation ? Plasma proteins.

[46] Direct product sequestration of extracellular proteins from microbial batch cultures can be achieved by continuous or intermittent broth recycle through an external extractive loop. Here, we describe the development of a fluidisable, mixed mode adsorbent, designed to tolerate increasing ionic strength (synonymous with extended productive batch cultures). This facilitated operations for the integrated recovery of an extracellular acid protease from cultures of Yarrowia lipolytica. Mixed mode adsorbents were prepared using chemistries containing hydrophobic and ionic groups. Matrix hydrophobicity and titration ranges were matched to the requirements of integrated protease adsorption. A single expanded bed was able to service the productive phase of growth without recourse to the pH adjustment of the broth previously required for ion exchange adsorption. This resulted in increased yields of product, accompanied by further increases in enzyme specific activity. A step change from pH 4.5 to 2.6, across the isoelectric point of the protease, enabled high resolution fixed bed elution induced by electrostatic repulsion. The generic application of mixed mode chemistries, which combine the physical robustness of ion-exchange ligands in sanitisation and sterilisation procedures with a selectivity, which approaches that of affinity interactions, is discussed.

[47] The mixture of phosphopeptides enriched from proteome samples are very complex. To reduce the complexity it is necessary to fractionate the phosphopeptides. However, conventional enrichment methods typically only enrich phosphopeptides but not fractionate phosphopeptides. In this study, the application of strong anion exchange (SAX) chromatography for enrichment and fractionation of phosphopeptides was presented. It was found that phosphopeptides were highly enriched by SAX and majority of unmodified peptides did not bind onto SAX. Compared with Fe3+ immobilized metal affinity chromatography (Fe3 + −IMAC), almost double phosphopeptides were identified from the same sample when only one fraction was generated by SAX. SAX and Fe3 + −IMAC showed the complementarity in enrichment and identification of phosphopeptides. It was also demonstrated that SAX have the ability to fractionate phosphopeptides under gradient elution based on their different interaction with SAX adsorbent. SAX was further applied to enrich and fractionate phosphopeptides in tryptic digest of proteins extracted from human liver tissue adjacent to tumorous region for phosphoproteome profiling. This resulted in the highly confident identification of 274 phosphorylation sites from 305 unique phosphopeptides corresponding to 168 proteins at false discovery rate (FDR) of 0.96 %.

[48] Mixed-mode hydrophilic interaction/cation-exchange chromatography (HILIC/CEX) is a novel high-performance technique which has excellent potential for peptide separations. Separations by HILIX/CEX are carried out by subjecting peptides to linear increasing salt gradients in the presence of high levels of acetonitrile, which promotes hydrophilic interactions overlaid on ionic interactions with the cation-exchange matrix. In the present study, HILIC/CEX has been compared to reversed-phase liquid chromatography (RP-HPLC) for separation of mixtures of diastereomeric amphipathic alpha-helical peptide analogues, where L- and D-amino acid substitutions were made in the centre of the hydrophilic face of the amphipathic alpha-helix. Unlike RP-HPLC, temperature had a substantial effect on HILIC/CEX of the peptides, with a rise in temperature from 25 to 65 degrees C increasing the retention times of the peptides as well as improving resolution. Our results again highlight the potential of HILIC/CEX as a peptide separation mode in its own right as well as an excellent complement to RP-HPLC.

[49] Separation of polar compounds on polar stationary phases with partly aqueous eluents is by no means a new separation mode in LC. The first HPLC applications were published more than 30 years ago, and were for a long time mostly confined to carbohydrate analysis. In the early 1990s new phases started to emerge, and the practice was given a name, hydrophilic interaction chromatography (HILIC). Although the use of this separation mode has been relatively limited, we have seen a sudden increase in popularity over the last few years, promoted by the need to analyze polar compounds in increasingly complex mixtures. Another reason for the increase in popularity is the widespread use of MS coupled to LC. The partly aqueous eluents high in ACN with a limited need of adding salt is almost ideal for ESI. The applications now encompass most categories of polar

63. Hirs CHW, Stein WH et al (1951) Chromatography of proteins. Ribonuclease. J Am Chem Soc 73(4):1893
64. Hjertén S (1964) The preparation of agarose spheres for chromatography of molecules and particles. Biochim Biophys Acta (BBA) – Special Sect Biophys Subjects 79(2):393–398[50]
65. Hjertén S, Mosbach R (1962) "Molecular-sieve" chromatography of proteins on columns of cross-linked polyacrylamide. Anal Biochem 3 (2):109–118[51]
66. Hjertén S, Rosengren J et al (1974) Hydrophobic interaction chromatography: the synthesis and the use of some alkyl and aryl derivatives of agarose. J Chromatogr A 101(2):281–288[52]
67. Hong P, Koza S et al (2012) Size-exclusion chromatography for the analysis of protein biotherapeutics and their aggregates. J Liq Chromatogr Relat Technol 35(20):2923–2950[53]
68. Huang P, Jin X et al (1999) Use of a mixed-mode packing and voltage tuning for peptide mixture separation in pressurized capillary electrochromatography with an ion trap storage/reflectron time-of-flight mass spectrometer detector. Anal Chem 71(9):1786–1791[54]
69. Ibrahim MEA, Lucy CA (2013) Stationary phases for Hilic. Hydrophilic interaction chromatography, Wiley, 43–85[55]

compounds, charged as well as uncharged, although HILIC is particularly well suited for solutes lacking charge where coulombic interactions cannot be used to mediate retention. The review attempts to summarize the ongoing discussion on the separation mechanism and gives an overview of the stationary phases used and the applications addressed with this separation mode in LC.

[50] A method is described for preparation of spherical agarose or agar grains, to be used as bed material for chromatographic "sieving" of molecules and particles. Due to a comparatively great hardness of these grains, they give high flow rates even if they are made small in order to increase the resolving power of the column.

[51] Columns packed with cross-linked polyacrylamide have been used for chromatographic separation of high molecular weight substances, especially proteins. These columns also allow separation of large molecules from small ones, for instance proteins from amino acids, peptides, salts. There is a positive correlation between the molecular size of a protein and its Rf value.

[52] Aliphatic and aromatic alcohols in the form of glycidyl ethers have been coupled to agarose gels. These neutral agarose derivatives, which thus contain hydrophobic substituents, have been used as adsorbents in hydrophobic interaction chromatography. The coupling yield and the degree of substitution have been determined for one aliphatic and one aromatic model substance. Different fractionation problems require different degrees of hydrophobicity of the substituents. To "tailor make" gels, the hydrophobicity can be varied in small steps by the use of aliphatic alcohols of different chain length. The agarose derivatives described have been used for the purification of proteins, demonstrated with a plasma fractionation, viruses (STNV) and even whole cells (baker's yeast). Under suitable experimental conditions, the interactions can be very mild (enzyme activities have been recovered in a 50–100 % yield). Enzyme reactors with a high capacity can be prepared in a simple manner by applying the enzyme solution at any pH on to a suitable hydrophobically interacting bed. As the enzymes are not covalently linked to the bed, they can easily be recovered in the free form. Contrary to ion-exchange chromatography, the adsorption in hydrophobic interaction chromatography decreases with a decrease in ionic strength and temperature.

[53] In recent years, the use and number of biotherapeutics has increased significantly. For these largely protein-based therapies, the quantitation of aggregates is of particular concern given their potential effect on efficacy and immunogenicity. This need has renewed interest in size-exclusion chromatography (SEC). In the following review we will outline the history and background of SEC for the analysis of proteins. We will also discuss the instrumentation for these analyses, including the use of different types of detectors. Method development for protein analysis by SEC will also be outlined, including the effect of mobile phase and column parameters (column length, pore size). We will also review some of the applications of this mode of separation that are of particular importance to protein biopharmaceutical development and highlight some considerations in their implementation.

[54] A mixed-mode (reversed-phase/anion-exchange) stationary phase has been used as the capillary column packing for investigation of the separation of peptide mixtures in pressurized capillary electrochromatography (pCEC). This stationary phase contains both octadecylsilanes and dialkylamines. The amine groups of the stationary phase determine the charge density on the surface of the packing and can produce a strong and constant electroosmotic flow (EOF) at low pH. A comparison was made in terms of the capability of separating tryptic digests between the mixed-mode phase and C18 reversed phase. In addition, the constant EOF enabled the tuning of the retention and the selectivity of the separation by adjusting the mobile phase pH from 2 to 5. Furthermore, the magnitude and the polarity of the electric voltage were demonstrated to greatly influence the elution profiles of the peptides in pCEC. An ion trap storage/reflectron time-of-flight mass spectrometer was used as an on-line detector in these experiments due to its ability to provide rapid and accurate mass detection of the sample components eluting from the separation column.

[55] Literature and research on hydrophilic interaction liquid chromatography (HILIC) has increased dramatically in recent years. This has been accompanied by a correspondingly rapid increase in stationary phases developed for HILIC. This chapter first discusses all classes of stationary phases used in HILIC mode in terms of chemistry,

70. Ikegami T, Tomomatsu K et al (2008) Separation efficiencies in hydrophilic interaction chromatography. J Chromatogr A 1184(1–2):474–503[56]
71. Intoh A, Kurisaki A et al (2009) Separation with zwitterionic hydrophilic interaction liquid chromatography improves protein identification by matrix-assisted laser desorption/ionization-based proteomic analysis. Biomed Chromatogr 23(6):607–614[57]
72. Irvine GB (2001) Determination of molecular size by size-exclusion chromatography (gel filtration). Current protocols in cell biology, Wiley[58]
73. Irvine GB (2003) High-performance size-exclusion chromatography of peptides. J Biochem Biophys Methods 56(1–3):233–242[59]
74. Jacobs JM, Mottaz HM et al (2003) Multidimensional proteome analysis of human mammary epithelial cells. J Proteome Res 3(1):68–75[60]

available trade names, and representative applications. The classes of stationary phases include underivatized silica phase, derivatized silica phase, and nonsilica phases. Important characteristics of some selected commercial HILIC phases are summarized in a table. The table classifies HILIC phases according to their chemical nature. Then, the chapter compares these HILIC phases in terms of efficiency, retention, and selectivity.

[56] Hydrophilic interaction chromatography (HILIC) is important for the separation of highly polar substances including biologically active compounds, such as pharmaceutical drugs, neurotransmitters, nucleosides, nucleotides, amino acids, peptides, proteins, oligosaccharides, carbohydrates, etc. In the HILIC mode separation, aqueous organic solvents are used as mobile phases on more polar stationary phases that consist of bare silica, and silica phases modified with amino, amide, zwitterionic functional group, polyols including saccharides and other polar groups. This review discusses the column efficiency of HILIC materials in relation to solute and stationary phase structures, as well as comparisons between particle-packed and monolithic columns. In addition, a literature review consisting of 2006–2007 data is included, as a follow up to the excellent review by Hemström and Irgum.

[57] Comprehensive proteomic analyses necessitate efficient separation of peptide mixtures for the subsequent identification of proteins by mass spectrometry (MS). However, digestion of proteins extracted from cells and tissues often yields complex peptide mixtures that confound direct comprehensive MS analysis. This study investigated a zwitterionic hydrophilic interaction liquid chromatography (ZIC-HILIC) technique for the peptide separation step, which was verified by subsequent MS analysis. Human serum albumin (HSA) was the model protein used for this analysis. HSA was digested with trypsin and resolved by ZIC-HILIC or conventional strong cation exchange (SCX) prior to MS analysis for peptide identification. Separation with ZIC-HILIC significantly improved the identification of HSA peptides over SCX chromatography. Detailed analyses of the identified peptides revealed that the ZIC-HILIC has better peptide fractionation ability. We further demonstrated that ZIC-HILIC is useful for quantitatively surveying cell surface markers specifically expressed in undifferentiated embryonic stem cells. These results suggested the value of ZIC-HILIC as a novel and efficient separation method for comprehensive and quantitative proteomic analyses.

[58] Size-exclusion or gel filtration chromatography is one of the most popular methods for determining the sizes of proteins. Proteins in solution, or other macromolecules, are applied to a column with a defined support medium. The behavior of the protein depends on its size and that of the pores in the medium. If the protein is small relative to the pore size, it will partition into the medium and emerge from the column after larger proteins. Besides a protein's size, this technique can also be used for protein purification, analysis of purity, and study of interactions between proteins. In this unit protocols are provided for size-exclusion high-performance liquid chromatography (SE-HPLC) and for conventional gel filtration, including calibration of columns (in terms of the Stokes radius) using protein standards.

[59] Gel filtration on soft gels has been employed for over 40 years for the separation, desalting and molecular weight estimation of peptides and proteins. Technical improvements have given rise to high-performance size-exclusion chromatography (HPSEC) on rigid supports, giving more rapid run times and increased resolution. Initially, these packings were more suitable for the separation of proteins than of peptides, but supports that operate in the fractionation range <10,000 Daltons (Da) are now available. In this report, HPSEC is described in relation to its application to peptides, especially regarding purification, estimation of molecular weight and study of molecular associations.

[60] Recent multidimensional liquid chromatography MS/MS studies have contributed to the identification of large numbers of expressed proteins for numerous species. The present study couples size exclusion chromatography of intact proteins with the separation of tryptically digested peptides using a combination of strong cation exchange and high resolution, reversed phase capillary chromatography to identify proteins extracted from human mammary epithelial cells (HMECs). In addition to conventional conservative criteria for protein identifications, the confidence levels were additionally increased through the use of peptide normalized elution times (NET) for the liquid chromatographic separation step. The combined approach resulted in a total of 5838 unique peptides identified covering 1574 different proteins with an estimated 4 % gene coverage of the human genome, as annotated by the National Center for Biotechnology Information (NCBI). This database provides a baseline for comparison against variations in other genetically and environmentally perturbed systems.

75. Jane I (1975) The separation of a wide range of drugs of abuse by high-pressure liquid chromatography. J Chromatogr A 111(1):227–233
76. Jiang W, Fischer G et al (2006) Zwitterionic stationary phase with covalently bonded phosphorylcholine type polymer grafts and its applicability to separation of peptides in the hydrophilic interaction liquid chromatography mode. J Chromatogr A 1127 (1–2):82–91[61]
77. Jiang W, Irgum K (2001). Synthesis and evaluation of polymer-based zwitterionic stationary phases for separation of ionic species. Anal Chem 73 (9):1993–2003[62]
78. Jiang W, Irgum K (2002) Tentacle-type zwitterionic stationary phase prepared by surface-initiated graft polymerization of 3-[N,N-Dimethyl-N-(Methacryloyloxyethyl)- ammonium] propanesulfonate through peroxide groups yethered on porous silica. Anal Chem 74(18):4682–4687[63]
79. Johansson BL, Belew M et al (2003) Preparation and characterization of prototypes for multi-modal

Proteins identified were categorized based upon intracellular location and biological process with the identification of numerous receptors, regulatory proteins, and extracellular proteins, demonstrating the usefulness of this application in the global analysis of human cells for future comparative studies. Keywords: human ? HMEC ? multidimensional ? liquid chromatography ? proteome ? global ? Size exclusion.

[61] A novel phosphorylcholine type zwitterionic stationary phase was synthesized by graft polymerization of 2-methacryloyloxyethyl phosphorylcholine onto the surface of porous silica particles. The resulting material possesses both negatively charged phosphoric acid and positively charged quaternary ammonium groups, which renders it a low net charge over a wide pH range. The composition of the surface grafts were determined by elemental analysis and solid state NMR, and the surface charge (zeta-potential) in different buffer solutions were measured using photon correlation spectroscopy. Separation of several peptides was investigated on packed columns in the hydrophilic interaction liquid chromatography (HILIC) separation mode. It was shown that small peptides can be separated based on hydrophilic interaction and ionic interaction between the stationary phase and analyte. The organic solvent composition, the pH and the salt concentration of the eluent have strong effects on the retention time. Compared to native silica before grafting, the newly synthesized zwitterionic material gave more stable retention times for basic peptides over pH range 3–7 due to elimination of the dissociation of silanol groups.

[62] Three different zwitterionic functional stationary phases for chromatography were synthesized on the basis of 2-hydroxyethyl methacrylate (HEMA) polymeric particles. Two synthesis routes, producing materials designated S300-ECH-DMA-PS or S300-TC-DMA-PS, involved activation of the hydroxyl groups of the HEMA material with epichlorohydrin or thionyl chloride, respectively, followed by dimethylamination and quaternizing 3-sulfopropylation with 1,3-propane sultone. The third route was accomplished by attaching methacrylate moieties to the HEMA through a reaction with methacrylic anhydride, followed by graft photopolymerization of the zwitterionic monomer 3-[N, N-dimethyl-N-(methacryloyloxyethyl)ammonium] propanesulfonate, initiated by benzoin methyl ether under 365-nm light. According to elemental analyses, both the S300-ECH-DMA-PS and S300-TC-DMA-PS materials appeared to have overall charge stoichometries close to unity, whereas the grafted material, S300-MAA-SPE, seemed to carry an excess of anion exchange sites in addition to the zwitterionic groups. Yet all three zwitterionic stationary phases were capable of separating inorganic anions and cations simultaneously and independently using aqueous solutions of perchloric acid or perchlorate salts as eluent, albeit with markedly different selectivities. On the S300-TC-DMA-PS and S300-MAA-SPE materials, the retention times increased for cations and decreased for anions with increasing eluent concentration, whereas with the S300-ECH-DMA-PS material, the retention times of both anions and cations decreased with increasing eluent concentration. These results demonstrate the importance of choosing appropriate synthesis conditions in order to prepare covalently bonded zwitterionic separation materials with an acceptable charge balance.

[63] A novel stationary phase with tentacle-type zwitterionic interaction layer was synthesized by free radical graft polymerization of 3-[N,N-dimethyl-N-(methacryloyloxyethyl)ammonium]propanesulfonate (SPE) from the surface of Kromasil porous silica particles. The polymerization was initiated by thermal cleavage of tert-butylperoxy groups covalently attached to the particle surface, and the material therefore carries a tentacle-type polymeric interaction layer with 3-sulfopropylbetaine functional moieties. The composition of the surface graft was determined by elemental analysis, and the surface charge was measured using photon correlation spectroscopy. The measured zeta-potentials were close to 0 and nearly independent of pH, and the tentacle character of the interactive layers were evident from the lack of colloidal stability in the absence of salt (antipolyelectrolytic behavior) and a marked increase in column back-pressure when the concentration of perchloric acid or perchlorate salt was increased. The chromatographic properties were evaluated on columns packed with the functionalized material, and it was shown that this zwitterionic stationary phase could simultaneously and independently separate inorganic anions and cations using aqueous solutions of perchloric acid or perchlorate salts as eluents. The material was also capable of separating two acidic and three basic proteins in a single run, using gradient salt elution at constant pH.

separation aimed for capture of positively charged biomolecules at high-salt conditions. J Chromatogr A 1016(1):35–49[64]

80. Johansson BL, Belew M et al (2003) Preparation and characterization of prototypes for multi-modal separation media aimed for capture of negatively charged biomolecules at high salt conditions. J Chromatogr A 1016(1):21–33[65]

81. Jungbauer A. (2005) Chromatographic media for bioseparation. J Chromatogr A 1065(1):3–12[66]

[64] Several prototypes of aromatic (Ar) and non-aromatic (NoAr) cation-exchange ligands suitable for capture of proteins from high conductivity (ca. 30 mS/cm) mobile phases were coupled to Sepharoseâ„¢ 6 Fast Flow. These new prototypes of multi-modal cation-exchangers were found by screening a diverse library of multi-modal ligands and selecting cation-exchangers resulting in elution of test proteins at high ionic-strength. Candidates were then tested with respect to breakthrough capacity of bovine serum albumin (BSA), human IgG and lysozyme in buffers adjusted to a high conductivity. By applying a salt-step or a pH-step the recoveries were also tested. We have found that aromatic multi-modal cation-exchanger ligands based on carboxylic acids seem to be optimal for the capture of proteins at high-salt conditions. Experimental evidence on the importance of the relative position of the aromatic group in order to improve the breakthrough capacity at high-salt conditions has been found. It was also found that an amide group on the Î ± −carbon was essential for capture of proteins at high-salt conditions. Compared to a strong cation-exchanger such as SP Sepharose™ Fast Flow the best new multi-modal weak cation-exchangers have breakthrough capacities of BSA, human IgG and lysozyme that are 10â€“30 times higher at high-salt conditions. The new multi-modal cation-exchangers can also be used at normal cation-exchange conditions and with either a salt-step or a pH-step (to pH-values where the proteins are negatively charged) to accomplish elution of proteins. In addition, the functional performance of the new cation-exchangers was found to be intact after treatment in 1.0 M sodium hydroxide solution for 10 days. For BSA it was also possible to design cation-exchangers based on non-aromatic carboxyl acid ligands with high capacities at high-salt conditions. A common feature of these ligands is that they contain hydrogen acceptor groups close to the carboxylic group. Furthermore, it was also possible to obtain high breakthrough capacities for lysozyme and BSA of a strong cation-exchanger (SP Sepharoseâ„¢ Fast Flow) if phenyl groups were attached to the beads. Varying the ligand ratio (SP/Phenyl) could be used for optimizing the function of mixed-ligand ion-exchange media.

[65] Several prototypes of multi-modal ligands suitable for the capture of negatively charged proteins from high conductivity (28 mS/cm) mobile phases were coupled to Sepharose 6 Fast Flow. These new prototypes of multi-modal anion-exchangers were found by screening a diverse library of multi-modal ligands and selecting anion-exchangers resulting in elution of test proteins at high ionic strength. Candidates were then tested with respect to breakthrough capacity of BSA in a buffer adjusted to a high conductivity (20 mM Piperazine and 0.25 M NaCl, pH 6.0). The recovery of BSA was also tested with a salt step (from 0.25 to 2.0 M NaCl using 20 mM Piperazine as buffer, pH 6.0) or with a pH-step to pH 4.0. We have found that non-aromatic multi-modal anion-exchange ligands based on primary or secondary amines (or both) are optimal for the capture of proteins at high salt conditions. Furthermore, these new multi-modal anion-exchange ligands have been designed to take advantage not only of electrostatic but also hydrogen bond interactions. This has been accomplished through modification of the ligands by the introduction of hydroxyl groups in the proximity of the ionic group. Experimental evidence on the importance of the relative position of the hydroxyl groups on the ligand in order to improve the breakthrough capacity of BSA has been found. Compared to strong anion-exchangers such as Q Sepharoseâ„¢ Fast Flow the new multi-modal weak anion-exchangers have breakthrough capacities of BSA at mobile phases of 28 mS/cm and pH 6.0 that are 20â€“30 times higher. The new multi-modal anion-exchangers can also be used at normal anion-exchange conditions and with either a salt step or a pH-step to acidic pH can accomplish the elution of proteins. In addition, the functional performance of the new anion-exchangers was found to be intact after treatment in 1.0 M sodium hydroxide solution for 1 week. A number of multi-modal anion-exchange ligands based on aromatic amines exhibiting high breakthrough capacity of BSA have been found. With these ligands recovery was often found to be low due to strong non-electrostatic interactions. However, for phenol derived anion-exchange media the recovery can be improved by desorption at high pH.

[66] Bioseparation processes are dominated by chromatographic steps. Even primary recovery is sometimes accomplished by chromatographic separation, using a fluidized bed instead of a fixed bed. In this review, the action principles, features of chromatography media regarding physical and chemical properties will be described. An attempt will be made to establish categories of different media. Characteristics for bioseparation are the large pores and particle sizes. To achieve sufficient capacity for ultralarge molecules, such as plasmids or nanoparticles, such as viruses monoliths are the media of choice. In these media, the mass transport is accomplished by convection, and thus, the low diffusivity can be overcome. Common to all modern chromatography media is the fast operation. There are examples where a residence time of less then 3 min, is sufficient to reach the full potential of the adsorbent.

82. Kakhniashvili DG, Bulla LA et al (2004) The human erythrocyte proteome: analysis by ion trap mass spectrometry. Mol Cell Proteomics 3(5):501–509
83. Karlsson E, Hirsh I (2011) Ion exchange chromatography. Protein purification, Wiley, 93–133[67]
84. Kawachi Y, Ikegami T et al (2011) Chromatographic characterization of hydrophilic interaction liquid chromatography stationary phases: hydrophilicity, charge effects, structural selectivity, and separation efficiency. J Chromatogr A 1218(35):5903–5919[68]
85. Kawasaki T, Niikura M et al (1990) Fundamental study of hydroxyapatite high-performance liquid chromatography: II. Experimental analysis on the basis of the general theory of gradient chromatography. J Chromatogr A 515(0):91–123[69]
86. Kirkland JJ (1973) Porous silica microsphere column packings for high-speed liquid—liquid chromatography. J Chromatogr A 83(0):149–167[70]
87. Kirkland JJ, Truszkowski FA et al (2000) Superficially porous silica microspheres for fast high-performance liquid chromatography of macromolecules. J Chromatogr A 890(1):3–13[71]

[67] This chapter contains sections titled: * Introduction * The Ion Exchange Process * Charge Properties of Proteins * The Stationary Phase—The Ion Exchangers * Nonionic Interactions * The Mobile Phase: Buffers and Salts * Experimental Planning and Preparation * Chromatographic Techniques * Handling of Isolated Proteins * Hydroxyapatite Chromatography * Applications * Acknowledgments * References.

[68] Fourteen commercially available particle-packed columns and a monolithic column for hydrophilic interaction liquid chromatography (HILIC) were characterized in terms of the degree of hydrophilicity, the selectivity for hydrophilic-hydrophobic substituents, the selectivity for the regio and configurational differences in hydrophilic substituents, the selectivity for molecular shapes, the evaluation of electrostatic interactions, and the evaluation of the acidic-basic nature of the stationary phases using nucleoside derivatives, phenyl glucoside derivatives, xanthine derivatives, sodium p-toluenesulfonate, and trimethylphenylammonium chloride as a set of samples. Principal component analysis based on the data of retention factors could separate three clusters of the HILIC phases. The column efficiency and the peak asymmetry factors were also discussed. These data on the selectivity for partial structural differences were summarized as radar-shaped diagrams. This method of column characterization is helpful to classify HILIC stationary phases on the basis of their chromatographic properties, and to choose better columns for targets to be separated. Judging from the retention factor for uridine, these HILIC columns could be separated into two groups: strongly retentive and weakly retentive stationary phases. Among the strongly retentive stationary phases, zwitterionic and amide functionalities were found to be the most selective on the basis of partial structural differences. The hydroxyethyl-type stationary phase showed the highest retention factor, but with low separation efficiency. Weakly retentive stationary phases generally showed lower selectivity for partial structural differences.

[69] In hydroxyapatite (HA) chromatography, competition occurs between the sample molecule and ions from the buffer for adsorption onto the crystal surface of HA. The competition mechanism for several proteins and nucleoside phosphates was analysed on the basis of the general theory of gradient chromatography that has been established recently. It was concluded that the number, x' of adsorbing sites of HA that are covered by an adsorbed molecule, in general, tends to increase slowly with increase in molecular mass, but that the correlation between molecular mass and x' is weak. The conclusion is consistent with the deduction made earlier that the stereochemical structure of the local molecular surface (which is highly characteristic of a molecule, and is intimately related to the x' value) is discerned by the regular crystal surface structure of HA. The capacity factor, k'
, is argued on the basis of the competition model.

[70] A new column packing for high-performance liquid chromatography, porous microspheres of silica produced by the agglutination of colloidal silica particles, has recently been introduced for use in adsorption chromatography. The narrow-size range, relatively homogeneous pore structure and short diffusion path lengths of these <10-μ particles result in very high column efficiencies, and the relatively large, highly available surface area provides for high sample capacity. The microsphere packing displays retention and efficiency characteristics which are less dependent on water content than wide-pore silica gel. Columns of the microspheres may be prepared which are reproducible in chromatographic performance, using a simple high-pressure slurry-packing procedure. More than 10,000 theoretical plates have been obtained on a single 25-cm-long column of 5-μ microspheres at carrier velocities of about 0.7 cm/sec. Plate heights of about five particle diameters and more than thirty-six effective plates/sec have been demonstrated for solutes with capacity factors (k') in the 2–5 range. These columns may be connected in series using low-volume fittings with little loss in efficiency. Columns of the 5-μ particles appear to be limited by mobile phase mass transfer effects, contrasted to the stagnant mobile phase mass transfer limitations exhibited by similar 8- to 9-μ particles.

[71] Very fast reversed-phase separations of biomacromolecules are performed using columns made with superficially porous silica microsphere column packings ("Poroshell"). These column packings consist of ultrapure "biofriendly" silica microspheres composed of solid cores and thin outer shells with uniform pores. The excellent kinetic properties of these new column packings allow stable, high-resolution gradient chromatography of polypeptides, proteins, nucleic acids, DNA fragments, etc. in a fraction of the time required for conventional separations. Contrasted with <2-μm non-porous

88. Kirsch S, Muthing J et al (2009) On-line nano-HPLC/ESI QTOF MS monitoring of alpha2-3 and alpha2-6 sialylation in granulocyte glycosphingolipidome. Biol Chem 390(7):657–672[72]
89. Layne J (2002) Characterization and comparison of the chromatographic performance of conventional, polar-embedded, and polar-endcapped reversed-phase liquid chromatography stationary phases. J Chromatogr A 957(2):149–164[73]
90. Lea DJ, Sehon AH (1962) Preparation of synthetic gels for chromatography of macromolecules. Can J Chem 40(1):159–160
91. Lecchi P, Gupte AR et al (2003) Size-exclusion chromatography in multidimensional separation schemes for proteome analysis. J Biochem Biophys Methods 56(1–3):141–152[74]
92. Leitner A, Reischl R et al (2012) Expanding the chemical cross-linking toolbox by the use of multiple proteases and enrichment by size exclusion chromatography. Mol Cell Proteomics 11(3)[75]
93. Li J, Shao S et al (2008) Simultaneous determination of cations, zwitterions and neutral compounds using mixed-mode reversed-phase and cation-exchange high-performance liquid chromatography. J Chromatogr A 1185(2):185–193[76]

particles, Poroshell packings can be used optimally with existing equipments and greater sample loading capacities, while retaining kinetic (and separation speed) advantages over conventional totally porous particles.

[72] A novel glycosphingolipidomic protocol using nano-high performance liquid chromatography coupled on-line to electrospray ionization quadrupole time-of-flight mass spectrometry (ESI-QTOF-MS) focusing on the separation of isomeric ganglioside structures is described here. A highly efficient separation of alpha2-3- and alpha2-6-sialylated ganglioside species of different carbohydrate chain length was achieved on an HILIC-amido column, followed by sensitive flow-through ESI-QTOF-MS detection and unambiguous structural identification by tandem MS experiments. The protocol was applied to encompass the glycosphingolipidome of human granulocytes, where 182 distinct components could be clearly identified and assigned regarding the ganglioside type and the isomer distribution.

[73] We have evaluated and compared the performance of several conventional C18 phases with those possessing either a polar-endcapping group or a polar-embedded group within the primary alkyl ligand and found distinct differences in the chromatographic behavior among the three groups, as well as a high degree of variability within each group. The trend is for the polar-endcapped phases to display similar hydrophobic retention characteristics as the conventional C18 columns, but to express higher hydrogen bonding capacities and silanol activity. The polar-embedded phases displayed the opposite behavior, with a greatly reduced hydrophobic nature compared to the conventional and polar-endcapped C18 phases, and also a very much reduced silanol activity. Most interestingly, it appears that ionic or dipole interactions play a significant role in the overall retention behavior of the polar-embedded phases towards basic and acidic analytes.

[74] Size-exclusion chromatography (SEC) is a separation technique with a relatively low resolving power, compared to those usually utilized in proteomics. Therefore, it is often overlooked in experimental protocols, when the main goal is resolving complex biological mixtures. In this report, we introduce innovative multidimensional schemes for proteomics analysis, in which SEC plays a practical role. Liquid isoelectric focusing (IEF) was combined with SEC, and experimental results were compared to those obtained by two-dimensional polyacrylamide gel electrophoresis (2D-PAGE), well-established techniques relying upon similar criteria for separation. Additional experiments were performed to evaluate the practical contribution of SEC in multidimensional chromatographic separations. Specifically, we evaluated the combination of SEC and ion exchange chromatography in an analytical scheme for the mass spectrometric analysis of protein-extracts obtained from bacterial cultures grown in stable isotope enriched media. Experimental conditions and practical considerations are discussed.

[75] Chemical cross-linking in combination with mass spectrometric analysis offers the potential to obtain low-resolution structural information from proteins and protein complexes. Identification of peptides connected by a cross-link provides direct evidence for the physical interaction of amino acid side chains, information that can be used for computational modeling purposes. Despite impressive advances that were made in recent years, the number of experimentally observed cross-links still falls below the number of possible contacts of cross-linkable side chains within the span of the cross-linker. Here, we propose two complementary experimental strategies to expand cross-linking data sets. First, enrichment of cross-linked peptides by size exclusion chromatography selects cross-linked peptides based on their higher molecular mass, thereby depleting the majority of unmodified peptides present in proteolytic digests of cross-linked samples. Second, we demonstrate that the use of proteases in addition to trypsin, such as Asp-N, can additionally boost the number of observable cross-linking sites. The benefits of both SEC enrichment and multiprotease digests are demonstrated on a set of model proteins and the improved workflow is applied to the characterization of the 20S proteasome from rabbit and Schizosaccharomyces pombe.

[76] A novel mixed-mode reversed-phase and cation-exchange high-performance liquid chromatography (HPLC) method is described to simultaneously determine four related impurities of cations, zwitterions and neutral compounds in developmental Drug A. The commercial column is Primesep 200 containing hydrophobic alkyl chains with embedded acidic groups in H+ form on a silica support. The mobile phase variables of acid

94. Linden JC, Lawhead CL (1975) Liquid chromatography of saccharides. J Chromatogr A 105 (1):125–133[77]
95. Lindner H, Helliger W (2004) Hydrophilic interaction chromatography. HPLC of peptides and proteins. MI Aguilar, Springer, New York, 251, 75–88
96. Lindqvist B, Storgards T (1955) Molecular-sieving properties of starch. Nature 175(4455):511–512
97. Link AJ, Eng J et al (1999) Direct analysis of protein complexes using mass spectrometry. Nat Biotechnol 17(7):676–682[78]
98. Link AJ, Eng J et al (1999) Direct analysis of protein complexes using mass spectrometry. Nat Biotech 17 (7):676–682
99. Lork KD, Unger KK Solute retention in reversed-phase chromatography as a function of stationary phase properties: effect of n-alkyl chain length and ligand density.
100. Luo J, Zhou W et al (2013). Comparison of fully-porous beads and cored beads in size exclusion chromatography for protein purification. Chem Eng Sci 102:99–105[79]
101. Mant CT, Hodges RS (2008) Mixed-mode hydrophilic interaction/cation-exchange chromatography: separation of complex mixtures of peptides of varying charge and hydrophobicity. J Sep Sci 31 (9):1573–1584[80]

additives, contents of acetonitrile and concentrations of potassium chloride have been thoroughly investigated to optimize the separation. The retention factors as a function of the concentrations of potassium chloride and the percentages of acetonitrile in the mobile phases are investigated to get an insight into the retention and separation mechanisms of each related impurity and Drug A. Furthermore, the elution orders of the related impurities and Drug A in an ion-pair chromatography (IPC) are compared to those in the mixed-mode HPLC to further understand the chromatographic retention behaviors of each related impurity and Drug A. The study found that the positively charged Degradant 1, Degradant 2 and Drug A were retained by both ion-exchange and reversed-phase partitioning mechanisms. RI2, a small ionic compound, was primarily retained by ion-exchange. RI4, a neutral compound, was retained through reversed-phase partitioning without ion-exchange. Moreover, the method performance characteristics of selectivity, sensitivity and accuracy have been demonstrated to be suitable to determine the related impurities in the capsules of Drug A.

[77] The analysis of saccharides by liquid chromatography on an automated instrument is described. Conditions for the resolution and quantitation of fructose, glucose, sucrose, melibiose, raffinose, betaine and three kestose isomers as well as starch hydrolysates are given. Liquid chromatographic analysis equals the precision and accuracy of gas–liquid chromatographic analysis. Greater analysis flexibility and reduced sample preparation are important advantages over gas–liquid chromatographic analysis.

[78] We describe a rapid, sensitive process for comprehensively identifying proteins in macromolecular complexes that uses multidimensional liquid chromatography (LC) and tandem mass spectrometry (MS/MS) to separate and fragment peptides. The SEQUEST algorithm, relying upon translated genomic sequences, infers amino acid sequences from the fragment ions. The method was applied to the Saccharomyces cerevisiae ribosome leading to the identification of a novel protein component of the yeast and human 40S subunit. By offering the ability to identify >100 proteins in a single run, this process enables components in even the largest macromolecular complexes to be analyzed comprehensively.

[79] Size-exclusion chromatography (SEC) relies exclusively on intraparticle diffusion to separate solutes of different molecular sizes and shapes. Thus, its feed volume can only be a small fraction of the column volume. Much larger columns are required for SEC than other forms of liquid chromatography. Becasue of this, SEC often employs less expensive soft gels in large-scale applications to reduce costs. Excessive bed compression forces engineers to use pancake-shaped columns instead of more desirable slim columns during scale-up. Cored beads have impenetrable rigid cores that result in lower pressure drops and better pressure resistance. They also provide sharper peaks due to shortened radial distance for diffusion. Using a new general rate model for SEC with cored beads, this work demonstrated that cored beads performed better than fully-porous beads for myoglobin and ovalbumin separation through computer simulation. This theoretical work could encourge the research and product development of cored beads for large-scale SEC that has not been reported. © 2013 Elsevier Ltd.

[80] Mixed-mode hydrophilic interaction/cation-exchange chromatography (HILIC/CEX) was applied to the separation of two mixtures of synthetic peptide standards: (i) a 27-peptide mixture containing three groups of peptides (each group containing nine peptides of the same net charge of +1, +2 or +3), where the hydrophilicity/hydrophobicity of adjacent peptides within the groups varied only subtly (generally by only a single carbon atom); and (ii) peptide pairs with the same composition but different sequences, where the sole difference between the peptides was the position of a single amino acid substitution. HILIC/CEX is essentially CEX chromatography in the presence of high levels of organic modifier (generally ACN). The present study demonstrated the dramatic effect of increasing ACN concentration (optimum levels of 60–80 %, depending on the application) on the separation of both mixtures of peptides. The greater the charge on the peptides, the better the separation achievable by HILIC/CEX. In addition, HILIC/CEX separation of both the peptide mixtures used in the present study was shown to be superior to that of the more

102. Mant CT, Parker JMR et al (1987) Siz-exclusion high-performance liquid chromatography of peptides: requirement for peptide standards to monitor column performance and non-ideal behaviour. J Chromatogr A 397(0):99–112[81]
103. Marchand DH, Croes K et al (2005) Column selectivity in reversed-phase liquid chromatography: VII. Cyanopropyl columns. J Chromatogr A 1062 (1):57–64[82]
104. Marino K, Bones J et al (2010) A systematic approach to protein glycosylation analysis: a path through the maze. Nat Chem Biol 6(10):713–723
105. Martin AJ, Synge RL (1941) A new form of chromatogram employing two liquid phases: a theory of chromatography. 2. Application to the micro-determination of the higher monoamino-acids in proteins. Biochem J 35(12):1358–1368
106. Mauko L, Nordborg A et al (2011) Glycan profiling of monoclonal antibodies using zwitterionic-type hydrophilic interaction chromatography coupled with electrospray ionization mass spectrometry detection. Anal Biochem 408(2):235–241[83]
107. McCalley DV (2007) Is hydrophilic interaction chromatography with silica columns a viable alternative to reversed-phase liquid chromatography for the analysis of ionisable compounds?. J Chromatogr A 1171(1–2):46–55[84]

commonly applied RP-HPLC mode. Our results highlight again the efficacy of HILIC/CEX as a peptide separation mode in its own right as well as an excellent complement to RP-HPLC.

[81] A series of five synthetic peptide polymers with the sequence Ac-(G-L-G-A-K-G-A-G-V-G)n-amide, where n = 1–5, was employed to assess the resolving power of high-performance size-exclusion columns in peptide separations. The peptide standards showed great versatility in monitoring both ideal (no interactions of solutes with the column material) and non-ideal (hydrophobic and/or ionic interactions of solutes with the column material) size-exclusion behaviour in volatile and non-volatile mobile phases. The effectiveness of adding salts or organic solvents to overcome non-specific interactions of solutes with the column materials was well illustrated by the standards. In addition, the advantageous use of non-ideal size-exclusion behaviour was highlighted. The ability to predict the position and/or elution order of peptides during size-exclusion chromatography (SEC) requires peptides to be separated by a pure size-exclusion process. Although the peptide standards demonstrated similar ideal size-exclusion profiles in non-denaturing medium on all the columns studied this study suggested that, if the conformational character of a peptide protein mixture in a particular mobile phase is uncertain ideal size-exclusion behaviour is required, SEC should be carried out under highly denaturing conditions.

[82] Eleven cyanopropyl ("cyano") columns were characterized by means of a relationship developed originally for alkyl-silica columns. Compared to type-B alkyl-silica columns (i.e., made from pure silica), cyano columns are much less hydrophobic (smaller H), less sterically restricted (smaller S*), and have lower hydrogen-bond acidity (smaller A). Because sample retention is generally much weaker on cyano versus other columns (e.g., C8, C18), a change to a cyano column usually requires a significantly weaker mobile phase in order to maintain comparable values of k for both columns. For this reason, practical comparisons of selectivity between cyano and other columns (i.e., involving different mobile phases for each column) must take into account possible changes in separation due to the change in mobile phase, as well as change in the column.

[83] We present a new method for the analysis of glycans enzymatically released from monoclonal antibodies (MAbs) employing a zwitterionic-type hydrophilic interaction chromatography (ZIC–HILIC) column coupled with electrospray ionization mass spectrometry (ESI–MS). Both native and reduced glycans were analyzed, and the developed procedure was compared with a standard HILIC procedure used in the pharmaceutical industry whereby fluorescent-labeled glycans are analyzed using a TSK Amide-80 column coupled with fluorescence detection. The separation of isobaric alditol oligosaccharides present in monoclonal antibodies and ribonuclease B is demonstrated, and ZIC–HILIC is shown to have good capability for structural recognition. Glycan profiles obtained with the ZIC–HILIC column and ESI–MS provided detailed information on MAb glycosylation, including identification of some less abundant glycan species, and are consistent with the profiles generated with the standard procedure. This new ZIC–HILIC method offers a simpler and faster approach for glycosylation analysis of therapeutic antibodies.

[84] The separation of acidic, neutral and particularly basic solutes was investigated using a bare silica column, mostly under hydrophilic interaction chromatography (HILIC) conditions with water concentrations >2.5 % and with >70 % acetonitrile (ACN). Profound changes in selectivity could be obtained by judicious selection of the buffer and its pH. Acidic solutes had low retention or showed exclusion in ammonium formate buffers, but were strongly retained when using trifluoroacetic acid (TFA) buffers, possibly due to suppression of repulsion of the solute anions from ionised silanol groups at the low pH s s of TFA solutions of aqueous ACN. At high buffer pH, the ionisation of weak bases was suppressed, reducing ionic (and possibly hydrophilic retention) leading to further opportunities for manipulation of selectivity. Peak shapes of basic solutes were excellent in ammonium formate buffers, and overloading effects, which are a major problem for charged bases in RPLC, were relatively insignificant in analytical separations using this buffer. HILIC separations were ideal for fast analysis of ionised bases, due to the low viscosity of mobile phases with high ACN

108. McCalley DV (2013) Separation mechanisms in hydrophilic interaction chromatography. Hydrophilic interaction chromatography, Wiley, 1–41[85]
109. McDonald WH, Ohi R et al (2002) Comparison of three directly coupled HPLC MS/MS strategies for identification of proteins from complex mixtures: single-dimension LC-MS/MS, 2-phase MudPIT, and 3-phase MudPIT. Int J Mass Spectrom 219 (1):245–251[86]
110. McNulty DE, Annan RS (2008) Hydrophilic interaction chromatography reduces the complexity of the phosphoproteome and improves global phosphopeptide isolation and detection. Mol Cell Proteomics 7(5):971–980[87]
111. Mihailova A, Lundanes E et al (2006) Determination and removal of impurities in 2-D LC-MS of peptides. J Sep Sci 29(4):576–581[88]
112. Miller NT, Feibush B et al (1984) Wide-pore silica-based ether-bonded phases for separation of proteins by high-performance hydrophobic-interaction and size exclusion chromatography. J Chromatogr A 316(0):519–536[89]

content, and the favourable Van Deemter curves which resulted from higher solute diffusivities.

[85] Hydrophilic interaction chromatography (HILIC) is a technique that has become increasingly popular for the separation of polar, hydrophilic, and ionizable compounds, which are difficult to separate by reversed-phase (RP) chromatography due to their poor retention when RP is used. HILIC typically uses a polar stationary phase such as bare silica or a polar bonded phase, together with an eluent. This chapter considers in some detail the various mechanisms that contribute to HILIC separations. Contributory mechanisms are likely to be partition, adsorption, ionic interactions, and even hydrophobic retention depending on the experimental conditions.

[86] One of the most effective methods for the direct identification of proteins from complex mixtures without first having to resolve them by polyacrylamide gel electrophoresis is to separate proteolytically generated peptides by microcapillary HPLC and then collect data directly on the eluent using a tandem mass spectrometer. Multidimensional HPLC separation techniques provide access to even more complex mixtures of proteins. A set of techniques for multidimensional analysis was developed in our lab; collectively they are known as multidimensional protein identification technology (MudPIT). These strategies employ a biphasic column with a section of reversed phase (RP) material flanked by strong cation exchange (SCX) resin and allow for multidimensional separation of peptides. A variation on MudPIT adds an additional section of RP material behind the SCX and RP. This 3-phase column can be used for "online" desalting of the sample. We compare the analysis of a complex mixture of proteins purified by their association with bovine brain microtubules using a single-dimension LC-MS/MS column, a 2-phase (standard) MudPIT column, and a 3-phase MudPIT column. We find that the 3-phase MudPIT column yields a greater number of protein identifications for this test sample and allows data to be collected on a set of hydrophilic peptides not sampled using the 2-phase MudPIT column.

[87] The diversity and complexity of proteins and peptides in biological systems requires powerful liquid chromatography-based separations to optimize resolution and detection of components. Proteomics strategies often combine two orthogonal separation modes to meet this challenge. In nearly all cases, the second dimension is a reverse phase separation interfaced directly to a mass spectrometer. Here we report on the use of hydrophilic interaction chromatography (HILIC) as part of a multidimensional chromatography strategy for proteomics. Tryptic peptides are separated on TSKgel Amide-80 columns using a shallow inverse organic gradient. Under these conditions, peptide retention is based on overall hydrophilicity, and a separation truly orthogonal to reverse phase is produced. Analysis of tryptic digests from HeLa cells yielded numbers of protein identifications comparable to that obtained using strong cation exchange. We also demonstrate that HILIC represents a significant advance in phosphoproteomics analysis. We exploited the strong hydrophilicity of the phosphate group to selectively enrich and fractionate phosphopeptides based on their increased retention under HILIC conditions. Subsequent IMAC enrichment of phosphopeptides from HILIC fractions showed better than 99 % selectivity. This was achieved without the use of derivatization or chemical modifiers. In a 300-μg equivalent of HeLa cell lysate we identified over 1000 unique phosphorylation sites. More than 700 novel sites were added to the HeLa phosphoproteome.

[88] Problems occurring during operation of a 2-D LC-MS system for separation and identification of neuropeptides, such as contamination of the used salts and column bleed, are described. When using polysulfoethyl aspartamide, which is widely used as a strong cation exchange stationary phase in the first dimension, interfering peaks were observed in the second-dimension reversed-phase chromatograms. The observed peaks, found to be caused by column bleeding, had abundance above the threshold value and influenced the quality of the analyses. The origin of the peaks was verified and appropriate measures are proposed. Additionally, peaks caused by polyethylene glycols (PEGs), covering approximately 5 min of feasible chromatographic time in every fraction, were observed. The commercial ammonium formate salts used to prepare the first-dimension mobile phase were found to contain PEG impurities, and in subsequent work the salt solutions were prepared from formic acid and ammonia to avoid any additional contaminations.

[89] This paper examines the use of wide-pore silica-based hydrophilic ether-bonded phases for the chromatographic separation of proteins under mild elution conditions. In particular, ether phases of the following structure

113. Mohammed S, Heck AJR (2011) Strong cation exchange (SCX) based analytical methods for the targeted analysis of protein post-translational modifications. Curr Opin Biotechnol 22(1):9–16[90]
114. Molnar I (2002) Computerized design of separation strategies by reversed-phase liquid chromatography: development of DryLab software. J Chromatogr A 965(1–2):175–194[91]
115. Moody RT (1999) 3 – Zorbax porous silica microsphere columns for high-performance size exclusion chromatography. Column handbook for size exclusion chromatography. Cs Wu. San Diego, Academic Press, pp 75–92
116. Moore AW, Jorgenson JW (1995) Comprehensive three-dimensional separation of peptides using size exclusion chromatography/reversed phase liquid chromatography/optically gated capillary zone electrophoresis. Anal Chem 67(19):3456–3463
117. Motoyama A, Xu T et al (2007). Anion and cation mixed-bed ion exchange for enhanced multidimensional separations of peptides and phosphopeptides. Anal Chem 79(10):3623–3634[92]
118. Naidong W (2003) Bioanalytical liquid chromatography tandem mass spectrometry methods on

Si-(CH2)3-O-(CH2-CH2-O)n-R, where n 1, 2, 3 and R methyl, ethyl or n-butyl, have been prepared. These phases can be employed either in high-performance hydrophobic-interaction or size-exclusion chromatography, depending on mobile phase conditions. In the hydrophobic-interaction mode, a gradient of decreasing salt concentration, e.g., from 3 M ammonium sulfate (pH 6.0, 25 °C), yields sharp peaks with high mass recovery of active proteins. In this mode, retention can be controlled by salt type and concentration, as well as by column temperature. In the size-exclusion mode, use of medium ionic strength, e.g., 0.5 M ammonium acetate (pH 6.0) yields linear calibration of log (MW[n]) vs. retention volume. Even at 0.05 M salt concentration, no stationary phase charge effects on protein elution are observed. These bonded-phase columns exhibit good column-to-column reproducibility and constant retention for at least 5 months of continual use. Examples of the high-performance separation of proteins in both modes are illustrated.

[90] The multidimensional combination of strong cation exchange (SCX) chromatography and reversed phase chromatography has emerged as a powerful approach to separate peptides originating from complex samples such as digested cellular lysates or tissues before analysis by mass spectrometry, enabling the identification of over 10,000 s of peptides and thousands of proteins in a single sample. Although, such multidimensional chromatography approaches are powerful, the in-depth analysis of protein post-translational modifications still requires additional sample preparation steps, involving the specific enrichment of peptides displaying the targeted modification. Here, we describe how in particular SCX chromatography can be used for the targeted analysis of important post-translational modifications, such as phosphorylation and N-terminal acetylation. Compared to other methods, SCX is less labor-intensive and more robust, and therefore likely more easily adaptable to main-stream research laboratories.

[91] The development of DryLab software is a special achievement in analytical HPLC which took place in the last 16 years. This paper tries to collect some of the historical mile stones and concepts. DryLab, being always subject to change according to the needs of the user, never stopped being developed. Under the influence of an ever changing science market, the DryLab development team had to consider not just scientific improvements, but also new technological achievements, such as the introduction of Windows 1.0 and 3.1, and later Windows NT and 2000. The recent availability of new 32-bit programming tools allowed calculations of chromatograms to be completed more quickly so as to show peak movements which result for example from slight changes in eluent pH. DryLab is a great success of interdisciplinary and intercontinental cooperation by many scientists.

[92] Shotgun proteomics typically uses multidimensional LC/MS/MS analysis of enzymatically digested proteins, where strong cation-exchange (SCX) and reversed-phase (RP) separations are coupled to increase the separation power and dynamic range of analysis. Here we report an on-line multidimensional LC method using an anion- and cation-exchange mixed bed for the first separation dimension. The mixed-bed ion-exchange resin improved peptide recovery over SCX resins alone and showed better orthogonality to RP separations in two-dimensional separations. The Donnan effect, which was enhanced by the introduction of fixed opposite charges in one column, is proposed as the mechanism responsible for improved peptide recovery by producing higher fluxes of salt cations and lower populations of salt anions proximal to the SCX phase. An increase in orthogonality was achieved by a combination of increased retention for acidic peptides and moderately reduced retention of neutral to basic peptides by the added anion-exchange resin. The combination of these effects led to ?100 % increase in the number of identified peptides from an analysis of a tryptic digest of a yeast whole cell lysate. The application of the method to phosphopeptide-enriched samples increased by 94 % phosphopeptide identifications over SCX alone. The lower pKa of phosphopeptides led to specific enrichment in a single salt step resolving acidic phosphopeptides from other phospho- and non-phosphopeptides. Unlike previous methods that use anion exchange to alter selectivity or enrich phosphopeptides, the proposed format is unique in that it works with typical acidic buffer systems used in electrospray ionization, making it feasible for online multidimensional LC/MS/MS applications.

underivatized silica columns with aqueous/organic mobile phases. J Chromatogr B 796(2):209–224[93]
119. Naidong W, Shou W et al (2001) Novel liquid chromatographic–tandem mass spectrometric methods using silica columns and aqueous–organic mobile phases for quantitative analysis of polar ionic analytes in biological fluids. J Chromatogr B Biomed Sci Appl 754(2):387–399[94]
120. Nikolov ZL, Reilly PJ (1985) Retention of carbohydrates on silica and amine-bonded silica stationary phases: application of the hydration model. J Chromatogr A 325:287–293
121. Nogueira R, Lämmerhofer M et al (2005) Alternative high-performance liquid chromatographic peptide separation and purification concept using a new mixed-mode reversed-phase/weak anion-exchange type stationary phase. J Chromatogr A 1089 (1–2):158–169[95]
122. Nogueira R, Lubda D et al (2006) Silica-based monolithic columns with mixed-mode reversed-phase/weak anion-exchange selectivity principle for high-performance liquid chromatography. J Sep Sci 29(7):966–978[96]

[93] This review article summarizes the recent progress on bioanalytical LC–MS/MS methods using underivatized silica columns and aqueous/organic mobile phases. Various types of polar analytes were extracted by using protein precipitation (PP), liquid/liquid extraction (LLE) or solid-phase extraction (SPE) and were then analyzed using LC–MS/MS on the silica columns. Use of silica columns and aqueous/organic mobile phases could significantly enhance LC–MS/MS method sensitivity, due to the high organic content in the mobile phase. Thanks to the very low backpressure generated from the silica column with low aqueous/high organic mobile phases, LC–MS/MS methods at high flow rates are feasible, resulting in significant timesaving. Because organic solvents have weaker eluting strength than water, direct injection of the organic solvent extracts from the reversed-phase solid-phase extraction onto the silica column was possible. Gradient elution on the silica columns using aqueous/organic mobile phases was also demonstrated. Contrary to what is commonly perceived, the silica column demonstrated superior column stability. This technology can be a valuable supplement to the reversed-phase LC–MS/MS.

[94] Use of silica stationary phase and aqueous–organic mobile phases could significantly enhance LC–MS–MS method sensitivity. The LC conditions were compatible with MS detection. Analytes with basic functional groups were eluted with acidic mobile phases and detected by MS in the positive ion mode. Analytes with acid functional groups were eluted with mobile phases at neutral pH and detected by MS in the negative ion mode. Analytes poorly retained on reversed-phase columns showed good retention on silica columns. Compared with reversed-phase LC–MS–MS, 5–8-fold sensitivity increases were observed for basic polar ionic compounds when using silica columns and aqueous–organic mobile phase. Up to a 20-fold sensitivity increase was observed for acidic polar ionic compounds. Silica columns and aqueous–organic mobile phases were used for assaying nicotine, cotinine, and albuterol in biological fluids.

[95] This article describes a new complementary peptide separation and purification concept that makes use of a novel mixed-mode reversed-phase/weak anion-exchange (RP/WAX) type stationary phase. The RP/WAX is based on N-(10-undecenoyl)-3-aminoquinuclidine selector, which is covalently immobilized on thiol-modified silica particles (5 μm, 100 Å pore diameter) by radical addition reaction. Remaining thiol groups are capped by radical addition with 1-hexene. This newly developed separation material contains two distinct binding domains in a single chromatographic interactive ligand: a lipophilic alkyl chain for hydrophobic interactions with lipophilic moieties of the solute, such as in the reversed-phase chromatography, and a cationic site for anion-exchange chromatography with oppositely charged solutes, which also enables repulsive ionic interactions with positively charged functional groups, leading to ion-exclusion phenomena. The beneficial effect that may result from the combination of the two chromatographic modes is exemplified by the application of this new separation material for the chromatographic separation of the N- and C-terminally protected tetrapeptide N-acetyl-Ile-Glu-Gly-Arg-p-nitroanilide from its side products. Mobile phase variables have been thoroughly investigated to optimize the separation and to get a deeper insight into the retention and separation mechanism, which turned out to be more complex than any of the individual chromatography modes alone. A significant anion-exchange retention contribution at optimal pH of 4.5 was found only for acetate but not for formate as counter-ion. In loadability studies using acetate, peptide masses up to 200 mg could be injected onto an analytical 250 mm × 4 mm i.d. RP/WAX column (5 μm) still without touching bands of major impurity and target peptide peaks. The corresponding loadability tests with formate allowed the injection of only 25 % of this amount. The analysis of the purified peptide by capillary high-performance liquid chromatography (HPLC)-UV and HPLC–ESI-MS employing RP-18 columns revealed that the known major impurities have all been removed by a single chromatographic step employing the RP/WAX stationary phase. The better selectivity and enhanced sample loading capacity in comparison to RP-HPLC resulted in an improved productivity of the new purification protocol. For example, the yield of pure peptide per chromatographic run on RP/WAX phase was by a factor of about 15 higher compared to the standard gradient elution RP-purification protocol.

[96] This article describes the synthesis, chromatographic characterization, and performance evaluation of analytical (100 x 4.6 mm id) and semipreparative (100 x 10 mm id) monolithic silica columns with mixed-mode RP/weak

123. O'Gara JE, Wyndham KD (2006) Porous hybrid organic-inorganic particles in reversed-phase liquid chromatography. J Liquid Chromatogr Related Technol 29(7–8):1025–1045[97]

124. Opiteck GJ, Jorgenson JW et al (1997) Two-Dimensional SEC/RPLC coupled to mass spectrometry for the analysis of peptides. Anal Chem 69 (13):2283–2291[98]

125. Opiteck GJ, Ramirez SM et al (1998) Comprehensive two-dimensional high-performance liquid chromatography for the isolation of overexpressed proteins and proteome mapping. Anal Biochem 258 (2):349–361[99]

anion-exchange (RP/WAX) surface modification. The monolithic RP/WAX columns were obtained by immobilization of N-(10-undecenoyl)-3-aminoquinuclidine onto thiol-modified monolithic silica columns (Chromolith) by a radical addition reaction. Their chromatographic characterization by Engelhardt and Tanaka tests revealed slightly lower hydrophobic selectivities than C-8 phases, as well as higher polarity and also improved shape selectivity than RP-18e silica rods. The surface modification enabled separation by both RP and anion-exchange chromatography principles, and thus showed complementary selectivities to the RP-18e monoliths. The mixed-mode monoliths have been tested for the separation of peptides and turned out to be particularly useful for hydrophilic acidic peptides, which are usually insufficiently retained on RP-18e monolithic columns. Compared to a corresponding particulate RP/WAX column (5 microm, 10 nm pore diameter), the analytical RP/WAX monolith caused lower system pressure drops and showed, as expected, higher efficiency (e.g. by a factor of about 2.5 lower C-term for a tetrapeptide). The upscaling from the analytical to semipreparative column dimension was also successful.

[97] Abstract Reversed?phase chromatographic media have recently become available that are based on porous hybrid organic?inorganic particles. The present paper reviews hybrid particles that are made from organosilanes (organic moiety) and tetraalkoxysilanes (inorganic moiety). The hybrid particles are defined and classified within the context of a broader definition of hybrid materials. First syntheses and chromatographic evaluations are discussed for this class of hybrid packing materials. Publications are then described, which characterize two distinguishing chemical properties of hybrid particles vs. silica gel: 1) less acidic silanols, and 2) markedly longer lifetimes in alkaline mobile phases. These properties are achieved without sacrificing mechanical strength, as is found for fully organic particles, i.e., polymers, with the same chemical features. Literature reports are then reviewed that employ hybrid based reversed?phase column packings for HPLC. Topics covered include fundamental retention mechanism studies, methods development studies, and applications made possible with the hybrid based products. Further review is presented on the use of theses hybrid particles for UPLC. The hybrid particles afford good mechanical strength without sacrificing retention and loading capacity, as is found for non?porous particles. Applications employing hybrid based particles in the UPLC mode are then reported.

Reversed?phase chromatographic media have recently become available that are based on porous hybrid organic?inorganic particles. The present paper reviews hybrid particles that are made from organosilanes (organic moiety) and tetraalkoxysilanes (inorganic moiety). The hybrid particles are defined and classified within the context of a broader definition of hybrid materials. First syntheses and chromatographic evaluations are discussed for this class of hybrid packing materials. Publications are then described, which characterize two distinguishing chemical properties of hybrid particles vs. silica gel: 1) less acidic silanols, and 2) markedly longer lifetimes in alkaline mobile phases. These properties are achieved without sacrificing mechanical strength, as is found for fully organic particles, i.e., polymers, with the same chemical features. Literature reports are then reviewed that employ hybrid based reversed?phase column packings for HPLC. Topics covered include fundamental retention mechanism studies, methods development studies, and applications made possible with the hybrid based products. Further review is presented on the use of theses hybrid particles for UPLC. The hybrid particles afford good mechanical strength without sacrificing retention and loading capacity, as is found for non?porous particles. Applications employing hybrid based particles in the UPLC mode are then reported.

[98] A two-dimensional liquid chromatography system is described here which uses size exclusion liquid chromatography (SEC) followed by reversed phase liquid chromatography (RPLC) to separate the mixture of peptides resulting from the enzymatic digestion of a protein. A novel LC/LC interface, using two RPLC columns in parallel rather than storage loops, joins the two chromatographic dimensions. This new interface design permits the use of conventional analytical diameter HPLC columns, 7.8 mm for SEC and 4.6 mm for RPLC, making construction and maintenance of this system very easy. The reversed phase chromatography utilizes 1.5 ?m diameter, nonporous C-18 modified silica particles, which produce fast and efficient analyses. Following the high-resolution two-dimensional chromatographic separation, an electrospray mass spectrometer detects the peptide fragments. The mass spectrometer scans a 2000 m/z range to identify the analytes from their molecular weights. The analyses of tryptic digests of ovalbumin and serum albumin are each described.

[99] A two-dimensional liquid chromatographic system is described here which uses size-exclusion liquid chromatography (SEC) followed by reversed-phase liquid chromatography (RPLC) to separate the mixture of proteins resulting from the lysis ofEscherichia colicells and to isolate the proteins that they produce. The size-exclusion

126. Oyler AR, Armstrong BL et al (1996) Hydrophilic interaction chromatography on amino-silica phases complements reversed-phase high-performance liquid chromatography and capillary electrophoresis for peptide analysis. J Chromatogr A 724 (1–2):378–383[100]
127. Pabst M, Altmann F (2011) Glycan analysis by modern instrumental methods. Proteomics 11 (4):631–643[101]
128. Peng J, Elias JE et al (2003) Evaluation of multidimensional chromatography coupled with tandem mass spectrometry (LC/LC-MS/MS) for large-scale protein analysis: the yeast proteome. J Proteome Res 2(1):43–50[102]
129. Phillips HL, Williamson JC et al (2010) Shotgun proteome analysis utilising mixed mode (reversed phase-anion exchange chromatography) in conjunction with reversed phase liquid chromatography mass spectrometry analysis. Proteomics 10 (16):2950–2960[103]

---

chromatography can be conducted under either denaturing or nondenaturing conditions. Peaks eluting from the first dimension are automatically subjected to reversed-phase chromatography to separate similarly sized proteins on the basis of their various hydrophobicities. The RPLC also serves to desalt the analytes so that they can be detected in the deep ultraviolet region at 215 nm regardless of the SEC mobile phase used. The two-dimensional (2D) chromatograms produced in this manner then strongly resemble the format of stained 2D gels, in that spots are displayed on aX–Yaxis and intensity represents quantity of analyte. Following chromatographic separation, the analytes are deposited into six 96-well (576 total) polypropylene microtiter plates via a fraction collector. Interesting fractions are analyzed by matrix-assisted laser desorption ionization time-of-flight mass spectrometry (MALDI-TOF/MS) or electrospray mass spectrometry (ESI/MS) depending on sample concentration, which both yield accurate (2 to 0.02 %) molecular weight information on intact proteins without any additional sample preparation, electroblotting, destaining, etc. The remaining 97 % of a fraction can then be used for other analyses, such Edman sequencing, amino acid analysis, or proteolytic digestion and sequencing by tandem mass spectrometry. This 2D HPLC protein purification and identification system was used to isolate the src homology (SH2) domain of the nonreceptor tyrosine kinase pp60c-srcand β-lactamase, both inserted intoE. coli,as well as a number of native proteins comprising a small portion of theE. coliproteome.

[100] Hydrophilic interaction chromatography (HILIC) on amine bonded-phase silica columns provides separations of peptides that are complementary to those obtained with reversed-phase HPLC and free solution capillary electrophoresis. This is illustrated with the peptide drug atosiban and nine diastereomers. Moreover, one of the HILIC methods was suitable for coupling with electrospray mass spectrometry.

[101] The oligosaccharides attached to proteins or lipids are among the most challenging analytical tasks due to their complexity and variety. Knowing the genes and enzymes responsible for their biosynthesis, a large but not unlimited number of different structures and isomers of such glycans can be imagined. Understanding of the biological role of structural variations requires the ability to unambiguously determine the identity and quantity of all glycan species. Here, we examine, which analytical strategies – with a certain high-throughput potential – may come near this ideal. After an expose of the relevant techniques, we try to depict how analytical raw data are translated into structural assignments using retention times, mass and fragment spectra. A method's ability to discriminate between the many conceivable isomeric structures together with the time, effort and sample amount needed for that purpose is suggested as a criterion for the comparative assessment of approaches and their evolutionary stages.

[102] Highly complex protein mixtures can be directly analyzed after proteolysis by liquid chromatography coupled with tandem mass spectrometry (LC–MS/MS). In this paper, we have utilized the combination of strong cation exchange (SCX) and reversed-phase (RP) chromatography to achieve two-dimensional separation prior to MS/MS. One milligram of whole yeast protein was proteolyzed and separated by SCX chromatography (2.1 mm i.d.) with fraction collection every minute during an 80-min elution. Eighty fractions were reduced in volume and then re-injected via an autosampler in an automated fashion using a vented-column (100 μm i.d.) approach for RP-LC-MS/MS analysis. More than 162 000 MS/MS spectra were collected with 26 815 matched to yeast peptides (7537 unique peptides). A total of 1504 yeast proteins were unambiguously identified in this single analysis. We present a comparison of this experiment with a previously published yeast proteome analysis by Yates and colleagues (Washburn, M. P.; Wolters, D.; Yates, J. R., III. Nat. Biotechnol. 2001, 19, 242–7). In addition, we report an in-depth analysis of the false-positive rates associated with peptide identification using the Sequest algorithm and a reversed yeast protein database. New criteria are proposed to decrease false-positives to less than 1 % and to greatly reduce the need for manual interpretation while permitting more proteins to be identified.

[103] The 2-D peptide separations employing mixed mode reversed phase anion exchange (MM (RP-AX)) HPLC in the first dimension in conjunction with RP chromatography in the second dimension were developed and utilised for shotgun proteome analysis. Compared with strong cation exchange (SCX) typically employed for shotgun proteomic analysis, peptide separations using MM (RP-AX) revealed improved separation efficiency and increased peptide distribution across the elution gradient.

130. Polson A (1961) Fractionation of protein mixtures on columns of granulated agar. Biochim Biophys Acta 50(3):565–567
131. Popovici ST, Schoenmakers PJ (2005) Fast size-exclusion chromatography—Theoretical and practical considerations. J Chromatogr A 1099 (1–2):92–102[104]
132. Porath J (1960) Gel filtration of proteins, peptides and amino acids. Biochim Biophys Acta 39 (2):193–207[105]
133. Porath J, Flodin PER (1959) Gel filtration: a method for desalting and group separation. Nature 183 (4676):1657–1659
134. Porath J, Sundberg L et al (1973) Salting-out in amphiphilic gels as a new approach to hydrophobic adsorption. Nature 245(5426):465–466
135. Porsch B. (1993) Epoxy- and diol-modified silica: optimization of surface bonding reaction. J Chromatogr A 653(1):1–7[106]
136. Queiroz JA, Tomaz CT, et al[107]
137. Regnier FE, Noel R (1976) Glycerolpropylsilane bonded phases in the steric exclusion chromatography of biological macromolecules. J Chromatogr Sci 14(7):316–320[108]
138. Ricker RD, Sandoval LA (1996) Fast, reproducible size-exclusion chromatography of biological macromolecules. J Chromatogr A 743(1):43–50[109]

In addition, improved sample handling, with no significant reduction in the orthogonality of the peptide separations was observed. The shotgun proteomic analysis of a mammalian nuclear cell lysate revealed additional proteome coverage (2818 versus 1125 unique peptides and 602 versus 238 proteins) using the MM (RP-AX) compared with the traditional SCX hyphenated to RP-LC-MS/MS. The MM analysis resulted in approximately 90 % of the unique peptides identified present in only one fraction, with a heterogeneous peptide distribution across all fractions. No clustering of the predominant peptide charge states was observed during the gradient elution. The application of MM (RP-AX) for 2-D LC proteomic studies was also extended in the analysis of iTRAQ-labelled HeLa and cyanobacterial proteomes using nano-flow chromatography interfaced to the MS/MS. We demonstrate MM (RP-AX) HPLC as an alternative approach for shotgun proteomic studies that offers significant advantages over traditional SCX peptide separations.

[104] Fast SEC is a very interesting modification of conventional SEC. The need for it emerges from combinatorial chemistry and high-throughput experimentation, where high-speed analyses are required. The different approaches to change the speed of analysis are extensively described in this paper. Special attention is paid to the trade-off between analysis time and resolution and to the selection of optimal column lengths and flow rates. Simulations are used to design and to understand experiments. Integrity plots are constructed to judge the quality of various SEC systems. Fast separations in size-exclusion chromatography are found to be more favorable than suggested by conventional theory. The results are based on experimental data obtained for polystyrene using THF as mobile phase.

[105] 1. 1. Mixtures of proteins, peptides and amino acids can be fractionated by filtration through beds of dextran gel containing only small amounts of carboxylic groups. 2. 2. Group separations are readily achieved. In highly cross-linked dextran proteins and large peptides move together ahead of amino acids. In dextran gels of low degree of cross-linking peptides and even proteins may be retained on the columns, so that a fractionation of substances within these groups may be obtained. 3. 3. Basic peptides and amino acids move slowly through the gels in certain basic solvents such as 1 M pyridine and faster in acidic solvents such as 1 M acetic acid. For acidic peptides and amino acids the influence of the solvents mentioned appears to be the reverse of that for the basic compounds. 4. 4. Aromatic substitution has a marked effect on the migration through the gels. The relative speed of dinitrophenylated amino acids is highly dependent on the buffer used. Such influence of the buffer was not noticed for phenylalanine, tyrosine and tryptophan, although these compounds are retarded to a different extent. 5. 5. When the columns are properly prepared, symmetrical distribution of each compound is always obtained. 6. 6. The column capacity is very high compared to other similar column methods (chromatography and zone electrophoresis). 7. 7. The reproducibility is very good. 8. 8. The gels are easily regenerated in the columns and may be used daily over a period of months without detectable deterioration.

[106] The 3-glycidyloxypropyltrimethoxysilane-silica bonding reaction was investigated. The carbon and bonded epoxide content after the bonding reaction and.

[107] In this article, an overview of hydrophobic interaction chromatography (HIC) of proteins is given. After a brief description of protein hydrophobicity and hydrophobic interactions, we present the different proposed theories for the retention mechanism of proteins in HIC. Additionally, the main parameters to consider for the optimization of fractionation processes by HIC and the stationary phases available were described. Selected examples of protein fractionation by HIC are also presented.

[108] Glycerolpropylsilane bonded phases have been found to control the adsorption and/or denaturation of proteins and nucleic acids on controlled porosity glass supports. The bonded-phase thickness is 18-19A while the amount of glycerol moiety varies from 80 to 150 mumoles/g depending on support pore diameter. It has been demonstrated that carbohydrate bonded supports may be used in the chromatography of proteins, nucleic acids, and polysaccharides.

[109] The size-dependent separation of biological macromolecules can be effectively carried out using size-exclusion chromatography (SEC) on silica-based

139. Roumeliotis P, Unger KK (1981) Assessment and optimization of system parameters in size exclusion separation of proteins on diol-modified silica columns. J Chromatogr A 218(0):535–546[110]
140. Ruhaak LR, Hennig R et al (2010) Optimized workflow for preparation of apts-labeled n-glycans allowing high-throughput analysis of human plasma glycomes using 48-channel multiplexed CGE-LIF. J Proteome Res 9(12):6655–6664[111]
141. Salisbury JJ (2008) Fused-core particles: a practical alternative to sub-2 micron particles. J Chromatogr Sci 46(10):883–886[112]
142. Sandra K, Moshir M et al (2008) Highly efficient peptide separations in proteomics: Part 1. Unidimensional high performance liquid chromatography. J Chromatogr B 866(1–2):48–63[113]

HPLC columns. For this technique to be successful, appropriate methods should be chosen. This paper presents practical guidelines for the development of reproducible SEC methods based upon optimized sample volume, flow-rate, column length and use of mobile phase conditions that reduce non-ideal SEC behavior – parameters often ignored in SEC. Adjustment of these parameters often results in more accurate elution times for proper molecular-mass determination, sharper peaks for improved resolution and shorter run times for increased throughput. In general, sample volume and flow-rate should be kept to a minimum for optimal resolution in SEC. Increasing column length improves resolution and may be achieved by placing columns in tandem. In addition, adjustment of the mobile phase conditions can significantly enhance resolution. However, the results are difficult to predict because the sample plays a major role in this interaction, as does the column packing. When possible, mobile phase ionic strength and pH should be altered until the peak(s) of interest elute at the expected time and with good peak shape. Finally, use of smaller-diameter columns (i.e., 4.6 mm rather than 9.4 mm) and small-diameter packing (4.5 μm) particles are also briefly discussed. The principles described here are demonstrated, using antibodies and a number of standard proteins under a variety of SEC conditions.

[110] On diol-modified silica columns the retention of proteins is governed by a size exclusion effect, but superimposed on this are some secondary effects, i.e., ionic and diol-ligand interactions which can be controlled and adjusted reproducibly by varying the eluent composition. The eluent composition also affects the column efficiency and peak shape. Both dependences can be employed to obtain a better resolution of proteins than can be expected from size exclusion alone.

[111] High-throughput methods for oligosaccharide analysis are required when searching for glycan-based biomarkers. Next to mass spectrometry-based methods, which allow fast and reproducible analysis of such compounds, further separation-based techniques are needed, which allow for quantitative analysis. Here, an optimized sample preparation method for N-glycan-profiling by multiplexed capillary gel electrophoresis with laser-induced fluorescence detection (CGE-LIF) was developed, enabling high-throughput glycosylation analysis. First, glycans are released enzymatically from denatured plasma glycoproteins. Second, glycans are labeled with APTS using 2-picoline borane as a nontoxic and efficient reducing agent. Reaction conditions are optimized for a high labeling efficiency, short handling times, and only limited loss of sialic acids. Third, samples are subjected to hydrophilic interaction chromatography (HILIC) purification at the 96-well plate format. Subsequently, purified APTS-labeled N-glycans are analyzed by CGE-LIF using a 48-capillary DNA sequencer. The method was found to be robust and suitable for high-throughput glycan analysis. Even though the method comprises two overnight incubations, 96 samples can be analyzed with an overall labor allocation time of 2.5 h. The method was applied to serum samples from a pregnant woman, which were sampled during first, second, and third trimesters of pregnancy, as well as 6 weeks, 3 months, and 6 months postpartum. Alterations in the glycosylation patterns were observed with gestation and time after delivery.

[112] The benefits of sub-2 micron particle size columns have been widely researched and published. The use of these columns on ultrahigh-pressure liquid chromatography (UHPLC) instrumentation may lead to increased efficiencies and higher throughput. However, these instruments may not be readily available to the pharmaceutical chemist. Within the past year, a practical alternative has been introduced which offers increased efficiencies, but at conventional HPLC pressure limitations. These particles are called fused-core particles and are comprised of a 1.7- micron solid core encompassed by a 0.5-micron porous silica layer (dp = 2.7 micron). The goal for this research was to test these columns for efficiency and robustness utilizing a mixture of Torcetrapib and its relative impurities. Our results indicate that excellent theoretical plates (~14,000) were achievable for run times less than 5 min. Compared to the Waters Acquity particles, the fused-core particles achieved approximately 80 % of the efficiency but with half the observed backpressure. Our robustness results concluded that these separations were reproducible for at least 500 injections while the % RSD for retention time, theoretical plates, peak asymmetry, and resolution was found to be less than 1 %.

[113] Sample complexity and dynamic range constitute enormous challenges in proteome analysis. The back-end technology in typical proteomics platforms, namely mass spectrometry (MS), can only tolerate a certain complexity, has a limited dynamic range per spectrum and is very sensitive towards ion suppression. Therefore, component overlap has to be minimized for successful mass spectrometric analysis and subsequent protein identification and quantification. The present review describes the advances that have been made in liquid-based separation

143. Saraswat M, Musante L et al (2013) Preparative purification of recombinant proteins: current status and future trends. BioMed Res Int 2013:2018
144. Selkirk, C. (2004). Ion-exchange chromatography. Protein purification protocols. P Cutler, Humana Press, 244, 125–131
145. Selman MHJ, Hemayatkar M et al (2011) Cotton HILIC SPE microtips for microscale purification and enrichment of glycans and glycopeptides. Anal Chem 83(7):2492–2499[114]
146. Selman MHJ, McDonnell LA et al (2010) Immunoglobulin G glycopeptide profiling by matrix-assisted laser desorption ionization Fourier transform ion cyclotron resonance mass spectrometry. Anal Chem 82(3):1073–1081[115]
147. Selman MHJ, Niks EH et al (2010) IgG Fc N-Glycosylation changes in lambert-eaton myasthenic syndrome and myasthenia gravis. J Proteome Res 10(1):143–152[116]
148. Shaltiel S, Er-El Z (1973) Hydrophobic chromatography: use for purification of glycogen synthetase. Proc Natl Acad Sci 70(3):778–781[117]

techniques with focus on the recent developments to boost the resolving power. The review is divided in two parts; the first part deals with unidimensional liquid chromatography and the second part with bi- and multidimensional liquid-based separation techniques. Part 1 mainly focuses on reversed-phase HPLC due to the fact that it is and will, in the near future, remain the technique of choice to be hyphenated with MS. The impact of increasing the column length, decreasing the particle diameter, replacing the traditional packed beds by monolithics, amongst others, is described. The review is complemented with data obtained in the laboratories of the authors.

[114] Solid-phase extraction microtips are important devices in modern bioanalytics, as they allow miniaturized sample preparation for mass spectrometric analysis. Here we introduce the use of cotton wool for the preparation of filter-free HILIC SPE microtips. To this end, pieces of cotton wool pads (approximately 500 μg) were packed into 10 μL pipet tips. The performance of the tips was evaluated for microscale purification of tryptic IgG Fc N-glycopeptides. Cotton wool HILIC SPE microtips allowed the removal of salts, most nonglycosylated peptides, and detergents such as SDS from glycoconjugate samples. MALDI-TOF-MS glycopeptide profiles were very repeatable with different tips as well as reused tips, and very similar profiles were obtained with different brands of cotton wool pads. In addition, we used cotton HILIC microtips to purify N-glycans after N-glycosidase F treatment of IgG and transferrin followed by MALDI-TOF-MS detection. In conclusion, we establish cotton wool microtips for glycan and glycopeptide purification with subsequent mass spectrometric detection.

[115] Immunoglobulin G (IgG) fragment crystallizable (Fc) glycosylation is essential for Fc-receptor-mediated activities. Changes in IgG Fc glycosylation have been found to be associated with various diseases. Here we describe a high-throughput IgG glycosylation profiling method. Sample preparation is performed in 96-well plate format: IgGs are purified from 2 ?L of human plasma using immobilized protein A. IgGs are cleaved with trypsin, and the resulting glycopeptides are purified by reversed-phase or hydrophilic interaction solid-phase extraction. Glycopeptides are analyzed by intermediate pressure matrix-assisted laser desorption ionization Fourier transform ion cyclotron resonance mass spectrometry (MALDI-FTICR-MS). Notably, both dihydroxybenzoic acid (DHB) and α-cyano-4-hydroxycinnamic acid (CHCA) matrixes allowed the registration of sialylated as well as nonsialylated glycopeptides. Data were automatically processed, and IgG isotype-specific Fc glycosylation profiles were obtained. The entire method showed an interday variation below 10 % for the six major glycoforms of both IgG1 and IgG2. The method was found suitable for isotype-specific high-throughput IgG glycosylation profiling from human plasma. As an example we successfully applied the method to profile the IgG glycosylation of 62 human samples.

[116] N-glycosylation of the immunoglobulin Fc moiety influences its biological activity by, for example, modulating the interaction with Fc receptors. Changes in IgG glycosylation have been found to be associated with various inflammatory diseases. Here we evaluated for the first time IgG Fc N-glycosylation changes in well-defined antibody-mediated autoimmune diseases, that is, the neurological disorders Lambert-Eaton myasthenic syndrome and myasthenia gravis, with antibodies to muscle nicotinic acetylcholine receptors or muscle-specific kinase. IgGs were purified from serum or plasma by protein A affinity chromatography and digested with trypsin. Glycopeptides were purified and analyzed by MALDI-FTICR?MS. Glycoform distributions of both IgG1 and IgG2 were determined for 229 patients and 56 controls. We observed an overall age and sex dependency of IgG Fc N-glycosylation, which was in accordance with literature. All three disease groups showed lower levels of IgG2 galactosylation compared to controls. In addition, LEMS patients showed lower IgG1 galactosylation. Notably, the galactosylation differences were not paralleled by a difference in IgG sialylation. Moreover, the level of IgG core-fucosylation and bisecting N-acetylglucosamine were evaluated. The control and disease groups revealed similar levels of IgG Fc core-fucosylation. Interestingly, LEMS patients below 50 years showed elevated levels of bisecting N-acetylglucosamine on IgG1 and IgG2, demonstrating for the first time the link of changes in the level of bisecting N-acetylglucosamine with disease.

[117] A homologous series of Î´-aminoalkylagaroses [Sepharose-NH(CH2)nNH2] that varied in the length of their hydrocarbon side chains was synthesized. This family of agaroses was used for a new type of

149. Shi Y, Xiang R et al (2004) The role of liquid chromatography in proteomics. J Chromatogr A 1053(1–2):27–36[118]
150. Simpson DC, Ahn S et al (2006) Using size exclusion chromatography-RPLC and RPLC-CIEF as two-dimensional separation strategies for protein profiling. Electrophoresis 27(13):2722–2733[119]
151. Snyder LR, Wrisley L et al (2006). Computer-aided optimization. HPLC made to measure, Wiley-VCH Verlag GmbH & Co. KGaA, 565–623[120]
152. Strege MA, Stevenson S et al (2000) Mixed-mode anion – cation exchange/hydrophilic interaction liquid chromatography – electrospray mass spectrometry as an alternative to reversed phase for small molecule drug discovery. Anal Chem 72 (19):4629–4633

chromatography, in which retention of proteins is achieved mainly through lipophilic interactions between the hydrocarbon side chains on the agarose and accessible hydrophobic pockets in the protein. When an extract of rabbit muscle was subjected to chromatography on these modified agaroses, the columns with short arms (n = 2 and n = 3) excluded glycogen synthetase (EC 2.4.1.11), but the enzyme was retained on Î'-aminobutyl-agarose (n = 4), from which it could be eluted with a linear NaCl gradient. Higher members of this series (e.g., n = 6) bind the synthetase so tightly that it can be eluted only in a denatured form. A column of Î´-aminobutyl-agarose, which retained the synthetase, excluded glycogen phosphorylase (EC 2.4.1.1), which in this column series and under the same conditions requires side chains 5-(or 6)-carbon-atoms long for retention. Therefore, it is possible to isolate glycogen synthetase by passage of muscle extract through Î´-aminobutyl-agarose, then to extract phosphorylase by subjecting the excluded proteins to chromatography on Î´-aminohexyl-agarose (n = 6). On a preparative scale, the synthetase (I form) was purified 25- to 50-fold in one step. This paper describes some basic features and potential uses of hydrophobic chromatography. The relevance of the results presented here to the design and use of affinity chromatography columns is discussed.

[118] Proteomics represents a significant challenge to separation scientists because of the diversity and complexity of proteins and peptides present in biological systems. Mass spectrometry as the central enabling technology in proteomics allows detection and identification of thousands of proteins and peptides in a single experiment. Liquid chromatography is recognized as an indispensable tool in proteomics research since it provides high-speed, high-resolution and high-sensitivity separation of macromolecules. In addition, the unique features of chromatography enable the detection of low-abundance species such as post-translationally modified proteins. Components such as phosphorylated proteins are often present in complex mixtures at vanishingly small concentrations. New chromatographic methods are needed to solve these analytical challenges, which are clearly formidable, but not insurmountable. This review covers recent advances in liquid chromatography, as it has impacted the area of proteomics. The future prospects for emerging chromatographic technologies such as monolithic capillary columns, high temperature chromatography and capillary electrochromatography are discussed.

[119] Bottom-up proteomics (analyzing peptides that result from protein digestion) has demonstrated capability for broad proteome coverage and good throughput. However, due to incomplete sequence coverage, this approach is not ideally suited to the study of modified proteins. The modification complement of a protein can best be elucidated by analyzing the intact protein. 2-DE, typically coupled with the analysis of peptides that result from in-gel digestion, is the most frequently applied protein separation technique in MS-based proteomics. As an alternative, numerous column-based liquid phase techniques, which are generally more amenable to automation, are being investigated. In this work, the combination of size-exclusion chromatography (SEC) fractionation with RPLC-Fourier-transform ion cyclotron resonance (FTICR)-MS is compared with the combination of RPLC fractionation with CIEF-FTICR-MS for the analysis of the Shewanella oneidensis proteome. SEC-RPLC-FTICR-MS allowed the detection of 297 proteins, as opposed to 166 using RPLC-CIEF-FTICR-MS, indicating that approaches based on LC-MS provide better coverage. However, there were significant differences in the sets of proteins detected and both approaches provide a basis for accurately quantifying changes in protein and modified protein abundances.

[120] This chapter contains sections titled: * Computer-Facilitated HPLC Method Development Using DryLab® Software Introduction HistoryTheoryDryLab Capabilities DryLab OperationMode ChoicesPractical Applications of DryLab® in the LaboratoryConclusions * References ChromSword® Software for Automated and Computer-Assisted Development of HPLC Methods Introduction Off-Line ModeOn-Line ModeChromSword® VersionsExperimental Set-Up for On-Line ModeMethod Development with ChromSword®Off-Line Mode (Computer-Assisted Method Development)On-Line Mode – Fully Automated Optimization of Isocratic and Gradient Separations Software Functions for AutomationHow Does the System Optimize Separations?Conclusion * References Multifactorial Systematic Method Development and Optimization in Reversed-Phase HPLC Introduction and Factorial ViewpointStrategy for Partially Automated Method DevelopmentComparison of Commercially Available Software Packages with Regard to Their Contribution to Factorial Method DevelopmentDevelopment of a New System for Multifactorial Method Development Selection of Stationary PhasesOptimizing Methods with HEUREKAEvaluation of Data with HEUREKAConclusion and Outlook * References.

153. Štulík K, Pacáková V et al (1997) Stationary phases for peptide analysis by high performance liquid chromatography: a review. Anal Chim Acta 352 (1–3):1–19[121]
154. Sun K, Sehon AH (1965) The use of polyacrylamide gels for chromatography of proteins. Can J Chem 43 (4):969–976
155. Cirkovic VelickovicT, JO, Mihajlovic L (2012) Separation of amino acids, peptides, and proteins by ion exchange chromatography. Ion exchange technology II: *applications*. ML Dr.Inamuddin. Netherlands, Springer, Dordrecht
156. Tanaka H, Zhou X et al (2003) Characterization of a novel diol column for high-performance liquid chromatography. J Chromatogr A 987(1–2):119–125[122]
157. Toll H, Oberacher H et al (2005). Separation, detection, and identification of peptides by ion-pair reversed-phase high-performance liquid chromatography-electrospray ionization mass spectrometry at high and low pH. J Chromatogr A 1079 (1–2):274–286[123]
158. Tolstikov VV, Fiehn O (2002). Analysis of highly polar compounds of plant origin: combination of hydrophilic interaction chromatography and electrospray ion trap mass spectrometry. Anal Biochem 301(2):298–307[124]

[121] A survey is given of modern stationary phases employed in high performance liquid chromatography (HPLC) analysis of peptides. The physico-chemical properties of peptides and their consequences for the selection and optimization of the separation system are briefly discussed, followed by a summary of the approaches to the selection and characterization of stationary phases. The properties and applicability of various stationary phases are then critically reviewed, including aspects such as size-exclusion, ion-exchange, reversed-phase, hydrophobic-interaction, affinity and chiral systems, as well as some specialized separation techniques. Emphasis is placed on the most recent literature.

[122] For the investigation of a diol phase (Inertsil Diol column) in hydrophilic interaction chromatography, urea, sucrose and glycine were used as test compounds. The chromatographic conditions were investigated for optimal column efficiency. The column temperature used in common reversed-phase liquid chromatography could also be used for the separation and the flow-rate should be adjusted to 0.3-0.5 ml/min to optimize column efficiency. It is suggested that the velocity of the hydrophilic interaction is slower than the hydrophobic interaction in RPLC. The addition of trifluoroacetic acid is effective for the retention of glycine, but ineffective for urea and sucrose. The diol phase exhibited sufficient chemical stability even if exposed to water in high percentage, and could be applied with isocratic elution for the separation/analysis of amino acids and glucose.

[123] Bioactive peptides and tryptic digests of various proteins were separated under acidic and alkaline conditions by ion-pair-reversed-phase high-performance liquid chromatography (RP-HPIPC) in 200 μm I.D. monolithic, poly(styrene-divinylbenzene)-based capillary columns using gradients of acetonitrile in 0.050 % aqueous trifluoroacetic acid, pH 2.1, or 1.0 % triethylamine-acetic acid, pH 10.6. Chromatographic performances with mobile phases of low and high-pH were practically equivalent and facilitated the separation of more than 50 tryptic peptides of bovine serum albumin within 15–20 min with peak widths at half height between 4 and 10 s. Neither a significant change in retentivity nor efficiency of the monolithic column was observed during 17-day operation at pH 10.6 and 50 °C. Upon separation by RP-HPIPC at high-pH, peptide detectabilities in full-scan negative-ion electrospray ionization mass spectrometry (negESI-MS) were about two to three times lower as compared to RP-HPIPC at low-pH with posESI-MS detection. Tandem mass spectra obtained by fragmentation of deprotonated peptide ions in negative ion mode yielded interpretable sequence information only in a few cases of relatively short peptides. However, in order to obtain sequence information for peptides separated with alkaline mobile phases, tandem mass spectrometry (MS/MS) could be performed in positive ion mode. The chromatographic selectivities were significantly different in separations performed with acidic and alkaline eluents, which facilitated the fractionation of a complex peptide mixture obtained by the tryptic digestion of 10 proteins utilizing off-line, two-dimensional RP-HPIPC at high pH × RP-HPIPC at low pH and subsequent on-line identification by posESI-MS/MS.

[124] The primary goal of metabolomic analysis is the unbiased relative quantification of every metabolite in a biological system. A number of different metabolite-profiling techniques must be combined to make this possible. Here we report the separation and analysis of highly polar compounds in a proof of concept study. Compounds were separated and analyzed using hydrophilic interaction liquid chromatography (HILIC) coupled to electrospray ionization (ESI) mass spectrometry. Two types of HILIC microbore columns (Polyhydroxyethyl A and TSK Gel Amide 80) were compared to normal phase silica HPLC columns. The best separations of standards mixtures and plant samples were achieved using the Amide 80 stationary phase. ESI enabled the detection of both positively and negatively charged metabolites, when coupled to a quadrupole ion trap mass spectrometer using continuous polarity switching. By stepwise mass spectrometric fragmentation of the most intense ions, unknown compounds could be identified and then included into a custom mass spectrometric library. This method was used to detect oligosaccharides, glycosides, amino sugars, amino acids, and sugar nucleotides in phloem exudates from petioles of fully expanded Cucurbita maxima leaves. Quantitative analysis was performed using external standards. The detection limit for stachyose was 0.5 ng per injection

159. Tran BQ, Hernandez C et al (2010) Addressing trypsin bias in large scale (phospho)proteome analysis by size exclusion chromatography and secondary digestion of large post-trypsin peptides. J Proteome Res 10(2):800–811[125]
160. van Deemter JJ, Zuiderweg FJ et al (1956) Longitudinal diffusion and resistance to mass transfer as causes of nonideality in chromatography. Chem Eng Sci 5(6):271–289[126]
161. Verhaar LAT, Kuster BFM (1982). Contribution to the elucidation of the mechanism of sugar retention on amine-modified silica in liquid chromatography. J Chromatogr A 234(1):57–64[127]
162. Wagner K, Miliotis T et al (2002) An automated on-line multidimensional HPLC system for protein and peptide mapping with integrated sample preparation. Anal Chem 74(4):809–820[128]
163. Walshe M, Kelly MT et al (1995) Retention studies on mixed-mode columns in high-performance liquid chromatography. J Chromatogr A 708(1):31–40[129]

(Amide 80). The concentration of stachyose in investigated phloem samples was in the range of 1–7 mM depending on the plant.

[125] In the vast majority of bottom-up proteomics studies, protein digestion is performed using only mammalian trypsin. Although it is clearly the best enzyme available, the sole use of trypsin rarely leads to complete sequence coverage, even for abundant proteins. It is commonly assumed that this is because many tryptic peptides are either too short or too long to be identified by RPLC/MS/MS. We show through in silico analysis that 20–30 %? of the total sequence of three proteomes (Schizosaccharomyces pombe, Saccharomyces cerevisiae, and Homo sapiens) is expected to be covered by Large post-Trypsin Peptides (LpTPs) with Mr above 3000 Da. We then established size exclusion chromatography to fractionate complex yeast tryptic digests into pools of peptides based on size. We found that secondary digestion of LpTPs followed by LC/MS/MS analysis leads to a significant increase in identified proteins and a 32–50 % relative increase in average sequence coverage compared to trypsin digestion alone. Application of the developed strategy to analyze the phosphoproteomes of S. pombe and of a human cell line identified a significant fraction of novel phosphosites. Overall our data indicate that specific targeting of LpTPs can complement standard bottom-up workflows to reveal a largely neglected portion of the proteome.

[126] The mechanisms of band broadening in linear, nonideal chromatography are examined. A development is presented of a rate theory for this process, wherein nonideality is caused by: • axial molecular diffusion; • axial eddy diffusion; • finiteness of transfer coefficient. The correspondence with the plate theory is given, so that the results can also be expressed in heights equivalent to a theoretical plate. The plate theory has been extended to the case of a finite volume of feed; the requirement for this feed volume to be negligible has been examined and a method is presented for evaluating concentration profiles obtained with a larger volume of feed. An analysis is given of experimental results, whereby the relative contributions to band broadening for various cooperating mechanisms could be ascertained.

[127] Liquid chromatography of reducing or non-reducing sugars results in single peaks on amine-modified silica with acetonitrile—water as eluent. In spite of the two anomeric forms of the reducing sugars, single peaks can be obtained because mutarotation is fast under these conditions. The bonded amine groups catalyse the mutarotation in such a way that triethylamine added to eluent has not influence. The separation of the sugars is the result of their partition between two liquid phases, because the composition of the stationary liquid phase appears to be much richer in water than the eluent.

[128] A comprehensive on-line two-dimensional 2D-HPLC system with integrated sample preparation was developed for the analysis of proteins and peptides with a molecular weight below 20 kDa. The system setup provided fast separations and high resolving power and is considered to be a complementary technique to 2D gel electrophoresis in proteomics. The on-line system reproducibly resolved ~ 1000 peaks within the total analysis time of 96 min and avoided sample losses by off-line sample handling. The low-molecular-weight target analytes were separated from the matrix using novel silica-based restricted access materials (RAM) with ion exchange functionalities. The size-selective sample fractionation step was followed by anion or cation exchange chromatography as the first dimension. The separation mechanism in the subsequent second dimension employed hydrophobic interactions using short reversed-phase (RP) columns. A new column-switching technique, including four parallel reversed-phase columns, was employed in the second dimension for on-line fractionation and separation. Gradient elution and UV detection of two columns were performed simultaneously while loading the third and regenerating the fourth column. The total integrated workstation was operated in an unattended mode. Selected peaks were collected and analyzed off-line by MALDI-TOF mass spectrometry. The system was applied to protein mapping of biological samples of human hemofiltrate as well as of cell lysates originating from a human fetal fibroblast cell line, demonstrating it to be a viable alternative to 2D gel electrophoresis for mapping peptides and small proteins.

[129] The retention properties of a column prepared by mixing together strong cation exchange (SCX) and reversed-phase (C18) packing materials were investigated using a range of test solutes. The column was found to exhibit chromatographic properties characteristic of both phases. The effects of changes in eluent composition, buffer ion, ionic strength and pH on the capacity factors of different compounds were determined. The dual nature of the retention mechanism allowed the retention of ionisable molecules to be adjusted by altering the composition of the aqueous component of the mobile phase

164. Wang X, Emmett MR et al (2010) Liquid chromatography electrospray ionization Fourier transform ion cyclotron resonance mass spectrometric characterization of N-linked glycans and glycopeptides. Anal Chem 82(15):6542–6548[130]
165. Wang X, Li W et al (2005) Orthogonal method development using hydrophilic interaction chromatography and reversed-phase high-performance liquid chromatography for the determination of pharmaceuticals and impurities. J Chromatogr A 1083(1–2):58–62[131]
166. Washburn MP, Wolters D et al (2001) Large-scale analysis of the yeast proteome by multidimensional protein identification technology. Nat Biotechnol 19 (3):242–247[132]
167. Weiss J, Jensen D (2003) Modern stationary phases for ion chromatography. Anal Bioanal Chem 375 (1):81–98
168. Westermeier R, Naven T et al (2008) Liquid chromatography techniques. Proteomics in practice, Wiley-VCH Verlag GmbH & Co. KGaA, 151–213[133]
169. Wohlgemuth J, Karas M et al (2010) Enhanced glyco-profiling by specific glycopeptide enrichment and complementary monolithic nano-LC (ZIC-HILIC/RP18e)/ESI-MS analysis. J Sep Sci 33 (6–7):880–890[134]

while those of compounds uncharged over the pH range investigated remained unaffected. Results were compared those obtained on a C18 column and it was found that the acidic and weakly basic compounds had higher capacity factors on this column whereas strongly basic compounds had higher capacity factors on the mixed-mode column.

[130] We combine liquid chromatography, electrospray ionization, and Fourier transform ion cyclotron resonance mass spectrometry (LC ESI FT-ICR MS) to determine the sugar composition, linkage pattern, and attachment sites of N-linked glycans. N-linked glycans were enzymatically released from glycoproteins with peptide N-glycosidase F, followed by purification with graphitized carbon cartridge solid-phase extraction and separation over a TSK-Gel Amide80 column under hydrophilic interaction chromatography (HILIC) conditions. Unique glycopeptide compositions were determined from experimentally measured masses for different combinations of glycans and glycopeptides. The method was validated by identifying four peptides glycosylated so as to yield 12 glycopeptides unique in glycan composition for the standard glycoprotein, bovine alpha-2-HS-glycoprotein. We then assigned a total of 137 unique glycopeptide compositions from 18 glycoproteins from fetal bovine serum, and the glycan structures for most of the assigned glycopeptides were heterogeneous. Highly accurate FT-ICR mass measurement is essential for reliable identification.

[131] A hydrophilic interaction chromatography (HILIC) method has been developed and validated as a secondary or orthogonal method complementary to a reversed-phase HPLC (RP-HPLC) method for quantitation of a polar active pharmaceutical ingredient and its three degradation products. The HILIC method uses a diol column and a mobile phase consisting of acetonitrile/water and ammonium chloride. The compounds of interest show significant differences in retention behaviors with the two very different chromatographic systems, which are desired in developing orthogonal methods. The HILIC method is validated and has met all validation acceptance criteria for the support of drug development activities.

[132] We describe a largely unbiased method for rapid and large-scale proteome analysis by multidimensional liquid chromatography, tandem mass spectrometry, and database searching by the SEQUEST algorithm, named multidimensional protein identification technology (MudPIT). MudPIT was applied to the proteome of the Saccharomyces cerevisiae strain BJ5460 grown to mid-log phase and yielded the largest proteome analysis to date. A total of 1484 proteins were detected and identified. Categorization of these hits demonstrated the ability of this technology to detect and identify proteins rarely seen in proteome analysis, including low-abundance proteins like transcription factors and protein kinases. Furthermore, we identified 131 proteins with three or more predicted transmembrane domains, which allowed us to map the soluble domains of many of the integral membrane proteins. MudPIT is useful for proteome analysis and may be specifically applied to integral membrane proteins to obtain detailed biochemical information on this unwieldy class of proteins.

[133] This chapter contains sections titled: * Basic Principles of Important Liquid Chromatography Techniques Ion Exchange ChromatographyReversed Phase ChromatographyAffinity ChromatographyGel Filtration * Strategic Approach and General Applicability * Liquid Chromatography Techniques and Applications in Proteome Analysis Peptide Separation2DLC Peptide SeparationAffinity Chromatography and LC-MS/MSProtein Pre-fractionation * Practical Considerations and Application of LC-based Protein Pre-fractionation Sample Extraction and PreparationExperimental SetupIon Exchange Chromatography and Protein Pre-fractionationReversed Phase Chromatography and Protein Pre-fractionationFraction Size and Number of Fractions * Critical Review and Outlook.

[134] Dedicated and specific sample preparation and adequate chromatographic resolution prior to MS are necessary for comprehensive and site-specific glycosylation analysis to compensate for high heterogeneity of protein glycosylation, low-abundance of specific glycoforms and ion-suppression effects caused by coelution of other peptides. This article describes a scheme for glycopeptide profiling, which comprises HILIC batch enrichment followed by complementary HILIC and RP-LC in 1-D and 2-D approaches. For reproducible and sensitive nano-

170. Wolters DA, Washburn MP et al (2001) An automated multidimensional protein identification technology for shotgun proteomics. Anal Chem 73 (23):5683–5690[135]
171. Wyndham KD, O'Gara JE et al (2003) Characterization and evaluation of C18 HPLC stationary phases based on ethyl-bridged hybrid organic/inorganic particles. Anal Chem 75(24):6781–6788[136]
172. Xia HF, Lin DQ et al (2008) Preparation and evaluation of cellulose adsorbents for hydrophobic charge induction chromatography. Ind Eng Chem Res 47 (23):9566–9572[137]
173. Xie S, Svec F et al (1997) Rigid porous polyacrylamide-based monolithic columns containing butyl methacrylate as a separation medium for the rapid hydrophobic interaction chro-

LC/ESI-MS analysis, we used ZIC-HILIC and RP18e monolithic silica capillaries and assessed their retention characteristics and complementarity for glycopeptide separations. The experiments revealed that pre-enrichment of glycopeptides in combination with LC employing both phases considerably improves site-specific elucidation of glycosylation heterogeneity. Zwitterionic hydrophilic interaction liquid chromatography showed high capability to separate glycopeptides by their glycan composition, which coeluted on RP18e. By varying solvent conditions, retention can be well tuned, and efficient separations were achieved even in absence of any additives like salt or formic acid. RP18e facilitated glycopeptide separations with high peak capacity based on peptide sequence and degree of sialylation. Implementing both orthogonal and complementary phases in 1-D and 2-D LC setups was shown to significantly increase the number of different identified glycoforms and possesses great potential for comprehensive glycoproteomics approaches.

[135] We describe an automated method for shotgun proteomics named multidimensional protein identification technology (MudPIT), which combines multidimensional liquid chromatography with electrospray ionization tandem mass spectrometry. The multidimensional liquid chromatography method integrates a strong cation-exchange (SCX) resin and reversed-phase resin in a biphasic column. We detail the improvements over a system described by Link et al. (Link, A. J.; Eng, J.; Schieltz, D. M.; Carmack, E.; Mize, G. J.; Morris, D. R.; Garvik, B. M.; Yates, J. R., III. Nat. Biotechnol. 1999, 17, 676?682) that separates and acquires tandem mass spectra for thousands of peptides. Peptides elute off the SCX phase by increasing pI, and elution off the SCX material is evenly distributed across an analysis. In addition, we describe the chromatographic benchmarks of MudPIT. MudPIT was reproducible within 0.5 % between two analyses. Furthermore, a dynamic range of 10?000 to 1 between the most abundant and least abundant proteins/peptides in a complex peptide mixture has been demonstrated. By improving sample preparation along with separations, the method improves the overall analysis of proteomes by identifying proteins of all functional and physical classes.

[136] The characterization and evaluation of three novel 5-?m HPLC column packings, prepared using ethyl-bridged hybrid organic/inorganic materials, is described. These highly spherical hybrid particles, which vary in specific surface area (140, 187, and 270 m2/g) and average pore diameter (185, 148, and 108 Å), were characterized by elemental analysis, SEM, and nitrogen sorption analysis and were chemically modified in a two-step process using octadecyltrichlorosilane and trimethylchlorosilane. The resultant bonded materials had an octadecyl surface concentration of 3.17?3.35 ?mol/m2, which is comparable to the coverage obtained for an identically bonded silica particle (3.44 ?mol/m2) that had a surface area of 344 m2/g. These hybrid materials were shown to have sufficient mechanical strength under conditions normally employed for traditional reversed-phase HPLC applications, using a high-pressure column flow test. The chromatographic properties of the C18 bonded hybrid phases were compared to a C18 bonded silica using a variety of neutral and basic analytes under the same mobile-phase conditions. The hybrid phases exhibited similar selectivity to the silica-based column, yet had improved peak tailing factors for the basic analytes. Column retentivity increased with increasing particle surface area. Elevated pH aging studies of these hybrid materials showed dramatic improvement in chemical stability for both bonded and unbonded hybrid materials compared to the C18 bonded silica phase, as determined by monitoring the loss in column efficiency through 140-h exposure to a pH 10 triethylamine mobile phase at 50 °C.

[137] Hydrophobic charge induction chromatography (HCIC) has been proven to be an efficient technique for antibody purification. Several HCIC adsorbents were prepared with macroporous cellulose?tungsten carbide composite beads (Cell-TuC) as the matrix. First, the cellulose beads were activated by allyl bromide (AB) or divinyl sulfone (DVS), and then they were coupled with three types of mercaptoheterocyclic groups?4-mercapto-ethyl-pyridine hydrochloride (MEP), 2-mercapto-1-methyl-imidazole (MMI), and 2-mercapto-benzimidazole (MBI)?as the HCIC ligands. Four types of HCIC adsorbents were obtained, labeled Cell-TuC-AB-MEP, Cell-TuC-DVS-MEP, Cell-TuC-DVS-MMI, and Cell-TuC-DVS-MBI. The activation and coupling conditions were optimized for high ligand density. The isotherm adsorption of immunoglobulin of egg yolk (IgY) on four HCIC adsorbents were investigated. High adsorption capacities of IgY could be obtained for all four adsorbents at pH 7, and low adsorption of IgY at pH 4 and of bovine serum albumin (BSA) at pH 7 was observed, which indicates that the HCIC adsorbents prepared have a potential application for antibody purification.

matography of proteins. J Chromatogr A 775 (1–2):65–72[138]

174. Yang Y, Geng X (2011). Mixed-mode chromatography and its applications to biopolymers. J Chromatogr A 1218(49):8813–8825[139]
175. Yon RJ (1972) Chromatography of lipophilic proteins on adsorbents containing mixed hydrophobic and ionic groups. Biochem J 126(3):765–767
176. Yon RJ (1974) Enzyme purification by hydrophobic chromatography: an alternative approach illustrated in the purification of aspartate transcarbamoylase from wheat germ (short communication). Biochem J 137(1):127–130[140]
177. Yon RJ (1981) Versatility of mixed-function adsorbents in biospecific protein desorption: accidental affinity and an improved purification of aspartate transcarbamoylase from wheat germ. Anal Biochem 113(2):219–228[141]
178. Yon RJ, Simmonds RJ (1975). Protein chromatography on adsorbents with hydrophobic and ionic groups. Some properties of N-(3-carboxypropionyl) aminodecyl-sepharose and its interaction with wheat-germ aspartate transcarbamoylase. Biochem J 151(2):281–290[142]

---

[138] Macroporous poly(acrylamide-co-butyl methacrylate-co-N,N′-methylenebisacrylamide) monoliths containing up to 15 % butyl methacrylate units have been prepared by direct polymerization within the confines of HPLC columns. The hydrodynamic and chromatographic properties of these 50 mm × 8 mm I.D. columns – such as back pressure at different flow-rates, effect of percentage of hydrophobic component in the polymerization mixture, effect of salt concentration on the retention of proteins, dynamic loading capacity, and recovery – were determined under conditions typical of hydrophobic interaction chromatography. Using the monolithic column, five proteins were easily separated within only 3 min.

[139] Mixed-mode chromatography is a type of chromatography in which a chromatographic stationary phase interacts with solutes through more than one interaction mode. This technique has been growing rapidly because of its advantages over conventional chromatography, such as its high resolution, high selectivity, high sample loading, high speed, and the ability to replace two conventionally corresponding columns in certain circumstances. In this work, some aspects of the development of mixed-mode chromatography are reviewed, such as stationary phase preparation, combinations of various separation modes, separation mechanisms, typical applications to biopolymers and peptides, and future prospects.

[140] Two adsorbents containing similar numbers of hydrocarbon (C(10)) chains but different numbers of carboxyl groups were made by chemical modification of Sepharose. The use of these adsorbents to purify proteins, under conditions where hydrophobic adsorption is partly resisted by electrostatic repulsion, is illustrated in the purification of aspartate transcarbamoylase (EC 2.1.3.2) from wheat germ.

[141] Under appropriate experimental conditions (usually but not invariably including low ionic strength) wheat germ aspartate transcarbamoylase can be specifically desorbed by the substrate, carbamoyl phosphate, from hydroxyapatite, from N-(3-carboxypropionyl) aminooctyl-Sepharose, from 10-carboxydecylamino-Sepharose, from Cibacron Blue F3GA-Sepharose, and from Coomassie Blue R250-Sepharose. Experimental evidence suggests that (a) the enzyme is adsorbed at heterogeneous sites on each column, only some of which are susceptible to substrate-specific desorption; (b) in none of these cases is the initial adsorption essentially biospecific, i.e., these are not cases of classical affinity chromatography; (c) in the case of 10-carboxydecylamino-Sepharose, and therefore presumably also in the other cases, the desorption is biospecific, i.e., involves the formation of the catalytically significant enzyme-carbamoyl phosphate complex. Substrate-specific desorption in these cases appears to derive from accidental affinity between, on the one hand, clusters of active (ionic, hydrophobic, aromatic, etc.) groups on the protein and, on the other, complementary clusters on the adsorbent, some of these interactions being perturbed when the ligands binds to the protein. Biospecific desorption from 10-carboxydecylamino-Sepharose has been incorporated as the sole chromatographic step in a new, 8000-fold purification of the enzyme. It is suggested that biospecific desorption from essentially nonbiospecific adsorbents could explain some published purifications currently described as "affinity chromatography".

[142] 1. The charge state of two derivatives of Sepharose prepared by the CNBr activation method were studied by acid-base titration and by ion-exchange chromatography. Dodecyl-Sepharose exhibited cationic groups (21mumol/ml of settled gel; pKa = 9.6) that were tentatively assigned to the coupling isourea group. 2. CPAD-Sepharose [N-(3-carboxypropionyl)aminodecyl-Sepharose] has anionic (carboxyl) groups (pKa = 4.5) and cationic groups (pKa = 9.6) in roughly equal concentrations (e coupling group. CPAD-Sepharose is slightly negatively charged at pH 7.0 and substantially negatively charged at pH 8.5. 3. The pKa values of dodecyl-Sepharose and CPAD-Sepharose are unaffected by a 100-fold increase in the concentration of KCl. 4. CPAD-Sepharose has considerable affinity for wheat-germ aspartate transcarbamoylase at pH 8.5 when the adsorbent and enzyme are both negatively charged. The interaction involves the C10 chain but is relatively moderate compared with C10 chains associated only with positive charge. 5. Desorption of the enzyme adsorbed to CPAD-Sepharose can be achieved by raising the pH to increase the electrostatic repulsion, or by introducing the detergent sodium deoxycholate. Acetone and butan-1-ol also weaken the adsorption at pH 8.5. 6. High concentrations of sodium acetate or sodium phosphate induced the enzyme to bind more tightly to CPAD-Sepharose. 7. These results are discussed in terms of a

179. Yoshida T (2004) Peptide separation by hydrophilic-interaction chromatography: a review. J Biochem Biophys Methods 60(3):265–280[143]
180. Zauner G, Deelder AM et al (2011) Recent advances in hydrophilic interaction liquid chromatography (HILIC) for structural glycomics. Electrophoresis 32(24):3456–3466[144]
181. Zhao G, Dong XY et al (2009) Ligands for mixed-mode protein chromatography: principles, characteristics and design. J Biotechnol 144 (1):3–11[145]
182. Zhao G, Peng G et al (2008) 5-Aminoindole, a new ligand for hydrophobic charge induction chromatography. J Chromatogr A 1211(1â€"2):90–98[146]
183. Zhou H, Di Palma S et al (2012). Toward a comprehensive characterization of a human cancer cell phosphoproteome. J Proteome Res 12(1):260–271[147]

'repulsion-controlled' model or hydrophobic chromatography.

[143] Recent developments in the separation of peptides by high-performance liquid chromatography (HPLC) using polar sorbents with less polar eluents are summarized in this review. This separation mode is now commonly referred to as Hydrophilic-Interaction Chromatography (HILIC). The retention mechanism and chromatographic behavior of polar solutes under HILIC conditions are studied on TSKgel Amide-80 columns, which consist of carbamoyl groups bonded to a silica gel matrix, using a mixture of acetonitrile (MeCN)–water containing 0.1 % trifluoroacetic acid (TFA). Some applications are given in peptide field using Hydrophilic-Interaction Chromatography.

[144] This review presents recent progress in employing hydrophilic interaction liquid chromatography (HILIC) for glycan and glycopeptides analysis. After an introduction of this technique, the following themes are addressed: (i) implementation of HILIC in large-scale studies for analyzing the human plasma N-glycome; (ii) the use of HILIC UPLC (ultrahigh pressure liquid chromatography) for fast high-resolution runs and its successful application with online MS for glycan and glycopeptide analysis; (iii) high-throughput profiling using HILIC solid-phase extraction in combination with MS detection; (iv) HILIC sample preparation for CE and CGE; (v) the latest glycoproteomic approaches implementing HILIC separation; (vi) future perspectives of HILIC including its use in large-scale glycoproteomics studies such as the analysis of entire glycoproteomes at the glycopeptide level.

[145] Mixed-mode chromatography is a chromatographic method that utilizes more than one form of interactions between the stationary phase and the solutes in a feed stream. Compared with other types of chromatography, mixed-mode chromatography is advantageous in its salt-independent adsorption, facile elution by charge repulsion, and unique selectivity. Hence, it has already proved beneficial for the separation of proteins as well as other purposes. In this article, mixed-mode ligands for protein purification have been reviewed. These ligands usually have an aliphatic or aromatic group as the hydrophobic moiety and an amino, carboxyl or sulfonic group as the ionic moiety. Heterocyclic groups are good ligand candidates for their unique hydrophobicity and dissociation property. Hydrogen bonding groups also have influences on the performance of mixed-mode adsorbents. These principles should be considered in the screening and design of mixed-mode ligands. Strategies for the design of synthetic affinity ligands, especially the bioinformatics and combinatorial methods, may be adopted for mixed-mode ligand design. More efforts are needed for the development of rational design and screening methods for mixed-mode protein ligands by sophisticated computational and experimental approaches.

[146] Hydrophobic charge induction chromatography (HCIC) is a mixed-mode chromatography that achieves high adsorption capacity by hydrophobic interaction and facile elution by pH-induced charge repulsion between the solute and ligand. This article reports a new medium, 5-aminoindole-modified Sepharose (AI-Sepharose) for HCIC. The adsorption equilibrium and kinetics of lysozyme and bovine serum albumin (BSA) to AI-Sepharose were determined by batch adsorption experiments at different conditions to provide insight into the adsorption properties of the medium. The influence of salt type on protein adsorption to AI-Sepharose corresponded with the trend for other hydrophobicity-related properties in literature. Both ligand density and salt concentration had positive influences on the adsorption of the two proteins investigated. The adsorption capacity of lysozyme, a basic protein, decreased rapidly when pH decreased from 7 to 3 due to the increase of electrostatic repulsion, while BSA, an acidic protein, achieved maximum adsorption capacity around its isoelectric point. Dynamic adsorption experiments showed that the effective pore diffusion coefficient of lysozyme remained constant at different salt concentrations, while that of BSA decreased with increased salt concentration due to its greater steric hindrance in pore diffusion. High protein recovery by adsorption at pH 7.10 elution at pH 3.0 was obtained at a number of NaCl concentrations, indicating that the adsorbent has typical characteristics of HCIC and potentials for applications in protein purification.

[147] Mass spectrometry (MS)-based phosphoproteomics has achieved extraordinary success in qualitative and quantitative analysis of cellular protein phosphorylation. Considering that an estimated level of phosphorylation in a cell is placed at well above 100?000 sites, there is still much room for improvement. Here, we attempt to extend the depth of phosphoproteome coverage while maintaining realistic aspirations in terms of available material, robustness, and instrument running time. We developed three strategies, where each provided a different balance between these three key parameters. The first strategy simply used enrichment by Ti4 + –IMAC followed by reversed chromatography LC-MS (termed

184. Zhou NE, Mant CT et al (1991) Comparison of silica-based cyanopropyl and octyl reversed-phase packings for the separation of peptides and proteins. J Chromatogr 548(1–2):179–193[148]
185. Zhu BY, Mant CT et al (1991). Hydrophilic-interaction chromatography of peptides on hydrophilic and strong cation-exchange columns. J Chromatogr 548(1–2):13–24[149]
186. Zhu BY, Mant CT et al (1992) Mixed-mode hydrophilic and ionic interaction chromatography rivals reversed-phase liquid chromatography for the separation of peptides. J Chromatogr A 594 (1–2):75–86[150]

1D). The second strategy incorporated an additional fractionation step through the use of HILIC (2D). Finally, a third strategy was designed employing first an SCX fractionation, followed by Ti4 + −IMAC enrichment and additional fractionation by HILIC (3D). A preliminary evaluation was performed on the HeLa cell line. Detecting 3700 phosphopeptides in about 2 h, the 1D strategy was found to be the most sensitive but limited in comprehensivity, mainly due to issues with complexity and dynamic range. Overall, the best balance was achieved using the 2D based strategy, identifying close to 17?000 phosphopeptides with less than 1 mg of material in about 48 h. Subsequently, we confirmed the findings with the K562 cell sample. When sufficient material was available, the 3D strategy increased phosphoproteome allowing over 22?000 unique phosphopeptides to be identified. Unfortunately, the 3D strategy required more time and over 1 mg of material before it started to outperform 2D. Ultimately, combining all strategies, we were able to identify over 16?000 and nearly 24?000 unique phosphorylation sites from the cancer cell lines HeLa and K562, respectively. In summary, we demonstrate the need to carry out extensive fractionation for deep mining of the phosphoproteome and provide a guide for appropriate strategies depending on sample amount and/or analysis time.

[148] The performance of a silica-based C8 packing was compared with that of a less hydrophobic, silica-based cyanopropyl (CN) packing during their application to reversed-phase high-performance liquid chromatography (linear trifluoroacetic acid-water to trifluoroacetic acid-acetonitrile gradients) of peptides and proteins. It was found that: (1) the CN column showed excellent selectivity for peptides which varied widely in hydrophobicity and peptide chain length; (2) peptides which could not be resolved easily on the C8 column were widely separated on the CN column; (3) certain mixtures of peptides and small organic molecules which could not be resolved on the C8 column were completely separated on the CN column; (4) impurities arising from solid-phase peptide synthesis were resolved by a wide margin on the CN column, unlike on the C8 column, where these compounds were eluted very close to the peptide product of interest: and (5) specific protein mixtures exhibited superior resolution and peak shape on the CN column compared with the C8 column. The results clearly demonstrate the effectiveness of employing stationary phases of different selectivities (as opposed to the more common optimization protocol of manipulating the mobile phase) for specific peptide and protein applications, an approach underestimated in the past.

[149] Hydrophilic-interaction chromatography (HILIC) was recently introduced as a potentially useful separation mode for the purification of peptides and other polar compounds. The elution order of peptides in HILIC, which separates solutes based on hydrophilic interactions, should be opposite to that obtained in reversed-phase chromatography, which separates solutes based on hydrophobic interactions. Three series of peptides, two of which consisted of positively charged peptides (independent of pH at pH less than 7) and one of which consisted of uncharged or negatively charged peptides (dependent on pH), and which varied in overall hydrophilicity/hydrophobicity, were utilized to examine the separation mechanism and efficiency of HILIC on hydrophilic and strong cation-exchange columns.

[150] Peptide separations based upon mixed-mode hydrophilic and ionic interactions with a strong cation-exchange column have been investigated. The peptide separations were generally achieved by utilizing a linear increasing salt (sodium perchlorate) gradient in the presence of acetonitrile (29–90 %, v/v) at pH 7. The presence of acetonitrile in the mobile phase promotes hydrophilic interactions with the hydrophilic stationary phase, these hydrophilic interactions becoming increasingly important to the separation process as the acetonitrile concentration is increased. At acetonitrile concentrations of 20–50 % (v/v) in the mobile phase, the peptides utilized in this study were eluted in order of increasing net positive charge, indicating that ionic interactions were dominating the separation process. Peptides with the same net positive charge were also well resolved by an hydrophilic interaction mechanism, being eluted in order of increasing hydrophilicity (decreasing hydrophobicity). At higher acetonitrile concentrations (70–90 %, v/v), column selectivity was changed dramatically, with hydrophilic interactions now dominating the separation process. Under these conditions, specific peptides may be eluted earlier or later than less highly charged peptides, depending upon their hydrophilic/hydrophobic character. This mixed-mode methodology was compared to reversed-phase liquid chromatography of the peptides at pH 2 and pH 7. The results of this comparison suggested that mixed-mode hydrophilic-ion-exchange chromatography on a strong cation-exchange column rivals reversed-phase liquid chromatography for peptide separations.

187. Zhu S, Zhang X et al (2012) Developing a strong anion exchange/RP (SAX/RP) 2D LC system for high-abundance proteins depletion in human plasma. Proteomics 12(23–24):3451–3463[151]
188. Zywicki B, Catchpole G et al[152]

[151] Human plasma is dominated by high-abundance proteins which severely impede the detection of low-abundance proteins. Unfortunately, now there is no efficient method for large-scale depletion of high-abundance proteins in human plasma. In this study, we developed a new strategy, strong anion exchange (SAX)/RP 2D LC system, which has potential for large-scale depletion of high-abundance proteins in human plasma. Separation gradients of the system were optimized to ensure an extensive separation of plasma proteins. Plasma was fractionated into 67 fractions by SAX. All these fractions were subjected a thorough separation by the 2D RPLC and 66 peaks with high UV absorption (>20 mAU) at 215 nm were collected. Proteins in these peaks were identified by LC-MS/MS analysis. Results showed that 83 proteins could be identified in these peaks, 68 among them were reported to be high- or middle-abundance proteins in plasma. All these proteins had definite retention times and were mapped in the 2D SAX-RP system, which resulted in accurate depletion of high-abundance proteins with ease. Our studies provide a convenient and effective method for large-scale depletion of high-abundance proteins and in-depth research in human plasma proteomics.

[152] Two rapid methods for highly selective detection and quantification of the two major glycoalkaloids in potatoes, α-chaconine and α-solanine, were compared for robustness in high-throughput operations for over 1000 analytical runs using potato tuber samples from field trials. Glycoalkaloids were analyzed using liquid chromatography coupled to tandem mass spectrometry in multiple reaction monitoring mode. An electrospray interface was used in the detection of glycoalkaloids in positive ion mode. Classical reversed phase (RP) and hydrophilic interaction (HILIC) columns were investigated for chromatographic separation, ruggedness, recovery, precision, and accuracy. During the validation procedure both methods proved to be precise and accurate enough in relation to the high degree of endogenous biological variability found for field-grown potato tubers. However, the RP method was found to be more precise, more accurate, and, more importantly, more rugged than the HILIC method for maintaining the analytes' peak shape symmetry in high-throughput operation. When applied to the comparison of six classically bred potato cultivars to six genetically modified (GM) lines engineered to synthesize health beneficial inulins, the glycoalkaloid content in potato peels of all GM lines was found within the range of the six cultivars. We suggest complementing current unbiased metabolomic strategies by validating quantitative analytical methods for important target analytes such as the toxic glycoalkaloids in potato plants.

Part II

Mass Spectrometry for Proteomics Analysis

Database Search Engines: Paradigms, Challenges and Solutions

6

Kenneth Verheggen, Lennart Martens, Frode S. Berven, Harald Barsnes, and Marc Vaudel

Abstract

The first step in identifying proteins from mass spectrometry based shotgun proteomics data is to infer peptides from tandem mass spectra, a task generally achieved using database search engines. In this chapter, the basic principles of database search engines are introduced with a focus on open source software, and the use of database search engines is demonstrated using the freely available SearchGUI interface. This chapter also discusses how to tackle general issues related to sequence database searching and shows how to minimize their impact.

Keywords

Peptide identification • Search engines • Shotgun proteomics • Sequence database searching

Abbreviations

PSM Peptide Spectrum Match
PTM Post-Translational Modification

6.1 Introduction

The raw output of modern mass spectrometers used in high throughput proteomics does not provide directly interpretable information

K. Verheggen • L. Martens
Department of Medical Protein Research, VIB, Ghent, Belgium

Department of Biochemistry, Faculty of Medicine and Health Sciences, Ghent University, Ghent, Belgium

F.S. Berven
Proteomics Unit, Department of Biomedicine, University of Bergen, Jonas Liesvei 91, N-5009 Bergen, Norway

KG Jebsen Centre for Multiple Sclerosis Research, Department of Clinical Medicine, University of Bergen, Bergen, Norway

Norwegian Multiple Sclerosis Competence Centre, Department of Neurology, Haukeland University Hospital, Bergen, Norway

H. Barsnes (✉) • M. Vaudel
Proteomics Unit, Department of Biomedicine, University of Bergen, Jonas Liesvei 91, N-5009 Bergen, Norway
e-mail: harald.barsnes@biomed.uib.no

H. Mirzaei and M. Carrasco (eds.), *Modern Proteomics – Sample Preparation, Analysis and Practical Applications*, Advances in Experimental Medicine and Biology 919, DOI 10.1007/978-3-319-41448-5_6

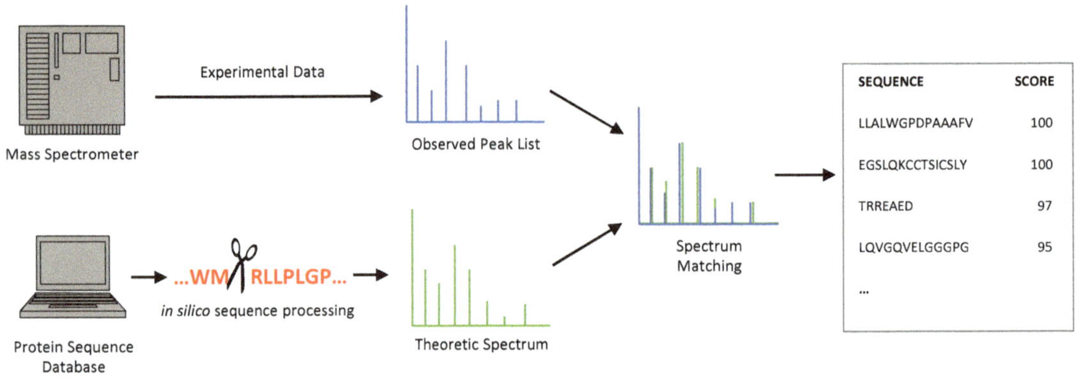

Fig. 6.1 Standard workflow for sequence database searches. The output from the mass spectrometer, consisting of experimental spectra, is compared to the theoretical spectra obtained from the peptides resulting from an *in silico* digestion of the proteins in the search database. The matching of these two types of spectra results in a list of peptide-to-spectrum matches (PSMs), each scored according to how well the peptide matches the spectrum

Table 6.1 A chronological (non-exhaustive) list of available sequence database search algorithms

Algorithm	Published	Free
SEQUEST [3]	1994	
Mascot [65]	1999	
X!Tandem [6]	2004	✓
OMSSA [10]	2004	✓
InSpect [66]	2005	✓
MyriMatch [7]	2007	✓
Crux [67]	2008	✓
MS-GF+ [9]	2010	✓
Tide [12]	2011	✓
MassWiz [68]	2011	✓
Andromeda [69]	2011	✓
Comet [11]	2012	✓
Byonic [70]	2012	
Peaks DB [71]	2012	
Morpheus [72]	2013	✓
MS Amanda [8]	2014	✓

Note that the Published column indicates when the manuscript describing the algorithm or search engine was published, and does not necessarily correspond to when it was made available to the community. The final column indicates if the algorithm is freely available

about the proteins originally present in the analyzed sample. In order to infer the protein composition of a sample, scientists rely on computational and statistical tools to process, control and interpret the data. An essential part of this process is the identification of peptides from tandem mass spectra, a task generally achieved by software referred to as (proteomics) search engines [1, 2].

As shown in Fig. 6.1, search engines compare the experimental spectra to theoretical protein sequences obtained from a protein sequence database. By performing an *in silico* processing of the theoretical sequences, the digestion and fragmentation of the actual experiment is mimicked by the software. Hence, theoretical spectra are generated and can then be compared to the experimentally obtained spectra. A match between an experimental and a theoretical spectrum is called a Peptide-to-Spectrum Match (PSM). The core output of a search engine is composed of a list of such candidate PSMs for every spectrum, with associated scores (typically reported as e-values) that provide an assessment of the quality of the match.

SEQUEST [3] was the first widely used algorithm implementing this technique, and was rapidly adopted in everyday lab practices. As listed in Table 6.1, several algorithms, commercial and academic, were subsequently made available to the community. In order to provide biologically meaningful results, these algorithms were also integrated into broader environments like the Trans-Proteomic Pipeline (TPP) [4] and OpenMS [5], where search engines can be combined with other tools, building complex proteomic workflows for protein inference, quantification and functional analysis.

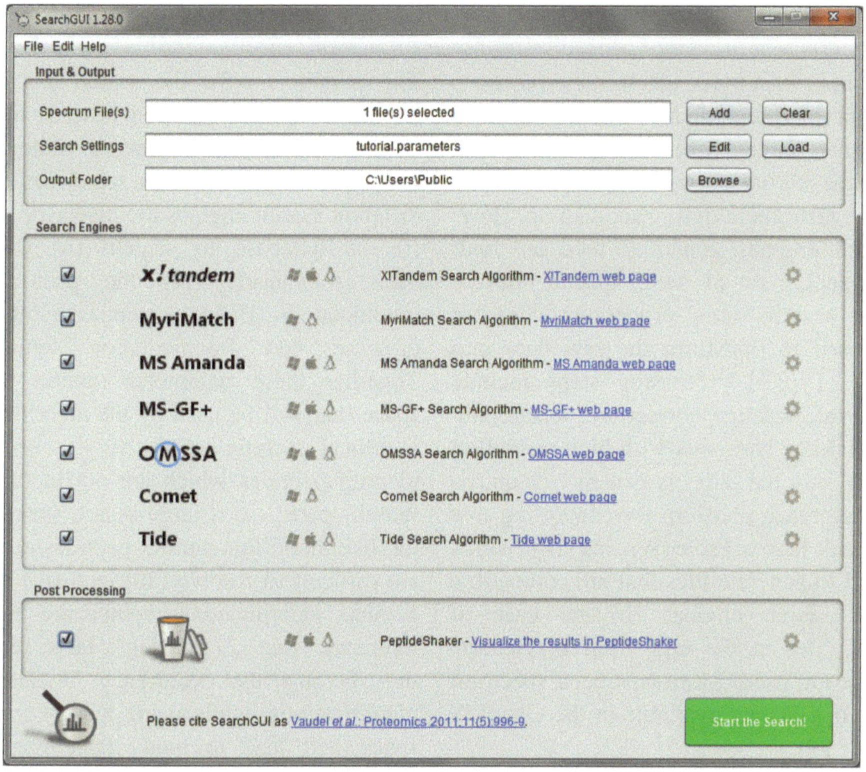

Fig. 6.2 SearchGUI main dialog. At the *top* the spectrum files to process, the general search settings and the output folder can be selected; in the *middle* the search engine(s) to use are chosen; and at the *bottom* optional post-processing to merge and view the results can be set up

In this chapter, we will demonstrate the use of search engines for peptide identification, with a focus on free and open source implementations. As an example, X!Tandem [6], MyriMatch [7], MS Amanda [8], MS-GF+ [9], OMSSA [10], Comet [11] and Tide [12], will be run *via* the SearchGUI [13] interface (http://compomics.github.io/projects/searchgui.html) – a user friendly framework to operate all seven of these command line algorithms. The main steps of the process will be detailed, highlighting potential pitfalls and available solutions. SearchGUI is open source software and does not require any installation except downloading and unzipping. Upon starting the tool, the dialog displayed in Fig. 6.2 appears, allowing the user to set up the desired search parameters and start the search.

The main dialog of SearchGUI consists of three sections: the top section, 'Input & Output' allows the input and output files, and the shared search settings used across all algorithms to be specified (more details below); in the 'Search Engines' section, the user can select which search engine(s) to use and to set search engine specific parameters (by clicking on the cogwheel next to the search engine in question); finally, the'Post Processing' section can be used to automatically run PeptideShaker [14] after the identification step. PeptideShaker will then import and merge the output from the different search engines, which generally gives better results than using only a single search engine [15]. Note however, that the post processing applied by a tool such as PeptideShaker is beyond the scope of this chapter and will not be detailed here. For further information on this topic, the interested reader is instead recommended to consult the extensive tutorial material on peptide and protein identification [16].

6.2 Spectrum Input

The raw output of a mass spectrometer consists of all spectra (MS1 and MS2) in a vendor specific binary format, along with various other data related to the chromatography and operational status of the instrument during acquisition. However, search engines generally take as input processed peak lists of MS2 spectra. Before starting the search, signal processing steps are therefore used to transform the raw data into peaks lists. Typical processing steps include noise removal, baseline correction, deisotoping and peak picking. Note that with high resolution instruments only the latter is generally required [17]. The reference platform for converting raw data into peak lists is ProteoWizard [18], which can be used to generate files that are compatible with most search engines. In the case of SearchGUI, files in the mgf (Mascot Generic File) format are used. More advanced spectrum processing options are available in the OpenMS platform [5, 19].

6.3 Search Settings

The parameters for the search can be set by clicking the 'Edit' button next to the 'Search Settings' field, which opens the dialog shown in Fig. 6.3. Here the common search settings for the different search engines are displayed, including: (i) the database to search; (ii) the allowed mass tolerances; (iii) the post-translational modifications (PTMs) to consider; and (iv) the protease and fragmentation settings used. Together these parameters define the search space that will be used by the algorithms, which is critical in three aspects: (i) it is impossible to identify peptides which are not included in the search space; (ii) a large search space increases the likelihood that similar peptides occur, which are difficult to resolve [20]; and (iii) ambiguous peptide identifications complicate the protein inference issue [21]. Using a large search space thus favours the occurrence of false positive identifications, while using a very small search space will lead to many false negatives and

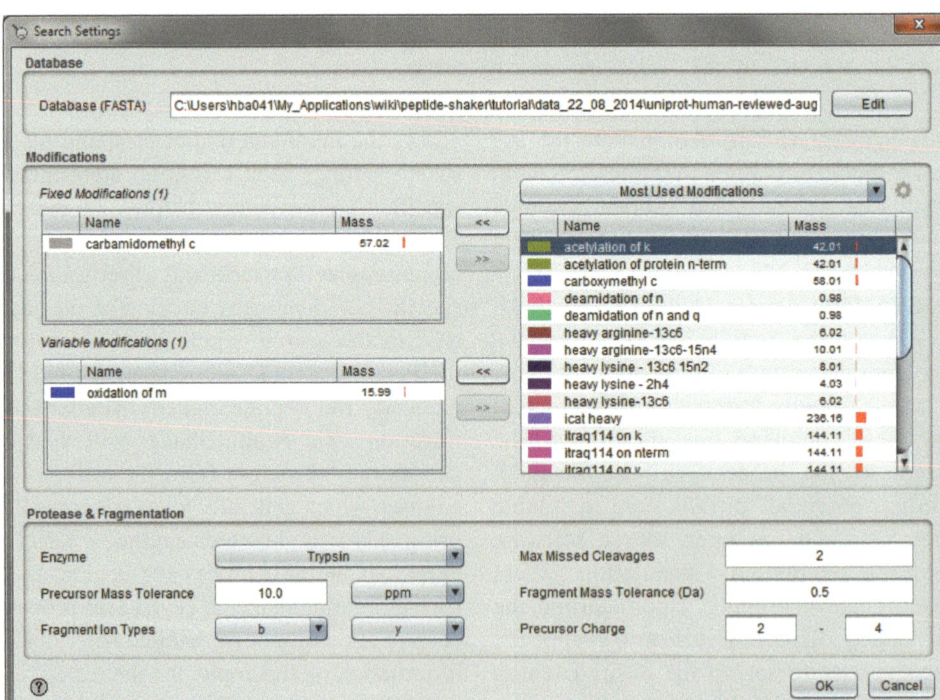

Fig. 6.3 SearchGUI search settings dialog. At the *top* the protein sequence database is selected; in the *middle* the modifications assumed to be in the sample are chosen (note the option to add modifications as either fixed or variable); and at the *bottom* the protease and fragmentation settings are inserted

unreliable scores for the reported scores. Finding the correct search settings is thus critical when using search engines.

6.4 Protein Databases

In order to match a spectrum to its theoretical counterpart, a protein sequence database is required. The database is in essence a list of all protein sequences that could presumably be found in the sample. Protein sequences can be obtained from online protein databases, generally in the text-based FASTA format. While specialized databases exist for specific species or pathologies (e.g. The Arabidopsis Information Resource [22], TAIR, for *Arabidopsis thaliana*, or TBDB [23] for Tuberculosis), generic resources for protein sequences such as the Universal Protein knowledgebase UniProtKB [24] (http://uniprot.org) and Ensembl [25] (http://ww.ensembl.org) have been established as well.

UniProt provides annotated sets of protein sequences, deduced from sequenced genomes for a large number of species, and consists of two main collections of sequences: Swiss-Prot and TrEMBL. Swiss-Prot contains manually annotated and reviewed protein sequences, while TrEMBL is automatically annotated and not yet manually curated. It is usually recommended to search against Swiss-Prot, as this ensures that the identifications are based on high quality protein information. If UniProt should contain no sequences for the organism under study, Ensembl can provide a useful alternative. In essence a nucleic acid sequence database that includes recently sequenced organisms, Ensembl also provides translated sequences in the form of protein databases.

In order to limit the search space, it is advised to tailor the set of sequences searched to those that are expected in the sample. This is achieved by restricting the search to the species of interest. Species specific sequence sets can be obtained from the UniProt website by selecting a specific taxonomy. However, it should be noted that this approach quickly becomes very complicated when working with poorly defined samples such as encountered in metaproteomics [26]. It is also important to note that protein databases are in constant development, and it is therefore crucial to clearly document the version of the database used for a given project (typically accompanied by the total number of sequences in that database), and to only compare results obtained with the same version of a given database.

Model organisms are well covered by UniProt. This is however not always the case for less characterized or strongly mutated organisms, where missing proteins can potentially be problematic. In such cases a related species with similar sequences is generally used. It is also worth mentioning that spectra from missing peptides are prone to generate false positive identifications [27], this is notably the case for contaminants which should be included in the list of searched proteins. This is especially important when searching non-human data, as minute amounts of human keratin, from hair or skin, often end up in the samples. If these are not filtered out as contaminants, the search engines may very well mistake them as evidence for proteins not actually in the sample [28]. A list of common contaminants can be found at the Global Proteome Machine [29] (GPM) website (http://www.thegpm.org/crap).

Although databases containing protein isoforms hold the promise for higher identification rates, the number of peptides identified is generally stable if not diminished, while the complexity of the subsequent protein inference step is dramatically increased [30]. Thus, using databases with high protein ambiguity results in increased number of proteins based on the same peptide sequences [31]. This ambiguity is particularly problematic in quantitative and functional analyses [32]. Thus, the option to include isoforms in the sequence database should be considered carefully, and the data resulting from such searches should be interpreted with due caution.

6.5 Post-Translational Modifications (PTMs)

Post-translational modifications (PTMs) can be categorized according to whether they occur *in vivo* or *in vitro*. Modifications in the first category are part of cellular mechanisms, for example, phosphorylation as a mechanism to activate proteins. Such natural modifications play an important part as control mechanisms for cellular regulation [33–35]. However, this category of modifications are often present in sub-stoichiometric amounts and are therefore unlikely to be found without prior enrichment [36]. Thus, the choice to include *in vivo* modifications in a search depends on whether the experiment actually targets these.

In vitro modifications are linked to intentional or unintentional modification due to sample handling and preparation, where the most widely encountered modifications are oxidation of methionine, an unintentional modification occurring due to the sample coming in contact with air, and the intentional protection of cysteine residues by alkylation after reduction of disulfide bonds. In the latter case, the modification is the result of a high yield chemical reaction that ensures that nearly all relevant sites will be modified. Another example where *in vitro* modifications are expected to occur in close to 100 % of the cases, is in label based quantification strategies such as SILAC, iTRAQ and TMT, where the incorporation or labelling efficiency is moreover typically verified (for a detailed example see [37]).

From the above, it is clear that PTMs are encountered in two different forms: (i) if a modification is expected to occur at (almost) all possible modification sites, it is referred to as a fixed (or static) modification, while (ii) a modification that is more unpredictable is called a variable (or dynamic) modification. It is important to note that fixed and variable modifications have a very different impact on the search space.

PTMs are identified by search engines *via* the mass shift they induce in the amino acid sequence at specific positions, e.g., at particular amino acids, or at the peptide or protein termini. For example, a peptide containing an oxidized methionine carries an extra oxygen atom. This means that all peptides containing an oxidized methionine will have their intact mass increased by approximately 16 Da. Moreover, each fragment ion that contains the modified methionine will also be affected in the same way. The search engine thus has to look for both versions of these peptides: the unmodified as well as the modified form. Given that this split into two distinct peptide forms has to be done for every methionine residue in a peptide, and given that peptides can contain more than one methionine, it should be clear that adding variable modifications has a dramatic impact on the size of the search space. And as already mentioned, increasing the search space also increases the likelihood of false positives. It is therefore generally good practice to evaluate the abundance of PTMs before including them in the search settings.

Fixed modifications on the other hand, do not impact the size of the search space, as a search engine can simply consider all potential modification sites as modified, effectively replacing the masses of the affected residues by their modified masses and eliminating the need to consider multiple alternatives for each peptide.

In SearchGUI, fixed and variable modifications are selected from a predefined list. But the list can be extended using the drop down menu above the table, and new modifications can be added by clicking on the cogwheel. Modifications are saved in a search engine independent structure [38] and can be reused in future searches.

6.6 Protease and Fragmentation

When digesting a sample using a specific protease, the peptides obtained abide to the enzyme cleavage rules. The leading protease in proteomics is trypsin [39]. Trypsin is commonly found in the digestive system of many vertebrates, and cleaves peptide chains at the carboxyl-terminal side of the amino acids lysine (K) and arginine (R), except when followed by proline (P). Due to this cleavage specificity, the amount of possible peptides is limited. Restricting the considered peptides to those fitting the tryptic cleavage rules dramatically reduces the search space, hence improving the search speed and reducing the number of false positives. However, note that using the cleavage rules as a filter sets a strong dependence on the quality of the digestion [40]. This factor is generally relaxed by allowing a given number of missed cleavages, thus accounting for the presence of sites which are not accessible to the enzyme [39]. Prediction tools exist that evaluate each potential cleavage site for missed cleavage [41–43] and are compared in [44], but these approaches have so far not been included in search engines.

It is also important to tailor the search space to the resolution of the mass spectrometer, both at the MS1 and MS2 levels; either in ppm (parts per million, tolerance relative to the precursor m/z) or in Dalton (absolute tolerance). The tolerances depend on the resolution of the measurements and can be optimized for a given setup [45]. Again, it should be mentioned that while relaxing the accuracy requirements, i.e., increasing the tolerances, may result in more peptides identified, it will in most cases also increase the number of false positives. Notably, for low resolution mass spectrometers, it is common practice to search with a wide tolerance and filter out the PSMs *a posteriori*, a method which can substantially increase the identification rate [46].

The charge states and ion types considered by the search engine should be adapted to the ionization technique and fragmentation type used, and can be done by setting the expected type of fragment ions and precursor charges. Fragment ion types generally consist of one forward ion (a, b or c; all containing the original amino-terminus of the peptide) and one rewind ion (x, y or z; all containing the original carboxyl-terminus), according to the nomenclature by Roepstorff and Fohlman [47]. Collision Induced Dissociation (CID) and Higher-energy Collisional Dissociation (HCD) generate mainly b and y ions, while Electron Capture Dissociation (ECD) and Electron Transfer Dissociation (ETD) yield mainly c and z ions. Setting the allowed precursor charge(s) depends on the ionization method used, with the defaults being +1 for Matrix Assisted Laser Desorption (MALDI) and +2, +3 and +4 for Electrospray Ionisation (ESI). Note that the charges encountered can differ based on the peptides present in solution and notably depending on the protease used for digestion, specific chemical modification [48], or by the mass spectrometer being tuned to target specific charges.

6.7 Search Engine Specific Settings

The search settings detailed above are common for all search engines. However, most search engines also have their own specific settings, allowing the user to customize the inner workings of the search engine algorithm. SearchGUI provides access to these search engine specific settings by clicking on the cogwheels located to the right of each search engine in the main dialog (see Fig. 6.2).

These advanced settings will not be described in detail here, and it is advised to refer to the documentation of the original algorithm before making any changes to the default values. Relevant options to inspect include the quick PTM searches options of X!Tandem and the related refinement procedure section. While the latter can increase the identification coverage by relaxing the search parameters in a so-called second pass search, it is known to bias the estimation of error rates [49]. It is also advised to verify the fragmentation method selected for MyriMatch, MS Amanda, and MS-GF+; and for the latter verify the selected detector and protocol. Also note that MS-GF+ does not take into account the provided MS2 tolerance, as it will optimize this setting internally [9].

6.8 Conclusion and Perspectives

One of the main challenges in peptide identification from mass spectrometry based shotgun proteomics data is the presence of false positive identifications. Their presence is controlled *a posteriori* in post processing software through the estimation of a False Discovery Rate (FDR), as reviewed in detail by Nesvizhskii [50]. The technique was pioneered by the PeptideProphet tool [51] using score distribution modelling. Subsequently, the target/decoy approach [52], relying on the inclusion of artificial, nonsensical sequences among the searched proteins was rapidly adopted in the field, providing more accurate error rate estimates [53]. These so-called decoy sequences can easily be generated and appended to the original protein sequences in SearchGUI when selecting the FASTA file in the search settings dialog.

Searching large datasets, or using a large search space containing, for example, different species or accounting for multiple modifications, quickly becomes impractical on standard desktop computers. Solutions have therefore been developed to speed up this process, including the use of distributed computing [54], graphical processing units (GPUs) [55], and the increasingly popular cloud computing [56–58]. Furthermore, by exploiting user friendly platforms for biological data processing such as Galaxy [59–61], powerful data analysis solutions are made available to every interested scientist.

Despite all this progress, database searching may not always be the method of choice for identifying peptides. For example, if no sequence database is available for the species under analysis, or if the search space cannot be reduced to a given species or a set of cleavage rules, search engines will not be of much use. In such cases, related approaches such as spectrum library searching [62] or *de novo* sequencing might be better alternatives [63]. Notably, the latter allows for mutation tolerant identification of proteins [64] and screening for unexpected modifications.

Acknowledgments K.V. acknowledges the support of Ghent University. L.M. acknowledges the support of Ghent University (Multidisciplinary Research Partnership "Bioinformatics: from nucleotides to networks") and the IWT SBO grant 'InSPECtor' (120025). H.B. is supported by the Research Council of Norway.

References

1. Mueller LN, Brusniak MY, Mani DR et al (2008) An assessment of software solutions for the analysis of mass spectrometry based quantitative proteomics data. J Proteome Res 7:51–61
2. Vaudel M, Sickmann A, Martens L (2010) Peptide and protein quantification: a map of the minefield. Proteomics 10:650–670
3. Eng J, McCormack AL, Yates JR III (1994) An approach to correlate tandem mass spectral data of peptides with amino acid sequences in a protein database. J Am Soc Mass Spectrom 5:976–989
4. Deutsch EW, Mendoza L, Shteynberg D et al (2010) A guided tour of the trans-proteomic pipeline. Proteomics 10:1150–1159
5. Sturm M, Bertsch A, Gropl C et al (2008) OpenMS – an open-source software framework for mass spectrometry. BMC Bioinf 9:163
6. Craig R, Beavis RC (2004) TANDEM: matching proteins with tandem mass spectra. Bioinformatics 20:1466–1467
7. Tabb DL, Fernando CG, Chambers MC (2007) MyriMatch: highly accurate tandem mass spectral peptide identification by multivariate hypergeometric analysis. J Proteome Res 6:654–661
8. Dorfer V, Pichler P, Stranzl T et al (2014) MS Amanda, a universal identification algorithm optimized for high accuracy tandem mass spectra. J Proteome Res 13:3679–3684
9. Kim S, Mischerikow N, Bandeira N et al (2010) The generating function of CID, ETD, and CID/ETD pairs of tandem mass spectra: applications to database search. Mol Cell Proteomics 9:2840–2852
10. Geer LY, Markey SP, Kowalak JA et al (2004) Open mass spectrometry search algorithm. J Proteome Res 3:958–964
11. Eng JK, Jahan TA, Hoopmann MR (2013) Comet: an open-source MS/MS sequence database search tool. Proteomics 13:22–24
12. Diament BJ, Noble WS (2011) Faster SEQUEST searching for peptide identification from tandem mass spectra. J Proteome Res 10:3871–3879
13. Vaudel M, Barsnes H, Berven FS et al (2011) SearchGUI: an open-source graphical user interface for simultaneous OMSSA and X!Tandem searches. Proteomics 11:996–999
14. Vaudel M, Burkhart JM, Zahedi RP et al (2015) PeptideShaker enables reanalysis of MS-derived proteomics data sets. Nat Biotechnol 33:22–24
15. Shteynberg D, Nesvizhskii AI, Moritz RL et al (2013) Combining results of multiple search engines in proteomics. Mol Cell Proteomics 12:2383–2393
16. Vaudel M, Venne AS, Berven FS et al (2014) Shedding light on black boxes in protein identification. Proteomics 14:1001–1005
17. Mancuso F, Bunkenborg J, Wierer M et al (2012) Data extraction from proteomics raw data: an evaluation of

nine tandem MS tools using a large Orbitrap data set. J Proteome 75:5293–5303
18. Kessner D, Chambers M, Burke R et al (2008) ProteoWizard: open source software for rapid proteomics tools development. Bioinformatics 24:2534–2536
19. Kohlbacher O, Reinert K, Gropl C et al (2007) TOPP – the OpenMS proteomics pipeline. Bioinformatics 23: e191–e197
20. Colaert N, Degroeve S, Helsens K et al (2011) Analysis of the resolution limitations of peptide identification algorithms. J Proteome Res 10:5555–5561
21. Nesvizhskii AI, Aebersold R (2005) Interpretation of shotgun proteomic data: the protein inference problem. Mol Cell Proteomics 4:1419–1440
22. Huala E, Dickerman AW, Garcia-Hernandez M et al (2001) The Arabidopsis Information Resource (TAIR): a comprehensive database and web-based information retrieval, analysis, and visualization system for a model plant. Nucleic Acids Res 29:102–105
23. Reddy TB, Riley R, Wymore F et al (2009) TB database: an integrated platform for tuberculosis research. Nucleic Acids Res 37:D499–D508
24. Apweiler R, Bairoch A, Wu CH et al (2004) UniProt: the Universal Protein knowledgebase. Nucleic Acids Res 32:D115–D119
25. Flicek P, Amode MR, Barrell D et al (2014) Ensembl 2014. Nucleic Acids Res 42:D749–D755
26. Muth T, Benndorf D, Reichl U et al (2013) Searching for a needle in a stack of needles: challenges in metaproteomics data analysis. Mol BioSyst 9:578–585
27. Knudsen GM, Chalkley RJ (2011) The effect of using an inappropriate protein database for proteomic data analysis. PLoS One 6:e20873
28. Ghesquiere B, Helsens K, Vandekerckhove J et al (2011) A stringent approach to improve the quality of nitrotyrosine peptide identifications. Proteomics 11:1094–1098
29. Craig R, Cortens JP, Beavis RC (2004) Open source system for analyzing, validating, and storing protein identification data. J Proteome Res 3:1234–1242
30. Martens L, Hermjakob H (2007) Proteomics data validation: why all must provide data. Mol Biosyst 3:518–522
31. Barsnes H, Martens L (2013) Crowdsourcing in proteomics: public resources lead to better experiments. Amino Acids 44:1129–1137
32. Vaudel M, Sickmann A, Martens L (2014) Introduction to opportunities and pitfalls in functional mass spectrometry based proteomics. Biochim Biophys Acta 1844:12–20
33. Venne AS, Kollipara L, Zahedi RP (2014) The next level of complexity: crosstalk of posttranslational modifications. Proteomics 14:513–524
34. Olsen JV, Mann M (2013) Status of large-scale analysis of post-translational modifications by mass spectrometry. Mol Cell Proteomics 12:3444–3452
35. Pawson T, Scott JD (2005) Protein phosphorylation in signaling – 50 years and counting. Trends Biochem Sci 30:286–290
36. Loroch S, Dickhut C, Zahedi RP et al (2013) Phosphoproteomics – more than meets the eye. Electrophoresis 34:1483–1492
37. Aasebo E, Vaudel M, Mjaavatten O et al (2014) Performance of super-SILAC based quantitative proteomics for comparison of different acute myeloid leukemia (AML) cell lines. Proteomics 14:1971–1976
38. Barsnes H, Vaudel M, Colaert N et al (2011) Compomics-utilities: an open-source Java library for computational proteomics. BMC Bioinf 12:70
39. Vandermarliere E, Mueller M, Martens L (2013) Getting intimate with trypsin, the leading protease in proteomics. Mass Spectrom Rev 32:453–465
40. Burkhart JM, Schumbrutzki C, Wortelkamp S et al (2012) Systematic and quantitative comparison of digest efficiency and specificity reveals the impact of trypsin quality on MS-based proteomics. J Proteome 75:1454–1462
41. Siepen JA, Keevil EJ, Knight D et al (2007) Prediction of missed cleavage sites in tryptic peptides aids protein identification in proteomics. J Proteome Res 6:399–408
42. Lawless C, Hubbard SJ (2012) Prediction of missed proteolytic cleavages for the selection of surrogate peptides for quantitative proteomics. OMICS 16:449–456
43. Fannes T, Vandermarliere E, Schietgat L et al (2013) Predicting tryptic cleavage from proteomics data using decision tree ensembles. J Proteome Res 12:2253–2259
44. Kelchtermans P, Bittremieux W, De Grave K et al (2014) Machine learning applications in proteomics research: how the past can boost the future. Proteomics 14:353–366
45. Vaudel M, Burkhart JM, Sickmann A et al (2011) Peptide identification quality control. Proteomics 11:2105–2114
46. Beausoleil SA, Villen J, Gerber SA et al (2006) A probability-based approach for high-throughput protein phosphorylation analysis and site localization. Nat Biotechnol 24:1285–1292
47. Roepstorff P, Fohlman J (1984) Proposal for a common nomenclature for sequence ions in mass spectra of peptides. Biomed Mass Spectrom 11:601
48. Thingholm TE, Palmisano G, Kjeldsen F et al (2010) Undesirable charge-enhancement of isobaric tagged phosphopeptides leads to reduced identification efficiency. J Proteome Res 9:4045–4052
49. Everett LJ, Bierl C, Master SR (2010) Unbiased statistical analysis for multi-stage proteomic search strategies. J Proteome Res 9:700–707
50. Nesvizhskii AI (2010) A survey of computational methods and error rate estimation procedures for peptide and protein identification in shotgun proteomics. J Proteome 73:2092–2123

51. Keller A, Nesvizhskii AI, Kolker E et al (2002) Empirical statistical model to estimate the accuracy of peptide identifications made by MS/MS and database search. Anal Chem 74:5383–5392
52. Elias JE, Gygi SP (2010) Target-decoy search strategy for mass spectrometry-based proteomics. Methods Mol Biol 604:55–71
53. Ma K, Vitek O, Nesvizhskii AI (2012) A statistical model-building perspective to identification of MS/MS spectra with PeptideProphet. BMC Bioinf 13(Suppl 16):S1
54. Verheggen K, Barsnes H, Martens L (2014) Distributed computing and data storage in proteomics: many hands make light work, and a stronger memory. Proteomics 14:367–377
55. Baumgardner LA, Shanmugam AK, Lam H et al (2011) Fast parallel tandem mass spectral library searching using GPU hardware acceleration. J Proteome Res 10:2882–2888
56. Trudgian DC, Mirzaei H (2012) Cloud CPFP: a shotgun proteomics data analysis pipeline using cloud and high performance computing. J Proteome Res 11:6282–6290
57. Muth T, Peters J, Blackburn J et al (2013) ProteoCloud: a full-featured open source proteomics cloud computing pipeline. J Proteome 88:104–108
58. Afgan E, Chapman B, Taylor J (2012) CloudMan as a platform for tool, data, and analysis distribution. BMC Bioinf 13:315
59. Giardine B, Riemer C, Hardison RC et al (2005) Galaxy: a platform for interactive large-scale genome analysis. Genome Res 15:1451–1455
60. Boekel J, Chilton JM, Cooke IR et al (2015) Multiomic data analysis using Galaxy. Nat Biotechnol 33:137–139
61. Goecks J, Nekrutenko A, Taylor J (2010) Galaxy: a comprehensive approach for supporting accessible, reproducible, and transparent computational research in the life sciences. Genome Biol 11:R86
62. Lam H (2011) Building and searching tandem mass spectral libraries for peptide identification. Mol Cell Proteomics 10(R111):008565
63. Allmer J (2011) Algorithms for the de novo sequencing of peptides from tandem mass spectra. Expert Rev Proteomics 8:645–657
64. Dasari S, Chambers MC, Slebos RJ et al (2010) TagRecon: high-throughput mutation identification through sequence tagging. J Proteome Res 9:1716–1726
65. Perkins DN, Pappin DJ, Creasy DM et al (1999) Probability-based protein identification by searching sequence databases using mass spectrometry data. Electrophoresis 20:3551–3567
66. Tanner S, Shu H, Frank A et al (2005) InsPecT: identification of posttranslationally modified peptides from tandem mass spectra. Anal Chem 77:4626–4639
67. Park CY, Klammer AA, Kall L et al (2008) Rapid and accurate peptide identification from tandem mass spectra. J Proteome Res 7:3022–3027
68. Yadav AK, Kumar D, Dash D (2011) MassWiz: a novel scoring algorithm with target-decoy based analysis pipeline for tandem mass spectrometry. J Proteome Res 10:2154–2160
69. Cox J, Neuhauser N, Michalski A et al (2011) Andromeda: a peptide search engine integrated into the MaxQuant environment. J Proteome Res 10:1794–1805
70. Bern M, Kil YJ, Becker C (2012) Byonic: advanced peptide and protein identification software. Curr Protoc Bioinf Chapter 13, Unit13 20
71. Zhang J, Xin L, Shan B et al (2012) PEAKS DB: de novo sequencing assisted database search for sensitive and accurate peptide identification. Mol Cell Proteomics 11:M111 010587
72. Wenger CD, Coon JJ (2013) A proteomics search algorithm specifically designed for high-resolution tandem mass spectra. J Proteome Res 12:1377–1386

7 Mass Analyzers and Mass Spectrometers

Anthony M. Haag

Abstract

Mass spectrometers are comprised of three main components: an ion source, a mass analyzer, and a detector. Ionization of the analyte occurs in the ion source and the resulting ions are counted at the detector. However, it is the mass analyzer that is responsible for determing the mass-to-charge ratio (*m*/*z*) of the ions (Jennings KR, Dolnikowski GG, Method Enzymol 193:37–61, 1990). Therefore, it is primarily the analyzer that allows the mass spectrometer to serve its primary goal – determining the mass of the analytes being measured. This becomes important in the field of molecular biology, where biomolecules may be of low molecular weight or often take on multiple charges (z) after ionization (Fenn JB, Mann M, Meng CK, Wong SF, Whitehouse CM, Science 246:64–71, 1989). For this reason, the choice of analyzer is dependant on the properties of the analyte after ionization and the requirements of the experiment being performed.

Keywords

Mass spectrometer • Mass analyzer • Quadrupole mass analyzer • Ion trap mass analyzer • Time-of-flight (TOF) mass analyzer • FT-ICR mass analyzer • Orbitrap mass analyzer • Tandem mass analyzer • Triple quadrupoles tandem mass analyzer • Q-TOF tandem mass analyzer • TOF/TOF tandem mass analyzer • Product ion scan • Precursor ion scan • Neutral loss scan • Selected reaction monitoring

7.1 Introduction

Mass spectrometers are comprised of three main components: an ion source, a mass analyzer, and a detector. Ionization of the analyte occurs in the ion source. The mass analyzer then resolves ions

A.M. Haag (✉)
Department of Pathology and Immunology, Baylor College of Medicine, Houston, TX 77030, USA

Texas Children's Microbiome Center, Texas Children's Hospital, Houston, TX 77030, USA
e-mail: Anthony.Haag@bcm.edu

H. Mirzaei and M. Carrasco (eds.), *Modern Proteomics – Sample Preparation, Analysis and Practical Applications*, Advances in Experimental Medicine and Biology 919, DOI 10.1007/978-3-319-41448-5_7

based on their mass-to-charge ratio (*m*/*z*) [1]. Ions most often impact a detector to produce a signal that is recorded. A mass spectrum is a plot of the relative abundance of ions against their *m*/*z*. It is primarily the analyzer that allows the mass spectrometer to serve its primary goal – determining the mass of the analyte being measured. Because analyzers only measure the *m*/*z* of ions, some amount of mass spectral interpretation is often required by the mass spectrometrist. This becomes important in the field of molecular biology, where biomolecules often take on multiple charges (*z*) after ionization [2]. For this reason, the *m*/*z* of a compound will often be a fraction of the actual mass (*m*) of the ion.

There are many different analyzer designs available. Along with their ability to resolve ions of different *m*/*z*, several analyzers are also capable of trapping and storing ions. Thus, such analyzers can function in a multitude of roles. The most common types of analyzers in commercial production include quadrupole, ion trap, time-of-flight (TOF), and Fourier transform analyzers (ion cyclotron [ICR] and Orbitrap), along with numerous combinations or hybrids of these analyzers. The choice of mass analyzer depends on a number of factors and experimental considerations. Such factors may include but are not limited to

1. The desired *m*/*z* range to be analyzed
2. The mass of the analyte
3. The required resolving power of the analyzer
4. The ability of the analyzer to interface with the ion source of the mass spectrometer
5. The limit of detection required

Because there is no single mass analyzer that is suitable for all applications, most laboratories will employ different mass spectrometers that utilize different analyzers. The most commonly utilized mass analyzers are discussed below.

7.2 Quadrupole

The quadrupole mass analyzer continues to be one of the most popular types of mass analyzer in use. Quadrupole mass analyzers are often employed in benchtop mass spectrometers due to their low cost, compact design, durability and reliability. For these reasons they have become the workhorse analyzer in the pharmaceutical industry. They are often used in tandem with each other such as in triple quadrupole mass spectrometers or with other mass analyzers such as time-of flight (TOF) [3].

A quadrupole analyzer is essentially a mass filter, due to its ability to discriminate and filter ions of different *m*/*z* [4]. Quadrupoles consist of four cylindrical or hyperbolic rods in parallel with each other (Fig. 7.1). Rods opposite each other are electrically connected together and a radio frequency (RF) potential is applied. A direct current (DC) potential is then superimposed over of the RF potential. The combination of RF and DC potential causes ions to oscillate as they pass through the quadrupole in the z-direction. Depending on the DC potential and frequency of the RF field, only ions of a particular *m*/*z* will have stable trajectories. Those ions that have unstable trajectories will collide into the rods and be filtered out. By varying the DC and RF potentials, ions of different *m*/*z* can be scanned or "filtered" through the quadrupoles [5].

A quadrupole or other multipole (hexapole or octupole) can also operate in an "RF-only" mode, in which the DC potential is reduced and only an RF potential is applied to the rods. This allows all ions to pass through the multipole, thereby transforming the quadrupole analyzer into a device for transmitting ions from one

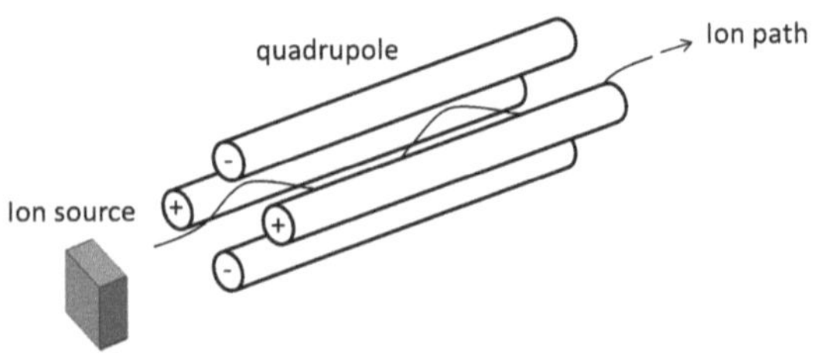

Fig. 7.1 A quadrupole mass analyzer consists of four metallic rods connected to both an RF and DC field. Ions entering the quadrupole will oscillate as they pass through the field between the quadrupole rods. Ions with stable oscillation trajectories will pass through while those that are unstable will collide with the rods

area of the mass spectrometer to another, such as moving ions from the ionization source into another analyzer. Thus, RF-only multipoles can act as ion transmission guides within a mass spectrometer where needed. RF-only multipoles can also act as collision cells for performing collision-induced dissociation (CID) [6]. When an inert gas is introduced into the collision cell and the RF-energy on the multipoles is increased, ions that are transmitted through the collision cell will undergo fragmentation via CID. By varying the RF-energy, the amount of ion fragmentation can be controlled.

As mentioned earlier, a major advantage of quadrupole mass spectrometers is their low cost and compact shape and size which makes them ideal for most laboratories. They are made by a variety of different manufacturers and have proven to be rugged and reliable for long periods of time, thus require little maintenance. They have excellent stability over long periods of time, thereby reducing the need for repeated calibrations. Because quadrupole analyzers have fast duty cycles and the need for a continuous flux of ions, they easily interface to both gas chromatography (GC) and liquid chromatography (LC) equipment [7, 8]. However, this makes quadrupole analyzers less suitable for pulsed ion sources such as matrix assisted laser desorption/ionization (MALDI). Also, quadrupole analyzers suffer from both limited mass ranges and poor resolution. This puts them at a disadvantage when analyzing large molecular weight compounds that may not form multiply charged ions or complex mixtures of compounds with similar masses.

7.3 Ion Trap

The ion trap mass analyzer is a modification of the quadrupole mass analyzer [9]. The 3D ion trap, also known as a Paul Trap, was the most common ion trap until the twenty-first century [10, 11]. Recently, the 2D linear ion trap has become more popular because of its numerous advantages over 3D traps in most commercially available equipment. The 3D traps consist of two hyperbolic electrode plates facing each other and a hyperbolic ring electrode placed in between them (Fig. 7.2). Using an oscillating RF field and a superimposed DC electric field, similar to that in quadrupoles, ions are trapped between the electrodes. In order to act as an analyzer, ions of different *m*/*z* are selectively ejected from the trap by varying the RF potential. The ejected ions are then registered at the detector. 2D traps, often referred to as linear traps, are equivalent to quadrupoles but a potential field is applied to each end of the quadrupole in order to trap the ions within the quadrupole itself. Ions can be selectively ejected either axially or radially depending on the design of the 2D trap.

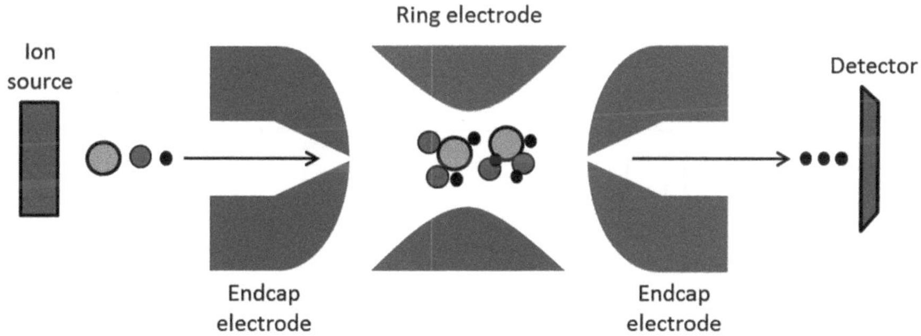

Fig. 7.2 In a 3D trap, ions enter a small opening in the endcap of one of the electrodes. An RF field is placed on the ring electrode, trapping the ions toward the center of the ion trap. The stability of ions within the trap is based on the RF frequency, the *m*/*z* of the ions, and the amplitude of the RF field. Ions may be selectively ejected at the opposite endcap electrode from which they entered by increasing the voltage of the RF field on the ring electrode

Because ion traps have the ability to accumulate ions over time, mass spectrometers that utilize them are known for their improved sensitivity. Much like their quadrupole counterpart, ion trap analyzers have the advantage of having a small and compact size, making them very affordable in most mass spectrometers. For this reason, they have played a major role in expanding the field of proteomics. Much of the early developments in identification of proteins in a complex mixture were performed on mass spectrometers utilizing ion trap analyzers [12, 13].

One of the biggest disadvantages to ion trap analyzers is their low resolving power. Even at slow scan speeds, ion trap analyzers (particularly 3D models) have only single unit mass resolution. With the advancement of other types of analyzers that have faster speed, better mass accuracy and superior resolution, there has been a shift away from performing proteomic analysis on mass spectrometers using only ion trap analyzers. However, due to their geometry, 2D analyzers still find widespread use in hybrid instruments, particularly those that utilize them as a precursor mass filter when performing tandem MS.

7.4 Time-of-Flight

Although Time-of-flight (TOF) analyzers have been around for some time, it has been the advent of MALDI ionization (which allowed for the easy analysis of large biomolecules) that propelled TOF analyzers to the forefront [14, 15]. TOF analyzers are the easiest to conceptualize as illustrated in (Fig. 7.3). In its simplest form, a TOF analyzer consists primarily of a flight tube and an acceleration grid that acts to accelerate a "packet" of ions from the ionization source to the MS detector [16]. Essentially, if two ions of different *m*/*z* are accelerated from the ion source with the same kinetic energy and allowed to drift through a field free region of the flight tube, then their arrival times at the detector will be different.

The equation for kinetic energy for any mass is

$$E_k = \frac{1}{2}mv^2$$

wherein E_k is the kinetic energy of the ion after acceleration, *m* is the mass of the ion and *v* is its velocity. Ion velocity remains constant after acceleration as it moves through the field free

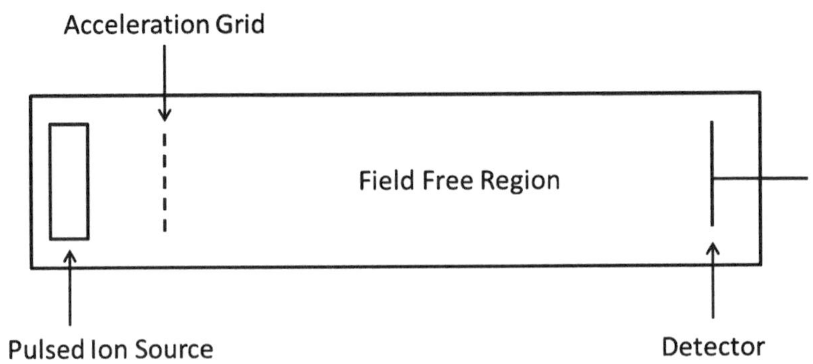

Fig. 7.3 Ions generated by the ion source are accelerated by placing a pulsed electric potential on the acceleration grid. The accelerated ions then drift through a field free region of a flight tube where they are separated based on their *m*/*z*. The greater the ion mass, the slower the drift through the flight tube. When an ion hits the detector, the mass spectrometer determines the time it took for the ion to drift through the flight tube. The drift time through the flight tube is proportional to the *m*/*z* of the ion

region of the flight tube. Because this velocity remains static, then velocity is given by

$$v = \frac{d}{t}$$

wherein d is the distance the ion travels and t is the time it takes for the ion to travel from its acceleration point to the detector. Substituting v into the kinetic energy equation results in

$$E_k = \frac{1}{2}m(d/t)^2$$

Solving for the mass of the ion yields

$$m = \frac{2E_k t^2}{d^2}$$

Because the initial kinetic energy (E_k) of the ions and the length of the flight tube (d) remain constant, mass is strictly a function of the time it takes for the ions to be detected after initial acceleration (time-of-flight).

TOF analyzers today also employ the use of a reflectron which reflects the ion path back in the direction of the ion source before being detected [17]. This allows for corrections in the small differences in initial kinetic energies of the ions that may occur during acceleration [18]. Other methods such as delayed extraction are also employed in order to increase resolution [19, 20]. After ions are formed, delayed extraction introduces a small delay (usually on the order of a few hundred nanoseconds) in the electric pulse of the acceleration grid before the ions are accelerated. This small delay allows the ions formed after ionization to equilibrate and have a more uniform average momentum before acceleration.

Due to the high ion transmission efficiencies of TOF analyzers, they can achieve the widest mass range of all mass analyzers. TOF analyzers allow for the separation of ions with masses of only a few Daltons to well over 100 kDa [21]. This makes them the analyzer of choice for observing singly charged high mass biomolecules such as proteins [22]. Because of their ability to simultaneously measure the masses of many peptides, TOF analyzers have been the most popular analyzer for performing peptide mass fingerprinting [23, 24]. Although new tandem analyzer configurations have allowed TOF analyzers to be interfaced ion sources that provide a continuous flux of ions, they have initially been employed with only pulsed ion sources such as MALDI.

7.5 FT-ICR

FT-ICR analyzers determine m/z by measuring the cyclotron frequency of ions in a fixed magnetic field [25]. Ions are first introduced into a Penning trap, a device similar to a 3D ion trap but using a magnetic field to trap ions rather than an electric field. The ions are injected into the magnetic field from the source as a "packet". The ions then experience a Lorentz force, which causes them to assume a circular motion in a plane perpendicular to the magnetic field (Fig. 7.4). The angular frequency, also known as the cyclotron frequency, is described by the equation

$$\omega_c = \frac{qB}{m}$$

where ω_c is the angular frequency of the ions in radians, m is the mass of the ion, q its charge and B is the strength of the magnetic field. However, because the ions are not in phase when initially introduced into the trap and typically have very small orbits, it is impossible to detect them. In order to detect these ions, they must be coherently excited to a larger radius within their plane of motion. This is accomplished by exciting the ions with a limited frequency sweep of a broadband RF field [26]. This excitation coherently places the ions in a higher cyclotron orbit, which allows them to be detected. As the ions are detected over time by the receiver plates, their signal intensity is digitized with respect to time and converted to a frequency spectrum via a Fourier transform. The cyclotron frequencies of the ions are proportional to their m/z.

One of the biggest advantages of FT-ICR analyzers are their very high mass accuracy and

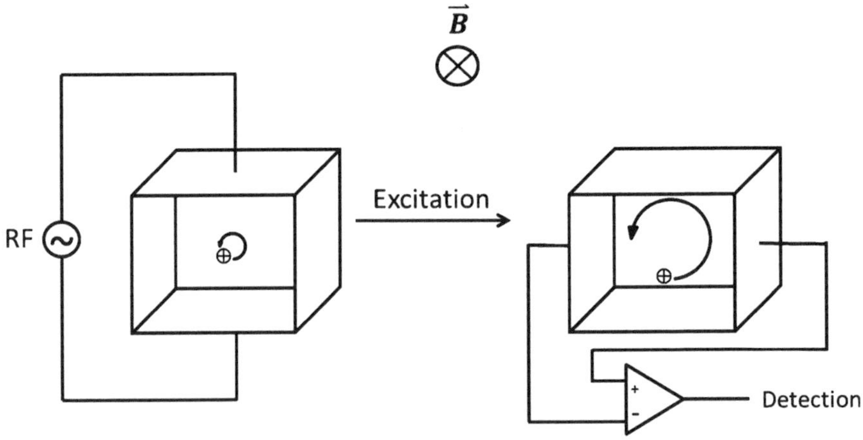

Fig. 7.4 Ions are injected into a magnetic field for which they then undergo a small cyclotron frequency perpendicular to the magnetic field. A brief broadband RF pulse excites the ions into a larger and coherent cyclotron orbit. The circular motion of the ions in the magnetic field is detected by the receiver plates and a Fourier transform converts the signal to a frequency spectrum. The angular frequency of the ions is determined by their *m*/*z*

resolving power. One million resolution has been reported on instruments with magnetic field strengths as low as 1 T [27]. All aspects of FT-ICR improve with higher magnetic field: increased resolution, increased mass accuracy, increased number of ions that can be put in the cell, decreased ion coalescence, etc. Most commercially available FT-ICR analyzers operate in magnetic field ranges between 7 and 12 T. This high resolution and mass accuracy is very useful when determining the elemental composition of small molecules based on their "mass defect" [28]. For example, two compounds, one with the empirical formula C_6H_{12} and the other with the empirical formula C_5NH_{10}, both appear to have the same mass of 84 Da. However, when calculating their mass with very high precision, C_6H_{12} has an exact mass of 84.09389 Da and $C_5H_{10}N$ has an exact mass of 84.08131 Da, a difference of 0.01258 Da. This is due to slight differences in the binding energies in the nuclei of the carbon and nitrogen atoms, thus causing a slight shift in their atomic mass. Unlike many other analyzers with lower resolving power, FT-ICR analyzers have the ability to obtain empirical formulas directly from mass data.

Another major application for FT-ICR is in the field of proteomics where high mass accuracy is often required. In order to maintain isotopic resolution of large molecular weight ions with multiple charge states, very high resolution must be employed. For example, a 50 KDa protein, regardless of charge state, would require a resolution of 50,000 of the analyzer in order to observe isotopic peaks. In order to perform top-down sequencing of proteins, it is preferable to have isotopic resolution of the protein and its MS/MS products. Because of the resolving power of FT-ICR, entire large proteins can be sequenced and identified when performing tandem MS. Post-translational modifications within an isolated protein can also be identified without having to first perform chemical or enzymatic cleavage of the isolated protein as required in bottom-up approaches [29].

Due to the need for very strong magnetic fields, FT-ICR analyzers require the use of large superconducting magnets. This introduces two major problems. First, large magnet sizes require large amounts of lab space to be available. This may also include the need for high laboratory ceilings in order to perform maintenance. Second, superconducting magnets require liquid helium as a coolant in order for them to operate. The cost of liquid helium is high and often beyond the budget of many small laboratories. The initial cost of most FT-ICR instruments is also very high.

Mass spectrometers that utilize FT-ICR also suffer from slow scan speeds compared to other analyzers such as time-of-flight. This makes it impractical for many LC-tandem MS experiments, such as Multi-dimensional Protein Identification Technology (MudPIT), where many different co-eluting peptides need to be analyzed at very high scan rates in order to collect as much tandem MS data as possible.

7.6 Orbitrap

Similar to the FT-ICR analyzer, the orbitrap is also a type of analyzer that makes use of a Fourier transform to convert a signal, produced by ions oscillating in a trap, from the time domain to a frequency domain [30, 31]. Unlike FT-ICR analyzers, which use a magnetic field to induce oscillation in the ions, orbitrap analyzers use an electric field to induce these oscillations [32].

The orbitrap mass analyzer is composed of three main parts, an inner spindle electrode covered by two hollow outer concave electrodes facing each other. The two outer electrodes are separated by a thin ring of dielectric material (Fig. 7.5). A voltage potential is applied between the inner and outer electrodes, creating a linear electric field between them. Ions are introduced tangentially into the orbitrap as a "packet" between the inner and outer electrodes through a hole machined into one of the outer electrodes. Due to the electric field between the inner and outer electrodes, the ion packet is bent towards the inner electrode while the tangential velocity of the ions creates an opposing centrifugal force. At a specific potential between the inner and outer electrodes, the ions remain in a spiral path around the inner electrode. However, due to the conical shape of the electrodes, a harmonic axial oscillation in the ions is induced. The outer electrodes also act as receiver plates that detect the back and forth axial harmonic motion of the ions. This signal image is digitized and transformed from the time domain to the frequency domain. Similar to FT-ICR, the axial harmonic frequencies are proportional to the m/z of the ions.

One of the major advantages of the orbitrap analyzer is its high resolving power, resulting in its use as a replacement for FT-ICR analyzers for many applications, particularly those involving proteomics. In general, FT-ICR analyzers are superior to orbitraps in the low molecular weight range, thus making them ideal for low mass

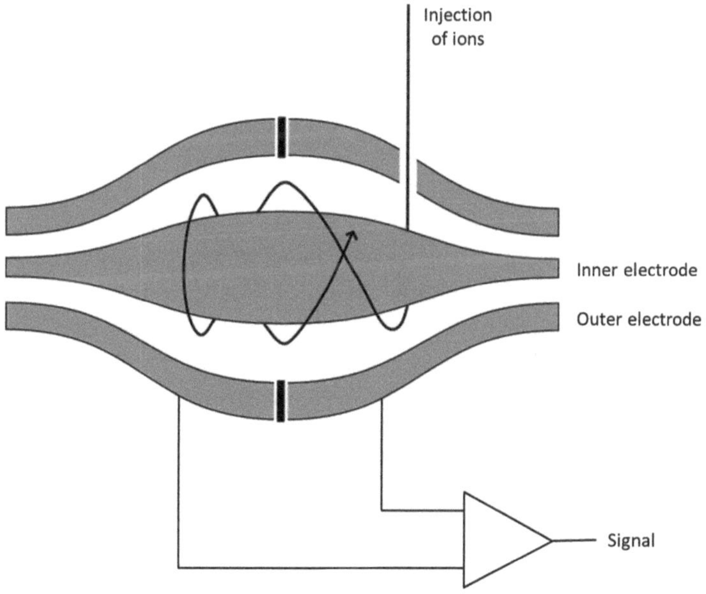

Fig. 7.5 In an orbitrap analyzer, ions enter through an opening in one of the outer electrodes. The entry of the ions is tangential to the inner electrode. At a particular potential between the inner and outer electrodes, the ions will continuously spin around the inner electrode. Ions will oscillate back and forth along the axis of the inner electrode. This oscillation is detected and transformed via a Fourier transform to obtain a mass spectrum

compounds. However, there is a fast decrease in the resolving power of FT-ICR analyzers at higher *m*/*z*. This decrease in resolving power of FT-ICR analyzers is inversely proportional with an increase in the *m*/*z* being measured. With orbitrap analyzers, this decrease in resolving power is inversely proportional to the square-root of the *m*/*z* being measured. Therefore, orbitrap analyzers often have better resolving power than FT-ICR analyzers at higher *m*/*z* [33]. This property can give the orbitrap an advantage when analyzing high molecular weight compounds such as proteins. Recently, there has been a move from bottom-up proteomic analysis to top-down analysis. Because top-down analysis requires very high resolving power, it was limited to FT-ICR analyzers and beyond the affordability of most MS labs. Orbitrap analyzers have been instrumental in overcoming this difficulty and have therefore pushed the advancement of top-down proteomics.

There are also a number of other advantages to orbitrap analyzers. Unlike the large size and operating costs of instruments utilizing FT-ICR, orbitrap instruments are much smaller and require very little maintenance. Orbitrap analyzers also do not use magnetic fields to operate, and therefore cryogenic refrigerants such as liquid helium are not necessary and operating costs are kept low. Although counterintuitive, the resolving power of the orbitrap analyzer is increased by the decrease in size of the analyzer. The main limitation to improved orbitrap design has been the tolerances needed in the machining process during manufacture. As machining processes improve, smaller orbitrap analyzers will no doubt continue to decrease the overall size of mass spectrometers that utilize them. Their smaller design will also allow for faster acquisition rates and higher resolution.

Improvements in orbirap analyzer design will continue to provide faster scan speeds and duty cycles. However, even with major improvements expected in the future, they will continue to be slower than that of TOF analyzers. This makes orbitrap analyzers potentially less ideal for performing MudPIT experiments where fast acquisition rates may outweigh the need for very high mass resolution or accuracy. Orbitrap analyzers are also very prone to space-charge effects and therefore the amount of ions entering the analyzer must be monitored by the MS software in order to trap a limited amount of ions.

7.7 Tandem Mass Analyzers

Mass spectrometers that utilize two or more mass analyzers consecutively are known as tandem mass spectrometers [34, 35]. Tandem MS analysis is the process by which the first analyzer is used to select ions of a particular *m*/*z* value, subject those ions to CID (as described in RF-only multipoles), and then analyze the resulting product ions using a second mass analyzer. CID is also sometimes referred to as collision-activated dissociation (CAD) and is a process by which ions are fragmented by colliding them with chemically inert gas (typically argon or nitrogen) at low pressure (~10^{-5} torr). The fragmentation occurs due to converting some of the kinetic energy from the collision of the analyte ion with inert gas atoms to internal energy of the ions, thus resulting in bond breakage of the analyte ion molecules [36]. These product ions formed from CID often provide information about the structure of the analyte molecules.

7.7.1 Triple Quadrupoles

Triple quadrupole mass spectrometers are one of the most commonly sold types of mass spectrometer and are one of the best examples of using analyzers in tandem [37]. In a triple quadrupole mass spectrometer, three sets of quadrupole analyzers are used in sequence (Fig. 7.6). The first analyzer is often referred to as Q1 and can scan across a range of *m*/*z* values or selectively filter ions of a selected *m*/*z*. Those ions that pass through Q1 then enter a second set of quadrupoles that are referred to as Q2. Unlike a quadrupole that operates as an analyzer, Q2 is used exclusively as a collision cell to fragment the selected ions from Q1. The product ions

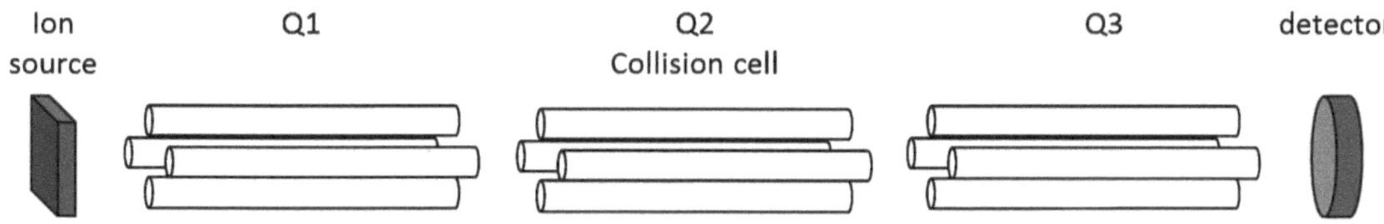

Fig. 7.6 A triple quadrupole mass spectrometer consists of three quadrupole analyzers used in tandem. Ions enter the first quadrupole, Q1. Ions that pass through Q1 enter into quadrupole Q2, where ions undergo CID. The resulting fragment ions are then filtered through the final quadrupole, Q3

Table 7.1 Table of different scan modes of triple quadrupole mass analyzers

	Q1	Q2	Q3
Product ion scan	Fixed *m/z*	CID	Scan full range
Precursor ion scan	Scan full range	CID	Fixed *m/z*
Neutral loss scan	Scan full range	CID	Scan full range Q3=Q1-Δm/z
Selected reaction monitoring	Fixed *m/z*	CID	Fixed *m/z*

formed in this process can then be either scanned through the final set of quadrupoles, Q3, to obtain a mass spectrum or Q3 can be fixed in order to monitor a particular ion. The combination of fixed or scanning modes of Q1 and Q3 determine the type of scan performed [38, 39]. The most common scan modes are described below and in Table 7.1.

Product ion scan – In a product ion scan, Q1 remains fixed such that only ions of a selected *m/z* are filtered through the quadrupole. These ions are then fragmented via CID through Q2. The resulting product ions are then scanned and analyzed in Q3. Once the product ions are recorded, Q1 can then fix on a new *m/z* and the process repeated. This technique is often used in order to determine structural information of specific analytes. For example, in a bottom-up proteomics approach, the sequence of many peptides eluting off a chromatographic column can be sequenced.

Precursor ion scan – In a precursor ion scan, Q1 is scanned across the entire *m/z* range of the analyzer. The precursor ions subsequently pass through Q2 for CID. However, Q3 is kept fixed such that only product ions of a specific *m/z* are filtered through the quadrupole. The mass chromatogram is plotted as the intensity of the ions exiting Q3 with respect to the *m/z* value that they originated from in Q1. In other words, precursor ion scanning allows one to determine the *m/z* of all precursor ions that have the same product ion. This is valuable in proteomics when one wants to identify all peptides that may have the same functional group. For example, performing a precursor ion scan at *m/z* = 216, a signature immonium ion for phosphotyrosine, allows one to selective identify peptides that may contain phosphotyrosine.

Neutral loss scan – A neutral loss scan is a technique to track ions before and after the loss of a neutral group. Both Q1 and Q3 are scanned simultaneously over the entire *m/z* range but with Q3 offset from the Q1 by an amount that corresponds to the loss of a neutral fragment from the ion. Using this method, all precursors that undergo the loss of the same neutral fragment can be monitored. Similar to precursor ion scanning, this technique can be a powerful tool for quickly and selectively identify peptides that are post-translationally modified such as those that have been phosphorylated. An example is in the identification of peptides with phosphorylated serine or threonine. Performing low energy CID on peptides that are phosphorylated will often result in the loss of phosphoric acid (H_3PO_4, m/z = 98) from the parent ion.

Selected reaction monitoring – Selected Reaction monitoring (SRM), sometimes referred to as multiple reaction monitoring (MRM), is a popular scanning technique for the quantification of compounds in a mixture. Q1 is fixed to allow only precursors of a particular *m/z* to filter through the quadrupole. CID then occurs in Q2 and all fragments sent to Q3. Q3 is fixed to only allow product ions of specific *m/z* to filter through. Thus, specific signature fragment ions originating from a compound of known mass can be monitored. This technique essentially allows for a single known compound to be monitored in real time. One caveat is that although the mass and potential *m/z* values for Q1 can be easily determined for a compound of interest, the *m/z* values of the product ions of that compound must be known prior to designing the SRM experiment. This can be solved but first performing a product ion scan of the compound of interest and recording the *m/z* of all product ions.

7.7.2 Q-TOF

Quadrupole analyzers prefer to operate efficiently when there is a continuous stream of ions from the ion source. However, TOF analyzers prefer a pulse or packet of ions. In order for the two analyzers to work in tandem, the TOF analyzer is placed in an orthogonal configuration after the quadrupole analyzer [40]. This configuration allows ions that are filtered through the quadrupole to be injected orthogonally into the TOF analyzer as a packet using a set of pusher and puller plates between the two analyzers (Fig. 7.7) [41].

Some of the biggest advantages to Q-TOF tandem analyzers are their higher mass accuracy, higher resolution and increased scan speed as compared to triple quadrupole mass analyzers and thus their ability to easily interface to liquid chromatography and perform very fast tandem MS. This allows many spectra to be acquired when there are many co-eluting compounds in a chromatographic run. Although the resolution of the data is not of the same quality as that when analyzed by an orbitrap or FT-ICR analyzer, it is far superior to that obtained by standard quadrupole or ion traps.

7.7.3 TOF/TOF

Time-of-flight/time-of-flight (TOF/TOF) is a method by which two TOF analyzers are used in tandem and CID is performed between the two TOF analyzers (Fig. 7.8) [42]. This allows one to perform tandem MS on biological compounds such as peptide and oligonucleotides that often are ionized by ionization methods such as MALDI [43]. Because of the speed at which TOF analyzers operate, sample analysis in both the MS and MS/MS level can be performed very rapidly.

In order to perform tandem MS in a TOF/TOF analyzer, very fast electronic switching must occur in a series of steps. First, ions of different *m/z* are separated through the flight tube based on their velocity. Second, ions of a particular *m/z* are

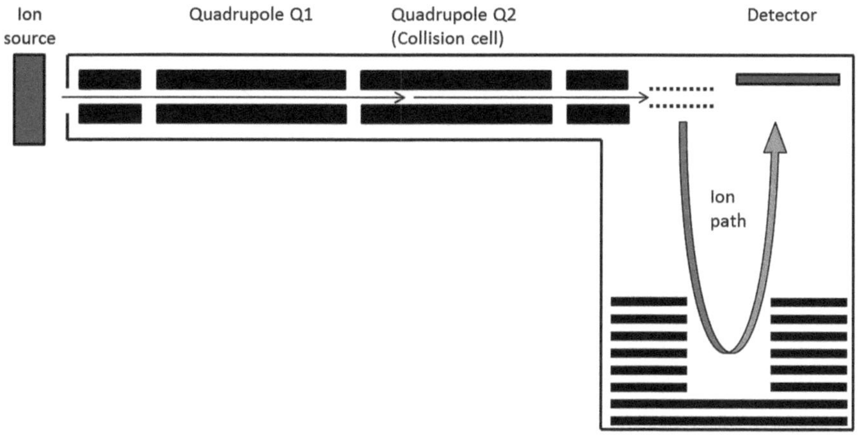

Fig. 7.7 The QTOF analyzer is a hybrid of triple quadrupole analyzer and a time-of-flight analyzer. It is analogous to a triple quadrupole system but with the exception that the last quadrupole is replaced by a time-of-flight analyzer

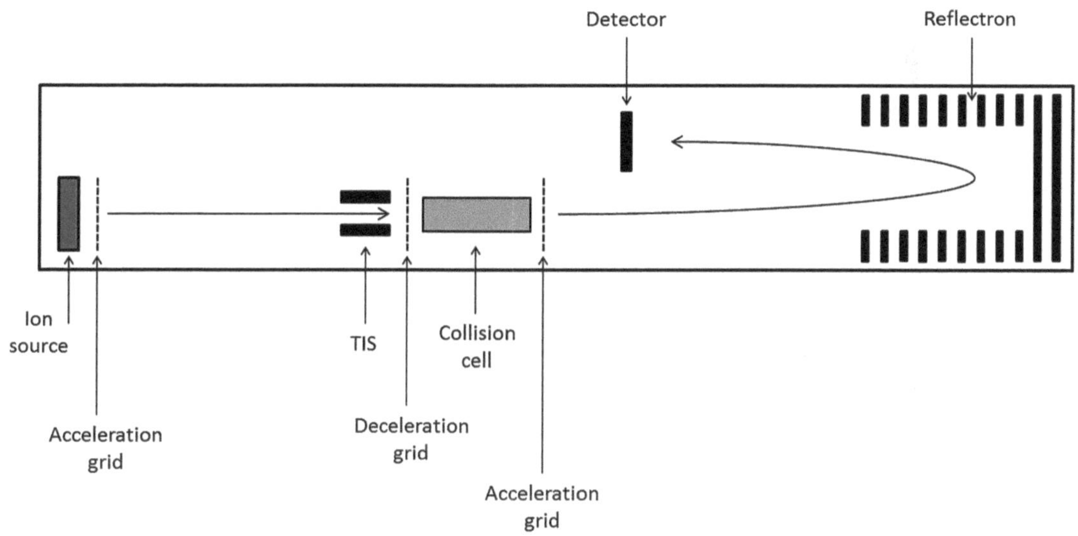

Fig. 7.8 TOF/TOF analyzers combine two TOF analyzers in tandem. Ions are accelerated in the first TOF. A timed ion selector allows ions of a particular m/z to pass. The selected ions are then decelerated before passing into a collision cell where they undergo CID. The resulting fragment ions are re-accelerated into the second TOF and their time-of-flight is measured to obtain a mass spectrum

selected while filtering out all others. This precursor ion selection is often performed using a Bradbury-Nielsen gate which is essentially a timed-ion-selector (TIS) that filters ions based on their arrival time to the gate [44]. Third, the selected precursor ions are then passed to a set of ion optics that de-accelerates them to a much slower velocity. Fourth, the ions then pass through a collision cell for CID. Finally, the product ions formed are re-accelerated into a second flight tube and analyzed. The fast analysis of this analyzer combination, combined with MALDI ion sources, make it ideal for the analysis of peptides from tryptic digests.

7.7.4 Other Tandem Analyzer Combinations

There are other combinations of mass analyzers which are far too numerous to list. In principle, the combination of mass analyzers, regardless of their type, allow the mass spectrometrist to perform tandem MS. The choice of combination is dependent on many necessary factors such as resolution, acquisition speed (duty cycle), mass accuracy, etc. For example, if high resolution of a product ion is required but not that of its precursor, the first analyzer may be a quadrupole or ion trap and the second analyzer an orbitrap or an FT-ICR. Newer instruments have a multitude of different analyzers that may be utilized in a number of different configurations. As newer combinations of analyzers continue, the variety of tandem MS methods will also continue to grow.

7.8 Other Analyzers

Although there are a number of other types of analyzers, those that have been described herein comprise the majority of analyzers currently used in mass spectrometers. There have been many other analyzers that were once popular but have been overtaken by the current selection of analyzers for a multitude of reasons. For example, magnetic sector analyzers were one of the first analyzers used in mass spectrometry. They can have high resolution (~200,000), good stability, and significant mass accuracy, but unfortunately suffer from their large size, low resolution for precursor ion selectivity, and slow scan speeds. For these reasons magnetic sector instruments have been less ideal for interfacing to LC. Other analyzers have found a niche market for a number of reasons. The QTRAP analyzer allows a triple-quad mass spectrometer to act as a quadrupole and linear ion trap tandem mass spectrometer. Although there are a number of advantages to this type of analyzer, the demand has not propelled it to the point that it has become one of the primary analyzers used in proteomics.

There is no doubt that newer analyzers will be developed along with improvements in current ones. These advancements will continue to push the limits of current mass spectrometry. Because of the complex nature of the proteomics field, the necessity for many different avenues of approach to problem solving by mass spectrometry will undoubtedly continue to grow.

References

1. Jennings KR, Dolnikowski GG (1990) Mass analyzers. Method Enzymol 193:37–61
2. Fenn JB, Mann M, Meng CK, Wong SF, Whitehouse CM (1989) Electrospray ionization for mass spectrometry of large biomolecules. Science 246:64–71
3. Chernushevich IV, Ens W, Standing KG (1999) Orthogonal-injection TOFMS for analyzing biomolecules. Anal Chem 71:452A–461A
4. Miller PE, Denton MB (1986) The quadrupole mass filter: basic operating concepts. J Chem Educ 63:617–622
5. March RE (1997) An introduction to quadrupole ion trap mass spectrometry. J Mass Spectrom 32:351–369
6. Hayes RN, Gross ML (1990) Collision-induced dissociation. Method Enzymol 193:237–263
7. Arpino PJ, Guiochon G (1979) LC/MS coupling. Anal Chem 51(7):692A–697A
8. Blakely CR, Vestal ML (1983) Thermospray interface for liquid chromatography/mass spectrometry. Anal Chem 55:750–754
9. Wong PSH, Cooks RG (1997) Ion trap mass spectrometry. Currentseparations.com 16(3)
10. Paul W, Steinwedel H (1953) Ein neues Massenspektrometer ohne Magnetfeld. Zeitschrift für Naturforschung A 8(7):448–450
11. Stafford GC, Kelley PE, Syka JEP, Reynolds WE, Todd JFJ (1984) Recent improvements in and applications of advanced ion trap technology. Int J Mass Spectrom Ion Process 60(1):85–98
12. Tong W, Link A, Eng JK, Yates JR (1999) Identification of proteins in complexes by solid-phase microextraction/multistep elution/capillary electrophoresis/tandem mass spectrometry. Anal Chem 71:2270–2278
13. Link AJ, Eng J, Schieltz DM, Carmack E, Mize GJ, Morris DR, Garvik BM, Yates JR (1999) Direct analysis of protein complexes using mass spectrometry. Nat Biotechnol 17:676–682
14. Karas M, Bachman D, Bagr U, Hillenkamp F (1987) Matrix-assisted ultraviolet laser desorption of non-volatile compounds. Int J Mass Spectrom Ion Process 78:53–68
15. Juhasz P, Roskey MT, Smirnov IP, Haff LA, Vestal ML, Martin SA (1996) Applications of delayed extraction matrix-assisted laser desorption ionization time-of-flight mass spectrometry to oligonucleotide analysis. Anal Chem 68:941–946
16. Mamyrin BA (2001) Time-of-flight mass spectrometry (concepts, achievements, and prospects). Int J Mass Spectrom 206:251–266
17. Cotter RJ (1999) The new time-of-flight mass spectrometry. Anal Chem 71:445A–451A

18. Mamyrin BA, Karataev VI, Shmikk DV, Zagulin VA (1973) The mass reflectron, a new nonmagnetic time-of-flight mass spectrometer with high resolution. Sov Phys – JETP 64:82–89
19. Vestal ML, Juhasz P, Martin SA (1995) Delayed extraction matrix-assisted laser desorption time-of-flight mass spectrometry. Rapid Commun Mass Spectrom 9(11):1044–1050
20. Juhasz P, Vestal ML, Martin SA (1997) On the initial velocity of ions generated by matrix-assisted laser desorption ionization and its effect on the calibration of delayed extraction time-of-flight mass spectra. J Am Soc Mass Spectrom 8:209–217
21. Karas M, Bahr U (1990) Laser desorption ionization mass spectrometry of large biomolecules. Trends Anal Chem 9(10):321–325
22. Hillenkamp F, Karas M (1990) Mass spectrometry of peptides and proteins by matrix assisted ultraviolet laser desorption/ionization. Method Ezymol 193:280–295
23. Pappin DJC, Hojrup P, Bleasby AJ (1993) Rapid identification of proteins by peptide-mass fingerprinting. Curr Biol 3:327–332
24. Thiede B, Höhenwarter W, Krah A, Mattow J, Schmid M, Schmidt F, Jungblut PR (2005) Peptide mass fingerprinting. Methods 35:237–247
25. Comisarow MB, Marshall AG (1974) Fourier transform mass Ion cyclotron resonance spectroscopy. Chem Phys Lett 25:282–283
26. Comisarow MB, Marshall AG (1974) Frequency-sweep Fourier transform ion cyclotron resonance spectroscopy. Chem Phys Lett 26:489–490
27. Gorshkov MV, Udseth HR, Anderson GA, Smith RD (2002) High performance electrospray ionization Fourier transform ion cyclotron resonance mass spectrometry at low magnetic field. Eur J Mass Spectrom 8:169–176
28. Marshall AG, Hendrickson CL, Jackson GS (1998) Fourier transform ion cyclotron resonance mass spectrometry: a primer. Mass Spectrom Rev 17:1–35
29. Bogdanov B, Smith RD (2005) Proteomics by FTICR mass spectrometry: top down and bottom up. Mass Spectrom Rev 24:168–200
30. Scigelova M, Hornshaw M, Giannakopulos A, Makarov A (2011) Fourier transform mass spectrometry. Mol Cell Proteomics 10(7): M111.009431. doi:10.1074/mcpM111.009431
31. Hu Q, Noll RJ, Li H, Makarov A, Hardman M, Cooks RG (2005) The orbitrap: a new mass spectrometer. J Mass Spectrom 40:430–443
32. Makarov A (2000) Electrostatic axially harmonic orbital trapping: a high-performance technique of mass analysis. Anal Chem 72:1156–1162
33. Zubarev RA, Makarov A (2013) Orbitrap mass spectrometry. Anal Chem 85:5288–5296
34. de Hoffmann E (1996) Tandem mass spectrometry: a primer. J Mass Spectrom 31:129–137
35. Yost RA, Boyd RK (1990) Tandem mass spectrometry: quadrupole and hybrid instruments. Method Enzymol 193:154–200
36. Cooks RG (1995) Collision-induced dissociation: readings and commentary. J Mass Spectrom 30:1215–1221
37. Yost RA, Enke CG (1979) Triple quadrupole mass spectrometry. Anal Chem 51(12):1251A–1264A
38. Yost RA, Enke CG (1978) Selected ion fragmentation with a tandem quadrupole mass spectrometer. J Am Chem Soc 100(7):2274–2275
39. Domon B, Aebersold R (2006) Mass spectrometry and protein analysis. Science 312:212–217
40. Morris HR, Paxton T, Dell A, Langhorne J, Berg M, Bordoli RS, Hoyes J, Bateman RH (1996) High sensitivity collisionally-activated decomposition tandem mass spectrometry on a novel quadrupole/orthogonal-acceleration time-of-flight mass spectrometer. Rapid Commun Mass Spectrom 10:889–896
41. Chernushevich IV, Loboda AV, Thomson BA (2001) An introduction to quadrupole-time-of-flight mass spectrometry. J Mass Spectrom 36:849–865
42. Medzihradszky KF, Campbell JM, Baldwin MA, Falick AM, Juhasz P, Vestal ML, Burlingame AL (2000) The characteristics of peptide collision-induced dissociation using a high-performance MALDI-TOF/TOF tandem mass spectrometer. Anal Chem 72:552–558
43. Vestal ML, Campbell JM (2005) Tandem time-of-flight mass spectrometry. Method Enzymol 402:79–108
44. Bradbury NE, Nielsen RA (1936) Absolute values of the electron mobility in hydrogen. Phys Rev 49:388–393

Top-Down Mass Spectrometry: Proteomics to Proteoforms

8

Steven M. Patrie

Abstract

This chapter highlights many of the fundamental concepts and technologies in the field of top-down mass spectrometry (TDMS), and provides numerous examples of contributions that TD is making in biology, biophysics, and clinical investigations. TD workflows include variegated steps that may include non-specific or targeted preparative strategies, orthogonal liquid chromatography techniques, analyte ionization, mass analysis, tandem mass spectrometry (MS/MS) and informatics procedures. This diversity of experimental designs has evolved to manage the large dynamic range of protein expression and diverse physiochemical properties of proteins in proteome investigations, tackle proteoform microheterogeneity, as well as determine structure and composition of gas-phase proteins and protein assemblies.

Keywords

Review • Proteomics • Proteoform • Protein • Mass spectrometry • Top-down • Bottom-up • Informatics • Electrospray ionization • Alternative splicing • SNPs • Post-translational modifications • Fourier transform • FT-ICR • Orbitrap • Time-of-flight • Quadrupole ion trap • Native MS • Data-dependent • Data-independent • Reversed-phase • Tandem MS • Charge reduction • Supercharging • Chromatography • Hydrogen/deuterium exchange • Label-free quantitation • SILAC • Membrane proteins

S.M. Patrie (✉)
Computational and Systems Biology & Biomedical Engineering Graduate Programs, University of Texas Southwestern Medical Center, Dallas, TX, USA

Department of Pathology, University of Texas Southwestern Medical Center, Dallas, TX, USA

Department of Bioengineering, University of Texas at Dallas, Richardson, TX, USA
e-mail: Steven.Patrie@UTSouthwestern.edu

H. Mirzaei and M. Carrasco (eds.), *Modern Proteomics – Sample Preparation, Analysis and Practical Applications*, Advances in Experimental Medicine and Biology 919, DOI 10.1007/978-3-319-41448-5_8

Abbreviations

2DGE	2D gel electrophoresis
μESI	micro electrospray ionization
ALS	acid labile surfactant
BU	bottom-up
CAD	collision activated dissociation
CF	chromatofocusing
CFGE	continuous flow gel elution
CID	collision induced dissociation
CZE	capillary zone electrophoresis
DB	database
DD	data-dependent
DI	data-independent
dMS	differential mass spectrometry
DT	drift tube
ECD	electron capture dissociation
ESI	electrospray ionization
ETD	electron transfer dissociation
FAIMS	field asymmetric ion mobility spectrometry
FT	Fourier transform
FTMS	Fourier transform mass spectrometry
GELFrEE	gel-eluted liquid fraction entrapment electrophoresis
HCD	higher-energy collision dissociation
HDL	high-density lipoprotein
HDX	hydrogen/deuterium exchange
HILIC	Hydrophilic interaction liquid chromatography
ICR	ion cyclotron resonance
IEC	Ion exchange chromatography
IEF	isoelectric focusing
IMAC	immobilized metal-ion affinity chromatography
IMP	Integral membrane protein
IM-MS	Ion mobility-mass spectrometry
IPG	immobilized pH gradient
ISD	in-source dissociation
ISD	in source dissociation (nozzle/skimmer dissociation)
IRMPD	infrared multiphoton dissociation
LC-MS	liquid chromatography mass spectrometry
LQT	linear quadrupole trap
MALDI	matrix assisted laser desorption ionization
MBP	myelin basic protein
MD	middle-down
MS/MS	tandem mass spectrometry
m/z	mass to charge ratio
nESI	nano electrospray ionization
pI	isoelectric point
PIITA	precursor ion independent top-down algorithm
PLRP	Polystyrene-divinylbenzene copolymers, commercialized by Agilent
PT	protein-platinum
PTM	post translational modification
Q	quadrupole
QIT	quadrupole ion trap
QIT	quadrupole ion-trap
RP	reversed-phase
SAX	strong anion exchange
SCX	strong cation exchange
SDS	sodium dodecyl sulfate
SEC	size exclusion chromatography
SID	surface-induced dissociation
S/N	signal to noise
TD	top-down
TDMS	top-down mass spectrometry
TOF	time-of-flight
TW	traveling-wave
UPLC	ultra-high performance LC
UVPD	ultraviolet photon dissociation
WAX	weak anion exchange
WCX	weak cation exchange
Zip-tips®	solid-phase capture/extraction tips, commercialized by Millipore

8.1 Introduction

Proteome diversity can substantially deviate from that predicted from the central dogma of biology [1] which states "the coded genetic information hard-wired into DNA is transcribed into individual transportable cassettes, composed of messenger RNA (mRNA); each mRNA cassette contains the program for synthesis of a particular protein (or small number of proteins)" (Fig. 8.1A). A "protein" is often a mixture of poly-peptidyl products (proteoforms [2]) that are molecularly similar, sharing appreciable amino acid homology, yet chemically distinct. The chemical heterogeneity occurs through mechanisms such as allelic alterations (e.g., single nucleotide polymorphisms (SNP), or mutations), alternative splicing, and post-translational modifications (PTMs). Regarding PTMs, hundreds are currently described and may manifest both enzymatically (e.g., phosphorylation) and non-enzymatically (e.g., carbonylation) at times coincident with translation (e.g. glycosylation) or after inter- and extra-cellular displacement. The existence of such diversity ensures that reasonably specific molecular tools are available for diverse jobs (e.g., protein trafficking, protein complex formation, membrane assembly, or signal cascades). Proteoform diversification may also impact disease pathobiology [3–5]; therefore, proteoform expression ratios are being actively investigated with the goal of translation to the clinic as diagnostic or prognostic markers.

Highlighted throughout this chapter is the tool development for protein and proteoform-level investigations that has occurred in the field of top-down mass spectrometry (TDMS) over the past few decades. This review includes discussion on advancements in sample processing, chromatography, MS and tandem MS (MS/MS or MS^n), and bioinformatics. The tenet that governs TD innovations is quite simple: *proteome and proteoform diversity is best interrogated when proteins are analyzed intact* (Fig. 8.1B). A key distinction between protein-

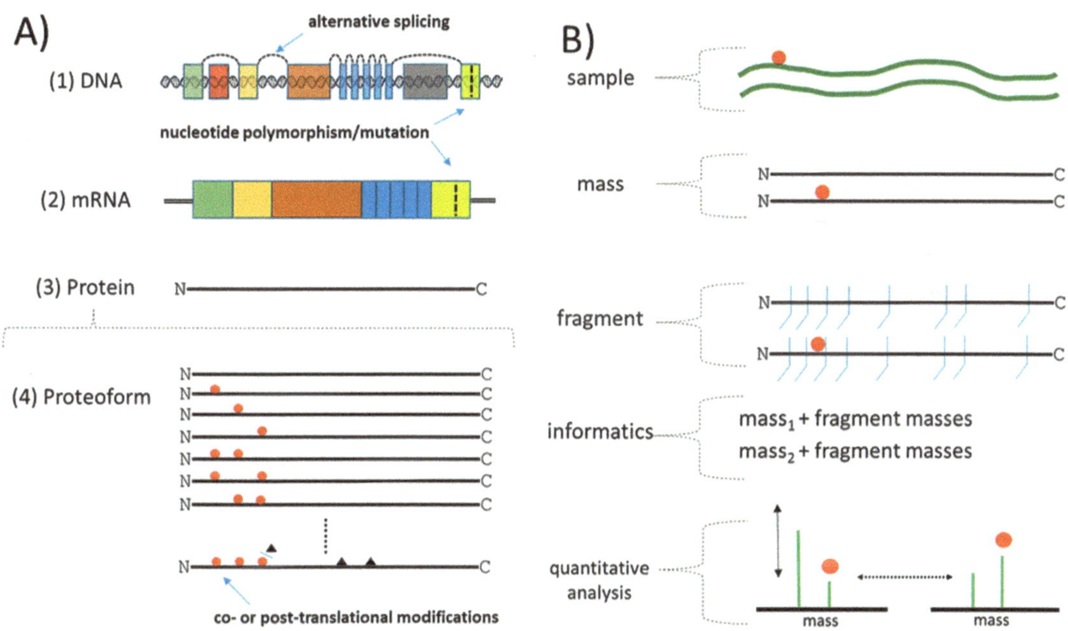

Fig. 8.1 Principles of proteome diversification and TDMS. (**A**) Schema highlights the information transfer from genome (DNA) through the transcriptome (messenger RNA) to the proteome (proteins). A protein often manifests as a collection of related "proteoforms" that derive from the same subset of genetic components, but are chemically diversified through polymorphisms, mutations, alternative splicing, and/or co- or post-translational modifications. (**B**) TD investigations will always measure the masses of the proteins/proteoforms in a sample. This is often followed by dissociation of the detected species in the mass spectrometer (e.g., MS/MS) and then informatics searches to identify the parent protein and localize positions of chemical microheterogeneity. Investigations may also seek to quantify between samples the expression changes of the protein or determine ratio changes between related proteoforms

centric vs. proteoform-centric studies is that while a protein centered study simply seeks to identify the proteins present within the samples, proteoform analysis attempts to localize all sources of molecular variation amongst related proteoforms. In addition, proteoform analyses seek to quantify expression changes at both the protein and proteoform levels between samples. These objectives largely set TD apart from bottom-up (BU) experiments where proteins are subjected to pre-analytical processing by proteases (e.g., trypsin) [6]. Elucidation of proteoform microheterogeneity is challenging with BU when the protein exhibits chemical diversity spatially distributed throughout the amino acid backbone [7]. An intermediate strategy between TD and BU called middle-down (MD) utilizes a limited digest at targeted amino acids to selectively cleave larger proteins into analytically manageable mid-size polypeptides [8]. While MD conceptually emulates BU, here it is considered a TD sample preparation method because it philosophically seeks to elucidate and quantify combinatorial products at the proteoform level.

8.2 Mass Spectrometry

8.2.1 Ionization

TDMS is performed on protein ions *in vacuo* (Fig. 8.2A). The gas-phase ions are generated by electrospray ionization (ESI) [9] or matrix assisted laser desorption ionization (MALDI) [10, 11]. ESI and MALDI are key in proteomics as they permit polypeptide analysis up to several hundred thousand Daltons (Da) without disrupting amino acid bonds, side-chain PTMs, or many non-covalent interactions in protein assemblies. Both ESI and MALDI have been applied in TD research; however, ESI is the workhorse largely because it seamlessly integrates liquid chromatography with mass spectrometry (LC-MS). ESI aerosolizes an analyte in solution using an atmospheric emitter held at a 1–4 kV voltage differential relative to the mass spectrometer's inlet. Generally, the ion formation occurs through steps of desolvation and Coulombic explosion of shrinking droplets, and surface evaporation of the charged analyte, all of which are facilitated by heated optics in the mass spectrometer inlet. ESI of a protein leads to multiple charge (z) states that sequentially differ by one charge and are observed in a *mass to charge* (m/z) spectrum (Fig. 8.2B). Molecular ions can be generated in protonated, cationized, or anionized states. Traditionally TD is performed in "positive ionization mode" where ESI, aided by the presence of acids in the sample (e.g., formic acid or acetic acid) generates multiply protonated states. The charge state may be assigned with the formula $(M + nH)^{n+}$, where n is the number of added protons (H^+) to a molecule with mass (M). Thus, the m/z is determined by $\frac{m}{z} = (M + n(mH^+))/n)$ where mH^+ is the mass of a proton (1.007277 Da).

Conventional ESI utilizes high flow rates (>1 μL/min) and a counter gas to facilitate desolvation. However, investigators often use micro-ESI (μESI) and nano-ESI (nESI) which exploit small inner diameter (ID) (1–50 μm ID) fused silica capillary emitters to introduce analyte at low flow rates (200–1000 nL/min and 10–200 nL/min, respectively) [12, 13]. This leads to smaller droplets for more efficient ion formation without the need for a desolvation gas. The limits of detection for proteins analyzed with μESI and nESI are usually in the nM to μM concentration range. Additionally, since most flow regimes applied in ESI are concentration sensitive [14], ESI is amenable to quantitative studies by overall spectral counts (intensity) in the m/z spectrum.

Despite the utility of ESI, the removal of excess salt (e.g. sodium or potassium), sample buffers (e.g. Tris(hydroxymethyl)aminomethane (Tris) or phosphate buffered saline (PBS)), detergents (e.g. sodium dodecyl sulfate (SDS)) or plasticizers from samples is critical because they create spectral artifacts and contribute to chemical noise in the ESI solution

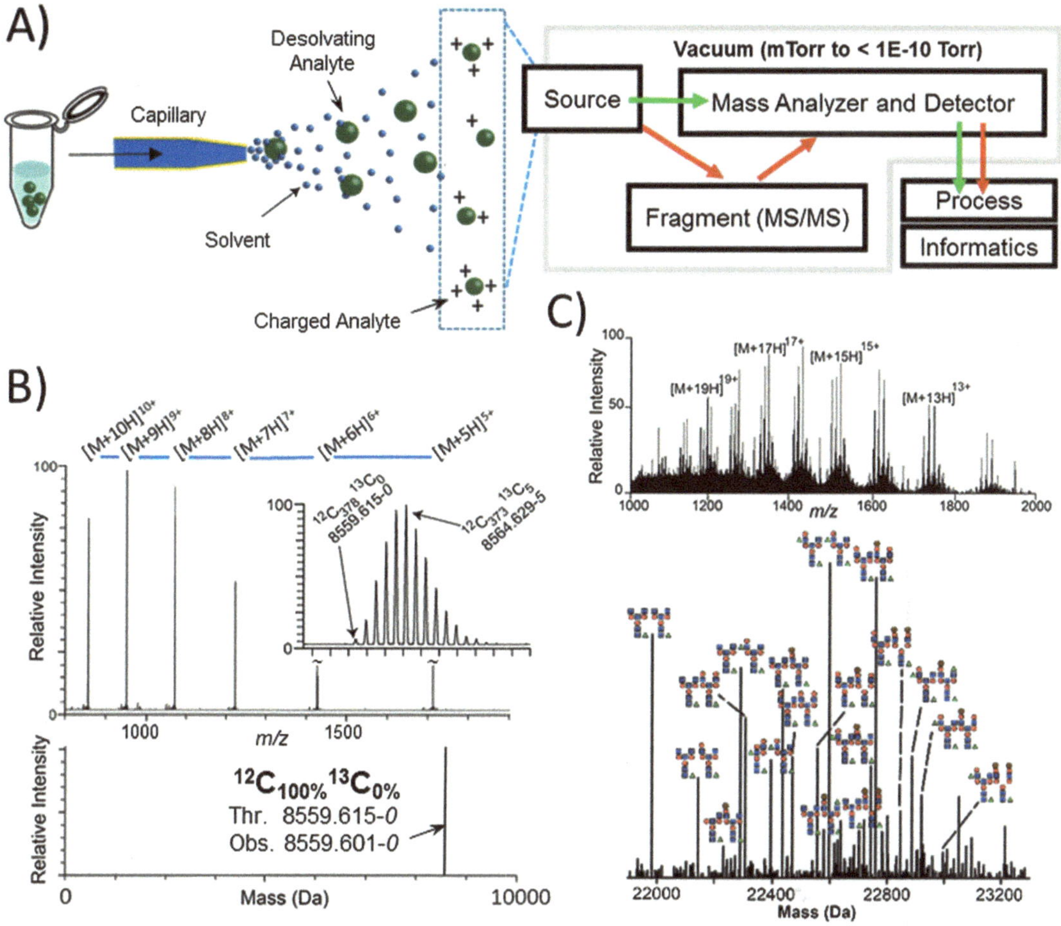

Fig. 8.2 Overview of ESI and MS data processing. (**A**) TD often applies ESI to aerosolize and ionize proteins suspended in solution. A mass to charge (*m*/*z*) spectrum is acquired for the parent protein and during subsequent fragmentation events. (**B**) A representative *m*/*z* spectrum (*top*) and corresponding zero charge mass spectrum (*bottom*) for bovine ubiquitin. The inset in the *m*/*z* spectrum highlights with sufficient instrument resolving power carbon-12 (^{12}C) carbon-13 (^{13}C) isotopologues for each charge state may be observed (*see* Sect. 8.3.4). Deconvolution of isotopologue distributions to monoisotopic masses often leads to low part per million mass accuracies. The "– #" indicates the isotopologue for which a mass is reported. (**C**) Representative *m*/*z* (*top*) and corresponding zero charge mass spectrum (*bottom*) for di-N-glycosylated lipocalin-type prostaglandin d synthase. The data highlights significant spectral complexity often associated with proteoform-level investigations, in this case multiple glycoproteoforms present at different ratios. Adapted with permission from Ref. [106]. Copyright 2014 by John Wiley & Sons, Inc

[15]. Therefore, successful implementation of ESI in proteomics often includes scaling LC columns to small diameter fused silica capillary columns (50–150 μm ID, 5–25 cm length) with integrated ESI emitters [16]. Of note, ESI is often performed with microfluidic devices [17] or robots [18] which facilitate assembly line processing for both "offline" direct infusion of individual proteins and "online" LC-MS experiments on complex mixtures. When applied with direct infusion the devices permit spectral averaging for improved spectral signal to noise (S/N) [18].

8.2.2 Mass Analyzers

To separate ions by their *m*/*z*, mass analyzers use electric or magnetic fields to apply a force that lead to both mass-dependent (Newton's second law) and ion-dependent (Lorentz force law) accelerations. Detection typically occurs in units of time, frequency, or current. The analyzer classes most relevant to TD include: (1) quadrupole mass filters (Q), (2) quadrupole ion-trap (QIT) in the design of 3D cylindrical-hyperbolic rings or 2D linear traps with quadrupole rods (LQT), (3) time-of-flight (TOF), and (4) Fourier transform MS (FTMS) which includes ion cyclotron resonance (FT-ICR) and Orbitrap. While analyzer performance characteristics vary, (Table 8.1) a notable distinction between analyzers comes in terms of processing duty cycle (scan rate) vs. spectral resolving power ($m/\Delta m_{50\%}$) and mass accuracy. The resolving power is typically determined from the minimum peak width (Δm_{fwhm}) at a set *m*/*z* value (e.g. 400 *m*/*z*) which permits comparison of different instrument types. Scanning instruments such as Q, QIT, and TOF provide a high scan rate (millisecond) that is useful for high-throughput MS/MS and reliable quantitative sampling during LC-MS. [19]. However, the higher acquisition rate may reduce resolving power and consequently mass accuracy (*see* Sect. 8.3.4). FTMS uses high-field superconducting magnets (FT-ICR) [20] or high-electric fields (Orbitrap) [21] to trap ions prior to frequency-based detection. FTMS is often performed at lower scan rate (100–1000's ms) which serves to dramatically improve spectral resolving power. On a well calibrated instrument, high spectral resolving power permits precise mass measurements at part-per-million or part-per-billion mass accuracy [22, 23]. As described in more detail below, the benefits of high resolving power and mass accuracy include the use of accurate mass tags to discriminate between elemental/chemical compositions of species in a database [24, 25], the resolution of metabolically incorporated isotopic labels for quantitation (see Sect. 8.4.2) [26], and the discrimination of PTMs with similar mass (e.g., O-phosphorylation, 79.96633 Da, vs. -O-sulfonation, 79.95682 Da) [27].

A notable feature of modern instruments is that they often combine mass analyzers in tandem (e.g., Q-TOF, QqQ, QqQ-FTMS, QIT-FTMS). Hybrids serve to improve dynamic range for continuous ion sources (such as ESI) [28, 29], aid selected enrichment of specific species in a sample [18, 30] and permit the parallel processing of high-resolving power scans in the FTMS with lower-resolving power MS/MS events in a separate QIT [31].

FTMS has largely dominated the TD field. Orbitraps, which do not require superconducting magnets, have been broadly accepted for routine LC-MS applications [32], while FT-ICR has been widely applied in detailed proteoform investigations where the highest resolving power and mass accuracies are required [33, 34]. However, continued innovation has decreased the performance gaps between

Table 8.1 Mass analyzers: Typical characteristics of mass analyzers used for top-down experiments

Mass analyzer	Ionization	Resolving power	Spectral duty cycle(s)	Upper m/z range	Mass accuracy (ppm)
QIT/LTQ	ESI, MALDI	1–3000	0.02–0.2	2000–3000	100–250
Q-TOF	ESI, MALDI	10,000–50,000	<0.01	>100,000	5.0–15.0
Orbitrap	ESI, MALDI	15,000–250,000	0.01–1.0	20,000–50,000	2.0–10.0
FT-ICR	ESI, MALDI	15,000–5,000,000	0.1–5.0	50,000	0.5–5.0

Abbreviations: *FT-ICR* Fourier transform ion cyclotron resonance, *LTQ* linear trap quadrupole, *QIT* quadrupole ion trap, *Q-TOF* quadrupole time-of-flight analyzers resulting in a broad application of TD on other classes of instrument [35–37].

8.2.3 Tandem Mass Spectrometry (MS/MS, MS^n)

Mass spectrometers apply a variety of fragmentation techniques (Table 8.2) that either have slow (10^{-5}–1 s) or fast ($<10^{-8}$ s) activation timeframes [37, 38]. The different techniques produce distinct types of terminal fragment ions [38, 39] that are annotated by the Roepstorff, Fohlman, and Biemann nomenclature (Fig. 8.3, right) [40, 41]. In TDMS, thermal or vibrational heating of the amino acid backbone occurs through collisions with gas or photons from a laser. Examples include resonant-based collision induced dissociation (CID) [39, 42], non-resonant higher-energy collisional dissociation (HCD) or collision activated dissociation (CAD) [43], infrared multiphoton dissociation (IRMPD) [44], and ultraviolet photon dissociation (UVPD) [45, 46]. These methods predominately cleave the polypeptide at the weakest amide bonds leading to "b" (N-terminal) and "y" (C-terminal) fragments; although the rapid activation by UVPD (~10^{-15} s) results in relatively random backbone cleavage that leads to complex spectra presenting most terminal fragment types. Electron capture dissociation (ECD) [47] and electron transfer dissociation (ETD) [48] cleave the backbone N-Cα bonds to form "c" (N-terminal) and "z·" (C-terminal) fragment ions. Addition of the electron onto the polypeptide forms a radical cation that rapidly cleaves the backbone. Consequently, ECD and ETD spectra are fragment rich due to the random cleavage events. In contrast with most other methods [49], ECD and ETD do not eject labile side-chain PTMs (such as phosphorylation or glycosylation) which increases their utility for localizing PTMs along the protein's backbone [50, 51].

In high-throughput "omics" investigations, the automation of MS and MS/MS events may be divided into data-dependent (DD) vs. data-independent (DI) acquisitions (Fig. 8.4) [30, 52]. These balance highly selective MS/MS events on a single spectral target (DD) vs. throughput with parallel fragmentation of co-existing species in a spectrum (DI). A protein's gas-phase structure and charge number, as well as the choice of MS/MS method influences the number and position of fragmentation events for a protein. Therefore, in DD acquisitions, MS/MS methods are often exploited in parallel to improve the total number of identifications across a proteome. Alternatively, MS/MS methods may be used

Table 8.2 Fragmentation techniques: Comparison of common *in vacuo* dissociation methods

Technique	Fragmentation	Mechanisms (cleavage site)	Special equipment	Automation
CID[a]	Collision	Resonant excitation[b] (b, y)	Ion trap	DD
CAD/HCD	Collision	Non-resonant excitation[c] (b, y)	Collision cell	DD, DI
ECD	Electron	Electron transfer (c, z)	Heated cathode	DD
ETD	Electron	Radical transfer (c, z)	Chemical ionization source	DD
IRMPD	Photon	Direct excitation[d] (b, y)	CO_2 laser	DD, DI
ISD	Collision	Non-resonant excitation[c] (b, y)	MS inlet (nozzle/skimmer)	DI
SID	Collision	Non-resonant excitation[c] (b, y)[e]	Metal surface	DD, DI
UVPD	Photon	Direct excitation[d] (a, b, c, x, y, z)	Excimer laser	DD, DI

Abbreviations: *CID* collision-induced dissociation, *CAD* collision assisted dissociation, *ECD* electron capture dissociation, *ETD* electron transfer dissociation, *HCD* higher-energy collisional dissociation, *IRMPD* infrared multiphoton dissociation, *ISD* in source dissociation (nozzle/skimmer dissociation), *SID* surface induced dissociation, *UVPD* ultraviolet photo-dissociation

[a]Called sustained off-resonance irradiation collision-activated (induced) dissociation (SORI-CAD) when performed in FT-ICR

[b]Application of radio frequencies to excite/dissociate to increase kinetic energy of trapped ions

[c]Application of DC potentials to accelerate ions into a high pressure region or surface

[d]Introduction of single or multiple photons to trapped ions

[e]Commonly results in ejection of macromolecular assemblies (see native MS)

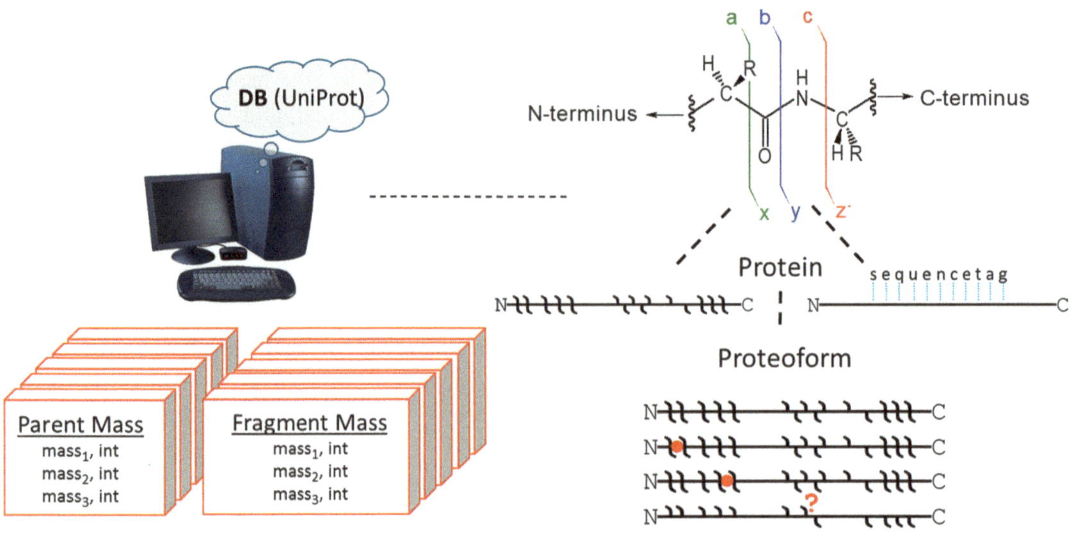

Fig. 8.3 Protein identification: A set of *in silico* informatics tools are used to process raw MS data, as well as, identify and characterize the proteoforms present. Deconvoluted spectra are typically provided as lists of parent masses and/or associated fragment masses. Fragmentation data often supports the generation of amino acid sequence tags that are searched against databases for proteins that contain the tag within its amino acid sequence. Alternatively, since fragment ions often contain either the N- and C- terminus, observed fragment masses are searched against theoretical terminal fragment ion masses for each protein in a database

sequentially (MS^n) on product ions to improve the resolution of proteoform characterization [18, 30, 49, 53–55]. These experiments typically use inclusion and exclusion criteria via decision-tree methods to automatically target or adjust fragmentation variables that may be influenced by the precursors mass or charge (e.g., activation times and energy levels) [18, 56, 57]. DI acquisitions have been applied to study proteins >25 kDa where low resolving power broadband spectrum are obtained followed by coincident dissociation of all components [32]. Alternatively, segmentation of the *m*/*z* range into multiple ~30–80 Δ *m*/*z* windows may be applied to enhance the dynamic range for intact precursor measurement and improve the sensitivity of fragment ion detection [18, 58].

8.2.4 Data Analysis and Informatics

Algorithms and software tools are available through instrument vendors or online to support automated spectral deconvolution of low and high resolution data [59–61]. Generally, the *z* for a protonated ion in a spectrum is readily derived from spacing between adjacent charge states: $z_{\frac{m}{z}i} = \frac{m}{z}_{i+1} / \left(\frac{m}{z}_{i+1} - \frac{m}{z}_{i}\right)$. Instruments that attain sufficient spectral resolving power may also separate peaks into isotopologues that predominately reflects the natural variation of carbon-12 (^{12}C) and carbon-13 (^{13}C) in the polypeptide (Fig. 8.2B, inset). Here, *z* is derived by counting the number of isotopes in a single *m*/*z* unit or by way of the Δ*m*/*z* difference between adjacent isotopes ($z = 1/\Delta m/z_{iso1\text{-}iso2}$). When reporting from high-resolving power datasets either the monoisotopic mass ($^{12}C_{100}{}^{13}C_0$, i.e., polypeptide containing only ^{12}C) or most abundant isotope ($^{12}C_{100\text{-}n}{}^{13}C_n$) mass is given, contrasting lower resolution approaches where the average molecular mass for all unresolved isotopes is reported.

As highlighted above, MS/MS serves a key role in identifying proteins and differentiating proteoforms. For example, the masses of fragments may be used for *de novo* sequence analysis. Here, a series of fragment ions differing

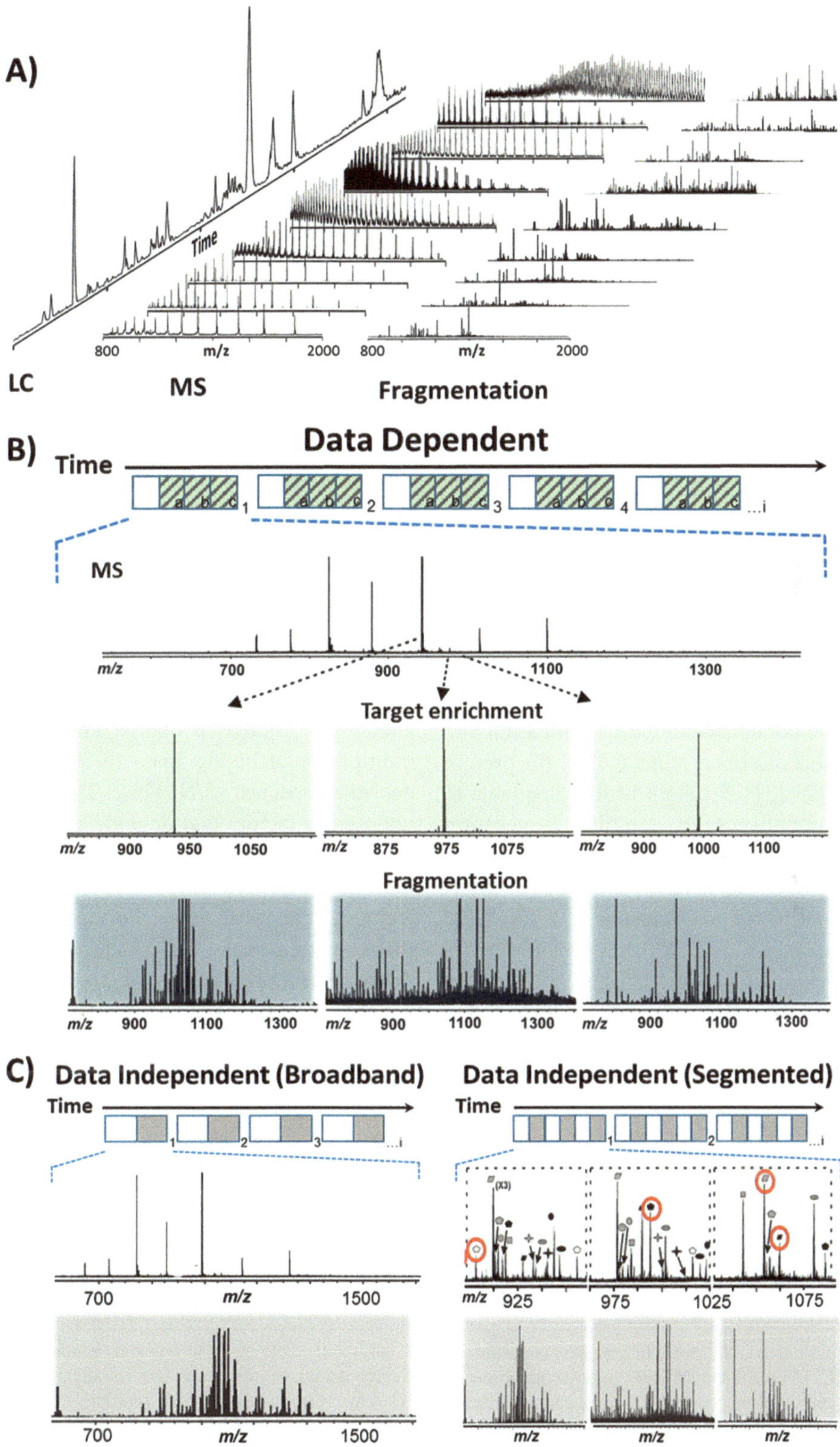

Fig. 8.4 Overview of data-dependent (DD) vs. data-independent (DI) fragmentation. (**A**) "Omics" investigations by TD typically include reversed-phased LC where proteins are automatically interrogated over time with DD or DI fragmentation. (**B**) In DD analysis, automated selection, enrichment, and fragmentation of

in masses equal to that of distinct amino acids (i.e., *sequence tag)* are searched against the protein database for matches with the same consensus sequence (Fig. 8.3) [62–64]. An alternative approach to protein identification is to correlate predicted terminal fragment ions of proteins to those observed in the MS/MS spectrum. For individual spectra, diverse open access resources are available for assignment of tandem MS data to target sequences (e.g., PROWL [65], BUPID [66], and MASH Suite [67]). Similarly, Kelleher and coworkers have created the ProSight series of search engines (e.g., web-based ProSight PTM (free), ProSight Lite (free), and ProSightPC 3.0 (commercialized by ThermoFisher Scientific) [68, 69]. ProSight uses a Poisson model to determine the significance of an identification made from MS/MS fragment matches [70]. The probability of a random identification is dependent upon the experimental mass accuracy and the number of fragments that match a protein in a database relative to the total number of fragments observed. Various scoring metrics are available to determine confidence in the identification and estimate false discovery rates (FDR) for protein identifications [32]. ProSight is also amenable to assigning confidence in complex proteoform studies [71]. For example, in work on histone H4, Pesavento et al. performed *in silico* "shotgun" annotation of all plausible H4 PTMs to create a database for MS/MS searches [34].

Other informatics approaches have also been created for TD. For example, Pevzner and coworkers created a spectral alignment algorithm that identifies protein forms presenting with concomitant PTMs. [72] In a follow-up report, the algorithm, MS-Align-E, was used to characterize histone H4 proteoforms, proving particularly useful for proteoform assignments in the absence of highly annotated databases [73]. The precursor ion independent top-down algorithm (PIITA) cross-correlates deconvoluted MS^2 spectra to theoretical MS [2] spectra in a manner similar to the SEQUEST algorithm used in ion peptide studies [74]. After identification, PIITA uses any difference observed between the observed precursor mass and theoretical precursor mass to identify and locate PTMs. Preliminary work with PIITA characterized 154 proteins at <1 % FDR from *Salmonella typhimurium* membrane extracts [74]. BIG-Mascot was created to extend the working mass range of the peptide-based search engine MASCOT (Matrix Science) [75]. Initial examples highlighted the identification of protein variants from 8 to 669 kDa through a combination of accurate mass tags and/or MS/MS events.

8.3 Chromatography

MS on intact proteins presents significant challenges. Increased charge and isotopologue multiplicity at higher mass (>20 kDa) quickly decrease spectral S/N [76]. This compounds upon other factors that degrade signal including charge competition between different proteins during ESI, protein solubility, and biological or technical chemical noise. Fortunately, many of these issues can be overcome by chromatographic preprocessing (e.g., molecular weight cutoff filters, dialysis, or immunoprecipitation). For complex mixtures, the observational capacity of the workflow is also increased by multidimensional steps which fractionate proteins by orthogonal physicochemical properties (e.g., size, isoelectric point (p*I*), hydrophobicity, and polarity). Many of these tools are briefly discussed here.

Fig. 8.4 (continued) individual charge states is performed on-the-fly. (**C**) In broadband DI (*left*), mass or *m*/*z* information is not used as a pre-selection criteria and all spectral partners (different charge states or different proteins) are simultaneously fragmented. Segmented DI (*right*) serves as an intermediate between DD and broadband DI because fragmentation events occur sequentially on enriched *m*/*z* windows that often contain multiple charge states of more than one protein. Adapted with permission from Ref. [52]. Copyright 2004 by Elsevier

8.3.1 Reversed Phase Separations

Whether offline, online, or via solid-phase capture/extraction tips (products such as Millipore's ZipTips), reversed-phase (RP) chromatography plays a pivotal role in TDMS. RP separations are mediated by the strength of the interaction between the hydrophobic domains of proteins and the non-polar stationary phase (Fig. 8.3A). Strong interactions permit simple sample clean-up from polar salts and buffers, as well as, support mixture partitioning by gradient elution with increased organic solvent concentration over time. In contrast with peptide-based analysis, when using conventional porous RPLC resins proteins often elute from the RP columns over several minutes. The poor peak capacity originates from structural and proteoform-level diversity, column precipitation, and poor diffusion/mass transfer characteristics through porous media. This has prompted the application of different resin architectures or chemistries for improved resolution in TD studies. Straightforward adjustments include the use of small C(n)-alkyl chain ($n = 4$ vs 18) resins which decrease the strength of protein and surface interactions. Larger pore size (1000 Å vs. 120–300 Å) has also been shown to improve peak capacity for larger proteins, as well as, increase of the organic mobile phase eluotropic strength (e.g., isopropanol) improves solubility of hydrophobic proteins [77]. Additionally, extension of conventional resins to ultra-high performance LC (UPLC) permits protein separations at high back pressure (400–1600 bar) [78]. For example, Ansong et al. used an UPLC system with 5 μm particle sizes packed into 80 cm long columns and extended gradients (~4 h) to identify 563 small to mid-sized polypeptides (and 1665 proteoforms) from lysates of *Salmonella typhimurium* [79].

Novel RP resin architectures have also been pursued. For example, polystyrene-divinylbenzene copolymers (e.g. Agilent's PLRP-S media) provide good mechanical and chemical stability under acidic/basic pH extremes and at elevated column temperatures (e.g. 50–65 °C). The use of higher temperatures enhances adsorption/desorption kinetics, lowers the mobile-phase viscosity, and helps to denature proteins. Vellaichamy et al. used PLRP in capillary-columns and found a ~2–3× improvement in resolution and spectral S/N versus conventional porous silica [77]. Monolithic stationary phases composed of a cross-linked network of mesoporous material (e.g., polymer, silica, and organic-silica hybrid monoliths) also provide good mass transfer characteristics for proteins. Eeltinik et al. created a 200 μm × 250 mm capillary polymer monolithic column and showed peak capacities >600 for complex mixtures studies [80]. Monolithic separations have also been applied in TD investigations on milk proteome [81] and the characterization of 19S and 20S proteasome proteins [82]. In a separate approach, Roth et al. showed superficially porous resins, consisting of a solid core with <1 μm porous outer shell packed into capillary columns, yield protein elution in <10 s, with 10^4 quantitative dynamic range, and attomole detection limits for standard proteins and lysates on heart tissue or cell cultures [16]. Finally, predictions of lower plate height minimum in Van Deemter plots suggests that extension of RPLC to <2 μm particles will improve chromatographic resolution [83]. Wu et al. showed that 0.5 μm diameter nonporous silica particles in capillary columns limit eddy diffusion and create a "slip flow" phenomenon along capillary walls that enhances flow and decreased velocity distributions of analyte [84]. When applied in TD studies on *Escherichia coli*, a peak capacity of 750 was observed for a 60 min gradient.

8.3.2 Ion Exchange

Ion exchange chromatography (IEC), such as weak anion exchange (WAX), weak cation exchange (WCX), strong anion exchange (SAX), strong cation exchange (SCX), and immobilized metal-ion affinity chromatography (IMAC), separates proteins based upon charge-charge interactions between a protein and a charged resin [85]. A step-wise or linear gradient increase in counter ion concentration (supplied by salts or changes in pH) helps to elute proteins, often under non-denaturing conditions. Combined with RPLC, both online and offline IEC support multidimensional TD experiments. For example, Shrama et al. used WAX prefractionation and online RPLC and detected 715 intact proteins from a *Shewanella oneidensis* MR-1 cell lysate [86]. Roth et al. used WAX/RPLC in a 2D workflow and analyzed >600 proteoforms from human primary leukocytes harvested from leukoreduction filters [87]. Similarly, SAX as a first dimension separation technique has been applied for integrated TD and BU studies on E. coli [82, 83]. These studies highlight the complementarity of the approaches, with small and larger proteins often over-represented for TD and BU, respectively [88, 89].

8.3.3 Hydrophilic Interactions

Hydrophilic interaction liquid chromatography (HILIC) separates polypeptides via a normal- or polar-stationary phase in the presence of a less polar mobile phase [90]. In HILIC, the stationary-phase is typically primed with water to form a hydrophilic shell prior to addition of the organic mobile phase. Separation is achieved by partitioning the polypeptides between the hydrophilic and hydrophobic layers with gradient elution by increasing water concentration over time, resulting in elution based on hydrophilicity. When the stationary phase is supplied by IEC columns, the added ionic interactions provide added selectivity. For example, HILIC on a WCX column has found widespread use for the sub-fractionation of histone proteoforms in TD and MD studies. Garcia et al. utilized WCX-HILIC offline with subsequent RPLC and direct infusion by μESI to identify 150 and 42 different proteoforms on histone H3.2 and H4, respectively [33, 91]. Young and coworkers extended WCX-HILIC to capillary-based columns for MD applications [4]. They created a "saltless" pH gradient for direct coupling to the mass spectrometer, and characterized >200 H3.2 and 70 H4 species from 1 μg of material in 2 h. Similarly, Tian, et al. recently created an online multidimensional histone fractionation system that automatically fractionated ~7.5 μg of major histone family members by RPLC prior to metal-free WCX/HILIC-MS/MS analysis [92, 93]. They identified 105 H4, 110 H2B, 77 H2A, and 416 H3 proteoforms in a single run.

8.3.4 Capillary Zone Electrophoresis & Isoelectric Focusing

Solution-based electrophoretic approaches have received considerable attention in TD studies owing to their high separation efficiencies. Online approaches with capillary zone electrophoresis (CZE) have been used to characterize proteins from microorganisms, biofluids, protein-ligand interactions, biopharmaceuticals and dietary proteins [94–96]. For example, Sun et al. created an electrokinetically pumped "sheath-flow" ESI-CZE–MS interface, providing proof of concept for TD on standard proteins and *Mycobaterium marinum* secretome [97, 98]. Li et al. created a similar apparatus as part of a multidimensional scheme that size sorted proteins into discrete molecular mass windows prior to CZE-ESI-MS/MS, identifying 30 proteins from 30 to 80 kDa from *Pseudomonas aeruginosa* [99]. Haselberg et al. created a "sheathless" CZE ESI interface for the characterization of 18 and 74 glycoproteoforms of recombinant human interferon-β and human erythropoietin respectively [100]. Han et al. applied a similar approach as part of an RPLC-CZE TD workflow characterizing ~300 proteoforms from 270 ng of protein from *Pyrococcus furiosus*, as well as proteins in the Dam1 complex from *Saccharomyces cerevisiae* [101].

Chromatofocusing (CF) and isoelectric focusing (IEF) separate proteins based upon their isoelectric point (p*I*). CF exploits a pH gradient on IEC columns and has been applied in studies on methanosarcinides [18], membrane proteins [102] and cancer cells [103]. Similarly, a variety of IEF systems (e.g., Rotofor, Free Flow Electrophoresis, IsoelectriQ, OFFGEL and Zoom IEF) have been applied for first-dimension fractionation of intact proteins (0.05–5 mg) [104]. Each system has different routes for generation of the pH gradient (e.g. carrier ampholytes and/or immobilized pH gradient membranes). For example, Zhang et al. recently used off-gel IEF prior to LC-MS to characterize hundreds of proteoforms in heart sarcomeres ranging from 5 to 230 kDa in mass [105, 106]. Here, protein separation occurs in solution via voltage-driven migration through an immobilized p*H* gradient (IPG) (Fig. 8.5A). Jensen et al. created a totally solution-based capillary IEF (cIEF) procedure for measurements of the *E. coli* and *D. radiodurans* proteomes, characterizing up ~ 1000 proteoforms from a total injection of ~ 300 ng [107].

8.3.5 Size-Based Separations

Size/mass separations are an attractive option to overcome S/N bias associated with measuring intact proteins over broad mass ranges [76]. While 2D gel electrophoresis (2DGE) has unsurpassed peak capacity for proteins from 5 to 250 kDa [108], poor duty cycle for gel elution has largely prevented its use in TD. Alternatively, solution-based size exclusion chromatography (SEC) and continuous flow gel elution (CFGE) have been routinely exploited in TD. In SEC, proteins migrate through a porous polymeric column and are separated by their hydrodynamic volume with larger proteins eluting before smaller ones [109]. Examples of SEC in TD include the characterization of lumen proteins from *Arabidopsis thaliana* [110], membrane protein complexes in *Synechocystis sp. PC6803* [111], degradation products of biopharmaceuticals [112], and structural studies on amyloid beta oligomers [113]. Chen and Ge recently reported that a novel ultrahigh pressure SEC approach utilizing MS compatible elution buffers permitted MS analysis of proteins from 6 to 669 kDa in <7 min [114].

CFGE separates proteins on a tubular polyacrylamide gel electrophoresis column with eluting proteins fraction collected by increasing protein mass over time. Meng et al. used a preparative CFGE apparatus with an acid labile surfactant (ALS) to fractionate low mass yeast proteome in ~5 kDa windows prior to offline RPLC and μESI [115]. Use of ALS instead of SDS allowed direct processing of PAGE fractions without precipitation. Tran et al. further refined this approach creating a

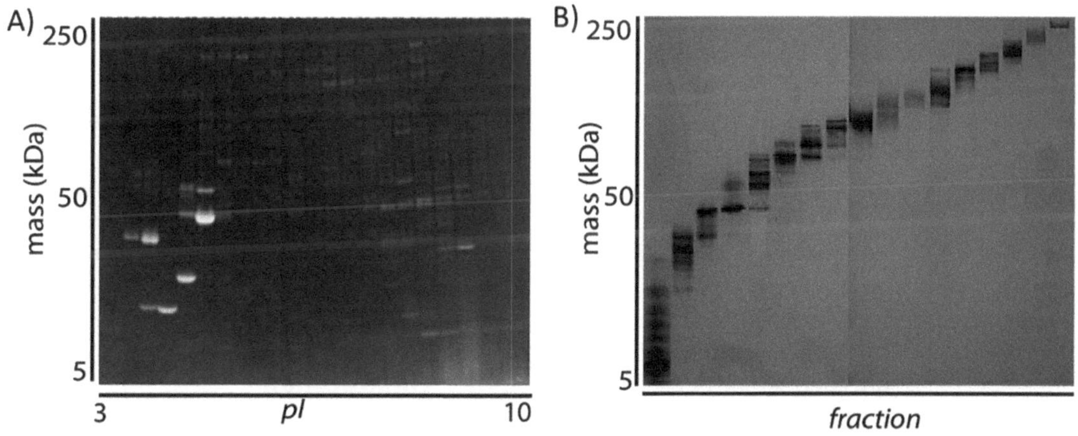

Fig. 8.5 (**A**) SDS/PAGE analysis of mouse heart myofibrils fractionated by off-gel IEF with a 3–10 immobilized pH gradient. Adapted with permission from Ref. [105]. Copyright 2013 by American Chemical Society. (**B**) SDS/PAGE analysis of yeast proteome fractionated by GELFrEE mass separation. Adapted with permission from Ref. [116]. Copyright 2009 by American Chemical Society

gel-eluted liquid fraction entrapment electrophoresis (GELFrEE) [116] technique that exploits smaller tube dimensions with short resolving gels [116]. GELFrEE could reproducibly separate μg to milligram levels of material from 5 to 250 kDa with high recovery (Fig. 8.5B) [70, 77], and has been used in recent multidimensional TD investigations. For example, Kelleher and coworkers combined IEF, GELFrEE, and LC-MS/MS into a 4D dimension TD workflow that provided a theoretical 4D peak capacity >100,000. The work represents the largest dynamic range for TD on mammalian cell lysate reported to date, as well as, highlights that in TD proteomics investigations, the number of observed proteoforms will typically exceed the number of identified proteins by ˜3× [32, 116]. Scaling of CFGE to microfluidic devises is also showing potential for ultrafast size based separations. For example, Root et al. fully resolved various standard proteins through a 75 μm ID, 25 cm long fused-silica capillary coated with a poly-(N-hydroxyethylacrylamide) polymer in <3 min. [117]

8.4 Frontiers

8.4.1 Comprehensive Proteoform Studies

As highlighted above, the transition to multidimensional separations and online LC-MS/MS has dramatically increased proteome coverage for complex mixtures and detailed proteoform investigations on target proteins [32, 79]. In targeted studies, the investigations into the histone family members (H1, H2a, H2b, H4 and H3) (see Sect. 8.4.3) exemplify the utility of top-down for the examination of extreme complexity associated with heterogeneous modification states [4, 33, 34, 91, 118–120]. Extreme proteoform diversification is not limited to epigenetic regulators. For example, Zhang et al. used TD to characterize 12 PTMs at 11 different sites on myelin basic protein (MBP), an intrinsically disordered protein the myelin sheath [121]. They found diverse PTM classes, including N-terminal and internal acetylation, mono- and dimethylatation deamination, deamidation, and phosphorylation. Similar investigations have been performed on clinical biomarkers including transthyretin and hemoglobin variants [122–126], studies into the deamidation kinetics in ribonuclease A [127] and beta-2-microglobulin [128], and hundreds of nitration and oxidation events on calmodulin [129]. Additionally, Ge and coworkers have used TD to monitor phosphorylation on diverse myofibril proteins in the context of chronic heart failure [130–132]. In infectious disease research, Burnaevskiy et al. characterized a novel N-terminal demyristoylation and coincident amidation event on ADP-ribosylation factor (ARF)1p and (ARF)2p by *Shigella flexneri* virulence factor invasion plasmid antigen J (IpaJ) [133].

TD is also being applied for the evaluation of microheterogeneity associated with protein glycosylation [112, 134–140]. For example, Bourgoin-Voillard et al. used CID, IRMPD, ETD and ECD to characterize fragmentation dynamics of intact RNAse B and its bound N-glycans [66]. In the characterization of two isoforms of prostate specific antigen the Association of Bimolecular Resource Facilities (ABRF) determined that TD quantified glycoproteoforms with the same reliability as conventional peptide-*N*-glycosidase F (PNGase F) release of N-glycans procedures [141, 142]. In another example, Zhang et al. used off-gel IEF to fractionate total cerebrospinal fluid (CSF) proteins prior to LC-MS. [105, 106] The approach permitted the generation of virtual 2D gels (p*I* vs mass) that resolved >200 di-n-glycosylated lipocalin-type prostaglandin d synthase glycoproteoforms directly from the CSF-proteome [106].

8.4.2 Quantitation

TD quantitation methods have largely mirrored those used in bottom-up [143]. Label-free quantitation (LFQ) offer a distinct advantage for comparison of clinical samples where metabolic-labeling is not possible and chemical-labeling costs may be too restrictive. For example, Gucinski and Boyne examined the impact of multiple charge states, isotopologues, and resolving power on linearity of TD LFQ on protein therapeutics, highlighting that absolute quantitation by standard curves, as well as relative quantitation between analyte and internal standards, is readily possible [144]. For complex mixtures, Julka et al. combined ultraviolet detection with 2D SAX LC-MS to quantify proteins spiked into *E. coli*, demonstrating good linearity ($R^2 > 0.99$) over a ten-fold concentration range [145]. In biomarker discovery, Mazur et al. applied an automated differential MS (dMS) infrastructure to examine changes between patient expression in human high-density lipoprotein (HDL) [139, 146]. More recently, Ntai extended LFQ to multidimensional TD workflows that sought to characterize proteome and proteoform-level effects of deleting histone deacetylase (rpd3) in budding yeast. [147] Numerous other investigations have also evaluated the reliability of TD for proteoform ratio determination [33, 87, 91, 118, 119, 148], often finding that the ratios of isobaric proteoforms, for example resulting from PTM positional isomers, must be determined from fragment ion ratios [118, 148, 149].

Metabolic labeling by SILAC (*S*table *I*sotope *L*abeling by *A*mino Acid in *C*ell culture) has found use for proteins isolated from cell cultures [86, 150, 151]. For example, Veenstra et al. introduced isotopically labeled leucine into proteins in *E. coli* characterizing expression changes via CIEF and FT-ICR [152]. Parks et al. used WAX and RPLC separations with FT-ICR to characterize $^{14}N/^{15}N$ metabolic labels on histidine, leucine, and tryptophan to determine expression ratios on 231 metabolically labeled protein pairs in *S. cerevisiae* [153]. Similar work by Pesavento et al. monitored changes to histone H4K20 methylation during the HeLa cell cycle progression [154]. Collier et al. also applied metabolic labeling for quantitative comparison of hundreds of proteins from *Aspergillus flavus* and human embryonic stem cells [150, 155]. More recently, Rhoads et al. created a novel neutron-encoded mass signature strategy which labeled proteins in yeast cultures with either $^{13}C_6{}^{15}N_2$-lysine or $^{2}H_8$-lysine [26]. The work highlighted a mass difference of 0.036 Da between the isotopologues, that was not distinguished during low resolving power scans, but could be discriminated upon acquiring a high-resolution scan, permitted quantitation based upon isotopologue ratios. The strategy has potential for TD metabolic-labeling studies because it permits multiplexing without increasing spectral congestion [26].

8.4.3 Biologics and Biosimilars

Protein-drugs are often modified by PTMs (such as N-glycosylation) and careful attention to the location, composition, or structure of these modifications is key to ensuring both biological efficacy and toxicity of biologics and biosimilars [156]. TD has been increasingly exploited to meet these regulatory challenges. For example, Boyne and coworkers applied TD to evaluate PTM-profiles on FDA-approved and unapproved filgrastim therapeutics [156, 157]. Additionally, they defined sequence variations for different forms of herring protamine sulfate and low abundant impurities in the complex drug product. Similar efforts have extended TD to antibody based drugs with investigators creating MS and MS/MS protocols for intact monoclonal IgGs (˜150 kDa) or on the two light chains (L_c, ˜25 kDa each) and two heavy chains (H_c, ˜50 kDa each) independently. For example, Mazur et al. used SEC with CID and ETD to characterize IgG impurities (e.g., proteolytic breakdown products) [158]. Studies on intact antibodies with Orbitrap and FT-ICR show that ETD and ECD can provide ˜33 % sequence coverage [159, 160]. Zhang et al. performed ISD on an intact antibody followed by CID of ISD fragments to improve sequence coverage to 46 % and 27 % for L_c and H_c, respectively [161]. LC–MS/MS have also been performed on individual L_c and H_c after offline disulfide bond reduction [162–164]. In a another novel approach, Nicolardi et al. extended reduction steps to online LCMS, using an inline electrochemical reduction cell to systematically release interchain disulfide bonds and disassemble L_c and H_c from the full IgG1 mAb [165].

8.4.4 Membrane Proteins

Integral membrane proteins (IMPs) play key roles in transmembrane signaling and are thought to constitute approximately a third of the proteome [102, 111]. IMPs have challenged both TD and BU workflows because of their amphipathic characteristics (harboring polar soluble domains and hydrophobic transmembrane domains) and heterogeneous PTMs [166]. Methods for solubilization of IMPs often include SDS or Triton X-100 [167, 168]. To overcome the workflow mismatch, precipitation by chloroform/methanol/water or extraction with acetone are commonly used [169]. Similarly, other agents (e.g., urea, sodium deoxycholate, or acid-labile surfactants) also enhance solubility of IMPs and are compatible with SEC and RPLC workflows [170]. Other investigations have also shown IMP solubility is enhanced through >80 % formic acid or organic solvents with strong eluotropic strength (e.g., isopropanol) [167–169, 171]. For example, Doucette et al. recently examined the effectiveness of acetone and chloroform/methanol/water precipitation for SDS removal and found that both provided >90 % recovery when resolubilization of the precipitated proteins was performed in cold (−21 °C) 80 % formic acid [172].

Recent investigations have demonstrated the scale at which polytopic IMPs can be interrogated by TD. For example, Carroll et al. used Q-TOF with CID to characterize several small proteolipids (1–4 transmembrane helices) and larger proteins (1–18 transmembrane helices) from bovine mitochondrial preparations [171]. LC-MS/MS with CAD or ETD was used to characterize eleven integral and five peripheral subunits of the 750 kDa Photosystem II (PSII) complex from the eukaryotic red alga, *Galdieria sulphuraria* [173], and numerous subunits of the cytochrome b6f complex from chloroplasts and cyanobacteria [174]. In another example, Catherman et al. enriched mitochondrial proteins from NCI H-1299 cells and used multidimensional separations to characterize over 300 IMPs with up to 12 transmembrane helixes [175]. Of note, CID has been shown to preferentially fragment in transmembrane domains [176]. Additionally, it complements ETD or ECD (with or without vibrational activation by IRMPD [177]) for high amino acid resolution of PTMs over variegated inter-, trans-, or intracellular domains (e.g., proteolytic processing, phosphorylation, disulfide bonds, cysteinylation, heme-modification, pyroglutamate, acetylation, amidation, formulation, N^6-retinylidene, etc.) [166, 174–176].

8.4.5 Hydrogen/Deuterium Exchange

Characterization of a protein's structure is important for understanding its function. As protein in solution unfold and refold, hydrogen bonds break and reform. The dynamics of these structural changes can be monitored though hydrogen/deuterium exchange (HDX) at solvent-exposed amides [178]. The conservation of HDX information in intact proteins via TD analysis offers distinct benefits over peptide-based assays where 10–50 % deuterium back-exchange often results during subsequent proteolytic processing. Key to the utility of TD in HDX is that ECD leads to low H/D "scrambling" along the amino acid backbone during gas-phase dissociation. Low scrambling preserves structural information associated with H/D positions [179, 180]. Wang et al. recently used HDX with ECD to gain conformer-specific information on non-native protein states of ubiquitin [181]. Similarly, Pan et al. performed HDX with short HPLC gradients to characterize the therapeutic protein interferon α2a, a cancer drug, and several variants [182]. Their methods provided insight into the protein's primary structure including identification of preferential oxidation on methionine residues that led to distinct PTM-induced structural changes. To minimize back-exchange over extended periods, Amon et al. recently devised a sub-zero cooled microchip for nESI showing <5 % back-exchange over 10 min [183]. Pan et al. also extended subzero temperature screens to HPLC runs showing ˜2 % back-exchange over 10 min elution gradients when carrying out structural comparability tests on intact antibodies [184].

8.4.6 Native MS and Protein Complex Studies

Studies on protein macromolecular assemblies in their native state are key to understanding protein function. Numerous TD protocols for "native" MS have emerged to complement conventional structural biology techniques (e.g., NMR and X-ray crystallography), providing information on the complex's protein composition, stoichiometry, spatial arrangements of subunits, assembly dynamics, or interactions with other ligands or metal ions [185]. To support these variegated efforts, researchers have developed various MS compatible buffers and additives that can preserve native assemblies during ESI and support both controlled disassembly of the complexes and denaturation of the proteins [185, 186].

A notable distinction between native MS and conventional protein MS is the relatively high *m*/*z* values observed for protein complexes as compared to individual denatured proteins. This is because charge incorporation during ESI on folded structures is largely restricted to surface sites. The need for high *m*/*z* sampling has made ESI-Q-TOF the mainstay for native MS because TOF permits sensitive analysis largely independent of *m*/*z*. For example, Zhou et al. performed native ESI on the 801 kDa chaperonin complex GroEL and used CID and surface-induced dissociation (SID) to show that ejection of highly charged monomers is the predominant dissociation pathway in CID while SID results in extensive dissociation into a wide variety of products (Fig. 8.6) [187]. Similarly, Snijder et al. recently used ESI-Q-TOF to study the 18 MDa capsid of bacteriophage HK97 [188], and Blackwell et al. used Q-TOF with CID and SID to dissociate the heterohexamer toyocamycin nitrile hydratase into monomers and subcomplexes [189]. These efforts highlight how the choice of dissociation mode and the extent of energy deposition dictates the degree of disassembly [178, 180, 187, 189].

The improved mass range associated with FTMS technologies (e.g., high field Orbitrap,

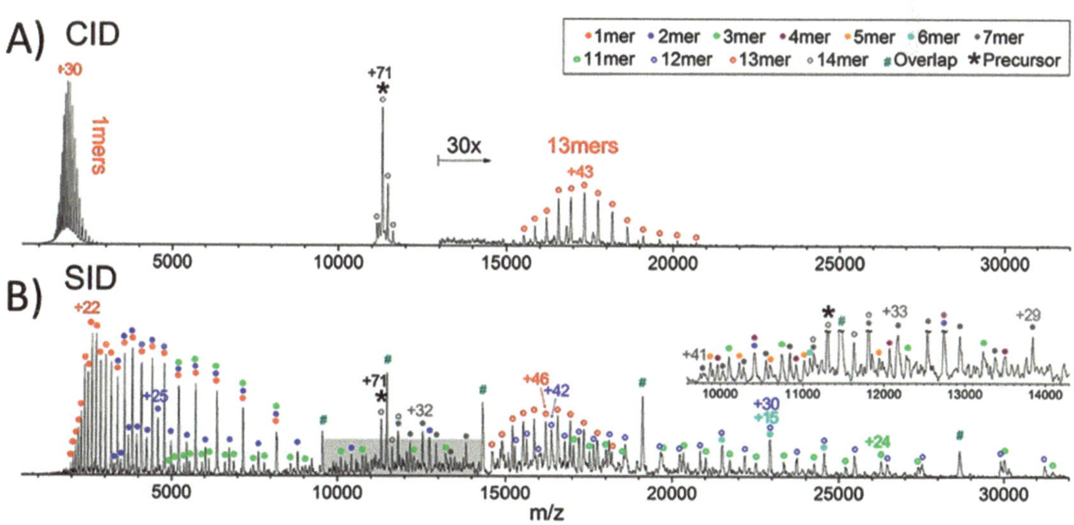

Fig. 8.6 Native MS. (**A**) Collision induced dissociation of the +71 GroEL tetradecamer. (**B**) Surface-induced dissociation of GroEL tetradecamer +71. The inset spectrum is a zoom-in view of the region shaded in the middle of the full SID spectrum. Charge states of several peaks discussed are selectively labeled with the corresponding colors of the dots in the legend. Adapted with permission from Ref. [187]. Copyright 2013 by American Chemical Society

superconducting magnets in excess of 15 T for FT-ICR, and expanded use of absorption-mode FT data-processing modes) [30, 190–197] has significant promise for native MS. For example, FT-ICR with CAD on complexes has been used to probe nucleotides and metal ligand-binding sites [198] while ETD and ECD were used to localize the position of bound ligands and the topology of protein complexes [199]. Similarly Li et al. used a 15 T FT-ICR-MS with an infinity trap for studies on yeast and horse liver alcohol dehydrogenase (147 kDa) [200]. Their approach permitted the isotopic resolution of a yeast ADH (yADH) tetramer (147 kDa) and showed ~40 % sequence coverage could be obtained directly from the native yADH tetramer complex when using a combination of ECD, ISD, CAD, and IRMPD methods. Also, Skinner et al. recently adapted GELFrEE for native state size separations followed by MS/MS on an Orbitrap to characterize protein complexes from mouse heart tissue and fundal secretome of *Trichoderma reesei* [201].

8.4.7 Ion Mobility

Ion mobility (IM) separates ions based on their collision cross section (ratio of size-to-charge) prior to MS analysis [202]. Structure-based separations derive from low-energy collisions with a neutral drift gas region in the presence of electrostatic or electrodynamic fields. Separations can occur either through space or temporal dispersion. Spatial dispersion is accomplished via differential mobility and field asymmetric ion mobility spectrometry (FAIMS), while time separations are accomplished with drift tube (DT) or traveling-wave (TW) formats.

Since IM occurs on the millisecond timeframe, it may be applied as an orthogonal post-LC separation strategy to improve dynamic range in proteomics. For example, IM has been combined with nano-LC-IM-MS/MS to evaluate charge-state specific fragmentation tendencies, with results generally showing MS/MS on high charge unfolded ions leads to improved sequence coverage [203]. IM has also been integrated post dissociation to reduce spectral congestion by resolving overlapping product ion series prior to MS analysis [204]. For example, Zinnel et al. created a hybrid CID-IM-MS strategy showing a 2–10× increased sequence coverage for various peptides or proteins [205].

In studies that probe gas-phase conformations of intact proteins, results on ubiquitin show that solution structure is consistent with ions produced by ESI as long as experimental conditions avoid thermal unfolding or Coulomb repulsion-induced unfolding at higher charge [206, 207]. Continued improvement in IM resolving power is also providing insight into conformational flexibility of proteins. For example, Clemmer and coworkers applied a frequency-based linear DT (overtone mobility spectrometry) to sample continuous ion sources at resolving powers >100 [208]. Additionally, Shvartsburg has improved FAIMS resolving power to ~400 through the application of elevated electric fields and hydrogen-rich collision gases. This approach separated charge states of standard proteins (up to 30 kDa) into over 20 gas-phase conformers per charge state [209]. Continued application of high resolving power IM is expected to facilitate the interrogation of isobaric proteoforms through differences in gas-phase structure [210], as well as complement native MS investigations by monitoring disassembly of heterogeneous complexes [211].

IM is also being applied in mainstream biopharmaceutical and biomarker research. Escribano et al. used IM and TDMS to analyze the specificity and structural behavior of several protein-platinum (Pt) metallodrug adducts and to determine the primary binding site(s) in dicarboxylate Pt compounds [212]. Bowers and coworkers applied IM to characterize factors leading to aberrant aggregation in neurodegeneration and neuroinflammation, such as amyloid beta and tau [213]. Young et al. expanded upon these efforts by establishing an high-throughput small molecule screen to identify inhibitors of amyloid aggregation [214]. In related work, Beveridge et al. proposed a generalized IM-MS framework that permits more accurate prediction of the extent of disorder in protein when compared to conventional charge hydropathy plots [215].

8.4.8 Charge Manipulation

Charge Reduction MS and MS/MS spectra are often populated by overlapping charge state distributions associated with simultaneous detection of proteins or fragment ions. The spectral congestion may lead to missed assignment of an ion's charge, particularly under conditions with inadequate resolving power [140]. Gas-phase "proton-transfer" (PT) reaction methods seek to reduce spectral congestion by converting multiply-charged ions into readily interpreted mono- and di-protonated species [216–219]. For example, McLuckey and coworkers have charge-reduced intact proteins with anion-based perfluorocarbons [193]. Their approach applies an electrodynamic QIT which enables frequency-based accumulation of specific charge reduced states [220]. In similar efforts, they have shown charge reduction on CID fragments provides an order of magnitude improvement in informatics scores [221, 222] (Fig. 8.7A). Efforts by Smith and coworkers have shown that novel [204] Po α particle and coronal discharge sources also permit controlled proton-reduction on mixtures of electrosprayed proteins [223, 224]. For omics investigations, Chi et al. have demonstrated that benzoate anions introduced post ETD enabled the identification of *E. coli* 70S ribosomal proteins during online LC-MS/MS on an LTQ [225]. Similar work by Huang et al. combined CID and ETD with ion/ion proton transfer reactions in a Q-TOF to characterize unknown proteins with novel PTMs from *E. coli* [36].

Super-Charging Enhancement of gas-phase charge also has benefits in TD, including improved resolving power at lower *m/z* for most mass analyzers, improved sensitivity of charge-sensitive detection (e.g., FTMS), and enhanced fragmentation efficiency due to a more unfolded gas-phase structure [226]. Therefore investigations have sought to identify agents that enhance protonation beyond that typically obtained by ESI. Early observations suggest that denaturing conditions in an ESI solution containing surfactants at low concentrations enhances charging (e.g., cationic, cetyl trimethylammonium bromide (CTAB) and zwitterionic, 3-(3-cholamidopropyl) dimethylammonio-1-propanesulfonate (CHAPS), and many nonionic saccharide-based detergents) [227]. More recently, Iavarone et al. showed that addition of glycerol and m-nitrobenzyl alcohol (m-NBA) can effectively "supercharge" proteins [228]. Other diverse organic compounds or common organic solvents (e.g., DMSO) also enhance charging in either direct infusion or LC-MS applications [186, 229, 230]. For example, Teo and Donald recently compared m-NBA, sulfolane, 2-nitroanisole, ethylene carbonate (EC) and propylene carbonate (PC) showing that the latter two promote charge states higher than the theoretical maximum predicted by proton-transfer reactivity (Fig. 8.7B) [226]. Williams and coworkers have recently introduced an electrothermal supercharging method that rapidly switches between native and denaturing conditions via changing ESI potentials, showing that inclusion of sulphate and phosphate anions increases protein charging from aqueous ammonium and sodium buffers [231, 232].

8.5 Concluding Remarks

Advancements in chromatography, MS, and informatics have made "birds-eye-view" (top-down) proteomics increasingly available to the masses. Key to the field's expansion is the high resolving power afforded by modern mass spectrometers which provides unparalleled clarity on the microheterogeneity that exists in a proteome at the proteoform-level. TD has proven to be an exceptionally powerful resource for

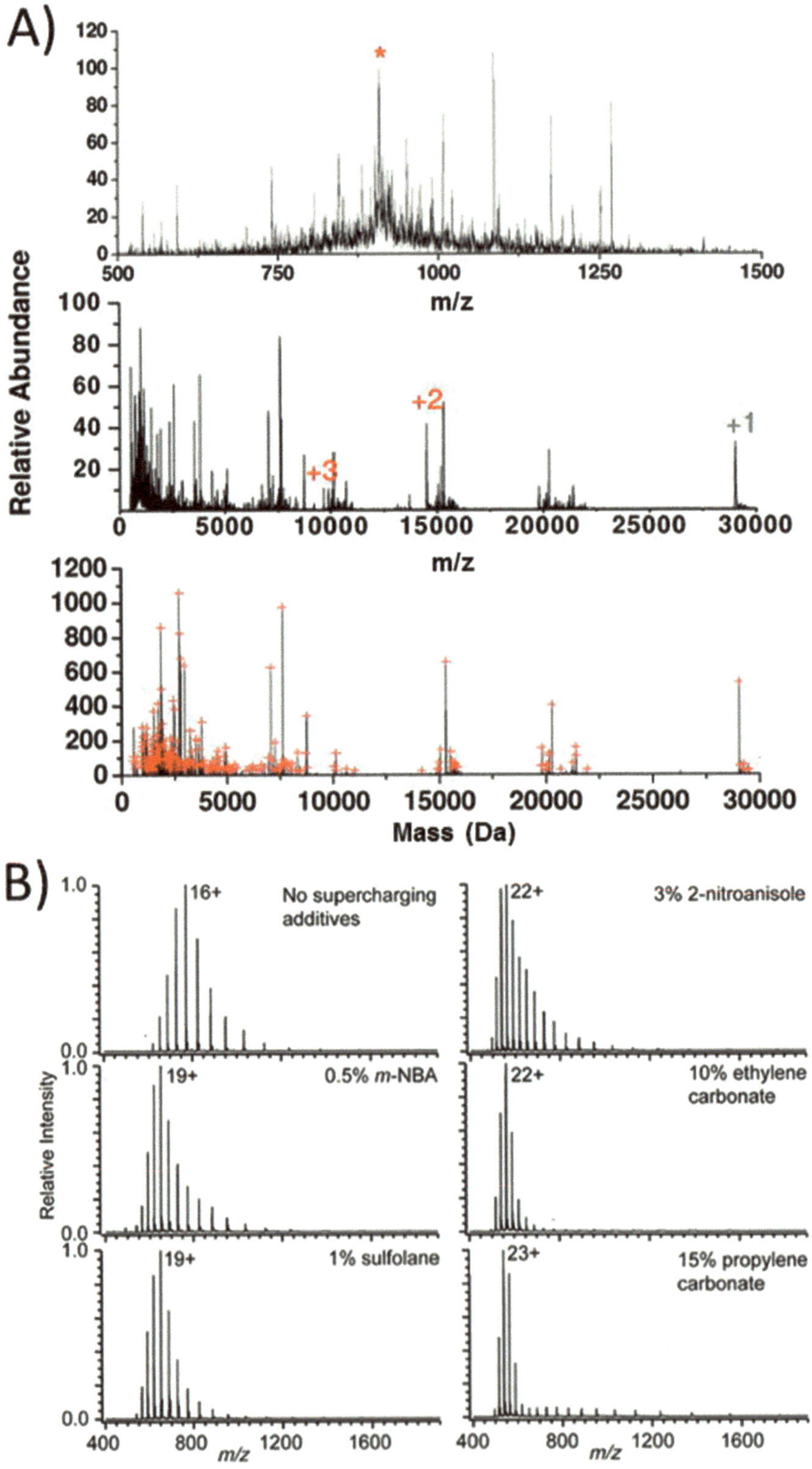

Fig. 8.7 Charge reduction and supercharging: (**A**) Collision activated dissociation of carbonic anhydrase $[M + 32H]^{32+}$ (*top*). Simultaneous CAD and ion/ion proton charge reduction reaction (*middle*) with the corresponding deconvoluted spectrum (*bottom*). Adapted with permission from Ref. [221]. Copyright 2009 by

hypothesis-driven research on defined protein targets. With technology advances showing duty-cycles and sensitivities that surpasses many conventional bioassays (gel-electrophoresis or western blots), it is easy to envision molecular biologists applying TD in their daily screens. While omic-level screens are still largely done in dedicated research labs, more widespread implementation is expected with the continued refinement of procedures, particularly informatics for comprehensive proteoform studies as well as the training of a new generation of researchers on TD protocols.

Acknowledgement This work was supported by the National Institute of General Medical Sciences of the National Institutes of Health under award number 1R01GM115739-01A1. Any opinions, findings, conclusions or recommendations expressed in this material are those of the authors and do not necessarily reflect the views of the National Institutes of Health. This work was also supported by the Multiple Sclerosis Society [PP-1503-04034], The Darrel K. Royal Research Fund for Alzheimer's Disease [48680-DKR], The Texas Alzheimer's Research and Care Consortium Investigator Grant Program [354091], and the UT System Neuroscience and Neurotechnology Research Institute [363027].

References

1. Lodish H, Berk A, Zipursky SL, and al., e (2000) Molecular cell biology, 4th edn. W. H. Freeman, New York
2. Smith LM, Kelleher NL, Proteomics CTD (2013) Proteoform: a single term describing protein complexity. Nat Methods 10:186–187
3. Janke C, Bulinski JC (2011) Post-translational regulation of the microtubule cytoskeleton: mechanisms and functions. Nat Rev Mol Cell Biol 12:773–786
4. Young NL, DiMaggio PA, Plazas-Mayorca MD, Baliban RC, Floudas CA, Garcia BA (2009) High throughput characterization of combinatorial histone codes. Mol Cell Proteomics 8:2266–2284
5. Jahn O, Tenzer S, Werner HB (2009) Myelin proteomics: molecular anatomy of an insulating sheath. Mol Neurobiol 40:55–72, 2758371
6. Aebersold R, Mann M (2003) Mass spectrometry-based proteomics. Nature 422:198–207
7. Meyer B, Papasotiriou DG, Karas M (2011) 100 % protein sequence coverage: a modern form of surrealism in proteomics. Amino Acids 41:291–310
8. Cannon J, Lohnes K, Wynne C, Wang Y, Edwards N, Fenselau C (2010) High-throughput middle-down analysis using an orbitrap. J Proteome Res 9:3886–3890, PMC2917504
9. Fenn JB, Mann M, Meng CK, Wong SF, Whitehouse CM (1989) Electrospray ionization for mass-spectrometry of large biomolecules. Science 246:64–71
10. Karas M, Hillenkamp F (1988) Laser desorption ionization of proteins with molecular masses exceeding 10,000 daltons. Anal Chem 60:2299–2301
11. Tanaka K, Waki H, Yutaka I, Akita S, Yoshida Y, Yoshida T (1988) Protein and polymer analyses up to m/z 100 000 by laser ionization time-of-flight mass spectrometry. Rapid Commun Mass Spectrom 2:151–153
12. Emmett MR, Caprioli RM (1994) Micro-electrospray mass spectrometry: ultra-high-sensitivity analysis of peptides and proteins. J Am Soc Mass Spectrom 5:605–613
13. Valaskovic GA, Kelleher NL, Little DP, Aaserud DJ, McLafferty FW (1995) Attomole-sensitivity electrospray source for large-molecule mass spectrometry. Anal Chem 67:3802–3805
14. Marginean I, Kelly RT, Prior DC, LaMarche BL, Tang K, Smith RD (2008) Analytical characterization of the electrospray ion source in the nanoflow regime. Anal Chem 80:6573–6579, PMC2692497
15. Cech NB, Enke CG (2001) Practical implications of some recent studies in electrospray ionization fundamentals. Mass Spectrom Rev 20:362–387
16. Roth MJ, Plymire DA, Chang AN, Kim J, Maresh EM, Larson SE, Patrie SM (2011) Sensitive and reproducible intact mass analysis of complex protein mixtures with superficially porous capillary reversed-phase liquid chromatography mass spectrometry. Anal Chem 83:9586–9592
17. Needham SR, Valaskovic GA (2015) Microspray and microflow LC-MS/MS: the perfect fit for bioanalysis. Bioanalysis 7:1061–1064
18. Patrie SM, Ferguson JT, Robinson DE, Whipple D, Rother M, Metcalf WW, Kelleher NL (2006) Top down mass spectrometry of <60-kDa proteins from Methanosarcina acetivorans using quadrupole FRMS with automated octopole collisionally activated dissociation. Mol Cell Proteomics MCP 5:14–25

Fig. 8.7 (continued) American Chemical Society. (**B**) ESI mass spectra of 45/54/1 methanol/water/acetic acid solutions containing 10 μM Cytochrome C and no supercharging additive, 0.5 % *m*-nitrobenzyl alcohol, 1 % sulfolane, 3 % *o*-nitroanisole, 10 % ethylene carbonate, or 15 % propylene carbonate. Adapted with permission from Ref. [225]. Copyright 2014 by American Chemical Society

19. Bielow C, Aiche S, Andreotti S, Reinert K (2011) MSSimulator: simulation of mass spectrometry data. J Proteome Res 10:2922–2929
20. Marshall AG, Hendrickson CL, Jackson GS (1998) Fourier transform ion cyclotron resonance mass spectrometry: a primer. Mass Spectrom Rev 17:1–35
21. Hu Q, Noll RJ, Li H, Makarov A, Hardman M, Graham Cooks R (2005) The orbitrap: a new mass spectrometer. J Mass Spectrom 40:430–443
22. Valeja SG, Kaiser NK, Xian F, Hendrickson CL, Rouse JC, Marshall AG (2011) Unit mass baseline resolution for an intact 148 kDa therapeutic monoclonal antibody by Fourier transform Ion cyclotron resonance mass spectrometry. Anal Chem 83:8391–8395
23. Williams DK, Muddiman DC (2007) Parts-Per-billion mass measurement accuracy achieved through the combination of multiple linear regression and automatic gain control in a Fourier transform Ion cyclotron resonance mass spectrometer. Anal Chem 79:5058–5063
24. Smith RD, Anderson GA, Lipton MS, Pasa-Tolic L, Shen Y, Conrads TP, Veenstra TD, Udseth HR (2002) An accurate mass tag strategy for quantitative and high-throughput proteome measurements. Proteomics 2:513–523
25. Conrads TP, Anderson GA, Veenstra TD, Pasa-Tolic L, Smith RD (2000) Utility of accurate mass tags for proteome-wide protein identification. Anal Chem 72:3349–3354
26. Rhoads TW, Rose CM, Bailey DJ, Riley NM, Molden RC, Nestler AJ, Merrill AE, Smith LM, Hebert AS, Westphall MS, Pagliarini DJ, Garcia BA, Coon JJ (2014) Neutron-encoded mass signatures for quantitative top-down proteomics. Anal Chem 86:2314–2319
27. Mao Y, Zamdborg L, Kelleher NL, Hendrickson CL, Marshall AG (2011) Identification of phosphorylated human peptides by accurate mass measurement alone. Int J Mass Spectrom 308:357–361
28. Senko MW, Hendrickson CL, Pasa-Tolic L, Marto JA, White FM, Guan S, Marshall AG (1996) Electrospray ionization Fourier transform ion cyclotron resonance at 9.4 T. Rapid Commun Mass Spectrom 10:1824–1828
29. Glish GL, Burinsky DJ (2008) Hybrid mass spectrometers for tandem mass spectrometry. J Am Soc Mass Spectrom 19:161–172
30. Patrie SM, Charlebois JP, Whipple D, Kelleher NL, Hendrickson CL, Quinn JP, Marshall AG, Mukhopadhyay B (2004) Construction of a hybrid quadrupole/Fourier transform ion cyclotron resonance mass spectrometer for versatile MS/MS above 10 kDa. J Am Soc Mass Spectrom 15:1099–1108
31. Michalski A, Damoc E, Lange O, Denisov E, Nolting D, Müller M, Viner R, Schwartz J, Remes P, Belford M, Dunyach J-J, Cox J, Horning S, Mann M, Makarov A (2012) Ultra high resolution linear Ion trap orbitrap mass spectrometer (orbitrap elite) facilitates top down LC MS/MS and versatile peptide fragmentation modes. Mol Cell Proteomics MCP 11, O111.013698
32. Tran JC, Zamdborg L, Ahlf DR, Lee JE, Catherman AD, Durbin KR, Tipton JD, Vellaichamy A, Kellie JF, Li MX, Wu C, Sweet SMM, Early BP, Siuti N, LeDuc RD, Compton PD, Thomas PM, Kelleher NL (2011) Mapping intact protein isoforms in discovery mode using top-down proteomics. Nature 480:254, U141
33. Garcia BA, Pesavento JJ, Mizzen CA, Kelleher NL (2007) Pervasive combinatorial modification of histone H3 in human cells. Nat Methods 4:487–489
34. Pesavento JJ, Kim YB, Taylor GK, Kelleher NL (2004) Shotgun annotation of histone modifications: a new approach for streamlined characterization of proteins by top down mass spectrometry. J Am Chem Soc 126:3386–3387
35. Ginter JM, Zhou F, Johnston MV (2004) Generating protein sequence tags by combining cone and conventional collision induced dissociation in a quadrupole time-of-flight mass spectrometer. J Am Soc Mass Spectrom 15:1478–1486
36. Huang TY, McLuckey SA (2010) Top-down protein characterization facilitated by ion/ion reactions on a quadrupole/time of flight platform. Proteomics 10:3577–3588
37. Madsen JA, Gardner MW, Smith SI, Ledvina AR, Coon JJ, Schwartz JC, Stafford GC, Brodbelt JS (2009) Top-down protein fragmentation by infrared multiphoton dissociation in a dual pressure linear Ion trap. Anal Chem 81:8677–8686
38. Sleno L, Volmer DA (2004) Ion activation methods for tandem mass spectrometry. J Mass Spectrom 39:1091–1112
39. Wells JM, McLuckey SA (2005) Collision-induced dissociation (CID) of peptides and proteins. Methods Enzymol 402:148–185
40. Roepstorff P, Fohlman J (1984) Proposal for a common nomenclature for sequence ions in mass spectra of peptides. Biomed Mass Spectrom 11:601
41. Biemann K (1988) Contributions of mass spectrometry to peptide and protein structure. Biomed Environ Mass Spectrom 16:99–111
42. Bean MF, Carr SA, Thorne GC, Reilly MH, Gaskell SJ (1991) Tandem mass spectrometry of peptides using hybrid and four-sector instruments: a comparative study. Anal Chem 63:1473–1481
43. Olsen JV, Macek B, Lange O, Makarov A, Horning S, Mann M (2007) Higher-energy C-trap dissociation for peptide modification analysis. Nat Methods 4:709–712
44. Little DP, Speir JP, Senko MW, O'Connor PB, McLafferty FW (1994) Infrared multiphoton dissociation of large multiply charged ions for biomolecule sequencing. Anal Chem 66:2809–2815
45. Shaw JB, Li WZ, Holden DD, Zhang Y, Griep-Raming J, Fellers RT, Early BP, Thomas PM,

Kelleher NL, Brodbelt JS (2013) Complete protein characterization using top-down mass spectrometry and ultraviolet photodissociation. J Am Chem Soc 135:12646–12651
46. Madsen JA, Boutz DR, Brodbelt JS (2010) Ultrafast ultraviolet photodissociation at 193 nm and its applicability to proteomic workflows. J Proteome Res 9:4205–4214
47. McLafferty FW, Horn DM, Breuker K, Ge Y, Lewis MA, Cerda B, Zubarev RA, Carpenter BK (2001) Electron capture dissociation of gaseous multiply charged ions by Fourier-transform ion cyclotron resonance. J Am Soc Mass Spectrom 12:245–249
48. Mikesh LM, Ueberheide B, Chi A, Coon JJ, Syka JE, Shabanowitz J, Hunt DF (2006) The utility of ETD mass spectrometry in proteomic analysis. Biochim Biophys Acta 1764:1811–1822
49. Hakansson K, Chalmers MJ, Quinn JP, McFarland MA, Hendrickson CL, Marshall AG (2003) Combined electron capture and infrared multiphoton dissociation for multistage MS/MS in a Fourier transform ion cyclotron resonance mass spectrometer. Anal Chem 75:3256–3262
50. Wuhrer M, Catalina MI, Deelder AM, Hokke CH (2007) Glycoproteomics based on tandem mass spectrometry of glycopeptides. J Chromatogr B Anal Technol Biomed Life Sci 849:115–128
51. Boersema PJ, Mohammed S, Heck AJ (2009) Phosphopeptide fragmentation and analysis by mass spectrometry. J Mass Spectrom 44:861–878
52. Patrie SM, Robinson DE, Meng F, Du Y, Kelleher NL (2004) Strategies for automating top-down protein analysis with Q-FTICR MS. Int J Mass Spectrom 234:175–184
53. Hakansson K, Cooper HJ, Emmett MR, Costello CE, Marshall AG, Nilsson CL (2001) Electron capture dissociation and infrared multiphoton dissociation MS/MS of an N-glycosylated tryptic peptic to yield complementary sequence information. Anal Chem 73:4530–4536
54. Catalina MI, Koeleman CA, Deelder AM, Wuhrer M (2007) Electron transfer dissociation of N-glycopeptides: loss of the entire N-glycosylated asparagine side chain. Rapid Commun Mass Spectrom 21:1053–1061
55. Wu SL, Huhmer AF, Hao Z, Karger BL (2007) On-line LC-MS approach combining collision-induced dissociation (CID), electron-transfer dissociation (ETD), and CID of an isolated charge-reduced species for the trace-level characterization of proteins with post-translational modifications. J Proteome Res 6:4230–4244, 2557440
56. Wenger CD, Boyne MT, Ferguson JT, Robinson DE, Kelleher NL (2008) Versatile online-offline engine for automated acquisition of high-resolution tandem mass spectra. Anal Chem 80:8055–8063
57. Rozman M, Gaskell SJ (2012) Charge state dependent top-down characterisation using electron transfer dissociation. Rapid Commun Mass Spectrom 26:282–286
58. Tipton JD, Tran JC, Catherman AD, Ahlf DR, Durbin KR, Lee JE, Kellie JF, Kelleher NL, Hendrickson CL, Marshall AG (2012) Nano-LC FTICR tandem mass spectrometry for top-down proteomics: routine baseline unit mass resolution of whole cell lysate proteins up to 72 kDa. Anal Chem 84:2111–2117
59. Zhang ZQ, Marshall AG (1998) A universal algorithm for fast and automated charge state deconvolution of electrospray mass-to-charge ratio spectra. J Am Soc Mass Spectrom 9:225–233
60. Horn DM, Zubarev RA, McLafferty FW (2000) Automated de novo sequencing of proteins by tandem high-resolution mass spectrometry. Proc Natl Acad Sci U S A 97:10313–10317
61. Liu X, Inbar Y, Dorrestein PC, Wynne C, Edwards N, Souda P, Whitelegge JP, Bafna V, Pevzner PA (2010) Deconvolution and database search of complex tandem mass spectra of intact proteins: a combinatorial approach. Mol Cell Proteomics MCP 9:2772–2782, 3101958
62. Frank A, Tanner S, Bafna V, Pevzner P (2005) Peptide sequence tags for fast database search in mass-spectrometry. J Proteome Res 4:1287–1295
63. Sheng QH, Xie T, Ding DF (2000) De novo interpretation of MS/MS spectra and protein identification via database searching. Sheng wu hua xue yu sheng wu wu li xue bao Acta biochimica et biophysica Sinica 32:595–600
64. Liu X, Dekker LJ, Wu S, Vanduijn MM, Luider TM, Tolic N, Kou Q, Dvorkin M, Alexandrova S, Vyatkina K, Pasa-Tolic L, Pevzner PA (2014) De novo protein sequencing by combining top-down and bottom-up tandem mass spectra. J Proteome Res 13:3241–3248
65. Beavis R, Fenyö D (2004) Finding protein sequences using PROWL. In: Current protocols in bioinformatics. Wiley, New York
66. Bourgoin-Voillard S, Leymarie N, Costello CE (2014) Top-down tandem mass spectrometry on RNase A and B using a Qh/FT-ICR hybrid mass spectrometer. Proteomics 14:1174–1184, 4095805
67. Guner H, Close PL, Cai WX, Zhang H, Peng Y, Gregorich ZR, Ge Y (2014) MASH suite: a user-friendly and versatile software interface for high-resolution mass spectrometry data interpretation and visualization. J Am Soc Mass Spectrom 25:464–470
68. Fellers RT, Greer JB, Early BP, Yu X, LeDuc RD, Kelleher NL, Thomas PM (2014) ProSight Lite: Graphical software to analyze top-down mass spectrometry data. Proteomics
69. LeDuc RD, Taylor GK, Kim YB, Januszyk TE, Bynum LH, Sola JV, Garavelli JS, Kelleher NL (2004) ProSight PTM: an integrated environment for protein identification and characterization by

top-down mass spectrometry. Nucleic Acids Res 32: W340–W345
70. Meng F, Cargile BJ, Miller LM, Forbes AJ, Johnson JR, Kelleher NL (2001) Informatics and multiplexing of intact protein identification in bacteria and the archaea. Nat Biotechnol 19:952–957
71. LeDuc RD, Fellers RT, Early BP, Greer JB, Thomas PM, Kelleher NL (2014) The C-score: a Bayesian framework to sharply improve proteoform scoring in high-throughput top down proteomics. J Proteome Res 13:3231–3240, 4084843
72. Frank AM, Pesavento JJ, Mizzen CA, Kelleher NL, Pevzner PA (2008) Interpreting top-down mass spectra using spectral alignment. Anal Chem 80:2499–2505
73. Liu XW, Hengel S, Wu S, Tolic N, Pasa-Tolic L, Pevzner PA (2013) Identification of ultramodified proteins using top-down tandem mass spectra. J Proteome Res 12:5830–5838
74. Tsai YS, Scherl A, Shaw JL, MacKay CL, Shaffer SA, Langridge-Smith PR, Goodlett DR (2009) Precursor ion independent algorithm for top-down shotgun proteomics. J Am Soc Mass Spectrom 20:2154–2166
75. Karabacak NM, Li L, Tiwari A, Hayward LJ, Hong P, Easterling ML, Agar JN (2009) Sensitive and specific identification of wild type and variant proteins from 8 to 669 kDa using top-down mass spectrometry. Mol Cell Proteomics 8:846–856
76. Compton PD, Zamdborg L, Thomas PM, Kelleher NL (2011) On the scalability and requirements of whole protein mass spectrometry. Anal Chem 83:6868–6874, 3165072
77. Vellaichamy A, Tran JC, Catherman AD, Lee JE, Kellie JF, Sweet SM, Zamdborg L, Thomas PM, Ahlf DR, Durbin KR, Valaskovic GA, Kelleher NL (2010) Size-sorting combined with improved nanocapillary liquid chromatography-mass spectrometry for identification of intact proteins up to 80 kDa. Anal Chem 82:1234–1244, 2823583
78. MacNair JE, Opiteck GJ, Jorgenson JW, Moseley MA 3rd (1997) Rapid separation and characterization of protein and peptide mixtures using 1.5 microns diameter non-porous silica in packed capillary liquid chromatography/mass spectrometry. Rapid Commun Mass Spectrom 11:1279–1285
79. Ansong C, Wu S, Meng D, Liu X, Brewer HM, Deatherage Kaiser BL, Nakayasu ES, Cort JR, Pevzner P, Smith RD, Heffron F, Adkins JN, Pasa-Tolic L (2013) Top-down proteomics reveals a unique protein S-thiolation switch in Salmonella Typhimurium in response to infection-like conditions. Proc Natl Acad Sci U S A 110:10153–10158, 3690903
80. Eeltink S, Wouters B, Desmet G, Ursem M, Blinco D, Kemp GD, Treumann A (2011) High-resolution separations of protein isoforms with liquid chromatography time-of-flight mass spectrometry using polymer monolithic capillary columns. J Chromatogr A 1218:5504–5511
81. Pierri G, Kotoni D, Simone P, Villani C, Pepe G, Campiglia P, Dugo P, Gasparrini F (2013) Analysis of bovine milk caseins on organic monolithic columns: An integrated capillary liquid chromatography-high resolution mass spectrometry approach for the study of time-dependent casein degradation. J Chromatogr A 1313:259–269
82. Lakshmanan R, Wolff JJ, Alvarado R, Loo JA (2014) Top-down protein identification of proteasome proteins with nanoLC-FT-ICR-MS employing data-independent fragmentation methods. Proteomics 14:1271–1282
83. Everley RA, Croley TR (2008) Ultra-performance liquid chromatography/mass spectrometry of intact proteins. J Chromatogr A 1192:239–247
84. Wu Z, Wei B, Zhang X, Wirth MJ (2014) Efficient separations of intact proteins using slip-flow with nano-liquid chromatography-mass spectrometry. Anal Chem 86:1592–1598, 3982985
85. Fekete S, Beck A, Veuthey JL, Guillarme D (2015) Ion-exchange chromatography for the characterization of biopharmaceuticals. J Pharm Biomed Anal 113:43
86. Sharma S, Simpson DC, Tolic N, Jaitly N, Mayampurath AM, Smith RD, Pasa-Tolic L (2007) Proteomic profiling of intact proteins using WAX-RPLC 2-D separations and FTICR mass spectrometry. J Proteome Res 6:602–610
87. Roth MJ, Parks BA, Ferguson JT, Boyne MT 2nd, Kelleher NL (2008) "Proteotyping": population proteomics of human leukocytes using top down mass spectrometry. Anal Chem 80:2857–2866, 2615201
88. Millea KM, Krull IS, Cohen SA, Gebler JC, Berger SJ (2006) Integration of multidimensional chromatographic protein separations with a combined "top-down" and "bottom-up" proteomic strategy. J Proteome Res 5:135–146
89. Bunger MK, Cargile BJ, Ngunjiri A, Bundy JL, Stephenson JL Jr (2008) Automated proteomics of E. coli via top-down electron-transfer dissociation mass spectrometry. Anal Chem 80:1459–1467
90. Buszewski B, Noga S (2012) Hydrophilic interaction liquid chromatography (HILIC)–a powerful separation technique. Anal Bioanal Chem 402:231–247, PMC3249561
91. Pesavento JJ, Bullock CR, LeDuc RD, Mizzen CA, Kelleher NL (2008) Combinatorial modification of human histone H4 quantitated by two-dimensional liquid chromatography coupled with top down mass spectrometry. J Biol Chem 283:14927–14937, 2397456
92. Tian Z, Zhao R, Tolic N, Moore RJ, Stenoien DL, Robinson EW, Smith RD, Pasa-Tolic L (2010) Two-dimensional liquid chromatography system for online top-down mass spectrometry. Proteomics 10:3610–3620, 3010896

93. Tian Z, Tolic N, Zhao R, Moore RJ, Hengel SM, Robinson EW, Stenoien DL, Wu S, Smith RD, Pasa-Tolic L (2012) Enhanced top-down characterization of histone post-translational modifications. Genome Biol 13:R86, 3491414
94. Haselberg R, de Jong GJ, Somsen GW (2011) Capillary electrophoresis-mass spectrometry for the analysis of intact proteins 2007–2010. Electrophoresis 32:66–82
95. Haselberg R, de Jong GJ, Somsen GW (2007) Capillary electrophoresis-mass spectrometry for the analysis of intact proteins. J Chromatogr A 1159:81–109
96. Simpson DC, Ahn S, Pasa-Tolic L, Bogdanov B, Mottaz HM, Vilkov AN, Anderson GA, Lipton MS, Smith RD (2006) Using size exclusion chromatography-RPLC and RPLC-CIEF as two-dimensional separation strategies for protein profiling. Electrophoresis 27:2722–2733
97. Sun L, Knierman MD, Zhu G, Dovichi NJ (2013) Fast top-down intact protein characterization with capillary zone electrophoresis-electrospray ionization tandem mass spectrometry. Anal Chem 85:5989–5995, 3770260
98. Zhao Y, Sun L, Champion MM, Knierman MD, Dovichi NJ (2014) Capillary zone electrophoresis-electrospray ionization-tandem mass spectrometry for top-down characterization of the Mycobacterium marinum secretome. Anal Chem 86:4873–4878, 4033641
99. Li Y, Compton PD, Tran JC, Ntai I, Kelleher NL (2014) Optimizing capillary electrophoresis for top-down proteomics of 30–80 kDa proteins. Proteomics 14:1158–1164, PMC4034378
100. Haselberg R, de Jong GJ, Somsen GW (2013) Low-flow sheathless capillary electrophoresis-mass spectrometry for sensitive glycoform profiling of intact pharmaceutical proteins. Anal Chem 85:2289–2296
101. Han X, Wang Y, Aslanian A, Fonslow B, Graczyk B, Davis TN, Yates JR 3rd (2014) In-line separation by capillary electrophoresis prior to analysis by top-down mass spectrometry enables sensitive characterization of protein complexes. J Proteome Res 13:6078–6086, 4262260
102. Whitelegge JP, Laganowsky A, Nishio J, Souda P, Zhang HM, Cramer WA (2006) Sequencing covalent modifications of membrane proteins. J Exp Bot 57:1515–1522
103. Yan F, Subramanian B, Nakeff A, Barder TJ, Parus SJ, Lubman DM (2003) A comparison of drug-treated and untreated HCT-116 human colon adenocarcinoma cells using a 2-D liquid separation mapping method based upon chromatofocusing PI fractionation. Anal Chem 75:2299–2308
104. Stoyanov A (2012) IEF-based multidimensional applications in proteomics: toward higher resolution. Electrophoresis 33:3281–3290
105. Zhang J, Roth MJ, Chang AN, Plymire DA, Corbett JR, Greenberg BM, Patrie SM (2013) Top-down mass spectrometry on tissue extracts and biofluids with isoelectric focusing and superficially porous silica liquid chromatography. Anal Chem 85:10377–10384
106. Zhang JM, Corbett JR, Plymire DA, Greenberg BM, Patrie SM (2014) Proteoform analysis of lipocalin-type prostaglandin D-synthase from human cerebrospinal fluid by isoelectric focusing and superficially porous liquid chromatography with Fourier transform mass spectrometry. Proteomics 14:1223–1231
107. Jensen PK, Pasa-Tolic L, Peden KK, Martinovic S, Lipton MS, Anderson GA, Tolic N, Wong KK, Smith RD (2000) Mass spectrometric detection for capillary isoelectric focusing separations of complex protein mixtures. Electrophoresis 21:1372–1380
108. Arentz G, Weiland F, Oehler MK, Hoffmann P (2015) State of the art of 2D DIGE. Proteomics Clin Appl 9:277–288
109. Uliyanchenko E (2014) Size-exclusion chromatography-from high-performance to ultra-performance. Anal Bioanal Chem 406:6087–6094
110. Zabrouskov V, Giacomelli L, van Wijk KJ, McLafferty FW (2003) New approach for plant proteomics – characterization of chloroplast proteins of Arabidopsis thaliana by top-down mass spectrometry. Mol Cell Proteomics 2:1253–1260
111. Whitelegge J (2005) Tandem mass spectrometry of integral membrane proteins for top-down proteomics. TrAC-Trends Anal Chem 24:576–582
112. Mazur Mt Fau – Seipert RS, Seipert Rs Fau – Mahon D, Mahon D Fau – Zhou Q, Zhou Q Fau – Liu T, Liu T (2012) A platform for characterizing therapeutic monoclonal antibody breakdown products by 2D chromatography and top-down mass spectrometry. AAPS J 14
113. Pan J, Han J, Borchers CH, Konermann L (2012) Structure and dynamics of small soluble Abeta (1–40) oligomers studied by top-down hydrogen exchange mass spectrometry. Biochemistry 51:3694–3703
114. Chen X, Ge Y (2013) Ultrahigh pressure fast size exclusion chromatography for top-down proteomics. Proteomics 13:2563–2566
115. Meng F, Cargile BJ, Patrie SM, Johnson JR, McLoughlin SM, Kelleher NL (2002) Processing complex mixtures of intact proteins for direct analysis by mass spectrometry. Anal Chem 74:2923–2929
116. Tran JC, Doucette AA (2009) Multiplexed size separation of intact proteins in solution phase for mass spectrometry. Anal Chem 81:6201–6209
117. Root BE, Zhang B, Barron AE (2009) Size-based protein separations by microchip electrophoresis using an acid-labile surfactant as a replacement for SDS. Electrophoresis 30:2117–2122
118. Garcia BA, Thomas CE, Kelleher NL, Mizzen CA (2008) Tissue-specific expression and post-translational modification of histone H3 variants. J Proteome Res 7:4225–4236

119. Boyne MT 2nd, Pesavento JJ, Mizzen CA, Kelleher NL (2006) Precise characterization of human histones in the H2A gene family by top down mass spectrometry. J Proteome Res 5:248–253
120. Dang X, Scotcher J, Wu S, Chu RK, Tolic N, Ntai I, Thomas PM, Fellers RT, Early BP, Zheng Y, Durbin KR, Leduc RD, Wolff JJ, Thompson CJ, Pan J, Han J, Shaw JB, Salisbury JP, Easterling M, Borchers CH, Brodbelt JS, Agar JN, Pasa-Tolic L, Kelleher NL, Young NL (2014) The first pilot project of the consortium for top-down proteomics: a status report. Proteomics 14:1130–1140, 4146406
121. Zhang C, Walker AK, Zand R, Moscarello MA, Yan JM, Andrews PC (2012) Myelin basic protein undergoes a broader range of modifications in mammals than in lower vertebrates. J Proteome Res 11:4791–4802, 3612544
122. Nepomuceno AI, Mason CJ, Muddiman DC, Bergen HR, Zeldenrust SR (2004) Detection of genetic variants of transthyretin by liquid chromatography-dual electrospray ionization Fourier-transform ion-cyclotron-resonance mass spectrometry. Clin Chem 50:1535–1543
123. Theberge R, Infusini G, Tong W, McComb ME, Costello CE (2011) Top-down analysis of small plasma proteins using an LTQ-orbitrap. Potential for mass spectrometry-based clinical assays for transthyretin and hemoglobin. Int J Mass Spectrom 300:130–142, 3098445
124. Edwards RL, Griffiths P, Bunch J, Cooper HJ (2012) Top-down proteomics and direct surface sampling of neonatal dried blood spots: diagnosis of unknown hemoglobin variants. J Am Soc Mass Spectrom 23:1921–1930
125. Mao P, Wang D (2014) Top-down proteomics of a drop of blood for diabetes monitoring. J Proteome Res 13:1560–1569, 3993886
126. Sarsby J, Martin NJ, Lalor PF, Bunch J, Cooper HJ (2014) Top-down and bottom-up identification of proteins by liquid extraction surface analysis mass spectrometry of healthy and diseased human liver tissue. J Am Soc Mass Spectrom 25:1953–1961, 4197381
127. Zabrouskov V, Han XM, Welker E, Zhai HL, Lin C, van Wijk KJ, Scheraga HA, McLafferty FW (2006) Stepwise deamidation of ribonuclease A at five sites determined by top down mass spectrometry. Biochemistry 45:987–992
128. Li X, Yu X, Costello CE, Lin C, O'Connor PB (2012) Top-down study of beta(2)-microglobulin deamidation. Anal Chem 84:6150–6157
129. Lourette N, Smallwood H, Wu S, Robinson EW, Squier TC, Smith RD, Pasa-Tolic L (2010) A top-down LC-FTICR MS-based strategy for characterizing oxidized calmodulin in activated macrophages. J Am Soc Mass Spectrom 21:930–939
130. Zabrouskov V, Ge Y, Schwartz J, Walker JW (2008) Unraveling molecular complexity of phosphorylated human cardiac troponin I by top down electron capture dissociation/electron transfer dissociation mass spectrometry. Mol Cell Proteomics 7:1838–1849
131. Zhang J, Guy MJ, Norman HS, Chen YC, Xu QG, Dong XT, Guner H, Wang SJ, Kohmoto T, Young KH, Moss RL, Ge Y (2011) Top-down quantitative proteomics identified phosphorylation of cardiac troponin I as a candidate biomarker for chronic heart failure. J Proteome Res 10:4054–4065
132. Dong X, Sumandea CA, Chen YC, Garcia-Cazarin ML, Zhang J, Balke CW, Sumandea MP, Ge Y (2012) Augmented phosphorylation of cardiac troponin I in hypertensive heart failure. J Biol Chem 287:848–857, 3256890
133. Burnaevskiy N, Fox TG, Plymire DA, Ertelt JM, Weigele BA, Selyunin AS, Way SS, Patrie SM, Alto NM (2013) Proteolytic elimination of N-myristoyl modifications by the Shigella virulence factor IpaJ. Nature 496:106–109, 3722872
134. Twine SM, Reid CW, Aubry A, McMullin DR, Fulton KM, Austin J, Logan SM (2009) Motility and flagellar glycosylation in clostridium difficile. J Bacteriol 191:7050–7062
135. Chamot-Rooke J, Rousseau B, Lanternier F, Mikaty G, Mairey E, Malosse C, Bouchoux G, Pelicic V, Camoin L, Nassif X, Dumenil G (2007) Alternative Neisseria spp. type IV pilin glycosylation with a glyceramido acetamido trideoxyhexose residue. Proc Natl Acad Sci USA 104:14783–14788
136. Twine SM, Paul CJ, Vinogradov E, McNally DJ, Brisson JR, Mullen JA, McMullin DR, Jarrell HC, Austin JW, Kelly JF, Logan SM (2008) Flagellar glycosylation in Clostridium botulinum. FEBS J 275:4428–4444
137. Wagner-Rousset E, Bednarczyk A, Bussat MC, Colas O, Corvaia N, Schaeffer C, Van Dorsselaer A, Beck A (2008) The way forward, enhanced characterization of therapeutic antibody glycosylation: comparison of three level mass spectrometry-based strategies. J Chromatogr B 872:23–37
138. Reid GE, Stephenson JL, McLuckey SA (2002) Tandem mass spectrometry of ribonuclease A and B: N-linked glycosylation site analysis of whole protein ions. Anal Chem 74:577–583
139. Mazur MT, Cardasis HL, Spellman DS, Liaw A, Yates NA, Hendrickson RC Quantitative analysis of intact apolipoproteins in human HDL by top-down differential mass spectrometry. Proc Natl Acad Sci USA 107:7728–7733
140. Reid GE, McLuckey SA (2002) 'Top down' protein characterization via tandem mass spectrometry. J Mass Spectrom 37:663–675
141. Friedman DB, Andacht TM, Bunger MK, Chien AS, Hawke DH, Krijgsveld J, Lane WS, Lilley KS, MacCoss MJ, Moritz RL, Settlage RE, Sherman NE, Weintraub ST, Witkowska HE, Yates NA, Turck CW (2011) The ABRF Proteomics Research Group studies: educational exercises for qualitative

and quantitative proteomic analyses. Proteomics 11:1371–1381
142. Leymarie N, Griffin PJ, Jonscher K, Kolarich D, Orlando R, McComb M, Zaia J, Aguilan J, Alley WR, Altmann F, Ball LE, Basumallick L, Bazemore-Walker CR, Behnken H, Blank MA, Brown KJ, Bunz SC, Cairo CW, Cipollo JF, Daneshfar R, Desaire H, Drake RR, Go EP, Goldman R, Gruber C, Halim A, Hathout Y, Hensbergen PJ, Horn DM, Hurum D, Jabs W, Larson G, Ly M, Mann BF, Marx K, Mechref Y, Meyer B, Moginger U, Neusubeta C, Nilsson J, Novotny MV, Nyalwidhe JO, Packer NH, Pompach P, Reiz B, Resemann A, Rohrer JS, Ruthenbeck A, Sanda M, Schulz JM, Schweiger-Hufnagel U, Sihlbom C, Song E, Staples GO, Suckau D, Tang H, Thaysen-Andersen M, Viner RI, An Y, Valmu L, Wada Y, Watson M, Windwarder M, Whittal R, Wuhrer M, Zhu Y, Zou C (2013) Interlaboratory study on differential analysis of protein glycosylation by mass spectrometry: the ABRF glycoprotein research multi-institutional study 2012. Mol Cell Proteomics: MCP 12:2935–2951, 3790302
143. Collier TS, Muddiman DC (2012) Analytical strategies for the global quantification of intact proteins. Amino Acids 43:1109
144. Gucinski AC, Boyne MT 2nd (2012) Evaluation of intact mass spectrometry for the quantitative analysis of protein therapeutics. Anal Chem 84:8045–8051
145. Julka S, Folkenroth J, Young SA (2011) Two dimensional liquid chromatography-ultraviolet/mass spectrometric (2DLC-UV/MS) analyses for quantitation of intact proteins in complex biological matrices. J Chromatogr B 879:2057–2063
146. Mazur MT, Fyhr R (2011) An algorithm for identifying multiply modified endogenous proteins using both full-scan and high-resolution tandem mass spectrometric data. Rapid Commun Mass Spectrom 25:3617–3626
147. Ntai I, Kim K, Fellers RT, Skinner OS, Smith AD, Early BP, Savaryn JP, LeDuc RD, Thomas PM, Kelleher NL (2014) Applying label-free quantitation to top down proteomics. Anal Chem 86:4961–4968
148. Pesavento JJ, Mizzen CA, Kelleher NL (2006) Quantitative analysis of modified proteins and their positional isomers by tandem mass spectrometry: human histone H4. Anal Chem 78:4271–4280
149. Thomas CE, Kelleher NL, Mizzen CA (2006) Mass spectrometric characterization of human histone H3: a bird's eye view. J Proteome Res 5:240–247
150. Collier TS, Sarkar P, Franck WL, Rao BM, Dean RA, Muddiman DC Direct comparison of stable isotope labeling by amino acids in cell culture and spectral counting for quantitative proteomics. Anal Chem 82:8696–8702
151. Waanders LF, Hanke S, Mann M (2007) Top-down quantitation and characterization of SILAC-labeled proteins. J Am Soc Mass Spectrom 18:2058–2064
152. Veenstra TD, Martinovic S, Anderson GA, Pasa-Tolic L, Smith RD (2000) Proteome analysis using selective incorporation of isotopically labeled amino acids. J Am Soc Mass Spectrom 11:78–82
153. Parks BA, Jiang L, Thomas PM, Wenger CD, Roth MJ, Boyne MT 2nd, Burke PV, Kwast KE, Kelleher NL (2007) Top-down proteomics on a chromatographic time scale using linear ion trap fourier transform hybrid mass spectrometers. Anal Chem 79:7984–7991, 2361135
154. Pesavento JJ, Yang H, Kelleher NL, Mizzen CA (2008) Certain and progressive methylation of histone H4 at lysine 20 during the cell cycle. Mol Cell Biol 28:468–486
155. Collier TS, Sarkar P, Rao B, Muddiman DC. Quantitative top-down proteomics of SILAC labeled human embryonic stem cells. J Am Soc Mass Spectrom 21:879–889
156. Levy MJ, Gucinski AC, Sommers CD, Ghasriani H, Wang B, Keire DA, Boyne MT 2nd (2014) Analytical techniques and bioactivity assays to compare the structure and function of filgrastim (granulocyte-colony stimulating factor) therapeutics from different manufacturers. Anal Bioanal Chem 406:6559–6567
157. Gucinski AC, Boyne MT 2nd (2014) Identification of site-specific heterogeneity in peptide drugs using intact mass spectrometry with electron transfer dissociation. Rapid Commun Mass Spectrom 28:1757–1763
158. Mazur MT, Seipert RS, Mahon D, Zhou Q, Liu T (2012) A platform for characterizing therapeutic monoclonal antibody breakdown products by 2D chromatography and top-down mass spectrometry. AAPS J 14:530–541, 3385834
159. Fornelli L, Damoc E, Thomas PM, Kelleher NL, Aizikov K, Denisov E, Makarov A, Tsybin YO (2012) Analysis of intact monoclonal antibody IgG1 by electron transfer dissociation Orbitrap FTMS. Mol Cell Proteomics: MCP 11:1758–1767, 3518117
160. Mao Y, Valeja SG, Rouse JC, Hendrickson CL, Marshall AG (2013) Top-down structural analysis of an intact monoclonal antibody by electron capture dissociation-Fourier transform ion cyclotron resonance-mass spectrometry. Anal Chem 85:4239–4246
161. Zhang Z, Shah B (2007) Characterization of variable regions of monoclonal antibodies by top-down mass spectrometry. Anal Chem 79:5723–5729
162. Bondarenko PV, Second TP, Zabrouskov V, Makarov AA, Zhang Z (2009) Mass measurement and top-down HPLC/MS analysis of intact monoclonal antibodies on a hybrid linear quadrupole ion trap-Orbitrap mass spectrometer. J Am Soc Mass Spectrom 20:1415–1424
163. Ren D, Pipes GD, Hambly D, Bondarenko PV, Treuheit MJ, Gadgil HS (2009) Top-down N-terminal sequencing of Immunoglobulin subunits with

electrospray ionization time of flight mass spectrometry. Anal Biochem 384:42–48
164. Liu H, Gaza-Bulseco G, Chumsae C (2009) Analysis of reduced monoclonal antibodies using size exclusion chromatography coupled with mass spectrometry. J Am Soc Mass Spectrom 20:2258–2264
165. Nicolardi S, Deelder AM, Palmblad M, van der Burgt YE (2014) Structural analysis of an intact monoclonal antibody by online electrochemical reduction of disulfide bonds and Fourier transform ion cyclotron resonance mass spectrometry. Anal Chem 86:5376–5382
166. Souda P, Ryan CM, Cramer WA, Whitelegge J (2011) Profiling of integral membrane proteins and their post translational modifications using high-resolution mass spectrometry. Methods 55:330–336
167. Whitelegge J, Halgand F, Souda P, Zabrouskov V (2006) Top-down mass spectrometry of integral membrane proteins. Expert Rev Proteomics 3:585–596
168. Whitelegge JP (2005) Sequencing covalent modifications of membrane proteins. Comp Biochem Phys A 141:S249
169. Schindler PA, Van Dorsselaer A, Falick AM (1993) Analysis of hydrophobic proteins and peptides by electrospray ionization mass spectrometry. Anal Biochem 213:256–263
170. Whitelegge JP, Zhang H, Aguilera R, Taylor RM, Cramer WA (2002) Full subunit coverage liquid chromatography electrospray ionization mass spectrometry (LCMS+) of an oligomeric membrane protein: cytochrome b(6)f complex from spinach and the cyanobacterium Mastigocladus laminosus. Mol Cell Proteomics: MCP 1:816–827
171. Carroll J, Altman MC, Fearnley IM, Walker JE (2007) Identification of membrane proteins by tandem mass spectrometry of protein ions. Proc Natl Acad Sci U S A 104:14330–14335, 1952138
172. Doucette AA, Vieira DB, Orton DJ, Wall MJ (2014) Resolubilization of precipitated intact membrane proteins with cold formic acid for analysis by mass spectrometry. J Proteome Res 13:6001–6012
173. Thangaraj B, Ryan CM, Souda P, Krause K, Faull KF, Weber AP, Fromme P, Whitelegge JP (2010) Data-directed top-down Fourier-transform mass spectrometry of a large integral membrane protein complex: photosystem II from Galdieria sulphuraria. Proteomics 10:3644–3656, 3517113
174. Ryan CM, Souda P, Bassilian S, Ujwal R, Zhang J, Abramson J, Ping P, Durazo A, Bowie JU, Hasan SS, Baniulis D, Cramer WA, Faull KF, Whitelegge JP (2010) Post-translational modifications of integral membrane proteins resolved by top-down Fourier transform mass spectrometry with collisionally activated dissociation. Mol Cell Proteomics: MCP 9:791–803, 2871414
175. Catherman AD, Durbin KR, Ahlf DR, Early BP, Fellers RT, Tran JC, Thomas PM, Kelleher NL (2013) Large-scale top-down proteomics of the human proteome: membrane proteins, mitochondria, and senescence. Mol Cell Proteomics: MCP 12:3465–3473, 3861700
176. Catherman AD, Li M, Tran JC, Durbin KR, Compton PD, Early BP, Thomas PM, Kelleher NL (2013) Top down proteomics of human membrane proteins from enriched mitochondrial fractions. Anal Chem 85:1880–1888, 3565750
177. Zabrouskov V, Whitelegge JP (2007) Increased coverage in the transmembrane domain with activated-ion electron capture dissociation for top-down Fourier-transform mass spectrometry of integral membrane proteins. J Proteome Res 6:2205–2210
178. Kaltashov IA, Bobst CE, Abzalimov RR (2009) H/D exchange and mass spectrometry in the studies of protein conformation and dynamics: is there a need for a top-down approach? Anal Chem 81:7892–7899
179. Abzalimov RR, Kaplan DA, Easterling ML, Kaltashov IA (2009) Protein conformations can be probed in top-down HDX MS experiments utilizing electron transfer dissociation of protein ions without hydrogen scrambling. J Am Soc Mass Spectrom 20:1514–1517
180. Pan J, Han J, Borchers CH, Konermann L (2008) Electron capture dissociation of electrosprayed protein ions for spatially resolved hydrogen exchange measurements. J Am Chem Soc 130:11574–11575
181. Wang G, Kaltashov IA (2014) Approach to characterization of the higher order structure of disulfide-containing proteins using hydrogen/deuterium exchange and top-down mass spectrometry. Anal Chem 86:7293–7298, 4144750
182. Pan J, Borchers CH (2014) Top-down mass spectrometry and hydrogen/deuterium exchange for comprehensive structural characterization of interferons: implications for biosimilars. Proteomics 14:1249–1258
183. Amon S, Trelle MB, Jensen ON, Jorgensen TJ (2012) Spatially resolved protein hydrogen exchange measured by subzero-cooled chip-based nanoelectrospray ionization tandem mass spectrometry. Anal Chem 84:4467–4473
184. Pan JX, Zhang SP, Parker CE, Borchers CH (2014) Subzero temperature chromatography and top-down mass spectrometry for protein higher-order structure characterization: method validation and application to therapeutic antibodies. J Am Chem Soc 136:13065–13071
185. Lorenzen K, van Duijn E (2010) Native mass spectrometry as a tool in structural biology.In John EC et al (ed) Current protocols in protein science. Chapter 17, Unit17 12
186. Lomeli SH, Peng IX, Yin S, Loo RR, Loo JA (2010) New reagents for increasing ESI multiple charging of proteins and protein complexes. J Am Soc Mass Spectrom 21:127–131, 2821426
187. Zhou M, Jones CM, Wysocki VH (2013) Dissecting the large noncovalent protein complex GroEL with

surface-induced dissociation and ion mobility-mass spectrometry. Anal Chem 85:8262–8267
188. Snijder J, Rose RJ, Veesler D, Johnson JE, Heck AJR (2013) Studying 18 MDa virus assemblies with native mass spectrometry. Angew Chem Int Ed 52:4020–4023
189. Blackwell AE, Dodds ED, Bandarian V, Wysocki VH (2011) Revealing the quaternary structure of a heterogeneous noncovalent protein complex through surface-induced dissociation. Anal Chem 83:2862–2865, 3343771
190. Belov ME, Damoc E, Denisov E, Compton PD, Horning S, Makarov AA, Kelleher NL (2013) From protein complexes to subunit backbone fragments: a multi-stage approach to native mass spectrometry. Anal Chem 85:11163–11173
191. Ahlf DR, Compton PD, Tran JC, Early BP, Thomas PM, Kelleher NL (2012) Evaluation of the compact high-field orbitrap for top-down proteomics of human cells. J Proteome Res 11:4308–4314
192. Beu SC, Blakney GT, Quinn JP, Hendrickson CL, Marshall AG (2004) Broadband phase correction of FT-ICR mass spectra via simultaneous excitation and detection. Anal Chem 76:5756–5761
193. Makarov A, Denisov E, Lange O (2009) Performance evaluation of a high-field Orbitrap mass analyzer. J Am Soc Mass Spectrom 20:1391–1396
194. Schaub TM, Hendrickson CL, Horning S, Quinn JP, Senko MW, Marshall AG (2008) High-performance mass spectrometry: Fourier transform ion cyclotron resonance at 14.5 Tesla. Anal Chem 80:3985–3990
195. Scigelova M, Hornshaw M, Giannakopulos A, Makarov A (2011) Fourier transform mass spectrometry. Mol Cell Proteomics 10:M111 009431, 3134075
196. Xian F, Hendrickson CL, Blakney GT, Beu SC, Marshall AG (2010) Automated broadband phase correction of Fourier transform Ion cyclotron resonance mass spectra. Anal Chem 82:8807–8812
197. Dyachenko A, Wang G, Belov M, Makarov A, de Jong RN, van den Bremer ET, Parren PW, Heck AJ (2015) Tandem native mass-spectrometry on antibody-drug conjugates and submillion Da antibody-antigen protein assemblies on an orbitrap EMR equipped with a high-mass quadrupole mass selector. Anal Chem 87:6095–6102
198. Yin S, Loo JA (2010) Elucidating the site of protein-ATP binding by top-down mass spectrometry. J Am Soc Mass Spectrom 21:899–907
199. Zhang H, Cui W, Wen J, Blankenship RE, Gross ML (2010) Native electrospray and electron-capture dissociation in FTICR mass spectrometry provide top-down sequencing of a protein component in an intact protein assembly. J Am Soc Mass Spectrom 21:1966–1968, 2991543
200. Li H, Wongkongkathep P, Van Orden SL, Ogorzalek Loo RR, Loo JA (2014) Revealing ligand binding sites and quantifying subunit variants of noncovalent protein complexes in a single native top-down FTICR MS experiment. J Am Soc Mass Spectrom 25:2060–2068
201. Skinner OS, Do Vale LH, Catherman AD, Havugimana PC, Sousa MV, Compton PD, Kelleher NL (2015) Native GELFrEE: a New separation technique for biomolecular assemblies. Anal Chem 87:3032–3038
202. May JC, Goodwin CR, McLean JA (2015) Ion mobility-mass spectrometry strategies for untargeted systems, synthetic, and chemical biology. Curr Opin Biotechnol 31:117–121, PMC4297680
203. Sowell RA, Koeniger SL, Valentine SJ, Moon MH, Clemmer DE (2004) Nanoflow LC/IMS-MS and LC/IMS-CID/MS of protein mixtures. J Am Soc Mass Spectrom 15:1341–1353
204. McKenna T (2007) Top-down sequencing using the SynaptHigh Definition Mass Spectrometry™(HDMS™) System. Nat Methods| Application Notes
205. Zinnel NF, Pai PJ, Russell DH (2012) Ion mobility-mass spectrometry (IM-MS) for top-down proteomics: increased dynamic range affords increased sequence coverage. Anal Chem 84:3390–3397
206. Wyttenbach T, Bowers MT (2011) Structural stability from solution to the Gas phase: native solution structure of ubiquitin survives analysis in a solvent-free Ion mobility-mass spectrometry environment. J Phys Chem B 115:12266–12275
207. Shi HL, Pierson NA, Valentine SJ, Clemmer DE (2012) Conformation types of ubiquitin [M + 8H] (8+) ions from water:methanol solutions: evidence for the N and A states in aqueous solution. J Phys Chem B 116:3344–3352
208. Ewing MA, Conant CR, Zucker SM, Griffith KJ, Clemmer DE (2015) Selected overtone mobility spectrometry. Anal Chem 87:5132–5138
209. Shvartsburg AA (2014) Ultrahigh-resolution differential ion mobility separations of conformers for proteins above 10 kDa: onset of dipole alignment? Anal Chem 86:10608–10615
210. Shvartsburg AA, Zheng YP, Smith RD, Kelleher NL (2012) Ion mobility separation of variant histone tails extending to the "middle-down" range. Anal Chem 84:4271–4276
211. Cui W, Zhang H, Blankenship RE, Gross ML (2015) Electron-capture dissociation and ion mobility mass spectrometry for characterization of the hemoglobin protein assembly. Protein Sci 24:1325
212. Escribano E, Madurga S, Vilaseca M, Moreno V (2014) Ion mobility and Top-down MS complementary approaches for the structural analysis of protein models bound to anticancer metallodrugs. Inorg Chim Acta Part B 423:60–69
213. Do TD, Economou NJ, Chamas A, Buratto SK, Shea JE, Bowers MT (2014) Interactions between amyloid-beta and Tau fragments promote aberrant aggregates: implications for amyloid toxicity. J Phys Chem B 118:11220–11230

214. Young LM, Saunders JC, Mahood RA, Revill CH, Foster RJ, Tu L-H, Raleigh DP, Radford SE, Ashcroft AE (2015) Screening and classifying small-molecule inhibitors of amyloid formation using ion mobility spectrometry–mass spectrometry. Nat Chem 7:73–81
215. Beveridge R, Covill S, Pacholarz KJ, Kalapothakis JMD, MacPhee CE, Barran PE (2014) A mass-spectrometry-based framework to define the extent of disorder in proteins. Anal Chem 86:10979–10991
216. McLuckey SA, Glish GL, Van Berkel GJ (1991) Charge determination of product ions formed from collision-induced dissociation of multiply protonated molecules via ion/molecule reactions. Anal Chem 63:1971–1978
217. Abzalimov RR, Kaltashov IA (2010) Electrospray ionization mass spectrometry of highly heterogeneous protein systems: protein ion charge state assignment via incomplete charge reduction. Anal Chem 82:7523–7526
218. Hassell KM, LeBlanc YC, McLuckey SA (2011) Chemical noise reduction via mass spectrometry and ion/ion charge inversion: amino acids. Anal Chem 83:3252–3255, 3084898
219. Pitteri SJ, McLuckey SA (2005) Recent developments in the ion/ion chemistry of high-mass multiply charged ions. Mass Spectrom Rev 24:931–958
220. McLuckey S (2009) Peptide and protein Ion/Ion reactions in electrodynamic Ion traps: tools and methods. In: Lipton M, Paša-Tolic L (eds) Mass spectrometry of proteins and peptides. Humana Press, New York, pp 395–412
221. McLuckey SA, Reid GE, Wells JM (2002) Ion parking during ion/ion reactions in electrodynamic ion traps. Anal Chem 74:336–346
222. Liu J, Huang TY, McLuckey SA (2009) Simultaneous transmission mode collision-induced dissociation and ion/ion reactions for top-down protein identification/characterization using a quadrupole/time-of-flight tandem mass spectrometer. Anal Chem 81:2159–2167, 2667222
223. Frey BL, Lin Y, Westphall MS, Smith LM (2005) Controlling gas-phase reactions for efficient charge reduction electrospray mass spectrometry of intact proteins. J Am Soc Mass Spectrom 16:1876–1887, 1489883
224. Scalf M, Westphall MS, Smith LM (2000) Charge reduction electrospray mass spectrometry. Anal Chem 72:52–60
225. Chi A, Bai DL, Geer LY, Shabanowitz J, Hunt DF (2007) Analysis of intact proteins on a chromatographic time scale by electron transfer dissociation tandem mass spectrometry. Int J Mass Spectrom 259:197–203
226. Teo CA, Donald WA (2014) Solution additives for supercharging proteins beyond the theoretical maximum proton-transfer limit in electrospray ionization mass spectrometry. Anal Chem 86:4455–4462
227. Loo RRO, Dales N, Andrews PC (1994) Surfactant effects on protein-structure examined by electrospray-ionization mass-spectrometry. Protein Sci 3:1975–1983
228. Iavarone AT, Jurchen JC, Williams ER (2001) Supercharged protein and peptide ions formed by electrospray ionization. Anal Chem 73:1455–1460, 1414801
229. Valeja SG, Tipton JD, Emmett MR, Marshall AG (2010) New reagents for enhanced liquid chromatographic separation and charging of intact protein ions for electrospray ionization mass spectrometry. Anal Chem 82:7515–7519, 2932825
230. Iavarone AT, Jurchen JC, Williams ER (2000) Effects of solvent on the maximum charge state and charge state distribution of protein ions produced by electrospray ionization. J Am Soc Mass Spectrom 11:976–985
231. Cassou CA, Williams ER (2014) Anions in electrothermal supercharging of proteins with electrospray ionization follow a reverse Hofmeister series. Anal Chem 86:1640–1647, PMC3983018
232. Sterling HJ, Cassou CA, Susa AC, Williams ER (2012) Electrothermal supercharging of proteins in native electrospray ionization. Anal Chem 84:3795–3801, PMC3328611

Part III

Bioinformatic Tools for Proteomics data Analysis and Interpretation

Platforms and Pipelines for Proteomics Data Analysis and Management

9

Marius Cosmin Codrea and Sven Nahnsen

Abstract

Since mass spectrometry was introduced as the core technology for large-scale analysis of the proteome, the speed of data acquisition, dynamic ranges of measurements, and data quality are continuously improving. These improvements are triggered by regular launches of new methodologies and instruments.

Keywords

Bioinformatics • Proteomics data processing • Protein identification • Protein quantification • Data processing pipeline • Trans proteomic pipeline • OpenMS pipeline • The Central Proteomics Facilities Pipeline (CPFP) • MaxQuant pipeline • Scaffold pipeline • Sorcerer pipeline • IPA/IP2 pipeline

Abbreviations

FDR	False Discovery Rate
GO	Gene Ontology
GUI	Graphical User Interface
I/O	input, output
iTRAQ	Isobaric tags for relative and absolute quantitation
M/Z	mass-to-charge
PTM	Post-Translational Modification
RT	retention time
SILAC	Stable isotope labeling by amino acids in cell culture
SRM	Selected Reaction Monitoring
TB	terra byte
TPP	Trans-Proteomics Pipeline

M.C. Codrea • S. Nahnsen (✉)
Quantitative Biology Center (QBiC), University of Tübingen, Auf der Morgenstelle 10, 72076 Tübingen, Germany
e-mail: sven.nahnsen@uni-tuebingen.de

9.1 Introduction

Since mass spectrometry was introduced as the core technology for large-scale analysis of the proteome, the speed of data acquisition, dynamic ranges of measurements, and data quality are continuously improving. These improvements are triggered by regular launches of new methodologies and instruments.

H. Mirzaei and M. Carrasco (eds.), *Modern Proteomics – Sample Preparation, Analysis and Practical Applications*, Advances in Experimental Medicine and Biology 919, DOI 10.1007/978-3-319-41448-5_9

A consequence of higher throughput and performance is the increased size and complexity of the data. Mass spectrometry studies using the latest technology can readily generate datasets in the TB range (e.g., [1, 2]). These datasets contain millions of spectra that need to be converted into biological insights, and manual completion of this task is prohibitive. Hence, the importance of bioinformatics tools for the analysis of proteomics data is growing rapidly. Besides efficient implementations of the required features to meet these demands, there is a strong need for flexible and user-friendly interfaces. Software tools in the field of proteomics need to provide a broad operability to allow for a scalable integration for the complete workflow. The processing and analysis software should be usable by biologists, instrument technicians and computer scientists alike.

From a computational point of view, algorithms for the identification and quantification of peptides and proteins from a selection of mass spectra are at the core of the workflow. For the remainder of this chapter, these fundamental tasks are summarized as *data processing*. Following data processing, proteomics studies usually require bioinformatics *data analysis*, which comprises statistical assessment of quantitative data, functional enrichment, visualization and the integration with other -OMICS technologies. The underlying data in proteomics can be derived from targeted, data-independent acquisition, or shotgun proteomics experiments [3]. We will mainly focus on tools for shotgun proteomics in this chapter. In shotgun proteomics (also referred to as bottom-up proteomics), proteins are enzymatically digested to peptides and separated using chromatography systems, most frequently reversed-phase chromatography using a high-performance liquid chromatographer (HPLC). The eluents of the chromatographic separation are ionized (e.g., via Matrix assisted laser desorption/ionization or more commonly via electrospray ionization) and online injected into the mass spectrometer. A mass spectrometry experiment is frequently set up as a two stage mass measurement. The first mass measurement records mass-to-charge ratios and intensities of the entire peptide in a survey scan mass spectrum (MS1). At the same time, a certain number of peptide ions are automatically selected for fragmentation based on their occurring intensity – this method is called data-dependent acquisition. Different methods for peptide fragmentation have been established (e.g., collision-induced fragmentation (CID), higher-energy collisional dissociation (HCD), electron-transfer dissociation (ETD) etc.). Following fragmentation, the resulting product ions are measured and recorded in a fragment ion or tandem MS spectrum (MS/MS or MS2). This measurement can take place consecutively in the same mass analyzer (tandem MS in time) or on a hybrid instrument with an additional mass analyzer (tandem MS in space). Classically, the MS2 spectra are used to identify peptides using database mapping, while the MS1 spectra allow one to estimate their relative quantities [4]. Such proteomics experiments can lead to hundreds of gigabytes that need automated bioinformatics data management, processing and analysis tools.

Bioinformatics tools comprise a diverse selection of approaches for the processing, analyzing and managing of mass spectrometry-based proteomics data. The underlying architecture varies from monolithic commercial applications to free- and open source software libraries.

Common to all tools that enable the complete proteomics processing and analysis workflow are algorithmic solutions to search tandem MS spectra against a protein database, as well as methods for the statistical post-processing of identification results and quantification. The development of tandem MS search engines has been a topic in computational proteomics research since the emergence of the field in the early 1990s. These developments include Sequest [5], Mascot [6], OMSSA [7] or Andromeda [8] among many others (for a comprehensive review, see [9]). As a second step of the computational identification, the raw search results are subject to post-processing tools, such as PeptideProphet [10] or Percolator [11].

Computational tools that provide the analysis workflow of proteomics data are available from the instrument vendors or bioinformatics firms.

Oftentimes, they are also available as open source and freeware applications that are frequently developed as bioinformatics research projects. The applicability of such tools ranges from single PCs to high performance compute clusters. While many tools are implemented in an operating system-independent fashion, some tools are available for one operating system only. With the advent of very complex and data-intense experiments, scalability is slowly becoming an important factor to consider when choosing an adequate software suit or data analysis strategy. While it is difficult to draw a clear-cut functional grouping of software tools and platforms, there is a conceptual difference among the available tools: monolithic and modular software tools.

Monolithic applications are usually very user-friendly and come with a reduced complexity for their application. Most of the monolithic software platforms in proteomics [12] are easy to deploy and their usage is facilitated through an intuitive graphical user interface (GUI). The most prominent examples for monolithic analyses software/platforms in proteomics are MaxQuant [13] and commercial packages, such as proprietary software from the instrument vendors. A drawback of large monolithic applications is that the entire codebase is packaged and software maintenance is a very advanced task that can partly only be done by the developers. Most importantly for the data-intense field of proteomics, monolithic applications are difficult to scale for high-throughput data processing and customization. Adding new functionalities is also very difficult, if not impossible.

Classically, the counterpart of monolithic applications is a modularized collection of software tools. Prominent examples in proteomics include the Trans Proteomic Pipeline (TPP) [14] and OpenMS [15]. Comparative details for these and others can be found in a recent benchmarking paper [16].

Modularized software applications group individual parts of the whole functionality into separate modules that may also have stand-alone functionality. Even individual algorithms can be refactored into modules and thereby reduce the complexity of the tasks. A general advantage of this approach is the maintainable codebase and facilitated development. Most community-wide software projects have an underlying modularized design. Due to the separation of individual tasks, the modularized design is more scalable compared to the one-codebase, monolithic application. On the other hand, the refactoring of tasks and algorithms frequently introduces an I/O overload.

For proteomics data processing and analysis, the user has the option to choose between different software solutions. While the solutions can be clearly associated with either a monolithic or modular architectures, the choice between these two options is not trivial. The technical challenges in computational proteomics were recently discussed in a series of reviews [17]. Valuable insights of the state-of-the-art in open source libraries for proteomics data analysis can be found in [18]. This chapter briefly introduces the technical background of the different solutions and provides means for the selection of the most suited tool for a given dataset. The choice of the software to analyze research data is obviously at the forefront of factors to contribute to the success of research projects. It is important to carefully decide the most suitable tool. The main questions each user of proteomics software will need to address are: What is the level of expertise in software application and development? Is it enough to use existing tools or will it be required to implement additional functionality? What degree of flexibility (for both developers and users) is needed? What is the size of data that needs to be processed? Do I have enough hardware resources to run my analysis in-house and is my software compatible with these resources and the data that will be generated?

9.2 Material and Methods

As described above, common to all software tools for the analysis of shotgun proteomics data are processing strategies for data I/O, for peptide and protein identification as well as different quantification techniques. We describe the nodes of such a processing workflow as depicted in Fig. 9.1. We use the modular workflow description for illustration, but the underlying functionality is applicable for monolithic implementations as well.

9.2.1 Data I/O

To process or analyze any mass spectrometry data, it is essential to have access to the required information. Most instruments write their own binary files in proprietary formats and accessing the content is only possible if software tools can use the necessary libraries encoding this functionality. The content and the structure of the data files varies among the proprietary formats (e.g., ThermoFisher Scientific *.raw; ABI/Sciex *.wiff; Agilent *.d). Due to the continuous development of technology and the addition of new features, these formats are frequently updated. Software tools that analyze MS data produce and require additional information than the raw spectra and general acquisition settings, which are encoded in the raw files.

Fortunately, there is a viable community effort towards the implementation of common, open standard file formats and as result of such an effort, the HUPO Proteomics Standards Initiative (PSI, a wide variety of XML-based open formats) has been introduced [20].

For example, mzML is the current standard format for storing MS data (i.e., spectra) whereas mzIdentML, mzTab and mzQuantML formats are for storing analysis results (peptide/protein identification and quantification). The usage of these standard formats in proteomics software development is illustrated in a recent tutorial [21]. The emerging mz5 format [22] has been introduced as an alternative to the XML encoding, which is more compact and efficient file format, since it avoids the heavy load of XML tags (http://www.hdfgroup.org/HDF5/). Despite the strong need to streamline data analysis in open formats, software tools are still lagging behind. Open formats are not only indispensable for a sustainable proteomics research, but also to facilitate software development and maintenance. They encourage reproducible data analysis, enhance data sharing and enable benchmarking of analysis algorithms.

The ProteoWizard suite of proteomics data tools [23, 24] provides the "msConvert" utility for converting between common mass spectrometer file formats to mzML and mzXML.

While Fig. 9.1 points out the conversion step as a node, it should be noted that not all tools and pipelines require this step, and can read the binary formats directly.

9.2.2 Signal Processing

Appropriate pre-processing of the mass spectra can significantly improve the quality of the results. The most common methods include: (1) filtering or "denoising"; (2) baseline correction, which eliminates systematic trends; (3) normalization; (4) peak detection;

Modern high-resolution instruments [25, 26] have made the raw data signal processing steps much simpler than years ago. It is no longer a critical step in the workflow, but is occasionally needed.

9.2.3 Feature Finding

Mass spectrometers measure eluting analytes over a certain period of time, resulting in the analyte's elution profile. Furthermore, the elemental compositions of the molecular species give rise to isotopic patterns. Integrating over the elution profile and the isotopic pattern, all signals for one analyte can be summed up and peptide feature intensities can be derived [27]. Figure 9.2 illustrates the distribution of the isotopic peaks of a peptide feature with charge two. Feature finding in this case, aims at automatically collect all the individual peaks that are visible along the m/z and the RT access. When electrospray ionization is used, peptides are frequently observed with different charges (e.g., $z = +1$, $z = +2$, $z = +3$) and therefore peaks

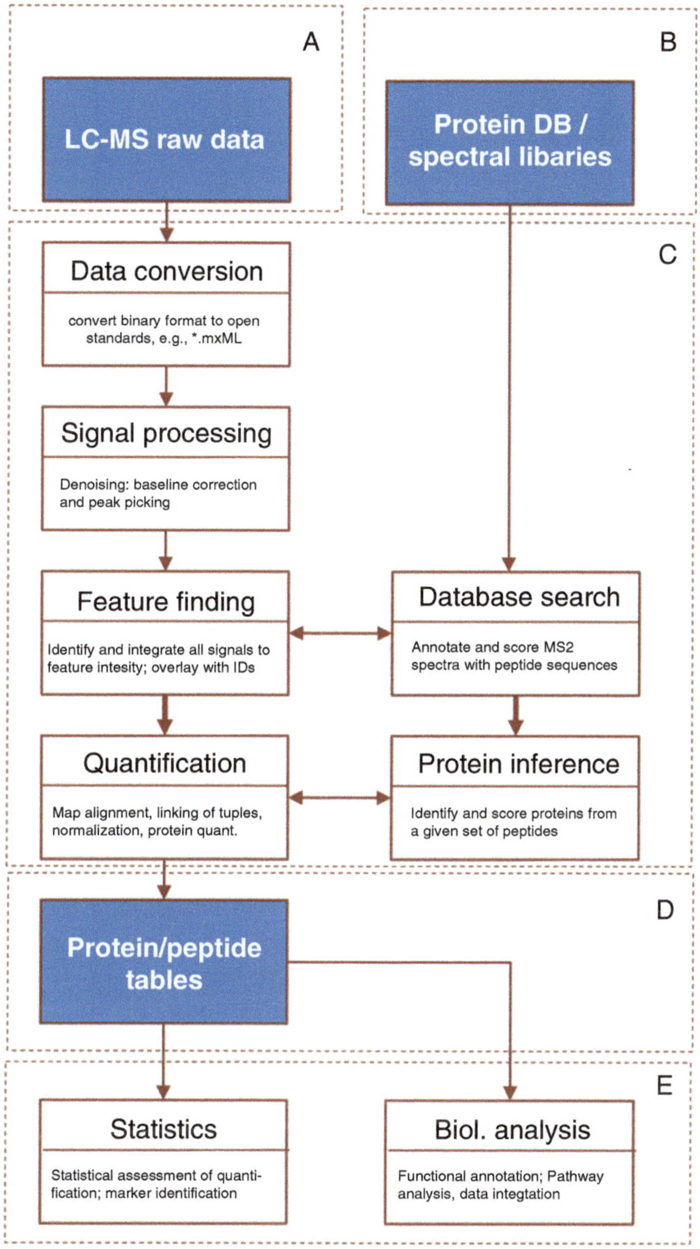

Fig. 9.1 A typical workflow for the analysis of shotgun proteomics data. (**a**) shall exemplify the data acquired during the LC-MS runs, (**b**) corresponds to the *a priori* needed knowledge about the samples, e.g., which organisms are analyzed. Nodes outlined in (**c**) are summarized as data processing tools; a detailed description of these tasks is outlined below. The data processing workflow generates an output of peptide and protein IDs (**d**) along with their differential or absolute quantification. The format of this output varies from pipeline to pipeline and ranges from tsv tables to graphical output and since recently also to the open standard exchange format, mzTAB [19]. (**e**) After completion of the data processing, the resulting information is subject to data analysis. These analyses involve the statistical assessment of differential protein expression and/or biomarker identification (especially in clinical studies). Performing functional annotation or enrichment analyses facilitates the biological interpretation of high-throughput proteomics data. With the emergence of multi-OMICS data, the biological analysis frequently also involve data integration steps

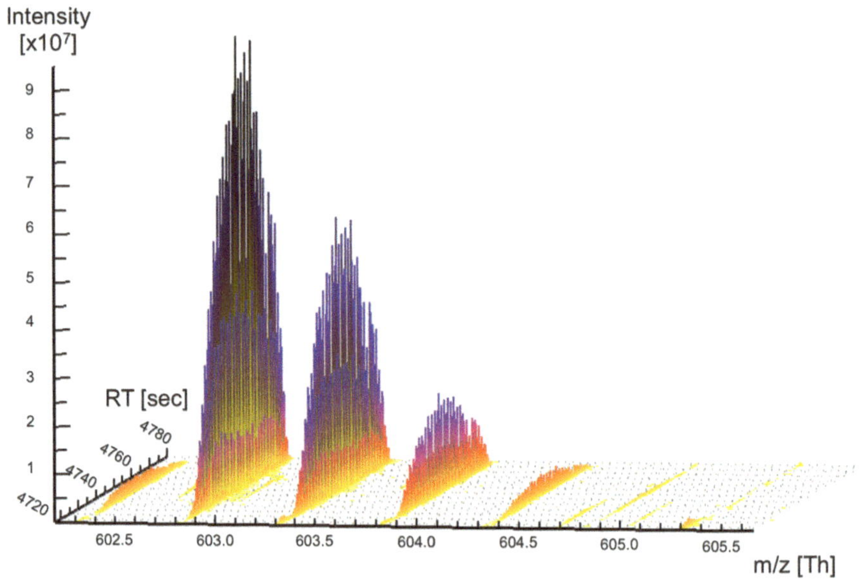

Fig. 9.2 A peptide feature captures all information of an eluting peptide. Here a doubly charged feature with its monoisotopic ion measured at 602.8 Th. The whole peptide elutes over 1 min from the column

can be observed at m/1, m/2 and m/3, where m is the mass of the analyte. Charge determination aims at identifying peak groups that can explain the presence of an analyte at a certain charge. For MS1 peaks that are selected for MS2 fragmentation, this is referred to as "precursor charge". Tools that implement the functionality of merging different charge states of the same peptide are called charge state deconvolution tools. Such tools assemble quantities of the different charge state features into a single *decharged* peptide feature.

For accurate quantification, algorithmic tools are needed to find the isotopic groups and combine the corresponding peaks into features, and more importantly, collate the individual peak intensities into single feature intensities and optionally perform charge state deconvolution. Independent of the nature of the data (with or without labels), feature finding is the most crucial step in any quantification workflow.

9.2.4 Identification

Peptide Identification Classically, in a tandem MS setup, MS2 spectra are used to identify peptides by matching the fragment spectra against the theoretical spectra derived from the target protein database (i.e., known sequences). In brief, using database searching, peptide sequences are assigned to these spectra as outlined above. Following database searches, statistical assessment is required to distinguish correct from incorrect identifications, most commonly using the target/decoy strategy [28]. Using this strategy, reversed, shuffled or randomized protein sequences can serve as negative controls and thus help to estimate the overall FDR.

Protein Inference Following peptide identification, parent protein identity can be inferred from its daughter peptides. In particular, due to false discoveries on the peptide level as well as the fact that peptides can map to multiple proteins, protein inference with accurate error estimation remains a difficult problem. Continual release of tools providing solutions for protein inference attempt to alleviate this problem. Protein Prophet [29] is one of the most widely used tools but other tools include MAYU [30] or more recent BP-Quant [31] that specifically addresses the problem of alternative splicing and other proteoforms – which is the core challenge in protein inference.

9.2.5 Quantification

Protein quantification can be done by attaching labels to proteins or peptides, chemically [32] or metabolically [33]. Label-free strategies are another approach for protein quantification. For a comprehensive review of quantification strategies, we refer to [34]. The choice of the data analysis strategy obviously depends on the methods that were used for the generation of the data. For the quantitative analysis of labeled data many dedicated tools are available [35, 36]. Recently, label-free quantification is gaining increasing interest due to the practical simplicity for data generation and the expansion of software applications for quantifying label-free data [4, 37]. Most of the libraries and computational frameworks provide algorithmic solutions to a wide range of quantitative data. Besides the feature-based quantification, an alternative approach to assess differential peptide and protein quantities include methods that are summarized as spectral counting. These methods rely on the counting of the number of MS/MS events that can be associated with a certain protein and thereby allow for differential quantification. Spectral counting methods have been comprehensively reviewed in [38].

9.2.6 Alignment and Normalization of Multiple Runs

Quantitative studies usually involve a panel of samples reflecting the experimental design of interest. While recent advances in large-scale multiplexing [39] can reduce the number of runs significantly, moderately large experiments (e.g., for clinical studies) still require multiple injections into the LC-MS setup. If analytes need to be quantified across multiple LC-MS runs, the unavoidable technical variability need to be corrected. Algorithmic solutions for this are summarized as *map alignment* with a map referring to the 2D-space (retention time (RT) vs. mass-to-charge (m/z) of an LC-MS run. The variations in the m/z is marginal in latest instrumentation, the RT dimension, however, can be quite variable within one experiment; these variabilities can even be non-linear [27]. All tools discussed within this chapter provide algorithmic solutions that account for these variabilities (see [40]) for the algorithm as implemented in TOPP [41]). A broader overview of different alignment tools can be found in [42].

After the accurate identification of features, calculation of individual feature intensities, the next step in the quantification node needs to account for systematic biases that have been introduced during the sample preparation and/or the measurements. This procedure is commonly referred to as map normalization. Normalization steps also include an optional assessment of biological variation in biological replicates – such analyses, however, should be done with caution, since biological variation is an inherent property of any biological system. Nonetheless, normalization is beneficial for any quantitative set-up, but is essential for label-free analyses due to a much higher technical variation in comparison to labeled data, where multiple samples are measured in the same run.

9.2.7 Statistical Analysis

Various quality-control measures such as mass error or charge distribution during an LC run can readily give a diagnostic on the dataset [43]. Moreover, simple descriptive statistics, scatter plots, clustering or PCA plots can be informative in their own right. However, in non-trivial experiments, statistical analysis has to match the original experimental design and therefore expert knowledge is often needed to choose the right package and method [Chap. 11].

9.3 Tools and Platforms

The following section outlines the most commonly used software solutions in proteomics. This section details the underlying functionality

Table 9.1 Summary of the software tools and platforms for proteomics data analysis

Tool	License	Interface	[b]Current version	[c]Platforms	File formats	URL
TPP	Open-source	Command-line	2.4.2	W, L	mz(ML|XML)	http://tools.proteomecenter.org
	GPL v. 2.0 and LGPL	Web				
OpenMS	Open-source	Command-line	1.11	W, L, M	mz(ML|XML|Data)	http://open-ms.sourceforge.net/
		TOPP				
CPFP	Open-source	Web	2.1.1	L, M	mz(ML|XML)	http://cpfp.sourceforge.net/
	CDDL[a]				MySQL	
MaxQuant	Freeware	GUI	1.5.2.8.	W	Thermo.RAW	www.maxquant.org
					mzXML	
Scaffold	commercial	GUI	4.4.1	W, L, M	Major vendor formats	www.proteomesoftware.com
Sorcerer	commercial	Web	Visit URL	L	Major vendor formats	www.sagenresearch.com
IPA	commercial	Web	IP 2	L	ms1, ms2, mzXML,	www.integratedproteomics.com
					DTASelect	

[a]CDDL: OSI approved Common Development and Distribution License
[b]December 2014
[c]Operating systems: *W* Windows, *L* Linux, *M* Mac OS

and applicability of the tools and if available, it points the user to the resources and provides information on the licensing for the individual software applications. Table 9.1 summarizes the major properties of the individual tools.

9.3.1 Trans Proteomic Pipeline (Open Source)

The Trans-Proteomics Pipeline (TPP) is one of the most mature suites of software tools for the analysis of LC-MS/MS data [14]. The tools cover all the steps in a shotgun proteomics analysis workflow from raw data conversion to protein-level identification and quantification [44], (See Fig. 9.1). Within TPP, particular emphasize has been put into statistical validation of the identifications. Typically, peptide identifications from different search engines are validated by PeptideProphet [10] and refined and merged with iProphet [45]. Subsequently, protein inference is performed with ProteinProphet [29] and results provided at different false discovery rates (FDR).

TPP is shipped with Comet [46] and X!TANDEM [47] search engines but the currently supported engines are: SEQUEST [5]; MSGF+ [48]; Inspect [49]; OMSSA [7]; MyriMatch [50]; Mascot [6].

In addition to the traditional database (sequence) search, TPP provides the SpectraST tool as an alternative approach [51]. SpectraST is a spectral library building and searching tool wherein: (1) Previously observed and identified peptide MS/MS spectra are compiled and stored into "spectral libraries" and (2) Newly observed spectra to be identified are matched against the entire target spectral library. This approach shows great potential in complementing and/or substituting the classical sequence searching. TPP includes ASAPRatio [52] and XPRESS [53] for relative abundances of proteins from ICAT-reagent labeled data. iTRAQ and TMT labeled samples can be analyzed and quantified with the *Libra* TPP module.

TPP offers a web-based GUI (called Petunia), which gives access to the tools and data in a visual environment as an alternative to the command-line interface.

9.3.2 OpenMS (Open Source)

OpenMS has been designed as a software framework for mass spectrometry [15]. As such, it provides data structures and algorithms to rapidly design and assemble analysis pipelines. OpenMS is developed in the C++ programming language and its code is freely available under the 3-clause BSD license at https://github.com/OpenMS/OpenMS. Besides the core structures, OpenMS is shipped with TOPP, the OpenMS Proteomics Pipeline [41], which is a collection of precompiled building blocks that can be chained together to form production-ready processing pipelines. Both the library and the TOPP tools are available for all major operating system (iOS, Windows and Linux). The TOPP tools include nodes for data handling, raw data signal processing, peptide and protein identification as well as for the quantification of peptides and proteins using different labeling strategies (e.g., isobaric labeling or SILAC) or label-free [4]. OpenMS tools and the resulting constructed pipelines can easily be executed on high-performance computing clusters using different workflow systems, e.g., Galaxy [54], thus providing a scalable solution for large-scale data centers.

Furthermore, the OpenMS framework also provides tools for the visualization of MS raw data and analysis results, as well as a pipeline designing tool, the OpenMS Proteomics Pipeline Assistant, TOPPAS [55]. Using TOPPAS, the user can intuitively build customized processing pipelines.

9.3.3 CPFP (Web-Based Freeware)

The Central Proteomics Facilities Pipeline (cpfp.sourceforge.net) is in essence a web based wrapper around TPP tools, various search engines (Mascot [6], OMSSA [7], and X!TANDEM [47]) and a MySQL back-end for storing spectra and results [56]. It is primarily suited for core facilities, providing an easy to use web interface to upload data, trigger workflows and browse the results.

The analysis pipeline covers identification, quantitation and validation of peptides and proteins [14].

9.3.4 MaxQuant (Freeware)

MaxQuant is a proteomics software application designed for quantitative analysis of LC-MS/MS data [13]. It is a freely available, closed-source (written in C# using the.NET Framework), monolithic, and Windows only application (www.maxquant.org). Its algorithms are particularly tailored for high-resolution data such as Thermo Orbitrap and FT.

MaxQuant provides all steps for a shotgun proteomics analysis workflow (See Fig. 9.1) organized into the "Quant" module, Andromeda search engine [8] and the "Identify" module.

It can provide: (1) protein identification, (2) Feature-based label-free quantification and (3) quantification for SILAC, TMT and iTRAQ-labeled samples.

In essence, the user has to choose the raw data files, the target database and make the appropriate parameter settings (e.g., SILAC labels, mass tolerances). These can be done via the GUI, which is a typical desktop application interface. The workflow then runs all the intermediate necessary steps (e.g., feature detection, MS/MS searches, filtering, protein assembly and quantification) in what appears to the user as a single analysis run.

MaxQuant is equipped with a built-in "Viewer" for data inspection and browsing the results. Recently, the authors recommend the "Perseus" framework (www.perseus-framework.org/) for subsequent statistical analysis of MaxQuant output.

9.3.5 Scaffold (Commercially Available)

Scaffold (Proteome Software, Portland OR, USA, www.proteomesoftware.com) is a feature-rich software suite to assist in analysis, visualization, quantification, annotation and validation of complex LC-MS/MS experiments. It supports a wide variety of search engines: Mascot [6], MascotDistiller, MatrixScience (London, UK, http://www.matrixscience.com/distiller.html), Proteome Discoverer, Thermo Fisher (Bremen, Germany, http://www.thermoscientific.com), Spectrum Mill (Agilent, Santa Clara, USA, http://www.chem.agilent.com/), SEQUEST [5], IdentityE/PLGS (Waters, Manchester, UK, www.waters.com), OMSSA [7], X!TANDEM [47] and MaxQuant/Andromeda [8, 13]. Validation is achieved by the Peptide Prophet/Protein Prophet algorithms [14] with an enhanced protein grouping method [57]. It supports label free quantitation (MS1 precursor intensity as well as MS2 spectral counting). Scaffold Q+ and Q + S can perform iTRAQ, TMT and SILAC based quantitation. Basic statistics like t-test, ANOVA or Kruskal Wallis test are included and offer built-in differential expression analysis. Data can be filtered using various criteria like peptide/protein probabilities (FDR), search engine scores, expression values (fold change), etc. Scaffold maintains multiple GO annotation databases and allows GO filtering as well as categorical GO term quantitation. Visualization spans from raw MS/MS spectra to peptides and proteins coverage, differential expression, modifications, GO annotation, Venn diagrams, as well as intensity scatterplots and quantitation charts.

Quality control visualization includes: search engines scatterplot comparisons, ROC plots for sensitivity/specificity, error estimates and randomized permutation calculation.

Scaffold PTM module provides PTM site localization and probability, motif validation and sequence visualization and filtering.

9.3.6 Sorcerer (Commercially Available)

Sorcerer (Sage-N Research, Milpitas CA, USA, www.sagenresearch.com) platforms include tightly integrated hardware & software solutions. The Enterprise solution is customizable, scalable and provides very high-throughput aggregate analysis and integrated optimized storage. SORCERER™ 2 is also a fully integrated data analysis system particularly tailored for labs with moderate throughput (or high, but not continuous throughput). SORCERER-V (standing for virtual), is a scaled down, yet complete platform packed into a virtual machine that can run on a regular modern PC. This is offered as an entry-level product for scientists to explore and start building data analysis and data-mining platforms.

SORCERER analysis is accessible via Scaffold or the Trans-Proteomics Pipeline.

9.3.7 IPA/IP2 (Commercially Available)

Integrated Proteomics Pipeline (IP2, Integrated Proteomics Applications, Inc. San Diego, CA, USA, www.integratedproteomics.com) provides complete solutions for proteomics data analysis. The core methods for protein identification, quantification, filtering and analysis are Sequest/ProLuCID [5] (http://fields.scripps.edu/downloads.php), DTASelect2 [58, 59], and Census [60]. It uses internal file formats: ms1 and ms2 (RawExtract, http://fields.scripps.edu/downloads.php) [61]. It has a project-oriented web interface and features GO analysis and PTM analysis as well as basic statistics support.

9.4 Technical Aspects and Data Dissemination

9.4.1 Computation and Data Management

While the software tools can, in principle, run on a single PC, the true computational throughput and performance is achieved when running on computer clusters or on clouds, in parallel. One can, for example, run MaxQuant on a regular PC or even on a modern laptop for projects with a small number of samples. Obviously, this strategy does not scale well with respect to the ever accumulating data and the expected turnaround times. Dedicated hardware and advanced tools and IT knowledge are needed to securely store, back-up and retrieve the data.

9.4.2 Resources and Repositories

The major freely accessible resources of protein sequence, annotation, functional information are: UniProt (Universal Protein Resource, www.uniprot.org); Ensembl (www.ensembl.org) and NCBI (The National Center for Biotechnology Information www.ncbi.nlm.nih.gov).

Modern proteomics journals require that upon publication of articles, the data (both raw and processed) to be made publicly available. The ProteomeXchange consortium (www.proteomexchange.org) has emerged as the primary coordinator of the main existing proteomics repositories. The current repository of choice for tandem MS/MS datasets is PRIDE (www.ebi.ac.uk/pride) [62] and for Selected Reaction Monitoring (SRM) datasets, the PASSEL component of PeptideAtlas (www.peptideatlas.org/passel/) [63].

The consortium collects, centralizes and disseminates the raw data, processed data (as published by the contributors) as well as the essential metainformation about the dataset (e.g., species, tissue, genetic background or health state).

9.5 Conclusion

The growing variety of proteomics software tools and platforms can only reflect an increasing interest in the field of proteomics and its expected impact on life and medical sciences. The application areas of the described tools range from specialized experiments to generic solutions for the most commonly performed experiments. Libraries and their modular building blocks primarily fulfill the latter functionality. Specialized applications are most frequently covered by monolithic stand-alone applications.

Therefore, the choice of appropriate software tools and platforms is primarily driven by the needs and capabilities of one's lab. Commercial, turn-key solutions can obviously cut down in the necessary IT support, while setting-up, developing and maintaining a custom, open-source platform may require extensive IT and bioinformatics knowledge and assistance.

Small proteomics labs or labs without dedicated bioinformatics staff may opt for pre-assembled, GUI-based applications, to avoid IT overhead. More advanced or large-scale facilities will require higher flexibility and scalability in their bioinformatics applications and infrastructure.

Since the technological development in proteomics is an on-going process, it can be anticipated that new software applications will emerge and that proteomics software development will remain a vivid field of bioinformatics.

References

1. Kirkwood KJ et al (2013) Characterization of native protein complexes and protein isoform variation using size-fractionation-based quantitative proteomics. Mol Cell Proteomics 12(12):3851–3873
2. Kim MS et al (2014) A draft map of the human proteome. Nature 509(7502):575–581
3. Domon B, Aebersold R (2010) Options and considerations when selecting a quantitative proteomics strategy. Nat Biotechnol 28(7):710–721
4. Weisser H et al (2013) An automated pipeline for high-throughput label-free quantitative proteomics. J Proteome Res 12:1628
5. Eng JK, McCormack AL, Yates JR (1994) An approach to correlate tandem mass spectral data of peptides with amino acid sequences in a protein database. J Am Soc Mass Spectrom 5(11):976–989
6. Perkins DN et al (1999) Probability-based protein identification by searching sequence databases using mass spectrometry data. Electrophoresis 20 (18):3551–3567
7. Geer LY et al (2004) Open mass spectrometry search algorithm. J Proteome Res 3(5):958–964
8. Cox J et al (2011) Andromeda: a peptide search engine integrated into the MaxQuant environment. J Proteome Res 10(4):1794–1805
9. Eng JK et al (2011) A face in the crowd: recognizing peptides through database search. Mol Cell Proteomics 10(11):R111.009522
10. Keller A et al (2002) Empirical statistical model to estimate the accuracy of peptide identifications made by MS/MS and database search. Anal Chem 74 (20):5383–5392
11. Kall L et al (2007) Semi-supervised learning for peptide identification from shotgun proteomics datasets. Nat Methods 4(11):923–925

12. Zhang R et al (2010) Evaluation of computational platforms for LS-MS based label-free Quantitative-Proteomics: a global view. J Proteomics Bioinform 3:260–265
13. Cox J, Mann M (2008) MaxQuant enables high peptide identification rates, individualized p.p.b.-range mass accuracies and proteome-wide protein quantification. Nat Biotechnol 26(12):1367–1372
14. Deutsch EW et al (2010) A guided tour of the trans-proteomic pipeline. Proteomics 10(6):1150–1159
15. Sturm M et al (2008) OpenMS – an open-source software framework for mass spectrometry. BMC Bioinf 9:163
16. Hoekman B et al (2012) msCompare: a framework for quantitative analysis of label-free LC-MS data for comparative biomarker studies. Mol Cell Proteomics 11:M111.015974
17. Aebersold R (2011) Editorial: from data to results. Mol Cell Proteomics 10(11):E111 014787
18. Perez-Riverol Y et al (2014) Open source libraries and frameworks for mass spectrometry based proteomics: a developer's perspective. Biochim Biophys Acta 1844(1 Pt A):63–76
19. Griss J et al (2014) The mzTab data exchange format: communicating mass-spectrometry-based proteomics and metabolomics experimental results to a wider audience. Mol Cell Proteomics 13(10):2765–2775
20. Deutsch EW (2012) File formats commonly used in mass spectrometry proteomics. Mol Cell Proteomics 11(12):1612–1621
21. Gonzalez-Galarza FF et al (2014) A tutorial for software development in quantitative proteomics using PSI standard formats. Biochim Biophys Acta 1844 (1 Pt A):88–97
22. Wilhelm M et al (2012) mz5: space- and time-efficient storage of mass spectrometry data sets. Mol Cell Proteomics 11(1):O111 011379
23. Kessner D et al (2008) ProteoWizard: open source software for rapid proteomics tools development. Bioinformatics (Oxford, England) 24(21):2534–2536
24. Chambers MC et al (2012) A cross-platform toolkit for mass spectrometry and proteomics. Nat Biotechnol 30(10):918–920
25. Olsen JV et al (2009) A dual pressure linear ion trap orbitrap instrument with very high sequencing speed. Mol Cell Proteomics 8(12):2759–2769
26. Kelstrup CD et al (2014) Rapid and deep proteomes by faster sequencing on a benchtop quadrupole ultra-high-field orbitrap mass spectrometer. J Proteome Res 13(12):6187–6195
27. Nahnsen S et al (2013) Tools for label-free peptide quantification. Mol Cell Proteomics 12(3):549–556
28. Elias JE, Gygi SP (2007) Target-decoy search strategy for increased confidence in large-scale protein identifications by mass spectrometry. Nat Methods 4 (3):207–214
29. Nesvizhskii AI et al (2003) A statistical model for identifying proteins by tandem mass spectrometry. Anal Chem 75(17):4646–4658
30. Reiter L et al (2009) Protein identification false discovery rates for very large proteomics data sets generated by tandem mass spectrometry. Mol Cell Proteomics 8(11):2405–2417
31. Webb-Robertson BJ et al (2014) Bayesian proteoform modeling improves protein quantification of global proteomic measurements. Mol Cell Proteomics 13 (12):3639–3646
32. Gygi SP et al (1999) Quantitative analysis of complex protein mixtures using isotope-coded affinity tags. Nat Biotechnol 17(10):994–999
33. Ong S-E et al (2002) Stable isotope labeling by amino acids in cell culture, SILAC, as a simple and accurate approach to expression proteomics. Mol Cell Proteomics 1(5):376–386
34. Bantscheff M et al (2007) Quantitative mass spectrometry in proteomics: a critical review. Anal Bioanal Chem 389(4):1017–1031
35. Liao Z et al (2012) IsoQuant: a software tool for stable isotope labeling by amino acids in cell culture-based mass spectrometry quantitation. Anal Chem 84 (10):4535–4543
36. Wen B et al (2014) IQuant: an automated pipeline for quantitative proteomics based upon isobaric tags. Proteomics 14(20):2280–2285
37. Cox J et al (2014) Accurate proteomewide label-free quantification by delayed normalization and maximal peptide ratio extraction, termed MaxLFQ. Mol Cell Proteomics 13(9):2513–2526
38. Lundgren DH et al (2010) Role of spectral counting in quantitative proteomics. Expert Rev Proteomics 7 (1):39–53
39. Dephoure N, Gygi SP (2012) Hyperplexing: a method for higher-order multiplexed quantitative proteomics provides a map of the dynamic response to rapamycin in yeast. Sci Signal 5(217):rs2
40. Lange E et al (2007) A geometric approach for the alignment of liquid chromatography-mass spectrometry data. Bioinformatics (Oxford, England) 23(13): i273–i281
41. Kohlbacher O et al (2007) TOPP-the OpenMS proteomics pipeline. Bioinformatics (Oxford, England) 23 (2):e191–e197
42. Lange E et al (2008) Critical assessment of alignment procedures for LC-MS proteomics and metabolomics measurements. BMC Bioinf 9:375
43. Walzer M et al (2014) qcML: an exchange format for quality control metrics from mass spectrometry experiments. Mol Cell Proteomics 13(8): 1905–1913
44. Keller A, Shteynberg D (2011) Software pipeline and data analysis for MS/MS proteomics: the trans-proteomic pipeline. Methods Mol Biol 694: 169–189
45. Shteynberg D et al (2011) iProphet: multi-level integrative analysis of shotgun proteomic data improves peptide and protein identification rates and error estimates. Mol Cell Proteomics MCP 10(12):M111 007690

46. Eng JK, Jahan TA, Hoopmann MR (2013) Comet: an open-source MS/MS sequence database search tool. Proteomics 13(1):22–24
47. Craig R, Beavis RC (2004) TANDEM: matching proteins with tandem mass spectra. Bioinformatics (Oxford, England) 20(9):1466–1467
48. Kim S, Pevzner PA (2014) MS-GF+ makes progress towards a universal database search tool for proteomics. Nat Commun 5:5277
49. Tanner S et al (2005) InsPecT: Identification of posttransiationally modified peptides from tandem mass spectra. Anal Chem 77(14):4626–4639
50. Tabb DL, Fernando CG, Chambers MC (2007) MyriMatch: highly accurate tandem mass spectral peptide identification by multivariate hypergeometric analysis. J Proteome Res 6(2):654–661
51. Lam H et al (2008) Building consensus spectral libraries for peptide identification in proteomics. Nat Methods 5(10):873–875
52. Li XJ et al (2003) Automated statistical analysis of protein abundance ratios from data generated by stable-isotope dilution and tandem mass spectrometry. Anal Chem 75(23):6648–6657
53. Han DK et al (2001) Quantitative profiling of differentiation-induced microsomal proteins using isotope-coded affinity tags and mass spectrometry. Nat Biotechnol 19(10):946–951
54. Goecks J et al (2010) Galaxy: a comprehensive approach for supporting accessible, reproducible, and transparent computational research in the life sciences. Genome Biol 11(8):R86
55. Junker J et al (2012) TOPPAS: a graphical workflow editor for the analysis of high-throughput proteomics data. J Proteome Res 11(7):3914–3920
56. Trudgian DC et al (2010) CPFP: a central proteomics facilities pipeline. Bioinformatics 26(8):1131–1132
57. Searle BC (2010) Scaffold: a bioinformatic tool for validating MS/MS-based proteomic studies. Proteomics 10(6):1265–1269
58. Tabb DL, McDonald WH, Yates JR 3rd (2002) DTASelect and Contrast: tools for assembling and comparing protein identifications from shotgun proteomics. J Proteome Res 1(1):21–26
59. Cociorva D, Tabb LD, Yates JR (2007) Validation of tandem mass spectrometry database search results using DTASelect. Curr Protoc Bioinformatics Chapter 13: p. Unit 13.4
60. Park SK et al (2008) A quantitative analysis software tool for mass spectrometry-based proteomics. Nat Methods 5(4):319–322
61. McDonald WH et al (2004) MS1, MS2, and SQT-three unified, compact, and easily parsed file formats for the storage of shotgun proteomic spectra and identifications. Rapid Commun Mass Spectrom 18(18):2162–2168
62. Vizcaino JA et al (2009) A guide to the proteomics identifications database proteomics data repository. Proteomics 9(18):4276–4283
63. Deutsch EW, Lam H, Aebersold R (2008) PeptideAtlas: a resource for target selection for emerging targeted proteomics workflows. EMBO Rep 9(5):429–434

Tandem Mass Spectrum Sequencing: An Alternative to Database Search Engines in Shotgun Proteomics

10

Thilo Muth, Erdmann Rapp, Frode S. Berven, Harald Barsnes, and Marc Vaudel

Abstract

Protein identification *via* database searches has become the gold standard in mass spectrometry based shotgun proteomics. However, as the quality of tandem mass spectra improves, direct mass spectrum sequencing gains interest as a database-independent alternative. In this chapter, the general principle of this so-called *de novo* sequencing is introduced along with pitfalls and challenges of the technique. The main tools available are presented with a focus on user friendly open source software which can be directly applied in everyday proteomic workflows.

Keywords

de novo identification • Mass spectrum sequencing • Quality control • Visualization

T. Muth • E. Rapp
Max Planck Institute for Dynamics of Complex Technical Systems, Magdeburg, Germany

glyXera GmbH, Magdeburg, Germany

F.S. Berven
Proteomics Unit, Department of Biomedicine, University of Bergen, Jonas Liesvei 91, N-5009 Bergen, Norway

KG Jebsen Centre for Multiple Sclerosis Research, Department of Clinical Medicine, University of Bergen, Bergen, Norway

Norwegian Multiple Sclerosis Competence Centre, Department of Neurology, Haukeland University Hospital, Bergen, Norway

Abbreviations

PSM	Peptide Spectrum Match
FDR	False Discovery Rate
m/z	Mass over Charge
BLAST	Basic Local Alignment Search Tool
PTM	Post-Translational Modification

H. Barsnes (✉) • M. Vaudel
Proteomics Unit, Department of Biomedicine, University of Bergen, Jonas Liesvei 91, N-5009 Bergen, Norway
e-mail: harald.barsnes@biomed.uib.no

H. Mirzaei and M. Carrasco (eds.), *Modern Proteomics – Sample Preparation, Analysis and Practical Applications*, Advances in Experimental Medicine and Biology 919, DOI 10.1007/978-3-319-41448-5_10

10.1 Introduction

In the early days of shotgun proteomics, three paradigms emerged for the computational derivation of peptide sequences from tandem mass spectra as replacement to chemical strategies like Edman degradation [1]:

(i) Spectral matching as inherited from small molecule analyses approaches [2]
(ii) Automated sequencing of spectra as exemplified by the SEQPEP algorithm [3]
(iii) Making use of the growing protein sequence databases to restrain the search space [4], giving birth to the first search engines, like the pioneer algorithms SEQUEST [5] and MOWSE (later employed in Mascot) [6, 7].

Mass spectrum sequencing is generally termed *de novo* peptide identification by opposition to database search engines since it does not rely on *a priori* knowledge about the possible peptide sequences. By definition, it consists of building a sequence from the spectrum fragment ions, by chaining peaks separated by amino acid characteristic masses. While in the ideal case all fragment ions can be assigned and a full peptide sequence from the N-terminus to the C-terminus built, peptides generally fragment unevenly rendering the detection of some fragment ions unlikely or even impossible [8]. As a result, peaks are generally missing and full sequence assignments are usually not possible, introducing sequence ambiguities or gaps due to missing fragments. Conversely, the occurrence of a high number of peaks typically observed for multiply charged precursors, neutral losses and noisy spectra makes the sequencing practically impossible.

The success of *de novo* identification thus strongly relies on the quality of the fragmentation and resolution of the mass spectrometer, and is computationally demanding. Three decades ago, the resolution of mass spectrometers and the computational speed made spectral libraries searching and direct sequencing challenging for everyday lab practices, and search engines were thus rapidly established as the gold standard for shotgun protein identification. Nowadays however, the enhanced fragmentation quality, sub-ppm resolution of modern mass spectrometers, and increased computational speed and parallelization of computers make it realistic to (re)introduce mass spectrum sequencing.

Thereby, peptides are inferred from the spectra in an unbiased way, independently from any database, thus providing the unique potential to identify protein isoforms, mutated sequences, and unexpected modifications. However, this advantage comes at the cost of high computational complexity and challenging protein inference.

In this chapter, we will present the different paradigms of *de novo* identification and the algorithmic implementations. We will also illustrate how mass spectrum sequencing can be integrated in standard proteomic workflows *via* user-friendly interfaces. Finally, we will discuss the remaining challenges for a complete integration of mass spectrum sequencing in everyday practices.

10.2 Paradigms and Algorithms

Two main paradigms emerged in mass spectrum sequencing: tag-based approaches, and complete sequence *de novo* identification algorithms. While the latter attempts to derive the entire peptide sequence from the spectrum, the tag approach only partially identifies the peptide *via* a sequence tag of a few amino acids. The rationale behind the tag approach is that while complete fragment ion coverage is rare, spectra generally present a series of a few high intense peaks providing a high quality tag which can be used for further identification. Mann and Wilm pioneered tag sequencing suggesting that these "islands" of sequence ions present valuable information complementary to database search results [9]. The approach was applied for non-sequenced organisms in MultiTag [10], and

Table 10.1 Software available for mass spectrum sequencing

Sequencing type	Name	Publication year	Number of citations (Total/ Average/Trend)	Free	Maintained
Tag-based	MultiTag [10]	2003	88/7.33/↘	Yes	No
Tag-based	GutenTag [11]	2003	175/14.58/↘	Yes	Yes
Tag-based	DirecTag [12]	2008	32/4.57/↗	Yes	Yes
Full *de novo*	Lutefisk [20]	1997	238/13.22/↘	Yes	No
Full *de novo*	SeqMS [21]	2000	36/2.40/↘	Yes	No
Full *de novo*	Sub [22]	2003	59/4.92/↘	Yes	?
Full *de novo*	NovoHMM [23]	2005	73/7.30/↗	Yes	No
Full *de novo*	Audens [24]	2005	37/3.70/↘	Yes	?
Full *de novo*	MSNovo [25]	2007	38/4.75/↗	Yes	?
Full *de novo*	Vonode [26]	2010	11/2.2/→	Yes	?
Full *de novo*	GenoMS [27]	2010	14/2.80/↗	Yes	?
Full *de novo*	CompNovo [28]	2009	18/3.00/→	Yes	No
Full *de novo*	MetaSPS [29]	2013	N/A	Yes	?
Full *de novo*	pNovo + [30]	2013	N/A	Yes	?
Full *de novo*	PepNovo + [31]	2005	280/28.0/↗	Yes	No
Full *de novo*	UniNovo [32]	2013	N/A	Yes	No
Full *de novo* coupled with database search	PEAKS [33, 42]	2003	364/30.33/↗	No	Yes
Full *de novo* coupled with database search	Bionics [34]	2007	64/8.00/↗	No	Yes
Peptide assembly	TagRecon [39]	2010	30/6.00/↗	Yes	Yes
Sequence similarity	FASTA [38]	1988	9457/350.56/↘	Yes	?
Sequence similarity	BLAST [37]	1990	38,684/1548.08/↗	Yes	?
Sequence similarity	PepExplorer [41]	2014	N/A	Yes	?
Protein inference	MSDA [40]	2014	N/A	Yes	Yes
Protein inference	IdPicker [44]	2007	136/17.00/↗	Yes	Yes
Graphical interface	BumberDash		N/A	Yes	Yes
Graphical interface	DeNovoGUI [47]	2013	N/A	Yes	Yes

The table lists the software mentioned in this book chapter, classified by use case, and provides the publication year and corresponding reference. For tools published earlier than 2013, the number of citations according to Thomson Reuters™ Web of Science™ is given. Finally, the table indicates whether the software is free and maintained. Whenever a tool could not be found, it was marked as not maintained. Note that the number of citations is solely given as an indicator of the tool usage. The total and average numbers of citations per year are given, as well as the trend for the last 3 years' citation average relative to the global average: ↗ increasing number of citations, → stable number of citations, and ↘ decreasing number of citations

in generic tools for mass spectrum sequencing by the Tabb lab: GutenTag [11] and DirecTag [12], as listed in Table 10.1.

Tag algorithms have the advantage that they can be extremely fast, however, they are criticized for requiring clearly defined consecutive lists of amino acids, and to provide only limited information about the sequence. Figure 10.1 displays the distribution of Peptide Spectrum Matches (PSMs) according to the length of the longest tag which can be derived from its spectrum annotation in a standard shotgun proteomic run (a tryptic *HeLa* digest measured on a Q Exactive, data from [13]) obtained from the combination of five search engines (MS Amanda [14], MS-GF+ [15], Myrimatch [16], OMSSA [17], and X!Tandem [18]) using PeptideShaker (http://www.ncbi.nlm.nih.gov/pubmed/25574629). From the figure, it is clear that a tag of at least three amino acids can

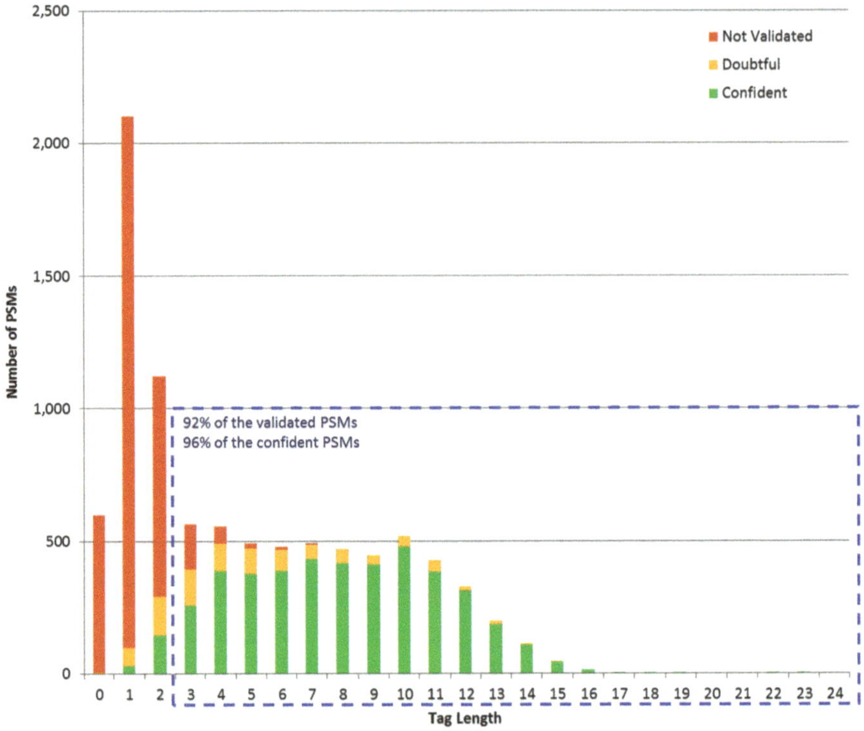

Fig. 10.1 Distribution of Peptide Spectrum Matches (PSMs) according to the length of the longest tag which can be derived from its spectrum annotation in a standard shotgun proteomic experiment obtained from the combination of five search engines (see text for details). PSMs are sorted into four categories: (i) Not Validated – PSMs which do not pass a 1 % False Discovery Rate (FDR) threshold, (ii) Doubtful – PSMs passing a 1 % FDR threshold but not the quality filters embedded in PeptideShaker, and (iii) Confident – PSMs passing a 1 % FDR threshold and the quality filters. The PSMs with a tag length of at least three is circled in blue, comprising 92 % of the validated PSMs and 96 % of the confident PSMs. When no combination of two annotated peaks separated by a single amino acid mass could be found, the PSM was categorized in the '0' category

be derived from 92 % of the PSMs validated at 1 % False Discovery Rate (FDR), and from 96 % of the PSMs passing the quality filters implemented in PeptideShaker. The potential identification rate of tag approaches thus appears to be comparable to search engines. However, it requires the intervention of a downstream algorithm to infer the complete peptide sequence, a point which will be touched upon in the following section.

Besides the tag-based approaches, several algorithms have been developed aiming at a complete sequencing of tandem mass spectra. As schematized in Fig. 10.2, the standard approach relies on a spectrum graph that consists of vertices and edges: the peaks of the spectrum are converted into vertices with attributed m/z values [19]. If the mass difference between two different peaks corresponds to the mass of an amino acid, possibly carrying a modification, an edge is drawn to connect the respective vertices. This procedure is repeated until a full path is found that connects the N-terminal with the C-terminal vertices. Additionally, these connections are scored, for example, based on the intensity of the peaks or the accuracy of the peak m/z matching. The *de novo* sequencing algorithms then try to find the path with the best score and this path is transferred back to a peptide sequence suggestion. Some algorithms also include peptide fragmentation models in order to provide statistical significance for the scoring: can a peak be explained by a predicted fragmentation rule or is it simply a random match?

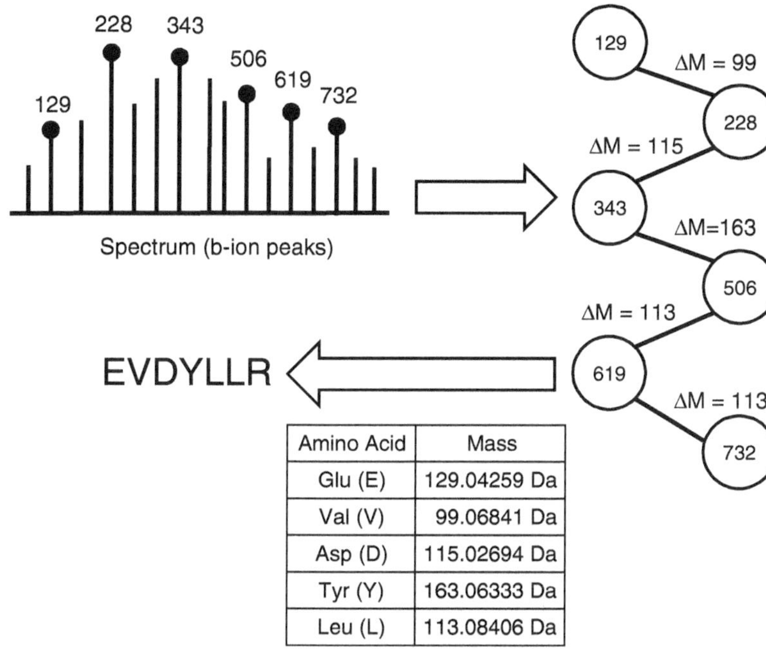

Amino Acid	Mass
Glu (E)	129.04259 Da
Val (V)	99.06841 Da
Asp (D)	115.02694 Da
Tyr (Y)	163.06333 Da
Leu (L)	113.08406 Da

Fig. 10.2 The spectrum *de novo* sequencing principle. The spectrum is converted into a spectrum graph, and the peptide sequence is then derived from the graph

De novo identification algorithms are available as both free and commercial software, see Table 10.1. One of the most popular pioneer algorithm is Lutefisk [20], which was followed by SeqMS [21], sub [22], NovoHMM [23], Audens [24], and MSNovo [25]. Vonode [26] and GenoMS [27] were subsequently specifically developed for proteogenomic studies. CompNovo [28], MetaSPS [29] and pNovo + [30] were developed for dataset presenting complementary multiple fragmentation techniques. Finally, we have the tools of the Pevzner group, PepNovo + [31] and UniNovo [32]. PEAKS [33] (http://www.bioinfor.com) and Bionic [34] (http://www.proteinmetrics.com) are among the most encountered commercial software tools supporting *de novo* sequencing strategies in their workflows. Most of these tools are indexed in the 'OMICS tools' platform [35] (http://omictools.com) maintaining links to the respective web pages.

10.3 Mapping *de novo* Sequences onto Protein Databases

The result of sequencing algorithms is a list of potential peptide sequences, possibly containing sequence gaps where the amino acid sequence could not be inferred. For most biological studies, however, protein level information is necessary to draw meaningful conclusions. Thus, these sequences, or partial sequences, are mapped to known protein sequences, for example, to UniProtKB [36] reference proteomes. The challenge is to provide relevant results in a reasonable time, without losing the hits not exactly matching the sequences in the database, as in the case of sequence mutations. The most frequently chosen option is to proceed with a sequence similarity search using the BLAST [37] (Basic Local Alignment Search Tool) or FASTA [38] algorithms available online, for example, from the UniProt website (http://uniprot.org). These approaches, however, lose the information of the precursor mass and thus do not take mass gaps into account. Moreover, these heuristics only resolve a limited set of mutations. Dedicated software has therefore been developed, such as TagRecon [39] for DirecTag results, MSDA [40] for PepNovo+, and PepExplorer [41] as a more generic solution supporting several algorithms. As a direct result, mass spectrum sequencing output can be readily interpreted, similar as for standard database search engine results.

In order to fully benefit from the advantages of both database search and mass spectrum sequencing, efforts have been put toward unifying the two approaches. This is for instance the case in the abovementioned commercial software (PEAKS [33] and Bionic [34]) where sequencing results are combined with database results [42]. The best representative of such efforts in academic freeware is IdPicker [43, 44], which, in a user-friendly interface, combines the strength of virtually any database search engine results (thanks to the standard mzIdentML format [45]), with the mass spectrum sequencing results of DirecTag combined with TagRecon, and the spectral matching results of Pepitome [46].

10.4 Using Sequencing Algorithms

Most of the algorithms presented above have to be run on the command line which may require additional technical expertise. In order to achieve the transfer of these algorithms to lab practices, it is therefore vital to provide user-friendly interfaces, together with respective teaching material [13], enabling the use of the tools and the inspection of the results without the need of advanced skills in the computer science domain. An example of such an interface is BumberDash (http://fenchurch.mc.vanderbilt.edu/software.php), which is dedicated to software from the Tabb group and allows operating the group's command line tools *via* a graphical user interface, before gathering the results in IdPicker. Notably, the Audens *de novo* algorithm also comes with a graphical interface.

Here, we present how to run the popular sequencing tools PepNovo + and DirecTag, as representatives for *de novo* sequencing and tag-based approaches, and inspect their results in a user-friendly interface called DeNovoGUI [47] (http://compomics.github.io/projects/denovogui.html) – an easy-to-use and open source software which does not require any specific installation. When starting the tool, the main dialog opens as displayed in Fig. 10.3.

Under 'Input & Output' located at the top of the interface, the user provides the peak list file (s) to analyze, the sequencing settings to use, and the output folder where the results will be saved. The results presented in this chapter are obtained from the example file included in DeNovoGUI, which can be accessed *via* the 'File' -> 'Load Example' menu. The sequencing settings can be edited by clicking on the 'Edit' button, opening the dialog shown in Fig. 10.4. This dialog allows for adjusting the general sequencing settings, such as mass tolerances, and algorithm specific settings. The user can also select post-

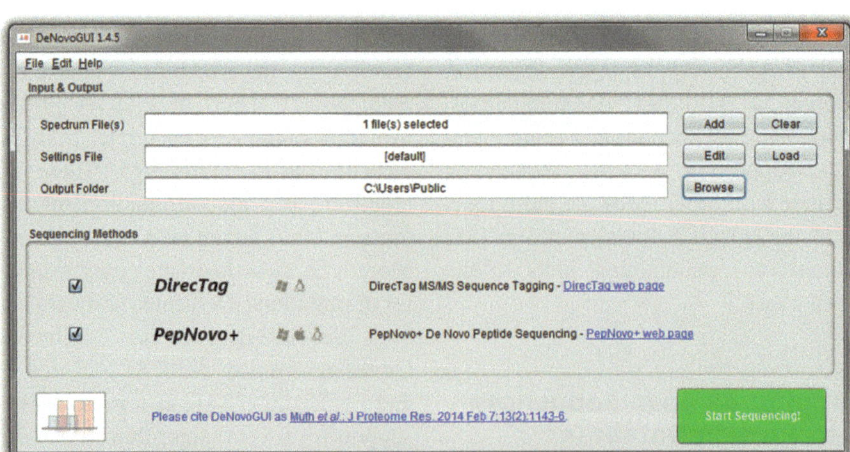

Fig. 10.3 The main DeNovoGUI dialog. The user provides the desired input, settings, and output for the tools to process at the *top*. The user then selects the algorithm(s) to operate, DirecTag and/or PepNovo+, and starts the sequencing

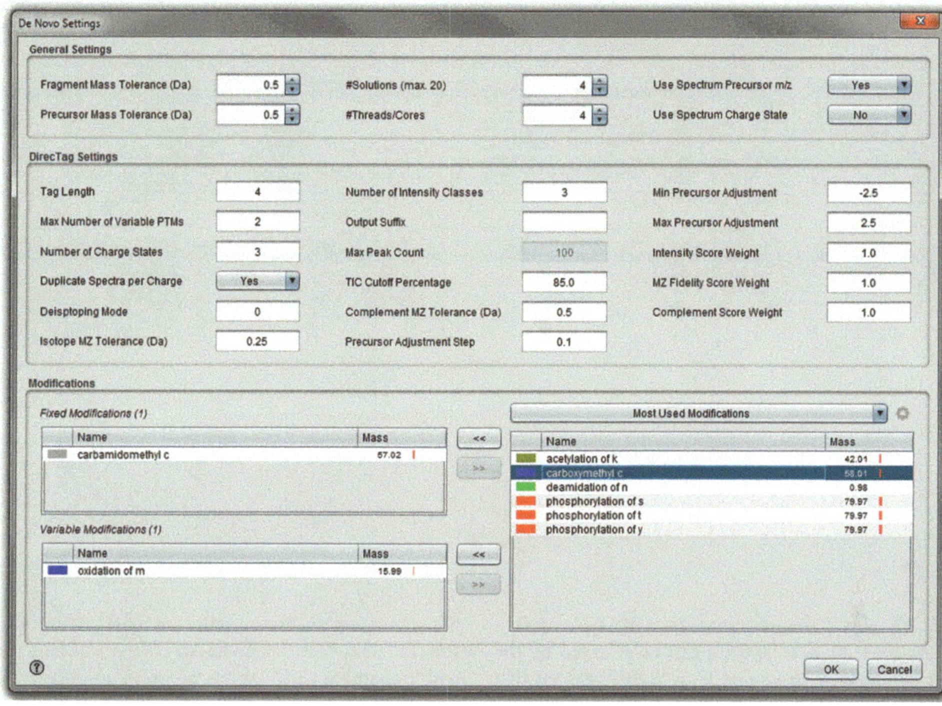

Fig. 10.4 This dialog allows editing the sequencing settings. General settings are listed at the *top*, notably including mass tolerances. These are followed by DirecTag specific settings, and finally post-translational modifications (PTMs). Above the table listing the PTMs, a drop down menu allows for displaying a more extended list of modifications and a cogwheel allows the creation of user-defined modifications

translational modifications (PTMs), and even add custom modifications using the compomics-utilities structure [48], by clicking the cogwheel above the table listing the PTMs.

In the 'Sequencing Methods' section of the main interface, the PepNovo + and the DirecTag algorithms can be selected. As soon as the settings and input files have been chosen, the mass spectrum sequencing can be started by clicking the 'Start Sequencing!' button. While the algorithms are running, the user is informed about the status of the sequencing and a progress bar is shown. When the sequencing has finished, the results are stored in the provided output folder (in the tools respective original formats), and the detailed results are parsed and displayed in the DeNovoGUI interface, as shown in Fig. 10.5. Note that previous sequencing result files can be opened directly *via* the 'File' -> 'Open' menu option.

At the top of the results display, all the input spectra are listed in the 'Query Spectra' table, and the *de novo* peptide sequences are shown for each selected spectrum in the 'De Novo Peptides' table. The 'Query Spectra' table displays information collected from the original spectra, such as title, precursor m/z, charge and identification state, while the 'De Novo Peptides' table shows details obtained from the *de novo* sequencing results on the selected spectrum: peptide sequence, precursor m/z and charge, terminal mass gaps and scores. Note that the mass gaps are annotated on the sequence as well, and that PTMs are indicated using a user customizable color coding. Finally, the last column allows for a direct online BLAST of the selected sequence.

At the bottom, the currently selected spectrum is displayed with the fragment ion annotation corresponding to the selected *de novo* peptide solution. A sequence overlay annotates the

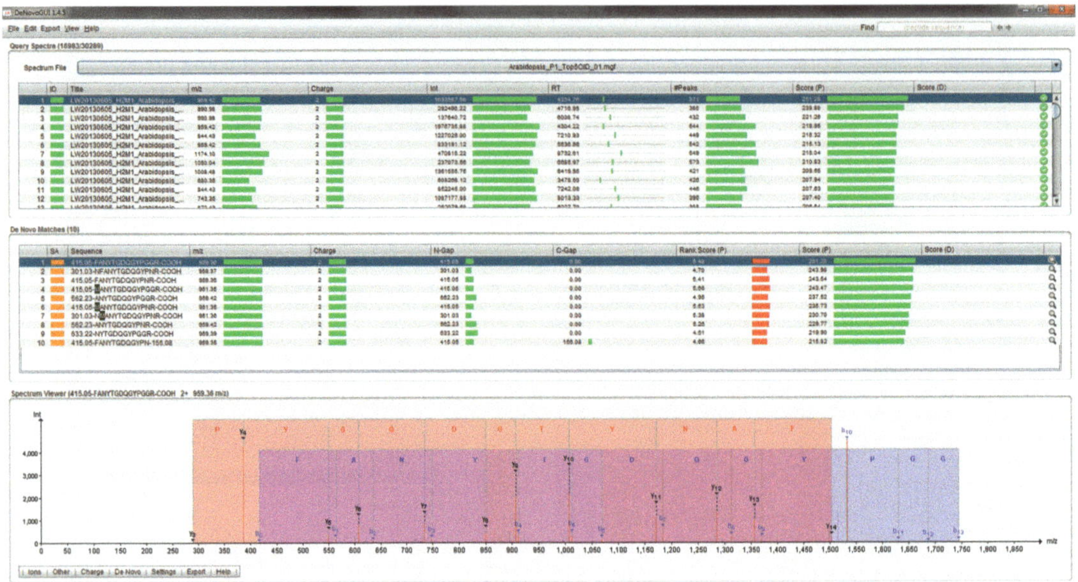

Fig. 10.5 Display of sequencing results in DeNovoGUI. At the *top*, the sequenced spectra can be selected by the user. The sequencing results of both algorithms on that file are listed in the middle table. At the bottom, the selected sequence is annotated on the selected spectrum with the amino acids annotated between the peaks

amino acids between fragment ion peaks. A menu under the spectrum allows customization of the spectrum annotation. Note that from the top menu and spectrum contextual menu, different export options are available, allowing the export of publication level illustrations, Microsoft Excel compatible tab separated tables, a simple matching to protein databases export, and an export of the whole dataset compatible with BLAST. Thus, the complete workflow from spectra sequencing, *via* interpretation of the *de novo* results, to the export of the results for further processing in other software, is supported within the same user-friendly framework.

10.5 Conclusion and Perspectives

Mass spectrum sequencing is a technique that is fully independent of an external protein database resource. This unbiased approach becomes more efficient and accurate as the quality of spectra produced by high accuracy and high resolution mass spectrometers increases. In principle, sequencing algorithms are able to retrieve previously unknown or mutated peptide sequences as well as unexpected PTMs. This approach can be used complementarily to database searches in fully integrated environments. One of the main issues barely touched upon in the literature, and beyond the scope of this chapter, is the evaluation of the quality of sequencing matches and the estimation of a reliable false discovery rate as done with database searches using the target/decoy strategy [49]. This can be especially challenging when evaluating matches containing sequence mutations.

Modern computational power and the use of computer clusters allow for a valuable integration of mass spectrum sequencing into any proteomics workflow. As mass spectrum sequencing performance is improving with better software and hardware optimizations, and is made easier to handle by relying on user-friendly interfaces, the application of this promising technique will surely increase in shotgun proteomic studies. Ideally, it should become integrated in standard proteomic workflows as an alternative, respectively, an add-on to conventional database search engines, which then would be able to provide improved identification coverage at controlled error rates.

Acknowledgements T.M. and E.R. acknowledge the support by Max Planck Society. H.B. is supported by the Research Council of Norway.

References

1. Edman P, Begg G (1967) A protein sequenator. Eur J Biochem 1:80–91
2. Martinsen DP, Song B-H (1985) Computer applications in mass spectral interpretation: a recent review. Mass Spectrom Rev 4:461–490
3. Johnson RS, Biemann K (1989) Computer program (SEQPEP) to aid in the interpretation of high-energy collision tandem mass spectra of peptides. Biomed Environ Mass Spectrom 18:945–957
4. Henzel WJ, Billeci TM, Stults JT et al (1993) Identifying proteins from two-dimensional gels by molecular mass searching of peptide fragments in protein sequence databases. Proc Natl Acad Sci U S A 90:5011–5015
5. Eng JK, McCormack AL, Yates JR (1994) An approach to correlate tandem mass spectral data of peptides with amino acid sequences in a protein database. J Am Soc Mass Spectrom 5:976–989
6. Pappin DJ, Hojrup P, Bleasby AJ (1993) Rapid identification of proteins by peptide-mass fingerprinting. Curr Biol 3:327–332
7. Perkins DN, Pappin DJ, Creasy DM et al (1999) Probability-based protein identification by searching sequence databases using mass spectrometry data. Electrophoresis 20:3551–3567
8. Barsnes H, Eidhammer I, Martens L (2011) A global analysis of peptide fragmentation variability. Proteomics 11:1181–1188
9. Mann M, Wilm M (1994) Error-tolerant identification of peptides in sequence databases by peptide sequence tags. Anal Chem 66:4390–4399
10. Sunyaev S, Liska AJ, Golod A et al (2003) MultiTag: multiple error-tolerant sequence tag search for the sequence-similarity identification of proteins by mass spectrometry. Anal Chem 75:1307–1315
11. Tabb DL, Saraf A, Yates JR 3rd (2003) GutenTag: high-throughput sequence tagging via an empirically derived fragmentation model. Anal Chem 75:6415–6421
12. Tabb DL, Ma ZQ, Martin DB et al (2008) DirecTag: accurate sequence tags from peptide MS/MS through statistical scoring. J Proteome Res 7:3838–3846
13. Vaudel M, Venne AS, Berven FS et al (2014) Shedding light on black boxes in protein identification. Proteomics 14:1001–1005
14. Dorfer V, Pichler P, Stranzl T et al (2014) MS Amanda, a universal identification algorithm optimized for high accuracy tandem mass spectra. J Proteome Res 13:3679–3684
15. Kim S, Gupta N, Pevzner PA (2008) Spectral probabilities and generating functions of tandem mass spectra: a strike against decoy databases. J Proteome Res 7:3354–3363
16. Tabb DL, Fernando CG, Chambers MC (2007) MyriMatch: highly accurate tandem mass spectral peptide identification by multivariate hypergeometric analysis. J Proteome Res 6:654–661
17. Geer LY, Markey SP, Kowalak JA et al (2004) Open mass spectrometry search algorithm. J Proteome Res 3:958–964
18. Craig R, Beavis RC (2004) TANDEM: matching proteins with tandem mass spectra. Bioinformatics 20:1466–1467
19. Chen T, Kao MY, Tepel M et al (2001) A dynamic programming approach to de novo peptide sequencing via tandem mass spectrometry. J Comput Biol 8:325–337
20. Taylor JA, Johnson RS (1997) Sequence database searches via de novo peptide sequencing by tandem mass spectrometry. Rapid Commun Mass Spectrom 11:1067–1075
21. Fernandez-de-Cossio J, Gonzalez J, Satomi Y et al (2000) Automated interpretation of low-energy collision-induced dissociation spectra by SeqMS, a software aid for de novo sequencing by tandem mass spectrometry. Electrophoresis 21:1694–1699
22. Lu B, Chen T (2003) A suboptimal algorithm for de novo peptide sequencing via tandem mass spectrometry. J Comput Biol 10:1–12
23. Fischer B, Roth V, Roos F et al (2005) NovoHMM: a hidden Markov model for de novo peptide sequencing. Anal Chem 77:7265–7273
24. Grossmann J, Roos FF, Cieliebak M et al (2005) AUDENS: a tool for automated peptide de novo sequencing. J Proteome Res 4:1768–1774
25. Mo L, Dutta D, Wan Y et al (2007) MSNovo: a dynamic programming algorithm for de novo peptide sequencing via tandem mass spectrometry. Anal Chem 79:4870–4878
26. Pan C, Park BH, McDonald WH et al (2010) A high-throughput de novo sequencing approach for shotgun proteomics using high-resolution tandem mass spectrometry. BMC Bioinf 11:118
27. Castellana NE, Pham V, Arnott D et al (2010) Template proteogenomics: sequencing whole proteins using an imperfect database. Mol Cell Proteomics 9:1260–1270
28. Bertsch A, Leinenbach A, Pervukhin A et al (2009) De novo peptide sequencing by tandem MS using complementary CID and electron transfer dissociation. Electrophoresis 30:3736–3747
29. Guthals A, Clauser KR, Frank AM et al (2013) Sequencing-grade de novo analysis of MS/MS triplets (CID/HCD/ETD) from overlapping peptides. J Proteome Res 12:2846–2857
30. Chi H, Chen H, He K et al (2013) pNovo+: de novo peptide sequencing using complementary HCD and ETD tandem mass spectra. J Proteome Res 12:615–625

31. Frank A, Pevzner P (2005) PepNovo: de novo peptide sequencing via probabilistic network modeling. Anal Chem 77:964–973
32. Jeong K, Kim S, Pevzner PA (2013) UniNovo: a universal tool for de novo peptide sequencing. Bioinformatics 29:1953–1962
33. Ma B, Zhang K, Hendrie C et al (2003) PEAKS: powerful software for peptide de novo sequencing by tandem mass spectrometry. Rapid Commun Mass Spectrom 17:2337–2342
34. Bern M, Cai Y, Goldberg D (2007) Lookup peaks: a hybrid of de novo sequencing and database search for protein identification by tandem mass spectrometry. Anal Chem 79:1393–1400
35. Henry VJ, Bandrowski AE, Pepin AS et al (2014) OMICtools: an informative directory for multi-omic data analysis. Database J Biol Databases Curation 2014. Available from: http://www.ncbi.nlm.nih.gov/pubmed/25024350
36. Apweiler R, Bairoch A, Wu CH et al (2004) UniProt: the Universal Protein knowledgebase. Nucleic Acids Res 32:D115–D119
37. Altschul SF, Gish W, Miller W et al (1990) Basic local alignment search tool. J Mol Biol 215:403–410
38. Pearson WR, Lipman DJ (1988) Improved tools for biological sequence comparison. Proc Natl Acad Sci U S A 85:2444–2448
39. Dasari S, Chambers MC, Slebos RJ et al (2010) TagRecon: high-throughput mutation identification through sequence tagging. J Proteome Res 9:1716–1726
40. Carapito C, Burel A, Guterl P et al (2014) MSDA, a proteomics software suite for in-depth Mass Spectrometry Data Analysis using grid computing. Proteomics 14:1014–1019
41. Leprevost FV, Valente RH, Borges DL et al (2014) PepExplorer: a similarity-driven tool for analyzing de novo sequencing results. Mol Cell Proteomics 13 (9):2480–2489
42. Zhang J, Xin L, Shan B et al (2012) PEAKS DB: de novo sequencing assisted database search for sensitive and accurate peptide identification. Mol Cell Proteomics 11:M111.010587
43. Ma ZQ, Dasari S, Chambers MC et al (2009) IDPicker 2.0: improved protein assembly with high discrimination peptide identification filtering. J Proteome Res 8:3872–3881
44. Zhang B, Chambers MC, Tabb DL (2007) Proteomic parsimony through bipartite graph analysis improves accuracy and transparency. J Proteome Res 6:3549–3557
45. Jones AR, Eisenacher M, Mayer G et al (2012) The mzIdentML data standard for mass spectrometry-based proteomics results. Mol Cell Proteomics 11: M111.014381
46. Dasari S, Chambers MC, Martinez MA et al (2012) Pepitome: evaluating improved spectral library search for identification complementarity and quality assessment. J Proteome Res 11:1686–1695
47. Muth T, Weilnbock L, Rapp E et al (2014) DeNovoGUI: an open source graphical user interface for de novo sequencing of tandem mass spectra. J Proteome Res 13(2):1143–1146
48. Barsnes H, Vaudel M, Colaert N et al (2011) compomics-utilities: an open-source Java library for computational proteomics. BMC Bioinf 12:70
49. Elias JE, Gygi SP (2010) Target-decoy search strategy for mass spectrometry-based proteomics. Methods Mol Biol 604:55–71

Visualization, Inspection and Interpretation of Shotgun Proteomics Identification Results

11

Ragnhild R. Lereim, Eystein Oveland, Frode S. Berven, Marc Vaudel, and Harald Barsnes

Abstract

Shotgun proteomics is a high throughput technique for protein identification able to identify up to several thousand proteins from a single sample. In order to make sense of this large amount of data, proteomics analysis software is needed, aimed at making the data intuitively accessible to beginners as well as experienced scientists. This chapter provides insight on where to start when analyzing shotgun proteomics data, with a focus on explaining the most common pitfalls in protein identification analysis and how to avoid them. Finally, the move to seeing beyond the list of identified proteins and to putting the results into a bigger biological context is discussed.

Keywords

Protein identification • Visualization • Protein annotation • Validation

R.R. Lereim • F.S. Berven
Proteomics Unit, Department of Biomedicine, University of Bergen, Jonas Liesvei 91, N-5009 Bergen, Norway

KG Jebsen Centre for Multiple Sclerosis Research, Department of Clinical Medicine, University of Bergen, Bergen, Norway

Norwegian Multiple Sclerosis Competence Centre, Department of Neurology, Haukeland University, Bergen, Norway

E. Oveland
KG Jebsen Centre for Multiple Sclerosis Research, Department of Clinical Medicine, University of Bergen, Bergen, Norway

Norwegian Multiple Sclerosis Competence Centre, Department of Neurology, Haukeland University, Bergen, Norway

Department of Clinical Medicine, University of Bergen, Bergen, Norway

Abbreviations

PTM	Post-Translational Modification
PSM	Peptide Spectrum Match
FDR	False Discovery Rate
FNR	False Negative Rate
GO	Gene Ontology
PI	Protein Inference

M. Vaudel • H. Barsnes (✉)
Proteomics Unit, Department of Biomedicine, University of Bergen, Jonas Liesvei 91, N-5009 Bergen, Norway
e-mail: harald.barsnes@biomed.uib.no

H. Mirzaei and M. Carrasco (eds.), *Modern Proteomics – Sample Preparation, Analysis and Practical Applications*, Advances in Experimental Medicine and Biology 919, DOI 10.1007/978-3-319-41448-5_11

11.1 Background

In shotgun proteomics, thousands of proteins are digested into peptides prior to mass spectrometry, and the generated MS/MS spectra are matched to theoretical peptides from a protein sequence database using dedicated algorithms [1]. These matches, termed Peptide Spectrum Matches (PSMs), are scored and ranked, and the best match per spectrum is used as the peptide candidate to infer the proteins. By design, shotgun proteomics thus investigates peptides and not proteins, a fact that gives rise to numerous computational difficulties when trying to figure out which peptide belongs to which protein [2, 3]. A task made even more complicated by the existence of post-translational modifications (PTMs). In order to avoid the most common pitfalls when analyzing shotgun proteomics data, a basic understanding of the computational and statistical methods is needed.

Inherent to the shotgun approach, the matching between theoretical and experimental spectra will generate false positives, *i.e.*, a wrong match passing the validation threshold. The control of the share of false positive matches, the False Discovery Rate (FDR), and its optimization are an important focus of the data interpretation [4, 5]. Generally, so-called decoy sequences are included in the database, which are used to match the experimental data to the theoretical data [6]. The distribution of decoy hits is then used to evaluate the quality of the identifications as reviewed in detail elsewhere [4, 7].

As multiple proteins can share one or several peptide sequences, a PSM can end up being used as evidence for the wrong protein, referred to as the protein inference problem [3]. To ensure correct identification, protein inference cases may have to be inspected manually. Furthermore, the localization of a PTM in a protein can be of biological importance. But the exact location is often difficult to assess based on mass spectrometry data alone. Algorithms exist to estimate the quality of the localization [8], but statistics regarding PTM localization still ought to be further evaluated to ensure correct identification.

The goal of most shotgun experiments is to put the results into a bigger biological context, often by relating the results to information available in protein annotation databases, *e.g.*, related to genes, protein functions, protein structures or biological pathways [9, 10]. There are several software solutions aimed at visualizing and interpreting shotgun proteomic data (http://www.ncbi.nlm.nih.gov/pubmed/25504833), and setting the findings into a bigger biological context, such as the freely available MaxQuant [11], or commercial alternatives like ProteomeDiscoverer (Thermo Scientific, Thermo Fisher Scientific Inc.) or Scaffold (Proteome Software, Inc.).

In this chapter, the open source and freely available analysis software PeptideShaker (http://www.ncbi.nlm.nih.gov/pubmed/25574629) will be used as an example to show what can be achieved *via* the use of such software packages. The main concepts and knowledge should however be transferable to most shotgun analysis tools.

11.2 Shotgun Proteomics Data

In PeptideShaker, a new project can be created based on the results from multiple identification algorithms, plus a set of identification parameters, a protein sequence database, and one or more spectrum files in the standard mgf format (http://www.matrixscience.com/help/data_file_help.html#GEN). If the search has not already been performed, it is possible to use multiple search engines *via* SearchGUI [12] and open the data in PeptideShaker. Public and private datasets stored in the PRIDE database [13], made available *via* the ProteomeXchange consortium [14], can also be reanalyzed via the "PRIDE Reshake" feature. In this chapter, the dataset made available by the developers (ProteomeXchange accession PXD000674) will be used. The dataset can be loaded by clicking "Open Example" in the PeptideShaker Welcome Dialog. For detailed tutorials on project creation, tool usage and proteomics identification in general, please see http://compomics.com/bioinformatics-for-proteomics [15].

11.3 Getting an Overview

After loading the data, the PeptideShaker "Overview" tab displays the combined search engine identification results at the protein, peptide and PSM level (Fig. 11.1). Three linked tables are used to represent the proteins, peptides and PSMs, meaning that selecting a protein displays the identified peptides of that protein, similarly, selecting a peptide displays corresponding PSMs. In addition to the tables (http://www.ncbi.nlm.nih.gov/pubmed/25422159), the tab includes visualization of the PSMs in a spectrum viewer [16] and a display of the protein sequence coverage. The spectrum viewer allows for inspection of the quality of the PSMs, while the sequence coverage at the bottom shows the location of the identified peptides in a linear representation of the protein, with the selected peptide in blue. Notably, PTMs identified for the protein are also mapped onto the sequence using user-defined color coding. A targeted search for specific proteins or peptides is facilitated by a search box in the upper right corner. In short, the "Overview" tab quickly gives the user an overview of the search result and provides direct interaction with the data. Additional tabs in the upper right corner can be used to further investigate different aspects of the shotgun proteomics result.

11.4 Protein Inference

A protein is identified either by peptides that can only derive from that specific protein, so-called unique peptides, or by peptides that can derive from several distinct proteins, so-called shared or degenerate peptides. The latter is often due to protein isoforms (or more generally proteoforms), but the proteins can also be unrelated. When a group of proteins cannot be distinguished by a unique peptide, a so-called ambiguity group is created [3], and a representative protein is chosen for the group (also sometimes referred to as a leading protein).

Fig. 11.1 PeptideShaker overview tab: (*1*) the search box allows for targeted investigation of proteins and peptides; (*2*) select other tabs for additional analysis and quality control; (*3*) the protein table displays details on the proteins identified in the dataset; (*4*) the peptide table lists details on the peptides used to identify the selected protein; (*5*) the PSM table lists details on the PSMs used to identify the selected peptide; (*6*) the spectrum viewer displays the selected PSM; and (*7*) the sequence coverage of the selected protein is displayed in the sequence coverage panel

Related proteins may have similar functions, and unless the experiment focuses on a specific proteoform, having a group of related proteins will usually have limited impact on the outcome. Groups of unrelated proteins are however more problematic, given that they can lead to incorrect biological interpretations. Note that the ambiguity groups that are created, and the chosen leading protein, can be different when comparing different analysis software. In practical terms, this means that the same peptides can lead to different protein identifications depending on which algorithm is used. In PeptideShaker, the protein inference (PI) status is color coded in the protein and peptide tables, and clicking it displays details on the respective peptide to protein matching. It is here possible to change the protein representing the group and its PI status. However, this reduces the reproducibility, and all such changes should therefore be well grounded and documented.

11.5 Inspecting Spectrum Identifications

For proteins, peptides and PSMs, the quality of the identification is indicated in the confidence and validation columns. Low confidence often results from poor peptide to spectrum matching. The spectra for each peptide can be inspected either in the "Overview" tab or (in more detail) in the "Spectrum ID" tab. When selecting a spectrum in the "Overview" tab, the peaks that can be explained by the peptide fragmentation are annotated and outlined in red in the spectrum viewer. The spectrum viewer also includes several additional plots that can be used to manually investigate the PSM quality (Fig. 11.2).

A high quality PSM generally has clearly defined peptide fragment ion peaks, covering the most intense peaks in the spectrum with low mass errors, *i.e.*, the difference between the masses of the peaks in the experimental spectrum

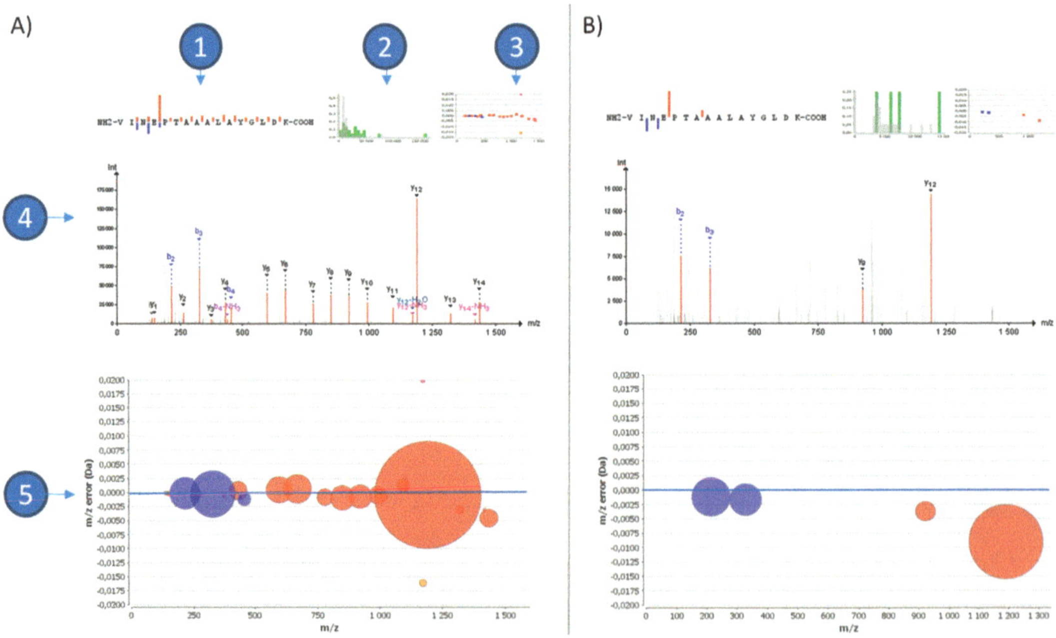

Fig. 11.2 PSM investigation in the PeptideShaker overview tab: (**a**) Example of a PSM classified as confident in PeptideShaker. (*1*) The most intense fragment ions identified and their intensities are illustrated with colored bars. (*2*) A histogram displays the intensities covered by the fragment ions (*green*), and the background intensities (*grey*). (*3*) The mass error is plotted against the m/z for every annotated fragment ion. (*4*) The PSM mass spectrum showing annotated peaks (*red*) and background peaks (*gray*). (*5*) A bubble plot of the fragment ion mass error plotted against the m/z, where the size of the bubble represents the peak intensity. (**b**) Example of a PSM classified as doubtful in PeptideShaker: few detected peptide fragment ions, most of the high intensity peaks are not detected, and the annotated peaks have a high mass error

and the theoretical spectrum [17]. Note that artifacts during mass spectrometry analysis can result in increasing mass error with increasing m/z values. Therefore, a high mass error is not necessarily due to a false PSM. High mass errors resulting from wrongly annotated spectra are typically sporadic, distributing with a great spread and no clear trend with increasing m/z. The distribution of the mass errors can be visualized by clicking the "Bubble Plot" option below the spectrum viewer in the "Overview" tab (Fig. 11.2a(5)).

As manual investigation of PSMs is time consuming, matches passing the statistical threshold are generally trusted. However, if a peptide or protein of interest is based on only spectra with low confidence, it is advised to further verify the presence of this peptide or protein.

11.6 Search Engine Performance

The performance of the identification algorithms, generally search engines, can be compared in the "Spectrum IDs" tab. A spectrum can be assigned to different peptides, by one or more algorithms. The matches for one particular spectrum can be viewed *via* the table at the top, where all the spectra generated by the mass spectrometer are listed. Selecting a spectrum displays the peptides inferred by each algorithm, showing search engine agreement (or disagreement), and the match retained by PeptideShaker. By selecting a PSM, the spectrum will be annotated accordingly, thus making it possible to compare the spectrum annotation for the different peptide candidates and inspect their validity.

As search engines have slightly different approaches, there will be differences in the numbers of annotated spectra. When the results of a given algorithm clearly deviate from the others, it may indicate that the given search engine under or over performed on the given experiment, *e.g.*, a specific search engine may perform best/worst using certain identification parameters or database types. It is important to verify the source of such differences to avoid any bias in the final result. The individual search engine performance can also be compared to the combined result generated by the analysis software. If the combined result is impaired by one of the search engines one should consider excluding it, as this may improve the overall identification rate.

11.7 Validating the Identifications

Shotgun analysis software statistically validates the identification results at a user-defined false discovery rate (FDR) threshold, and provides a confidence level illustrating the quality of the match. Both FDR and confidence are generally estimated using the target/decoy approach [6]. For an extensive description on how these error rates are calculated, see Nesvizhskii et al. [4]. Matches can also be further validated by automated expert inspection of the matches [18].

Importantly, a 1 % FDR threshold indicates that there is an estimated amount of one false discovery per 100 validated entries. This means that even validated proteins might be false positives, which is important to keep in mind during the data analysis. The false negative rate (FNR) is also stated as an estimate of how many correct matches that are left out due to the FDR threshold.

The "Validation" tab provides an overview of the total number of validated entries, and the associated FDR and FNR levels. It allows the user to tune the statistical thresholds, balancing between sensitivity and specificity. Experiments requiring high quality results should be validated at a stringent FDR (typically 1 %), while experiments interested in a high identification coverage can tune the validation threshold toward FNR minimization.

In PeptideShaker the validation approach combines statistical validation and expert inspection, resulting in three color coded categories: (i) Confident – indicating that the statistical threshold was passed as well as the quality filters (green); (ii) Doubtful – indicating that the statistical threshold was passed but not the quality filters (yellow); and (ii) Not Validated – indicating that the statistical threshold was not passed (red). Clicking the icon representing the validation level (found in the rightmost columns in the tables in the "Overview" tab) opens a dialog with details on the match validation criteria.

The following default quality filters are employed at the PSM, peptide and protein level: (i) PSMs must have a low mass error and a high fragment ion sequence coverage; (ii) peptides have to be identified by at least two confident PSMs; and (iii) proteins have to be identified by at least two confident peptides and at least two confident PSMs. Together with the statistical validation, labeling the entries as confident or doubtful on the basis on these quality filters makes it easier to find out which identifications to trust.

11.8 Overall Quality Control

Analysis software collect statistics that allow for optimization of search parameters and evaluation of the success rate of the experiment. These quality control statistics can vary from software to software. The "QC Plots" tab in PeptideShaker displays quality metrics at the protein, peptide and PSM level, thus making it straightforward to evaluate the overall quality of the validated entries before continuing the analysis. At the protein level there are statistics on how many peptides the proteins are identified by, plus the distributions of the sequence coverage and protein lengths. The peptide QC plots include statistics on how many peptides that have missed cleavages (due to incomplete digestion), the number of PSMs the peptides are identified by, and the peptide lengths. The PSM statistics include the precursor mass error, and the precursor charge.

These metrics can be used as a measure of the success of the laboratory procedure and the parameters used during data analysis. For example, if the precursor mass deviation of the PSMs is large, tailoring the mass accuracy parameters might improve the number of identifications [5]. However, large mass errors may also indicate a calibration issue with the mass spectrometer [19]. If any of the quality metrics are unexpected, the source of the problem should be detected and eliminated before continuing the analysis.

11.9 Validating PTMs

Modified peptides are often much less abundant than unmodified peptides *in vivo*. Therefore, to detect modified peptides, the samples are commonly enriched for a certain type of PTM prior to mass spectrometry analysis. Searching for PTMs also results in several computational difficulties. First of all, the modified and unmodified peptides are most often counted as separate identifications. This can lead to an increased number of peptides for each protein, even though the protein sequence coverage remains unchanged. In practical terms, this means that proteins can pass the quality control filter of at least two confident peptides per protein on the basis of a single peptide sequence.

In order to accurately determine the PTM localization in a peptide, a large degree of the peptide fragments ought to be detected. The appearance of an m/z addition in all fragment ions past a certain point in the peptide sequence, named site determining ions [20], indicates the location of the PTM. However, as all peptide fragment ions are usually not identified, False Localization Rate (FLR), is typically higher than the False Discovery Rate (FDR), which does not account for PTM localization. In such cases the algorithms calculate the probability for the PTM localization at the modifiable residues within the peptide using PTM localization scores. In PeptideShaker, the popular A-score [20] and PhosphoRS [21] PTM probabilistic localization scores can be used complementarily to the D-score [22]. As a result, confidently localized PTMs are indicated on peptide sequences with a colored background while ambiguous sites are shown with a white background throughout the interface. For sites of interest, the relevant PSMs can be inspected by selecting them in the "Variable Modifications" table in the "Modifications" tab. Overlapping peptides, listed in the "Related Peptides" table with the same PTM localization can then also be used as an additional quality control. More details on PTM localization inspection can be found elsewhere [8, 23].

11.10 Biological Context

In shotgun analysis software, protein annotations can be used to further understand the identified proteins by linking to commonly used protein and gene knowledge databases. A detailed list of resources and specific tools can be found in dedicated reviews [10, 24]. The ones integrated in PeptideShaker are highlighted in the following sections, as a brief introduction to the annotation possibilities.

In the "Overview" tab in PeptideShaker, the protein accession numbers are linked to the UniProt knowledgebase [25], and the chromosome annotation is provided using Ensembl [26]. Clicking the protein accession number opens the UniProt web page for the given protein, while clicking the chromosome number displays the related Ensembl gene name as well as a list of Gene Ontology (GO) terms for that gene. This provides an easy access to the basic gene and protein information about the leading protein of the identified protein ambiguity group.

GO analysis can be conducted for the entire dataset in the "GO Analysis" tab. A subset of the available GO terms (a so-called GO Slim) is used to annotate the validated proteins in the dataset. The frequency of proteins annotated by each GO term is compared to the annotation frequency of the same term for the studied species in Ensembl. This can be used to see if the dataset has a significantly higher or lower frequency of proteins with gene information linked to a specific GO term (such as "aging", "cell division", etc.) in the selected organism. Information about a specific GO term can be accessed by clicking the GO identifier linked to the EBI QuickGO web service [27].

The "3D Structures" tab uses information from the Protein Data Bank (PDB) [28] to map the identified peptides and PTMs onto the 3D structure of the protein, displayed *via* Jmol [29]. By selecting specific peptides, the researcher can investigate their location on the protein structure. This function can be used to resolve PTM location conflicts [30], as PTMs located on the protein according to their function, typically at reactive sites at the surface. Additional information about the structures is available by clicking the PDB identifier.

Annotation can be collected manually, but this can be time consuming, and knowing which database to use is not always easy. The "Annotation" tab can be used to obtain annotations from several online databases and resources. This can be done for a single protein, or for the complete list of validated proteins, and includes pathway databases such as STRING [31] and Reactome [32], protein functional databases such as DAVID [33], protein interaction databases such as IntAct [34] and protein signature databases such as InterPro [35]. Finally, there are databases that collect information from multiple resources, such as DASty [36].

Finally, it is important to keep in mind that databases are not static entities, and change with the constant input from new literature. For this reason, the database version used for annotation should always be stated in the publication, and the quality of the data should also be carefully considered [37, 38].

11.11 Conclusions and Perspectives

Visualization of shotgun proteomics results allows the researcher to investigate both the identification algorithms performance and the quality of the experimental results. User-friendly and visual analysis software interfaces thus empower the experimentalists, allowing them to critically interpret their data using state of the art algorithms without demanding advanced knowledge in (bio)informatics.

The computational difficulties in interpreting and combining data from several search engines as highlighted in this chapter, show the importance of using high quality analysis software as a tool to interact with and understand proteomics data. Several software exist, with different ways of selecting leading proteins in PI ambiguity groups, using quality control filters, and algorithms for combining results from different search engines. For this reason, directly comparing results from different software ought to be done with caution.

To conclude, the more the researcher knows about the bioinformatics tools used for the analysis, the better the results of the analysis. However tempting, manual interference with the results should be done with the utmost caution, for both experimentalists and bioinformaticians, due to resulting in reduced reproducibility and the chance of introducing interpretation biases.

PeptideShaker allows the collecting of data from a single mass spectrometry run, and can also analyze several fractions together. However, comparing one project to another has to be done manually, by exporting the data and comparing them in programs such as Perseus (http://www.maxquant.org). Given that an increasing number of proteomics experiments aim at comparing different conditions measured in parallel, there is a strong need for a broader free interface allowing intuitive comparison of multiple projects, as available in commercial software.

Acknowledgments R.R.L, E.O. and F.S.B. acknowledge the support by Kjell Alme's Legacy for Research in Multiple Sclerosis, and the Kristian Gerhard Jebsen Foundation. H.B. is supported by the Research Council of Norway.

References

1. Aebersold R, Mann M (2003) Mass spectrometry-based proteomics. Nature 422:198–207
2. Duncan MW, Aebersold R, Caprioli RM (2010) The pros and cons of peptide-centric proteomics. Nat Biotechnol 28:659–664
3. Nesvizhskii AI, Aebersold R (2005) Interpretation of shotgun proteomic data: the protein inference problem. Mol Cell Proteomics 4:1419–1440
4. Nesvizhskii AI (2010) A survey of computational methods and error rate estimation procedures for peptide and protein identification in shotgun proteomics. J Proteomics 73:2092–2123
5. Vaudel M, Burkhart JM, Sickmann A et al (2011) Peptide identification quality control. Proteomics 11:2105–2114
6. Elias JE, Gygi SP (2010) Target-decoy search strategy for mass spectrometry-based proteomics. Methods Mol Biol 604:55–71
7. Vaudel M, Sickmann A, Martens L (2012) Current methods for global proteome identification. Expert Rev Proteomics 9:519–532
8. Chalkley RJ, Clauser KR (2012) Modification site localization scoring: strategies and performance. Mol Cell Proteomics 11:3–14
9. Barsnes H, Martens L (2013) Crowdsourcing in proteomics: public resources lead to better experiments. Amino Acids 44:1129–1137
10. Vizcaino JA, Mueller M, Hermjakob H et al (2009) Charting online OMICS resources: a navigational chart for clinical researchers. Proteomics Clin Appl 3:18–29
11. Cox J, Mann M (2008) MaxQuant enables high peptide identification rates, individualized p.p.b.-range mass accuracies and proteome-wide protein quantification. Nat Biotechnol 26:1367–1372
12. Vaudel M, Barsnes H, Berven FS et al (2011) SearchGUI: an open-source graphical user interface for simultaneous OMSSA and X!Tandem searches. Proteomics 11:996–999
13. Vizcaino JA, Cote RG, Csordas A et al (2013) The PRoteomics IDEntifications (PRIDE) database and associated tools: status in 2013. Nucleic Acids Res 41:D1063–D1069
14. Vizcaino JA, Deutsch EW, Wang R et al (2014) ProteomeXchange provides globally coordinated proteomics data submission and dissemination. Nat Biotechnol 32:223–226
15. Vaudel M, Venne AS, Berven FS et al (2014) Shedding light on black boxes in protein identification. Proteomics 14:1001–1005
16. Barsnes H, Vaudel M, Colaert N et al (2011) Compomics-utilities: an open-source Java library for computational proteomics. BMC Bioinf 12:70
17. Barsnes H, Eidhammer I, Martens L (2011) A global analysis of peptide fragmentation variability. Proteomics 11:1181–1188
18. Helsens K, Timmerman E, Vandekerckhove J et al (2008) Peptizer, a tool for assessing false positive peptide identifications and manually validating selected results. Mol Cell Proteomics 7:2364–2372
19. Olsen JV, de Godoy LM, Li G et al (2005) Parts per million mass accuracy on an Orbitrap mass spectrometer via lock mass injection into a C-trap. Mol Cell Proteomics 4:2010–2021
20. Beausoleil SA, Villen J, Gerber SA et al (2006) A probability-based approach for high-throughput protein phosphorylation analysis and site localization. Nat Biotechnol 24:1285–1292
21. Savitski MM, Lemeer S, Boesche M et al (2011) Confident phosphorylation site localization using the Mascot Delta Score. Mol Cell Proteomics 10: M110.003830
22. Vaudel M, Breiter D, Beck F et al (2013) D-score: a search engine independent MD-score. Proteomics 13:1036–1041

23. Olsen JV, Mann M (2013) Status of large-scale analysis of post-translational modifications by mass spectrometry. Mol Cell Proteomics 12:3444–3452
24. Vaudel M, Sickmann A, Martens L (2014) Introduction to opportunities and pitfalls in functional mass spectrometry based proteomics. Biochim Biophys Acta 1844:12–20
25. Apweiler R, Bairoch A, Wu CH et al (2004) UniProt: the Universal Protein knowledgebase. Nucleic Acids Res 32:D115–D119
26. Flicek P, Amode MR, Barrell D et al (2011) Ensembl 2011. Nucleic Acids Res 39:D800–D806
27. Binns D, Dimmer E, Huntley R et al (2009) QuickGO: a web-based tool for Gene Ontology searching. Bioinformatics 25:3045–3046
28. Sussman JL, Lin D, Jiang J et al (1998) Protein Data Bank (PDB): database of three-dimensional structural information of biological macromolecules. Acta Crystallogr D Biol Crystallogr 54:1078–1084
29. Herraez A (2006) Biomolecules in the computer: Jmol to the rescue. Biochem Mol Biol Educ 34:255–261
30. Vandermarliere E, Martens L (2013) Protein structure as a means to triage proposed PTM sites. Proteomics 13:1028–1035
31. von Mering C, Huynen M, Jaeggi D et al (2003) STRING: a database of predicted functional associations between proteins. Nucleic Acids Res 31:258–261
32. Croft D, O'Kelly G, Wu G et al (2011) Reactome: a database of reactions, pathways and biological processes. Nucleic Acids Res 39:D691–D697
33. da Huang W, Sherman BT, Lempicki RA (2009) Systematic and integrative analysis of large gene lists using DAVID bioinformatics resources. Nat Protoc 4:44–57
34. Kerrien S, Aranda B, Breuza L et al (2012) The IntAct molecular interaction database in 2012. Nucleic Acids Res 40:D841–D846
35. Hunter S, Jones P, Mitchell A et al (2012) InterPro in 2011: new developments in the family and domain prediction database. Nucleic Acids Res 40:D306–D312
36. Villaveces JM, Jimenez RC, Garcia LJ et al (2011) Dasty3, a WEB framework for DAS. Bioinformatics 27:2616–2617
37. Muller T, Schrotter A, Loosse C et al (2011) Sense and nonsense of pathway analysis software in proteomics. J Proteome Res 10:5398–5408
38. Khatri P, Sirota M, Butte AJ (2012) Ten years of pathway analysis: current approaches and outstanding challenges. PLoS Comput Biol 8, e1002375

Protein Inference

12

Zengyou He, Ting Huang, Can Zhao, and Ben Teng

Abstract

Protein inference is one of the most important steps in protein identification, which transforms peptides identified from tandem mass spectra into a list of proteins. In this chapter, we provide a brief introduction on this problem and present a short summary on the existing protein inference methods in the literature.

Keywords

Protein identification • Protein inference

12.1 Problem Statement and Challenges

Protein inference describes the process used to assemble identified peptides into a list of proteins that are believed to be present in a sample. The standard input of protein inference can be considered as a bipartite graph [1], as shown in Fig. 12.1. The two sets of nodes in this bipartite graph represent the identified peptides reported by the peptide identification algorithms and the candidate proteins, respectively. Additionally, each peptide vertex has a corresponding identification score or probability. Protein vertices are the candidate proteins that may be present in the sample. If the sequence of a protein vertex in the database contains the sequence of at least one peptide vertex, this protein is a candidate protein resulting in a connection between the peptide and the protein in the bipartite graph. The task of protein inference is to make a selection from the candidate proteins that best explains all the identified peptides.

The two biggest challenges in protein inference are how to tackle degenerate peptides and one-hit wonders. Degenerate peptides are peptides that are shared by multiple candidate proteins. It is difficult to distinguish from which protein any given degenerate peptide originated. One-hit wonders are proteins that match with only one identified peptide. Since current peptide identification algorithms are not perfect, this peptide may be discovered by chance and the

Z. He (✉) • C. Zhao • B. Teng
School of Software, Dalian University of Technology, Dalian, China
e-mail: zyhe@dlut.edu.cn

T. Huang
College of Computer and Information Science, Northeastern University, Boston, MA, USA

H. Mirzaei and M. Carrasco (eds.), *Modern Proteomics – Sample Preparation, Analysis and Practical Applications*, Advances in Experimental Medicine and Biology 919, DOI 10.1007/978-3-319-41448-5_12

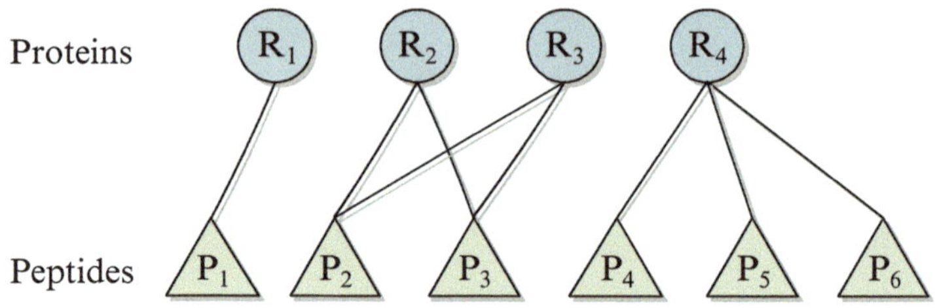

Fig. 12.1 The standard input of protein inference problem. Here protein R_1 is a one-hit wonder and peptides P_2 and P_3 are degenerate peptides

reliability of the one-hit wonders cannot be guaranteed. To address these problems, different protein inference algorithms and tools have been developed during the past decade.

12.2 Algorithms and Tools

The available methods for solving the protein inference problem can be categorized into two classes [1]: the bipartite graph model and the supplementary information model, as shown in Fig. 12.2. This classification is based on the different input information that inference algorithms used to assemble the identified peptides.

12.2.1 Bipartite Graph Model

The algorithms that belong to the bipartite graph model mainly use the information generated from bipartite graphs, such as peptide-protein relationships and peptide identification scores. They can be subdivided into three categories based on different models used: the parsimonious model, the statistical model and the optimistic model.

12.2.1.1 Parsimonious Model

Since protein inference algorithms are aimed at finding a subset of protein vertices that can cover all the identified peptides, it is natural to apply the parsimony principle (Occam's razor principle) to solve the inference problem. More precisely, the objective is to report a minimum subset of proteins that can "explain" all identified peptides. In practice, a greedy algorithm is often used to find the solution efficiently. The greedy algorithm typically works as follows: it first selects the protein that matches with the largest number of peptides and removes all its matching peptides from the identified peptide set; then it repeats the first step until the identified peptide set is empty. The selected proteins are considered to be present in the sample. For example, applying the parsimony principle to the sample in Fig. 12.1 will successively report proteins $\{R_4, R_3, R_1\}$ or $\{R_4, R_2, R_1\}$.

IDPicker [2, 3] is a typical method which uses the parsimony principle for protein inference. It reports the minimum protein identifications through a greedy algorithm. Meanwhile, in its latest version, IDPicker has been extended to integrate multiple peptide identification scores generated by different peptide identification methods.

DBParser [4], MassSieve [5] and LDFA [6] also employ the parsimony analysis to remove redundant protein identifications. But LDFA is a little different from the other parsimonious methods. It assigns the shared peptide to the corresponding protein according to peptide detectability, rather than the number of sibling peptides matched to the same protein. Peptide detectability is an intrinsic property of the peptide. It indicates the probability of detecting a peptide in a standard sample by a standard proteomics routine if its parent protein is present.

The methods that employ the parsimony principle have deterministic results and fast running speeds. They require very few parameters and thus are easy to use. However, only reporting the minimum number of proteins in any given sample may lead to the loss of useful information. For example, homologous proteins are likely to have the same set of identified peptides and they may all be present in the sample. Unfortunately, the parsimonious methods will

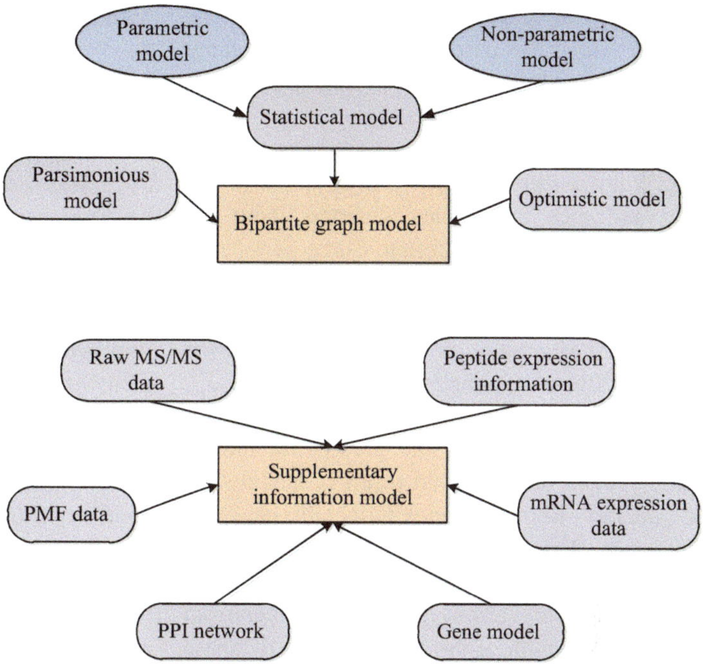

Fig. 12.2 The classification of protein inference methods

probably only report one of them. Moreover, each reported protein will not have a score and the probability that any one of the selected proteins is present in the sample is unknown.

12.2.1.2 Statistical Model

According to the assumptions made by these statistical methods, the methods can be divided into two categories: the *non-parametric model* and the *parametric model*.

The non-parametric model does not rely on the assumption that the data are drawn from a given parametric probability distribution.

ProteinProphet [7] is the most widely used method to solve the protein inference problem. ProteinProphet employs an iterative procedure to estimate protein probabilities. It first computes the protein probability as the probability that at least one identified peptide corresponding to the protein is correct, and then re-computes the peptide weight conditioned on the protein probabilities. The above iteration process continues until convergence. ProteinProphet also considers the number of sibling peptides in the scoring procedure to facilitate the assignment of degenerate peptides to the most likely protein. ProteinProphet is integrated into the popular Trans-Proteomic Pipeline software.

MSBayesPro [8] describes two Bayesian approaches to address the protein inference problem. The basic Bayesian model assumes that all the peptides have equal identification scores. Another advanced model incorporates the peptide identification scores into the Bayesian model. Moreover, MSBayesPro provides a Gibbs sampling algorithm to quickly approximate the protein posterior probabilities. MSBayesPro has two important features: (1) the use of peptide detectability; (2) the use of both identified and non-identified peptides. These salient features will help improve the identification accuracy.

ProteinLP [9] uses the joint probability that both a protein and its constituent peptide are present in the sample as the unknown variable to compute the protein probability. It first makes a mathematical transformation of such joint probability to obtain a new variable. Then, both the peptide probability and protein probability are represented as a formula that is built on the linear combination of these new variables. Finally, the protein inference problem is formulated as a linear programming problem. Since ProteinLP is based on linear programming (LP) model, it can be solved efficiently with existing LP software packages.

The parametric model first assumes that the data follows some form of probability distribution and then makes an inference about the parameters of the distribution. Since parametric methods make more assumptions than non-parametric methods, they may produce more accurate protein probability estimations if these additional assumptions are correct.

PROT_PROBE [10] is a typical method for protein inference using the parametric model. Each protein identification result is modeled as a random Bernoulli event which has two outcomes: a protein is either identified or not. The probability of the protein identification at each Bernoulli event is determined either from the relative length of the protein in the database (null hypothesis) or from the hyper-geometric probabilities of peptides (alternative hypothesis). By comparing the two distributions, the one that the protein belongs to is determined.

12.2.1.3 Optimistic Model

In contrast to the parsimonious model which reports the minimum list of protein identifications, optimistic model returns all potential proteins that meet some simple criterion. Two-peptide rule is a typical example of an optimistic model. It reports all the candidate proteins matching at least two peptides without any further filtering. For instance, applying the two-peptide rule to the example in Fig. 12.1 will report proteins R_2, R_3 and R_4 to the user.

DTASelect [11] also falls into the category of an optimistic model. In this method, a protein is regarded as being present in the sample if it matches a sufficient number of different peptides or at least one peptide that appears many times.

The optimistic model is simple to understand and easy to use. However, if the filtering condition is overly strict, some true protein identifications would be missed. Alternatively, if the filtering condition is overly liberal, the set of reported proteins would include too many false positives.

12.2.2 Supplementary Information Model

In the bipartite graph model, it is difficult to further improve the identification performance, no matter how ideal the algorithm is. This is because the input information of this model is limited. For example, proteins P_2 and P_3 are very difficult to be distinguished if only based on the information shown in Fig. 12.1. In order to improve the identification accuracy, some supplementary information can be incorporated into the protein inference process. Such supplementary information can facilitate identification of proteins that may not be identifiable with high confidence by MS/MS evidence alone. So far, there are six types of supplementary information that have been used: raw MS/MS data, peptide mass fingerprinting data, peptide expression profiles, protein interaction networks, mRNA expression data and gene models.

Raw MS/MS Data This data takes advantage of the raw MS/MS spectra information. Protein identification includes two steps: peptide identification and protein inference. This separation may lead to a significant loss of information during the protein inference. For example, suppose only the best-matched peptide is reported for each spectrum. For a particular spectrum, if this best-matched peptide is incorrect, then the information about the second-ranked, possibly correct peptide, is not available to protein inference algorithms. Thus, the raw MS/MS data model directly conducts protein inference from the raw spectra in order to obtain better identification results. HSM [12] is a typical protein inference method that utilizes raw MS/MS data. It is an integrated statistical model, which jointly assess the confidence of the peptides and proteins identified from raw MS/MS data.

Peptide Mass Fingerprinting Data There are two types of data for identifying proteins in the sample: single-stage MS data and MS/MS data. Shotgun proteomics is based on tandem mass spectrometry data. Peptide mass fingerprinting (PMF) is the identification method that utilizes single stage MS data. PMF assumes that every protein has a set of peptides and thus masses of these peptides can form its unique fingerprinting. PMF matches observed peptide masses with theoretical peptide masses to identify proteins. MS-based methods provide wider coverage than MS/MS-based method, while their identification accuracy is lower since MS data have less information than MS/MS data. It is a natural idea to combine MS data and MS/MS data in a unified model so that the identification performance can be improved. The PSC method [13] combines the MS data and MS/MS data together under a partial set covering model to identify the proteins in the sample.

Peptide Expression Information Peptide expression information, such as peptide intensity information, is widely used in label-free quantitative proteomics studies. Recently, it has been used to improve protein identifications. PIPER [14] assumes that peptides derived from the same protein should have similar expression profiles. Thus, according to the known peptide expression profiles, PIPER can filter out some identified proteins to obtain more accurate inference results.

Protein Interaction Network Most of the protein inference methods consider the candidate proteins independently. In fact, two or more proteins usually bind together to carry out the biological functions, which form the protein-protein interaction network (PPI network). That is, certain proteins are correlated with each other. Thus, it is reasonable to consider the protein-protein interaction information in the protein inference procedure. CEA [15] tries to revive the eliminated proteins by incorporating the protein interaction network. It assumes that a non-confident protein will become confident if it has a sufficient number of confident neighbor proteins in the PPI network. The model uses the relationships among proteins to adjust the identification results generated by other protein inference methods.

mRNA Expression Data mRNA expression information during transcription can be used to help estimate the protein probability as well. For example, MSpresso [16] re-calculates protein identification probabilities given their mRNA abundances.

Gene Model Compared to protein interaction network and mRNA expression data, it is easier to accurately and quickly obtain accurate gene information. A DNA segment can generate multiple proteins and these proteins are relevant. The existence of one protein may indicate that other proteins originating from the same gene are also present in the sample. The typical application of the gene model is Markovian Inference of Proteins and Gene Models (MIPGEM) [17]. It addresses the problem of protein and gene model inference through a probabilistic graphical model.

Different supplementary information models have their own characteristics. Methods that incorporate MS-related data (raw MS/MS data, PMF data and peptide expression data) can be applied to the analysis of any sample since such data are always available. In contrast, approaches that use other biological data can only work when the required supplementary information are available.

12.3 Validation for Protein Identifications

Since none of the protein inference algorithms are perfect, controlling the quality of inferred proteins is as important as developing protein inference algorithms. For a long time, the assessment of inferred proteins has been confused with the validation of peptide identifications. In fact, inferred proteins are more biologically relevant than identified peptides in a proteomics experiment. Therefore, it is vital to control the quality of the identification results at the protein-level. However, the accurate assessment of the confidence of protein identifications remains an open question. To date, several research efforts have been made to estimate the protein-level error rate in terms of false discovery rate (FDR).

On one hand, some methods rely on the use of decoy databases during FDR estimation. In these methods, the MS/MS spectra are first searched against a target-decoy database and then the number of false positive protein identifications is estimated according to the number of decoy entries. The naive target-decoy method and MAYU are two examples in this category. For the naive target-decoy method, FDR is calculated by doubling the ratio of the number of decoy proteins and the total number of protein identifications. MAYU [18] uses a more sophisticated statistical model to estimate the expected number of false positive protein identifications.

On the other hand, the decoy-free method evaluates the protein inference results without searching a decoy database. For instance, the method in [19] uses a random permutation method to estimate the confidence of each protein in terms of p-value and calculates the FDR from these p-values.

12.4 Conclusions

Researchers have proposed many solutions from different angles to tackle the protein inference problem. However, the performance of current available protein inference methods is still far from satisfactory in practice. Therefore, more research efforts are still needed towards this direction.

Acknowledgement This work was partially supported by the Natural Science Foundation of China under Grant No. 61003176 and the Fundamental Research Funds for the Central Universities of China (DUT14QY07).

References

1. Huang T, Wang J, Yu W et al (2012) Protein inference: a review. Brief Bioinform 13(5):586–614
2. Zhang B, Chambers MC, Tabb DL (2007) Proteomic parsimony through bipartite graph analysis improves accuracy and transparency. J Proteome Res 6 (9):3549–3557
3. Ma Z-Q, Dasari S, Chambers MC et al (2009) IDPicker 2.0: improved protein assembly with high discrimination peptide identification filtering. J Proteome Res 8(8):3872–3881
4. Yang X, Dondeti V, Dezube R et al (2004) DBParser: web-based software for shotgun proteomic data analyses. J Proteome Res 3(5):1002–1008
5. Slotta DJ, Mcfarland MA, Markey SP (2010) MassSieve: panning MS/MS peptide data for proteins. Proteomics 10(16):3035–3039
6. Alves P, Arnold RJ, Novotny MV et al (2007) Advancement in protein inference from shotgun proteomics using peptide detectability. Pac Symp Biocomput 12:409–420
7. Nesvizhskii AI, Keller A, Kolker E et al (2003) A statistical model for identifying proteins by tandem mass spectrometry. Anal Chem 75(17):4646–4658
8. Li YF, Arnold RJ, Li Y et al (2009) A Bayesian approach to protein inference problem in shotgun proteomics. J Comput Biol 16(8):1–11
9. Huang T, He Z (2012) A linear programming model for protein inference problem in shotgun proteomics. Bioinformatics 28(22):2956–2962
10. Sadygov RG, Liu H, Yates JR (2004) Statistical models for protein validation using tandem mass spectral data and protein amino acid sequence databases. Anal Chem 76(6):1664–1671
11. Tabb DL, Mcdonald H, Yates JR (2002) DTASelect and contrast: tools for assembling and comparing protein identifications from shotgun proteomics. J Proteome Res 1:21–26
12. Shen C, Wang ZH, Shankar G et al (2008) A hierarchical statistical model to assess the confidence of peptides and proteins inferred from tandem mass spectrometry. Bioinformatics 24(2):202–208
13. He Z, Yang C, Yu W (2011) A partial set covering model for protein mixture identification using mass spectrometry data. IEEE/ACM Trans Comput Biol Bioinform 8(2):368–380
14. Kearney P, Butler H, Eng K et al (2008) Protein identification and peptide expression resolver: harmonizing protein identification with protein expression data. J Proteome Res 7(1):234–244
15. Li J, Zimmerman LJ, Park B-H et al (2009) Network-assisted protein identification and data interpretation in shotgun proteomics. Mol Syst Biol 5:303
16. Ramakrishnan SR, Vogel C, Prince JT et al (2009) Integrating shotgun proteomics and mRNA expression data to improve protein identification. Bioinformatics 25(11):1397–1403
17. Gerstera S, Qelib E, Ahrensb CH et al (2010) Protein and gene model inference based on statistical modeling in k-partite graphs. Proc Natl Acad Sci U S A 107(27):12101–12106
18. Reiter L, Claassen M, Schrimpf SP et al (2009) Protein identification false discovery rates for very large proteomics datasets generated by tandem mass spectrometry. Mol Cell Proteomics 8(11):2405–2417
19. Teng B, Huang T, He Z (2014) Decoy-free protein-level false discovery rate estimation. Bioinformatics 30(5):675–681

Modification Site Localization in Peptides

13

Robert J. Chalkley

Abstract

There are a large number of search engines designed to take mass spectrometry fragmentation spectra and match them to peptides from proteins in a database. These peptides could be unmodified, but they could also bear modifications that were added biologically or during sample preparation. As a measure of reliability for the peptide identification, software normally calculates how likely a given quality of match could have been achieved at random, most commonly through the use of target-decoy database searching (Elias and Gygi, Nat Methods 4(3): 207–214, 2007). Matching the correct peptide but with the wrong modification localization is not a random match, so results with this error will normally still be assessed as reliable identifications by the search engine. Hence, an extra step is required to determine site localization reliability, and the software approaches to measure this are the subject of this part of the chapter.

Keywords

Modification site localization • False localization rate • Peak picking

13.1 Approaches

Site localization scoring approaches can be broken down broadly into two camps:

1. Those that make use of score/probability differences reported directly from the search engine that was used for peptide identification
2. Those that independently calculate a score based on an estimation of how likely a given site-determining peak may have been observed at random

I will give three examples of software employing each approach, but a more in-depth coverage of a wider range of tools has previously been published [2]. Table 13.1 summarizes approaches used by these six software tools, which will be described in more detail below.

R.J. Chalkley (✉)
Department of Pharmaceutical Chemistry, University of California San Francisco, San Francisco, CA 94143, USA
e-mail: chalkley@cgl.ucsf.edu

H. Mirzaei and M. Carrasco (eds.), *Modern Proteomics – Sample Preparation, Analysis and Practical Applications*, Advances in Experimental Medicine and Biology 919, DOI 10.1007/978-3-319-41448-5_13

Table 13.1 Comparison of site localization software

Software	Peak picking	Scoring: Probability or difference score	Representing ambiguity	Search engine results applied to
A-Score	N peaks per 100 Th	Probability	Reports best site; does not indicate best alternative location	Sequest. Can be applied to multiple search engines via Scaffold.
Mascot Delta Score	N peaks per 110 Th	Difference score	Reports best site; does not indicate best alternative location	Mascot
PhosphoRS	Variable number of peaks per 100 Th	Probability	Reports probability for all sites	Sequest and Mascot via ProteomeDiscoverer
PTM Score	N peaks per 100 Th	Probability	Reports probability for all sites	Andromeda (MaxQuant)
SLIP Score	N most intense peaks in each half of observed m/z range	Difference score	Lists all sites within score threshold	Protein Prospector
Variable modification localization score	25 peaks with highest S/N after precursor and isotope removal	Difference score	Lists all sites within score threshold	Spectrum Mill

Examples of search engine based site localization scoring include Mascot Delta Score [3], SLIP scoring in Protein Prospector [4] and variable modification localization scoring in Spectrum Mill [5]. These scores are automatically reported by Protein Prospector and Spectrum Mill, whereas Mascot Delta Score is calculated separately by software that processes the Mascot search result output. In each case the localization score is derived by determining a score or probability difference between the top scoring peptide / site combination and the next highest scoring match of the same peptide but with a different modification site localization. Spectrum Mill reports arbitrary scores for peptide identifications, and the resulting site localization is scored on the same scale. Mascot and Protein Prospector both report probability / expectation value scores. Site localization scores reported by these two software programs are derived from differences in reported probabilities for peptides identified with different modification site localizations. Scores are reported on a log10 scale, such that a score of 10 represents an order of magnitude difference and 20 represents two orders of magnitude difference in probability score. It is important to note that although these values are derived from probability scores they are not probability measures for site localization, so should be treated simply as arbitrary scores.

Examples of software that calculate scores based on estimating the probability of matching a peak at random include A-Score [6] (which is also available as part of Scaffold [7]), PTM Score [8] in MaxQuant and PhosphoRS [9] (which is also available in ProteomeDiscoverer [10]). Both A-Score and PTM Score treat observed masses as integer values (which is a reasonable step for low mass accuracy ion trap CID data, but less so for high mass accuracy fragmentation data), then calculate probabilities under the assumption that if, for example, four peaks per 100 m/z are considered, then the probability of randomly matching a peak is 4 in 100. In the case of A-Score the resulting output is converted into a score that is $-10 \times \log10(p)$, so 13 corresponds to 95 % confidence and 20 corresponds to 99 % confidence. In the case of PTM Score and PhosphoRS they invert their probabilities of random matching into probabilities of correct localization, then normalize values for all sites in the peptide so they sum to 1 (for a singly modified peptide); i.e. the peptide is definitely modified somewhere.

In order to be able to localize the site of modification in a peptide it is necessary to observe fragments formed by cleavage between two potential sites of modification; if such a fragment is not observed, then the site localization should be reported as ambiguous. Unfortunately, a given MS/MS spectrum will usually contain a mixture of fragments from the component of interest, but also 'background' ions derived from other co-isolated precursors, or maybe electrical noise. Hence, software needs to make a decision as to which peaks should be considered during scoring and site localization. This choice is normally made on the basis of intensity. Software could use a constant intensity threshold across the whole spectrum, as is the case for Spectrum Mill, but most split the spectrum into parts and then pick a certain number of the most intense peaks in each part. For example, Protein Prospector divides the observed mass range in half and then considers an equal number of peaks (as a default 20) per half. A-Score, PTM Score and Mascot split the spectrum into bins of m/z 100 (or m/z 110 in the case of Mascot), then use a constant number of peaks per m/z bin. PhosphoRS performs the same binning, but can vary the number of peaks used within each bin.

13.2 Assessing Performance

For benchmarking performance of software identifying peptides, the use of target-decoy database searching to calculate false discovery rates (FDRs) allows comparison of tools on a level playing field [1]. However, there is no equivalent approach that can be used to calculate a modification site false localization rate (FLR). The only practical way to do this is to produce datasets where the correct answers are known and hope scores have the same meaning when analyzing other data. Two approaches have been used to create such datasets.

The first is to create synthetic peptide libraries with known modification sites. The publications describing Mascot Delta Score [3] and PhosphoRS [9] created libraries of about 180 phosphopeptides for benchmarking their performance, and the former of these datasets was also used for evaluating A-Score [3] and SLIP scoring in Protein Prospector [4]. More recently a much larger synthetic phosphopeptide library of greater than 100,000 phosphopeptides was created using a limited number of seed sequences, then permuting the residues −1 and +1 from the modified residue, and this reference dataset was used for comparing PTM Score, PhosphoRS and Mascot Delta Score [11]. An interesting result from this comparison was the surprisingly high complementarity of tools; i.e. each tool reliably identified a different subset of sites, so combining multiple tools on a dataset could significantly increase the number of reliable site localizations.

The other approach employed for creating a dataset of known answers was to take data where there is only one potential site of modification, then measure how often there is an assignment to decoy amino acid residues [4]. Using proline and glutamine residues as decoy potential phosphorylation sites about 10,000 phosphopeptide spectra were assessed, from which false localization rates for a given score could be calculated for SLIP scoring in Protein Prospector.

13.3 Effect of Fragment Mass Accuracy

As previously stated, both A-Score and PTM-Score were designed for analysis of low mass accuracy fragmentation data and assume only unit mass accuracy. Hence, they do not make use of higher accuracy mass measurement. By narrowing the mass window bins; e.g. using 0.1 Da instead of 1 Da bins, then this information could be utilized, and this type of approach is what PhosphoRS does. By using narrow mass bins, for most windows it will be impossible to produce a peptide fragment within the given mass range, so the approach of assuming equal likelihood of matching a peak in all bins (and hence probability calculation) falls down. However, as final probabilities are all normalized to sum to 1, it is unclear how much of an issue this really is. Search engine site localization scoring will automatically make use of mass accuracy through the fragment mass tolerance during database searching, so may have advantages for analyzing high mass accuracy data.

13.4 Handling Ambiguity

It is rare to get fragmentation of all peptide bonds in a peptide, especially in CID or HCD fragmentation, which both have a strong sequence preference in bond cleavage [12]. ETD produces more even fragmentation, meaning you are more likely to be able to determine modification sites from this type of data [11], but there will still be spectra where a site cannot be reliably determined. Software programs handle this issue in different ways. In the case of A-Score and Mascot Delta Score, the software has to assign a site, even if the score is 0, and they do not indicate what the alternative modification location would be. Protein Prospector and Spectrum Mill employ localization score thresholds, and if the score is below this threshold then they report all residues that could be modified. In the case of PTM Score and PhosphoRS they report probabilities for all sites. An attractive feature of the Protein Prospector output is that it provides hyperlinks to annotated spectra where if there is localization ambiguity then it will annotate with both/all localizations and indicate which peaks (if any) are unique to one localization interpretation (Fig. 13.1). This feature is also accessible through a web interface to support data sharing and publication, using MS-Viewer [13].

13.5 Application

Phosphorylation is the post-translational modification that has seen the greatest need for site localization scoring due to the combination of it being heavily studied, but also because it can occur on several amino acids, most commonly serine, threonine and tyrosine residues, so there are often multiple potential sites of phosphorylation present in a given peptide. Indeed, several of the tools described in this chapter were specifically designed for phosphopeptide analysis, although there is no reason why they cannot be adapted for other PTMs. However, the user must appreciate that each software calculates

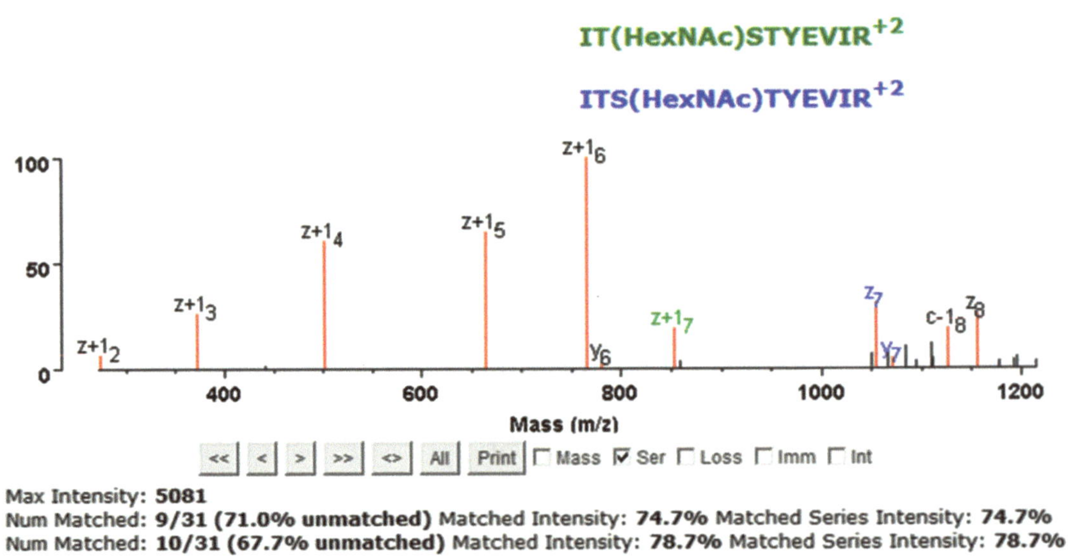

Fig. 13.1 Annotation of alternative site localization. Protein Prospector reported the site localization of the HexNAc sugar modification in this peptide as ambiguous between Thr-2 and Ser-3. Annotating the alternatives shows z ions consistent with both site localizations, suggesting this may be a mixture spectrum of the two different singly modified versions co-eluting

scores/probabilities under the assumption that only the residues specified could bear a particular modification. This becomes a complication when different modifications produce a similar or identical mass change. For example, lysine methylation is an important biological modification [14]. However, methylation of carboxylic acid residues can easily be introduced during sample handling (e.g. staining and storage of gels in solutions containing methanol and acid is common) and also many single amino acid substitutions such as serine to threonine, or valine to leucine or isoleucine produce the same mass shift. Hence, it is important to make sure software is considering any potential location for the modification. It is also worth remembering that during the peptide identification step itself the reliability of identifications of modified peptides may be lower than for unmodified because the search engine generally considers many more modified peptides than unmodified, leading to a higher proportion of the incorrect answers at a given FDR threshold being modified peptides [15].

13.6 Conclusions

Using modified peptide identifications from a search engine without any evaluation of site localization reliability produces many incorrect results. There are several tools that can evaluate site localization reliability, although in many cases the choice of tool is dictated by the search engine that was used for the peptide identification step, such that a user may not formally have any choice as to which they use. Reassuringly, these tools seem to perform reasonably at evaluating the reliability of the results they report at a given acceptance threshold, although they all clearly have many false negative results where they correctly identify the modification site but score the result below a confidence threshold. Hence, there is clear room for improvement in the performance of these tools.

Acknowledgement This work was supported by NIH NIGMS grant 8P41GM103481.

References

1. Elias JE, Gygi SP (2007) Target-decoy search strategy for increased confidence in large-scale protein identifications by mass spectrometry. Nat Methods 4 (3):207–214
2. Chalkley RJ, Clauser KR (2012) Modification site localization scoring: strategies and performance. Mol Cell Proteomics 11(5):3–14
3. Savitski MM, Lemeer S, Boesche M, Lang M, Mathieson T, Bantscheff M, Kuster B (2011) Confident phosphorylation site localization using the Mascot Delta Score. Mol Cell Proteomics 10(2): M110.003830
4. Baker PR, Trinidad JC, Chalkley RJ (2011) Modification site localization scoring integrated into a search engine. Mol Cell Proteomics 10(7):M111 008078
5. Spectrum Mill – Agilent Technologies Inc. Available from: http://www.chem.agilent.com/en-US/Products/software/chromatography/ms/spectrummillformasshunterworkstation/pages/default.aspx
6. Beausoleil SA, Villen J, Gerber SA, Rush J, Gygi SP (2006) A probability-based approach for high-throughput protein phosphorylation analysis and site localization. Nat Biotechnol 24(10):1285–1292
7. Scaffold – Proteome Software. Available from: http://www.proteomesoftware.com/products/ptm/
8. Olsen JV, Blagoev B, Gnad F, Macek B, Kumar C, Mortensen P, Mann M (2006) Global, in vivo, and site-specific phosphorylation dynamics in signaling networks. Cell 127(3):635–648
9. Taus T, Kocher T, Pichler P, Paschke C, Schmidt A, Henrich C, Mechtler K (2011) Universal and confident phosphorylation site localization using phosphoRS. J Proteome Res 10(12):5354–5362
10. Proteome Discoverer – Thermo Scientific. Available from: http://www.thermoscientific.com/en/product/proteome-discoverer-software.html
11. Marx H, Lemeer S, Schliep JE, Matheron L, Mohammed S, Cox J, Mann M, Heck AJ, Kuster B (2013) A large synthetic peptide and phosphopeptide reference library for mass spectrometry-based proteomics. Nat Biotechnol 31(6):557–564
12. Kapp EA, Schutz F, Reid GE, Eddes JS, Moritz RL, O'Hair RA, Speed TP, Simpson RJ (2003) Mining a tandem mass spectrometry database to determine the trends and global factors influencing peptide fragmentation. Anal Chem 75(22):6251–6264
13. Baker PR, Chalkley RJ (2014) MS-viewer: a web-based spectral viewer for proteomics results. Mol Cell Proteomics 13(5):1392–1396
14. Moore KE, Gozani O (2014) An unexpected journey: lysine methylation across the proteome. Biochim Biophys Acta 1839(12):1395–1403
15. Chalkley RJ (2013) When target-decoy false discovery rate estimations are inaccurate and how to spot instances. J Proteome Res 12(2):1062–1064

Useful Web Resources

14

Andre Bui and Maria D. Person

Abstract

An increasing number of web resources are available for aiding in proteomics research. Databases contain repositories of proteins and associated information. A recent article by Chen et al. (Genomics Proteomics Bioinformatics 13(1):36–39, 2015) evaluates a number of MS-based proteomics repositories containing MS and expression data, including repositories devoted to cataloguing high confidence post-translational modifications. Many sites have tools developed by research labs that are shared with the community and online tutorials and videos for learning how to use the tools. This chapter contains a selection of web sites useful for proteomics analyses but is by no means comprehensive. Using a search engine such as Google is the easiest way to find the sites using the name given below.

Keywords

Proteomics • Informatics • Web resources • Mass spectrometry • Protein identification • Databases

An increasing number of web resources are available for aiding in proteomics research. Databases contain repositories of proteins and associated information. A recent article by Chen et al. [1] evaluates a number of MS-based proteomics repositories containing MS and expression data, including repositories devoted to cataloguing high confidence post-translational modifications. Many sites have tools developed by research labs that are shared with the community and online tutorials and videos for learning how to use the tools. This chapter contains a selection of web sites useful for proteomics analyses but is by no means comprehensive. Using a search engine such as Google is the easiest way to find the sites using the name given below.

ExPASy Launched by the Swiss Institute of Bioinformatics (SIB), the Expert Protein Analysis System (ExPASy) is a bioinformatics resource portal that collects various web resources and

A. Bui • M.D. Person (✉)
Proteomics Facility, Institute for Cellular and Molecular Biology and College of Pharmacy, The University of Texas at Austin, Austin, TX, USA
e-mail: mperson@austin.utexas.edu

H. Mirzaei and M. Carrasco (eds.), *Modern Proteomics – Sample Preparation, Analysis and Practical Applications*, Advances in Experimental Medicine and Biology 919, DOI 10.1007/978-3-319-41448-5_14

repositories for protein and proteomics analyses. Proteomics and protein specific applications include various molecular weight calculators, software for 2-D PAGE analysis, sequence alignment tools, and protein structure and modeling tools.

UniProt UniProt is a repository of proteome sequence databases for multiple species stored in FASTA formats that can be used across various proteomics software packages for analysis. A European consortium of EMBL, SIB, and PIR maintains it. The UniProt databases contain both manually annotated and reviewed Swiss-Prot databases and TrEMBL databases, which are automatically annotated and unreviewed. Individual protein information can also be accessed, which includes function, protein localization, commonly associated PTMs, sequence and sequence similarity, and structural information, along with references. Full proteomes for each species can be downloaded as FASTA files for use in database search algorithms and these include all proteins validated, either with or without the isoforms.

14.1 Human Proteome Resources

There are a number of species-specific genomics resources [2], and 2014 saw publication of papers detailing the most in-depth exploration of the human proteome by mass spectrometry methods [3, 4]. Over 10,000 gene products were detected in both studies using high resolution FT-MS data, and covering all major organs as well as several cell types. However, the extensive datasets did not detect every protein and there is controversy over the accurate calculation of false discovery rates that have a major impact on the number of identified proteins. In 2015, the antibody based complementary Human Protein Atlas was completed [5].

neXtProt As an elaboration of the annotation seen in UniProt Swiss-Prot entries, neXtProt is limited to annotating and organizing data on human proteins from SIB and GeneBio. Information on function, expression, interaction, localization, proteomics, structures, GO terms and medical implications is presented.

Human Protein Atlas This site contains a spatial proteomics atlas resolved at the single cell level with images from immunohistochemical (IHC) staining for predicted proteins from a consortium of investigators in Sweden and India. The proteins are organized by tissue using a gene centric approach. There are atlases for 44 tissues and 46 cell lines, staining from 20 types of cancer, with the first complete draft released in 2015 containing 13 million annotated images, all provided in an accessible platform. Abundance estimates are made based on RNA-seq FPKM values, including coverage plots of the reads, while protein expression is visualized with tissue IHC. While there is RNA evidence for all the predicted protein encoding genes, the antibody evidence exists for 83 % of the predicted proteins, about 17 K proteins. Tissue enriched and enhanced proteins are identified, housekeeping, secretome, membrane, regulatory, isoform, cancer and druggable proteomes are defined and explored. The antigen peptide library of protein epitope signature tags (PrEST) has enabled defined standards for LC-MS/MS based quantitation. From these, ratios between RNA and protein levels have been defined for a subset of proteins in specific cell lines. The primary caveat with the use of antibody-based data is cross reactivity, estimated at causing 25–50 % of IHC staining. However, extensive validation of antibodies is performed, including checking if paired antibody pairs display identical behavior in human tissue and using genetic methods like siRNA and CRISPER on cell lines. A subcellular proteome atlas is being developed for release in 2016 to detail protein expression in subcellular compartments.

Human Proteome Map This LC-MS/MS based resource provides a graphical overview of expression levels in multiple tissues. These are the results of a large-scale project on 17 adult and seven fetal tissues, and six hematopoietic cells. The proteins in this database represent over 17,000 genes. Proteotypic peptide sequences are given for each protein to facilitate targeted studies.

ProteomicsDB An LC-MS/MS based database of human proteins with coverage of 93 % of predicted proteins, over 18,000 proteins represented. You can search for individual proteins and find projects where this protein has been identified and the proteotypical peptide for the protein. Tissue expression levels are displayed, and RNA-Seq data has been incorporated in a heat map display format.

14.2 Protein Identification and Quantitation

Mascot Matrix Science has an extensive collection of educational materials on the protein identification process as well as help for one of the most commonly used, platform independent database search algorithms for protein identification.

ProteinProspector Developed by the UCSF Mass Spectrometry Facility, ProteinProspector is a suite of platform independent tools designed for MS based proteomics experiments, ranging from initial experimental design to data processing and database searching. Some of the more useful toolsets are highlighted here:

MS-Digest — a tool for performing *in silico* digests of proteins and protein databases using various available proteases, with parameters supplied by the user.
MS-Viewer — Allows for the annotation and visualization of MS/MS spectra from database searches.
DB-Stat — Mines for statistical information from supplied FASTA databases, such as total number of entries, entries within a selected molecular weight range, mass of the longest protein, and other desired information.

Trans Proteomic Pipeline The Trans Proteomic Pipeline (TPP) is a collection of freeware software tools for MS/MS based proteomics analysis originating from the Ruedi Aebersold lab. The workflow of the TPP is designed to be platform independent, handling vendor specific file formats by converting to the universal standard file format. The search engine of choice performs database searching, while peptide and protein validations are handled by the PeptideProphet and ProteinProphet toolsets. Quantitation and visualization modules are also available, making the TPP an all-in-one freeware platform for tandem mass spectrometry based protein identification experiments.

Maxquant The site contains tools for protein identification in Andromeda, for protein quantitation by stable isotope labeled and label free methods, and for statistical analysis and visualization in Perseus developed by the Jurgen Cox lab. They have given particular attention to accurate false discovery rate calculations, using q values to determine protein FDR rates that are more stringent than the PSM or peptide FDRs are given by other programs. Tools for learning the software include videos and help sites, and a summer course is held every year to master the software.

pFind Studio pFind Studio is a freeware suite of programs designed for computational analysis of MS/MS based proteomics experiments from the Chinese Academy of Sciences. Included in the package are:

pFind: protein identification software
pLink: designed for the analysis of cross-linked peptides and SUMOylation
pNovo: De novo peptide sequencing using various fragmentation methods
pLabel: spectral annotation software for visualizing spectral matches in MS/MS results

Skyline Skyline is a platform independent, freely available application designed for chromatography-based quantitation using MS1 and MS2 ion intensities developed by the Michael MacCoss lab. Skyline originated as a toolset for the development of SRM/MRM assays for triple quadrupole instruments, but has expanded to include parallel reaction monitoring (PRM), DIA, and targeted DDA methods. Alongside its method development features, Skyline offers the ability to QC individual peptides for quantitation, a robust toolset for visualizing the results of individual quantitative MS runs, and the ability to export customized reports depending on the experimental requirements. Tutorials show how to use the software, and extensive visualization is possible for manual quality control. Webinars are held as new tools are developed.

Panorama Also developed by the MacCoss lab, Panorama is freely available software that is designed to act as a repository server application for Skyline targeted proteomics experiments. Panorama is designed to be a platform for sharing and organizing data in the Skyline format for easy visualization and access. Information such as results, spectral annotations, or spectral library chromatograms can be shared and accessed amongst collaborators for easy access.

CRAPome The Contaminant Repository for Affinity Purification (CRAPome) is an annotated database provided by a collaboration between the Alexey Nesvizhskii and Anne-Claude Gingras labs which contains negative controls considered as common protein contaminants in mass spectrometry experiments. The CRAPome also contains software available for the analysis of data generated from tandem MS experiments against the CRAPome database.

14.3 Protein Modifications

Luciphor2 Luciphor2 is an enhanced version of the original Luciphor. While the original implementation focused solely on site localization and scoring of phosphorylation events on individual peptides, Luciphor2 expands this to include any PTM. Using a JAVA based implementation, the advantage of Luciphor2 over the original is not only its ability to evaluate any PTM but also score results from any search tool. Luciphor2 can process PeptideProphet XML files derived from the TPP or tab-delimited files with scores from any protein search engine.

ProteomeScout A compendia of information on PTMs from six large databases and additional experiments created by the Neagle lab. Query with protein name to get information about modifications, binding partners, mutations, domains and structural elements.

ProSight Lite Developed for top-down analysis of protein sequences with fragment matching for a variety of fragment types by the Neil Kelleher lab at Northwestern University. ProSight Lite is available for free download. Using deconvoluted MS and MS/MS data acquired from intact proteins, protein modifications can be mapped.

14.4 Protein Interactions

String-DB A database of protein-protein interactions maintained by a European consortium including the University of Copenhagen, EMBL and University of Zurich. The information is culled from many sources including experimental evidence from pull down mass spectrometry experiments, co-expression of transcripts, genomic context and retrieval from the literature. The interactions are displayed graphically with protein balls connected by lines showing the type or confidence of the evidence. The protein structures from Protein Data Bank are accessed with a single click, making this site a great source for graphics.

DAVID A source for functional classification of user data based on enrichment of GO terms and other annotation metrics from SAIC-Frederick.

14.5 Orbitrap Information

PlanetOrbitrap A website designed to act as an umbrella and informational repository for the Thermo Orbitrap family of instruments. Included is a science library that has access to peer-reviewed scientific papers, application notes and technical guides, poster presentations from conferences, product support notes, and webinars for the application of the Orbitrap to various experimental needs. A community forum is also available for members to interact and get troubleshooting and tips from the wide network of Thermo Orbitrap users.

Acknowledgments Thanks to Michelle Gadush for website suggestions. This work was supported by CPRIT grant RP110782 and the UT system proteomics core facility network grant.

References

1. Chen T, Zhao J, Ma J, Zhu Y (2015) Web resources for mass spectrometry-based proteomics. Genomics Proteomics Bioinformatics 13(1):36–39. doi:10.1016/j.gpb.2015.01.004. Review
2. Tang B, Wang Y, Zhu J, Zhao W (2015) Web resources for model organism studies. Genomics Proteomics Bioinformatics 13(1):64–68. doi:10.1016/j.gpb.2015.01.003. Review
3. Kim MS, Pinto SM, Getnet D, Nirujogi RS, Manda SS, Chaerkady R, Madugundu AK, Kelkar DS, Isserlin R, Jain S, Thomas JK, Muthusamy B, Leal-Rojas P, Kumar P, Sahasrabuddhe NA, Balakrishnan L, Advani J, George B, Renuse S, Selvan LD, Patil AH, Nanjappa V, Radhakrishnan A, Prasad S, Subbannayya T, Raju R, Kumar M, Sreenivasamurthy SK, Marimuthu A, Sathe GJ, Chavan S, Datta KK, Subbannayya Y, Sahu A, Yelamanchi SD, Jayaram S, Rajagopalan P, Sharma J, Murthy KR, Syed N, Goel R, Khan AA, Ahmad S, Dey G, Mudgal K, Chatterjee A, Huang TC, Zhong J, Wu X, Shaw PG, Freed D, Zahari MS, Mukherjee KK, Shankar S, Mahadevan A, Lam H, Mitchell CJ, Shankar SK, Satishchandra P, Schroeder JT, Sirdeshmukh R, Maitra A, Leach SD, Drake CG, Halushka MK, Prasad TS, Hruban RH, Kerr CL, Bader GD, Iacobuzio-Donahue CA, Gowda H, Pandey A (2014) A draft map of the human proteome. Nature 509(7502):575–581. doi:10.1038/nature13302. PubMed PMID: 24870542; PubMed Central PMCID: PMC4403737
4. Wilhelm M, Schlegl J, Hahne H, MoghaddasGholami A, Lieberenz M, Savitski MM, Ziegler E, Butzmann L, Gessulat S, Marx H, Mathieson T, Lemeer S, Schnatbaum K, Reimer U, Wenschuh H, Mollenhauer M, Slotta-Huspenina J, Boese JH, Bantscheff M, Gerstmair A, Faerber F, Kuster B (2014) Mass-spectrometry-based draft of the human proteome. Nature 509(7502):582–587. doi:10.1038/nature13319
5. Uhlén M, Fagerberg L, Hallström BM, Lindskog C, Oksvold P, Mardinoglu A, Sivertsson Å, Kampf C, Sjöstedt E, Asplund A, Olsson I, Edlund K, Lundberg E, Navani S, Szigyarto CA, Odeberg J, Djureinovic D, Takanen JO, Hober S, Alm T, Edqvist PH, Berling H, Tegel H, Mulder J, Rockberg J, Nilsson P, Schwenk JM, Hamsten M, von Feilitzen K, Forsberg M, Persson L, Johansson F, Zwahlen M, von Heijne G, Nielsen J, Pontén F (2015) Proteomics. Tissue-based map of the human proteome. Science 347(6220):1260419. doi:10.1126/science.1260419

Mass Spectrometry-Based Protein Quantification

15

Yun Chen, Fuqiang Wang, Feifei Xu, and Ting Yang

Abstract

Quantification of individual proteins and even entire proteomes is an important theme in proteomics research. Quantitative proteomics is an approach to obtain quantitative information about proteins in a sample. Compared to qualitative or semi-quantitative proteomics, this approach can provide more insight into the effects of a specific stimulus, such as a change in the expression level of a protein and its posttranslational modifications, or to a panel of proposed biomarkers in a given disease state. Proteomics methodologies, along with a variety of bioinformatics approaches, are a major tool in quantitative proteomics. As the theory and technological aspects underlying the proteomics methodologies will be extensively described in Chap. 20, and protein identification as a prerequisite of quantification has been discussed in Chap. 17, we will focus on the quantitative proteomics bioinformatics algorithms and software tools in this chapter. Our goal is to provide researchers and newcomers a rational framework to select suitable bioinformatics tools for data analysis, interpretation, and integration in protein quantification. Before doing so, a brief overview of quantitative proteomics is provided.

Keywords

Protein quantification bioinformatics • Quantitative signal processing • SAPRatio • MAXquant • Progenesis QI • APEX • Trans-Proteomic Pipeline (TPP) • IsobariQ and Iquant • Targeted proteomics • PeptideAtlas • Skyline • ATAQS

15.1 Brief Introduction of Quantitative Proteomics

Despite a number of recent developments in proteomics-associated technologies, such as

Y. Chen (✉) • F. Wang • F. Xu • T. Yang
School of Pharmacy, Nanjing Medical University, 818 Tian Yuan East Road, Nanjing 211166, China
e-mail: ychen@njmu.edu

H. Mirzaei and M. Carrasco (eds.), *Modern Proteomics – Sample Preparation, Analysis and Practical Applications*, Advances in Experimental Medicine and Biology 919, DOI 10.1007/978-3-319-41448-5_15

two-dimensional polyacrylamide gel electrophoresis (2D-PAGE) and protein microarrays [1], mass spectrometry (MS)-based proteomics remains an essential technique for quantitative proteome analyses. Our focus here is on MS-based proteomics.

In MS-based proteomics, two fundamental approaches are currently employed: top-down and bottom-up proteomics. In top-down proteomics, intact proteins or large fragments are subjected to mass spectrometry. Bottom-up proteomics relies on proteolytic peptides, which are generated by enzymatic digestion of proteins. Of several such strategies that have been developed, all involve the digestion of proteins into peptides, typically with trypsin, followed by chromatographic separation, ionization and mass spectrometric analysis of the complex peptide samples. Due to the protein size limitation (<50 kDa) and the general reduction of sensitivity by one order of magnitude in top-down proteomics [2], bottom-up proteomics is more commonly used. Currently, bottom-up proteomics has three types of classifications:

1. Relative and absolute quantification (according to the information they can provide)
2. Label-based and label free proteomics (according to the underlying methodology)
3. Discovery and targeted proteomics (according to the pre-selected range of proteins) (Table 15.1)

Targeted proteomics (hypothesis-driven proteomics) is a new concept in protein quantification for highly selective and high-throughput analysis of one or more target proteins, corresponding to discovery (shotgun) proteomics [3].

Relative quantification compares the specific protein level in different samples, with results being expressed as a relative fold change of protein abundance, whereas absolute quantification is the determination of the exact amount or mass concentration of a protein, for example, in units of ng/mL of a plasma biomarker or in mol/cell of a cellular protein. Both relative and absolute quantification can be achieved using isotopic/isobaric labeling or label-free strategies. Stable isotope labeling typically compares naturally abundant stable isotope peptides to physicochemically identical peptides with atoms enriched in a heavy stable isotope at the MS level. Isobaric labeling allows peptides or proteins to be labeled with isobaric reagents and is usually detected at the MS/MS level [4]. Labeling technologies include *in vivo* labeling *via* metabolic incorporation or *in vitro* labeling *via* chemical reactions. Metabolic incorporation such as SILAC (stable isotope labeling with amino acids in cell culture) introduces stable isotope labeled amino acids in cells. Chemical reactions such as iTRAQ or TMT incorporate amine-specific isobaric tags onto sites such as the N-terminus, C-terminus, cysteines, lysines, and tyrosines. Alternatively, label-free quantification uses ion signal intensities acquired by the mass spectrometer (i.e., ion intensity measurement) or the number of spectra matched to peptides from a protein (i.e., spectral counting) as a proxy to assess the protein quantities within the sample.

Absolute quantification provides accurate protein amount information by spiking protein or peptide samples with known concentrations of heavy isotope labeled synthetic peptides. As an essential approach in absolute quantification, targeted proteomics has recently taken front stage in the proteomics community [5]. Targeted proteomics specifically refers to absolute quantification using selected/multiple reaction monitoring (SRM or MRM) on a triple quadrupole instrument and a stable isotope labeled internal standard [3]. Targeted proteomics strategies limit the number of proteins that are monitored and optimize the chromatography instrument tuning and acquisition methods to achieve the highest sensitivity and throughput for hundreds or thousands of samples. Discovery proteomics often requires large sample quantities and multi-dimensional fractionation, which diminishes sensitivity and throughput [6].

Table 15.1 Classification of proteomics

	Absolute quantification	Relative quantification	
Label-based	AQUA, SISCAPA	Metabolic	^{15}N, SILAC
	PSAQ	Chemical	ICAT, ICPL,
	Absolute SILAC		iTRAQ, TMT, IPTL,
	ELEX!Quant		DML, mTRAQ
	QCONcat	Enzymatic	^{18}O
Label-free	PAI, emPAI	Ion intensity (XIC)	
	APEX		
	Top3	Spectral counting	
	iBAQ		

15.2 Bioinformatic Tools in Quantitative Proteomics

Data analysis in MS-based proteomics is more challenging than for other high-throughput technologies and remains a principal bottleneck in proteomics [7]. There are a number of bioinformatics tools available in quantitative proteomics. These tools can be combined in various ways to generate different proteomics data analysis pipelines. Table 15.2 provides a summary of quantitative proteomics approaches, along with the associated software tools. Theoretically, all the approaches are achieved by providing/comparing the amounts of peptides and proteins. There are common challenges for calculating quantitative values at the protein and peptide level. Thus, these common issues will be first presented with bioinformatics solutions, followed by a selective description of several software tools for both relative and absolute quantification. As the majority of tools have been described in protein identification, we will focus on their quantitative features in data analysis. Additionally, targeted proteomics and its recently developed tools, such as Skyline and ATAQs, will be extensively illustrated due to their novelty in proteomics.

Taking previous reviews into account [8–10], Table 15.3 shows the details of most available software tools, including the supported instruments, free/commercial software, type of data, database search engine, requisite input files and software dependencies, as well as programming languages and supported operating systems. We hope that this information can provide a starting point for further reading or an initial guide for newcomers.

15.3 Common Issues in Proteomics Quantification

Data analysis in relative proteomics quantification generally includes raw data processing, followed by an ion chromatogram ratio calculation to infer the peptide abundance ratio, followed by relative protein abundances calculation using peptide ratios [11]. Absolute quantification is often performed in a manner similar to relative quantification. Absolute quantity of a peptide is calculated by comparing its ion intensity with the ion intensity of an identical chemically synthesized heavy isotope labeled peptide spiked in with known concentration as an internal standard. Many of the same problems encountered in relative quantification still occur in absolute quantification, and the existing software for relative quantification could be easily adapted for absolute quantitative purposes. Figure 15.1 is an overview of quantitative proteomics data analysis. Common computational and statistical issues will be interpreted below. Baseline subtraction, noise filtering, mass calibration, retention time alignment, and peak detection that are primarily employed in spectra processing and protein identification will also be briefly described for completeness.

Table 15.2 Proteomics strategies and associated software tools

Label-based	Metabolic	MSQuant	http://msquant.alwaysdata.net/
		Maxquant	http://maxquant.org/
		MFPaQ	http://mfpaq.sourceforge.net/
		OpenMS	http://open-ms.sourceforge.net/
		Proteome Discoverer	http://www.thermoscientific.com
		WARP-LC	http://www.bdal.com/products/software/warp-lc/
		PVIEW	http://compbio.cs.princeton.edu/pview/
		Elucidator	http://www.rosettabio.com/
		ASAPRatio	http://tools.proteomecenter.org/wiki/
		Mascot Distiller	http://Matrixscience.com/distiller.html
		Scaffold	http://www.proteomesoftware.com/products/
		Census	http://fields.scripps.edu/census/
		PEAKSQ	http://www.bioinfomaticssolutions.com/products/peaks/quantification.php
		MaXIC-Q	http://ms64.iis.sinica.edu.tw/MaXIC-Q_web/index.html
	Chemical	Mascot Distiller	http://Matrixscience.com/distiller.html
		MaXIC-Q	http://ms64.iis.sinica.edu.tw/MaXIC-Q_web/index.html
		Maxquant	http://maxquant.org/
		MFPaQ	http://mfpaq.sourceforge.net/
		OpenMS	http://open-ms.sourceforge.net/
		PeakQuant	http://www.medizinisches-proteom-center.de/software
		ProRata	http://code.google.com/p/prorata/
		Proteios	http://www.proteios.org/
		TPP	http://www.proteomecenter.org/software.php
		-Libra	http://tools.proteomecenter.org/wiki/
		VIPER	http://omics.pnl.gov/software/VIPER.php
		X-Tracker	http://www.x-tracker.info/
		Proteome Discoverer	http://www.thermoscientific.com
		WARP-LC	http://www.bdal.com/products/software/warp-lc/
		Mascot	http://www.matrixscience.com
		PVIEW	http://compbio.cs.princeton.edu/pview/
		IsobariQ	http://www.biotek.uio.no/research/thiede_group/software
		jTraqX	http://sourceforge.net/projects/protms
		IQuant	http://sourceforge.net/projects/iquant/
		Rover	http://genesis.ugent.be/rover/
		VEMS	http://www.portugene.com/software.html
		Elucidator	http://www.rosettabio.com/
		XPRESS	http://tools.proteomecenter.org/wiki/
		ASAPRatio	http://tools.proteomecenter.org/wiki/
		ZoomQuant	http://proteomics.mcw.edu/zoomquant.html
		PEAKSQ	http://www.bioinfomaticssolutions.com/products/peaks/quantification.php
		ProteinPilot	http://absciex.com/
		Multi-Q iTracker	http://ms.iis.sinica.edu.tw/Multi-Q/ http://www.cranfield.ac.uk/health/researchareas/bioinformatics/page6801.html
		MSQuant	http://msquant.alwaysdata.net/
		Scaffold Q +	http://www.proteomesoftware.com/
	Enzymatic	MSQuant	http://msquant.alwaysdata.net/
		ProRata	http://code.google.com/p/prorata/
		VIPER	http://omics.pnl.gov/software/VIPER.php
		ZoomQuant	http://proteomics.mcw.edu/zoomquant.html

(continued)

Table 15.2 (continued)

		Proteome Discoverer	http://www.thermoscientific.com
		WARP-LC	http://www.bdal.com/products/software/warp-lc/
		PVIEW	http://compbio.cs.princeton.edu/pview/
		Mascot Distiller	http://Matrixscience.com/distiller.html
		PEAKSQ	http://www.bioinfomaticssolutions.com/products/peaks/quantification.php
	Intensity-based	Maxquant	http://maxquant.org/
		MSQuant	http://msquant.alwaysdata.net/
		-SpecArray	http://tools.proteomecenter.org/wiki/
		Corra	http://tools.proteomecenter.org/wiki/index.php?title=Software:Corra
		Expression [E]	http://www.waters.com/waters/nav.htm?cid=10011719
		IDEAL-Q	http://ms.iis.sinica.edu.tw/IDEAL-Q/
		Mascot Distiller	http://www.matrixscience.com/distiller.html http://www.chem.agilent.com/en-
		MassHunter Mass Profiler	US/Products/software/chromatography/ms/masshunterprofiling/pages/default.aspx
		msBID	http://tools.proteomecenter.org/wiki/index.php?title=Software:msBID
		MsXelerator	http://www.msmetrix.com/
		Profile Analysis	http://www.bdal.com/products/software/profileanalysis/overview.html
		Progenesis LC-MS	http://www.nonlinear.com/
		ProteinQuant	http://www.ncgg.indiana.edu/
		ProtQuant	http://www.agbase.msstate.edu/cgi-bin/tools/index.cgi
		QuanLynx	http://www.waters.com/waters/nav.htm?locale=de_DE&cid=513662
		Refiner MS	http://www.genedata.com/products/expressionist/modules.html
		Peaks Q	http://www.bioinfomaticssolutions.com/products/peaks/quantification.php
		PVIEW	http://compbio.cs.princeton.edu/pview/
		VIPER	http://omics.pnl.gov/software/VIPER.php
		TOPP	http://open-ms.sourceforge.net/news.php
		SuperHirn	http://tools.proteomecenter.org/wiki/
		SIEVE	http://www.thermoscientific.com/
		PEPPeR	http://www.broadinstitute.org/cancer/software/genepattern/desc/proteomics
		msInspect	http://proteomics.fhcrc.org/CPL/msinspect/index.html
		Msight	http://www.expasy.org/MSight/
		DeCyder MS	http://www.gelifesciences.com/
	Spectral counting	APEX	http://pfgrc.jcvi.org/index.php/bioinformatics/apex.html
		ProteoIQ	http://bioinquire.com
		Abacus	http://abacustpp.sourceforge.net/
		emPAI (Mascot)	http://www.matrixscience.com/
		emPAICalc	http://empai.iab.keio.ac.jp/
		Scaffold 3	http://www.proteomesoftware.com/
Targeted proteomics	Skyline		http://proteome.gs.washington.edu/software/skyline
	MaxQuant		http://maxquant.org/
	ATAQS		http://tools.proteomecenter.org/ATAQS/ATAQS.html
	TIQAM		http://tools.proteomecenter.org/TIQAM/TIQAM.html
	MRMaid		http://138.250.31.29/mrmaid/
	SRMCollider		http://www.srmcollider.org/srmcollider/srmcollider.py
	MRMer		http://proteomics.fhcrc.org/CPL/MRMer.html

Table 15.3 Detailed information of selected software tools

Software	Version	Technique	Compatible search engine	Type of data	Instruments	Input files	Distribution	Language	Operation system (OS)
ASAP Ratio	–	ICPL, ICAT, SILAC	SEQUEST	MS^2	Any via mzXML or mzML/Thermo	Via T PP/pepXML (generated in the TPP)	Free	C	Windows, Linux OSX/Mac
APEX	1.1.0	Improvement of SC	SEQUEST, MASCOT and X!T andem	LC-MS/MS	Any via protXML files	.fasta, .oi, protXML	Free open source	Java	Windows, OSX, Linux
Census	1.72/2.3	15N, SILAC, iTRAQ, SC	–	MS^1/MS^2	Any via mzXML	MS1/MS2, DTASelect, mzXML, pepXML	Free	Java	Windows, OSX, Linux
Corra	v3.0	Intensity-based	SEQUEST	LC-MS	Any via mzXML	mzXML	Free	Java	Linux
IQuant	2.0.1	iTRAQ, TMT	MASCOT	MS^1/MS^2	Any via mzXML	mzXML	Free open source	Java	Windows XP, Ubuntu 8.04 - the Hardy Heron, and Mac OS × 10.4.11 platforms
IsobariQ	1.1	iTRAQ, TMT/IPTL	MASCOT	MS^1/MS^2	Any via mzXML	mzXML	Free	C++	windows
iTracker	1.1	iTRAQ	SEQUEST, MASCOT	MS^2	All machines that export noncentroided spectra and support. dta or .mgf file	.mgf, .dta	Free	Windows exe Perl	Windows, Linux
-Libra	–	iTRAQ, TMT	SEQUEST	–	Any via mzXML or mzML/Thermo	Via TPP/pepXML or summary.html files	Free	C	Windows, Linux, OSX
MapQuant	2.1.1	Label free image recognition	SEQUEST	LC-MS	Thermo, Waters	OpenRawSEQUEST/ mzXML and MQScript files	Free open source	Visual C++/C	Windows, Linux
MaxQuant	1.2.2.5/ 1.4.1.2	SILAC, ICPL, Label free/ Intensity-based	MASCOT	MS^1/MS^2	LTQ, Orbitrap, FT-ICR(Thermo)	.raw (Thermo)	Free	C#	Windows
mProphet	1.0.4.1	SRM, AQUA, QconCAT, PSAQ	–	LC-MS/MS	Any via mzXML	mzXML, .xls	Free open source	Perl/R	Windows, Linux

MRMaid	2.0	SRM, label-free	–	LC-MS/MS	–	.mgf, .pkl, or mzXML	Free	Java	Web-based
MRMer	–	SRM, label-free	–	LC-MS	Waters	mzXML	Free	Java	Windows, OSX, Linux
MSQuant	2.0b6	15N, SILAC, ICAT, ^{18}O, label-free/ Intensity-based	MASCOT	MS^1/MS^2	QSTAR (ABI), Q-ToF(Waters), LTQ, FT, Orbitrap (Thermo)	.raw (Thermo), .dat(Waters), .wiff(ABI)/MASCOT html and raw spectral file(.wiff, . raw and .dat supported)	Free open source	C# and VB .NET	Windows
Multi-Q	1.6.5.4	iTRAQ	SEQUEST, MASCOT and X!T andem	LC-MS/MS	Any via mzXM L/Applied Biosystems, Thermo,Waters & Bruker Daltonics	mzXML, .wiff (ABI)/ Vendor formats (.wiff, .raw) are converted into a reduced mzXML	Free	VB .NET	Windows, Web
OpenMS	1.8/ 1.11.1	iTRAQ, SILAC, labelfree	XTandem, SEQUEST, MASCOT, OMSSA	MS^1/MS^2	Any via mzXML or mzML	.dta, mzData, mzXML, mzML	Free open source	C++	Windows, Linux, OSX
PeakQuant	1.5.42	^{15}N, SILAC, iTRAQ	–	MS^1/MS^2	Any via mzXML	mzXML	Free	Java	Windows, Linux, OSX
PEPPeR	–	Label-free (Intensity-based)	–	MS^1/MS^2	Any via mzXML	mzXML	Free	Perl	Windows
Progenesis QI	1.0	Label-free	MASCOT	LC-MS	Any	mzXML, mzML and NetCDF	Commercial	–	–
ProRata	1.0	^{15}N, SILAC, ICAT, ^{18}O	SEQUEST	LC-MS	Any via mzXML/ Thermo	mzXML/mzXML and DTASelect output file	Free open source	C++	Windows, Linux
Proteios	2.16/2.19	iTRAQ, TMT	Mascot, XITandem, OMSSA	MS^1/MS^2	Any via mzXML	mzML	Free open source	Java	Windows, Linux, OSX
QUIL	–	^{18}O, ICAT	SEQUEST	LC-MS	LCQ, LTQ, FT-ICR (Thermo)	–	Available on request/ Free	Visual C+ +	Windows
Qupe	–	^{15}N	–	LC-MS	LTQ, Orbitrap (Thermo)	mzXML	Web	Java	Web-based
Skyline	1.2/2.5	SRM, label-free	–	LC-MS	Any via mzXML	mzXML, pepXML	Free open source	C#	Windows

(continued)

Table 15.3 (continued)

Software	Version	Technique	Compatible search engine	Type of data	Instruments	Input files	Distribution	Language	Operation system (OS)
STEM	–	^{18}O	MASCOT	LC-MS	Waters	.pkl (ProteinLynx, Waters)/MASCOT .dat file and .raw file	Free	–	Windows
TPP	4.5.0	ICAT, SILAC, iTRAQ	Any	MS^1/MS^2		mzXML, mzML	Free	–	Windows, Linux, OSX
VIPER	3.48.456/3.49	^{18}O, ICAT/Intensity-based	–	MS^1/MS^2	Any via mzML	.pek, .CSV, .mzXML,. mzData, .raw(Thermo)	Free open source	–	Windows
XPRESS	–	ICPL, ICAT, SILAC, ^{14}N/^{15}N	SEQUEST	–	Any via mzXML or mzML/Thermo	Via T PP/pepXM L (generated in the TPP)	Free	C	Windows, Linux, OSX
X-Tracker	1.3	iTRAQ, ^{15}N	–	MS^1/MS^2	Any via mzML	mzML, mzIdentML	Free open source	Java	Windows, Linux, OSX,
ZoomQuant	–	^{18}O	SEQUEST	LC-MS	LTQ (Thermo)	.raw (Thermo)/Uses various formats(i) Xcaliber raw file (ii) SEQUEST.out file Processed to internal.colon file	Free	Perl	Windows, Linux, OSX

Some information was cited from [9]

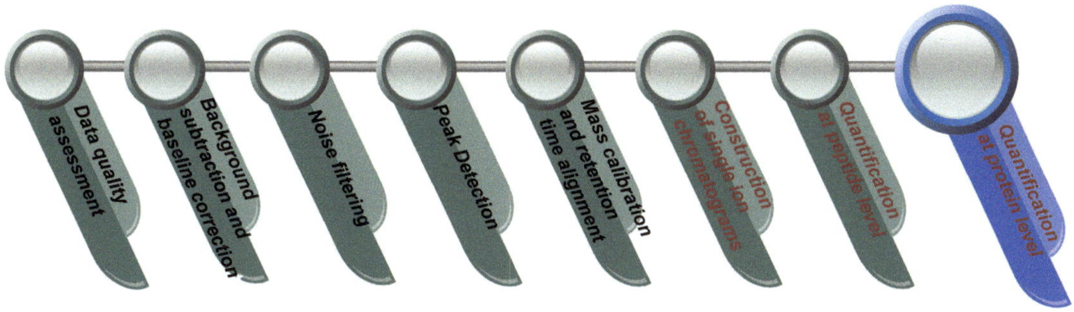

Fig. 15.1 General software workflow for quantitative proteomics

15.3.1 Data Quality Assessment

The proteomics process is sensitive to changes in sample preparation and spectra collection, especially for label-free approaches. Extreme caution must be used to maintain the same sample preparation protocol throughout the experiments. Introduction of any systematic bias into the data collection and sample handling will significantly impact the result, even if sophisticated bioinformatics tools are used [12]. This topic has been extensively investigated by Hilario et al. [13].

15.3.2 Background Subtraction and Baseline Correction

The baseline signal must be subtracted from the raw spectrum because the detector may overestimate the number of ions arriving at its surface, especially in low-molecular-weight regions. Recently, this type of correction has no longer been necessary, and the general assumption has become that commercial mass spectrometry instrument software will remove the background signal automatically [14].

15.3.3 Noise Filtering

Two types of errors are often present in experimental data, systematic error and random error [15]. Systematic error can be caused by a variety of factors, such as drift of calibration constants with time or temperature which can be easily prevented by routine calibration [14]. Random error, also called noise, can be divided into low and high frequency noise. The aim of the noise filtering step is to remove the random noise in the mass spectra and to enhance the signal to noise ratio (S/N). In general, noise filtering is performed before identifying peptide peaks. The selected features left after the removal of the noise are often called peaks. There are several methods to remove the noise and select the peaks, including (1) filter methods, (2) wrapper methods, and (3) embedded methods [16]. The filter methods are most commonly used, and a wavelet filter has been reported to perform best among these filters (e.g., average filter, Savitzky-Golay filter, Gaussian filter, Kaiser window, and wavelet based filters) [17].

15.3.4 Peak Detection

After noise filtering, the charge state is defined by analyzing the isotope distribution, and peak overlap is also resolved. The algorithms and tools associated with isotope and charge state deconvolution have been reviewed elsewhere [18, 19]. The peaks are detected using methods including the isotopic cluster identification method by Horn et al. [13], the local maximum peak detection method by Yasui et al. [14, 15], or the mean-spectrum undecimated discrete wavelet transform-based peak detection method by Morris et al. [16]. The resulting spectral information is subsequently subjected to database searching. Commercial search engines such as

MASCOT, SEQUEST and Phenyx support many relevant instruments and their fragmentation methods. It is possible to perform analyses with publicly available tools as well, for instance, VEMS v3.0. The most common databases used in searching are NCBI's Entrez Protein, RefSeq, IPI, Swiss-Prot, UniProt, and TrEMBL [2]. Once peptide identification has been deemed acceptable, the identified peptide information is used to locate specific peptide elution time in quantification applications.

15.3.5 Mass Calibration and Retention Time Alignment

Software provides advanced systems for mass calibration. The mass dimension rarely requires calibration, and the data alignment can conveniently be reduced to a simpler problem of aligning the retention time dimension [20]. A variety of alignment approaches have been suggested, including dynamic time warping, correlation optimized warping, parametric time warping, and peak alignment [21]. Good alignment is especially required for label-free quantification. According to the experimental design, the alignment can be performed before or after peptide identification.

15.3.6 Construction of Single Ion Chromatograms

The steps described above are also employed in protein identification. The following workflow including peptide ion chromatogram extraction, quantification at the peptide level and quantification at the protein level represents the major concept in quantitative proteomics.

Determination of the correct start and end points of peptide elution peaks in chromatograms is crucial for accurate and precise quantification results. The ion chromatogram extraction process is a computationally intensive step in quantitative analysis. Strictly speaking, an updated peak profile is reconstructed using the identified peptide information. There are three possible m/z values that can be used to define the m/z value for a given peptide in order to integrate the signal and extract its ion chromatogram: the experimentally observed m/z reported by the instrument's software, the experimental m/z reported by the search engine (which may differ), and the exact theoretical m/z calculated from the sequence in a given ion charge state [8]. Currently, most software tools use the theoretical peptide sequence mass to determine the m/z value. Using the determined m/z value, a single ion chromatogram can be extracted. However, the construction method for the ion chromatogram varies amongst the software tools [8]. Some tools construct several single ion chromatograms (from multiple charge states) and average them, whereas some others construct only one ion chromatogram from the most abundant precursor ion.

15.3.7 Quantification at Peptide Level

After defining peptide elution peaks, the next step is to calculate peptide abundances for the light and/or heavy peptides. There are several different algorithms to calculate peptide abundances – peak area, least squares regression and principal component analysis [22]. The peak area approach calculates the area of peaks. In the least squares regression approach, the peak profiles of light and heavy peptides are converted into a scatterplot based on their ion intensities. The slope of the regression provides a measure of the background-subtracted ratio, the intercept provides a measure of the ratio of the two backgrounds, and the correlation coefficient provides a measure of the ratio quality [23]. Principal component analysis generates a similar scatterplot of ion intensities from both light and heavy peptides, and calculates two principal components and their values. The slope of the first principal component indicates the peptide abundance ratio. The criteria of peak area integration depend on the S/N of the chromatogram, the chromatogram peak shape, and even individual users' biases.

15.3.8 Quantification at Protein Level

If there is more than one peptide ratio for a target protein, the individual peptide values must be combined in some fashion. There are three different approaches for calculation [8, 22]. One approach is to calculate the mean or median of all peptide measurements, fitting the experimental values to a normal distribution [24]. The second method is a weighted average in which peptides with given weights, based on scores such as the quality or standard deviation, are used to derive a protein abundance ratio. The third approach is to calculate the protein ratio from an estimated likelihood function and the significance of the protein ratio is also related to the maximum of the likelihood function [25]. Sometimes, protein ratio calculation is complicated because several identified peptides are not unique to the target protein and may occur in other proteins. These peptides that cannot be used to estimate the final protein ratio should be removed as outliers. Nesvizhskii and Aebersold have provided a review of resolving multipeptide/protein issues [26]. Statistical tests (e.g., *t* test or ANOVA) assign significance levels to ratio estimations and help to control error rates.

Many of the bioinformatics tools in quantitative proteomics follow the procedures described above. However, the algorithms implemented in these tools are used to correct potential artifacts created by the different proteomics approaches [8, 18]; thus, one software is not suitable for all quantitative strategies. In addition, their application is restricted by the nature of the experiment to be performed and available instrumentation. To further explain bioinformatics tools, we would like to provide more details about several selected tools in the next section of this chapter. These software packages were chosen because sufficient details of the implemented algorithms are available in the respective publications, whereas the others are "black box" designs tied to instrument vendors or are commercial products.

15.4 Selected Bioinformatics Tools

15.4.1 Automated Statistical Analysis of Protein Abundance Ratios (ASAPRatio)

ASAPRatio is currently applied for ICAT data analysis. The algorithms of ASAPRatio utilize Savitzky-Golay smoothing filter, statistics for weighted samples, and Dixon's test for outliers, to evaluate relative protein abundance ratios and their associated errors [24]. In the construction of single ion chromatograms, ASAPRatio considers signal integrated over three isotopic peaks, for each isotopic variant and for four charge states. Background subtraction and outlier removal are then performed prior to the calculation of an abundance ratio for each peptide. For each unique peptide, the abundance ratio is calculated for each observed charge state and then all valid abundance ratios from the different charge states are collected, weighted by the sum of the two corresponding elution peak areas. Ratios are averaged for individual peaks, and then over all peaks (using weights in both cases). If there are more than three ion ratios, a final 'unique peptide ratio' is produced for each peptide. If there is more than one peptide ratio for a particular protein, ASAPRatio use a weighted mean of the peptide ratios to calculate the protein ratio, using estimated errors for the peptide ratios [8].

15.4.2 MAXquant

Maxquant produced by Mattias Mann's group was developed based on the MSQuant and has similar properties with MSQuant [9]. Since it is designed for analyzing large mass spectrometric data sets, Maxquant is more suitable to high resolution data generated by the Thermo Orbitrap and FT mass spectrometers. This software supports label-free methods in addition to SILAC and ICPL (Isotope-coded protein label). Peaks are detected in each MS scan by fitting a Gaussian peak shape to the three central raw data points [27]. Using correlation analysis and graph theory, MaxQuant detects peaks, isotope clusters and SILAC peptide pairs as three-dimensional objects in *m*/*z*, elution time and signal intensity space. It currently uses Mascot to generate peptide candidates for MS/MS spectra. The subsequent analysis includes robust processing and filtering for peptide mass accuracy and false discovery rate (FDR) thresholds at protein and peptide level [28]. Protein ratios are calculated as the median of all peptide ratios and can be normalized to correct for unequal protein amounts.

15.4.3 Progenesis QI

Progenesis QI software package from Nonlinear Dynamics, Waters is a software solution for label-free quantification. Ion intensities are employed to provide quantification. Progenesis QI is capable of processing a large number of replicates, and has an accessible graphical user interface allowing users to view their MS data in two- or three-dimensional (2D or 3D) maps to verify if peptides have been quantified accurately. The peptide outlines mark the boundaries of each peptide isotope. The peptide abundance is the sum of the intensities within the isotope boundaries. To obtain the protein abundance, the sum of all unique normalized peptide ion abundances for a specific protein on each run is calculated. Furthermore, several post-processing software (i.e., the Progenesis Post-Processor) extends the application of Progenesis QI to label-based quantification by embedding Progenesis QI in the analysis of stable isotope labeling data and top3 pseudo-absolute quantification [29]. The validated quantification range has been reported in the 2–1000 fmol.

15.4.4 APEX (Absolute Protein Expression)

APEX is a modified spectral counting technique. Spectral counting techniques typically infer the relative quantity of a protein by dividing the number of MS-identified tryptic peptides derived form that protein by the total number of MS-identified peptides [30]. However, this technique is confounded by peptide physicochemical properties, affecting MS detection and resulting in each peptide having a different detection probability. In APEX, machine-learning algorithms are used to predict weighting factors for each peptide-spectrum match (PSM) based on the predicted properties of the peptide. The spectral count is weighted accordingly and used to calculate the protein abundance. The user-supplied normalization factor, typically an estimate of total protein concentration, converts the relative abundance values into absolute terms. Thus, quantification results over basic spectral counting can be improved.

15.4.5 Trans-Proteomic Pipeline (TPP)

TPP, a collection of software tools, is instrument-independent and supports commonly used proteomics workflows [31]. Importantly, the pipeline uses open, standard data formats and calculates estimates of sensitivity and error rates, thus allowing for meaningful data exchange. TPP relies on, and integrates in its workflow, external search engines (e.g., peptide identification (SEQUEST, Mascot, COMET, PeptideProphet and X!Tandem) and protein identification (ProteinProphet)) [32, 33]. Quantification analysis tools such as ASAPRatio described above and Libra can also be used in the pipeline flow for peptide and protein quantification. Due to its special usage, we will give more details here based on an example data of Tandem Mass Tags (TMT) 6-plex labeling (named as dataset HuN9; searched using Mascot 2.1 against human protein sequences of UniProt 2013_12). TMT is an isobaric compound that allows peptides from up to six samples to be identified and quantified in a single experiment. The intensities of six reporter ions are used for the quantification of peptides in different samples. Figure 15.2 shows representative mass spectra of a surrogate peptide used for quantification.

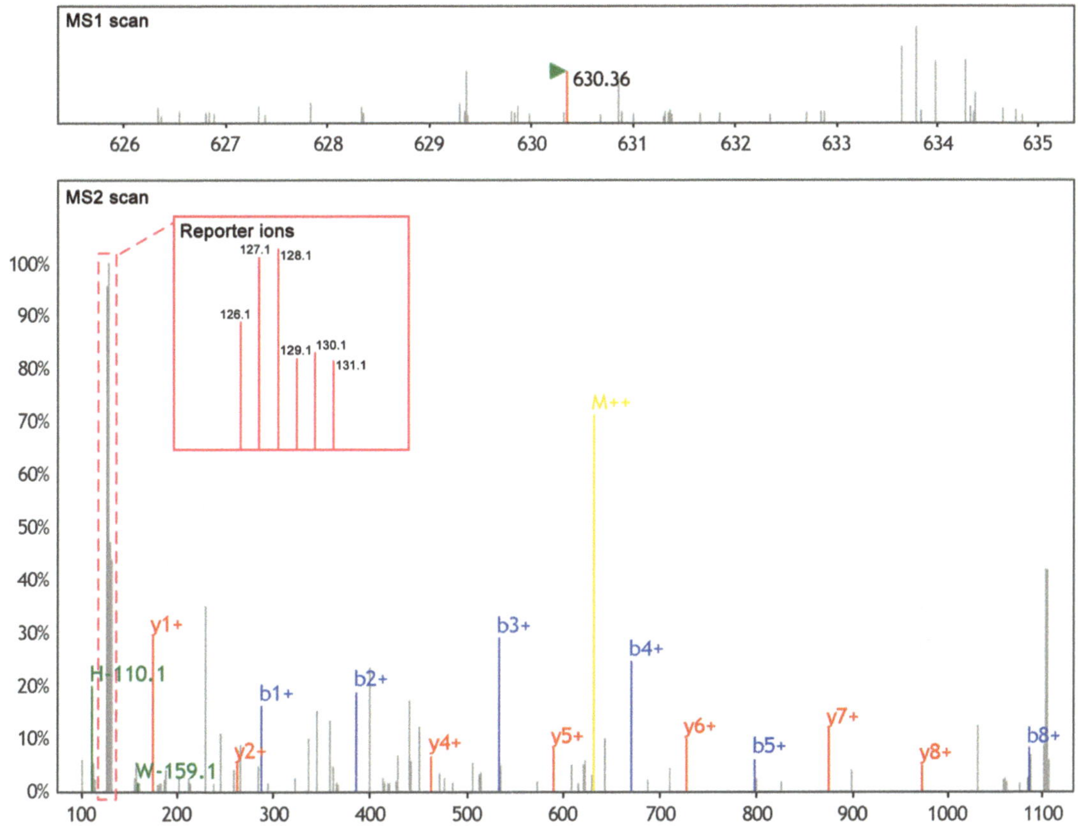

Fig. 15.2 Representative parent and product ion spectra of GVFHQTVSR, one of the four unique peptides of protein P12109 (Gene name: COL6A1, collagen alpha-1(VI) chain)

We used TPP to interpret the search result of HuN9. After loading the Mascot result, the TPP pipeline uses PeptideProphet and ProteinProphet to validate the identification of peptides and proteins [34]. With Libra, each peptide channel was normalized by the sum of peptides' channels. The values that deviated from the average by more than two sigma were removed. The protein level of each sample (labeling) was calculated as the median of normalized values of the corresponding ions of peptides. Using two peptides of protein P12109 (GVFHQTVSR and GDEGPPGSEGAR) as quantifiers, Fig. 15.3 shows the raw intensities of peptides (A), normalized values for peptides (B) and final relative expression levels of protein using individual peptides (C). The dataset HuN9 is designed as a 3 versus 3 sample comparison (Group 1: TMT126-128 samples; Group 2: TMT129-131 samples). This protein shows a significant decrease in group 2 (average fold change 1.7, p value < 0.001).

15.4.6 IsobariQ and Iquant

As isobaric labeling is more efficient in peptide labeling and, thus, is more widely used to find differently expressed proteins between two samples from different physiological or pathological states, the recently developed tools IQuant and IsobariQ are also discussed here, followed by their comparison using the same data set in TPP.

IsobariQ was developed in C++ for the windows platform and released under the GNU General Public License version 3. The statistical language R and the server Rserve must be

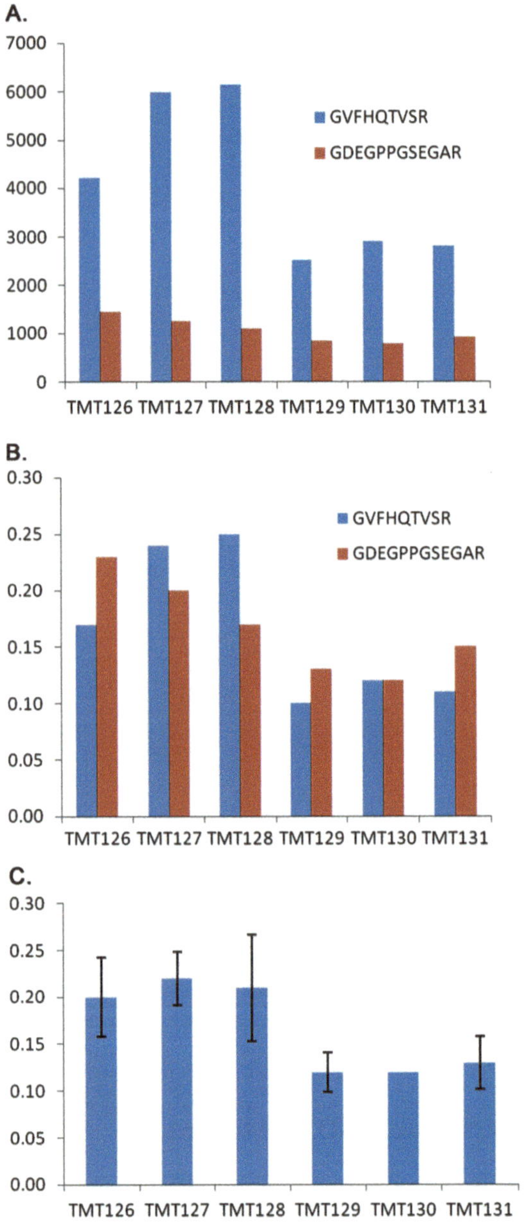

Fig. 15.3 Raw intensities (**a**) and normalized values (**b**) of surrogate peptides and relative quantitative values (**c**) of P12109

installed separately. The user can choose between three different types of normalization: (1) Division by median, (2) Variance stabilizing normalization (VSN), (3) Division by channel sum (reporter ions only). Once all the peptides are successfully quantified and normalized, the protein ratios are calculated as the median of the individual peptide ratios (reporter ions) or pooled standard deviation of all quantification points (IPTL). The user can select which peptides are included in the calculation of protein values. IsobariQ uses both z-statistics, similar to how MaxQuant treats SILAC data, to address the significance of a protein ratio. The Benjamini-Hochberg method is applied for ratio correction.

IQuant is implemented in JAVA and R, provides a GUI as well as a command-line interface, and works on both Windows and Linux system [35]. It integrates Mascot Percolator and advanced statistical algorithms to process the mass data. The abundance of reporter ion can be normalized through VSN and median-based approach. The VSN method has also been adapted in IsobariQ. Non-unique peptides and outlier peptide ratios are removed prior to quantitative calculation. The weight approach is employed to evaluate the ratios of protein quantity based on reporter ion intensities [36]. To estimate the statistical significance of the protein quantitative ratios, IQuant adopts the permutation test, a non-parametric approach. For each protein, IQuant as IsobariQ provides a significance evaluation that is corrected for multiple hypothesis testing by the Benjamini-Hochberg method [37].

Both IQuant and IsobariQ perform the analysis based on the Mascot identification results. They both need the identification result of Mascot ("dat" file) as input. IQuant further needs the fasta file of a sequence database for protein inference. The key steps shared by these two tools in quantitative proteomics are: tag impurity correction, peptide quantification, peptide ratios normalization, and protein quantification. As described above, they both require installation of R software for performing VSN [11]. VSN provides a robust variant of the maximum-likelihood estimator for differential expression, which was originally used for microarray data. Both tools provide peptide normalization based on division by median. And IsobariQ additionally supports normalization by channel sum. For protein quantification, the protein ratios were finally calculated as the median values of the individual peptide ratios in IsobariQ, whereas IQuant employed a weight approach to evaluate the ratios of protein [12].

To further illustrate the different normalization method performances, the previous HuN9 data set was processed. As shown in Table 15.4, the normalization by median-based approach produced similar result to the theoretical prediction in this study. However, VSN is recommended by IQuant as default method to solve the issue of heterogeneity of variance among peaks.

We further compared the quantification results of IsobariQ and IQuant using the example file provided by IQuant (File name: tte-1-1.dat; iTRAQ8Plex labeling). IQuant quantified more proteins than IsobariQ, likely due to its use of MascotPercolator to improve protein identification (Fig. 15.4a). We also calculated the ratios (116/114) of 495 common proteins using VSN and median normalization methods in IQuant and IsobariQ, respectively. The quantification results showed good correlations between these two tools, with a Spearman coefficient of 0.977 (Fig. 15.4b).

There are many excellent software tools available, and there is no consensus on how to calculate protein and peptide abundances. The selection of these tools is partly restricted by the nature of the experiment and available instrumentation, as well as the type of information the end-user is looking for.

15.5 Targeted Proteomics by SRM

Due to the special experimental design and data analysis of targeted proteomics, this approach is discussed individually here. In a targeted analysis for protein quantification, liquid chromatography coupled on-line to SRM (LC/SRM) assays are developed to detect fragment ion signals from proteolytic peptides driven from target proteins [38, 39]. The ion mass of the precursor peptide is filtered through in the first mass analyzer (Q1), while a peptide fragment ion, generated by collision-induced dissociation in Q2, is filtered through in the third mass analyzer (Q3). The precursor/product ion *m*/*z* pair, referred to as the transition, is used to yield the chromatogram. The area under the curve of the chromatogram provides a quantitative measurement for each desired peptide and target protein. As Method of the Year 2012, LC/MS/MS-based targeted proteomics allows researchers to quantify proteins with high sensitivity, high selectivity and wide dynamic ranges [40]. Low-abundance proteins of interest are not ignored as they are in discovery proteomics. If the retention time of the surrogate peptides is used as a constraint in data acquisition (scheduled SRM), several hundred peptides can be quantified during a single LC/MS/MS run [41, 42].

The most critical step in the establishment of a targeted proteomics assay is the selection of proteolytic peptides that (1) are unique to a candidate protein, (2) would ionize efficiently, (3) are completely digested (carry no miscleavages) and (4) can generate high-quality SRM (high S/N). Given these criteria that must be met for each transition, designing SRM assays for a protein can be time-consuming, and the workload increases rapidly as the number of target proteins is increased. To streamline this process, freely available software resources have been developed. There are primarily two opposite approaches, theoretical and experimental, either using *in silico* prediction by various algorithms or based on spectral evidence using existing mass

Table 15.4 Comparison of different methods for peptide normalization

Normalization	Ratio 127/126	Ratio 128/126	Ratio 129/126	Ratio 130/126	Ratio 131/126
NONE	0.95	0.90	0.93	0.96	0.91
SUM (Isobaric Q)	1.01	1.01	1.01	1.01	1.01
Median(Isobaric Q)	1.00	1.00	1.00	1.00	1.00
VSN(IQuant and Isobaric Q)	0.98	1.04	1.00	1.00	1.00

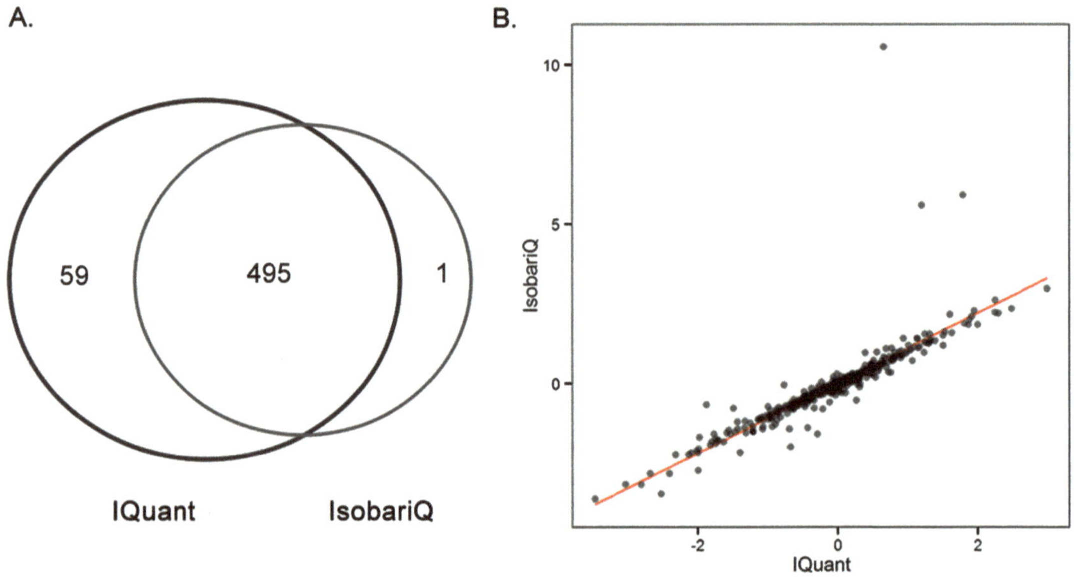

Fig. 15.4 Comparison of the quantification results of IsobariQ and IQuant. (**a**) The number of proteins obtained from each software. (**b**) The correlation of the protein ratios obtained from IQuant and IsobariQ. The PSMs were filtered with a q-value equal to or less than 0.01, and only proteins with equal to or more than two peptides were used for further analysis

Table 15.5 Predictive computational models for peptide selection

Prediction method	Website
ESP predictor	http://www.broadinstitute.org/cancer/software/genepattern/modules/ESPPredictor.htm
STEPP	http://cbb.pnnl.gov/portal/software/stepp.html
Peptide sieve (PAGE-ESI)	http://tools.proteomecenter.org/wiki/index.php?title=Software:PeptideSieve
Peptide detectability	http://darwin.informatics.indiana.edu/applications/PeptideDetectabilityPredictor/

spectral data from either public repositories or in-house experiments (e.g., spectra recorded during global discovery experiments). In a theoretical SRM design, it is possible to predict which peptides and product ions are most appropriate for SRM protein quantification by several computational tools (Table 15.5). However, it should be noted that the mechanisms of proteolysis, ionization, and fragmentation are not yet sufficiently well understood to produce accurate models for best SRM transition predictions. The current models can only assist to select high-responding peptides, particularly in the absence of experimental data. For example, ESP predictor considered 550 physicochemical properties to model the peptide response. For each physicochemical property, it computes the property value by averaging over all amino acids in each peptide. The software simulates a peptide response using the Random Forest algorithm. As reported previously, the ESP predictor can achieve a success rate of 89 % at selecting one or more high-responding peptides per protein on average [43]. There are some empirical criteria for peptide selection, which may be helpful at the primary stage of quantitative proteomics and are listed (Table 15.6).

The experimental approach uses experimentally obtained peptide spectra as evidence, and several software tools have been developed to extract the necessary information from those spectra to build SRM assays. Publicly available spectral repositories include PRIDE, GPMDB, PeptideAtlas and others (Table 15.7). The

Table 15.6 Empirical criteria for peptide selection in SRM

Necessary condition	1. The amino acid sequence of the peptide is unique for a target protein
	2. Length between 6 and 16 amino acids
	3. No posttranslational modifications and no single nucleotide polymorphism
	4. NO methionine or cysteine residues are included
	5. For membrane protein, No transmembrane region
	6. For trypsin digestion, NO continuous sequence of arginine or lysine residues (RR, KK, RK, KR) occurs in the digestion region
	7. For trypsin digestion, the peptide does NOT include a proline residue at the Cterminal side of an arginine or lysine residue (RP or KP) in the digestion region
Additional condition	1. No histidine residue
	2. Containing one of leucine, isoleucine, valine, alanine or proline residue
	3. Hydrophobic amino acids should comprise less than 40% of the peptide

Table 15.7 Public proteomics spectral repositories

Proteomics repositories	Website
PeptideAtlas	http://www.peptideatlas.org/speclib/
GPMDB	ftp://ftp.thegpm.org/projects/xhunter/libs/
PRIDE	http://www.ebi.ac.uk/pride
NIST	http://www.peptideatlas.org/speclib/
MacCoss	http://proteome.gs.washington.edu/software/bibliospec/documentation/libs.html

software tools are Targeted Identification for Quantitative Analysis by Multiple reaction monitoring (TIQAM), MRMer, SRMCollider, MaRiMba, MRMaid, Skyline and ATAQS as listed in Table 15.2, or commercial with the software platforms provided by mass spectrometer vendors (e.g., SRM Workflow software (based on SIEVE), Pinpoint and P3 predictor (Thermo Scientific), mTRAQ-reagent-based MRMPilot software and multiple reaction monitoring initiated detection and sequencing (MIDAS) Workflow Designer (Applied Biosystems), VerifyE and TargetLynx™ Application Manager (Waters), MassHunter Optimizer (Agilent Technologies)) [44]. Commercial software tools are not freely accessible and the algorithms are generally not published. This chapter focuses, therefore, on the freely available, platform-independent informatics resources for SRM transition design.

The initial software packages for SRM assay development were often single-user packages, such as MRMaid and MaRiMba, and they were limited in their specific scope. Newer software packages, such as Skyline and ATAQS, aim to integrate the entire targeted proteomic workflow and may be more comprehensive. We will give a simple example, re-analyzing data from recently published papers from our laboratory to explore the applicability of PeptideAtlas and quantitative capability of Skyline and ATAQS.

15.5.1 PeptideAtlas

Current proteomics repositories have been based on shotgun proteomics data. Databases such as PeptideAtlas are candidates that have the potential to handle SRM data. Features have been added to PeptideAtlas to leverage shotgun data in support of SRM experiment design [17]. An SRM-specific section of PeptideAtlas, SRMAtlas, has been created as a combined catalog of best-available transitions selected from either PeptideAtlas shotgun data, data collected for whole proteome synthetic tryptic peptide libraries [18], published validated transitions, and theoretical transition prediction approaches. SRMAtlas encompasses four levels of information and an algorithm (PeptideAtlas Best-transition Selection Tool (PABST)) to intelligently merge the levels with a weighting and scoring technique to provide ranked lists of peptides and transitions for all proteins for a species. Using these data, generic SRM measurements can be set up for protein quantification. PeptideAtlas SRM Experiment Library (PASSEL) is an active repository for SRM experimental data acquired in real-world studies. Different from SRMAtlas, this repository is specifically designed to store and present SRM experimental data in a publicly accessible manner [45].

Using HSP27 (P02786) recently investigated in our laboratory as an example [46], we can obtain a bulk list of recommended peptides and transitions and the desired user-settable parameters. Each transition was listed with its attributes, including final score and the source of the peptide as depicted in Fig. 15.5. The list may also be constrained to a subset of these classes, for example, to only optimize the transitions selected from a real QQQ spectrum. This table may be optimal for us to use in the quantification of real samples. In our study, the doubly charged ion VSASPLLYTLIEK was most abundant, and the corresponding transition of *m*/*z* 717.2 → 1089.8 had the greatest S/N in LC/SRM. Notably, only proteins with trypsin digestion can be processed using PeptideAtlas in the current version.

It is important to be judicious in the use of tryptic peptides in SRM assay development because public MS/MS spectra databases often lack information about the experimental methods and MS instrumentation used to obtain these spectra [47]. The predicted chromatographic and mass spectrometric behavior of peptides are not always sufficiently accurate to omit the need for experimental verification. While spectra generated on a triple-quadrupole instrument are often preferred, when not available, consensus ion trap spectra are often used as a substitute in many cases [48, 49]. Thus, the ion peak intensity ranking in a library is usually different from that provided by experiments (Table 15.8).

The peptide and transition information of proteins may also be queried programmatically by other software *via* web service interfaces, as described in detail at the SRMAtlas access help page (http://www.srmatlas.org/doc/webServiceAccess.php). This capability enables users to import the results from SRMAtlas directly into SRM bioinformatics tools such as ATAQs and Skyline.

15.5.2 Skyline

Skyline is a Windows client application for targeted proteomics method creation and quantitative data analysis. It is open source and freely available [50]. Skyline can not only establish an initial set of peptides and transitions but can also allow us to further refine and optimize these initial instrument methods after experimental runs.

Skyline supports all major publicly available spectral libraries. New spectral libraries can also be built, for example, post-translational modifications (PTMs) that are unavailable in public libraries. Skyline can support peptide and transition picking both *in silico* and from spectral libraries automatically. Peptide settings include the following: presence or absence of specific residues (including heavy amino acids), enzyme, peptide length, and charge states. Transition settings include the following: collision energy (CE) and declustering (set to instrument vendor-specific values if necessary), product ion *m*/*z* greater than the precursor, and monoisotopic or average masses. Retention time (RT) can also be predicted *ab initio* using a selection of "calculators", such as SSRCalc [51]. Matching spectra are shown in a graph pane with ion peak intensity ranking expressed in both the graph and document tree (Fig. 15.6). Several empirical criteria are valuable and provided for the creation of a new targeted assay, for instance, start with more transitions than required, prefer singly charged y-ions, etc. The resulting list of transitions can be exported in MS-vendor-specific formats, such as Agilent, Thermo, and Waters, so they can easily be scheduled in MS for quantitative monitoring later. Finally, empirical measurements in the experimental context are performed. After acquired SRM data are imported, subsequent method refinement is

ISB Home

SRMAtlas Home | Search | All Builds | Current Build | Queries | SRMAtlas | PTPAtlas | Submission | SWATH/DIA

Query Transitions | Transition Lists | SRMAtlas Builds | PASSEL Experiments | PASSEL Data

Form | Results

Data Access
Search SRM Assays
View SRM Builds

Background
Background
Data Contributors
Publications
Downloads
External Links
Contacts
SRM/MRM Assays
SRM/MRM Glossary

Related Resources
Peptide Atlas
PASSEL
SWATH Atlas
PASS

LOGIN

Institute for Systems Biology

hide column descriptions

Q1/Q3 transition pairs for SRM experiments

Protein	Protein Name/Accession.
External Links	Links to information about this protein in other resources
Pre AA	Preceding (towards the N terminus) amino acid
Sequence	Amino acid sequence of detected pepide, including any mass modifications.
Fol AA	Following (towards the C terminus) amino acid
Adj SS	Final suitablity score, the greater of ESS and PSS, adjusted by PABST weightings.
Source	Source from which transitions were obtained
Q1_mz	Mass to charge ratio of Q1 (precursor) peptide ion.
Q1_chg	Charge on Q1 (precursor) peptide ion.
Q3_mz	Mass to charge ratio of Q3 (fragment) ion.
Q3_chg	Charge on Q3 (fragment) ion.
Ion	Ion-series designation for fragment ion (Q3).
Rank	PABST Transition rank
RI	Relative Intensity of peak in CID spectrum
SSRT	Sequence Specific Retention time provides a hydrophobicity measure for each peptide using the algorithm of Krohkin et al. Version 3.0 [more]
N_map	Number of proteins in target proteome to which peptide maps
IonTrap	Consensus spectrum from Ion Trap instrument(s)

Protein	External Links	Pre AA	Sequence	Fol AA	Adj SS	Source	Q1_mz	Q1_chg	Q3_mz	Q3_chg	Ion	Rank	RI	SSR
P02786		R	VEYHFLSPYVSPK	E	1.74	IonTrap	522.61	3	345.70	2	y6	1	10000	32.0
P02786		R	SAFSNLFGGEPLSYTR	F	1.65	IonTrap	873.43	2	736.40	1	y6	1	10000	37.6
P02786		K	GFVEPDHYVVVGAQR	D	1.62	IonTrap	836.93	2	1240.64	1	y11	1	10000	29.4
P02786		R	HVFWGSGSHTLPALLENLK	L	1.61	IonTrap	702.71	3	935.50	2	y17	1	10000	44.2
P02786		K	LAVDEEENADNNTK	A	1.59	IonTrap	781.35	2	591.27	1	y5	1	8471	10.6
P02786		K	VSASPLLYTLIEK	T	1.57	IonTrap	717.42	2	1089.66	1	y9	1	10000	41.2
P02786		K	HPVTGQFLYQDSNWASK	V	1.41	IonTrap	659.99	3	872.91	2	b15	1	10000	32.0
P02786		R	LTTDFGNAEK	T	1.40	IonTrap	548.27	2	881.40	1	y8	1	2491	15.5
P02786		R	AAAEVAGQFVIK	L	1.24	IonTrap	602.34	2	691.41	1	y6	1	10000	29.4
P02786		K	LTHDVELNLDYER	Y	1.23	IonTrap	808.90	2	701.83	2	y11	1	10000	28.9
P02786		R	LVYLVENPGGYVAYSK	A	1.16	IonTrap	886.47	2	941.47	1	y9	1	10000	32.3
P02786		R	SSGLPNIPVQTISR	A	1.16	IonTrap	734.91	2	800.46	1	y7	1	10000	27.2
P02786		R	YNSQLLSFVR	D	1.09	IonTrap	613.83	2	508.29	1	y4	1	8396	34.3
P02786		K	LAQMFSDMVLK	D	1.05	IonTrap	641.83	2	839.43	1	y7	1	10000	36.7
P02786		K	DENLALYVENQFR	E	1.02	IonTrap	805.89	2	955.46	1	y7	1	10000	30.9
P02786		K	VANAESLNAIGVLIYMDQTK	F	0.99	IonTrap	1075.56	2	1167.61	1	y10	1	10000	42.8
P02786		K	DSAQNSVIIVDK	N	0.98	IonTrap	644.84	2	474.29	1	y4	1	7919	22.0
P02786		K	ILNIFGVIK	G	0.97	IonTrap	508.83	2	790.48	1	y7	1	10000	40.4
P02786		R	QVDGDNSHVEMK	L	0.91	IonTrap	679.80	2	508.73	2	y9	1	411	10.3
P02786		K	EMGLSLQWLYSAR	G	0.85	IonTrap	777.39	2	496.25	1	y4	1	10000	38.2
P02786		K	LLNENSYVPR	E	0.83	IonTrap	602.82	2	272.17	1	y2	1	10000	21.0
P02786		K	LVHANFGTK	K	0.82	IonTrap	986.54	1	682.37	1	b6	1	5890	17.4
P02786		K	LDSTDFTGTIK	L	0.75	IonTrap	599.30	2	969.49	1	y9	1	2899	22.7
P02786		K	AVLGTSNFK	V	0.73	IonTrap	468.76	2	766.41	1	y7	1	10000	19.4
P02786		K	AFTYINLDK	A	0.72	IonTrap	542.79	2	866.46	1	y7	1	1037	27.1
P02786		K	SGVGTALLLK	L	0.71	IonTrap	479.80	2	715.47	1	y7	1	5815	29.4
P02786		R	EEPGEDFPAAR	R	0.67	IonTrap	609.28	2	480.23	2	y9	1	10000	16.2
P02786		R	DAWGPGAAK	S	0.20	IonTrap	872.43	1	757.40	1	y8	1	7533	13.5
P02786		R	LYWDDLK	R	0.10	IonTrap	476.74	2	676.33	1	y5	1	10000	29.2
P02786		R	NQLALATWTIQGAANALSGDVWDIDNEF	-	0.08	IonTrap	1517.23	2	1296.54	1	y11	1	10000	56.0
P02786		R	SIIFASWSAGDFGSVGATEWLEGYLSSLHLK	A	0.07	IonTrap	1110.22	3	1183.59	2	y22	1	10000	59.5
P02786		K	LTVSNVLK	E	0.06	IonTrap	437.27	2	659.41	1	y6	1	9391	21.2
P02786		R	MVTSESK	N	0.02	IonTrap	781.38	1	551.27	1	y5	1	6626	5.1

[Show fewer rows]

Fig. 15.5 Result of PABST processing for HSP27. The list of peptides is provided in reverse sorted order with the best peptides appearing first, i.e., those with the highest value in the "Adj SS" column

carried out based on these results to achieve a highly effective instrument method.

Skyline fully supports protein quantification, with dialogs for defining static and heavy isotope modifications (Fig. 15.7) and assigning them broadly or explicitly to individual peptides. After importing the results files, Skyline calculates ratios between the unlabeled peptide

Table 15.8 Comparison of the Atlas library rank and the experimental rank of transitions for the peptide of QLSSGVSEIR

Compound group	Compound name	Precursor ion	Product ion	Ion name	Library rank	Experimental rank
sp|P04792| HSPB1_HUMAN	QLSSGVSEIR	538.3	834.4	y6	2	**3**
sp|P04792| HSPB1_HUMAN	QLSSGVSEIR	538.3	660.4	y5	5	**4**
sp|P04792| HSPB1_HUMAN	QLSSGVSEIR	538.3	504.3	y4	1	**1**
sp|P04792| HSPB1_HUMAN	QLSSGVSEIR	538.3	417.2	y3	3	**5**
sp|P04792| HSPB1_HUMAN	QLSSGVSEIR	538.3	288.2	b5	4	**2**

The data were obtained using an Agilent Series 1200 HPLC system (Agilent Technologies, Waldbronn, Germany) and a 6410 Triple Quad LC/MS mass spectrometer (Agilent Technologies, Santa Clara, CA, USA)

and the labeled internal standard and provides direct editing of integration boundaries [50]. The comma separated value format can also be obtained for further analysis with statistical tools such as Excel and R.

15.5.3 ATAQS (Automated and Targeted Analysis with Quantitative SRM)

ATAQS is an integrated software platform that supports all stages of targeted, SRM-based proteomics experiments including target selection, transition optimization and post acquisition data analysis [52]. ATAQS is written in Java and provides a graphic user interface for the popular browser Mozilla Firefox. Different from Skyline, which is a desktop application with manual inspection for validation, ATAQS provides modules with algorithms that collectively support all steps of the SRM assay development and deployment workflow for targeted proteomic experiments. For example, mProphet in ATAQS can provide which transition group has higher validated score. ATAQS can be easily extended and customized by the user with the addition of user-implemented algorithms at any of the workflow steps. The software uses FireGoose to connect to various Web services [53]. Among these Web services are PeptideAtlas, TIQAM, PIPE2 (to generate a list of proteins to design a SRM assay as well as for various analysis tasks), and PABST. A peptide transition list was generated using PABST based on user-defined criteria (Fig. 15.8).

Because the ATAQS software is primarily designed to support multiple users at an institution, several points deserve attention. (1) Because a number of algorithms were implemented in ATAQS and it is an institution-wide computing resource, the installation requirement is higher compared to the others (e.g., Tomcat v 6 .0.26 or higher, Java 1.6, Ant v1.7.1, MySQL v 5.0.77, Firefox 3.6.x, Firegoose-0.8.259.xpi, Adobe Flash Player 10, R 2.11.1). (2) The protein identification could be more reliable. As stated by the developers, Skyline uses a single score (a hydrophobicity value from SSRCalc) to provide confidence in identification, whereas ATAQs performs two sequential selections based on two separate peptide detectability algorithms (Peptide Sieve and Peptide Detectability Predictor) and user-defined criteria (e.g., number of amino acids (peptide length), amino acid composition and uniqueness of sequence (does not map to more than one protein or one region in the genome). (3) Peak signals in ATAQS are smoothed by discrete Fourier transformation and integrated using mQuest, compared to Skyline using CRAWDAD algorithms for chromatogram retention time alignment, warping and peak integration [49]. (4) ATAQS requires the experimenters to convert the file to

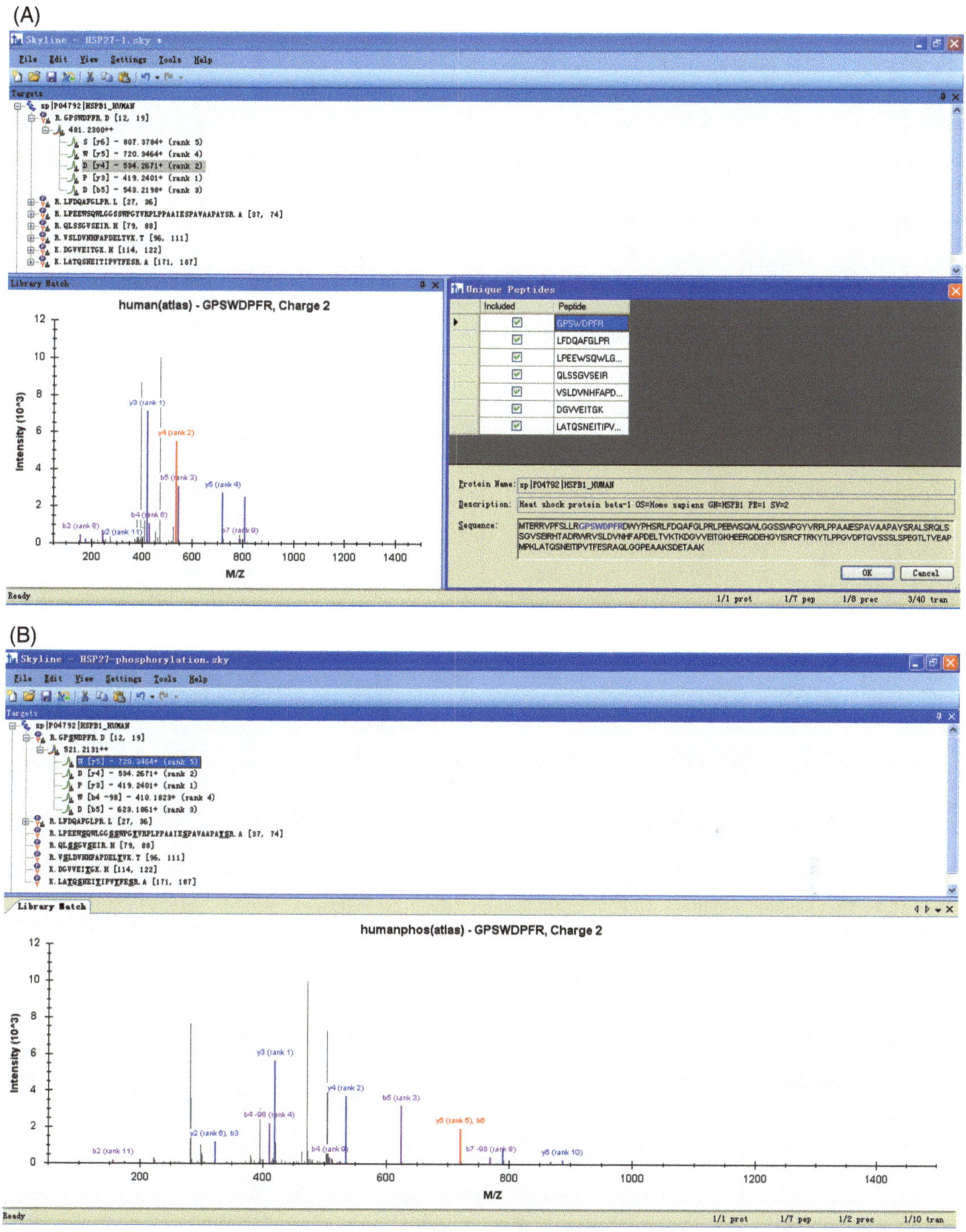

Fig. 15.6 The Skyline spectral library explorer showing spectral views of HSP27 (**a**) and its phosphorylation (**b**)

(A)

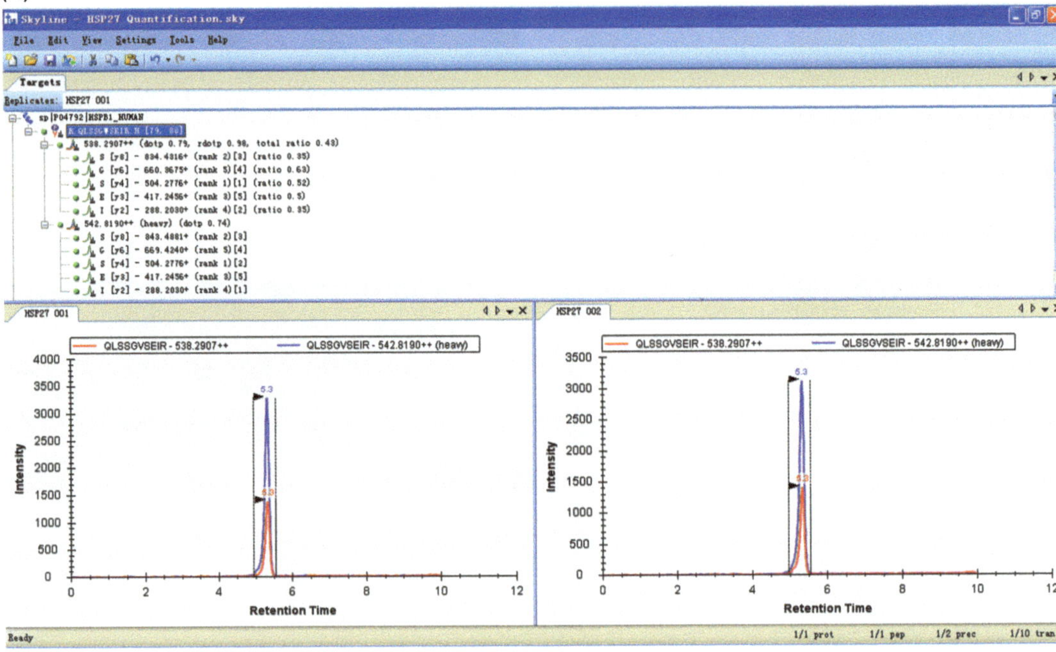

(B)

PeptideSequence	ProteinName	ReplicateName	PeptideRetentionTime	RatioLightToHeavy	Transition	AreaRatio
QLSSGVSEIR	sp|P04792|HSPB1_HUMAN	HSP27 S1	5.33	0.4262	S - y8+	0.3263+/-0.0303
QLSSGVSEIR	sp|P04792|HSPB1_HUMAN	HSP27 S1	5.33	0.4262	G - y6+	0.6576+/-0.0398
QLSSGVSEIR	sp|P04792|HSPB1_HUMAN	HSP27 S1	5.33	0.4262	S - y4+	0.4792+/-0.0513
QLSSGVSEIR	sp|P04792|HSPB1_HUMAN	HSP27 S1	5.33	0.4262	E - y3+	0.4818+/-0.0242
QLSSGVSEIR	sp|P04792|HSPB1_HUMAN	HSP27 S1	5.33	0.4262	I - y2+	0.3766+/-0.0354
QLSSGVSEIR	sp|P04792|HSPB1_HUMAN	HSP27 S2	5.33	0.8842	S - y8+	0.6836+/-0.0381
QLSSGVSEIR	sp|P04792|HSPB1_HUMAN	HSP27 S2	5.33	0.8842	G - y6+	1.3461+/-0.0781
QLSSGVSEIR	sp|P04792|HSPB1_HUMAN	HSP27 S2	5.33	0.8842	S - y4+	0.9876+/-0.0768
QLSSGVSEIR	sp|P04792|HSPB1_HUMAN	HSP27 S2	5.33	0.8842	E - y3+	0.9947+/-0.0452
QLSSGVSEIR	sp|P04792|HSPB1_HUMAN	HSP27 S2	5.33	0.8842	I - y2+	0.7836+/-0.0701

Fig. 15.7 The Skyline result of HSP27 quantification using the surrogate peptide QLSSGVSEIR and the corresponding [D_8]Val isotope-labeled internal standard. Spectral view (**a**) and exported area ratio result (**b**) are provided

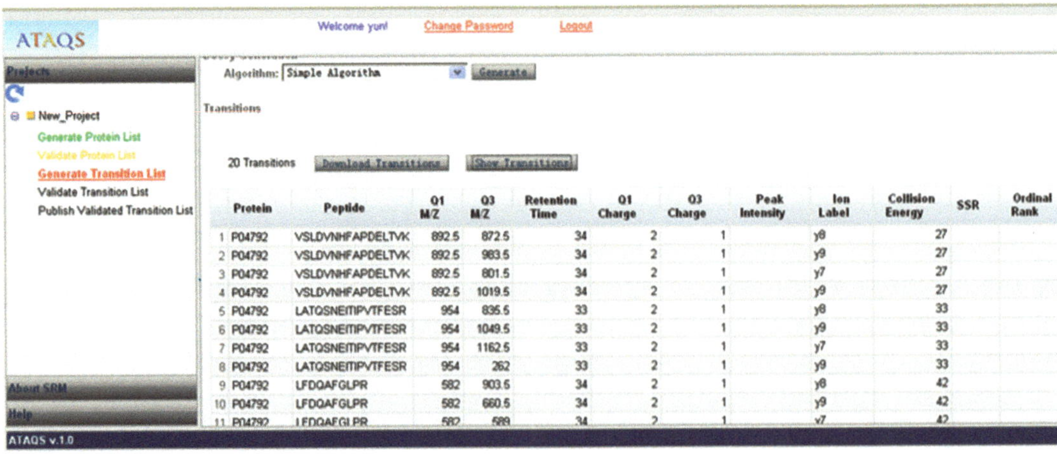

Fig. 15.8 The generated transition list in ATAQS

an open source format such as TraML, mzXML or mzML, whereas Skyline is the only open source program that can read all native vendor file formats.

15.6 Conclusions

As techniques for quantitative proteomics continue to grow, bioinformatics software tools are similarly expanding in number. Proteomics researchers are now faced with many tools to choose from, all with different advantages and disadvantages. Software that is developed for a particular type of mass spectrometer/method may be inadvertently or intentionally optimized for data from that instrument or using that proteomics approach and may be less well suited for more general use. Sometimes, the result is also easily influenced by the familiarity and expertise of the performers with the programs being processed. Thus, we cannot claim which tool is better. Currently, the rational way in quantitative proteomics is to select bioinformatics tools optimally suited to address the specific proteomics issue under consideration and the associated information.

References

1. Wang Y, Li H, Chen S (2010) Advances in quantitative proteomics. Front Biol 5:195–203
2. Jungblut PR (2014) The proteomics quantification dilemma. J Proteomics 107:98–102
3. Doerr A (2010) Targeted proteomics. Nat Methods 7:837–842
4. Nogueira FC, Palmisano G, Schwammle V, Campos FA, Larsen MR, Domont GB et al (2012) Performance of isobaric and isotopic labeling in quantitative plant proteomics. J Proteome Res 11:3046–3052
5. Yocum AK, Chinnaiyan AM (2009) Current affairs in quantitative targeted proteomics: multiple reaction monitoring-mass spectrometry. Brief Funct Genomic Proteomic 8:145–157
6. Colangelo CM, Chung L, Bruce C, Cheung KH (2013) Review of software tools for design and analysis of large scale MRM proteomic datasets. Methods 61:287–298
7. Patterson SD, Aebersold RH (2003) Proteomics: the first decade and beyond. Nat Genet 33 (Suppl):311–323
8. Lau KW, Jones AR, Swainston N, Siepen JA, Hubbard SJ (2007) Capture and analysis of quantitative proteomic data. Proteomics 7:2787–2799
9. Gonzalez-Galarza FF, Lawless C, Hubbard SJ, Fan J, Bessant C, Hermjakob H et al (2012) A critical appraisal of techniques, software packages, and standards for quantitative proteomic analysis. Omics 16:431–442
10. Lemeer S, Hahne H, Pachl F, Kuster B (2012) Software tools for MS-based quantitative proteomics: a brief overview. Methods Mol Biol 893:489–499
11. Park SK, Venable JD, Xu T, Yates JR 3rd (2008) A quantitative analysis software tool for mass spectrometry-based proteomics. Nat Methods 5:319–322
12. Matthiesen R, Datta S. Feature selection and machine learning with mass spectrometry data. In: Mass spectrometry data analysis in proteomics. Humana Press, pp 237–262
13. Hilario M, Kalousis A, Pellegrini C, Muller M (2006) Processing and classification of protein mass spectra. Mass Spectrom Rev 25:409–449
14. Matthiesen R, Matthiesen R. LC-MS spectra processing. In: Mass spectrometry data analysis in proteomics. Humana Press, pp 47–63
15. Mortensen P, Gouw JW, Olsen JV, Ong SE, Rigbolt KT, Bunkenborg J et al (2010) MSQuant, an open source platform for mass spectrometry-based quantitative proteomics. J Proteome Res 9:393–403
16. Azuaje F, Dopazo J (2005) Integrative data analysis and visualization: introduction to critical problems, goals and challenges. In: Data analysis and visualization in genomics and proteomics. Wiley, Hoboken, pp 1–9
17. Yang C, He Z, Yu W (2009) Comparison of public peak detection algorithms for MALDI mass spectrometry data analysis. BMC Bioinf 10:4
18. Matthiesen R (2007) Methods, algorithms and tools in computational proteomics: a practical point of view. Proteomics 7:2815–2832
19. Ryu SY (2014) Bioinformatics tools to identify and quantify proteins using mass spectrometry data. Adv Protein Chem Struct Biol 94:1–17
20. Matthiesen R, Azevedo L, Amorim A, Carvalho AS (2011) Discussion on common data analysis strategies used in MS-based proteomics. Proteomics 11:604–619
21. Vandenbogaert M, Li-Thiao-Te S, Kaltenbach HM, Zhang R, Aittokallio T, Schwikowski B (2008) Alignment of LC-MS images, with applications to biomarker discovery and protein identification. Proteomics 8:650–672
22. Lu B, Xu T, Park SK, McClatchy DB, Liao L, Yates JR 3rd (2009) Shotgun protein identification and quantification by mass spectrometry in neuroproteomics. Methods Mol Biol 566:229–259
23. MacCoss MJ (2005) Computational analysis of shotgun proteomics data. Curr Opin Chem Biol 9:88–94

24. Li XJ, Zhang H, Ranish JA, Aebersold R (2003) Automated statistical analysis of protein abundance ratios from data generated by stable-isotope dilution and tandem mass spectrometry. Anal Chem 75:6648–6657
25. Pan C, Kora G, McDonald WH, Tabb DL, VerBerkmoes NC, Hurst GB et al (2006) ProRata: a quantitative proteomics program for accurate protein abundance ratio estimation with confidence interval evaluation. Anal Chem 78:7121–7131
26. Nesvizhskii AI, Aebersold R (2005) Interpretation of shotgun proteomic data: the protein inference problem. Mol Cell Proteomics 4:1419–1440
27. Cox J, Mann M (2008) MaxQuant enables high peptide identification rates, individualized p.p.b.-range mass accuracies and proteome-wide protein quantification. Nat Biotechnol 26:1367–1372
28. Jagtap P, Bandhakavi S, Higgins L, McGowan T, Sa R, Stone MD et al (2012) Workflow for analysis of high mass accuracy salivary data set using MaxQuant and ProteinPilot search algorithm. Proteomics 12:1726–1730
29. Qi D, Brownridge P, Xia D, Mackay K, Gonzalez-Galarza FF, Kenyani J et al (2012) A software toolkit and interface for performing stable isotope labeling and top3 quantification using Progenesis LC-MS. Omics 16:489–495
30. Braisted JC, Kuntumalla S, Vogel C, Marcotte EM, Rodrigues AR, Wang R et al (2008) The APEX quantitative proteomics tool: generating protein quantitation estimates from LC-MS/MS proteomics results. BMC Bioinf 9:529
31. Pedrioli PG (2010) Trans-proteomic pipeline: a pipeline for proteomic analysis. Methods Mol Biol 604:213–238
32. Deutsch EW, Mendoza L, Shteynberg D, Farrah T, Lam H, Tasman N et al (2010) A guided tour of the trans-proteomic pipeline. Proteomics 10:1150–1159
33. Keller A, Eng J, Zhang N, Li XJ, Aebersold R (2005) A uniform proteomics MS/MS analysis platform utilizing open XML file formats. Mol Syst Biol 1:2005.0017
34. Nesvizhskii AI, Keller A, Kolker E, Aebersold R (2003) A statistical model for identifying proteins by tandem mass spectrometry. Anal Chem 75:4646–4658
35. Wen B, Zhou R, Feng Q, Wang Q, Wang J, Liu S (2014) IQuant: an automated pipeline for quantitative proteomics based upon isobaric tags. Proteomics 14:2280–2285
36. Breitwieser FP, Muller A, Dayon L, Kocher T, Hainard A, Pichler P et al (2011) General statistical modeling of data from protein relative expression isobaric tags. J Proteome Res 10:2758–2766
37. Benjamini Y, Drai D, Elmer G, Kafkafi N, Golani I (2001) Controlling the false discovery rate in behavior genetics research. Behav Brain Res 125:279–284
38. Doerr A (2011) Targeted proteomics. Nat Methods 8:43
39. Kelter G, Steinbach D, Konkimalla VB, Tahara T, Taketani S, Fiebig HH et al (2007) Role of transferrin receptor and the ABC transporters ABCB6 and ABCB7 for resistance and differentiation of tumor cells towards artesunate. PLoS One 2, e798
40. Aebersold R (2013) Method of the year 2012. Nat Methods 10:1
41. Lange V, Malmstrom JA, Didion J, King NL, Johansson BP, Schafer J et al (2008) Targeted quantitative analysis of Streptococcus pyogenes virulence factors by multiple reaction monitoring. Mol Cell Proteomics 7:1489–1500
42. Kiyonami R, Schoen A, Prakash A, Peterman S, Zabrouskov V, Picotti P et al (2011) Increased selectivity, analytical precision, and throughput in targeted proteomics. Mol Cell Proteomics 10:M110.002931
43. Fusaro VA, Mani DR, Mesirov JP, Carr SA (2009) Prediction of high-responding peptides for targeted protein assays by mass spectrometry. Nat Biotechnol 27:190–198
44. Cham Mead JA, Bianco L, Bessant C (2010) Free computational resources for designing selected reaction monitoring transitions. Proteomics 10:1106–1126
45. Farrah T, Deutsch EW, Kreisberg R, Sun Z, Campbell DS, Mendoza L et al (2012) PASSEL: the PeptideAtlas SRMexperiment library. Proteomics 12:1170–1175
46. Xu F, Yang T, Fang D, Xu Q, Chen Y (2014) An investigation of heat shock protein 27 and P-glycoprotein mediated multi-drug resistance in breast cancer using liquid chromatography-tandem mass spectrometry-based targeted proteomics. J Proteomics 108:188–197
47. Proc JL, Kuzyk MA, Hardie DB, Yang J, Smith DS, Jackson AM et al (2010) A quantitative study of the effects of chaotropic agents, surfactants, and solvents on the digestion efficiency of human plasma proteins by trypsin. J Proteome Res 9:5422–5437
48. Sherwood CA, Eastham A, Lee LW, Risler J, Vitek O, Martin DB (2009) Correlation between y-type ions observed in ion trap and triple quadrupole mass spectrometers. J Proteome Res 8:4243–4251
49. Prakash A, Tomazela DM, Frewen B, Maclean B, Merrihew G, Peterman S et al (2009) Expediting the development of targeted SRM assays: using data from shotgun proteomics to automate method development. J Proteome Res 8:2733–2739
50. MacLean B, Tomazela DM, Shulman N, Chambers M, Finney GL, Frewen B et al (2010) Skyline: an open source document editor for creating and analyzing targeted proteomics experiments. Bioinformatics 26:966–968
51. Krokhin OV, Craig R, Spicer V, Ens W, Standing KG, Beavis RC et al (2004) An improved model for prediction of retention times of tryptic peptides in ion pair reversed-phase HPLC: its application to protein peptide mapping by off-line HPLC-MALDI MS. Mol Cell Proteomics 3:908–919

52. Brusniak MY, Kwok ST, Christiansen M, Campbell D, Reiter L, Picotti P et al (2011) ATAQS: a computational software tool for high throughput transition optimization and validation for selected reaction monitoring mass spectrometry. BMC Bioinf 12:78

53. Shannon PT, Reiss DJ, Bonneau R, Baliga NS (2006) The Gaggle: an open-source software system for integrating bioinformatics software and data sources. BMC Bioinf 7:176

Bioinformatics Tools for Proteomics Data Interpretation

16

Karla Grisel Calderón-González, Jesús Hernández-Monge, María Esther Herrera-Aguirre, and Juan Pedro Luna-Arias

Abstract

Biological systems function via intricate cellular processes and networks in which RNAs, metabolites, proteins and other cellular compounds have a precise role and are exquisitely regulated (Kumar and Mann, FEBS Lett 583(11):1703–1712, 2009). The development of high-throughput technologies, such as the Next Generation DNA Sequencing (NGS) and DNA microarrays for sequencing genomes or metagenomes, have triggered a dramatic increase in the last few years in the amount of information stored in the GenBank and UniProt Knowledgebase (UniProtKB). GenBank release 210, reported in October 2015, contains 202,237,081,559 nucleotides corresponding to 188,372,017 sequences, whilst there are only 1,222,635,267,498 nucleotides corresponding to 309,198,943 sequences from Whole Genome Shotgun (WGS) projects. In the case of UniProKB/Swiss-Prot, release 2015_12 (December 9, 2015) contains 196,219,159 amino acids that correspond to 550,116 entries. Meanwhile, UniProtKB/TrEMBL (release 2015_12 of December 9 2015) contains 1,838,851,8871 amino acids corresponding to 555,270,679 entries. Proteomics has also improved our knowledge of proteins that are being expressed in cells at a certain time of the cell cycle. It has also allowed the identification of molecules forming part of multiprotein complexes and an increasing number of posttranslational modifications (PTMs) that are present in proteins, as well as the variants of proteins expressed.

K.G. Calderón-González • M.E. Herrera-Aguirre
J.P. Luna-Arias (✉)
Departamento de Biología Celular, Centro de Investigación y de Estudios Avanzados del Instituto Politécnico Nacional (Cinvestav-IPN), Av. Instituto Politécnico Nacional 2508, Col. San Pedro Zacatenco, Gustavo A. Madero, C.P. 07360 Ciudad de México, Mexico
e-mail: jpluna@cell.cinvestav.mx; jpluna@cinvestav.mx; jpluna2003@gmail.com

J. Hernández-Monge
Instituto de Física, Universidad Autónoma de San Luis Potosí, Av. Manuel Nava 6, Zona Universitaria, C.P. 78290 San Luis Potosí, S.L.P., Mexico

H. Mirzaei and M. Carrasco (eds.), *Modern Proteomics – Sample Preparation, Analysis and Practical Applications*, Advances in Experimental Medicine and Biology 919, DOI 10.1007/978-3-319-41448-5_16

Keywords
Proteomics data interpretation • Interactome mapping • Gene Ontology • STRING • MINT • IntAct • HPRD • BioGRID • PIPs • MPIDB • TAIR • PANTHER • DAVID • KEGG • IPA

Biological systems function via intricate cellular processes and networks in which RNAs, metabolites, proteins and other cellular compounds have a precise role and are exquisitely regulated [1]. The development of high-throughput technologies, such as the Next Generation DNA Sequencing (NGS) and DNA microarrays for sequencing genomes or metagenomes, have triggered a dramatic increase in the last few years in the amount of information stored in the GenBank and UniProt Knowledgebase (UniProtKB). GenBank release 210, reported in October 2015, contains 202,237,081,559 nucleotides corresponding to 188,372,017 sequences, whilst there are only 1,222,635,267,498 nucleotides corresponding to 309,198,943 sequences from Whole Genome Shotgun (WGS) projects. In the case of UniProKB/Swiss-Prot, release 2015_12 (December 9, 2015) contains 196,219,159 amino acids that correspond to 550,116 entries. Meanwhile, UniProtKB/TrEMBL (release 2015_12 of December 9 2015) contains 1,838,851,8871 amino acids corresponding to 555,270,679 entries. Proteomics has also improved our knowledge of proteins that are being expressed in cells at a certain time of the cell cycle. It has also allowed the identification of molecules forming part of multiprotein complexes and an increasing number of post-translational modifications (PTMs) that are present in proteins, as well as the variants of proteins expressed.

Considering that human cells contain between 20,000 and 30,000 protein-encoding genes and possibility that there could be approximately four alternative splice variants for each gene [2], the total number of proteins that could be expressed at a certain time would range between 80,000 and 120,000. Moreover, guessing four PTMs in each protein, then, the total number of proteins in a cell would range between 320,000 and 480,000. However, when we consider the more than 400 different PTMs that have been found [3] the number of proteins in a cell would easily grow to more than one million.

Proteins do not function alone; they usually carry their function by interacting with one or more partners. The main goal of the protein-protein interaction map is to catalogue interactions and to define the interactome. These interactions are currently determined using a vast array of technologies, including yeast two hybrid systems, tag-fusion proteins for the identification of interacting proteins, co-immunoprecipitation, chemical crosslinking, phage display, FRET (Fluorescence Resonance Energy Transfer), SPR (Surface Plasmon Resonance), tandem affinity purification, protein microarrays, protein domains, etc. Many of these techniques, if not all, use mass spectrometry and non-redundant gene and protein databases as the main tools for the identification of peptides and proteins. Many of the cellular protein-protein interaction networks have been catalogued and a number of interactome databases have been established. There are several protein-protein interaction databases freely available via World Wide Web that can be used to determine the putative functions of a protein based on its direct or indirect interactions. Protein-protein interaction maps in these databases are, in general, based on the information published, mostly in PubMed. In this section, we describe some of the most important databases available, including STRING, MINT, IntAct, HPRD, BioGRID, PIPs, MPIDB and TAIR. Furthermore, additional tools such as

Gene Ontology, PANTHER, DAVID, KEGG, and IPA, among others, have been developed to facilitate data mapping into these databases. We are certain that these tools will be useful in understanding the intricate interactions and functions of proteins in cells.

16.1 Gene Ontology

Many proteins are conserved through evolution and consequently share the same functions. However, the systems of nomenclature for genes and proteins stay divergent despite repeated evaluation of gene similarities by experts [4]. In order to tackle this challenge, the Gene Ontology (GO) consortium was created. The aim of the GO project is to provide a structured vocabulary to define specific biological domains that describe gene products in different organisms [5]. GO project began in 1998 as a collaborative effort between three organism databases: FlyBase (*Drosophila*), the Mouse Genome Informatics (MIG) project and the *Saccharomyces* Genome Database (SGD). The GO Consortium has been continuously growing due to the deposition of several animal, microbial and plant genome databases [6], as well as the recent addition of ontology areas, such as cell cycle and cilia-related terms, as well as multicellular organism processes [7]. By using these ontologies, it is possible to graph structures that comprise cellular components, molecular functions, biological processes, and the relationships between them in a species-independent manner [7]. In other words, GO is divided in two modules, the ontologies, called GO ontology, which includes defined terms and their relationships, and the GO annotations, which covers gene products and defined terms [8]. The GO annotation is generated either by a curator or automatically through predictive methods (95 % by this method).

The gene ontology relationships are developed like a tree, depicting a hierarchy from more general terms to more specific ones. Terms are linked by three possible relationships: "is_a", "part_of", and "positively regulates/negatively regulates". The "is_a" is a simple relationship between a class and a subclass. The "part_of" relationship is more complex than the former. C is part of D means that whenever C is present, it always belongs to D; for instance, an organelle (C) is always part of a cell (D), but not all cells have the same organelles. In the GO website (http://geneontology.org), a variety of browsers provide visualization and query capabilities for GO. For example, the AMIGO browser provides a web interface for searching and displaying ontologies, term definitions and associated annotated gene products for diverse organism databases [6]. The GO Online SQL (Structured Query Language) Environment (GOOSE) for AmiGO 2, allows users to freely enter SQL queries in the GO database. On the other hand, the PANTHER Classification System, that is further described next, provides enrichment analysis tools for GO.

16.2 PANTHER

PANTHER (Protein ANalysis THrough Evolutionary Relationships) is a classification system that combines ontology, gene function, pathways and statistical tools. This classification system can analyze sequencing, gene expression, and proteomics data [9]. PANTHER is a large database of gene families developed as a resource for family and subfamily classification of proteins [10]. PANTHER has two main components: PANTHER library (PANTHER/LIB) and PANTHER index (PANTHER/X). PANTHER library is a collection of protein families and subfamilies represented as phylogenetic trees assembled using Hidden Markov statistical models (HMMs) and a multiple sequence alignment algorithm (MSA) (Fig. 16.1a) [9–12]. PANTHER index is a set of ontological abbreviated terms that describe the function of proteins in biological processes or molecular functions [10–12]. In addition,

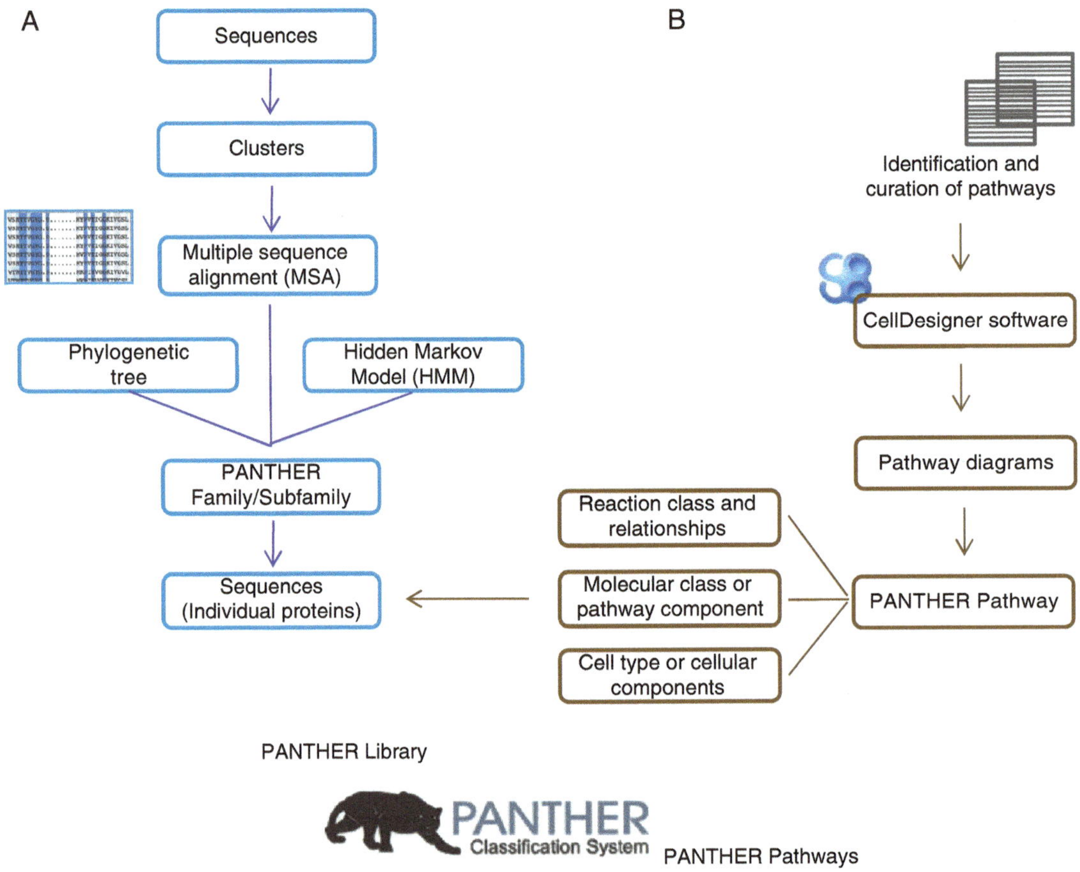

Fig. 16.1 PANTHER data overview. PANTHER has two main modules: (**a**) PANTHER Library which is a collection of families and subfamilies of proteins. This library is constructed from a selection of sequences built into clusters. These clusters are then used to generate multiple sequence alignments (MSA), phylogenetic trees, and statistical HMMs. (**b**) PANTHER Pathways are built using literature databases related to pathway components or a particular molecular class. Then, pathways are drawn and curated by expert curators using the CellDesigner software. Pathways are built based on molecular class or pathway component, reaction class and relationships, and cell type or cellular components. The pathway component is a link between various PANTHER modules

PANTHER has a Pathway module, in which the pathways are represented as a diagram generated with CellDesigner software (Fig. 16.1b) [13]. This module uses a defined vocabulary to describe pathways and their components, including pathway class and components, molecular class, reaction class, reaction relationships, cell type, and cellular components [14, 15]. PANTHER pathways are related to protein sequences in the PANTHER/LIB and, therefore, are also connected with families/subfamilies and HMM analysis (Fig. 16.1) [9, 10, 12]. Pathways are created and annotated by expert curators, according to evidence found in the literature. Moreover, pathways can be curated with the Pathway curation software (http://curation.pantherdb.org/) [14, 15]. Some of the pathways included in the PANTHER database are Cell cycle, DNA replication, General transcription regulation, Glycolysis, Tricarboxylic acid cycle, among others (http://www.pantherdb.org/pathway/

pathwayList.jsp). The PANTHER database contains the following information:

1. Genes (104 genomes; 1,424,953 total genes; 1,026,421 genes in PANTHER families with phylogenetic trees, MSA and HMMs)
2. Families (11,928 families and 83,190 subfamilies)
3. Pathways (177 pathways, 3092 pathway components, 2447 sequences related to pathways, and 2447 references captured for the pathways)
4. Ontologies (550 terms in PANTHER GO slim, 257 terms corresponding to biological process, 70 cellular components, and 223 molecular functions; 243 terms of protein class; 41,603 terms used in GO database annotations, including 9942 molecular functions, 27,852 biological processes, and 3809 cellular component terms (http://www.pantherdb.org/data).

The main window in PANTHER is composed of two main toolbars. The first one contains different links to individual topics (Fig. 16.2, items *1–5*), as well as an option for registration, login and contact (Fig. 16.2, items *6–8*). The second toolbar contains different options for data analysis, including gene list analysis, browse, sequence search, cSNP scoring, and keyword

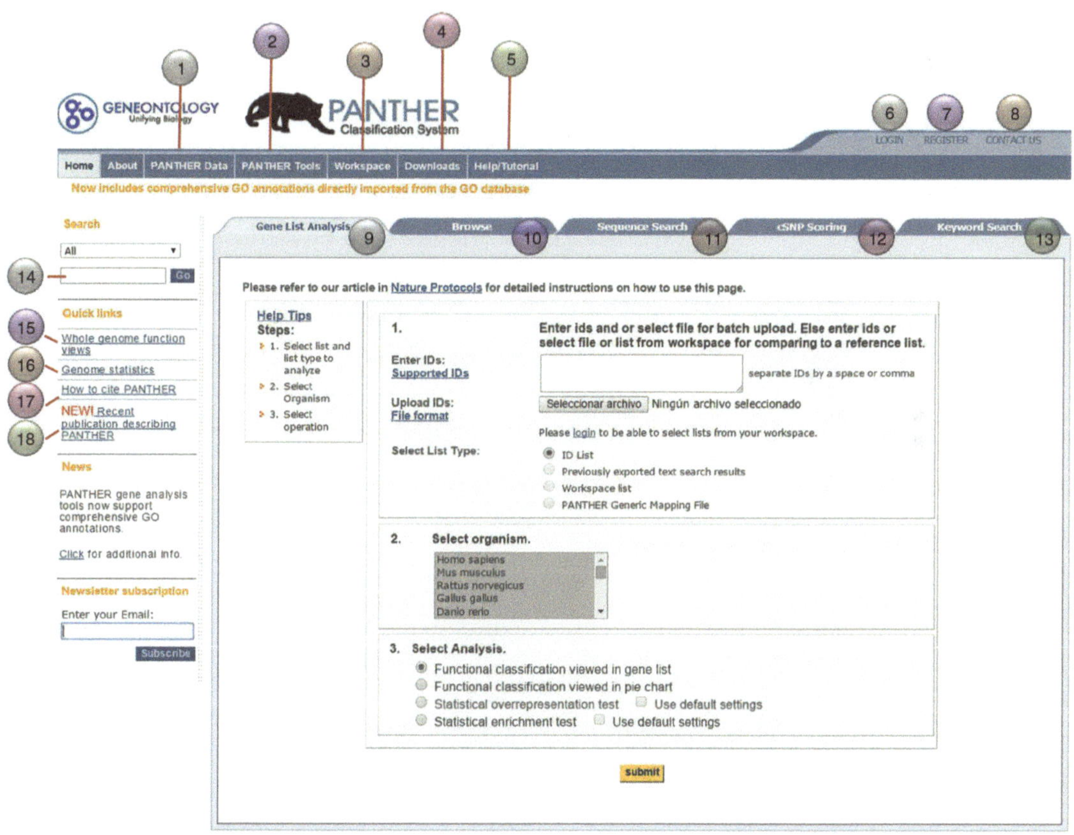

Fig. 16.2 PANTHER Classification System website. The main window in PANTHER contains two main toolbars. The first toolbar on top has links to different options inclduing: (*1*) PANTHER data, (*2*) PANTHER tools, (*3*) workspace, (*4*) downloads and (*5*) help/tutorial, and a section for (*6*) registration, (*7*) login, and (*8*) contact. The second toolbar, right under the first one, is for data analysis: (*9*) Gene list analysis, (*10*) browse, (*11*) sequence search, (*12*) cSNP scoping, and (*13*) keyword search. PANTHER also includes: (*14*) Quick keyword search, (*15*) whole genome function views, (*16*) genome statistics, (*17*) publications, and (*18*) recent publications describing PANTHER [16]

search (Fig. 16.2, items *9–13*). In addition, PANTHER has a panel for keyword search and quick links (Fig. 16.2, items *14–18*) [16]. In the analysis of list of genes or proteins, different functional classification views can be obtained, including gene list, bar or pie charts. Also, genes or proteins can be statistically analyzed through an enrichment test or a statistical overrepresentation test [17]. The PANTHER Ontology Browser also called PANTHER Prowler, browses and retrieves results (e.g. molecular functions, biological process, cellular component, protein class, pathway, and species) for input data related to ontology terms, such as genes and families [11, 17]. The PANTHER HMM sequence-scoring (sequence search) tool, can be used to search and compare protein sequences with the HMMs of PANTHER library. The top hit HMM can be observed in the results page, which also contains a statistical value for significance [17]. The Evolutionary Analysis of Coding SNPS (cSNP scoring) tool estimates the probability of a specific amino-acid change [17]. The keyword search tool can be used to obtain a variety of information, such as genes, families, pathways, and ontology terms for the protein of interest. However, we will focus on the generation of graphs for proteins classified in different categories.

16.3 PANTHER Gene List Analysis

To perform a gene list analysis using the PANTHER website (http://pantherdb.org), go to the toolbar gene list analysis (Fig. 16.3) and enter the

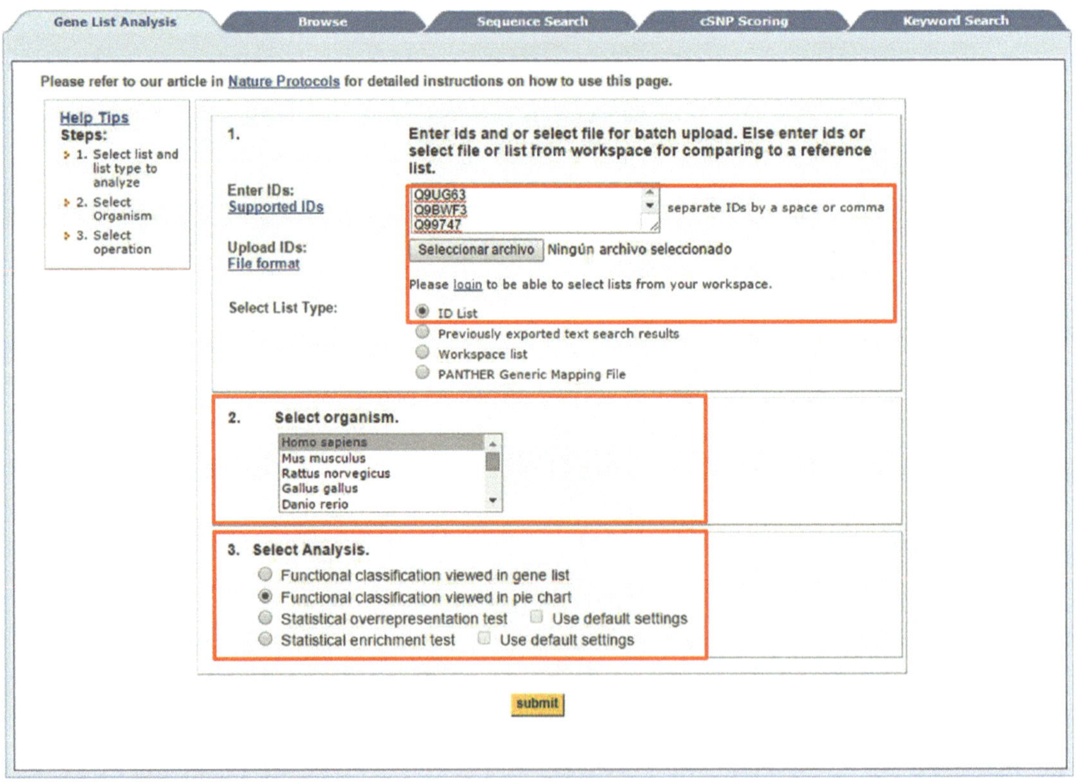

Fig. 16.3 Procedure to perform gene list analysis in PANTHER. The *red* section denotes the three primordial steps: (1) Enter the IDs of proteins to be analyzed, (2) select the organism, and (3) select the type of analysis to be performed

IDs of the genes or proteins in your list (Ensembl, Ensembl_PRO, Ensembl_TRS, Gene ID, Gene symbol, GI, HGNC, IPI, UniGene, UniProtKB ID) into the window, separating IDs by a space or comma. IDs can also be uploaded as a txt file. Then select the list type for query data (i.e. ID List, Previously exported gene list, Workspace list or PANTHER Generic Mapping File) and the organism of interest for analysis. In our example, we selected "ID list" and "*Homo sapiens*". Afterward, choose the type of analysis you like to perform. For example, we selected the "functional classification" viewed as a pie chart. Finally, click on the submit key (Fig. 16.3). In the results webpage, genes can be classified according to Molecular Function, Biological Process, Cellular Component, Protein Class, and Pathway (Fig. 16.4a). The chart obtained for a certain process can change for other processes. In addition, pie charts can be changed to bar charts and vice versa (Fig. 16.4b). The list of genes obtained in each ontological classification can be exported as a txt file. Classification categories may also contain different subcategories. When the cursor is located over a category in a chart, a message containing the following information will be displayed: Category name and its corresponding identifier, number of genes included from your list, the corresponding percentage of gene hits against the total number of identified genes, and the percentage of gene hits against the total number function hits (Fig. 16.4a). When a subcategory is selected, the corresponding gene list will be displayed (Fig. 16.5). As an example, we classified a list of overexpressed proteins in common between Luminal A (MCF7 and T47D) and Claudin-low (MDA-MB-231) breast cancer cells lines, which were recently described by Calderón-González et al. [18]. These proteins were categorized into Molecular functions and Cellular components (Fig. 16.4). In the first category, the most representative processes were: Binding and Catalytic activity with 25 and 21 genes, respectively (Figs. 16.4a and 16.5a). For Cellular component classification, categories with the higher number of genes were: Cell part (14 genes) and Macromolecular complex (10 genes) (Fig. 16.4b).

16.4 DAVID

The Database for Annotation, Visualization, and Integrated Discovery (DAVID) was developed in 2003 to address the emerging challenges posed by the post-genomic era [19]. DAVID, as well as other tools for the analysis of large gene lists, is based on the principle of gene enrichment that are functionally related to an altered gene/protein (generated by high throughput technologies). These enriched genes might potentially cooperate within a determined group and/or biological process [20]. DAVID is composed of the DAVID knowledgebase and five annotation tools:

1. DAVID Functional Annotation
2. DAVID Gene Functional Classification
3. DAVID Gene ID Conversion
4. DAVID Gene Name Viewer
5. NIAID Pathogen Annotation Browser.

The DAVID Knowledgebase is constructed around the "DAVID Gene Concept", which include tens of millions of gene/protein identifiers from several major public databases. This data concentration eliminates annotation redundancy among different resources and allows the organization of gene identifiers into more than 40 functional classification categories, e.g. Ontology (more than 40 million records), Protein-protein interactions (more than four millions), Disease gene associations (9000), Pathways (above 50,000), Functional categories (more than 6.9 millions), etc. [21].

DAVID Gene Functional Classification: This tool is useful for the exploration of large lists of genes into more feasible modules ordered according to their functional relationships. These functionally organized modules are very useful in processing large amounts of information, switching from a gene centric analysis to a module-centric analysis [21].

DAVID Functional Annotation Tool Suite: The Functional Annotation Tool Suite displays three ways for combining results: Functional Annotation Clustering, Functional Annotation Chart and Functional Annotation Table. The

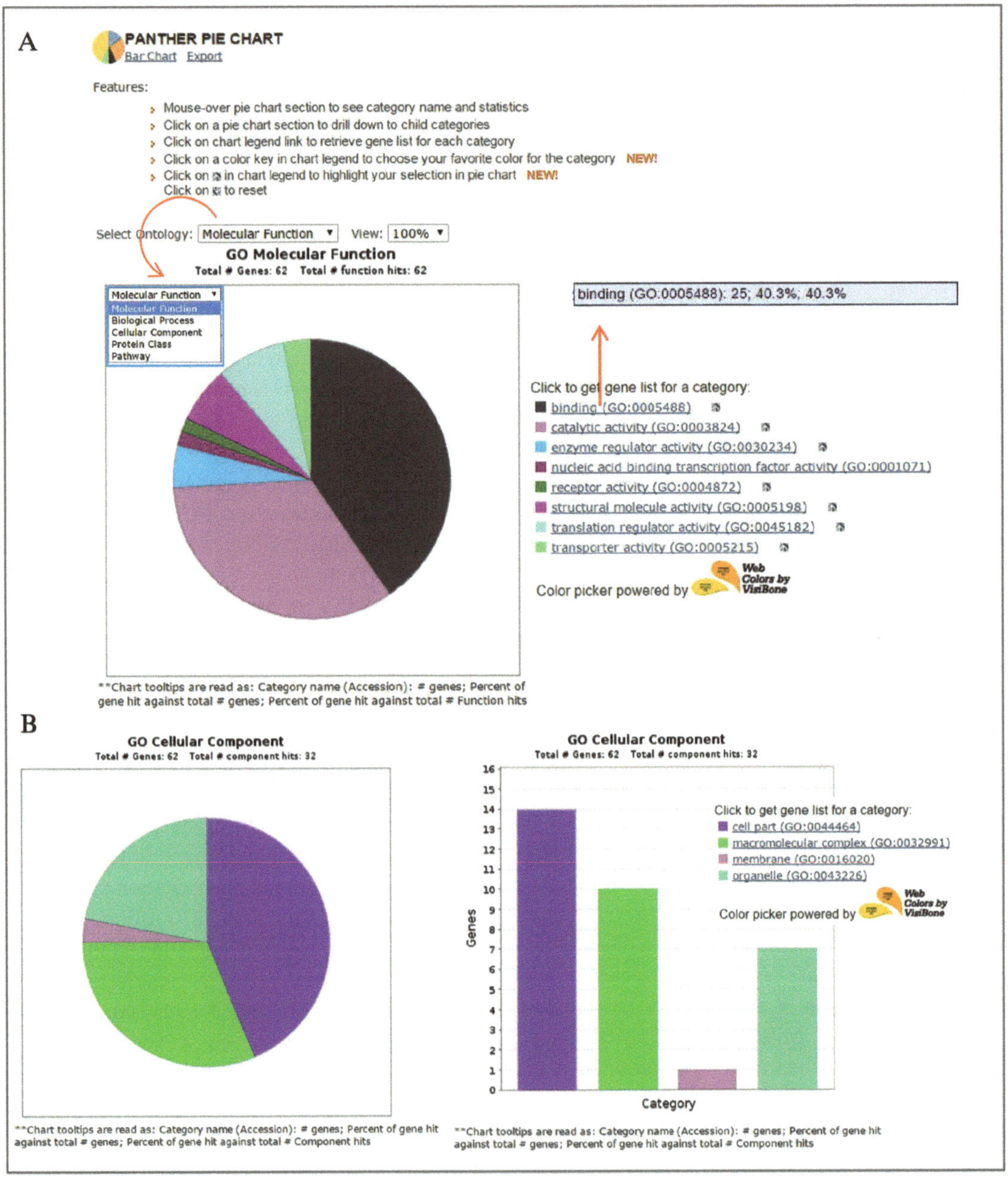

Fig. 16.4 Functional classification of proteins up-regulated in both Luminal A (MCF7 and T47D) and Claudin-low (MDA-MB-231) breast cancer cells lines. The proteins were classified into (**a**) Biological Processes and (**b**) Cellular Components. Figure shows the change of pie chart to a bar graphic as well

Functional Annotation Clustering tool allows the user to group genes depending on the degree of their functional association. It is performed with a novel algorithm that measures relationships among annotation terms. This process is useful to eliminate the redundant relationships that exist in many-genes-to-many-terms cases (i.e. when one gene is associated with many different redundant terms and one term is associated with many genes) [21]. Additional features of this

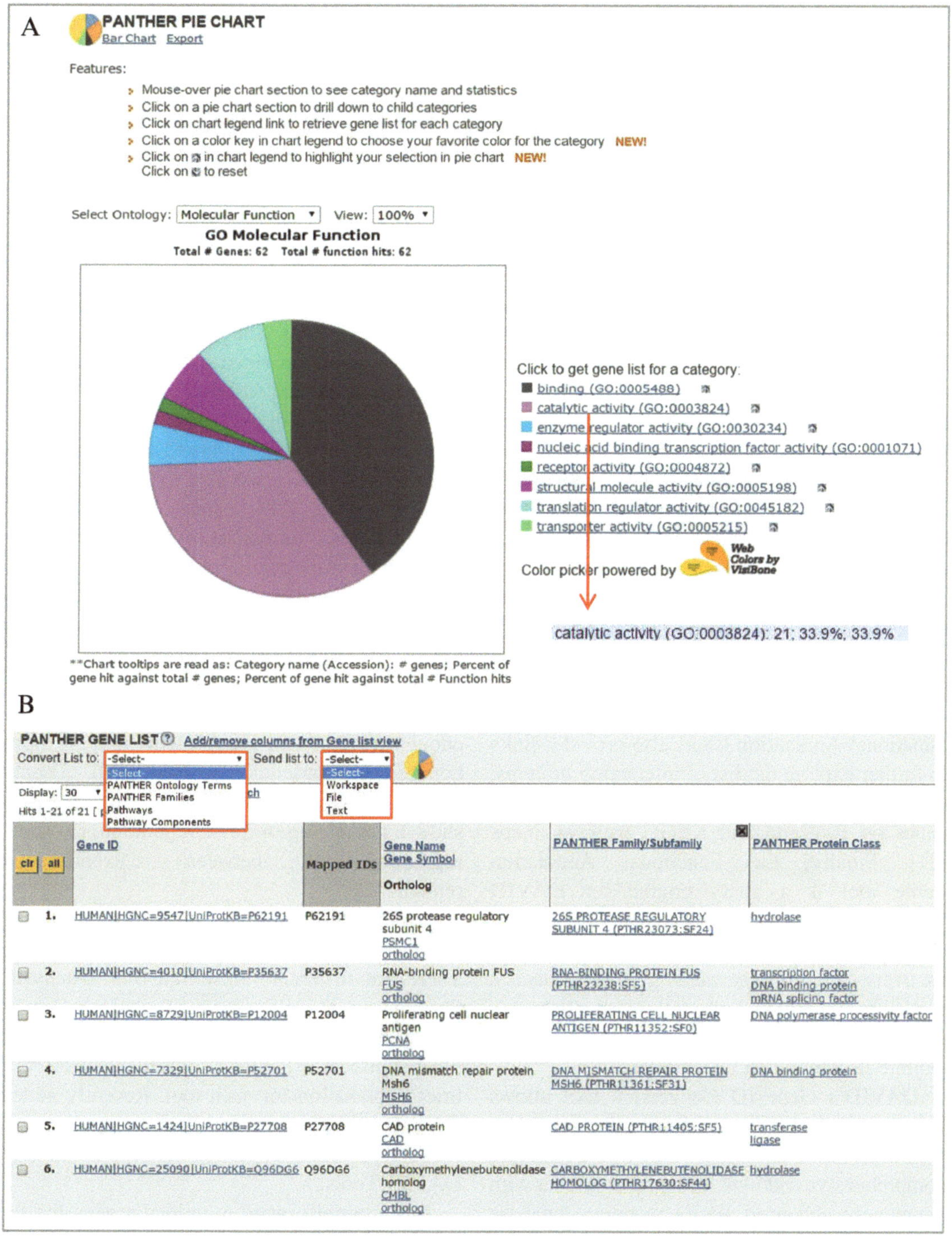

Fig. 16.5 Classification of Biological Processes for proteins up-regulated in both Luminal A (MCF7 and T47D) and Claudin-low (MDA-MB-231) breast cancer cells lines (**a**) Biological processes pie chart displaying different categories of processes, e.g. Metabolic Processes. (**b**) List of genes involved in the selected Metabolic Processes

clustering tool is the ability to rank the importance of annotation groups with an enrichment score (EASE scores) that uses the geometric mean of all the enrichment p-values of each annotation term in the group; the annotation clustering tool provides a link to a 2-D viewer for related gene-term relationships, allowing a fast way to focus on the genes that have common annotation terms [22]. On the other hand, The Functional Annotation Chart tool can be used to get the typical gene-GO term enrichment analysis (similar to other tools) to identify the most relevant (overrepresented) biological terms associated with a given gene list. However, DAVID offers extended annotation coverage in comparison to other enrichment analysis tools. The enhanced annotation coverage includes not only the GO terms but more than 40 annotation categories, such as protein-protein interactions, protein functional domains, disease associations, bio-pathways, sequence features, gene tissue expression, etc. This tool is helpful to identify enriched annotation terms associated with the gene list of interest in a linear tabular text format. Similar to the Annotation Clustering Tool, the Functional Annotation Chart also provides links to further explore the list of interacting proteins, link gene-disease associations and visualize genes on BioCarta and KEGG pathway maps [21]. Finally, the Functional Annotation Table tool is a query engine for DAVID Knowledgebase without statistical probes. It delivers annotation information in a table format for every gene from the users' gene list. This is a particularly useful tool when users want to have a closer look of some specific interesting genes and explore its annotation information.

DAVID's Gene ID Conversion tool allows conversion of user's input gene or gene product identifiers from any type to another in a more comprehensive and high throughput manner with a uniquely enhanced ID-ID mapping database leveraging heterogeneous annotations [23].

DAVID's Gene Name Viewer is another tool useful to quickly attach meaning to a list of gene IDs, translating them into their corresponding gene names. Thus, before proceeding to an in-depth analysis, researchers can quickly have an overview of gene names to gain insight into their biological system and have *a priori* general idea of interesting processes that might be involved.

DAVID's NIAID Pathogen Browser: The National Institute of Allergy and Infectious Diseases (NIAID) has defined three categories of priority pathogens, A, B and C. These pathogens are important for biodefense purposes and have become attractive study subjects because of the increasing research funding available to study them. The DAVID NIAID Pathogen Browser is provided as a support tool for researchers that would like to explore the biology of the priority pathogens types. For example, one may choose the word "anthrax" and type the key word "toxin", the result is a list of genes from *Bacillus anthracis* that matches to the typed key word. This tool may assist researchers in understanding the biology of a priority pathogen if the gene list retrieved from the DAVID NIAID Pathogen Browser is further analyzed by one of DAVID's Bioinformatics Resources [21].

Analysis of gene lists: To carry out an optimal gene list analysis, the list should; (1) have enough number of genes/proteins ranging from hundreds to thousands (e.g. 100–2000), (2) only include genes with statistical significance that show a notable up or down regulation, (3) show reproducibility between experimental replicas [22].

DAVID bioinformatics resources website is organized in two main toolbars (Fig. 16.6). There are different links, like Start Analysis, Shortcut to DAVID Tools, Technical Center, among others on top. On the left side, there are other shortcuts to DAVID Tools that also offers a brief explanation for each tool. Recently added DAVID NIAID Pathogen Annotation Browser tool can be found on the top menu in shortcut to DAVID Tools.

It is straightforward to upload a gene list for DAVID bioinformatics analysis (Fig. 16.7a). Firstly, go to https://david.ncifcrf.gov/gene2gene.jsp and select Start analysis. On the left side choose upload in the list manager, then: (1) Copy/paste the gene lists to be analyzed into box A; a text file or a gene IDs list can also be

Fig. 16.6 DAVID Bioinformatic Resources Website. This website has two main toolbars. The toolbar on the top has links to: (*1*) Start Analysis, (*2*) Shortcut to DAVID Tools, (*3*) Technical Center, (*4*) Downloads and APIs, (*5*) Terms of Service, (*6*) Why David, and (*7*) About Us. And the toolbar on the left side (*8*) has links to Tools that offer a brief explanation for each of DAVID's tool. Additionally, in (*2*) we can find the recently added tool NIAID Pathogen Annotation Browser (*9*)

uploaded in box B, (2) Choose the corresponding gene identifier type for your input gene IDs; alternatively use the ID conversion tool to seek (or convert) the correct gene identifier, (3) Select the type of list you are submitting, either gene list or gene background. The general guideline is to set up a pool of genes as population background. This usually includes all the genes that could be possibly detected (e.g. all the probes included in a particular DNA microarray). Since most of the studies are done in a genome-wide scale, there is no need to set a background (default background is the entire genome), (4) Submit the List. The different analysis suites are displayed (Fig. 16.7b) that will be applied to the submitted gene list shown on the left (highlighted in the Gene List Manager) (Fig. 16.7b). By clicking Start Analysis, users can go back at any time to upload another gene list or to access any analytical tool suite of interest.

In this section, a couple of examples are presented to showcase a few of the tools from David's toolbox that are most widely used using gene lists corresponding to proteins down regulated in both Luminal A (MCF7 and T47D) and Claudin-low (MDA-MB-231) breast cancer cell lines studied by Calderón-González et al. [18]. Selecting Functional Annotation Tool (Fig. 16.7b), results in Annotation Summary Results, which displays the number and percentage of genes (from the submitted gene list) involved in different GO categories

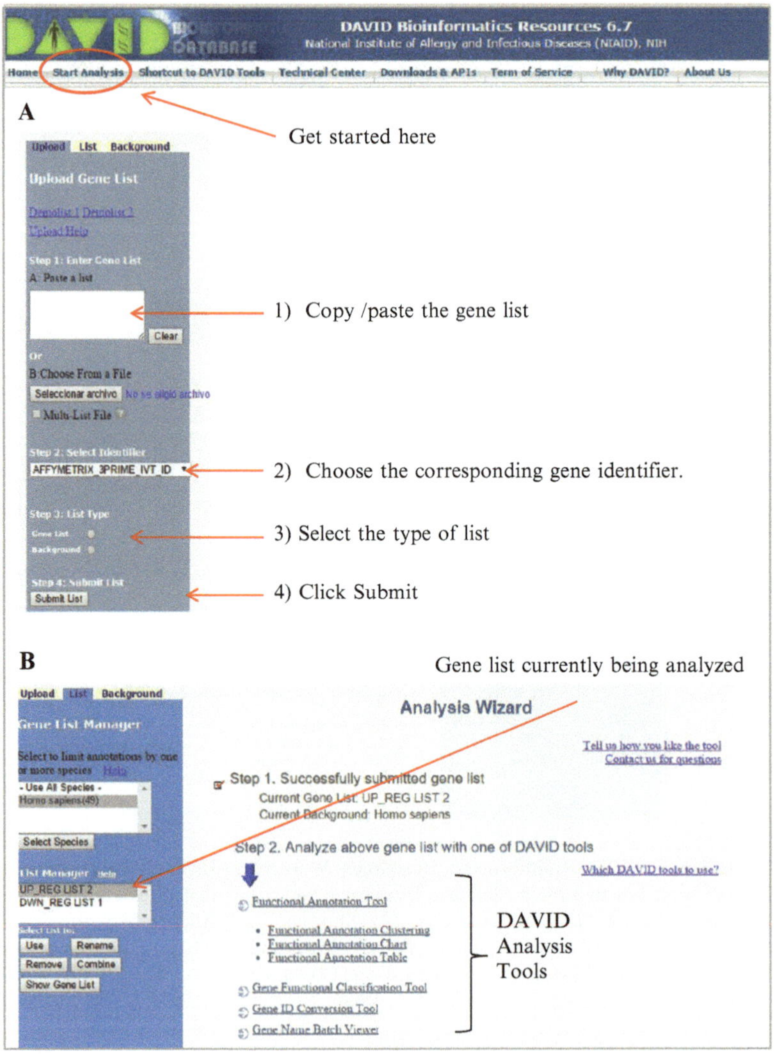

Fig. 16.7 Uploading data into David's gene list manager. (**a**) On the left side; (*1*) Upload a gene list, (*2*) Choose the corresponding gene identifier, (*3*) Select the type of list, either gene list or gene background, (*4*) Submit the gene list. (**b**) Once the user has submitted the gene list, the Analysis Wizard shows the shortcuts for the different DAVID Analysis tools

(Fig. 16.8). In each category, users can click on Chart to obtain an individual chart report for the selected category. Users can choose a number of categories for further analysis in the Combined Annotation Tools (Fig. 16.8). A table divided in several annotation clusters will be obtained by clicking on Annotation Clustering Tool. Every annotation cluster is formed by a group of terms from functionally related genes. Taken all together, the chance to identify a biological significance increases (Fig. 16.9). The degree of similarity between annotations is measured by Kappa statistics. This tool also provides a link to generate a 2D-view map that allows a fast way to associate genes that have common annotation terms.

From this very specific gene list, we observed an enriched group of genes involved in mitochondrial function. Noteworthy, the high correlation of this result in comparison with other tools previously explored. Since the submitted gene list corresponds to down-regulated genes in a proteomic approach, this result suggests that MCF7, T47D and MDA-MB231 breast cancer

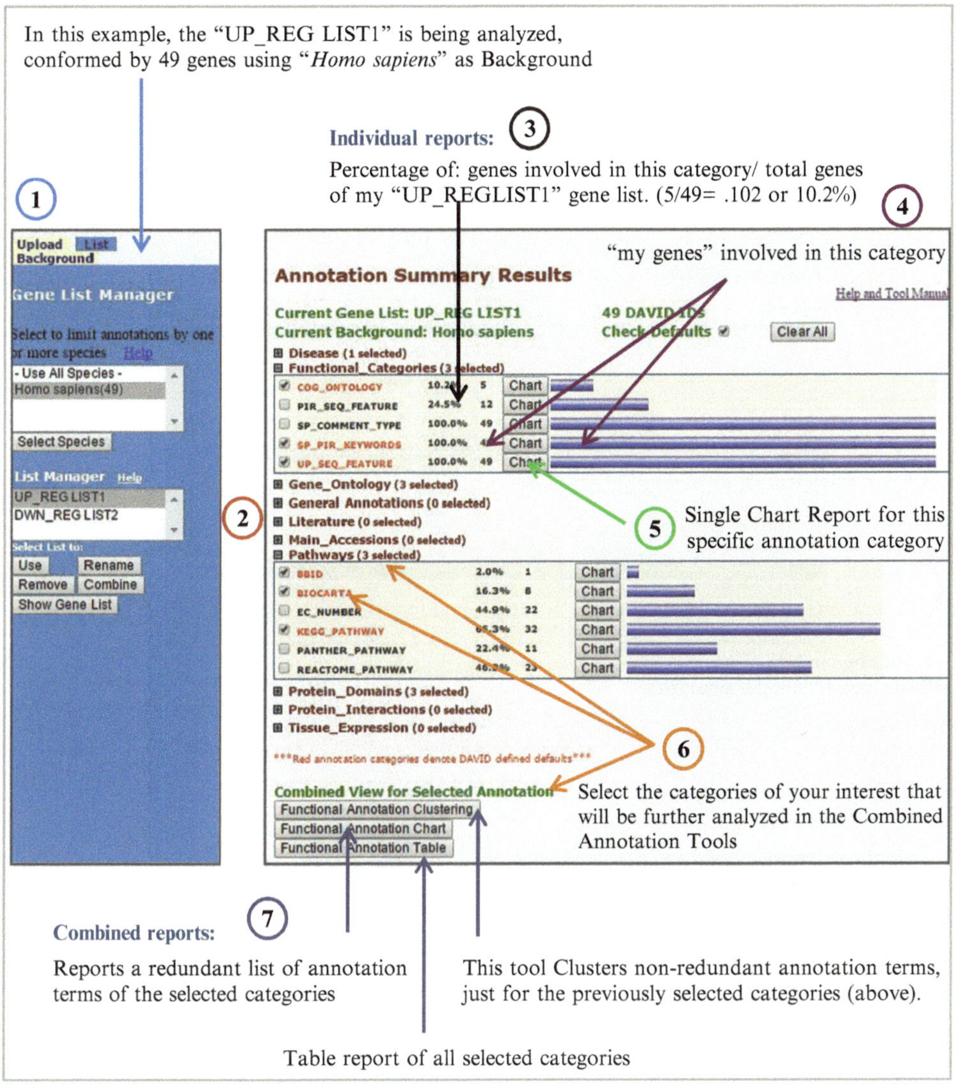

Fig. 16.8 Functional Annotation Tool Suite. (*1*) Gene List Manager showing the list that is being analyzed. (*2*) Annotation Summary results displaying different categories: (*3*) the number and (*4*) percentage of genes involved. (*5*) Clicking on this box will generate a chart report of functional categories. (*6*) The user can choose the number of categories to be considered for further analysis in the Combined Annotation Tools (*7*) by checking the check boxes next to each category

cell lines have an impaired mitochondrial function in comparison to the MCF10A control cell line.

For instance, NADH-coenzyme Q reductase, 3,2 trans-enoyl-Coenzyme A isomerase, cytochrome c oxidase, and malate dehydrogenase are some of the encoding genes that had a high EASE SCORE and are involved in the mitochondrial inner membrane function.

16.5 KEGG

The Kyoto Encyclopedia of Genes and Genomes (KEGG) is a database resource designed for understanding and interpreting biological systems using high-throughput data [24–26]. KEGG is composed of 17 databases organized into four categories:

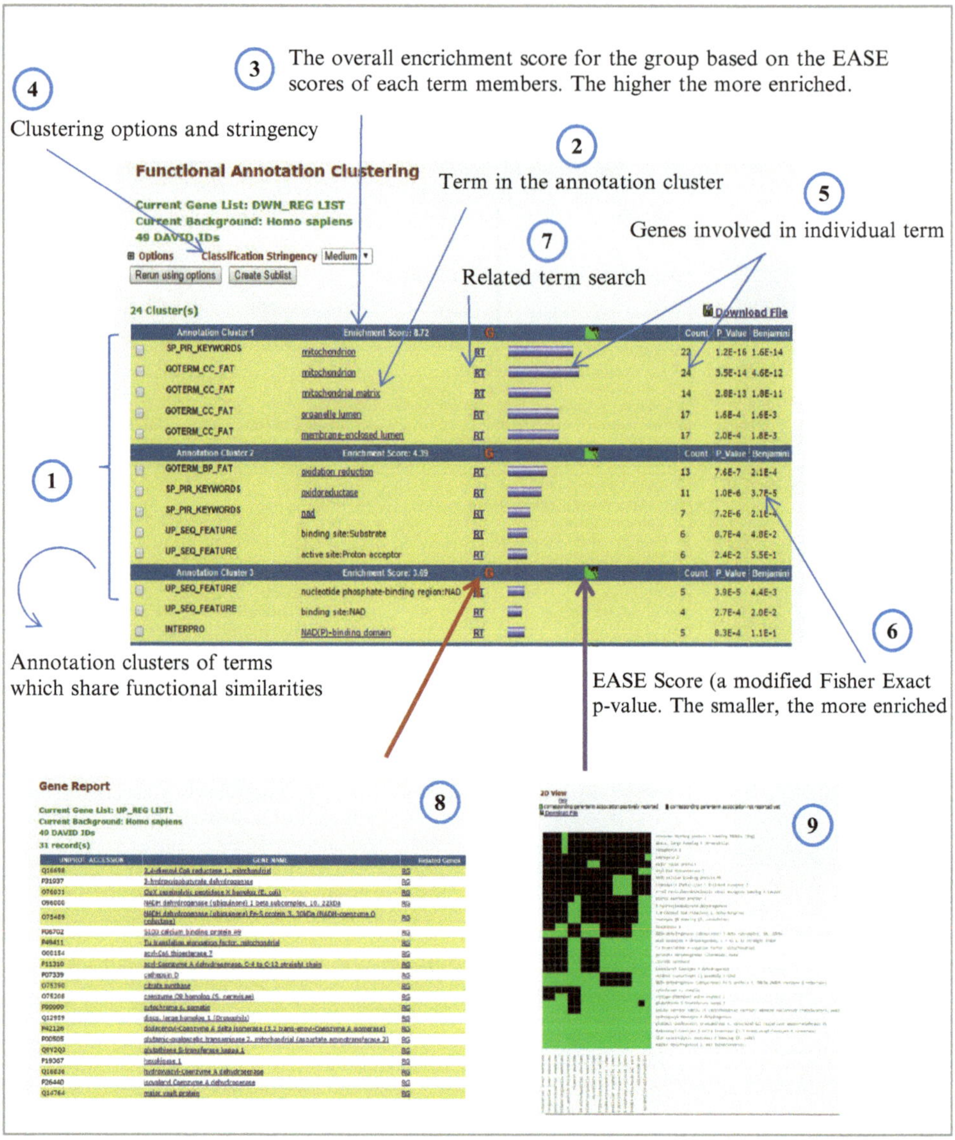

Fig. 16.9 An example of the Functional Annotation Clustering Tool. This image shows the results obtained by searching the "DWN_REG LIST". The search results show three clusters (*1*) each categorized further according to different terms (*2*). The clusters are ranked according to their enrichment score (*3*) and stringency (*4*). Genes involved in each term are shown for each cluster (*5*), as well as the EASE score (*6*) and the related term search (*7*). The links to obtain the gene list in each annotation cluster (*8*) and a 2D-Map View (*9*) are provided

1. Systems information: KEGG PATHWAY (pathway maps), KEGG BRITE (functional hierarchies and table files) and KEGG MODULE (Pathway, structural complex, functional set and signature modules). These databases are manually created using published literature
2. Genomic information: KEGG ORTHOLOGY (orthology (KO) groups), KEGG GENOME (complete genomes), KEGG GENES (gene catalogs) KEGG SSDB (sequence similarity database for genes), DGENES (draft genomes) and MGENES (metagenomes). The information about genes and genomes is

obtained from different databases, such as RefSeq (prokaryotes, eukaryotes, plasmids and viruses), GenBank (prokaryotes), and PubMed (addendum: collection of manually created protein sequences entry)
3. Chemical information, also called KEGG LIGAND: KEGG COMPOUND (metabolites and other small molecules), KEGG GLYCAN (glycans), KEGG REACTION (biochemical reactions), KEGG RPAIR (reactant pairs), KEGG RCLASS (reaction class), and KEGG ENZYME (enzyme nomenclature)
4. Health information commonly called KEGG MEDICUS: KEGG DISEASE (human diseases), KEGG DRUG (drugs), KEGG DGROUP (drug groups), KEGG ENVIRON (crude drugs and health related substances), JAPIC (drug labels in Japan) and DailyMed (links to drug labels in USA) [26].

The annotation system in KEGG is based on the correlation between functional information and orthologous groups (KEGG Orthology or KO) through the assignment of KO identifiers (K number). This information is stored in the KO database and is independent of the KEGG GENE database that contains completely sequenced genomes [26]. The KO system is essential for connecting the genomic information with systemic functional information resulting in the conversion of genes to K numbers, leading to an automatic reconstruction of KEGG PATHWAYS and other networks [26, 27]. Currently, KEGG has more than 4000 complete genomes annotated with the KO system [26].

KEGG has several analysis tools:

1. KEGG Mapper which is the interface used for KEGG Mapping. This is composed of KEGG BRITE, MODULE, and PATHWAY mapping tools, which map genes, proteins, small molecules, etc. (also called objects) into all brite functional hierarchies, modules and pathways maps, respectively [28]
2. KEGG Atlas is a graphical interface to navigate the global integrated maps in KEGG. Maps available are Metabolism (Biosynthesis of amino acids, Biosynthesis of secondary metabolites, Carbon metabolism, Degradation of aromatic compounds, Fatty acid metabolism, Microbial metabolism in diverse environments, and 2-Oxocarboxylic acid metabolism) and Cancer pathway [29]
3. BlastKOALA: KOALA is defined as KEGG Orthology And Links Annotation. BlastKOALA is used for the annotation of completely sequenced genomes. This tool utilizes the Pangenomes database
4. GhostKOALA: this tool is designed by the metagenome annotation and it uses the Pangenomes and Viruses databases [26, 27], (5) BLAST/FASTA performs searches of similar sequences
5. SIMCOMP searches for similar chemical structures

Pathway Maps Analysis To map proteins of interest into Pathways, go to the KEGG website (http://www.genome.jp/kegg/) and on the Data-oriented entry points, click on the KEGG PATHWAY key (Fig. 16.10). In the Pathway Mapping menu, select the mapping tool of interest: Search Pathway, Search&Color Pathway or Color Pathway. As an example, the up and down-regulated proteins found common between Luminal A (MCF7 and T47D) and Claudin-low (MDA-MB-231) breast cancer cells lines from Calderón-González et al. were analyzed with the Search&Color Pathway tool [18]. - Up-regulated proteins were colored in red, whilst down-regulated polypeptides were presented in green (Fig. 16.11). To perform this analysis, an organism must be selected first by clicking on the org key, after which a new window is displayed to find the three to four KEGG organism code. Type the desired organism in the window and then click on select. In this example, *H. sapiens* has the hsa code. The next step is to introduce IDs in UniProtKB format, followed by the word red or green as mentioned before. Other compatible ID formats are KEGG-Identifiers, NCBI-GeneID and NCBI-ProteinID. Alternatively, a file containing IDs can be uploaded. To perform

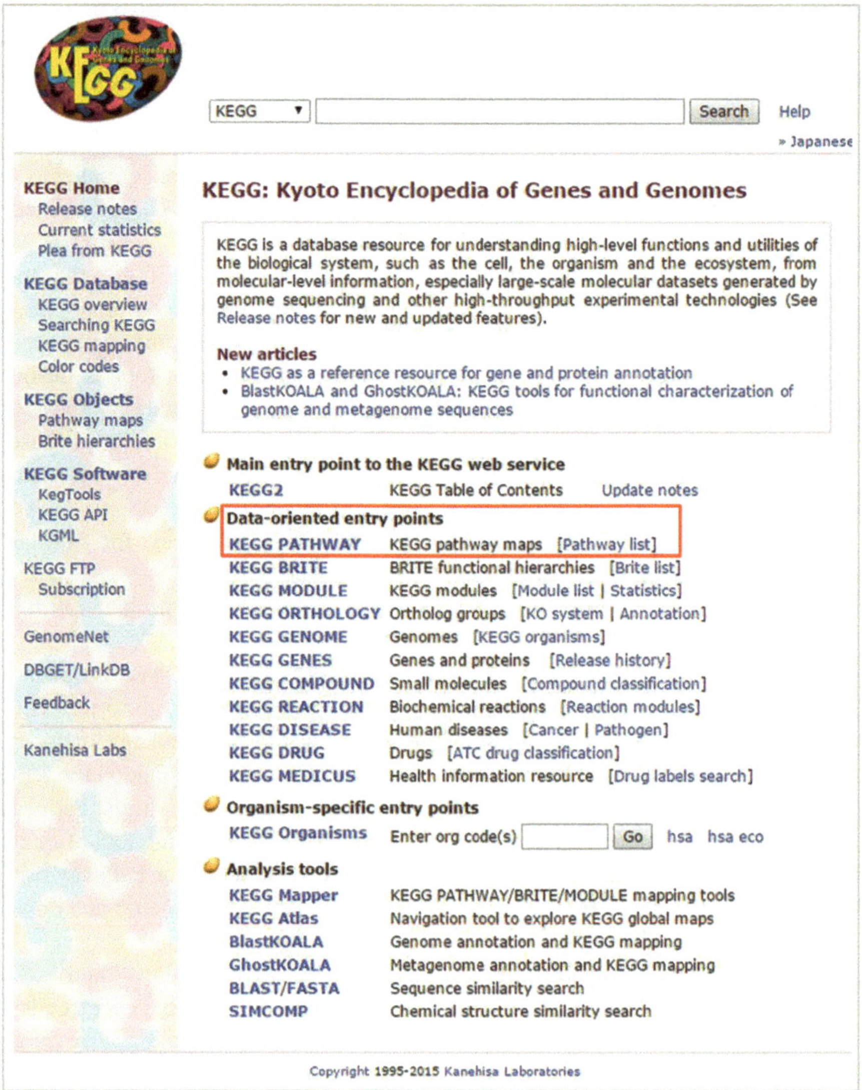

Fig. 16.10 KEGG website. This image shows the different links provided in KEGG's website, including KEGG Home, KEGG Database, KEGG Objects, KEGG Software, among others. The website also provides several tools for the data analysis including KEGG Mapper, KEGG Atlas, BlastKOALA, Ghost KOALA, BLAST/FASTA, SIMCOMP. KEGG Pathway modules are highlighted in a *red box*

the search, the following options were selected; (1) to include aliases and (2) to display objects not found in the search (Fig. 16.12a). The result window shows a list of pathways where proteins were mapped, as well as a list of protein IDs that were not found (Fig. 16.12a). A list of proteins found in each pathway, including their UniProtKB IDs and KEGG *H. sapiens* database codes is also displayed (Fig. 16.12b). Clicking a particular UniProtKB ID will display the information for the selected ID (Fig. 16.13a). On the other hand, if the code of the *H. sapiens* organism in KEGG is selected, a new window containing KEGG information about that protein, including Gene name, Disease, KEGG Orthology, Structure, Motifs in the protein, and Pathways, among other information will be displayed (Fig. 16.13b). Finally, when a certain pathway is selected, an

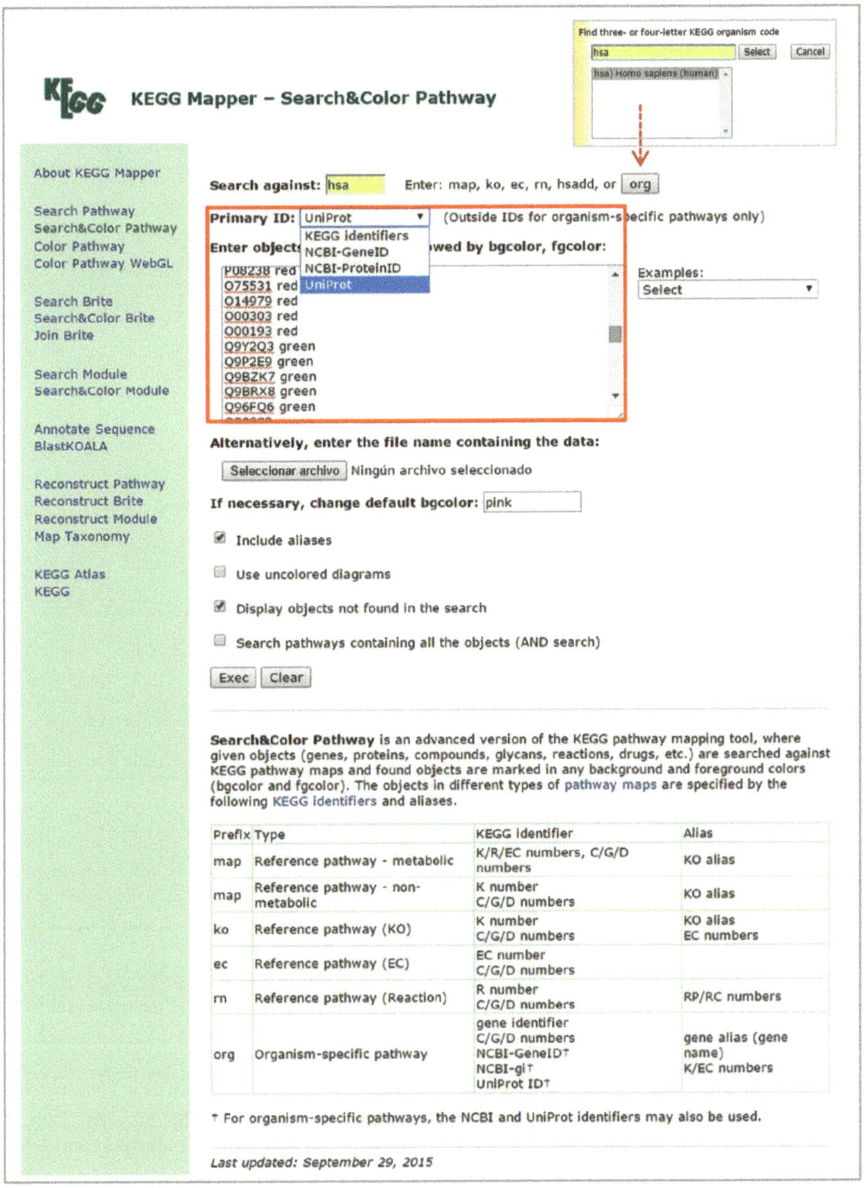

Fig. 16.11 KEGG pathway mapping tool. This image shows the general procedure for mapping proteins in Search & Color Pathway module. The format of IDs as well as the organism need to be selected. Protein accession numbers are followed with the word *red* or *green* to highlight up- or downregulated proteins, respectively

image is generated where up- or down-regulated proteins are highlighted in red or green respectively (Fig. 16.14). In the case of the breast cancer cell line, most quantified proteins mapped to metabolic processes, with 22 polypeptides [5 - up-regulated (↑) and 17 down-regulated (↓)]: ↓3HIDH, ↑ SAHH3, ↓ IVD (Amino acid metabolism), ↑ CMBL (Hydrolase), ↓ CISY (Carbon metabolism, 2-Oxocarboxylic acid metabolism, biosynthesis of amino acids, carbohydrate metabolism), ↓ AL1A3 (Carbohydrate metabolism, amino acid metabolism, metabolism of other amino acids, xenobiotics biodegradation and metabolism, chemical carcinogenesis),

Fig. 16.12 Search & Color Pathway result. (**a**) A list of proteins that were not found are shown at the *top*. The list of different pathways is also displayed with the number of proteins involved. (**b**) Two examples of proteins involved in RNA transport and DNA replication processes

↓ AATM (Carbon metabolism, 2-Oxocarboxylic acid metabolism, biosynthesis of amino acids, amino acid metabolism, fat digestion and absorption), ↓ HCDH (Fatty acid metabolism, carbohydrate metabolism, lipid metabolism, amino acid metabolism), ↓ HXK1 (Carbon metabolism, carbohydrate metabolism, biosynthesis of other secondary metabolites, HIF-1 signaling pathway, insulin signaling pathway, carbohydrate digestion and absorption, central carbon metabolism in cancer, endocrine and metabolic diseases), ↓ ACADM (Carbon metabolism, fatty acid metabolism, carbohydrate metabolism, lipid metabolism, amino acid metabolism, metabolism of other amino acids, PPAR signaling pathway), ↑ METK2 (Biosynthesis of amino acids, amino acid metabolism), ↓ MDHM (Carbon metabolism, carbohydrate metabolism, amino acid metabolism), ↓ NDUBA, ↓ NDUS3 (Energy metabolism, neurodegenerative diseases,

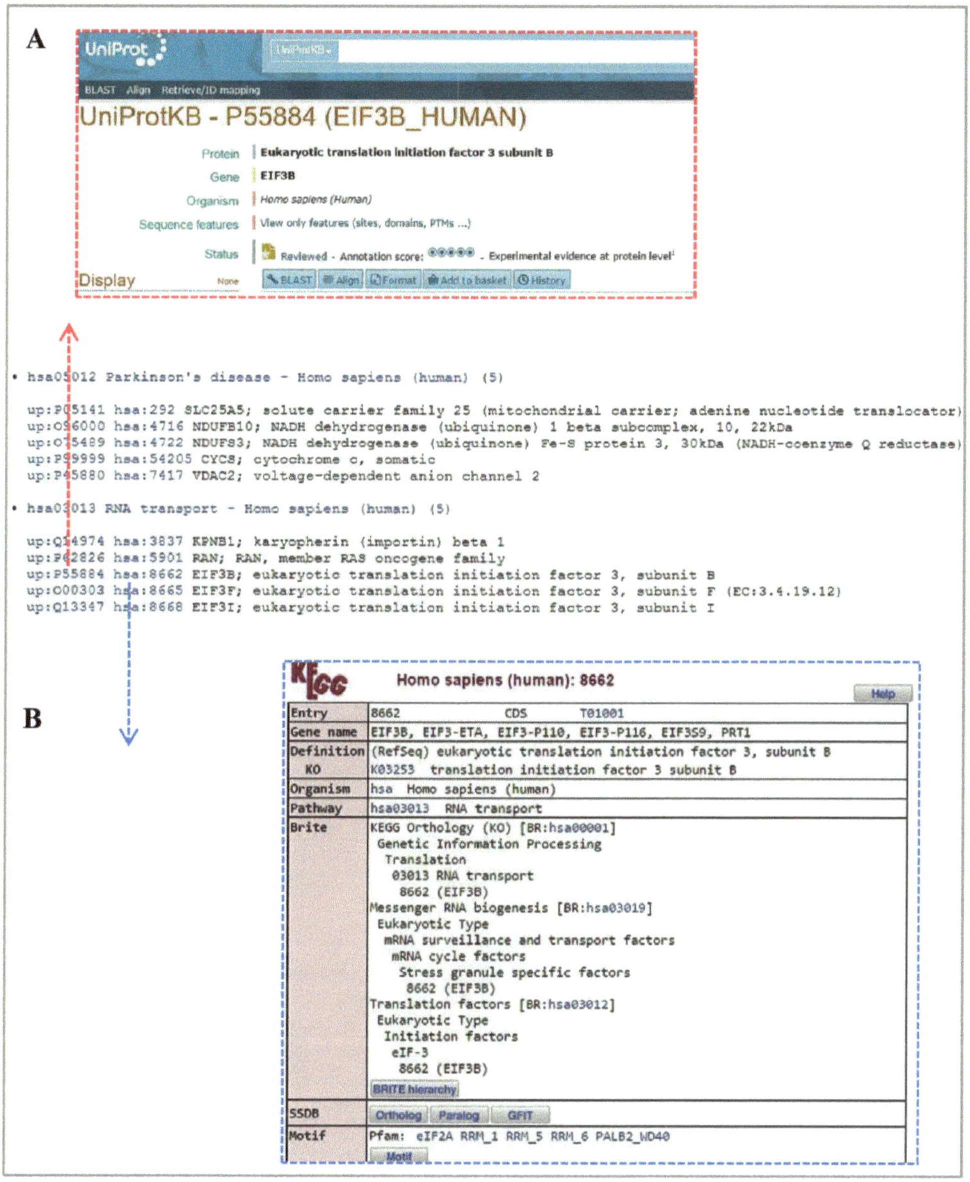

Fig. 16.13 Additional information for proteins in KEGG Database. The proteins displayed in each pathway have a link to additional information: (**a**) UniProtKB website and (**b**) KEGG database

endocrine and metabolic diseases), ↓ DHB12 (Fatty acid metabolism, lipid metabolism), ↓ ODPB (Carbon metabolism, carbohydrate metabolism, HIF-1 signaling pathway, glucagon signaling pathway, central carbon metabolism in cancer), ↑ PGAM1 (Carbon metabolism, biosynthesis of amino acids, carbohydrate metabolism, amino acid metabolism, glucagon signaling pathway, central carbon metabolism in cancer), ↓ CYC (Energy metabolism, cellular processes, pathways in cancer, neurodegenerative diseases, cardiovascular diseases, endocrine and metabolic diseases, infectious diseases), ↓ RPN1 (Glycan biosynthesis and metabolism, folding, sorting

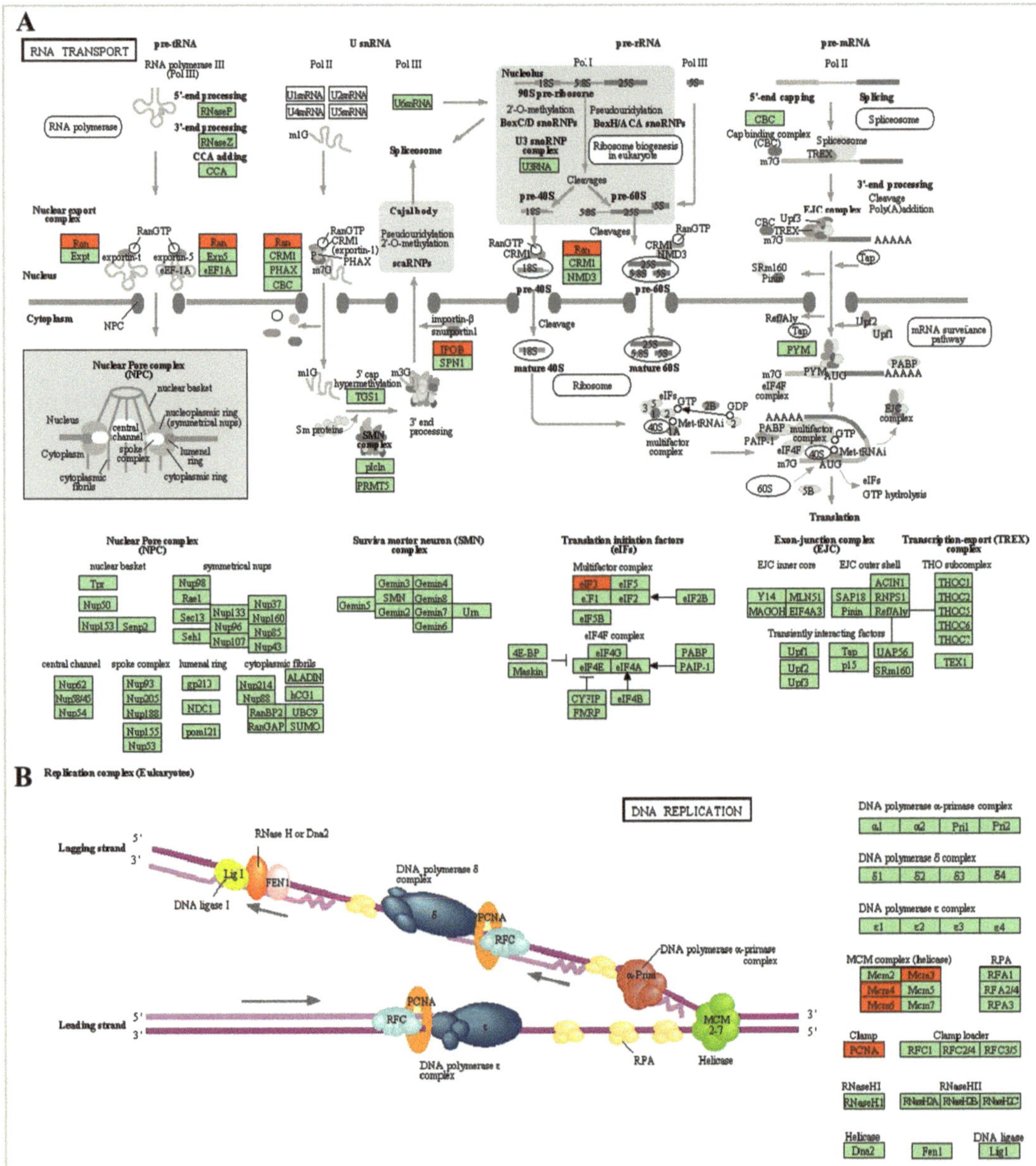

Fig. 16.14 Proteins mapped into KEGG PATHWAYS. Polypeptides found up- or down-regulated in both Luminal A (MCF7 and T47D) and Claudin-low (MDA-MB-231) breast cancer cell lines were submitted to KEGG mapping. Some of the processes found to be affected are, (**a**) RNA transport process, and (**b**) DNA replication process. Up-regulated proteins are colored in *red* and down-regulated proteins are in *green*

and degradation), ↓ NLTP (Lipid metabolism, cellular processes, PPAR signaling pathway), ↓ SPEE (Amino acid metabolism, metabolism of other amino acids), ↑ PYR1(Nucleotide metabolism, amino acid metabolism). Others mapped pathways were: RNA transport with 5 proteins ↑ IMB1, ↑ RAN, ↑ EIF3B, ↑ EIF3F, ↑ EIF3I) (Fig. 16.14a) and DNA replication with 4 polypeptides involved (↑MCM3, ↑ MCM4, ↑ MCM6, ↑ PCNA) (Fig. 16.14b).

16.6 Ingenuity Pathway Analysis (IPA)

Ingenuity Pathway Analysis (IPA, QIAGENs Redwood City, www.qiagen.com/ingenuity) is a software application platform developed for analysis, understanding, integration and interpretation of biological data [30]. Ingenuity can analyze data acquired using platforms such as microarrays, proteomics, metabolomics, etc. IPA uses the QIAGEN's Ingenuity Knowledge Base in which contents extracted from articles, biomedical literature, reviews, internally curated knowledge, and other sources are structured into Ontology terms. The information in this platform are categorized into several knowledgebases:

1. Ingenuity expert information, including Ingenuity expert findings and Ingenuity expert assist findings
2. Ingenuity supported third party information including MicroRNA-mRNA interactions (miRecords, TarBase, TargetScan)
3. Protein-Protein Interactions including BIND, cognia, DIP, Interactome studies, MINT, and MIPS
4. Additional sources: An open access database of genome-wide association results, BIOGRID, Breast cancer information core (BIC), Catalogue of somatic mutations in cancer (COSMIC), Chemical Carcinogenesis Research Information System (CCRIS), ClinicalTrials.gov, ClinVar, DrugBank, GO, GVK Biosciences, Hazardous Substances Data Bank (HSDB), HumanCyc, IntAct, miRBase, Mouse Genome Database (MGD), Obesity Gene Map Database, and Online Mendelian Inheritance in Man (OMIM).

The principal components of IPA suite are

1. Core Analyze
2. IPA-Tox
3. IPA-Biomarker
4. IPA-Metabolomics (Fig. 16.15)

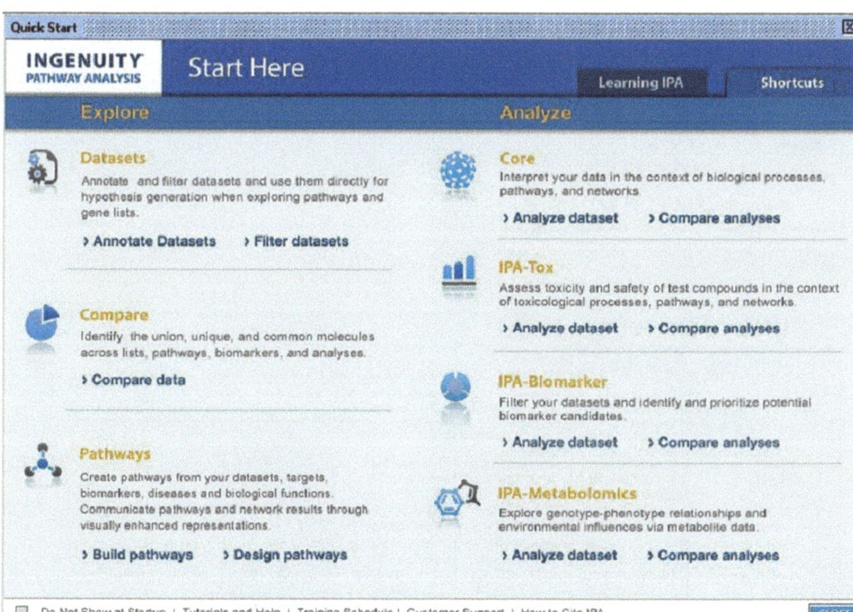

Fig. 16.15 The main page of Ingenuity Pathway Analysis suit. All functions are listed via in two main tabs, Learning IPA, and shortcuts. The shortcut tab contains the dataset- and pathway options, as well as different analysis options, including Core, IPA-Tox, IPA-Biomarker and IPA-Metabolomics

Core Analyze consists of classified data sets mapped into biological processes, networks and pathways. IPA-Tox module includes data classified in the context of toxicological processes. In this tool the toxicity and safety of compounds is evaluated. IPA-Tox keeps track of the biological processes that are related to compound toxicity at various biochemical and molecular levels. IPA-Biomarker tool is used to identify and prioritize potential biomarker candidates. The selection of these putative biomarkers is based on their biological characteristics. Finally, the fourth application IPA-Metabolomics, is able to analyze metabolomics data, which are then contextualized into biological insights (metabolism and cell physiology).

IPA supports several types of identifiers including Affymetrix, Affymetrix SNP ID, Agilent, CAS registry number, CodeLink, dbSNP, Ensembl, GenBank, Entrez gene, Gene Symbol-mouse, Gene Symbol-rat and Gene symbol—Human (Hugo/HGNC), GenPept, GI number, Human Metabolome Database (HMDB), Illumina, Ingenuity, International Protein Index, KEGG, Life Technologies (Applied Biosystems), miRBase (mature), miRBase (stemloop), PubChem CID, RefSeq, UCSC hg18 and 19, UniGene and UniProtKB/Swiss-Prot accession number. The confidence reported by IPA are either experimentally determined or theoretically predicted. Some tissues and cell lines covered by IPA include tissue and primary cells from nervous and other organ systems and cell lines from breast cancer, cervical, central nervous system (CNS), colon, hepatoma, immune, kidney, leukemia, lung, lymphoma, macrophage, melanoma, myeloma, neuroblastoma, osteosarcoma, ovarian, pancreatic, prostate and teratocarcinoma model systems. Mutations covered include functional effect, inheritance mode, translation impact, unclassified mutation, zygosity and wild type.

IPA analysis core protocol: To use IPA, a license needs to be purchased but one can use a trial version for a limited period of time. To perform an analysis in IPA, first an analysis dataset need to be created (Fig. 16.16). To create an analysis dataset, go to Annotate datasets

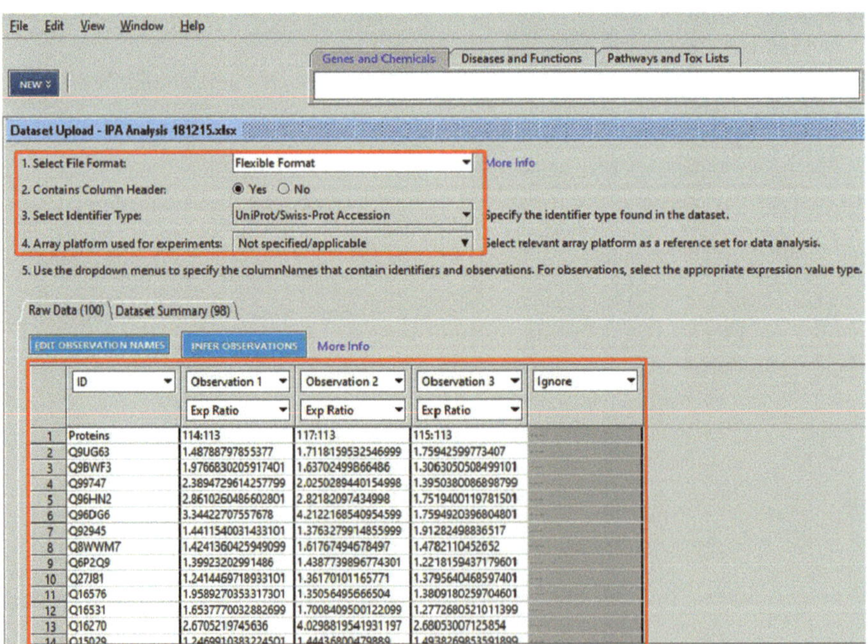

Fig. 16.16 Creation of a dataset with the IPA software. *Red* rectangles spotlight the basic steps to perform an analysis for a dataset

option in the IPA window (Fig. 16.15), select the file you wish to analyze and save the file. For illustration purposes, we analyzed proteins differentially expressed in common in Luminal A (MCF7 and T47D) and Claudin-low (MDA-MB-231) breast cancer cell lines from Calderón-González et al. [18]. It is necessary to specify the following information for the data that you wish to analyze:

1. File format: Flexible format
2. Column header: Yes
3. Identifier type: UniProt/Swiss-Prot accession
4. Array platform: In this case, it does not apply

Then the observation names must be edited, specifying the ID of proteins; in our case, the observation option 1 was selected (114:113. MCF7/MCF 10A), 2 (117:113. T47D/MCF 10A), 3 (115:113 MDA-MB-231/MCF10A), according to data number. Finally, the quantitative data format must be specified, which in our case we chose Exp Ratio (Fig. 16.16).

To carry out IPA Core analyses, we first uploaded the dataset previously created and then specified the parameters according to the goals of our study. The IPA platform gives different options to filter the data. We filtered the parameters for breast cancer disease as follows:

1. General settings: Ingenuity knowledge base (genes only). Considering direct and indirect relationships
2. Networks: 25 interaction networks with 35 molecules per interactome. Include endogenous chemicals (default parameters)
3. Data sources: All
4. Confidence: All
5. Species: Human with stringent filter
6. Tissues and cell lines: Mammary gland as organ and all breast cancer cell lines of database
7. Mutations: All.

At the end of the page, cutoff values are selected. We focused on up- and down-regulated proteins (Fig. 16.17). The statistical significance was determined by Fisher´s Exact Test, for which the p-value cutoff was set at 0.05. As a result of this analysis, we obtained three summary results, one for each observation. Then, we performed a Core Comparison Analysis. This analysis was performed using the following option (Core: Compare analysis). The procedure also requires

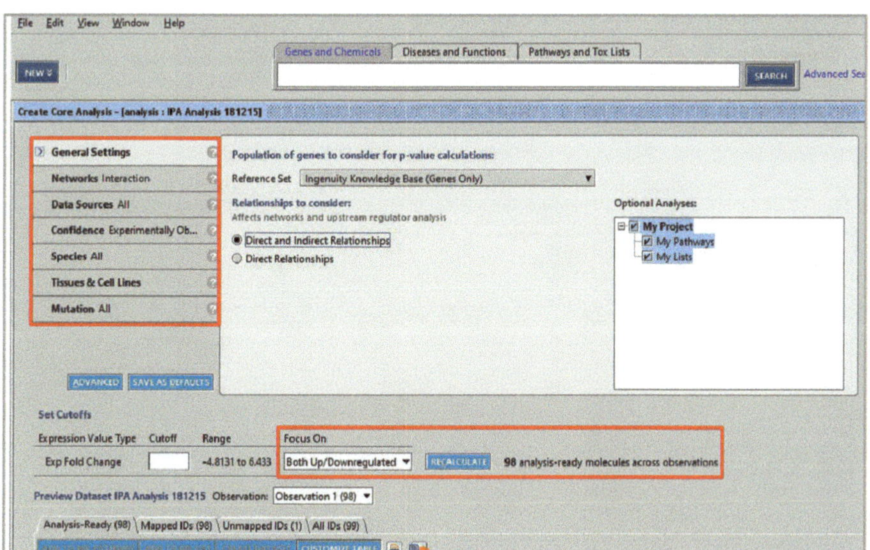

Fig. 16.17 Core parameters needed for IPA analysis. Figure shows the different parameters that need to be set to perform and delimit a Core Analysis. In this case the analysis was focus on breast cancer disease

selecting files for comparison. The summary results for all observation are reported in a single file. The Core Analysis result window shows different tool bars:

1. Canonical Pathways (Chart and HeatMap)
2. Upstream Analysis (Table and HeatMap)
3. Diseases & Functions (Chart and HeatMap)
4. Regulator effects (Table)
5. Networks (Networks for each observation or overlapping networks)
6. Molecules (Tables).

We focused our analysis on canonical pathway result obtained as a chart (Fig. 16.18a) or a HeatMap (Fig. 16.18b). In both cases, the number of up- and down-regulated proteins and their statistical probability were reported. Some of the

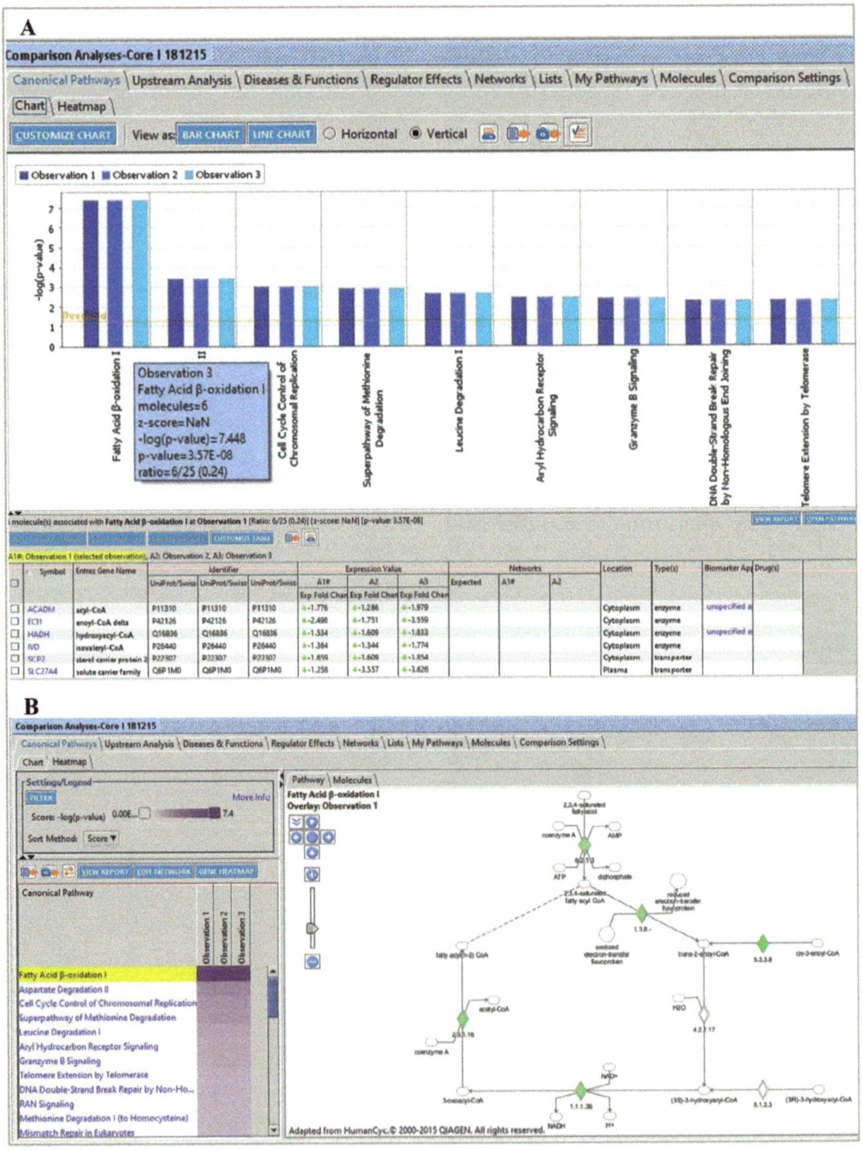

Fig. 16.18 Classification of proteins found up- or down-regulated in both Luminal A and Claudin-Low breast cancer cell lines into canonical pathways with IPA software. The result can be displayed as (**a**) Bar chart or (**b**) Heatmap

processes affected were: Fatty acid oxidation I (↓ACADM, ↓ECI1, ↓HADH, ↓IVD, ↓SCP2, ↓SLC27A4 with a p-value 3.57×10^{-8}), aspartate degradation II (↓GOT2 and ↓MDH2, p-value of 3.78×10^{-4}), cell cycle control of chromosomal replication (↑MCM3, ↑MCM4 and ↑MCM6, p-value 1.01×10^{-3}), telomere extension by telomerase (↑XRCC5 and ↑XRCC6, p-value 5.44×10^{-3}), and protein and ubiquitination pathway (HSP90AB1, ↑PSMA3, ↑PSMC1, ↑PSMD2, ↓PSMD3, and ↑PSMD7, p-value 8.65×10^{-3}).

Diseases functions are divided into two categories, Diseases and Bio Functions and Tox Functions. We only obtained the first category. We found the affected processes to be:

1. Cell-to-cell signaling and interaction: Formation of focal adhesions (↓CTNND1 and ↑STMN1, p-value 1.30×10^{-3})
2. Cellular assembly and organization: Formation of focal adhesions (↓CTNND1 and ↑STMN1, p- value 2.39×10^{-2}) and polymerization of microtubules (↑STMN1, p-value 2.39×10^{-2})
3. Cellular function and maintenance: Formation of focal adhesions (↓CTNND1 and ↑STMN1, p-value 1.30×10^{-3}) and polymerization of microtubules (↑STMN1, p-value 2.39×10^{-2})
4. Cell death and survival: Anoikis (↓CTNND1 and ↑ILK, p-value 3.99×10^{-3}) and cytotoxicity of breast cancer cell lines (↓RELA, p-value 3.17×10^{-2})
5. Drug metabolism: Synthesis and oxidation of tretinoin (↓ALDH1A3, p-value 8.02×10^{-3})
6. Cellular development: Epithelial-mesenchymal transition of breast cancer cell lines (↑ILK and ↑STMN1, p-value 4.45×10^{-2}) among other processes

The interactome data obtained in three separate experiments were processed resulting in identification of two principal networks related to: (1) Cellular development, cellular growth and proliferation, cellular movement, cell death and survival, and cancer, with a score of 19 and 14 molecules involved (↓ALDH1A3, ↓CTSD, ↓DLG1, ↓EZR, ↑FUS, ↑ILK, ↑KPNB1, ↓MVP, ↓RELA, ↓S100A8, ↑SET, ↓SLC25A5, ↑XRCC5 and ↑XRCC6) (Fig. 16.19a). (2) Cell death and survival, cellular development, DNA replication, recombination and repair, cancer and hereditary disorder obtained 12 proteins (↑ABCF2, ↑CAD, ↓CTNND1, ↓CYCS, ↑HSP90AB1, ↓LGALS3BP, ↑MAT2A, ↑MCM6, ↑MSH6, ↑NUMA1, ↑PCNA, ↑SNRPG) with a score of 15 (Fig. 16.19b). Proteins in red and green represent the up- and down- regulated proteins, respectively. Small molecules are shown in gray color to highlight their relationship with our proteins. Created Networks can be exported to IPA pathway for subcellular localization and decoration of network with organelles and backgrounds.

16.7 Biomarkers Module

To perform biomarker filtration, we used the Biomarkers module. As a first step in using the Biomarker module, we selected the analysis dataset function and choose a dataset created previously. Next we chose the following parameters:

1. Species: Human
2. Tissues and cell lines: mammary gland as organ and breast cancer cell lines
3. Molecules: All
4. Diseases: Cancer
5. Biofluids: All
6. Biomarkers: All biomarkers application (diagnosis, disease progression, efficacy, not applicable, prognosis, response to therapy, safety and unspecified application) and breast disease (breast cancer, breast carcinoma, ductal carcinoma, ductal carcinoma in situ, infiltrating ductal breast carcinoma, infiltrating lobular breast carcinoma, invasive ductal breast cancer, lobular breast cancer, mammary neoplasm, metastasic breast cancer) (Fig. 16.20a).

We then ran the analysis, saved the results, and performed a comparative analysis on our

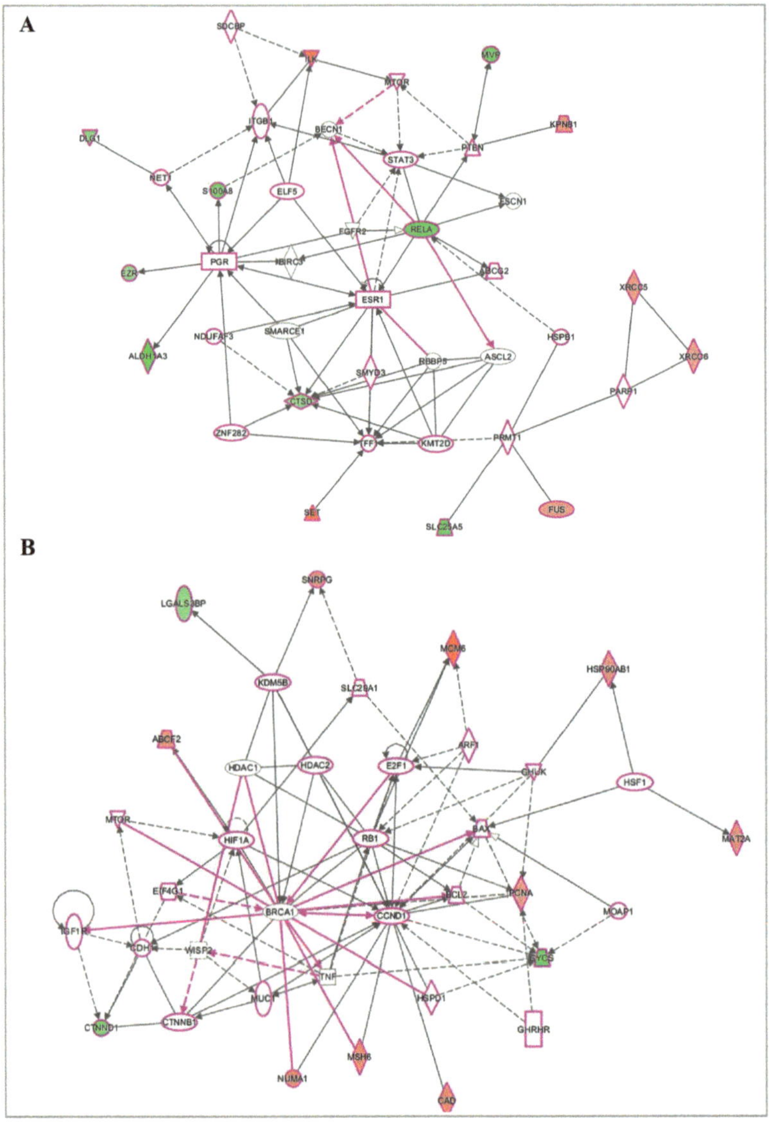

Fig. 16.19 IPA Networks of proteins found up- or down-regulated in both Luminal A and Claudin-Low breast cancer cell lines. The up- and down-regulated proteins are represented by molecules in *red* and *green* color, respectively. (**a**) Interactome related to cellular development, cellular growth and proliferation, cellular movement, cell death and survival, and cancer. (**b**) Interactome involved in cell death and survival, cellular development, DNA replication, recombination and repair, cancer and hereditary disorder

datasets. In this analysis, we had three datasets to compare (Fig. 16.20b) and only considered proteins found in all three datasets. We found four candidate biomarkers common between the luminal A and Claudin-low cells falling into different biomarker application categories: unspecified application (↑KHSRP protein found in nucleus and ↓S100A8 with cytoplasmic localization), diagnosis, efficacy (↓RELA localized in nucleus and ↑STMN1 found in cytoplasm) RELA was also found related to the drug NF-kappa B decoy (Fig. 16.21). All proteins were found in blood and all are related to cancer; however, they are not unique to this disease, as they are found in other diseases.

16.8 Protein-Protein Interactions Databases

16.8.1 STRING

STRING (Search Tool for the Retrieval of Interacting Genes/Proteins) is a database of

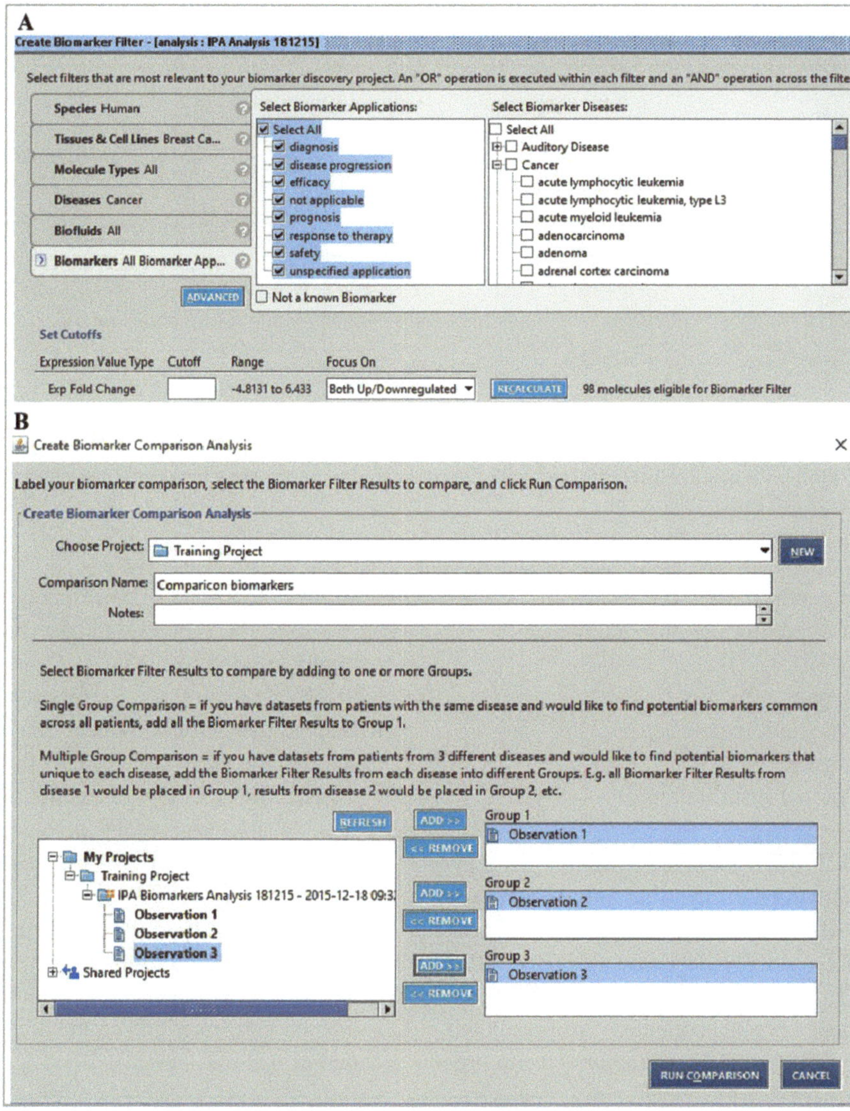

Fig. 16.20 Filter parameters for biomarker analysis in IPA software. (**a**) Creating a filter for putative biomarkers. (**b**) Comparison analysis between all observations (MCF7, T47D and MDA-MB-231)

known and predicted protein interactions [31]. This database was developed by the Center for Protein Research (CPR), The European Molecular Biology Laboratory (EMBL), The Swiss Institute of Bioinformatics (SIB), The University of Copenhagen (KU), The Technische Universität Dresden (TUD), and The Universität Zürich (UZH). STRING version 10.0 has 9,643,763 proteins from 2031 organisms. The main objective of this database is to integrate, predict and unify several protein-protein interactions [31, 32]. Associations between proteins can be physical (direct) or functional (indirect). The functional associations are defined as the interaction between two proteins that participate or contribute in the same cellular process or metabolic pathway, as well as other functional processes [32–34].

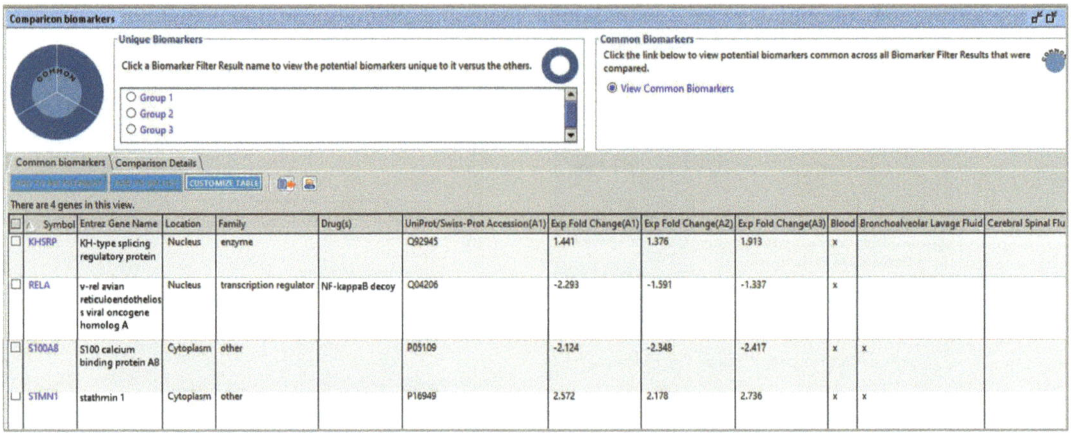

Fig. 16.21 Result of biomarker filter. Figure shows the four common biomarkers between. Luminal A and Claudin-low breast cancer cell lines

STRING database uses the following type of information to predict possible interaction:

1. Genomic data
2. High throughput experiments
3. Co-expression
4. Data extracted from literature

STRING import knowledge about protein-protein interactions from other databases such as IntAct, MINT, BioGRID, Reactome, KEGG, BIND, HPRD, DIP, NCI-Nature Pathway Interaction, GO, and EcoCyc [33]. In addition, STRING has a large collection of predicted interactions that are produced *de novo* using prediction algorithms [33, 35]. *De novo* predictions are made using genomic context such as conserved genomic neighborhood, gene fusion events, and co-occurrence of genes across the genome [34]. STRING also performs searches for genes with similar transcriptional response through a variety of conditions (co-expression) [33]. Information extracted from literature is another source used to extract protein association information from. In this case, STRING obtains information from all abstracts in PubMed database directly [36]. Finally, STRING assigns a probabilistic confidence score to all associations obtained through comparison of the association predictions against a reference database. STRING uses the KEGG database because this is manually curated [32, 37].

STRING website is composed of two components, the first component deals with protein analysis and the second covers the platforms (Fig. 16.22). The window of results displays the networks of protein-protein associations. The resulting interactome is represented by connecting lines. Each one of these lines represents different types of evidence. Networks can be viewed in three forms:

1. Evidence view in which connections are color coded as follows, neighborhood (green), gene fusion (red), co-occurrence (blue), co-expression (black), experiments (purple), database (light blue), text mining (yellow), and homology (gray)
2. Confidence view in which the thickness of connecting lines correlates with the strength of the associations
3. Interaction view in which the type of interactions is color coded as follows; activation (brilliant green), inhibition (red), binding (blue), phenotype (brilliant blue), catalysis (purple), posttranslational modifications (lilac), reaction (black) and expression (olive green)

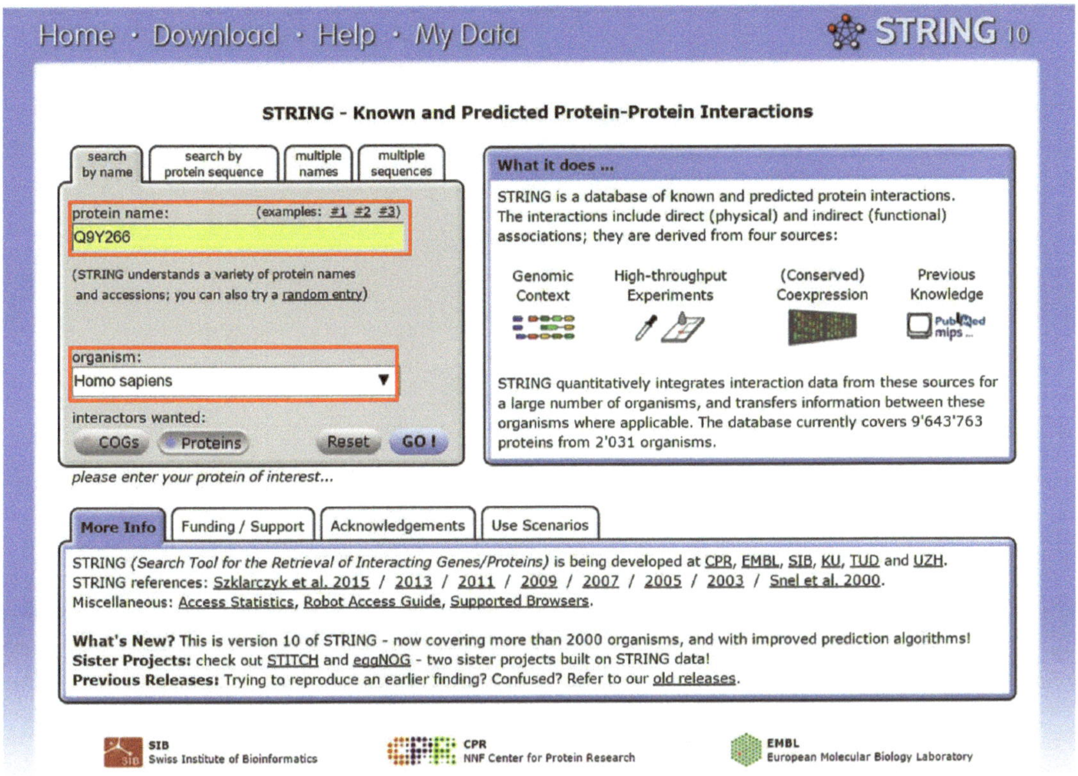

Fig. 16.22 STRING window view. The STRING webpage has different options to perform interaction analysis. The search can be done by the name of the protein or a protein sequence. The analysis can be performed for multiple proteins in the same way. In addition, the main page has various tabs with information about this platform

STRING has also an interactive view. In this option the network can by reordered by moving the proteins in the network. In advanced option, the network can be enriched into a GO Biological Processes, GO Molecular functions, GO Cellular components, KEGG Pathways, PFAM domains, INTERPRO domains, and Protein- Protein interactions. In each enrichment category, a new window is displayed containing a list of interactors, which contains different processes, the number of proteins involved as well as a p-value.

16.8.2 Protein-Protein Interaction Networks

To determine the protein-protein interaction of overexpressed NUDC protein exclusively found in Claudin-low breast cancer cell line [18], we accessed the STRING website http://string-db.org/.

To generate a network of protein interactions, a list (one or more) of protein names, accession number, or sequence, as well as the organism or species they originated from, need to be specified (Fig. 16.22). At the bottom of the result window there is a parameter box. The options in the parameter box are used to select the active prediction algorithm. The confidence score as well as the number of interactors can be adjusted as well (Fig. 16.23). The interactome can be seen according to evidence (Fig. 16.24a), confidence (Fig. 16.24b) and action (Fig. 16.24c). In each network, a score is generated according to each protein's interaction evidence. In addition, a brief description for each protein is also displayed (Fig. 16.24). NUDC protein is associated with PAFAH1B1

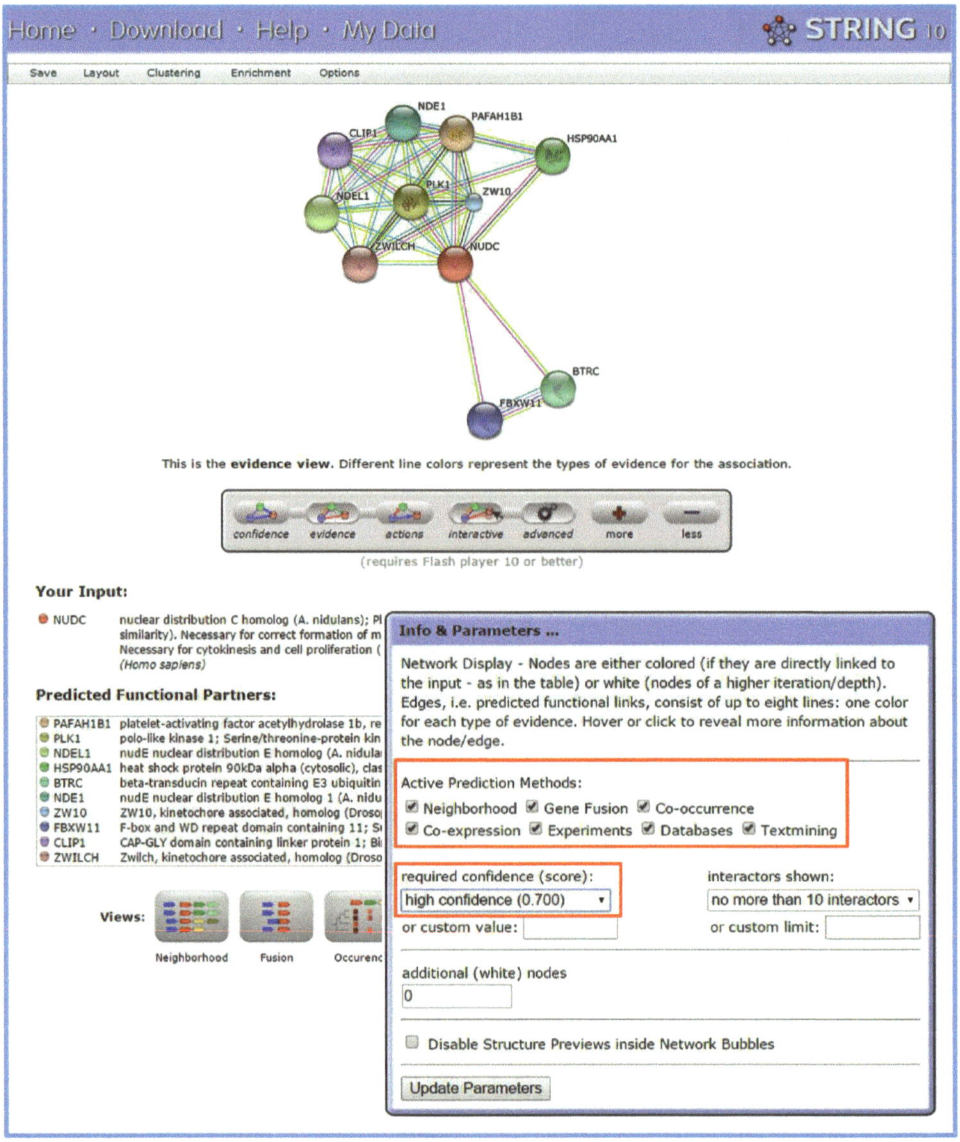

Fig. 16.23 STRING results view. A window containing different parameters is shown at the *bottom*. The active prediction methods as well as the confidence of the interactions in the network can be selected in this window

(platelet-activating factor acetylhydrolase 1b), PLK1 (polo-like kinase 1), NDEL1 (nudE nuclear distribution E homolog (A. nidulans)-like 1), HSP90AA1 (heat shock protein 90 kDa alpha), BTRC (beta-transducin repeat containing E3 ubiquitin protein ligase), NDE1 (nudE nuclear distribution E homolog 1 (A. nidulans)), ZW10 (ZW10, kinetochore associated, homolog (Drosophila), FBXW11 (F-box and WD repeat domain containing 11), CLIP1 (CAP-GLY domain containing linker protein 1) and ZWILCH (Zwilch, kinetochore associated, homolog (Drosophila)). All interactions have more than 0.90 score. In

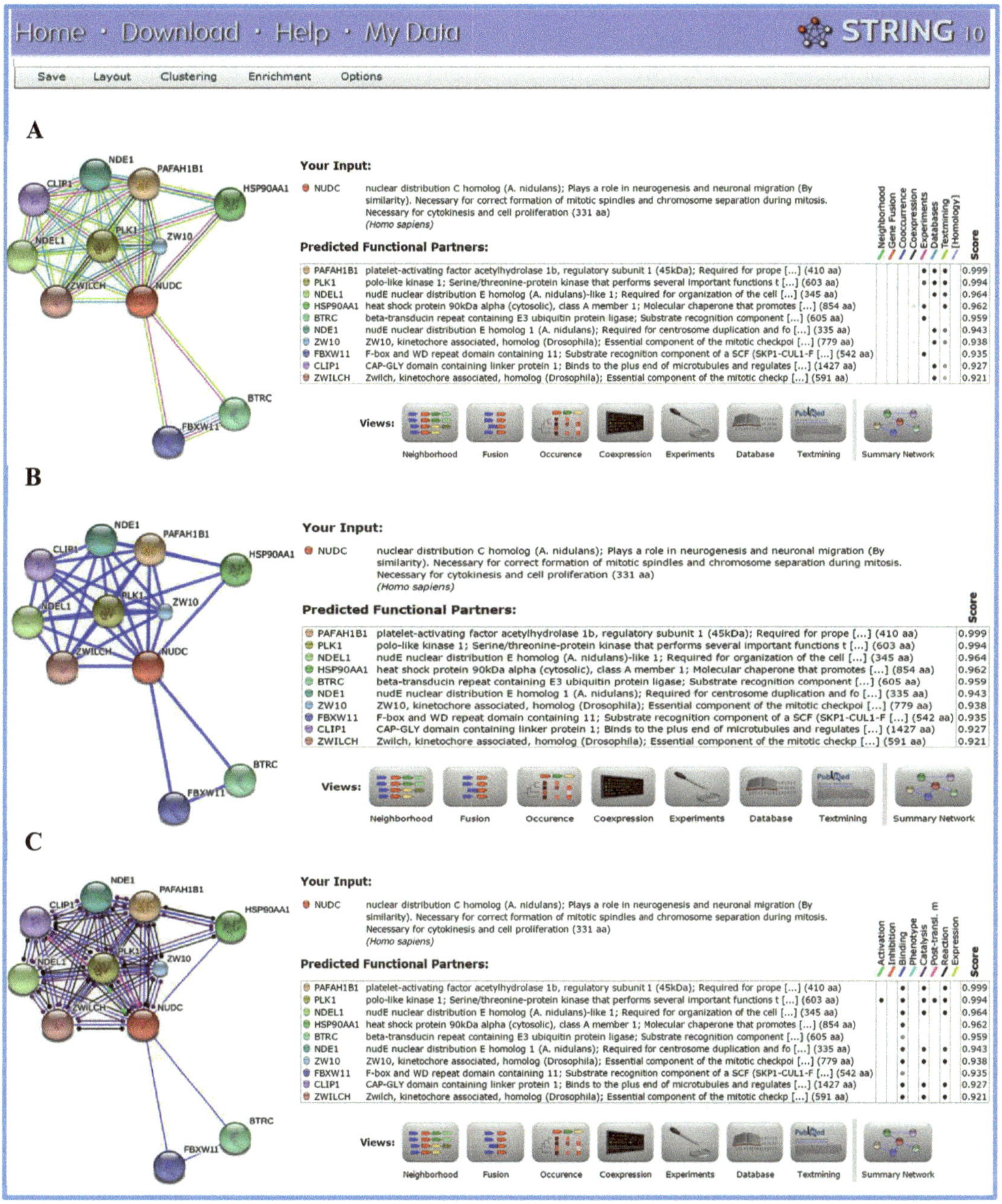

Fig. 16.24 Interaction network of NUDC protein. This polypeptide is overexpressed exclusively in Claudin-low breast cancer cell line. The interactome can be seen in three options. (**a**) Evidence view, where the color lines represent the diverse evidences of interactions: *Green*, neighborhood; *red*, gene fusion; *blue*, co-occurrence; *black*, co-expression; *purple*, experiments; *light blue*, database; *yellow*, text mining; *gray*, homology. (**b**) Confidence view where thicker lines represent stronger associations. (**c**) Interaction view, where the different modes of action are represented by different colors. *Brilliant green*, activation; *red*, inhibition; *blue*, binding; *brilliant blue*, phenotype; *purple*, catalysis, lilac, PTMs; *black*, reaction; *olive green*, expression. The three view modes provide a score of the different evidence of interaction

addition, the network was enriched into GO Biological Processes. Processes showed Enrichment with statistical significance were:

1. Mitotic prometaphase (4.940×10^{-13})
2. Mitotic anaphase (8.089×10^{-12})
3. Mitotic M phase (6.309×10^{-11})
4. M phase (6.309×10^{-11})
5. Mitotic cell cycle phase (4.300×10^{-10})
6. Cell cycle phase (4.300×10^{-10})

All processes mentioned above have at least eight proteins involved. We selected the cell cycle phase process as an example. The proteins enriched in this process are shown in color red (Fig. 16.25a). We selected the interacting proteins NUDC and ZW10 as examples to extract interaction information. ZW10 was selected because it is an essential component of the mitotic checkpoint that prevents cells from prematurely exiting mitosis. The evidence supporting the functional link between these two proteins are the following:

1. Co-expression (putative homologs are co-expressed in other species, score 0.065)
2. Association in curated database (score 0.900)
3. Co-mentioned in PubMed abstracts (score 0.285)

Also putative homologs are mentioned together in other species (score 0.192). The combined score is 0.938. There is also activity evidence, such as catalysis (score 0.900), binding (score 0.900) and reaction (score of 0.900) that support the interaction between these two proteins (Fig. 16.25b). For proteins selected in a network, STRING displays a window with information about their 3D structure, as well as links to Ensembl, GeneCards, KEGG, Nextprot and UniProt. Also, STRING can show the protein sequence and the sequence of its homologs in organisms stored in STRING. NUDC has three 3D structures obtained from Protein DataBase (PDB) (Fig. 16.25c). As mentioned above, STRING can perform network analysis for multiple proteins as well. We performed an interactome analysis for the up- and down-regulated proteins common in Luminal A (MCF7 and T47D) and Claudin-low (MDA-MB-231) breast cancer cells lines [18]. In this case, we used the highest confidence (0.900) possible to generate our interaction network. The network has several interaction nodes related to:

1. Energy metabolism
2. Translation
3. Proteasome
4. Replication and repair
5. Transcription

Red and green arrows indicate up- and down-regulated proteins, respectively (Fig. 16.26).

16.8.3 MINT

The Molecular INTeraction database or MINT is an open source protein-protein interaction database developed at the Università degli Studi di Roma Tor Vergata that has been experimentally verified [38, 39]. The webpage can be found at http://mint.bio.uniroma2.it/mint/Welcome.do (Fig. 16.27). The current version of MINT database (November 2015) contains 241,458 interactions, corresponding to 35,553 proteins and 5554 PMIDS (PubMed unique identifiers). Species included are *Drosophila melanogaster, Saccharomyces cerevisiae, Caenorhabditis elegans*, mammals and viruses, with mammal databases being the main datasets. Evidences for protein-protein interactions include association studies, co-localization, direct interactions, interactions in form of complexes, enzymatic reactions, and high throughput studies. Protein-protein interactions have been identified by a number of methods including co-immunoprecipitation with either anti-bait or anti-tag antibodies, fluorescence microscopy, peptide arrays, protein arrays, pull down experiments, SPR, tandem affinity isolation, two hybrid arrays, two hybrid pooling, and two hybrid systems, etc. Additionally, the MINT database is freely available for academic and commercial users.

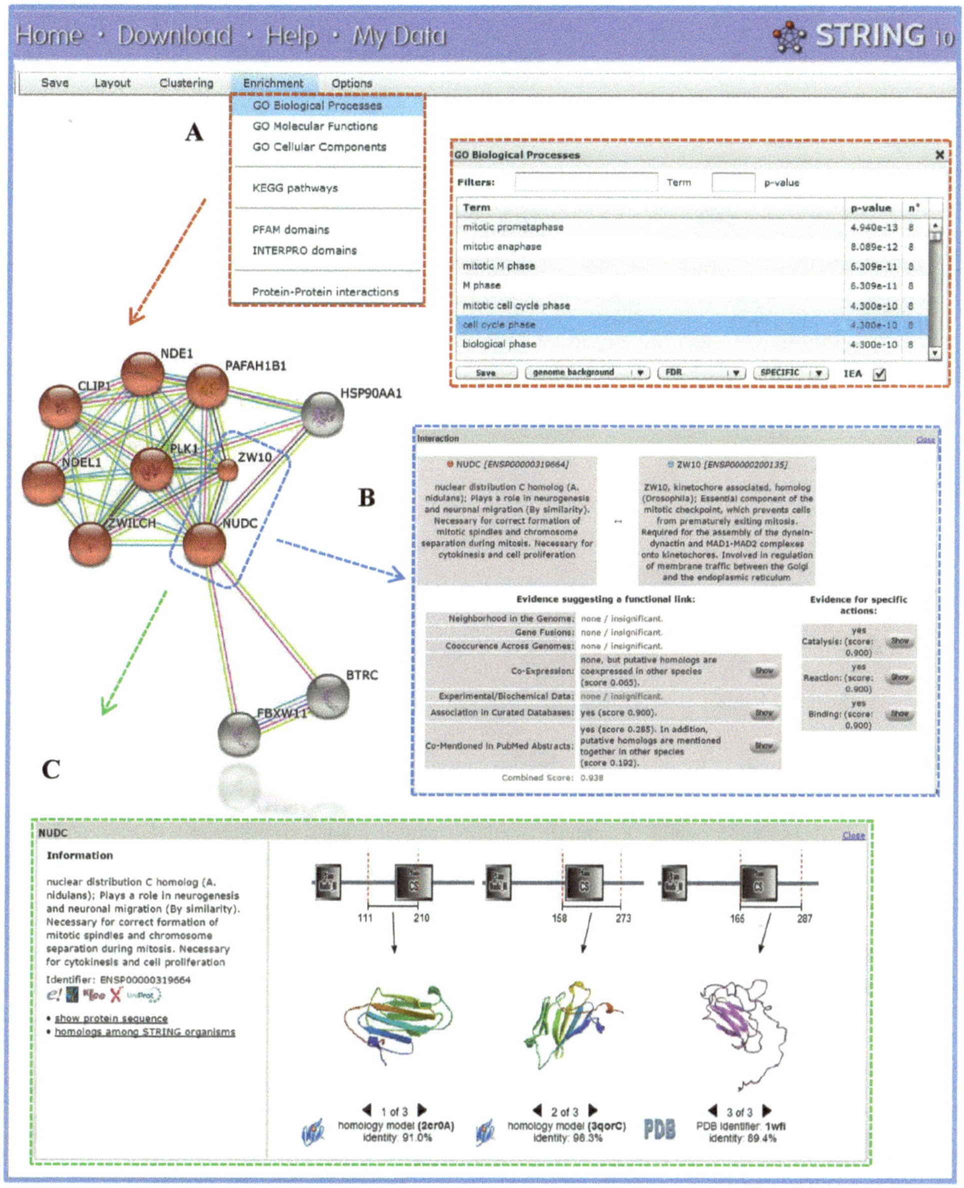

Fig. 16.25 Interaction network of NUDC overexpressed protein found exclusively in Claudin- low breast cancer cell line. STRING platform provides different information for the generated network. (**a**) Network enrichment for GO Biological Processes. The proteins in *red* which have a statistical significance (*p*-value) are involved in cell cycle phase. (**b**) Evidence supporting interaction between NUDC and ZW10. (**c**) 3D protein structure information

There are three additional databases available via MINT website including HomoMINT, Domino, and VirusMINT. The first one is an inferred network for human; the second is specialized in domain-peptide interactions, and the last is a protein-protein interaction database specialized on viruses.

Protein interaction searches in MINT database (Fig. 16.28a) can be carried out using PubMed ID, D.O.I, or author's name. Alternatively, this

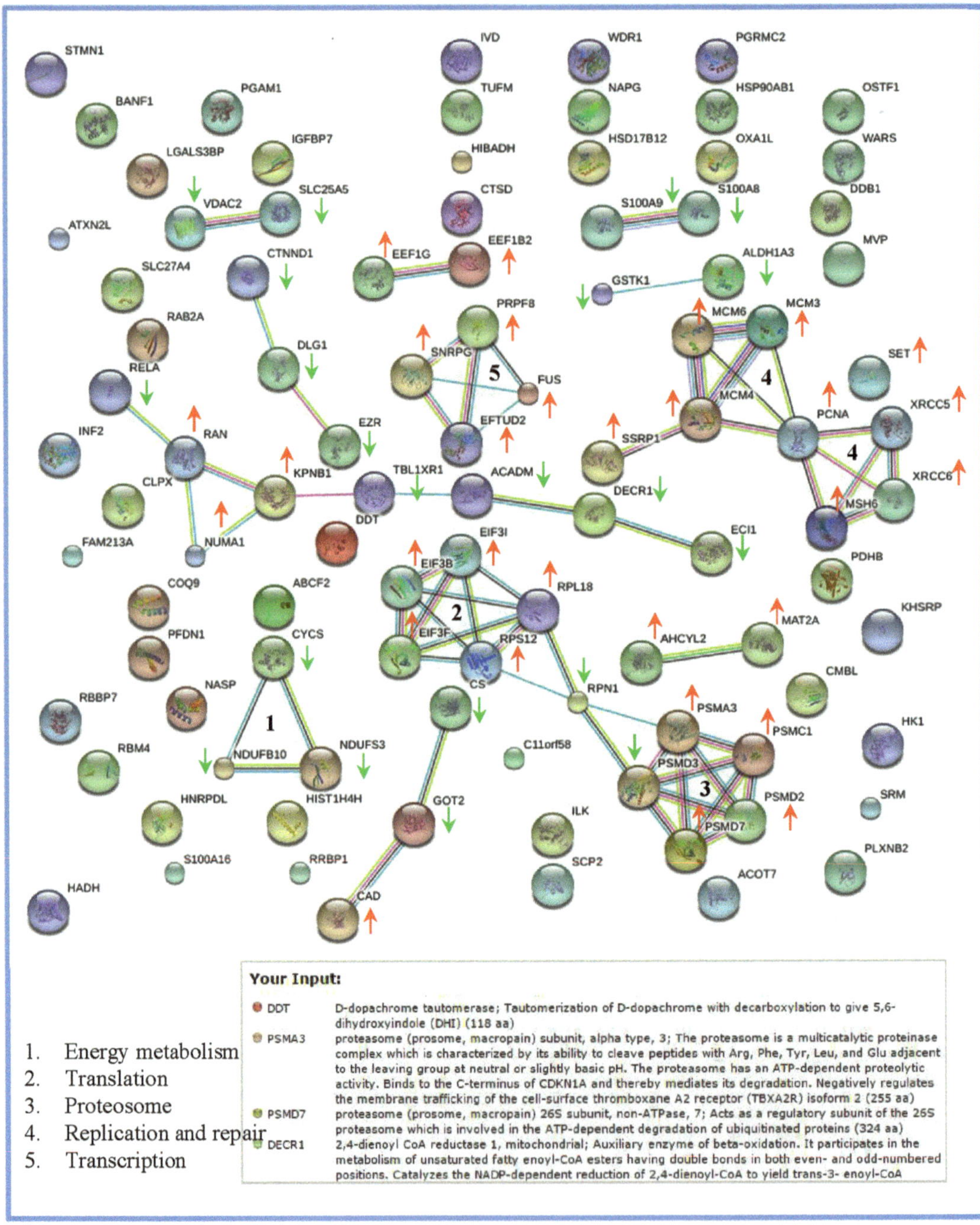

Fig. 16.26 STRING interaction network of proteins found up- or down-regulated in both Luminal A (MCF7 and T47D) and Claudin-low (MDA-MB-231) breast cancer cell lines. This list has interaction nodes related to: (*1*) Energy metabolism, (*2*) Translation, (*3*) Proteosome degradation, (*4*) Replication and repair, (*5*) Transcription. Colored lines represent different evidence of interaction: *Green*, neighborhood; *red*, gene fusion; *blue*, co-occurrence; *black*, co-expression; *purple*, experiments; *light blue*, database; *yellow*, text-mining; *gray*, homology. *Red arrows* indicate up-regulation and *green arrows* down-regulation. A box with information about some proteins is also shown

Fig. 16.27 Homepage of the Molecular INTeraction database, MINT

database can be searched against protein or gene name, protein accession number (Protein AN) or keywords. Protein accession numbers recognized by MINT search engine are FlyBase, Ensembl, Human Identified Gene Encoded Large Protein Analyzed database (HUGE), Nematode database (WormBase), OMIM, REACTOME pathway database, the *Saccharomyces* Genome Database (SGD), and Universal Protein Resource Knowledgebase (UniProtKB).

To demonstrate how MINT database works, we selected the vesicle-fusing ATPase NSF (P46459) for analysis. This protein is part of a set of proteins that were found overexpressed in several breast cancer cell lines [18]. To follow our analysis, click on the Search tab and type P46459 (Fig. 16.28, arrow 1) and then select the organism (Fig. 16.28, arrow 2) and then press the Search key (Fig. 16.28, arrow 3). Results show certain information for the queried protein including its ID, species, synonyms, domains found in query, a link to its role in diseases, its gene ontology, references covering the target protein, prediction of its modular domain interactions (ADAN), and its orthologs in MINT database (Fig. 16.28). Results also display a window containing a list of molecules interacting with the target according to MINT database, evidence for each interaction and a global score for each interaction (Fig. 16.28).

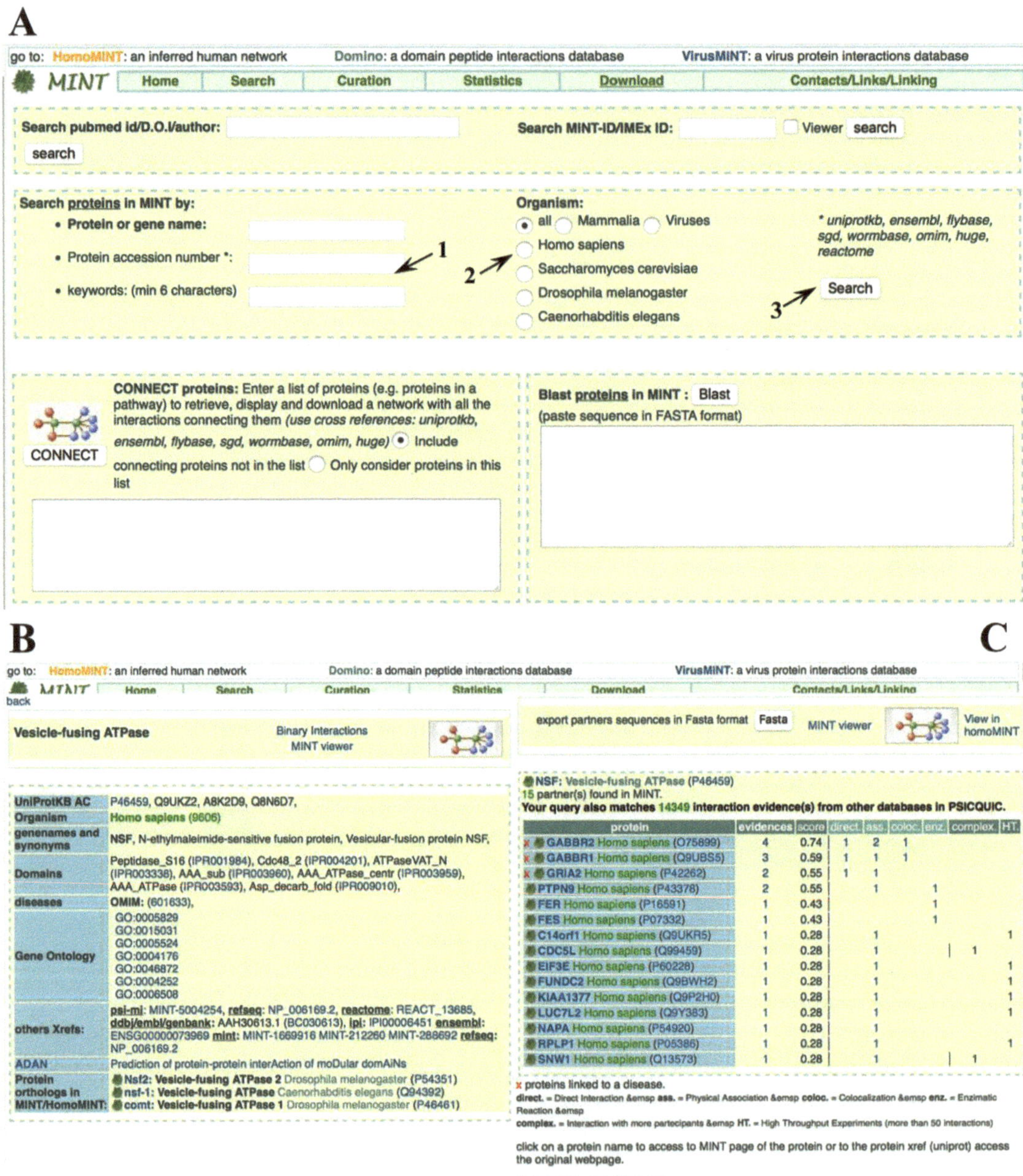

Fig. 16.28 MINT search webpage. (**a**) Search in MINT can be performed using: (*1*) Gene or protein name, Protein ID or keywords and the species of interest or the whole database, (*2*) Protein sequence in FASTA format, (*3*) a list of proteins. (**b**, **c**) Result of a query for vesicle-fusing ATPase NSF from Homo sapiens (UniProtKB/Swiss-Prot ID P46459). (**c**) List of NSF interactors are shown

Clicking on the MINT viewer will generate a list of interactions that are displayed as a function of score threshold. For each partner, a number showing evidence for interaction is shown (Fig. 16.29). As an example, we clicked on number 4 and a new window appeared showing the partner name, ID, and techniques used to determine the interaction, as well as a PubMed identifier containing this information (Fig. 16.29).

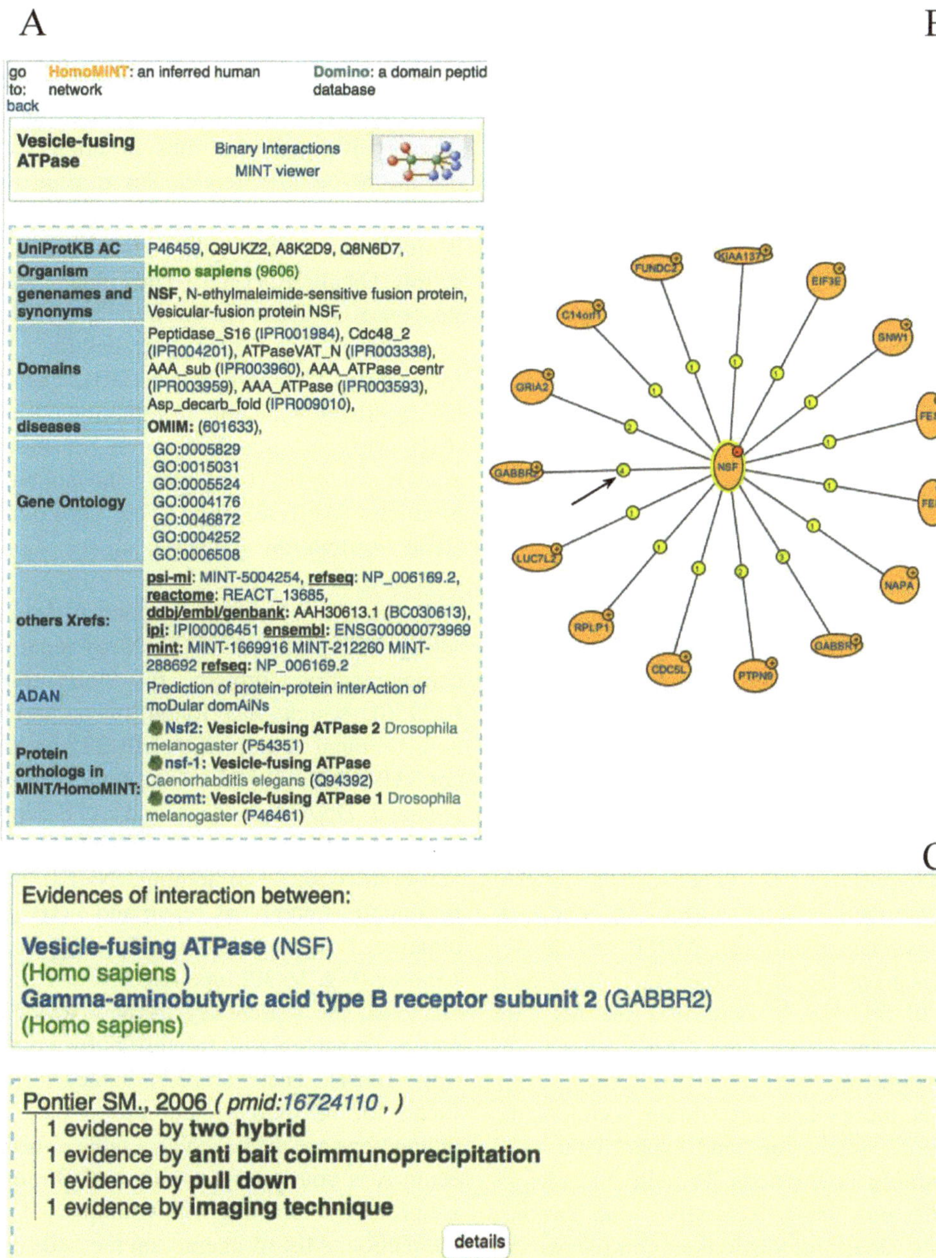

Fig. 16.29 Binary interactions of the N-ethylmaleimide-sensitive fusion protein NSF viewed in MINT database. (**a**) Basic information queried for NSF. (**b**) Binary interaction map of NSF with 15 interactors found in MINT database. (**c**) Selecting number 4 in (**b**), a new window is displayed showing the name of the corresponding interactor (GABBR2, Gamma-aminobutyric acid type B receptor subunit 2) and the experimental methods used to determine this interaction, as well as the PMID ID for the publication describing it

16.8.4 IntAct

IntAct is a database of protein-protein interactions, as well as a suite of analytical tools at The European Bioinformatics Institute (EBI), which is part of the European Molecular Biology Laboratory (EMBL) [40, 41]. All information has been curated by experts at the IntAct team.

This freely available database can be accessed through its webpage http://www.ebi.ac.uk/intact/.

As of November 26th, 2015 this database had registered 355,819 interactions, which included 89,340 interactors (proteins) described in 36,864 experiments, 13,892 PMIDs, and 564,831 binary interactions. Methods used for the determination of protein-protein interactions include tandem affinity purification, anti-tag co-immunoprecipitation, two hybrid systems, pull down experiments, two hybrid arrays, anti-bait co-immunoprecipitation, two hybrid pooling approach, and co-sedimentation, among others. The source of information mainly comes from human (42.5 %), various *S. cerevisiae* strains (22.8 %), *Mus musculus* (11.3 %), and *D. melanogaster* (8.1 %). Other species included are *Escherichia coli, C. elegans, A. thaliana, Campylobacter jejuni*, etc. MINT and IntAct databases have recently joined their individual efforts to optimize resources as the MIntAct project, thus avoiding duplication of activities [42].

IntAct model has three main components, interactions, interactors, and experiments used to determine interactions. Protein interactions are inferred using scientific publications, including binary interactions or complexes. An interactor can be defined as a biological molecule (mainly a protein) involved in a specific interaction. An interaction is not circumscribed to binary interactions only; it also includes interactions with more partners identified in the experiment performed, e.g. precipitation of multi-protein complexes. Search in IntAct database can be performed in different ways, including name of gene, protein, RNA or chemical compound, or UniProtKB, ChEBI (Chemical Entities of Biological Interest), RNA Central, PMID or IMEx (International Molecular Exchange) IDs. The principal page of IntAct (Fig. 16.30) contains links to other websites the might be of interest. These sites include MINT, UniProtKB, The Swiss Institute of Bioinformatics (SIB), The Interologous Interaction Database (I2D), The Innate Immune Response Database (Innate Database), Molecular Connections, The Extracellular Matrix Interactions Database (MatrixDB), The Modular Approach to Cellular Functions Resource (MB Info), a curated resource for functional analysis of agricultural plant and animal gene products (AgBase), and The cardiovascular Gene Annotation database at the London's Global University (UCL).

As an example of the function of IntAct, we selected the protein XRCC6 (X-ray repair cross-complementing protein 6, UniProtKB ID P12956), which was found overexpressed in both Luminal A and MDA-MB-231 breast cancer cell lines [18]. This protein is a single-stranded DNA-dependent and ATP-dependent 3′–5′ DNA helicase involved in DNA non-homologous end joining (NHEJ) required for double-strand break repair and V(D)J recombination. To reproduce our analysis, in the search window (Fig. 16.30) type XRCC6 or P12956 ID and push the search key. A new window will appear on screen with the results for your query (Fig. 16.31). There are 324 binary interaction found for XRCC6 protein up to date. These interactions are displayed as a table, where molecule A is your query or bait, and B molecules are proteins interacting with your query. For each interaction, a list of interaction methods used for the determination of such interactions is shown, their corresponding IDs, and the source database as well. When you click on the interactors tab, a new page will be shown containing a list of all interactors, showing the type of interactor, the number of interactions described, a link to access the description in UniProtKB, and a description of the interaction (Fig. 16.32). More information, including interactions described, the

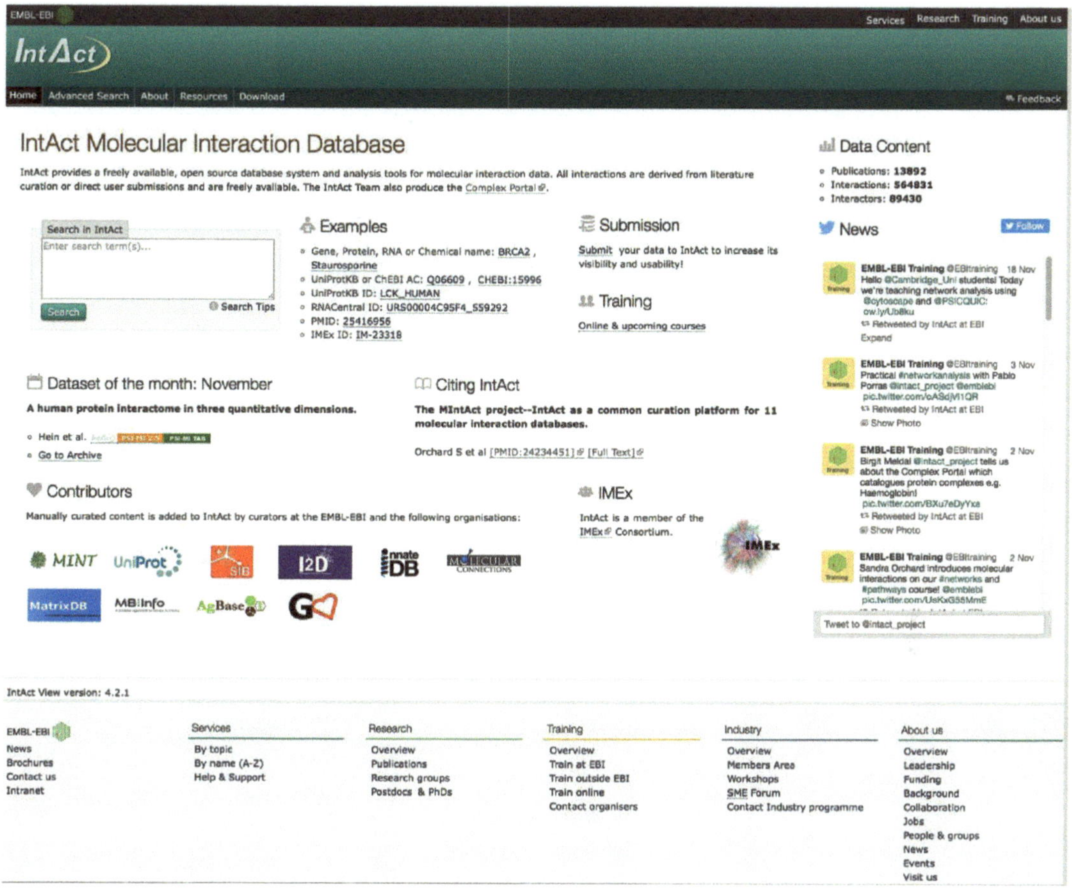

Fig. 16.30 Homepage of the IntAct Molecular Interaction Database

chromosome location in Ensembl webpage, the mRNA expression for interactor in the Expression Atlas webpage, and pathways is displayed when interactors are searched separately. The map of interactions for your query can be displayed in three layouts, force directed (Fig. 16.33), radial (Fig. 16.34) or circle (Fig. 16.35). In all cases, you can zoom in the graph with the tool window at the bottom.

Search can also be performed for a list of identifiers. The result will be more complex as all interactions for each member of your list will be shown. As an example, we only show the graph for ten proteins overexpressed in Luminal A and MDA-MB-231 breast cancer cell lines [18], where a total of 1101 binary interactions were found in database (Figs. 16.36, 16.37 and 16.38).

16.8.5 HPRD

The Human Protein Reference Database (HPRD) is a free web resource containing information of human proteins, including an information summary for each protein, their PTMs, protein-protein interactions, expression levels in tissues, mRNA and protein sequences, non-protein interactions, alternate names, participation in diseases, and domains found in proteins. All the information stored in this database is curated by a group of expert biologists from the Pandey Lab at Johns Hopkins University and the Institute of Bioinformatics in Bangalore, India [43]. The current version of HPRD is 9. It contains information for 30,047 proteins, 41,327 protein-protein interactions, 93,710 PTMs, 112,158

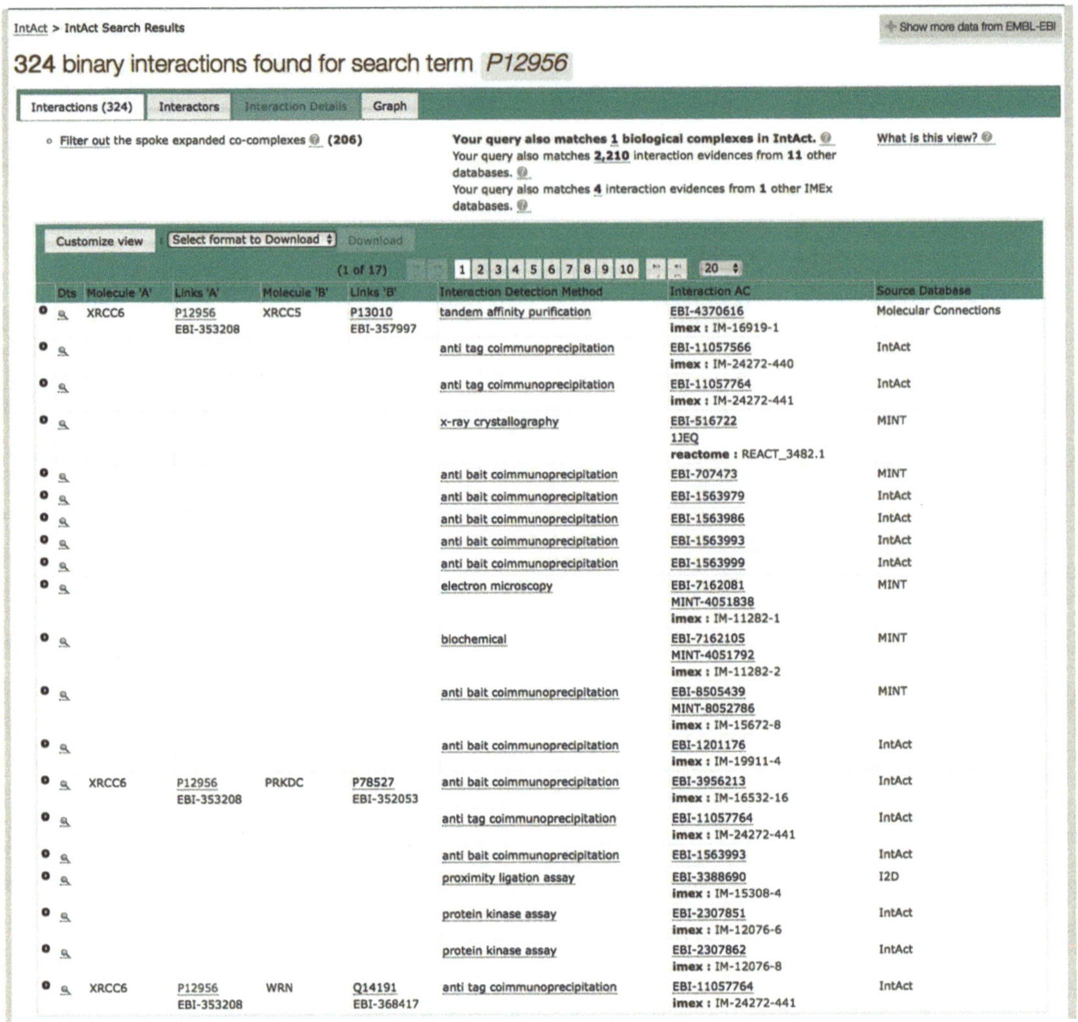

Fig. 16.31 List of binary interactions found for XRCC6 (the X-ray repair cross-complementing protein 6 from *Homo sapiens*, UniProtKB/Swiss-Prot ID P12956) in IntAct database. A total of 324 interactions were found for this protein

sites of protein expression, 22,490 sites of intracellular localization, 470 domains, and 453,521 PMIDs. In addition, two other applications have been recently added, the PhosphoMotif Finder and NetPath resources, which allow the identification of phosphorylation motifs for known kinases/phosphatases and binding motifs for phospho serine/threonine or phospho tyrosine in a compendium of signaling pathways in humans [43].

To perform a search, click on the Query key, type your query and push the Search button on the upper left part on screen (Fig. 16.39, arrow). There are several options for a query, including Protein Name, Accession Number (RefSeq, GenBank, OMIM, UniProtKB and Entrez Gene Name), HPRD identifier, Gene Symbol, Chromosome locus, Molecular Class (e.g. Nuclease, Serine Proteinase, Translation Regulatory protein, Glycosylase, etc.), PTMs (e.g. ADP

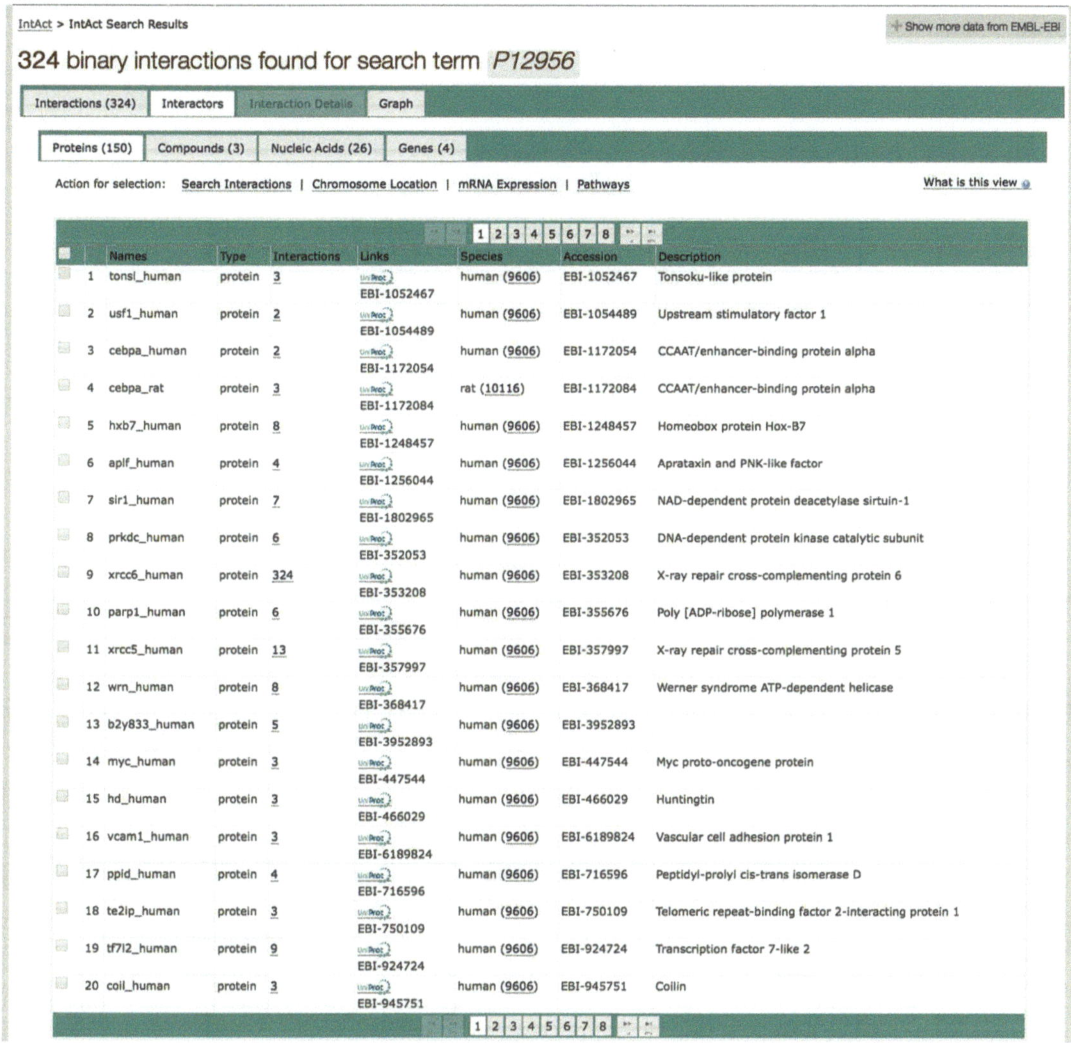

IntAct > IntAct Search Results

Show more data from EMBL-EBI

324 binary interactions found for search term P12956

Interactions (324) | Interactors | Interaction Details | Graph

Proteins (150) | Compounds (3) | Nucleic Acids (26) | Genes (4)

Action for selection: Search Interactions | Chromosome Location | mRNA Expression | Pathways

What is this view

1 2 3 4 5 6 7 8

	Names	Type	Interactions	Links	Species	Accession	Description
1	tonsl_human	protein	3	EBI-1052467	human (9606)	EBI-1052467	Tonsoku-like protein
2	usf1_human	protein	2	EBI-1054489	human (9606)	EBI-1054489	Upstream stimulatory factor 1
3	cebpa_human	protein	2	EBI-1172054	human (9606)	EBI-1172054	CCAAT/enhancer-binding protein alpha
4	cebpa_rat	protein	3	EBI-1172084	rat (10116)	EBI-1172084	CCAAT/enhancer-binding protein alpha
5	hxb7_human	protein	8	EBI-1248457	human (9606)	EBI-1248457	Homeobox protein Hox-B7
6	aplf_human	protein	4	EBI-1256044	human (9606)	EBI-1256044	Aprataxin and PNK-like factor
7	sir1_human	protein	7	EBI-1802965	human (9606)	EBI-1802965	NAD-dependent protein deacetylase sirtuin-1
8	prkdc_human	protein	6	EBI-352053	human (9606)	EBI-352053	DNA-dependent protein kinase catalytic subunit
9	xrcc6_human	protein	324	EBI-353208	human (9606)	EBI-353208	X-ray repair cross-complementing protein 6
10	parp1_human	protein	6	EBI-355676	human (9606)	EBI-355676	Poly [ADP-ribose] polymerase 1
11	xrcc5_human	protein	13	EBI-357997	human (9606)	EBI-357997	X-ray repair cross-complementing protein 5
12	wrn_human	protein	8	EBI-368417	human (9606)	EBI-368417	Werner syndrome ATP-dependent helicase
13	b2y833_human	protein	5	EBI-3952893	human (9606)	EBI-3952893	
14	myc_human	protein	3	EBI-447544	human (9606)	EBI-447544	Myc proto-oncogene protein
15	hd_human	protein	3	EBI-466029	human (9606)	EBI-466029	Huntingtin
16	vcam1_human	protein	3	EBI-6189824	human (9606)	EBI-6189824	Vascular cell adhesion protein 1
17	ppid_human	protein	4	EBI-716596	human (9606)	EBI-716596	Peptidyl-prolyl cis-trans isomerase D
18	te2ip_human	protein	3	EBI-750109	human (9606)	EBI-750109	Telomeric repeat-binding factor 2-interacting protein 1
19	tf7l2_human	protein	9	EBI-924724	human (9606)	EBI-924724	Transcription factor 7-like 2
20	coil_human	protein	3	EBI-945751	human (9606)	EBI-945751	Coilin

1 2 3 4 5 6 7 8

Fig. 16.32 List of binary interactions found for XRCC6 (the X-ray repair cross-complementing protein 6 from Homo sapiens, UniProtKB/Swiss-Prot ID P12956) in IntAct database. There are 150 proteins, three chemical compounds (XAV939, 15-deoxy-Delta(12,14)-prostaglandin J2 and Midostaurin), 26 nucleic acid molecules, and four genes (Klk3, kallikrein-related peptidase 3 encoding gene; Tmps2, Transmembrame protease serine 2). here only a list of 20 protein interactors is shown

Ribosylation, Glycation, Nitration, Sumoylation. Ubiquitination), Cellular Component, Domain Name, Motif, Expression Site, Length of Protein sequence, Molecular Mass, and Diseases (Fig. 16.40). To present an example, we searched NUMA1. Results are shown in Fig. 16.41. Information retrieved includes the name of protein (NUMA1 corresponds to the Nuclear mitotic apparatus protein 1, isoform 1), Molecular Class (Structural protein), Molecular Function (Structural molecule activity), and Biological Process (Cell growth and/or maintenance). Seven additional tabs are provided, which are Summary, Sequence, Interactions, External Links, Alternate Names, Diseases, PTMs, and Substrates. The General tab contains the

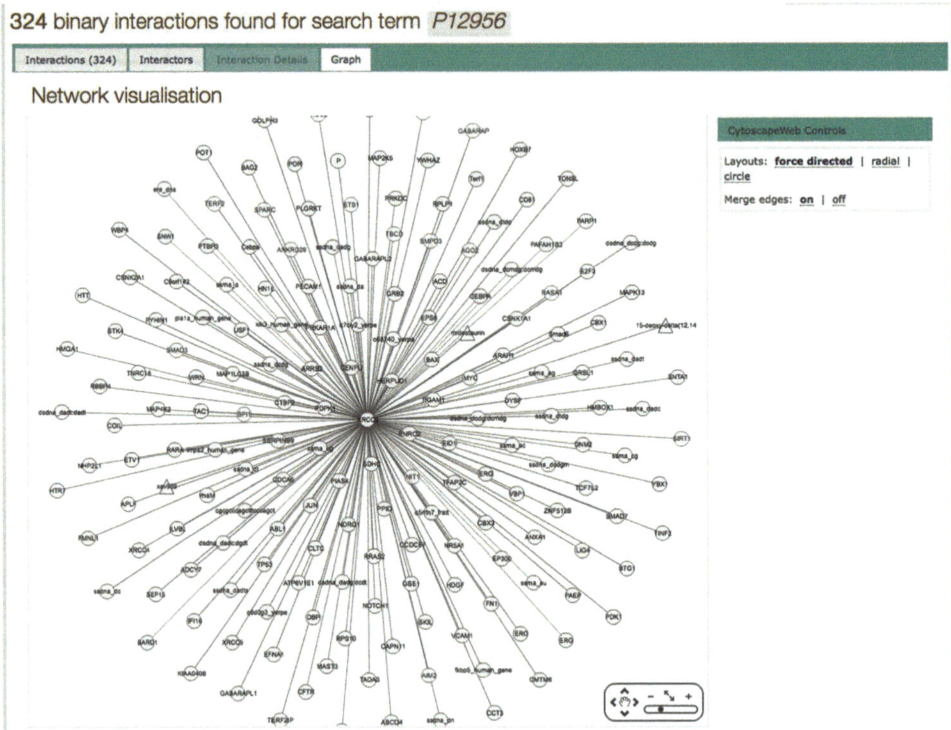

Fig. 16.33 Force-directed layout of the interaction map found for XRCC6 in IntAct database. XRCC6 protein is at the center of the map

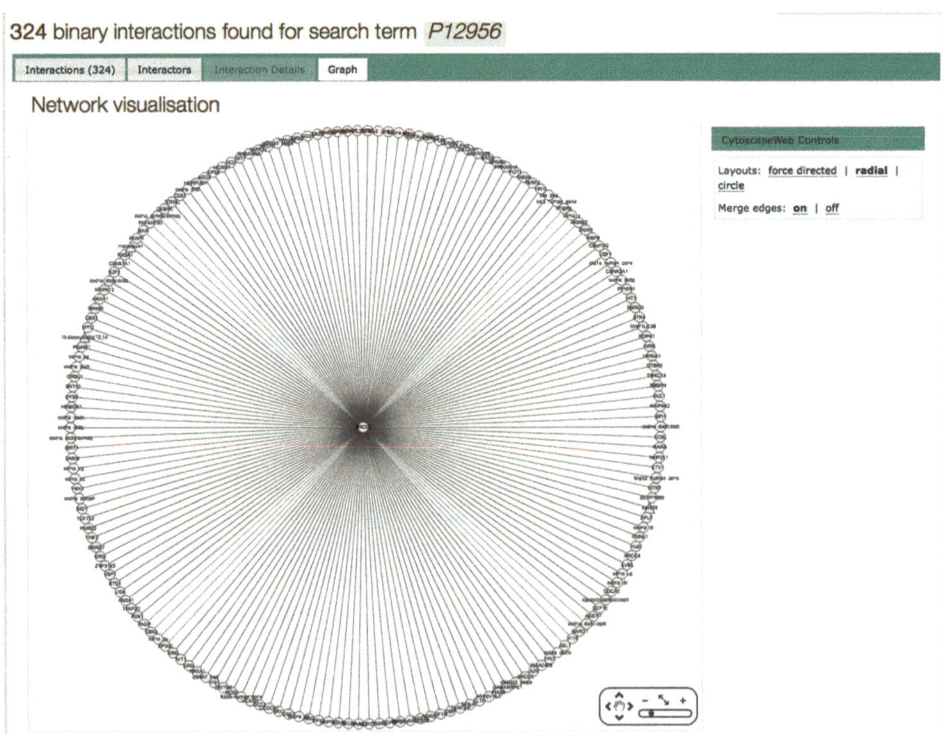

Fig. 16.34 Radial layout of the interaction map found for XRCC6 in IntAct database. XRCC6 protein query is at the center of the map

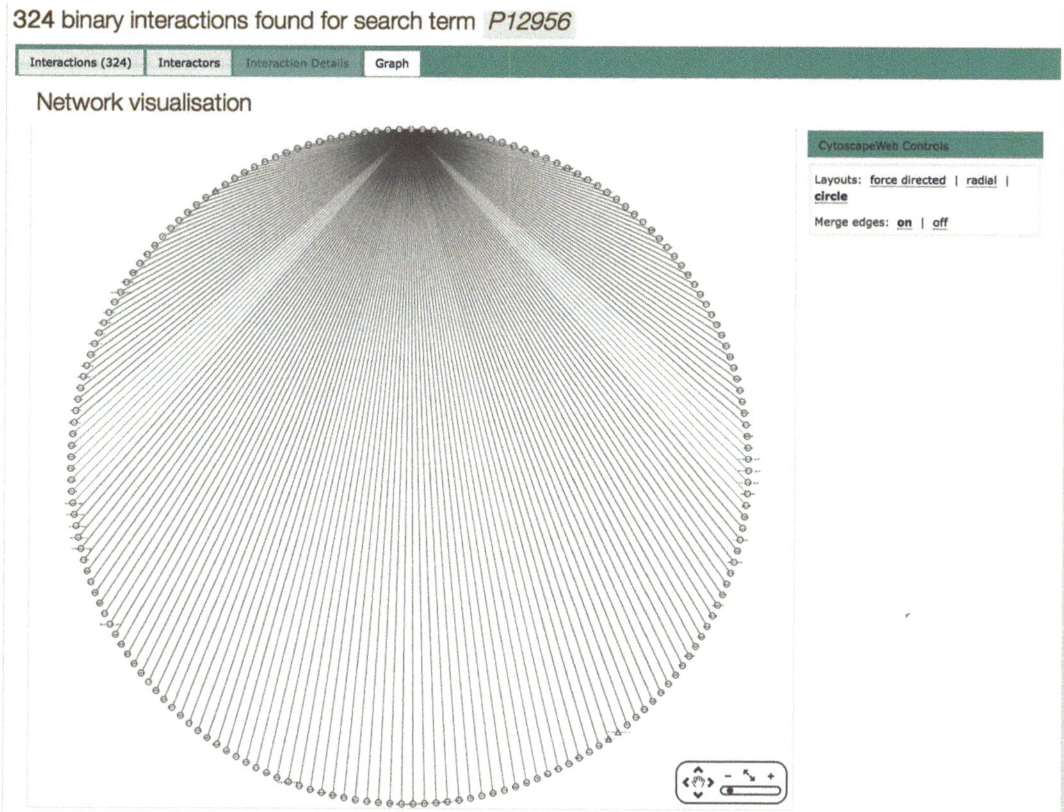

Fig. 16.35 Circle layout of the interaction map found for XRCC6 in IntAct database. XRCC6 protein query is located at the *top* of the map

corresponding HPRD ID 01236, Gene symbol NUMA1, Molecular Weight 238259 Da, Chromosome location 11q13, intracellular localization, domains and motifs, and sites of tissue gene expression (Fig. 16.41). The sequence of NUMA1 and its corresponding mRNA are obtained by clicking on Sequence tab (Fig. 16.42). A list of proteins that interact with NUMA1, and types of experiment and interactions (direct or in a complex) are shown in Fig. 16.43.

Alternatively, it is possible to search HPRD by browsing Molecule Class, Domains, Motifs, PTMs, and Localization by pushing the Browse key on the right of the main webpage (Fig. 16.39). Furthermore, access to Human Proteinpedia, Pathways, PhosphoMotif Finder, or downloading the complete HPRD are possible using the main menu.

16.8.6 BioGRID

The Biological General Repository for Interaction Datasets (BioGRID, http://thebiogrid.org), as many other protein-protein interactions databases, has as main goals to curate, organize and make it freely available. The funding partners of this important database are the National Institutes of Health (NIH), the

Fig. 16.36 Interaction map found for PSA3, SYWC, MCM4, SMAP, DDB1, EIF3, PYR1, MCM3, SSRP1 and METK2 proteins in IntAct database. Force directed layout of the network showing many more interactions that are contained in the IntAct database

Canadian Institutes of Health Research (CIHR), the Genome Canada, and GenomeQuébec. Many other institutions have joined efforts to BioGRID, including the Université de Montréal, Princeton University, Mount Sinai Hospital, University of Edinburgh, SGD, FlyBase, GeneDB, NCBI, WormBase, MaizeGDB, MINT, IntAct, String, MatrixDB, SIB, GO, UniProt, Reactome, Cytoscape, and many others that can be found in the BioGRID webpage. The current version of BioGRID database (3.4.131, December 2015) has information for several model organisms, including *A. thaliana*, *C. elegans*, *Candida albicans*, *Danio rerio*, *Dictyostellium discoideum*, *D. melanogaster*, *H. sapiens*, *Mus musculus*, *Neurospora crassa*, *Plasmodium falciparum*, *S. cerevisiae*, *Schizosaccharomyces pombe*, *Xenopus laevis*,

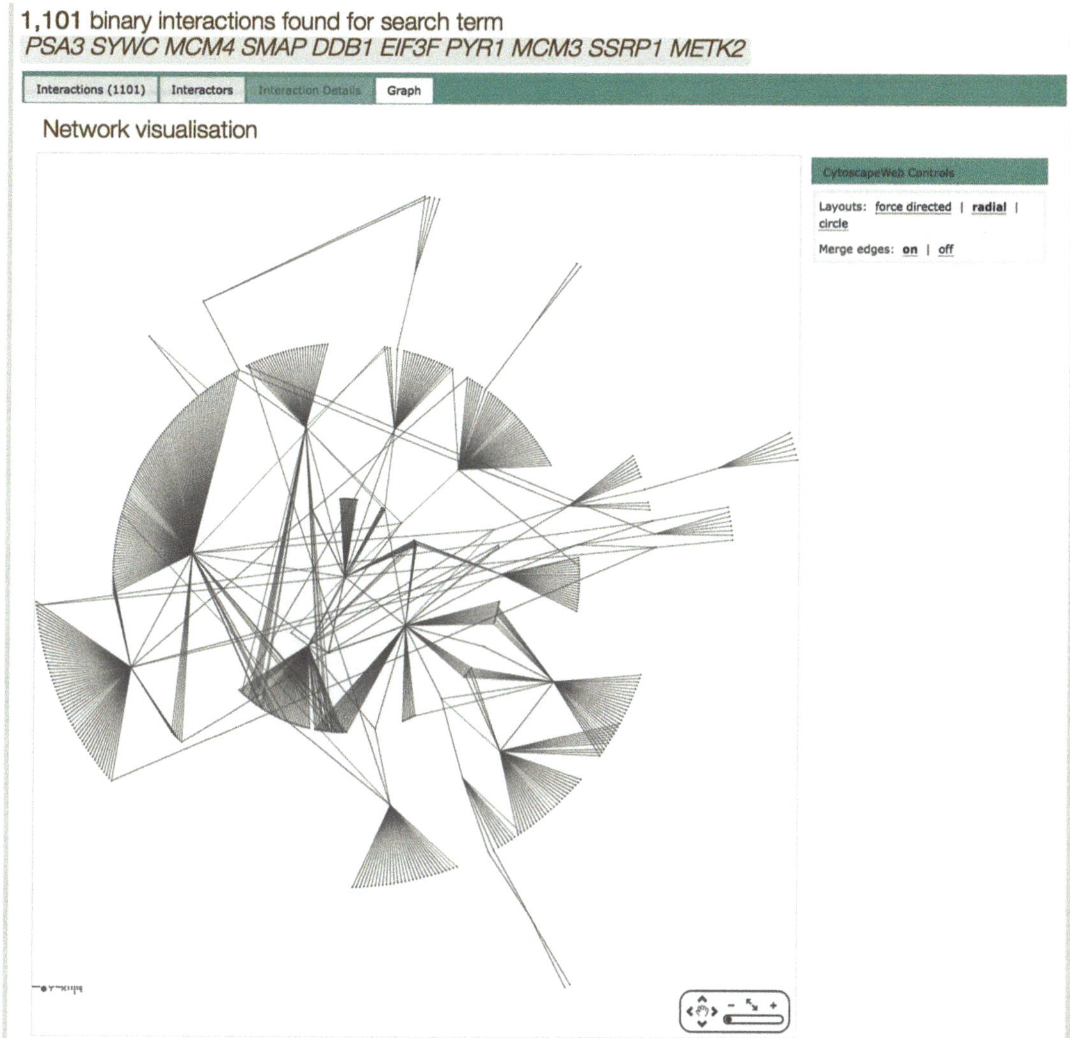

Fig. 16.37 Radial layout of the network found for PSA3, SYWC, MCM4, SMAP, DDB1, EIF3, PYR1, MCM3, SSRP1 and METK2 proteins in IntAct database

among other eukaryotic organisms. Furthermore, it has information of prokaryotic cells, such as *B. subtilis, E. coli, Mycobacterium tuberculosis, and Streptococcus pneumoniae.* Some viruses are included as well, e.g. Hepatitis C virus, Human Herpesvirus, Human Immunodeficiency virus, and Human Papillomavirus type 16 [44–46]. In its current version, the BioGRID database contains 749,213 non-redundant interactions, corresponding to 63,026 gene products and 45,623 unique publications. BioGRID database also includes 11,329 non-redundant interactions between 4851 unique chemical compounds and 2464 gene products accumulated from 8875 scientific publications. BioGRID also contains PTMs information. A total of 19,981 PTMs corresponding to 18,578 unassigned sites, 3165 unique proteins, 14,999 genes retrieved from 4317 publications are stored in this database.

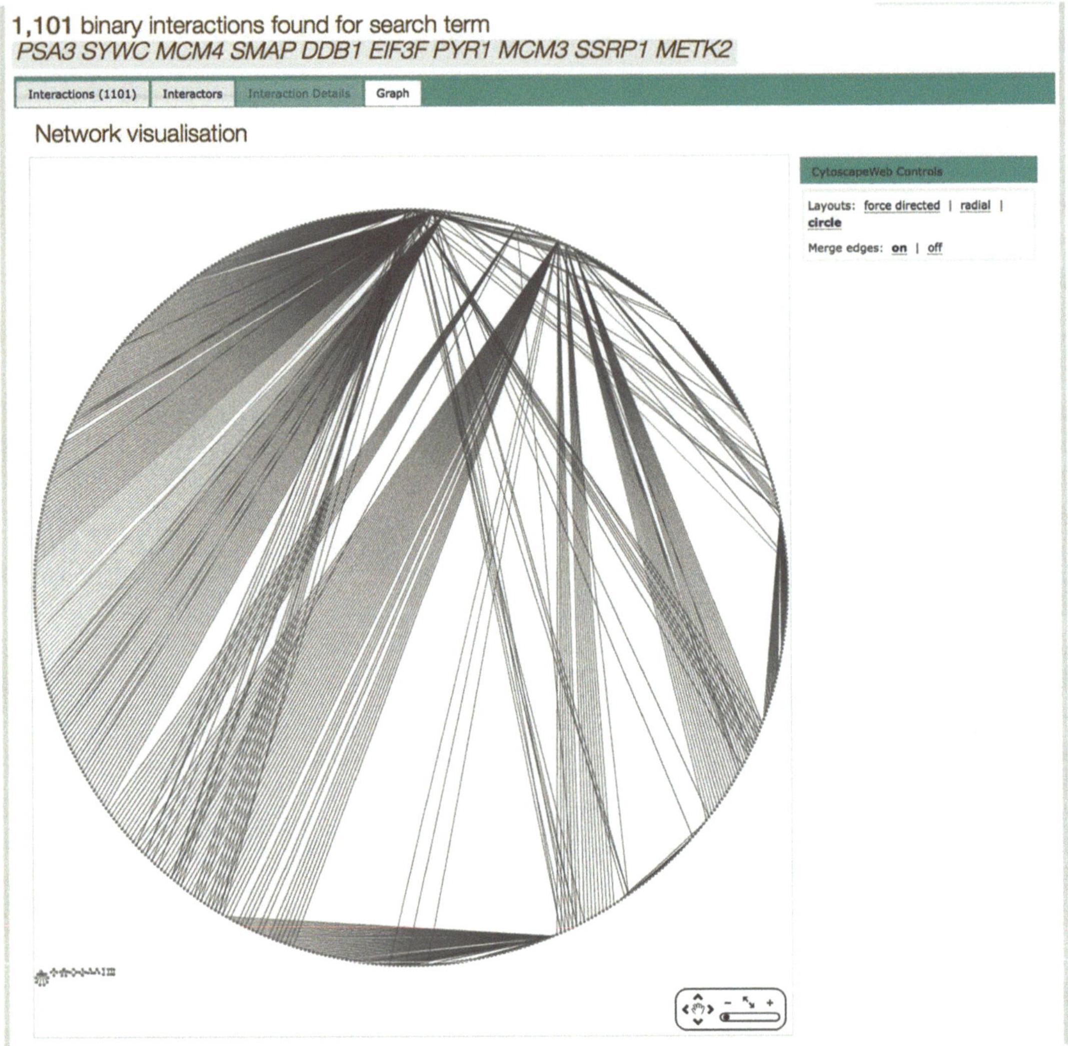

Fig. 16.38 Circle layout of the interaction map found for PSA3, SYWC, MCM4, SMAP, DDB1, EIF3, PYR1, MCM3, SSRP1 and METK2 proteins in IntAct database

To perform a search in BioGRID database, type your query (gene name, identifier or keywords) in the gene search window and select the species (Fig. 16.44). It is important to note that only one protein at a time can be searched. Alternatively, searches can be done by PubMed publication. However, searching of Multiple Genes or Publications will be available soon. As an example of a search, we selected the MCM6 protein, which was found overexpressed in both Luminal A and MDA-MB-231 breast cancer cell lines [18]. Results indicates that MCM6, the Minichromosome maintenance complex component 6, is involved in four GO Biological Processes:

1. DNA replication
2. DNA strand elongation involved in DNA replication
3. G1/S transition of mitotic cell cycle
4. Mitotic cell cycle

Fig. 16.39 Homepage of the Human Protein Reference Database HPRD

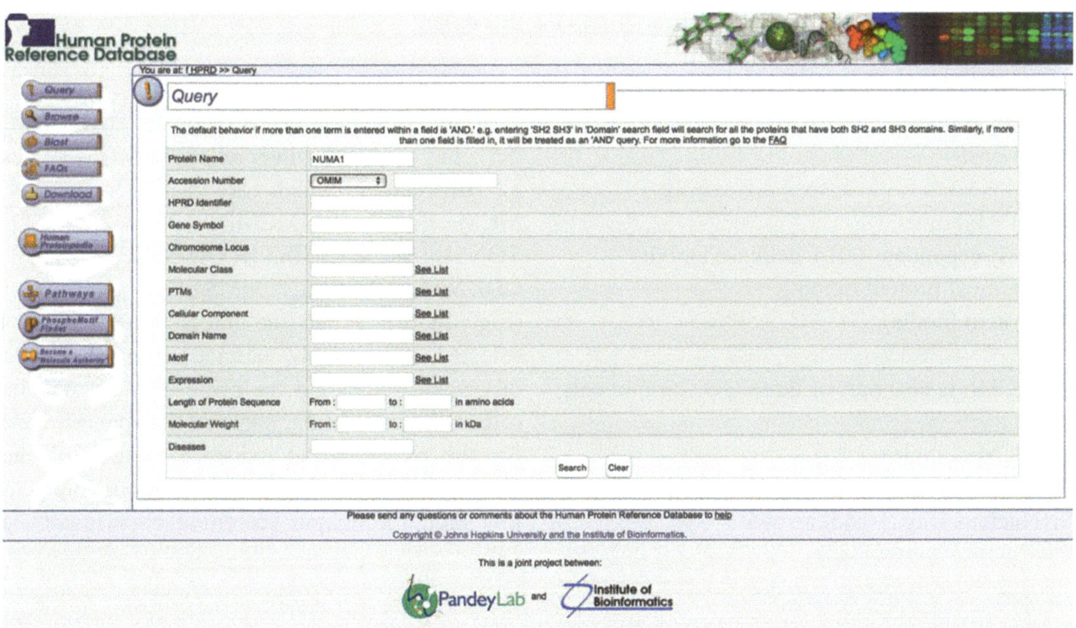

Fig. 16.40 Query webpage of the Human Protein Reference Database HPRD

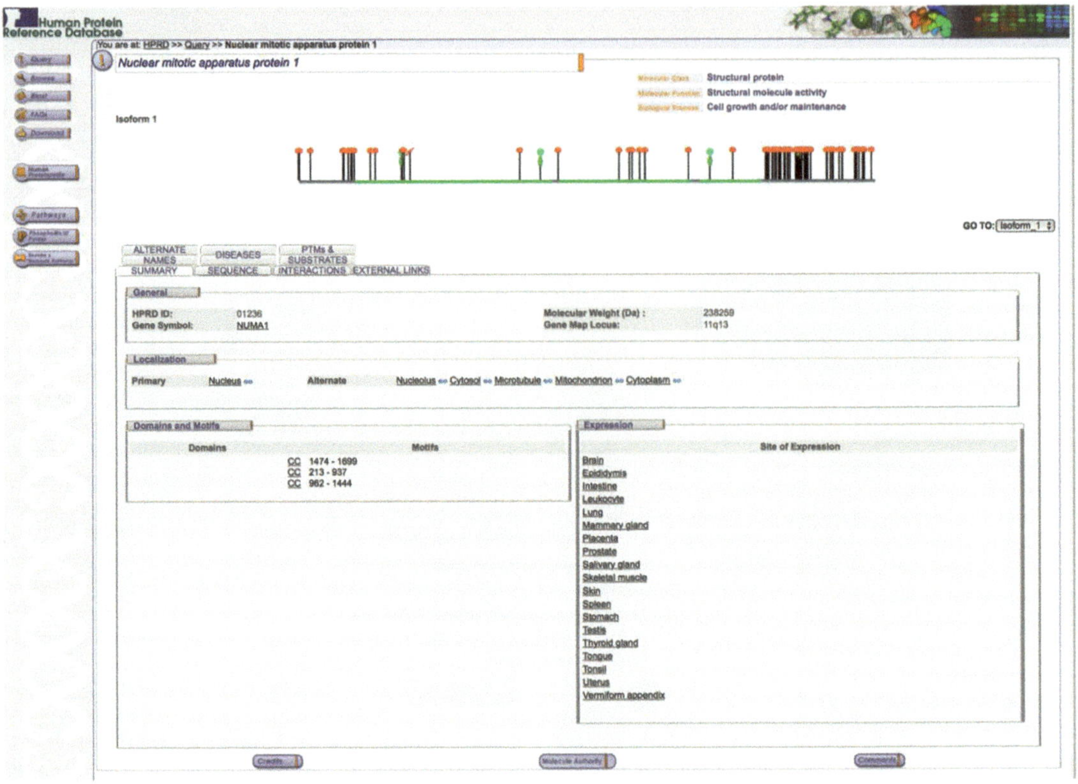

Fig. 16.41 HPRD query result for the Nuclear Mitotic Apparatus Protein 1, NUMA1. This screenshot shows a putative PTM map as well as a summary for NUMA1 indicating the chromosome localization, subcellular localization, domains, and tissues where the protein is expressed

This protein is also involved in four GO Functions:

1. ATP binding
2. ATP-dependent DNA helicase activity
3. Identical protein binding
4. Protein binding

MCM6 is also part of three GO Components:

1. MCM complex
2. Nucleoplasm
3. Nucleus (Fig. 16.45, arrows 1–3)

In order of significance according to the number of physical interactions, MCM6 has 82 interactors which are MCM2, MCM4, MCM7, MCM10, MCMBP, MCM3, CDT1, TONSL, MCM5, HIST1H4A, SSRP1, ASF1B, CDKN2A, ASF1A, MMS22L, and ING5 (Fig. 16.45). When the interactions option is selected, a list of 142 interactions are displayed on screen, indicating the name of interactor, its role in the interaction, name of the species, code for the experimental evidence, source of the dataset, whether interaction is from high or low high throughput screening experiments, a

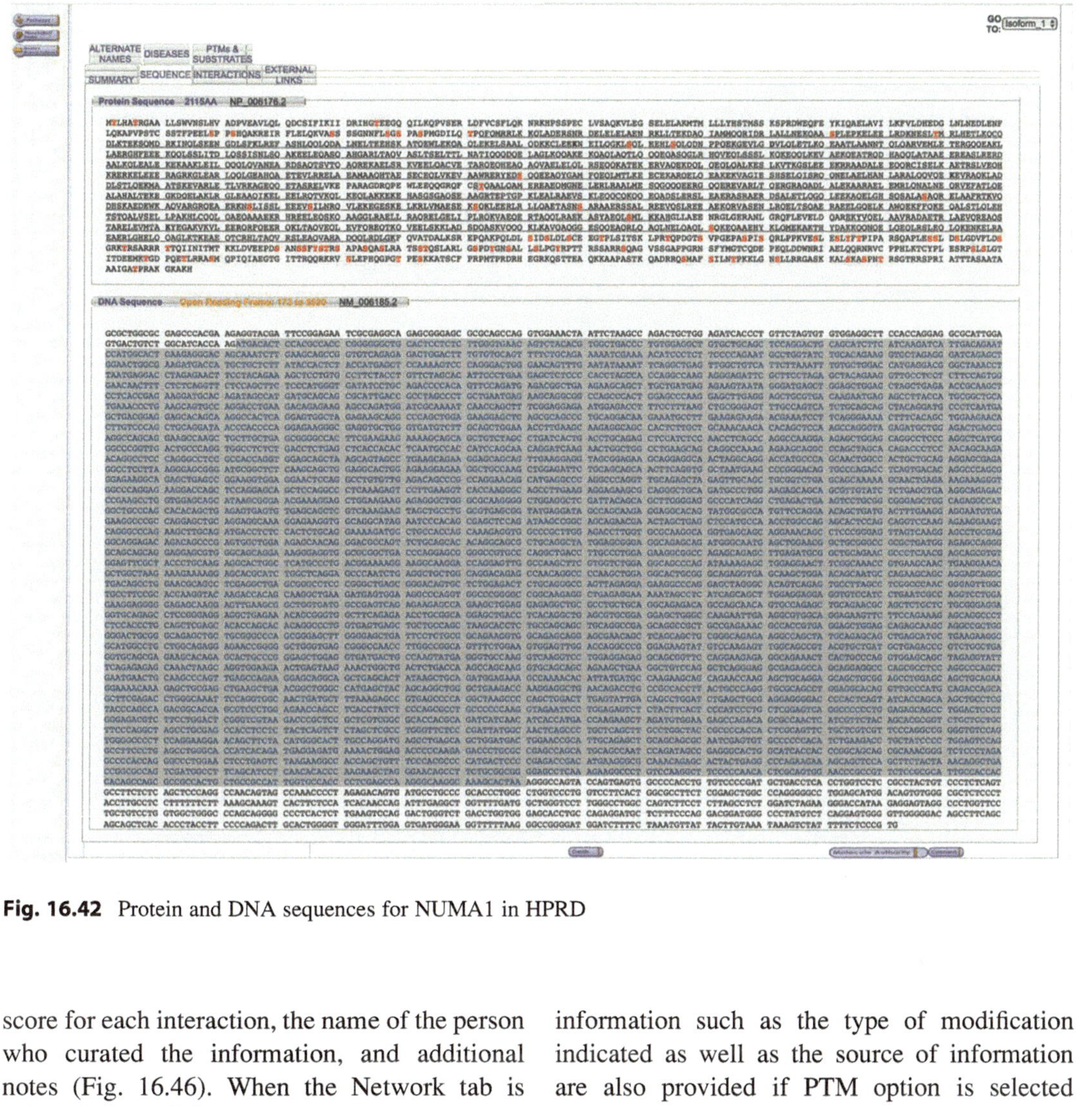

Fig. 16.42 Protein and DNA sequences for NUMA1 in HPRD

score for each interaction, the name of the person who curated the information, and additional notes (Fig. 16.46). When the Network tab is selected, three different layouts can be obtained: Concentric circles (Fig. 16.47), Single circle (Fig. 16.48), and Grid (Fig. 16.49). If the number of minimum evidence is changed to five for example, the number of interactions will drop (Fig. 16.50), thus reducing the complexity of the interaction map. When the PTM sites tab is selected, the amino acid sequence of the query is displayed and those residues with an identified PTM are highlighted in blue. Additional information such as the type of modification indicated as well as the source of information are also provided if PTM option is selected (Fig. 16.51). In the case of MCM6, there are 35 Lysine residues marked as ubiquitinated and two additional non-assigned PTMs (neddylation and sumoylation) (Fig. 16.52).

16.8.7 PIPs

The Human Protein-Protein Interaction Prediction (PIPs) is a specialized database containing a

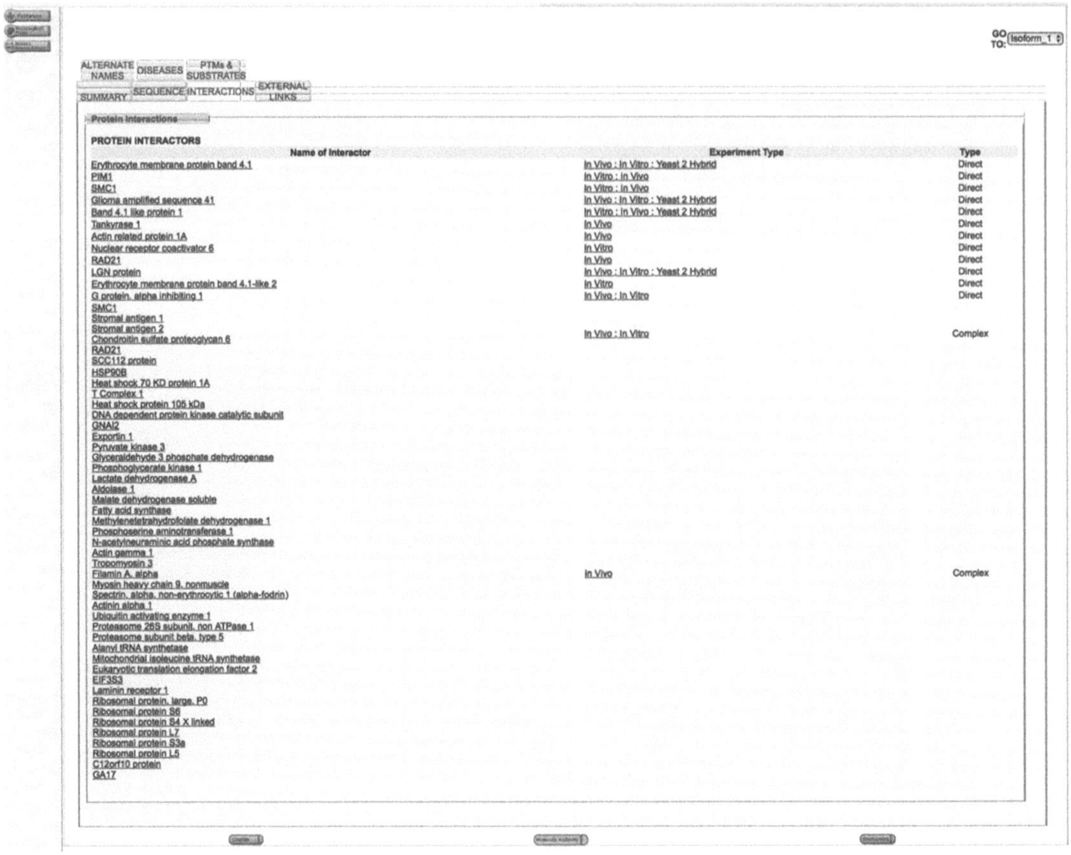

Fig. 16.43 List of protein interactors of NUMA1 queried in HPRD

catalogue of predicted human protein-protein interactions that have been probabilistically determined using a Bayesian model, which takes into account several modules: Expression, Orthology, Localization, Domain co-occurrence, PTMs co-occurrence, Disorder, and Transitive. Expression considers information from a number of gene expression profiles. Orthology uses the interactions that have been determined for orthologues from fly, human, worm and yeast. Localization is determined by using a human subcellular localization predictor (PSLT) in different subcellular compartments. Domain co-occurrence uses the information stored in InterPro (Protein sequence analysis and classification, http://www.ebi.ac.uk/interpro) and Pfam (Protein families, http://pfam.xfam.org) protein domain databases. PTM co-occurrence uses the information contained in HPRD and UniProtKB. Disorder refers to the prediction of intrinsic disorder of protein found in VLS2 prediction. Finally, Transitive is a module which involves the local topology of networks, considering all modules described above [47].

PIPs database is located at the University of Dundee and the current version (December 2015) contains 37,606 interactions with a score > 1.0, indicating a high probability of occurrence. To

Fig. 16.44 Homepage of the Biological General Repository for Interaction Databasets, BioGRID

perform a search, an ID in IPI, RefSeq or UniProtKB format must be entered in the search window. As an example, when TBP was used to initiate a query, results were displayed in several boxes each containing a number of interactions with a certain score. In this case, there are 65 interactions when a score value ≥ 1.0 was selected. For score values equal or larger than 2.5, 12.5, 25, 250, and 2500, there were 33, 15, 13, 7, and 3 interactions, respectively. When the number of interactions for a score ≥ 1.0 is selected, a list of interactors and the scores for each module used will be displayed on the screen.

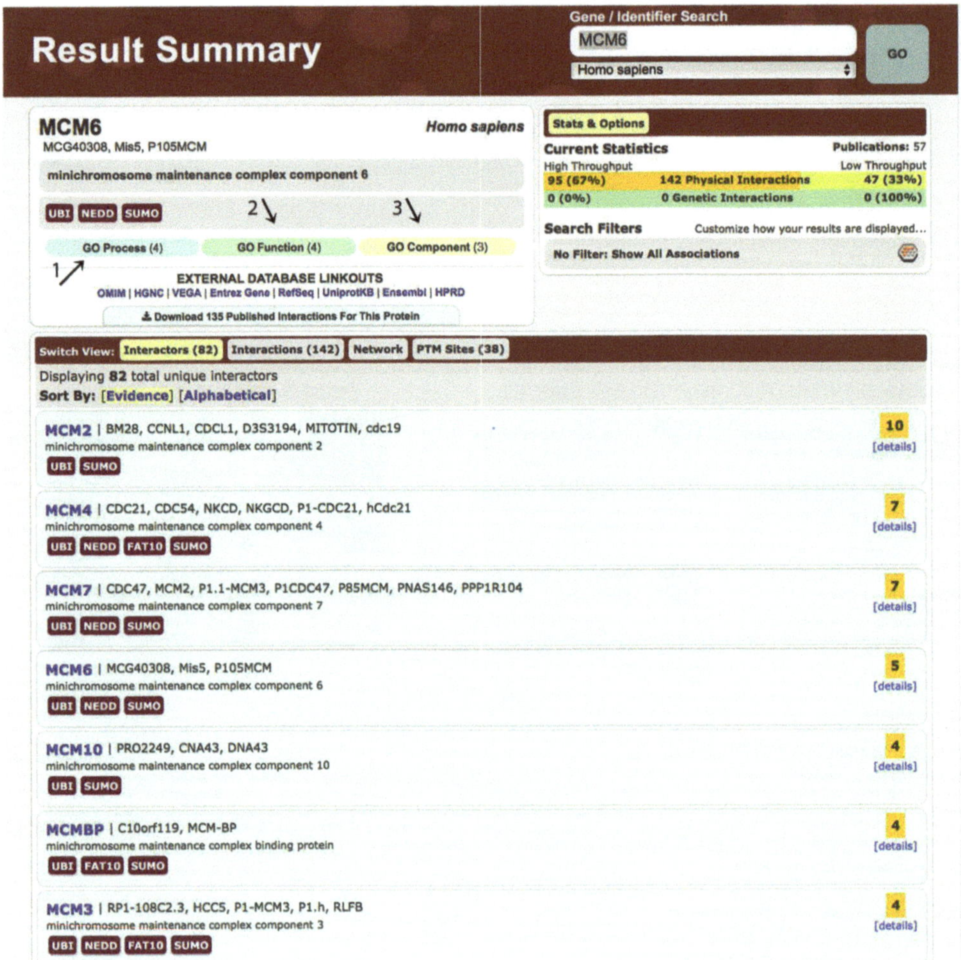

Fig. 16.45 Result summary for the Minichromosome Maintenance Complex Component 6, MCM6, queried in BioGRID. A total of 82 interactors were found in database

16.8.8 MPIDB

The Microbial Protein Interaction Database (MPIDB) at the Craig Venter Institute (http://jcvi.org/mpidb/about.php) is a database whose main goal is to gather information for all known protein interactions from microbial organisms [48]. The current version of MPIDB is 2009-11-18 and contains 24,295 interactions that have been experimentally determined for 250 species of bacteria. This number of interactions corresponds to 7810 proteins and 24,295 interactors. Like many other databases, MPIDB also imports information from other databases, including IntAct, Database of Interacting Proteins (DIP), The Biomolecular Interaction

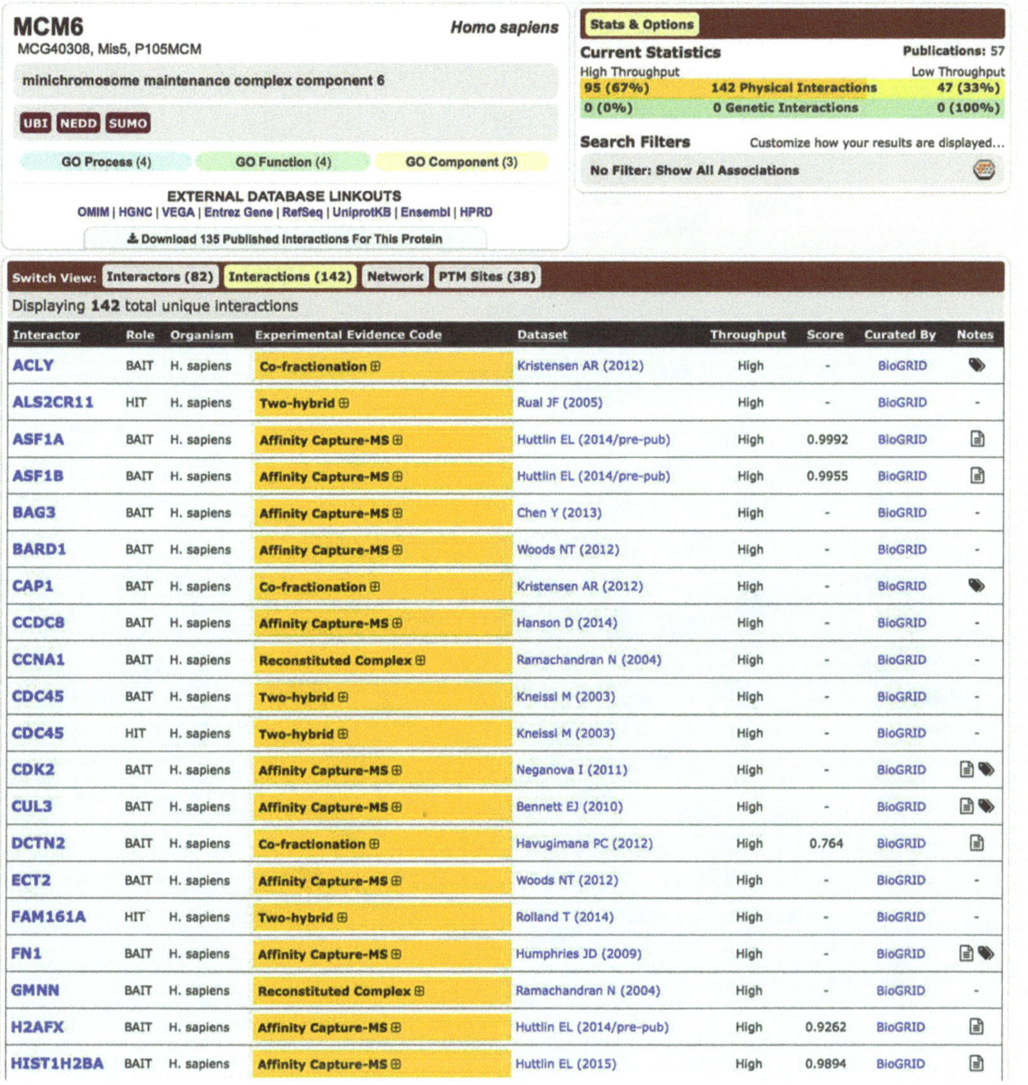
MCM6 *Homo sapiens*
MCG40308, Mis5, P105MCM
minichromosome maintenance complex component 6
UBI NEDD SUMO
GO Process (4) GO Function (4) GO Component (3)
EXTERNAL DATABASE LINKOUTS
OMIM | HGNC | VEGA | Entrez Gene | RefSeq | UniprotKB | Ensembl | HPRD
Download 135 Published Interactions For This Protein

Stats & Options
Current Statistics Publications: 57
High Throughput Low Throughput
95 (67%) 142 Physical Interactions 47 (33%)
0 (0%) 0 Genetic Interactions 0 (100%)
Search Filters Customize how your results are displayed...
No Filter: Show All Associations

Switch View: Interactors (82) Interactions (142) Network PTM Sites (38)
Displaying **142** total unique interactions

Interactor	Role	Organism	Experimental Evidence Code	Dataset	Throughput	Score	Curated By	Notes
ACLY	BAIT	H. sapiens	Co-fractionation	Kristensen AR (2012)	High	-	BioGRID	
ALS2CR11	HIT	H. sapiens	Two-hybrid	Rual JF (2005)	High	-	BioGRID	-
ASF1A	BAIT	H. sapiens	Affinity Capture-MS	Huttlin EL (2014/pre-pub)	High	0.9992	BioGRID	
ASF1B	BAIT	H. sapiens	Affinity Capture-MS	Huttlin EL (2014/pre-pub)	High	0.9955	BioGRID	
BAG3	BAIT	H. sapiens	Affinity Capture-MS	Chen Y (2013)	High	-	BioGRID	-
BARD1	BAIT	H. sapiens	Affinity Capture-MS	Woods NT (2012)	High	-	BioGRID	-
CAP1	BAIT	H. sapiens	Co-fractionation	Kristensen AR (2012)	High	-	BioGRID	
CCDC8	BAIT	H. sapiens	Affinity Capture-MS	Hanson D (2014)	High	-	BioGRID	-
CCNA1	BAIT	H. sapiens	Reconstituted Complex	Ramachandran N (2004)	High	-	BioGRID	-
CDC45	BAIT	H. sapiens	Two-hybrid	Kneissl M (2003)	High	-	BioGRID	-
CDC45	HIT	H. sapiens	Two-hybrid	Kneissl M (2003)	High	-	BioGRID	-
CDK2	BAIT	H. sapiens	Affinity Capture-MS	Neganova I (2011)	High	-	BioGRID	
CUL3	BAIT	H. sapiens	Affinity Capture-MS	Bennett EJ (2010)	High	-	BioGRID	
DCTN2	BAIT	H. sapiens	Co-fractionation	Havugimana PC (2012)	High	0.764	BioGRID	
ECT2	BAIT	H. sapiens	Affinity Capture-MS	Woods NT (2012)	High	-	BioGRID	-
FAM161A	HIT	H. sapiens	Two-hybrid	Rolland T (2014)	High	-	BioGRID	-
FN1	BAIT	H. sapiens	Affinity Capture-MS	Humphries JD (2009)	High	-	BioGRID	
GMNN	BAIT	H. sapiens	Reconstituted Complex	Ramachandran N (2004)	High	-	BioGRID	-
H2AFX	BAIT	H. sapiens	Affinity Capture-MS	Huttlin EL (2014/pre-pub)	High	0.9262	BioGRID	
HIST1H2BA	BAIT	H. sapiens	Affinity Capture-MS	Huttlin EL (2015)	High	0.9894	BioGRID	

Fig. 16.46 List of interactions found for MCM6 in BioGRID

Network Database (BIND) and MINT. Search can be performed using the name of a protein (UniProtKB ID or locus name) or by selecting species name. Results will be displayed as a table containing the UniProtKB ID, name of protein, interactor, loci of query and interactor, species for query and interactor and the number of evidences for such interaction.

16.8.9 TAIR

The Arabidopsis Information Resource (TAIR) at Phoenix Bioinformatics (https://www.arabidopsis.org) is a database of information for plant research model *A. thaliana*.

This database contains the whole *A. thaliana* genome sequence, analysis, structure and

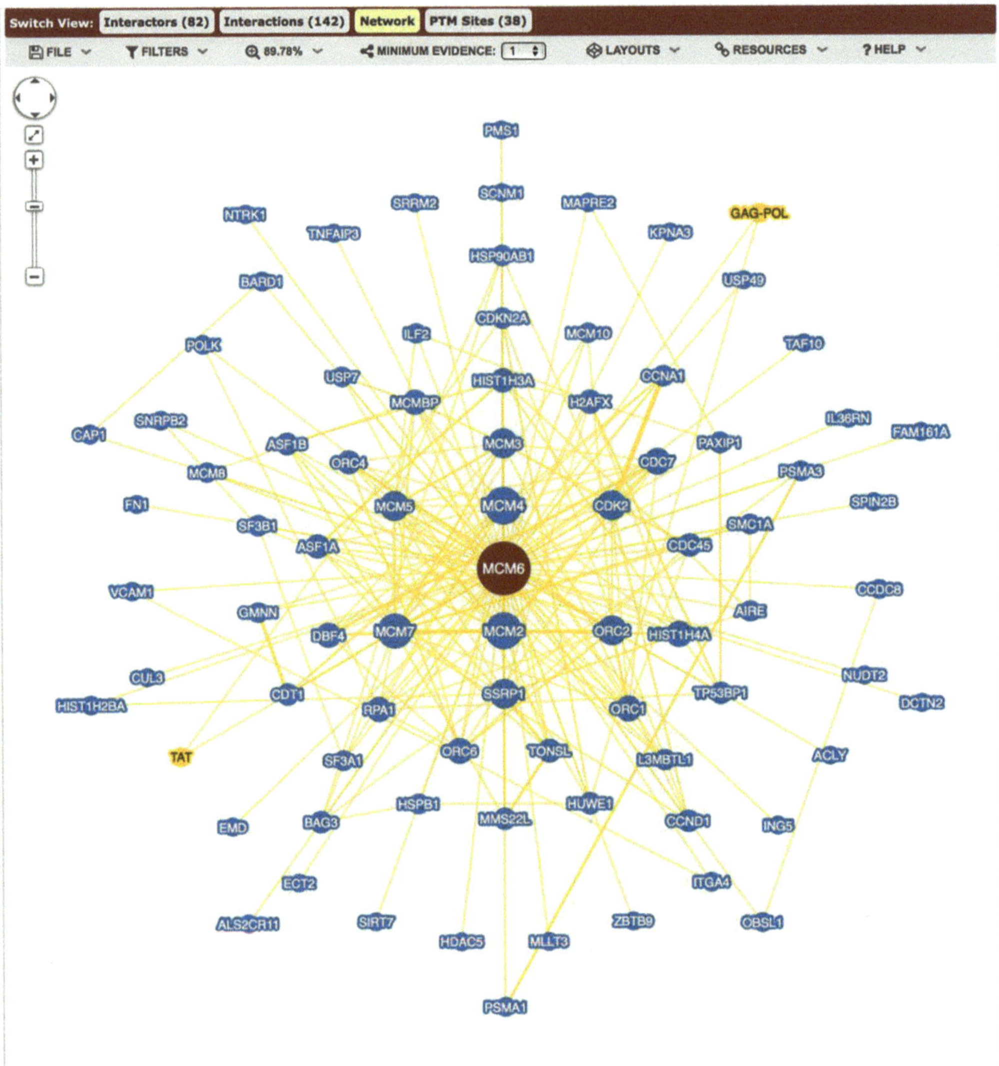

Fig. 16.47 Map of interactions for MCM6 in BioGRID database. Layout of interaction map is shown in concentric circles, where query protein is at the center

annotation of genes, information for all proteins encoded in its genome, data from gene expression experiments, genome maps, pathways, and other information useful to the scientific community [49]. Like other databases, experts from TAIR curate information using published experiments before entering them in this database. Search in TAIR can be performed in several ways: DNA/Clones, Ecotypes, Genes, Gene Ontology, Plant Ontology, Keywords, Locus, Markers, Microarray element, Microarray expression, People/Labs, Polymorphism/Alleles, Protein, Protocols, PMIDS, Seed/Germplasm, and Text. TAIR webpage also contains tools for

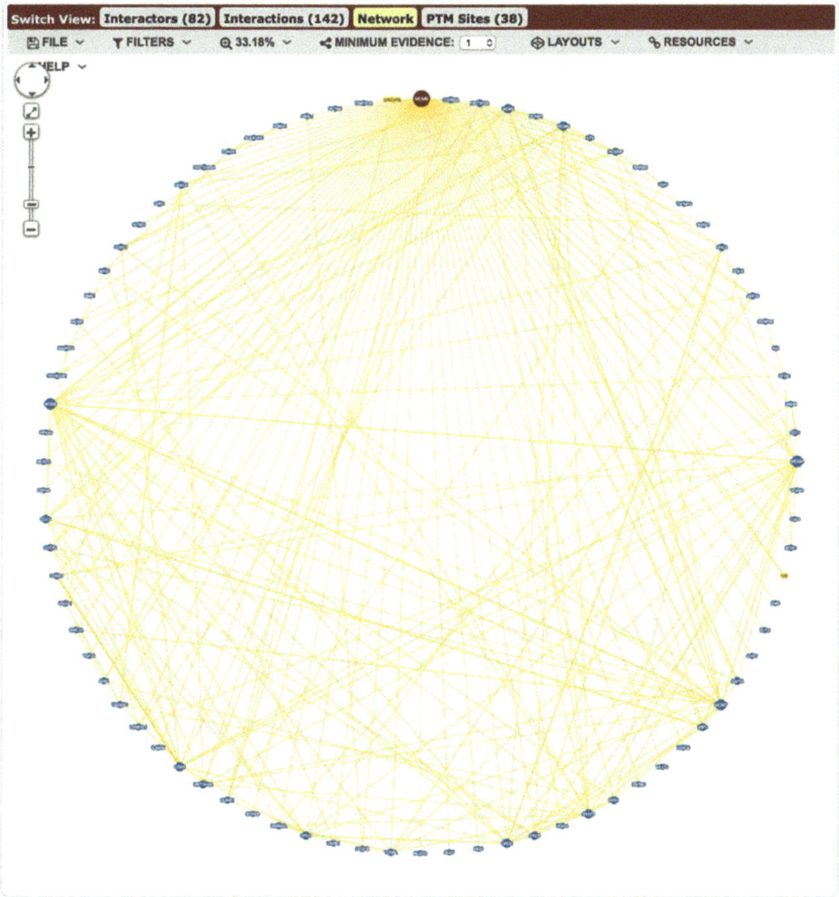

Fig. 16.48 Map of interactions for MCM6 in BioGRID database. Layout of interaction map is shown as a *single circle*, where MCM6 query protein is located at the *top* of the map

analysis of sequences, as well as viewers for maps and sequences. It is recommended to register in TAIR to download the whole genome sequence.

16.8.10 GeneCards

The Human Gene Database (GeneCards, http://www.genecards.org) is another useful database covering the human genome [50–53]. This database was created by scientists at the Weizmann Institute of Science and LifeMap Sciences. Search can be done using keywords, symbols, aliases, or identifiers. Information that can be retrieved from this database include:

1. Aliases for query
2. Links to HGNC (HUGO Gene Nomenclature Committee, http://www.genenames.org), Entrez Gene at NCBI, Ensembl (genome databases for vertebrates and other

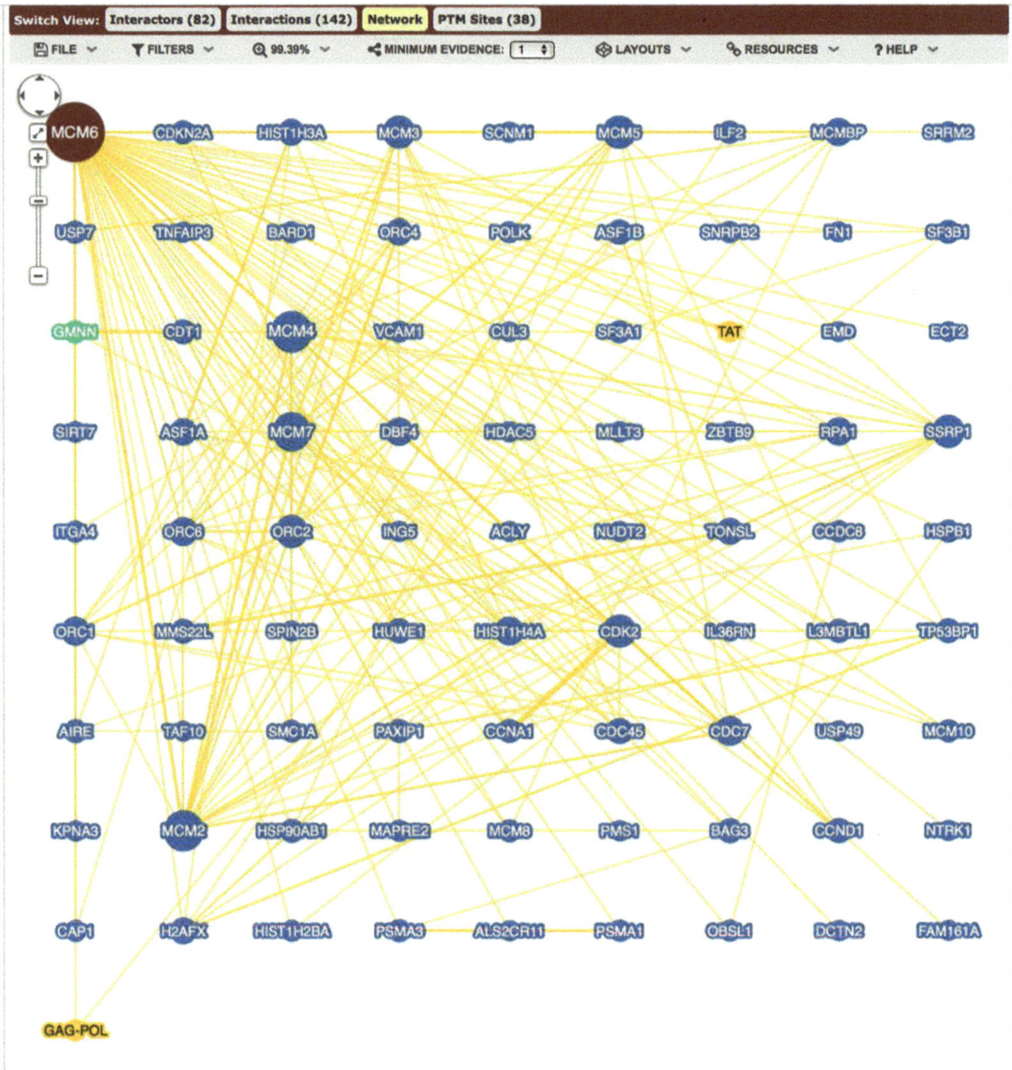

Fig. 16.49 Grid layout of the map of interactions for MCM6 in BioGRID database. MCM6 query protein is located at the *top left corner* of the map

eukaryotic species, http://www.ensembl.org/index.html), OMIM http://www.omim.org), and UniProtKB

3. Summaries of queries retrieved from different sources
4. Genomics data for query, including Regulatory Elements, Genomic location, Genomic region view, and RefSeq DNA sequence
5. Protein information such as Protein ID, Length in amino acids, Molecular Mass, Quaternary structure, Three dimensional structure from OCA (Brower-database for protein structure/function, http://oca.weizmann.ac.il/oca-docs/oca-home.html), Proteopedia (The free, collaborative D-encyclopedia of proteins & other

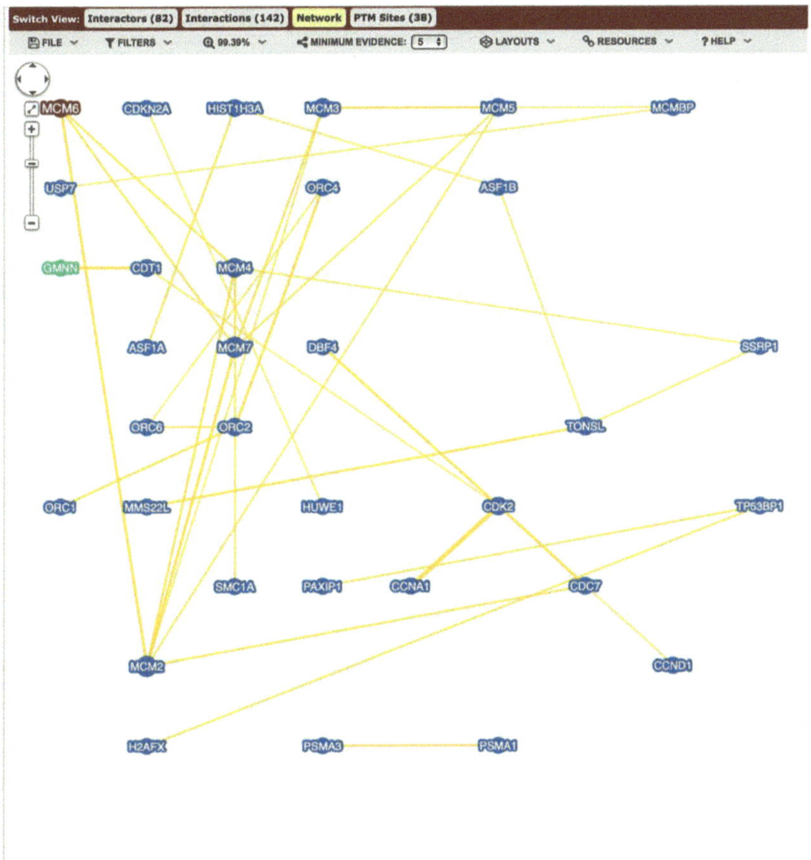

Fig. 16.50 Grid layout of the map of interactions for MCM6 in BioGRID database using a minimum value of 5 as evidence

molecules, http://proteopedia.org/wiki/index.php/Main_Page), Alternative splice forms, Data of protein expression in Proteomics DB (https://www.proteomicsdb.org/proteomicsdb/#overview), PaxDB (Protein Abundance Across Organisms, http://pax-db.org/#!home), MOPED (Multi-Omics Profiling Expression Database, https://www.proteinspire.org/MOPED/mopedviews/proteinExpressionDatabase.jsf), MaxQB (The MaxQuant DataBase, http://maxqb.biochem.mpg.de/mxdb/), and PTMs, (6) Domains in InterPro (Protein sequence, analysis and classification, http://www.ebi.ac.uk/interpro), ProtoNet (Automatic Hierarchical Classification of Proteins, http://www.protonet.cs.huji.ac.il/requested/cluster_card.php?global=protonet|no|6|61|lifetime|1|2|2&cluster=4023630&releaseid=6&firstEnterTimeClient=&blast=11053692|274977&clusteringNum=61)

6. Functions retrieved from UniProtKB, Enzyme Number; Gene Ontology; Phenotypes; Animal models for query; links to CRISPR products, miRNAs, siRNAs, shRNAs, clone products, etc.

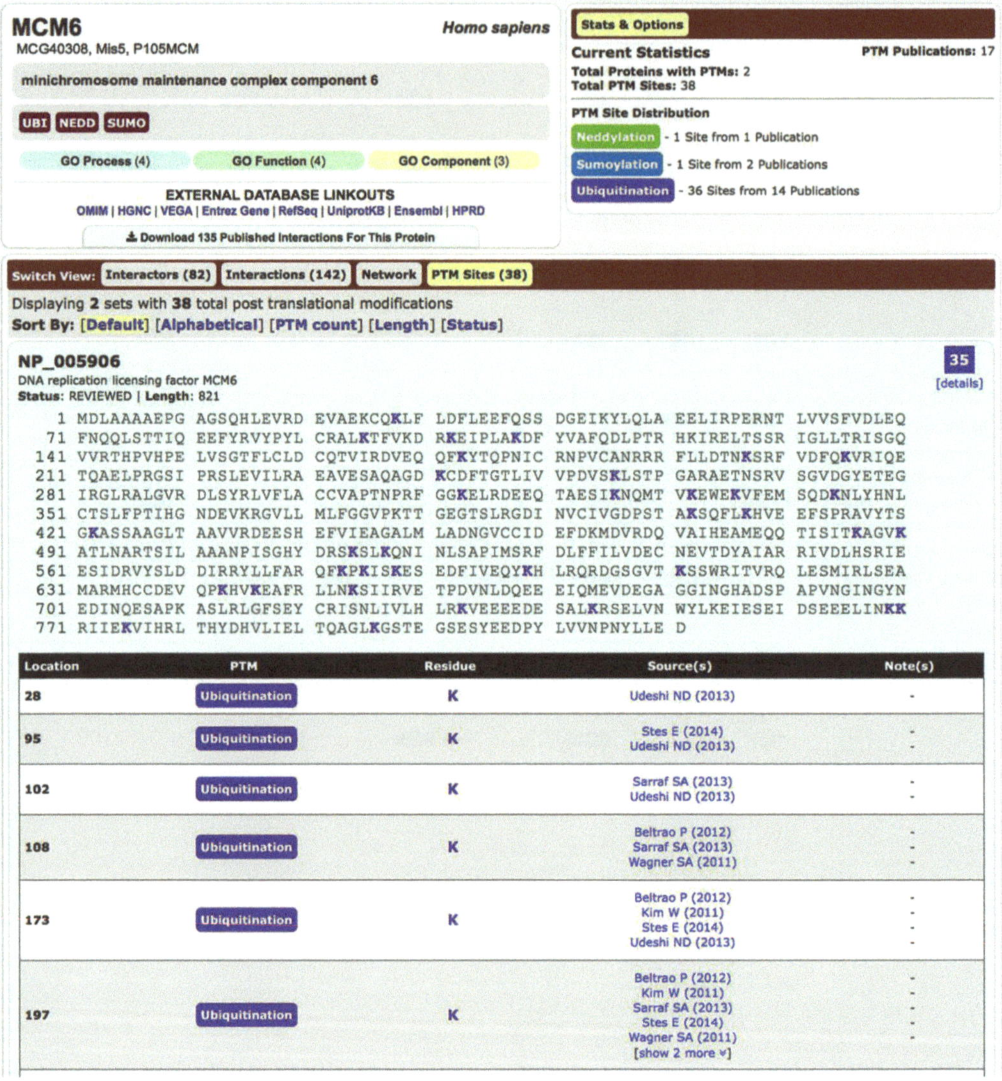

Fig. 16.51 PTMs reported for MCM6 in BioGRID database. There are a few sites shown to carry ubiquitination for MCM6. Reference is also provided

7. Localization of genes in chromosomes and subcellular location of proteins
8. Pathways
9. Drugs for query
10. Transcripts: Reference sequence (RefSeq), Enseml, Unigene Clusters
11. Expression in tissues: GeneAnalytics (http://geneanalytics.genecards.org/?utm_source=genecards&utm_medium=banner&utm_campaign=genecards&utm_content=banner_expression)
12. Orthologs
13. Paralogs
14. Variants
15. Disorders in MalaCards (The Humans Disease Database, http://www.malacards.org)
16. Publications

Fig. 16.52 PTMs reported for MCM6 in BioGRID database. Other PTMs are also shown in this figure for MCM6, including neddylation, sumoylation, as well as other ubiquitination sites

In addition, there are a lot of links to companies that might have products for the protein of interests, such as antibodies, immunofluorescence, animal models, silencing, etc.

Acknowledgements We thank the Instituto de Ciencia y Tecnología del Distrito Federal (ICyTDF), now renamed Secretaría de Ciencia, Tecnología e Innovación de la Ciudad de México (SECITI), for its support with the project ICyTDF-J.LA (CM-272/12-SECITI/033/2012), and Consejo Nacional de Ciencia y Tecnología (Conacyt) from Mexico, with the project number SALUD-2009-01-113674, both granted to Dr. Juan Pedro Luna Arias.

References

1. Kumar C, Mann M (2009) Bioinformatics analysis of mass spectrometry-based proteomics data sets. FEBS Lett 583(11):1703–1712

2. Su Z, Wang J, Yu J, Huang X, Gu X (2006) Evolution of alternative splicing after gene duplication. Genome Res 16(2):182–189
3. Twyman RM (2004) Principles of proteomics. Garland Biosciences/BIOS Scientific Publishers, Hampshire
4. Ashburner M, Ball CA, Blake JA, Botstein D, Butler H, Cherry JM, Davis AP, Dolinski K, Dwight SS, Eppig JT et al (2000) Gene ontology: tool for the unification of biology. The gene ontology consortium. Nat Genet 25(1):25–29
5. Gene Ontology Consortium (2001) Creating the gene ontology resource: design and implementation. Genome Res 11(8):1425–1433
6. Harris MA, Clark J, Ireland A, Lomax J, Ashburner M, Foulger R, Eilbeck K, Lewis S, Marshall B, Mungall C et al (2004) The Gene Ontology (GO) database and informatics resource. Nucleic Acids Res 32(Database issue):D258–D261
7. Gene Ontology C (2015) Gene ontology consortium: going forward. Nucleic Acids Res 43(Database issue): D1049–D1056
8. Rhee SY, Wood V, Dolinski K, Draghici S (2008) Use and misuse of the gene ontology annotations. Nat Rev Genet 9(7):509–515
9. Mi H, Muruganujan A, Casagrande JT, Thomas PD (2013) Large-scale gene function analysis with the PANTHER classification system. Nat Protoc 8 (8):1551–1566
10. Thomas PD, Campbell MJ, Kejariwal A, Mi H, Karlak B, Daverman R, Diemer K, Muruganujan A, Narechania A (2003) PANTHER: a library of protein families and subfamilies indexed by function. Genome Res 13(9):2129–2141
11. Thomas PD, Kejariwal A, Campbell MJ, Mi H, Diemer K, Guo N, Ladunga I, Ulitsky- Lazareva B, Muruganujan A, Rabkin S et al (2003) PANTHER: a browsable database of gene products organized by biological function, using curated protein family and subfamily classification. Nucleic Acids Res 31 (1):334–341
12. Mi H, Lazareva-Ulitsky B, Loo R, Kejariwal A, Vandergriff J, Rabkin S, Guo N, Muruganujan A, Doremieux O, Campbell MJ et al (2005) The PANTHER database of protein families, subfamilies, functions and pathways. Nucleic Acids Res 33(Database issue):D284–D288
13. Funahashi A, Jouraku A, Matsuoka Y, Morohashi M, Kikuchi N, Kitano H (2008) CellDesigner 3.5: a versatile modeling tool for biochemical networks. Proc IEEE 96(8):1254
14. Mi H, Guo N, Kejariwal A, Thomas PD (2007) PANTHER version 6: protein sequence and function evolution data with expanded representation of biological pathways. Nucleic Acids Res 35(Database issue): D247–D252
15. Mi H, Thomas P (2009) PANTHER pathway: an ontology-based pathway database coupled with data analysis tools. Methods Mol Biol 563:123–140
16. PANTHER User Manual (2015). http://pantherdb.org/help/PANTHER_user_manual.pdf
17. Mi H, Muruganujan A, Thomas PD (2013) PANTHER in 2013: modeling the evolution of gene function, and other gene attributes, in the context of phylogenetic trees. Nucleic Acids Res 41(Database issue):D377–D386
18. Calderon-Gonzalez KG, Valero Rustarazo ML, Labra-Barrios ML, Bazan-Mendez CI, Tavera-Tapia-A, Herrera-Aguirre M, Sanchez Del Pino MM, Gallegos-Perez JL, Gonzalez- Marquez H, Hernandez-Hernandez JM et al (2015) Data set of the protein expression profiles of Luminal A, Claudin-low and overexpressing HER2(+) breast cancer cell lines by iTRAQ labelling and tandem mass spectrometry. Data Brief 4:292–301
19. Dennis G Jr, Sherman BT, Hosack DA, Yang J, Gao W, Lane HC, Lempicki RA (2003) DAVID: database for annotation, visualization, and integrated discovery. Genome Biol 4(5):P3
20. da Huang W, Sherman BT, Lempicki RA (2009) Bioinformatics enrichment tools: paths toward the comprehensive functional analysis of large gene lists. Nucleic Acids Res 37(1):1–13
21. Huang DW, Sherman BT, Tan Q, Kir J, Liu D, Bryant D, Guo Y, Stephens R, Baseler MW, Lane HC et al (2007) DAVID bioinformatics resources: expanded annotation database and novel algorithms to better extract biology from large gene lists. Nucleic Acids Res 35(Web Server issue): W169–W175
22. da Huang W, Sherman BT, Lempicki RA (2009) Systematic and integrative analysis of large gene lists using DAVID bioinformatics resources. Nat Protoc 4(1):44–57
23. da Huang W, Sherman BT, Stephens R, Baseler MW, Lane HC, Lempicki RA (2008) DAVID gene ID conversion tool. Bioinformation 2(10):428–430
24. Kanehisa M, Goto S, Kawashima S, Okuno Y, Hattori M (2004) The KEGG resource for deciphering the genome. Nucleic Acids Res 32(Database issue): D277–D280
25. Kanehisa M, Goto S, Hattori M, Aoki-Kinoshita KF, Itoh M, Kawashima S, Katayama T, Araki M, Hirakawa M (2006) From genomics to chemical genomics: new developments in KEGG. Nucleic Acids Res 34(Database issue):D354–D357
26. Kanehisa M, Sato Y, Kawashima M, Furumichi M, Tanabe M (2015) KEGG as a reference resource for gene and protein annotation. Nucleic Acids Res 44:457
27. Kanehisa M, Sato Y, Morishima K (2015) BlastKOALA and GhostKOALA: KEGG tools for functional characterization of genome and metagenome sequences. J Mol Biol 428:726
28. Kanehisa M, Goto S, Sato Y, Furumichi M, Tanabe M (2012) KEGG for integration and interpretation of large-scale molecular data sets. Nucleic Acids Res 40(Database issue):D109–D114

29. Okuda S, Yamada T, Hamajima M, Itoh M, Katayama T, Bork P, Goto S, Kanehisa M (2008) KEGG Atlas mapping for global analysis of metabolic pathways. Nucleic Acids Res 36(Web Server issue): W423–W426
30. Chaiboonchoe A, Samarasinghe S, Kulasiri D, Salehi-Ashtiani K (2014) Integrated analysis of gene network in childhood leukemia from microarray and pathway databases. BioMed Res Int 2014:278748
31. von Mering C, Jensen LJ, Kuhn M, Chaffron S, Doerks T, Kruger B, Snel B, Bork P (2007) STRING 7--recent developments in the integration and prediction of protein interactions. Nucleic Acids Res 35 (Database issue):D358–D362
32. von Mering C, Jensen LJ, Snel B, Hooper SD, Krupp M, Foglierini M, Jouffre N, Huynen MA, Bork P (2005) STRING: known and predicted protein-protein associations, integrated and transferred across organisms. Nucleic Acids Res 33(Database issue):D433–D437
33. Jensen LJ, Kuhn M, Stark M, Chaffron S, Creevey C, Muller J, Doerks T, Julien P, Roth A, Simonovic M et al (2009) STRING 8--a global view on proteins and their functional interactions in 630 organisms. Nucleic Acids Res 37(Database issue):D412–D416
34. von Mering C, Huynen M, Jaeggi D, Schmidt S, Bork P, Snel B (2003) STRING: a database of predicted functional associations between proteins. Nucleic Acids Res 31(1):258–261
35. Harrington ED, Jensen LJ, Bork P (2008) Predicting biological networks from genomic data. FEBS Lett 582(8):1251–1258
36. Marcotte EM, Xenarios I, Eisenberg D (2001) Mining literature for protein-protein interactions. Bioinformatics 17(4):359–363
37. Szklarczyk D, Franceschini A, Kuhn M, Simonovic M, Roth A, Minguez P, Doerks T, Stark M, Muller J, Bork P et al (2011) The STRING database in 2011: functional interaction networks of proteins, globally integrated and scored. Nucleic Acids Res 39(Database issue):D561–D568
38. Zanzoni A, Montecchi-Palazzi L, Quondam M, Ausiello G, Helmer-Citterich M, Cesareni G (2002) MINT: a molecular INTeraction database. FEBS Lett 513(1):135–140
39. Licata L, Briganti L, Peluso D, Perfetto L, Iannuccelli M, Galeota E, Sacco F, Palma A, Nardozza AP, Santonico E et al (2012) MINT, the molecular interaction database: 2012 update. Nucleic Acids Res 40(Database issue):D857–D861
40. Hermjakob H, Montecchi-Palazzi L, Lewington C, Mudali S, Kerrien S, Orchard S, Vingron M, Roechert B, Roepstorff P, Valencia A et al (2004) IntAct: an open source molecular interaction database. Nucleic Acids Res 32(Database issue):D452–D455
41. Kerrien S, Alam-Faruque Y, Aranda B, Bancarz I, Bridge A, Derow C, Dimmer E, Feuermann M, Friedrichsen A, Huntley R et al (2007) IntAct--open source resource for molecular interaction data. Nucleic Acids Res 35(Database issue):D561–D565
42. Orchard S, Ammari M, Aranda B, Breuza L, Briganti L, Broackes-Carter F, Campbell NH, Chavali G, Chen C, del-Toro N et al (2014) The MIntAct project--IntAct as a common curation platform for 11 molecular interaction databases. Nucleic Acids Res 42(Database issue):D358–D363
43. Keshava Prasad TS, Goel R, Kandasamy K, Keerthikumar S, Kumar S, Mathivanan S, Telikicherla D, Raju R, Shafreen B, Venugopal A et al (2009) Human protein reference database--2009 update. Nucleic Acids Res 37(Database issue):D767–D772
44. Breitkreutz BJ, Stark C, Tyers M (2003) The GRID: the general repository for interaction datasets. Genome Biol 4(3):R23
45. Stark C, Breitkreutz BJ, Reguly T, Boucher L, Breitkreutz A, Tyers M (2006) BioGRID: a general repository for interaction datasets. Nucleic Acids Res 34(Database issue):D535–D539
46. Chatr-Aryamontri A, Breitkreutz BJ, Oughtred R, Boucher L, Heinicke S, Chen D, Stark C, Breitkreutz A, Kolas N, O'Donnell L et al (2015) The BioGRID interaction database: 2015 update. Nucleic Acids Res 43(Database issue):D470–D478
47. Scott MS, Barton GJ (2007) Probabilistic prediction and ranking of human protein-protein interactions. BMC Bioinf 8:239
48. Goll J, Rajagopala SV, Shiau SC, Wu H, Lamb BT, Uetz P (2008) MPIDB: the microbial protein interaction database. Bioinformatics 24(15):1743–1744
49. Lamesch P, Berardini TZ, Li D, Swarbreck D, Wilks C, Sasidharan R, Muller R, Dreher K, Alexander DL, Garcia-Hernandez M et al (2012) The Arabidopsis Information Resource (TAIR): improved gene annotation and new tools. Nucleic Acids Res 40 (Database issue):D1202–D1210
50. Rebhan M, Chalifa-Caspi V, Prilusky J, Lancet D (1997) GeneCards: integrating information about genes, proteins and diseases. Trends Genet 13(4):163
51. Safran MC-CV, Shmueli O, Rosen N, Benjamin-Rodrig H, Ophir R, Yanai I, Shmoish M, Lancet D (2003) The GeneCards family of databases: GeneCards, GeneLoc, GeneNote and GeneAnnot. In: Proceedings of the IEEE Computer Science Bioinformatics Conference CSB2003
52. Stelzer GHA, Dalah A, Rosen N, Shmoish M, Iny Stein T, Sirota A, Madi A, Safran M, Lancet D (2008) GeneCards: one stop site for human gene research. FISEB (ILANIT)
53. Harel A, Inger A, Stelzer G, Strichman-Almashanu L, Dalah I, Safran M, Lancet D (2009) GIFtS: annotation landscape analysis with GeneCards. BMC Bioinf 10:348

Part IV

Applications of Proteomics Technologies in Biological and Medical Sciences

Identification, Quantification, and Site Localization of Protein Posttranslational Modifications via Mass Spectrometry-Based Proteomics

17

Mi Ke, Hainan Shen, Linjue Wang, Shusheng Luo, Lin Lin, Jie Yang, and Ruijun Tian

Abstract

Posttranslational modifications (PTMs) are important biochemical processes for regulating various signaling pathways and determining specific cell fate. Mass spectrometry (MS)-based proteomics has been developed extensively in the past decade and is becoming the standard approach for systematic characterization of different PTMs on a global scale. In this chapter, we will explain the biological importance of various PTMs, summarize key innovations in PTMs enrichment strategies, high-performance liquid chromatography (HPLC)-based fractionation approaches, mass spectrometry detection methods, and lastly bioinformatic tools for PTMs related data analysis. With great effort in recent years by the proteomics community, highly efficient enriching methods and comprehensive resources have been developed. This chapter will specifically focus on five major types of PTMs; phosphorylation, glycosylation, ubiquitination/sumosylation, acetylation, and methylation.

Keywords

Posttranslational modifications (PTMs) • Phosphoproteomics • Glycoproteomics • Ubiquitinome and sumoylated proteome • Methylome • Acetylome • PTM enrichment • Metal oxide affinity chromatography (MOAC) • TiO_2 • IMAC • Immunoprecipitation • Peptide fractionation • Orthogonality • Lectin affinity • Hydrazine-based purification • Anti-k(GG) Ab • KMBD MBT-based purification • Lysine-methylation • Lysine-acetylation • Serial PTM enrichment • PTM mass spectrometry • CID • HCD • ETD • MS3

M. Ke • H. Shen • L. Wang • S. Luo • L. Lin • J. Yang
Department of Chemistry, South University of Science and Technology of China, 518055 Shenzhen, China

R. Tian (✉)
Department of Chemistry, South University of Science and Technology of China, 518055 Shenzhen, China

Shenzhen Key Laboratory of Cell Microenvironment, South University of Science and Technology of China, 518055 Shenzhen, China
e-mail: tian.rj@sustc.edu.cn

H. Mirzaei and M. Carrasco (eds.), *Modern Proteomics – Sample Preparation, Analysis and Practical Applications*, Advances in Experimental Medicine and Biology 919, DOI 10.1007/978-3-319-41448-5_17

17.1 Introduction

Posttranslational modification (PTMs) alter the biochemical properties of a protein by the addition of a chemical group to one or more of its amino acid residues. PTMs are extremely important for protein function as they can influence the activity state, stability, localization, turnover, and interactions with other proteins [1]. Up to now, more than 200 posttranslational modifications of proteins are known to occur physiologically. Analysis of these modifications is a great challenge, as biologically significant PTMs usually happen with low stoichiometry at low abundance. Since mass spectrometry (MS)-proteomic methodologies have demonstrated tremendous potential for quantitatively profiling various PTMs, increasing attention has been given to the development of powerful proteomic approaches to explore different PTMs in various biological systems [2].

In this chapter, we will provide a comprehensive review of MS-based proteomic analysis of different PTMs, including sample enrichment approaches, high-performance liquid chromatography (HPLC)-based fractionation approaches, MS detection techniques and bioinformatics methods. We will specifically focus our discussion on five major types of PTMs; phosphorylation, glycosylation, ubiquitination/sumosylation, acetylation, and methylation.

17.2 Biological Functions of PTMs

Posttranslational modifications (PTMs) are biochemical processes for modifying proteins with various chemical groups such as phosphate, glycan, methyl, acetyl, ubiquitin, etc. The majority of eukaryotic proteins (50–70 %) are modulated by different PTMs in space- and time-dependent manners, which are crucial for various biological functions. Over 200 types of PTMs have been identified so far, most of which are irreversible and lead to permanent changes in protein conformation and function [3]. With reversible phosphorylation as an example (Fig. 17.1), many key PTMs with indispensable biological functions are enzyme-dependent and reversible. These dynamic PTM modulations provide finely tuned biochemical mechanisms for the regulation of key physiological states in cells and their responses to the external environment. PTMs regulate the activity of proteins and also significantly expand the biological system complexity. Some of the major physiological roles for PTMs are summarized as follows:

1. PTMs can significantly change the three-dimensional structure of proteins, usually to modulate its function. A case in point is the hydroxylation of proline and lysine residues in collagens which stabilizes their coiled structure [4]
2. PTMs such as phosphorylation are key factors for regulating dynamic protein–protein interactions
3. The sub-cellular localization of some proteins is directed by PTMs. For example, proteins modified with glycosylphosphatidylinositol on cysteine are usually directed to the cellular membrane
4. Protein stability and half-life time are also modulated by PTMs. It is well-known that the K48-ubiquitination tag is a signal for protein degradation by proteasome pathway.

17.2.1 Phosphorylation

Phosphorylation is one of the most ubiquitous and important PTMs. It is a process that involves the kinase-regulated transportation of a phosphate group from ATP to specific amino acid residues, mainly serine (S), threonine (T), tyrosine (Y), and the recently discovered histidine (H). Reversible protein phosphorylation is a widely used modulation method which is utilized by both eukaryotic and prokaryotic organisms. Over 1/3 of proteins are modified by phosphorylation in eukaryotes at any given time [5, 6]. During signal transduction, upon the binding of a secreted ligand, the receptors are often phosphorylated, which subsequently activates

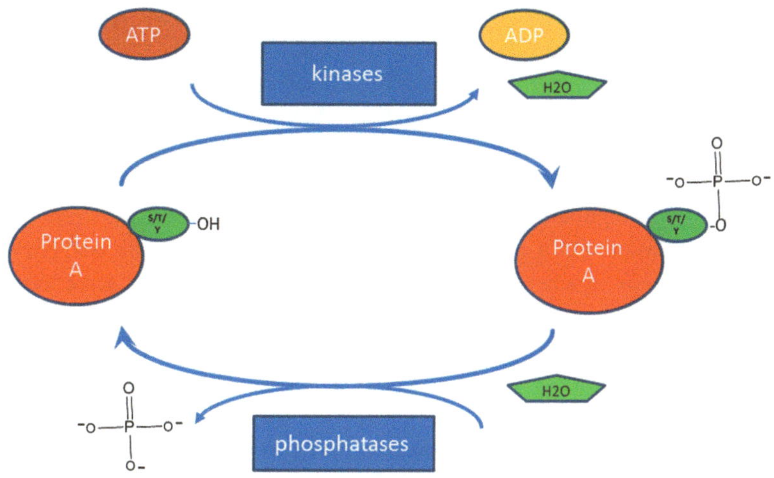

Fig. 17.1 Protein phosphorylation is a typical model for reversible and enzyme-dependent PTM. Phosphorylation is catalyzed by the "writer"-kinases and the dephosphorylation is performed by "eraser"-phosphatases, Protein phosphorylation mainly happens on specific amino acid residues including serine (S), threonine (T), and tyrosine (Y)

dynamic intracellular signaling pathways. It is currently known that many human diseases, for instance cancer and Alzheimer's disease (AD), result from abnormal phosphorylation of key functional proteins. For example, Tau protein holds more than 20 phosphorylation sites in AD which are rigorously modulated by various protein kinases, such as cycling dependent kinase 5 (CDK5), Mitogen-activated protein kinase (MAPK), etc. [7]

17.2.2 Glycosylation

Protein glycosylation is a process which involves the attachment of sugar moieties to proteins. Most of the proteins in the plasma membrane, endoplasmic reticulum (ER), and extracellular environment are glycosylated. Glycosylation is structurally the most complex PTM, hence, the mechanism of this process is highly-ordered, and its extent and complexity correlate closely with the level of evolution [8]. Different types of glycosylation have been well characterized, including N-linked glycosylation, O-linked glycosylation, C-mannosylation, glypiation (glycosylphosphatidylinositol anchor), etc. [9, 10]. Oligosaccharides vary in terms of the function and structure of their sugar residues (i.e. galactose, glucose, mannose, N-acetylgalactosamine, N-acetylglucosamine, fucose, sialic acid, etc.). Among them, N-linked and O-linked glycosylation are the most widely studied. N-linked glycans (essentially made up of two N-acetyl glucosamine and three mannose residues) always attached to asparagine (Asn) residues with the consensus sequence Asn-X-Ser/Thr (X represents any amino acid except proline) [11]. N-glycosylation is the most common glycosylation modification. O-glycosylated proteins also show their roles in various cellular functions, especially in cell metabolism. O-glycosylation occurs mainly on serine and threonine side chains, and sometimes can occur on oxidized forms of lysine and proline residues [12]. Glycosylation plays a role in many important cellular functions, such as cell adhesion, receptor activation, endocytosis, cell immune responses, etc. The sensitive recognition of protein-protein and cell-cell interactions, typically in the intracellular microenvironment, is the most well studied example of the functionality of protein glycosylation [13].

17.2.3 Ubiquitination and Sumoylation

Ubiquitin is a small (76 amino acids) polypeptide with that is usually characterized by two glycine residues (diGly) in the C-terminal domain. The ubiquitin peptide chain contains seven lysine residues at positions 6, 11, 27, 29, 33, 48 and 63, from the N- to the C-terminus, through which it can be attached to substrates [14]. Ubiquitin modification functions as a signaling mechanism, mainly by activating the protein degradation machinery. Attachment of ubiquitin to other proteins is catalyzed by an activating enzyme (E1), a ubiquitin-conjugating enzyme (E2) and a ubiquitin ligase (E3). This process plays significant roles in regulating cell apoptosis, transcription modulation, DNA repair, etc. In eukaryotes, over 30 % of newly synthesized proteins are degraded because of damage to their structure. Mono-ubiquitination regulates cellular functions by altering the activity of proteins, changing the binding affinity with other proteins, and transporting specific proteins to their site of activity [15]. A single ubiquitin moiety linked to a substrate is often a signal for additional linkages of ubiquitin molecules onto the existing ubiquitin, thus forming a polyubiquitin chain. Typically, K48-linked polyubiquitin chains target proteins for proteolysis. A chain of at least four ubiquitin molecules on a condemned protein can be recognized by the 26 s proteasome [16].

In addition to the well-studied ubiquitin system, later studies have uncovered other ubiquitin-like modifications including small ubiquitin-related modifier (SUMO) [17], Nedd8 [18], Atg8 [19], and ISG15 [20], etc. Unlike ubiquitination, sumoylation is a reversible and multifunctional modification, participating in many cellular signaling pathways. There are three well-characterized SUMO proteins in humans: SUMO1, SUMO2 and SUMO3. SUMO2 and SUMO3 share about 95 % sequence homology and have only few conspicuous functional differences. Conversely, the homology between SUMO2/3 and SUMO1 is only 44 % (Fig. 17.2) and they also have distinct target proteins.

17.2.4 Acetylation

Protein acetylation involves the introduction of an acetyl group into a polypeptide by replacing an active hydroxyl group. Lysine has emerged as the main acetylation site for many key functional proteins, such as histones, transcription regulators, and enzymes associated with glycolysis [21]. Like reversible phosphorylation, the acetylation process is catalyzed by a pair of enzymes, including histone acetyltransferase (HAT) and histone deacetylase (HDA). Acetylation regulates protein activity and can crosstalk with other PTMs such as phosphorylation and methylation in the dynamic control of transcription activity [22], cellular signaling [23], etc. Acetylation modification reduces the electrostatic attraction between histone 4 and the phospho-rich negatively charged DNA backbone, thereby loosening chromatin structure and resulting in increased transcription activity [22].

17.2.5 Methylation

Protein methylation is a common PTM that modifies proteins with methyl groups mostly on lysine (K) and arginine (R) amino acid residues. Arginine methylation is catalyzed mainly by two classes of arginine methyltransferases (PRMTs):

- Type I PRMTs including PRMT1, PRMT3, PRMT4, PRMT6, and PRMT8
- Type II PRMTs including PRMT5 and PRMT7

Both classes of enzymes can catalyze arginine monomethylation. Type I PRMTs can also add

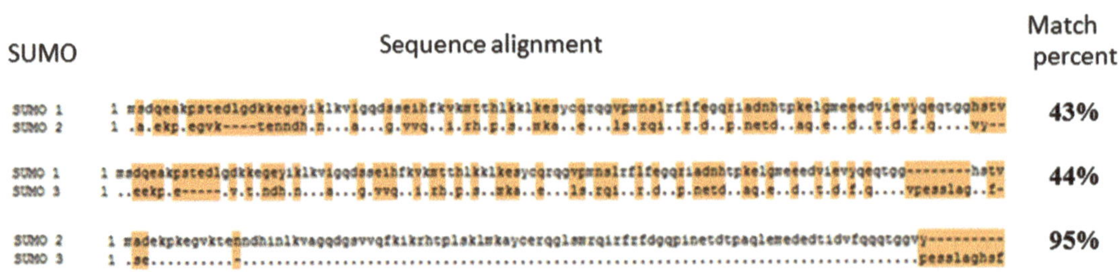

Fig. 17.2 Amino acid sequence alignment of SUMO1, 2 and 3 shows different degree of sequence distinction

methyl groups to arginine side chain forming asymmetric dimethyl arginine, while type II enzymes can further catalyze the symmetric dimethylation (Fig. 17.3a) [24]. Lysines can be mono-, di-, or trimethylated by lysine methyltransferases (KMTs), and this modification can also be reversed by demethylases (DMTs) (Fig. 17.3b) [25, 26]. Generally, methylation has been reported to regulate RNA processing, gene transcription, DNA damage repair, and signal transduction [27].

17.3 Proteomic Strategies for PTMs Analysis

The basic procedure for PTM analysis is roughly the same as the procedure used for the identification of proteins in 'classical' proteomics research. However, the PTMs analyses are generally more difficult for the following reasons:

1. The endogenously modified proteins only constitute a small fraction of the total protein numbers (low stoichiometry)
2. Since the covalent bond between the PTM and the amino acid side chain is typically labile, it is often difficult to maintain the peptides in their modified state during sample preparation and subsequent ionization in mass spectrometry
3. PTMs are frequently transient in the dynamic homeostasis of nature. Therefore, more effective sample preparation methods, more sensitive detection technology and more comprehensive data analysis strategies are needed in the analysis of PTMs.

17.3.1 Conventional Analysis Methods

Conventionally, PTMs analysis has been carried out by laborious biochemical approaches, including two-dimensional gel electrophoresis (2D-GE), western blot analysis, autoradiography and Edman sequencing.

2D-GE is a classical separation method that separates proteins on the basis of their isoelectric point and molecular weight [28]. 2D-GE has been successfully used to directly separate different types of modified proteins. For example, phosphorylation changes the charge of a protein and is often indicated by a horizontal trail of protein spots on a two-dimensional gel. Once the proteins have been isolated, a variety of detection techniques can be used in succession. The proteins in the gel can be unselectively visualized by staining the gel with coomassie blue or colloidal silver. The staining intensity of the gel spots roughly reflects the protein amount, providing information on the relative proportion of the various modified states.

The specific subsets of PTM-modified proteins present in the gel can also be selectively detected and visualized. This can be achieved by using a PTM-specific staining reagent to develop the gel or by using PTM-specific antibodies for western blotting, or by incorporating PTM-specific radiolabels into the proteins. For example, phosphoproteins can be selectively stained and visualized with phosphate-specific fluorescent probes (such as BO-IMI, in which a BODIPY dye is attached to a reactive imidazole group) [29]. Western blot analysis with antibodies against specific phosphorylation sites is widely used to detect different types of phosphoproteins (S, T, and Y) [30, 31]. Likewise, nitrated proteins can be detected with anti-nitrotyrosine antibodies. In these cases, the quality of the antibody, including specificity and sensitivity, is critical for the detection.

Autoradiography is an alternative detection technology that was widely used in the past, although it is sometimes expensive and hazardous. Proteins are labeled (*in vivo* or *in vitro*) with radioactive PTM precursors before extraction and separation, and subsequently visualized by autoradiography. A number of specific radiolabeling agents are available, such as $[^{32}P]$-phosphate or $[\gamma\text{-}^{32}P]$-ATP for

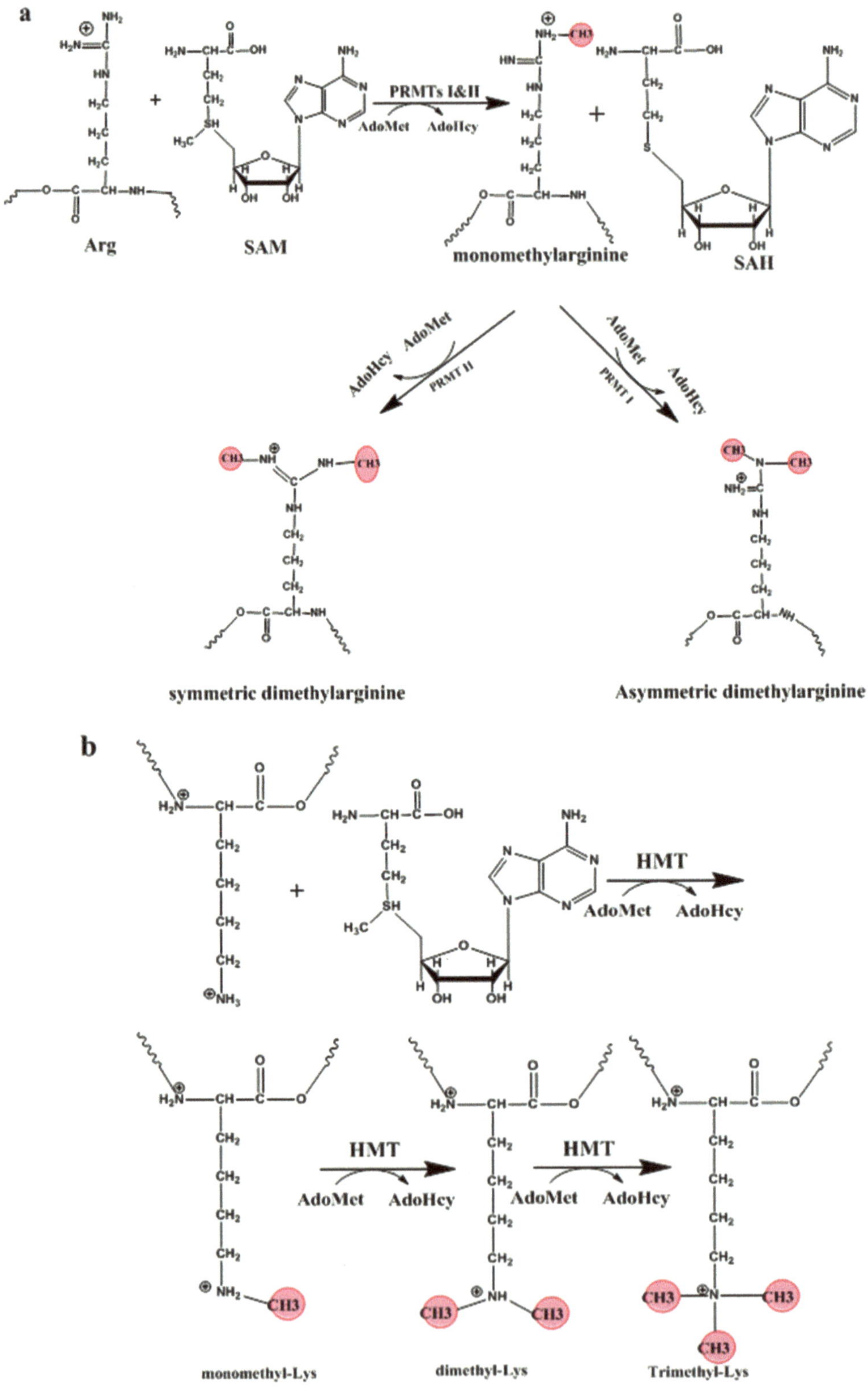

Fig. 17.3 Overview of arginine and lysine methylation. (**a**) Arginine methylation is catalyzed by protein arginine methyltransferases (PPRMTs) I or II (**a**) [24], of which PPRMT I can transfer a second methyl group to the same guanidino nitrogen amino of arginine, denoted as asymmetric dimethylation while PPRMT II catalyzes

phosphoproteins, [^{3}H]-inositol for GPI-anchored proteins, and [^{3}H]-myristyl for N-myristoylated proteins.

Edman degradation, the classical protein sequencing technique, can be used to locate modification sites. The proteolytic peptide fractions are applied to the sequencer and their amino acid sequence is determined. Modified amino acids become apparent for their absence or retention time shift in the corresponding sequencing cycle. Edman sequencing in combination with radiolabeling was once widely used for characterizing phosphorylation sites. For this, ^{32}P labeled proteins were digested into peptides and separated, and the candidate phosphorylation sites were identified by recording the cycle in which the radiolabeled amino acid was released [32].

While feasible, the traditional methods mentioned above suffer from various shortcomings. The 2D-GE separations are difficult to achieve when separating low abundance, acidic, basic, hydrophobic, very large, or very small proteins. Furthermore, this technique has reproducibility issues, a limited dynamic range and a low throughput, which hinders its application in the global characterization of PTMs. Antibody-based western blot analyses show poor performance in the detection of some types of PTMs due to steric hindrance of the recognition site. Autoradiography is hazardous and radio-isotopes of carbon and hydrogen are rather weak radio emitters, which makes it difficult to efficiently detect corresponding modified proteins (for example, ^{14}C or ^{3}H in the case of protein methylation and acetylation). Edman sequencing is tedious and requires massive amounts starting material and it has a lengthy analysis cycle. This is especially true when radiolabelling is involved, which limits its application in high-throughput studies.

Compared with those conventional analysis methods, mass spectrometry (MS) has emerged as a powerful technique to analyze PTMs due to its high efficiency, sensitivity, and selectivity. The extensive application of MS based proteomics in PTM analysis is the result of the development of effective enrichment strategies; faster, more sensitive MS detection technique and powerful bioinformatics methods. All these aspects will be described in detail in the subsequent paragraphs.

17.3.2 Enrichment of PTMs Prior to MS Analysis

PTMs are often found at sub-stoichiometric levels and represent a small proportion of all peptides present in a total cell lysate, which is why it requires enrichment/purification to improve their measurement prior to mass spectrometry identification. Table 17.1 summarizes the well-established enrichment methods for specific PTMs.

17.3.2.1 Phosphorylation

Enrichment

Phosphorylation is one of the most extensively studied PTMs due to its biological significance in cell signaling and regulation. Due to the sub-stoichiometric and highly dynamic nature of phosphorylation, large-scale studies of the phospho-proteome require sophisticated experimental workflows that primarily hinge upon achieving a highly efficiency, highly specific enrichment. Several affinity enrichment protocols have been established for enriching phosphorylated peptides from complex proteome digests such as cell lysates. These methods include metal oxide affinity chromatography

Fig. 17.3 (continued) the formation of symmetric dimethylarginine by adding the second methyl group to a different guanidine nitrogen atom of arginine. (**b**) Lysine methylation is catalyzed by the enzyme KMTs, usually are histone methyltransferases (HMTs) [25], adding one, two, or three methyls to the distinct guanidino nitrogen amino of lysine, forming monomethyl (Kme1), dimethyl (Kme2), or trimethyl (Kme3) lysine, respectively (Note: AdoMet, S-adenosylmethionine synthase; AdoH, *S*-Adenosyl-L-homocysteine, equals to SAH; PPRMT, protein arginine methyltransferase; KMT, lysine methyltransferase; SAM)

Table 17.1 PTMs MS shift and the reporter fragments observed by collision-based dissociation

PTMs type	Amino acid modified	Mass shift/Gross formula shift (stable in MS/MS fragmentation)	Diagnostic ions (specific fragment ions in MS/MS fragmentation)	Neutral loss (labile in MS/MS fragmentation)
Phosphorylation	Tyr	+79.9663 Da (HPO_3) [96, 97]	216.0426 Da (+) [98]	79.9663 (HPO_3) Da [99]
	Ser/Thr/Tyr		97 Da (−) [93]	97.976 (H_3PO_4) Da [97, 100]
			78.9591 Da (−) [97]	
			63 Da (−) [87]	
Glycosylation	N-linked (Asn)	>800 Da [83]	204.087 Da (+) [101]	203.079 Da (HexNAc) [102]
		Variable [96]		162.053 Da (Hexose) [102]
	O-linked (Ser/Thr)	+203.0793 Da (HexNAc) [101]	163.0606 Da (+) [101]	291.095 Da (Sialic acids) [102], 365.148 Da (HexHexNAc) [102]
		+162.0528 Da (Hexose) [101],	366.140 Da (+) [101]	
		+291.095 Da (Sialic acids) [102]	246.0977 Da (+) [101]	
		+365.148 Da (HexHexNAc) [102]	292.103 Da (+) [102]	
			274.093 Da (+) [102]	
Acetylation	Ser/Thr/Lys	+42.0105 Da (CH_3CO) [101]	126.0913 Da (+) [103]	n/a
			143.1179 Da (+) [103]	
Methylation	Lys/Arg	+14.01565 Da (CH_3) [96, 101]	71.06 Da (+), 46.06 Da (+) (Dimethylation) [104]	n/a
		+28.0313 Da (C_2H_6) [96]		
		+42.04695 Da (C_3H_9) [96]		
Ubiquitination	Lys	+114.043 Da (Gly-Gly) [105]	n/a	n/a

Note: (+): in positive-mode; (−): in negative-mode

(MOAC), immobilized metal ion affinity chromatography (IMAC), immunoprecipitation-based enrichment, and domain-based enrichment.

Metal Oxide Affinity Chromatography (MOAC)

MOAC represents one of the most commonly used strategies for phosphopeptide enrichment. This technique is based on the affinity that phosphate groups have towards metal oxides. Several metal oxides, including TiO_2 [33], ZrO_2 [34] and Nb_2O_5 [35], have been successfully used for this purpose. TiO_2 is the most popular MOAC substrate, with high enrichment efficiency and specificity. In a typical TiO_2-based MOAC procedure, the sample is mixed with an acidic buffer (e.g. 0.1 % (v/v) trifluoroacetic acid) to protonate acidic residues of non-phosphorylated peptides, preventing their adsorption to TiO_2. After a washing step, phosphopeptides are eluted from the TiO_2 column under alkaline conditions, such as ammonium bicarbonate at a pH of 9. Usually TiO_2-based MOAC enrichment suffers from low specificity due to the competitive binding of acidic amino residues (e.g. Glu and Asp) in non-phosphopeptides. Considerable efforts have been made to improve the specificity of this protocol by introducing competitive additives such as 2,5-hydroxybenzoic acid (DHB) [36], phthalic acid [37] and glutamic acid [38] into the loading buffers.

Immobilized Metal Ion Affinity Chromatography (IMAC)

IMAC is another widely used affinity purification technique for phospho-peptide enrichment. The affinity between phospho-peptides and IMAC resin is caused by electrostatic interactions between the negatively charged phosphate groups of phospho-peptides and the positively charged metal ions that are bound to a solid support via iminodiacetic acid (IDA) or nitriloacetic acid (NTA) ligands. Various metal ions have been tested for their efficiency in phosphorylated peptide enrichment, such as Fe^{3+}[39], Ti^{4+}[40], Zr^{4+}[41], Ga^{3+}[42], etc. The general procedure is similar to MOAC. First, the tryptic digest is dissolved in IMAC-binding buffer and loaded onto an IMAC column for incubation. Then nonphosphorylated peptides are removed by washing the resin with IMAC binding buffer. Phosphopeptides are then removed from beads at high pH or with phosphate salts. IMAC was first introduced by Anderson and Porathin in 1986 for the enrichment of phosphoproteins [43], and has been extensively improved by many other researchers. To reduce the nonspecific binding of nonphosphorylated acidic peptides to the IMAC resin, Ficarro et al. developed an technique that blocks the carboxylic groups that are present at the C-terminus of peptides and in acidic residues (i.e. Glu and Asp) by methyl-esterification [44]. Despite having increased specificity toward phosphopeptides, this approach suffers from incomplete reaction and side derivatization reactions which might complicate the MS identification. IMAC enrichment has a bias towards multi-phosphorylated peptides, which necessitates the implementation of complementary strategies such as SIMAC (sequential elution from IMAC) [45]. The SIMAC approach combines both Fe-IMAC and TiO_2 enrichment strategies for phospho-peptide enrichment in a consecutive manner. A typical SIMAC workflow starts with an IMAC enrichment to first capture multi-phosphorylated peptides. The flow-through and acid eluted fractions are then collected and subjected to TiO_2 enrichment, to capture most of the mono-phosphorylated peptides. Using such a strategy, Thingholm and coworker were able to double the identification of phosphorylation sites as compared with single TiO_2 enrichment [46].

A new type of IMAC approach with immobilized metal ions was developed by Zou et al. for high-efficient enrichment of phosphorylated peptides [47]. This resin, which uses Ti^{4+}, outperformed all other phosphopeptide enrichment methods that use other metal ions (Fe^{3+}-IMAC, Zr^{4+}-IMAC, TiO_2 and ZrO_2). The high specificity and efficiency of Ti^{4+}-IMAC is mainly due to the flexibility of the spacer arm that is linked to the polymer beads, and also to the specific interaction between the immobilized Ti^{4+} and the phosphate groups that prevents binding of acidic peptides [48].

Immunoprecipitation-Based Enrichment

Tyrosine phosphorylation often occurs at very low abundance and the occupancy is estimated at about 0.5 % of all human phosphorylation events with the majority occurring via serine (~90 %) or threonine (~10 %) residues [49]. Therefore, the aforementioned approaches are not well-suited for the study of tyrosine phosphorylation. Immunoprecipitation (IP) with immobilized antibodies against phosphotyrosine (pTyr) is a well-established strategy for the enrichment of pTyr carrying phosphopeptides. With the highly specific commercially available antibodies against pTyr (i.e., PY100), Rikova et al. identified 4551 phospho-tyrosinesites on 2700 different proteins and characterized tyrosine kinase signaling across 41 non-small cell lung cancer (NSCLC) cell lines and over 150 NSCLC tumors (Rikova, 2007, Cell). In addition to the pTyr-specific antibodies, substrates of RTKs such as Scr homology 2 (SH2) domains can also be used to enrich tyrosine-phosphorylated proteins. Using the SH2 domain of the adapter protein Grb2 (GST-SH2 fusion protein), Blagoev et al. identified 228 proteins. However, this approach is limited to those phosphotyrosine containing proteins that interact with the SH2-containing bait used in the assay.

IP approaches are not commonly used for phosphoserine and phosphothreonine enrichment, mainly because highly specific antibodies against pThr and pSer do not exist. Some studies have employed antibodies raised against the consensus motifs in phosphothreonine and phosphoserine peptides [50, 51]. However, yields of such approaches were relatively low, because those antibodies did not bind all pS/pT sites with the same efficiency.

Fractionation

Ion-Exchange Chromatography

Ion-Exchange Chromatography is a charge-based strategy for the enrichment of phosphopeptides according to the interaction between the negatively charged phosphate group and the Strong Cation Exchange (SCX) or Strong Anion Exchange (SAX) matrix. SCX chromatography has been one of the most popular fractionation strategies for sample complexity reduction in phosphoproteomics experiments [39]. The principle of SCX for fractionation of phosphopeptides is illustrated in Fig. 17.4. Under acidic conditions (e.g. pH 2.7), the N-terminal amino group and the C-terminal Lys/Arg residues of most tryptic phosphopeptides were protonized to have a net charge of 2^+, whereas mono-phosphopeptides have a charge state of only 1^+ due to the one unit of the attached negatively charged phosphate group (Left panel). The net charge of a phosphopeptide is decreased by one unit for each added phosphate group. This means that phosphopeptides have a decreased affinity (mono-phosphorylated) or no affinity (multi-phosphorylated) for the SCX media. [52].

In contrast to SCX chromatography, SAX chromatography tends to retain the negatively charged phosphopeptides more effectively than nonphosphorylated peptides [53]. SAX was shown to have a better selectivity for multiply phosphorylated peptides and was initially introduced to compensate one of the main issues associated with SCX, which is the relative inability to retain strongly acidic, negatively charged multi-phosphopeptides [54, 55]. Dai et al. have devised a multidimensional liquid chromatography (Yin-Yang MDLC) approach combining SCX and SAX to profile the phospho-proteome of mouse liver [54]. In this approach, protein digests were first loaded onto a SCX column. Flow through peptides from SCX were then collected and further loaded onto an SAX column. Both the SCX and SAX columns were eluted offline by a pH gradient to fractionate the phosphopeptides for following RP-LC/MS identification.

Reverse Phase Chromatography

Reverse phase chromatography (RPC) fractionation of protein/peptides is based on hydrophobic interactions of the protein/peptides with the RPC stationary phase. Theoretically, phosphopeptides are less retained by RP column and eluted earlier than the nonphosphorylated counterparts, due to their reduced hydrophobicity as a result of the attached phosphate groups. RPC is often used as a second dimension separation for phosphopeptide fractionation because of its superior separation efficiency and excellent compatibility with LC/MS. Despite its excellent ability to fractionate phosphopeptides, RPC is less commonly used for offline fractionation due to the lack of orthogonality with inline RPC LC/MS. For this reason, high-pH RPC was introduced by Gilar et al. as the first dimension of separation for peptide mixtures (Gilar 2005,

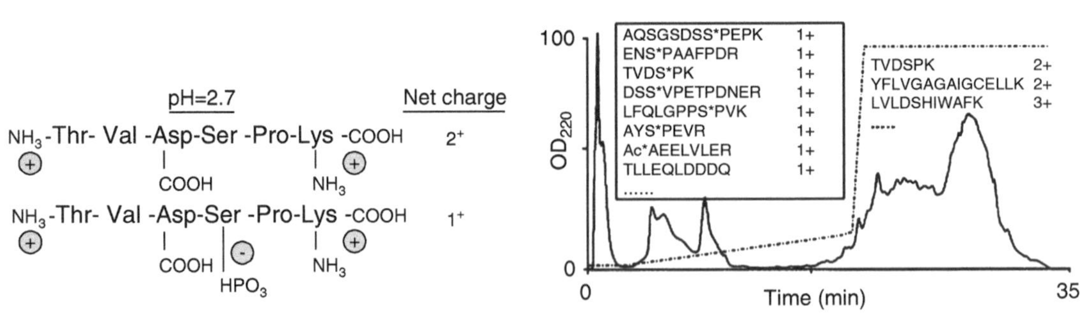

Fig. 17.4 Scheme for phosphopeptide enrichment by SCX chromatography. At pH 2.7, most peptides produced by trypsin proteolysis have a solution charge of 2, whereas phosphopeptides have a charge state of only 1 (*left panel*). SCX chromatography separation at pH 2.7 of a *HeLa* cell lysate after trypsin digestion. The dashed line indicates the salt gradient. Some identified peptides from the collected fractions are shown. Phosphorylation sites are denoted by an asterisk (*right panel*) (adopt from [52])

AC) showing excellent orthogonality with low-pH RPC, comparable to SCX-RP (Wang 2011, PROTEOMICS) for shotgun proteome analysis. The orthogonality of the high-pH RPC and low-pH RPC could be explained by the dramatic change in charge distribution within the peptide chain as a result of mobile-phase pH (Gilar 2005, AC). This approach was then adopted and refined by Zou et al. for global phosphopeptide analysis, which resulted in the identification of 30 % more peptides in mice liver compared to a conventional RPLC approach (Song 2010, AC).

HILIC/ERLIC

Hydrophilic interaction liquid chromatography (HILIC) and electrostatic repulsion hydrophilic interaction chromatography (ERLIC) are promising alternatives to ion-exchange and RP chromatography for the pre-fractionation and enrichment of phosphopeptides based on phosphopeptide hydrophilicity (polarity). HILIC uses a polar sorbent (e.g. TSK gel amide) to retain the highly hydrophilic phosphate groups. An organic containing loading buffer is used to promote hydrophilic interactions between phosphopeptides and the polar sorbent. Non-phosphorylated peptides, which are less hydrophilic, elute in the early fractions, followed by singly and multiply phosphorylated peptides in a gradient of increasing water. Alpert et al. first introduced ERLIC for the separation of phosphopeptides in 2008. This chromatography mode simultaneously uses hydrophilic interaction and electrostatic repulsion on a weak anion exchange (WAX) column to separate phosphopeptides [56]. When performed at low pH, non-phosphopeptides are protonized and are electrostatically repulsed by the WAX column, while the phosphopeptides, due to the presence of phosphate groups, are still negatively charged and electrostatically retained by the ERLIC column. With an increasing salt-gradient, phosphopeptides elute according to the number of phosphate groups, with monophosphorylated peptides eluting first.

Orthogonality in 2D-LC

Liquid chromatography (LC) has become the method-of-choice for the fractionation of peptides in complex mixtures due to its high resolving power and compatibility with downstream MS. By combining the resolving power of two orthogonal chromatography modes (2D-LC), complex peptide mixtures can be further simplified due to the increased resolution and higher peak capacity of the combined methods [54]. Gilar et al. comprehensively investigated the orthogonality of SCX, SEC, HILIC and RP for the 2D separation of defined peptides mixtures and showed that SCX-RP, HILIC-RP, and RP-RP (performed at high pH for the first dimension followed by low pH for the second dimension) provided the best combination in terms of orthogonality (Fig. 17.5).

The multidimensional combination of SCX and RPC has emerged as a powerful approach to separate phosphopeptides before analysis by mass spectrometry. By applying a multi-dimensional SCX-IMAC-RPC procedure, Gygi et al. were able to identify more than 5500 phosphoproteins with over 13,000 phosphorylation sites in mouse liver [57] and drosophila embryos [58]. McNulty et al. have demonstrated that HILIC could also be a good first dimension for the multidimensional separation of phosphopeptides by providing better orthogonality to the subsequent RPC than SCX. Using HILIC-RPC they were able to achieve higher coverage of the *Hela* phosphoproteome compared to SCX-RPC [59]. More recently, Song et al. established a new RPC-RPC approach for in depth phosphopeptides analysis [60]. They operated the first dimension of RPC separation at high pH (i.e. pH 10) and collected time-based fractions. They then pooled early fractions with late fractions that were collected in equal time intervals to decrease the total number of fractions before the second dimension RPLC-tandem mass spectrometry (MS/MS) at low pH. The resulting highly orthogonal 2D separation yielded 30 % more phosphopeptide identifications when compared to the conventional RPLC approach (Fig. 17.6).

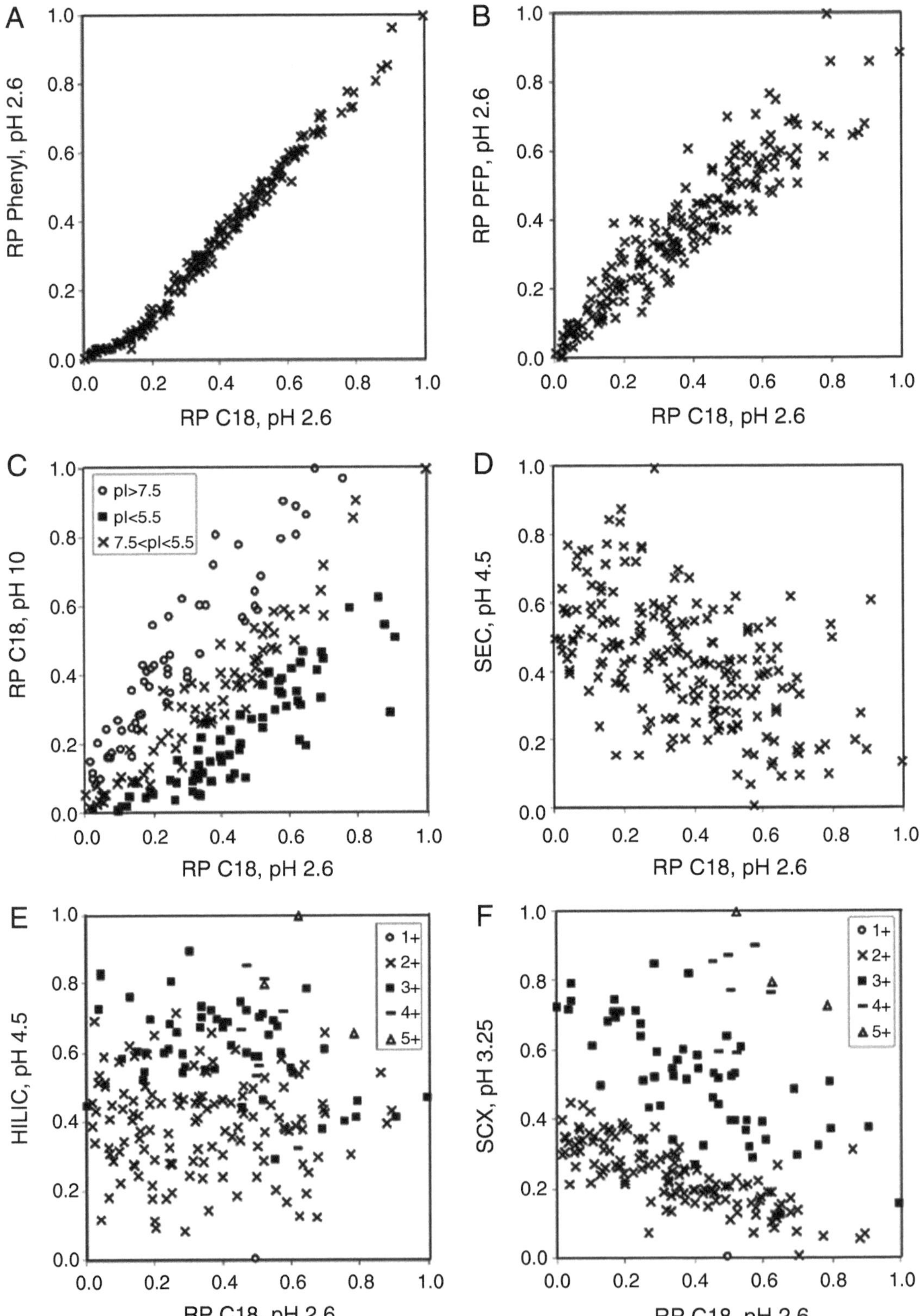

Fig. 17.5 Orthogonality of selected 2D-LC system

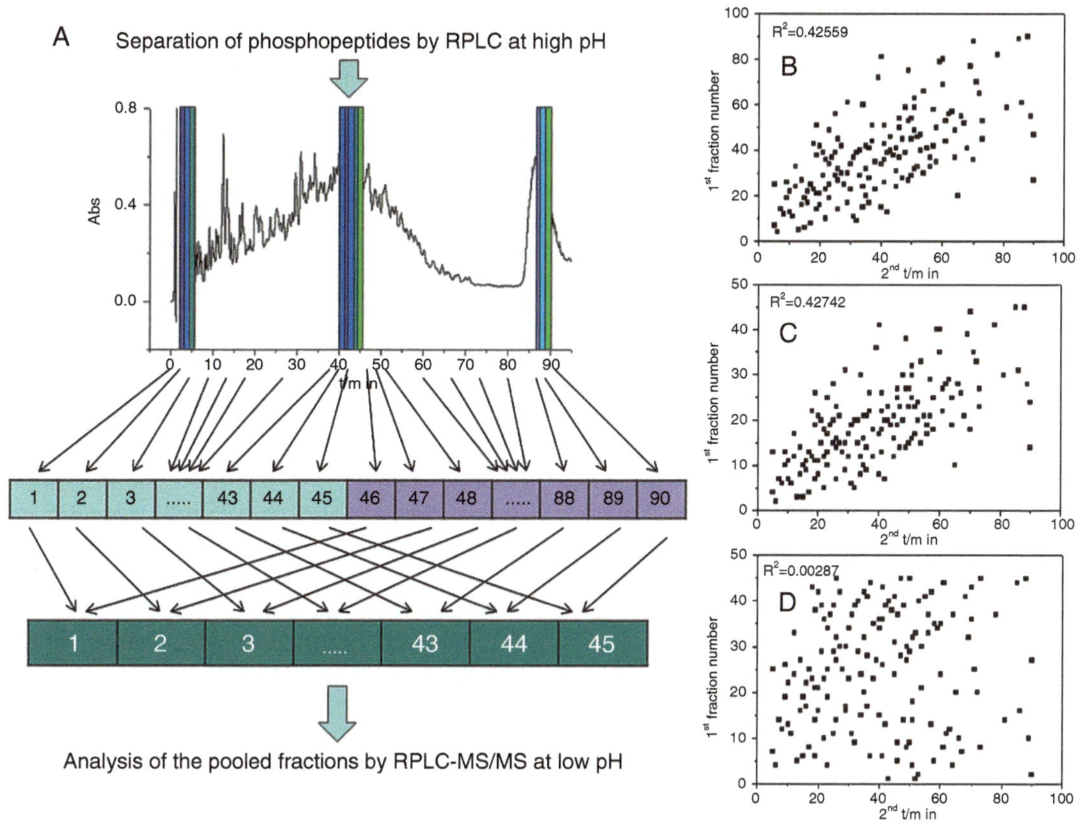

Fig. 17.6 The high pH RPC/low pH RPC approach with high orthogonality for separation of phosphopeptides. (**a**) The scheme for high pH RPC fractionation of phosphopeptides; (**b**) 2D retention plots for a hypothetical 2D separation of peptides, with 90 fractions; (**c**) reducing fraction number by pooling adjacent fractions; and (**d**) reducing fraction number by pooling equal interval fraction

Reducing Sample Complexity to Enhance Phosphoproteinome Coverage

Even with the most efficient and specific enrichment strategies, the PTM sample complexity will exceed the resolving power of state-of-the-art LC/MSMS systems. In order to reduce sample complexity and to increase the depth of PTM coverage, a combination of enrichment procedures and proper fractionation strategies is necessary. For instance, Trinidad et al. combined SCX fractionation with IMAC to study phosphorylation in mouse brain and reported a three-fold increase in phosphopeptide identification compared to SCX alone, demonstrating that a combination of fractionation and specific phosphopeptide enrichment is essential for large-scale phosphoproteomic studies. Peptide fractionation strategies such as SCX, SAX, HILIC and RPC conjugated with specific enrichment strategies (i.e. IMAC, TiO2, etc.) were comprehensively studied for their ability to reduce the sample complexity and to enhance the coverage of the PTM-ome. Gygi et al. have identified 5635 unique phosphorylation sites from 2328 proteins from mouse liver [57] and 13,720 different phosphorylation sites from 2702 proteins in developing *Drosophila* embryos [39] by applying a two-step phosphopeptide enrichment procedure consisting of SCX chromatography followed by IMAC. Similarly, Olsen et al. used SCX fractionation followed by TiO_2 enrichment was used to characterize the dynamics of human cell cycle phosphorylation [33, 61]. They detected 6600 phosphorylation sites from 2244 proteins in epidermal growth factor stimulated *HeLa* cancer cells

[33]. However, one important disadvantage associated with SCX fractionation is that phosphopeptides are not equally distributed in all fractions due to varying charge distributions amongst mono and polyphosphorylated peptides. To address this problem, McNulty et al. applied a HILIC-IMAC instead of the SCX-IMAC approach for fractionation of phosphopeptides, which resulted in higher coverage of *HeLa* cell phosphoproteome [59]. In this study, the authors also showed that the use of IMAC prior to a HILIC separation of phosphopeptides resulted in an increased contamination with non-phosphopeptides. The percentage of phosphopeptides increased to 99 % when performing IMAC on the HILIC fractions, which indicated the importance of a prefractionation step in reducing sample complexity and improving enrichment efficiency.

17.3.2.2 Glycosylation

N-Glycosylation

Lectin

Lectin affinity enrichment is an efficient strategy for glycoprotein/glycopeptide enrichment. Different types of lectins, immobilized on solid supports such as agarose or magnetic beads, are used to enrich glycoproteins/glycopeptides according to their glycan structures. The enrichment efficiency for different glycopeptides can be significantly increased by using lectins with broad specificities [62]. Alternatively, lectins with narrow specificity can be utilized as "structure specific affinity selectors". Concanavalin A (Con A), wheat germ agglutinin A (WGA), Peanut agglutinin (PNA) and aleuriaaurantia (AAA) are some of the most widely used lectins for enriching N-linked glycosylated proteins. ConA is a plant lectin that has high affinity for a series of high-mannose and hybrid-type N-glycans [63]. WGA recognizes N-acetylglucosamine and sialic acid residues while PNA is specific to T-antigen, which is commonly found in O-glycans [64]. AAA, on the other hand, shows broad specificity towards L-Fuc-containing glycans [47]. Figure 17.7 provides a detailed summary of different lectins that have been used for N-glycosylated proteins enrichment. The ability of different lectins to recognize specific glycosylation motifs was used to develop a multi-lectin affinity system that can achieve a

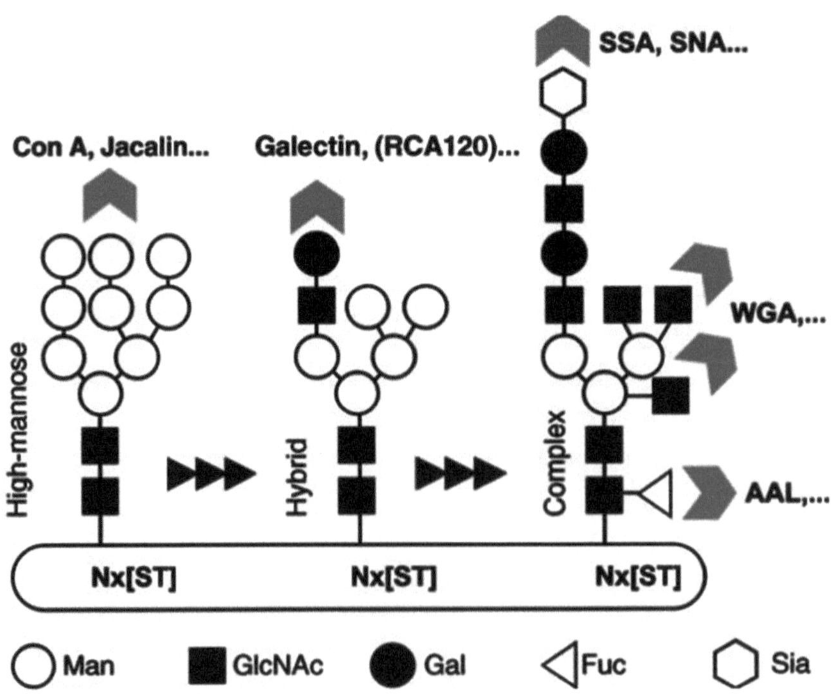

Fig. 17.7 N-linked glycans and their binding lectins [67]

comprehensive enrichment of glycoproteins from biological fluids [65]. Moreover, high-performance lectin affinity columns and microcolumns have been developed that can be used directly in line with LC-MS/MS systems [66].

Schematic illustration of various N-linked glycans attached to the polypeptide chain and several lectins with different binding specificity for the non-reducing end of the oligosaccharide. Lectins with affinity for specific oligosaccharides are denoted above or to the side of the chains (e.g., Con A, Jacalin...). Abbreviations: AAL, Aleuria aurantia agglutinin (lectin); RCA120, Ricinus communis agglutinin; SNA, Sambucus nigra agglutinin; SSA, Sambucus sieboldiana agglutinin; WGA, wheat germ agglutinin; Man, mannose; GlcNAc, N-acetylglucosamine; Gal, galactose; Fuc, fucose; Sia, sialic acid. Nx [ST] refers to a consensus tripeptide sequence for N-linked glycosylation.

Hydrazide Chemistry

Hydrazide chemistry, developed by Aebersold et al., is one of the most efficient techniques for N-linked glycopeptide enrichment [68]. As highlight in Fig. 17.8, glycoproteins are oxidized with sodium periodate to generate aldehydes in the carbohydrates, which then react with hydrazide groups immobilized on resin to form hydrazine bonds. After the removal of nonglycosylated peptides, the N-glycopeptides are selectively released from the resin by PNGase F cleavage for LC–MS analysis. In 2007 another group [69] modified this method to capture glycopeptides rather than glycoproteins to minimize sample

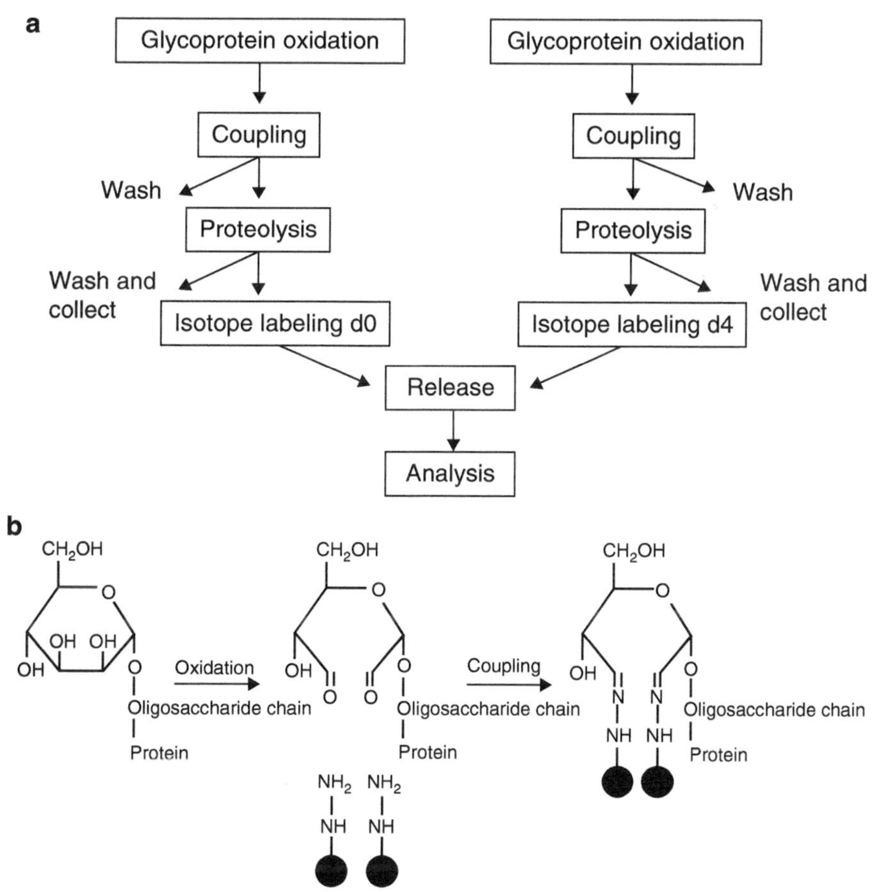

Fig. 17.8 Schematic diagram of quantitative analysis of N-linked glycopeptides [68]

loss and increase sensitivity. The development and application of this method has been well described in literature [70, 71].

Based on the hydrazide chemistry method described above, Wollscheid et al. [72] developed a cell surface capturing technology for labeling and enriching cell surface exposed N-glycoproteome before cell lysis.

(a) Strategy for quantitative analysis of glycopeptides. Proteins from two biological samples are oxidized and coupled to hydrazide resin. Nonglycosylated peptides are removed by proteolysis and extensive washes. The nonglycopeptides are isotope labeled by succinic anhydride carrying d0 or d4 tags. The beads are then combined and the isotopically tagged peptides are released by PNGase F and analyzed by LC-MS/MS. (b) Oxidation of a carbohydrate to an aldehyde followed by covalent coupling to hydrazide resin.

O-Glycosylation

The chemical/enzymatic photochemical cleavage (CEPC) method was used in O-GlcNAcylated peptide enrichment [73]. In this method (Fig. 17.8), O-GlcNAcylated peptides are first enzymatically labeled with azidogalactosamine (GalNAz). The free azido group in GalNAz is then conjugated to the alkyne group in a photocleavable biotin probe (PC-PEG-biotin-alkyne) through CuAAC. The biotinylated peptides are then enriched using avidin affinity chromatography, and subsequently released via photochemical cleavage. O-GlcNAc-modified peptides enriched by this method are tagged with a basic aminomethyltriazolacetylgalactosamine (AMTGalNAc) that facilitates ETD identification and site localization of O-GlcNAc–modified peptides [74].

A handful of complementary methods have been developed for the enrichment and identification of O-GlcNAcylation. Teo and coworker [73] obtained three antibodies capable of immunoprecipitating glycoproteins from HEK293T cell lysates. While each antibody captures a slightly different subset of targets, a total of 215 putatively O-GlcNAcglycosylated proteins were isolated and identified by shotgun mass spectrometry. Anonsen et al. [75] used the same strategy (combining antibody based enrichment with downstream MS analyses) to study the glycoproteome of *N. gonorrhoeae*.

17.3.2.3 Ubiquitination and Sumoylation

Tagging the Chain

Affinity tag based enrichment strategies are often used for ubiquitinome analysis. Typically, cells are transfected and ubiquitin is expressed with an epitope tag, such as a histidine tag or hemagglutinin (HA) tag at the N-terminus to facilitate subsequent affinity purification using nickel beads (for histidine tag) or an anti-epitope antibody. By using a yeast model system expressing 6xHis-tagged ubiquitin, Penget et al. [76], provided the first successful profiling of ubiquitinated proteins and ubiquitination sites using LC-MS/MS. Generally, a large percentage of proteins purified using a single-step ubiquitinome purification are not ubiquitinated (impurities include proteins with multiple histidine residues in a short sequence). In this case, tandem affinity tags for two-step purification were developed. Tagwerker et al. [77] described a fused a tandem histidine-biotin tag (HB-tag) strategy for two-step purification of the ubiquitinated proteome under fully denaturing conditions. The HB-tagged proteins were sequentially purified by Ni^{2+} chelate chromatography and streptavidin resins to greatly reduce the nonspecific proteins background.

Anti-k(GG)

Recently, a monoclonal antibody-based peptide-enrichment strategy has been developed for large-scale analysis of ubiquitination sites targeting dyglycine, anti-k(GG) moieties. This antibody can specifically target a diglycine adduct left at sites of ubiquitination after trypsin digestion with high efficiency. By using an anti-k (GG) antibody for enriching diglycine containing peptides, Guoqiang Xu and coworkers identify 374 diglycine-modified lysines on 236 ubiquitinated proteins in which 72 % of these proteins and 92 % of the ubiquitination sites were reported for the first time [78]. In another recent study by Kim et al. [79], more than 19,000 diGly-modified lysine residues from 5000 proteins were identified. This study proved the feasibility of global ubiquitinome profiling for the first time.

17.3.2.4 Methylation

KMBD MBT for Lysine Methylation

Recently a new strategy for enrichment of methylated proteins was introduced that relies on the affinity of naturally occurring 3xMBT domain repeats of L3MBTL1 for protein methylation. This affinity strategy was introduced as a universal method for detection and identification of proteins carrying a mono- or dimethylated lysine residue [33].

Antibody for Methylation

Protein methylation is a posttranslational modification that adds a single or multiple methyl group to the guanidino group of arginine or the primary amine of lysine residue side chains. Currently a high-throughput method for isolation and identification of lysine methylation does not exist due, in large part, to the lack of a specific antibody for methyl lysine. [26] Recently Michael's group [14] developed highly specific antibodies against methyl arginine and lysine motifs. These highly specific antibodies recognize monomethyl arginine; symmetric and asymmetric dimethyl arginine (sDMA and aDMA); and monomethyl, dimethyl, and trimethyl lysine motifs. These antibodies were used to enrich methyl peptides, over 1000 arginine methylation sites and 160 lysine methylation sites were identified, which is the most methylation sites identified in a single study to date. Other useful arginine methyl–specific antibodies have been developed, such as ASYM24 and ASYM25, which are specific for aDMA, and SYM10 and SYM11, which recognize sDMA [6].

17.3.2.5 Acetylation

Immunoprecipitation using monoclonal antibodies is the main enrichment strategy for acetylated lysine residues. Using this antibody enrichment strategy, a new study showed that more than 20 % of mitochondrial proteins are acetylated [80]. A global analysis of lysine acetylation using immunoprecipitation technique in a human cell line has recently identified 3600 sites on 1750 proteins [81].

17.3.2.6 Serial Enrichment of Different PTMs

More recently, a strategy for serial enrichments of different PTMs (SEPTM) from the same biological sample have been proposed by Mertins [82]. This approach enables the analysis of the phosphoproteome, ubiquitinome and acetylome from the same biological sample without decreasing the quality of each individual PTM. With their streamlined sequential use of IMAC (for phoshorylated peptides), K(GG)-specific antibodies (for ubiquitinated peptides) and K(Ac)-specific antibodies (for lysine-acetylated peptides) strategies, more than 20,000 phosphorylation sites, 15,000 ubiquitination sites, and 3000 acetylation sites were identified, of which 0.3 % of peptides contained different types of modifications. SEPTM approach, although in its infancy, might open a new avenue for systematic analysis various PTMs to study PTM crosstalk in cell signaling.

17.3.3 Mapping PTMs With Mass Spectrometry

Posttranslationally modified proteins are covalently modified with specific chemical groups. PTMs often occur at low stoichiometry and are often labile during mass spectrometry [83]. Due to these characteristics, global detection of PTMs requires mass spectrometers with high resolution, high scan speed, and high sensitivity [84].

The major goals of PTMs analysis are (i) identifying modified proteins, (ii) localizing modification sites on specific amino acids in protein sequence, (iii) measuring the stoichiometry of the modified sites, and (iv) accurately quantifying the dynamic changes of these covalent modifications. Achieving these goals require the right mass spectrometer with proper ionization, fragmentation, and detection technologies.

Since phosphorylation is one of the most important and well-studied PTMs in biological systems, we use it as an example to explain different ionization and fragmentation techniques that are commonly used to improve global phosphoproteome analysis.

17.3.3.1 Ionization Strategies

A successful detection of PTM by mass spectrometry is often challenging due to decreased peptide ionization efficiency as a result of PTM chemistry. For example, reduced ionization efficiency of phosphorylated peptides compared to their non-modified counterparts has been reported for both electrospray ionization (ESI) [85, 86] and matrix-assistant laser desorption ionization (MALDI) [85]. This is mainly due to the addition of a negatively charged phosphate group which reduces the ionization yield of phosphopeptides in positive-ion mode [86]. Negative-ion mode can yield more ions for phosphopeptides [87] but MS/MS scan for peptide sequencing still need to be done in the positive mode. Non-specific adsorption of phosphopeptides to stainless steel parts in the LC-MS system can also contribute to high detection limit [85, 88]. To compensate for reduced ionization efficiency and sensitivity, modified peptide pre-fractionation or enrichment steps are necessary.

17.3.3.2 Fragmentation Methods

Peptide PTM sites are usually detected according to the shift in the fragment ions m/z (Table 17.1), for example, a 80 Da mass shift reports the addition of HPO_3 group. Since a number of PTMs, such as phosphorylation and glycosylation, are labile during standard CID fragmentation, different types of fragmentation strategies have been developed for labile PTM analysis, including collision-based methods (CAD, HCD, MSA) and electron-based methods (ECD, ETD).

Collision-Induced Dissociation (CID)/ Collision-Activated Decomposition (CAD)

Collision-induced dissociation (CID) is also known as collision-activated decomposition (CAD). In standard CID/CAD fragmentation, protonated peptides collide with an inert neutral gas following an electric potential acceleration in the vacuum of the mass spectrometer. Non-modified peptides are generally fragmented at their backbone amide bonds which results in b- and y ladder ions that cover the peptide sequence from its N- and C-terminal respectively. However, due to the neutral loss of the phosphate group from phosphorylated peptides, the MS analysis of labile phosphorylation by CAD ionization is often challenging. As shown in Fig. 17.9a, the phosphate group is the preferred site for protonation and subsequent nucleophilic attack from a neighboring amide carbonyl group [89, 90]. This results in a dominant neutral loss peak, while sequence informative ions can rarely be observed as shown in Fig. 17.9b [91]. The extent of neutral loss depends on parameters, such as charge state, the chemical structure of the modified amino acid, the availability of mobile protons, peptide amino acid sequence, the amount of collision energy exerted in fragmentation and the type of mass spectrometer. Neutral loss is frequently observed in ion trap mass spectrometers that have lower collision energy and relatively longer activation time compared to QqQ or QTOF mass spectrometers. Moreover, the extent of neutral loss appears to also depend on the ratio of charge state versus number of basic amino acid residues [92]. When the ratio is higher, less neutral loss is observed. Since the mobile proton is available, the energy applied to charge-directed backbone fragmentation can be much lower. Phosphorylated tyrosine residues often lose HPO_3 (80 Da), while loss of H_3PO_4 (98 Da) is more observed in phosphorylated serine and threonine residues [93]. This is mainly due to the fact that the C-O bond of phosphorylated tyrosine residue is stronger than that of phosphorylated serine and threonine residues.

Apart from neutral loss issue, in ion trap instruments, gas-phase rearrangement of phosphate groups between different amino acid residues has been observed. This rearrangement complicates the correct and confident localization of phosphorylation sites. Palumbo et al. demonstrated that in gas phase and prior

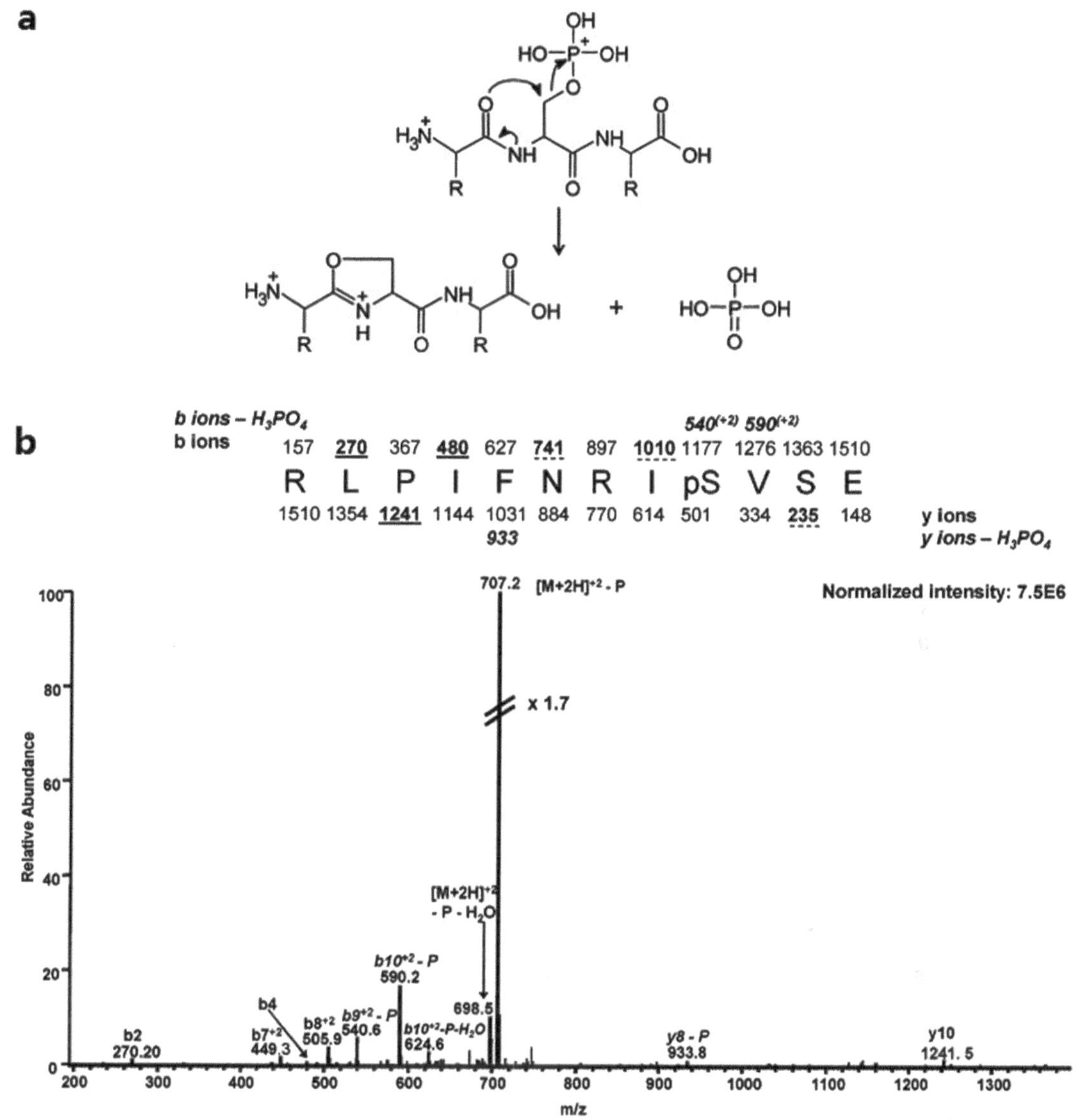

Fig. 17.9 Neutral loss in CAD fragmentation. (**a**) Fragmentation pattern with loss of phosphoric acid from a multiply protonated phosphopeptide by CAD (Adapted from [90]). (**b**) CAD spectrum of the $[M + 2H]^{2+}$ ion of RLPIFNRIpSVSE (*m/z* 756), dominated by neutral loss of phosphoric acid (Adapted from [94])

to fragmentation, phosphate groups can transfer to unmodified hydroxyl-containing amino acid residue [95].

MS3 Scan and Multistage Activation (MSA)

To obtain more sequence information for peptides carrying labile modifications with neutral loss in CID MS/MS, (MS/MS/MS) MS3 scan mode has been developed in ion trap mass spectrometers. MS3 scan is usually triggered in a data-dependent manner when a major neutral loss peak is observed. However, MS3 analysis may not yield an unambiguous phospho site localization due to the loss of H_3PO_4 prior to MS3 fragmentation. Additional complications in MS3 data interpretation may arise when the combined losses of HPO_3 from the phosphorylated residue and H_2O from another non-phosphorylated yields a neutral loss of 98 Da [95].

An alternative approach for MS3 scan is MSA scan, also termed pseudo MS3 scan, which uses a supplemental selective activation of the common neutral loss products and activation of the precursor ion simultaneously, and then records all the fragment ions [94]. Since the MSA spectrum contains both MS2 and data-dependent MS3 spectra, there is no need for MSA to isolate and fragment the neutral loss ions. A detailed comparison of the scan cycles for MS2, data-dependent neutral loss MS3 and MSA has been presented by Gygi et al. [91]. Compared with the MS2 and MS3 spectrum, MSA spectrum contains the relative intensity for both b/y-ions and b/y-98 ions without neutral loss peak. Successful MSA dissociation is usually performed on an ion trap mass spectrometer (LTQ), with relative low-mass resolution. However, it has been shown the standard MS2 scan outperforms MSA and MS3 when the experiments were performed on high-mass accuracy instrumentation (e.g. LTQ Orbitrap), in which accurate precursor mass determination is achieved in a high-throughput manner. The extra time needed to perform additional fragmentation in MSA or MS3 reduces the opportunities to sequence additional peptides [91].

High-Energy Collisional Dissociation (HCD)

High-energy collisional dissociation or higher-energy C-trap dissociation (HCD), also termed as "beam-type collisional activation", performs fragmentation in higher-energy collision dissociation cells with higher energy and shorter activation time as compared with ion trap CID. HCD tends to produce less neutral loss peaks and more sequence-specific fragment ions which predominantly contain y-ions and some b-ions. The proportion of b-ions is smaller than y-ions because they tend to be fragmented further to a-ions in HCD mode. With higher energy, HCD can also produce smaller species than ion trap CID, such as immonium ion to help identify specific modified residues [98, 106]. Since HCD fragmentation overcomes the "low mass cutoff" issue of ion trap fragmentation, it has been widely adapted for PTM analysis [98].

A detailed comparison of HCD and CID fragmentation of a synthetic phosphopeptide is shown in Fig. 17.10. In the low-mass region, HCD produced a clear a2/b2 ion pair, y1 and y2 ions, with relative higher abundance. Furthermore, the phosphotyrosine-specific immonium ion at *m/z* 216.0426 can be detected in HCD spectrum with high confidence. However, a full consensus has not been reached yet to determine whether a higher resolution, slower acquisition speed HCD-based strategy is better for modified peptide identification (e.g. phosphorylation) or a lower resolution, faster acquisition speed CID-based acquisition [106, 107].

Electron-Based Dissociations

Compared with CID, electron-based dissociation, such as electron capture dissociation (ECD) and electron transfer dissociation (ETD), yield sequence fragments while maintaining the modified group. In ECD, precursor ions are bombarded with near-thermal electrons (<0.2 eV). The basicity of the amide carbonyl oxygen can abstract a proton from an amino acid residue in the sequence. Then the N-Cα bond is dissociated with very low energy barriers, leading to c- and z- type ions. The process of peptide ion capturing an electron, which charge-reduces the peptide into a radical cation [108] (Fig. 17.11). In ETD process, fluoranthene radical-anions are used as reagents transferring an electron to a peptide with multiple charges. This reaction reduces the peptide charge by one, then triggers the peptide backbone fragmentation to produce a series of complementary c and z type fragment ions [109] (Fig. 17.11).

Figure 17.12 is a comparison between ETD and CAD analysis of a phosphopeptide [89]. The CAD spectrum (Fig. 17.12a) lacks sufficient fragmentation and cannot be matched to a correct sequence by database searching. In contrast, ETD fragmentation results in a more successful identification with near-complete backbone fragmentation (Fig. 17.12b). Proline, which frequently occurs at PTMs motif, does not cleave at its backbone amide bond due to its side chain

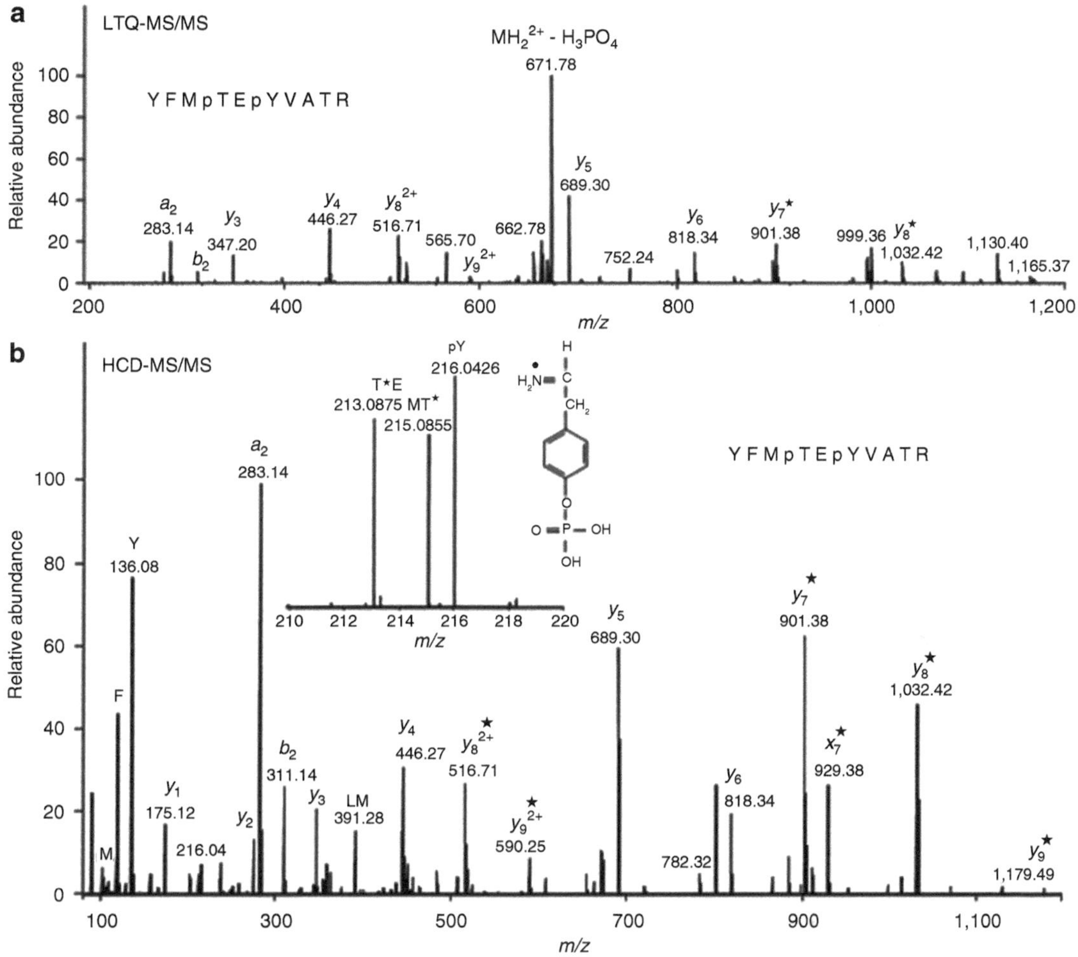

Fig. 17.10 Detection of a synthetic phosphopeptide containing the sequence YFMpTEpYVATR ("pY" indicates phosphotyrosine). (**a**) CID fragmentation by linear ion trap in Orbitrap mass spectrometer. (**b**) HCD fragmentation of the same phosphopeptide. Inset: close-up of the region with the phosphotyrosine-specific immonium ion at *m/z* 216.0426 (Adapted from [98])

ring structure and so the N–C bond N-terminal to the proline does not fragment by ETD [110]. And, as can be seen in Fig. 17.12b, the c and z-type ions at the N-terminal of proline are not detectable. But this limitation does not exist in CAD fragmentation (Fig. 17.12a). Furthermore, CAD was found to be more effective in fragmenting peptides containing lower net charge compared to ETD. Usually phosphoserine and phosphothreonine motifs with one or more basic residues fragment better by ETD. For large scale expansive identification of protein phosphorylation though, both CID and ETD should be applied complementary to increase the number of identifications and the coverage of peptide sequence [89, 111].

Several studies have confirmed that efficiency of ETD fragmentation is dependent on charge states >2 [89, 111] and the charge density (i.e. ratio of charge over number of amino acid residues) [93, 112]. Therefore, the peptides generated by Lys-C or trypsin with higher charge density will be excellent targets for ETD analysis. ETD can break the backbone randomly in longer peptides, such as peptides generated by the proteinase Lys C [113]. ETD can be

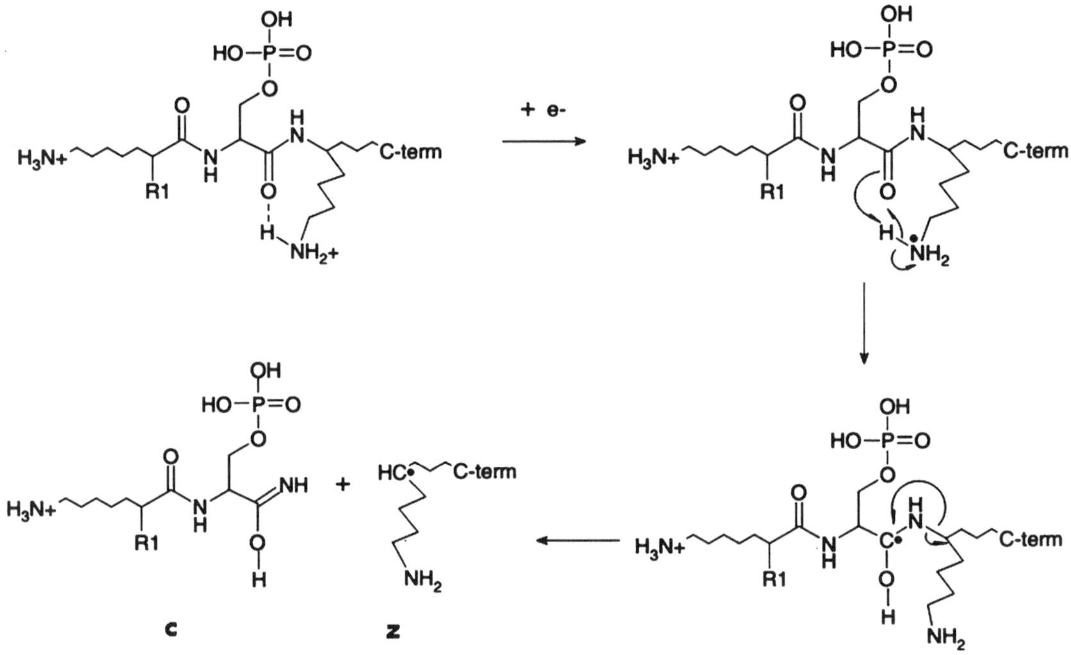

Fig. 17.11 Proposed ECD/ETD fragmentation mechanism of phosphorylated peptides (Adapted from [93])

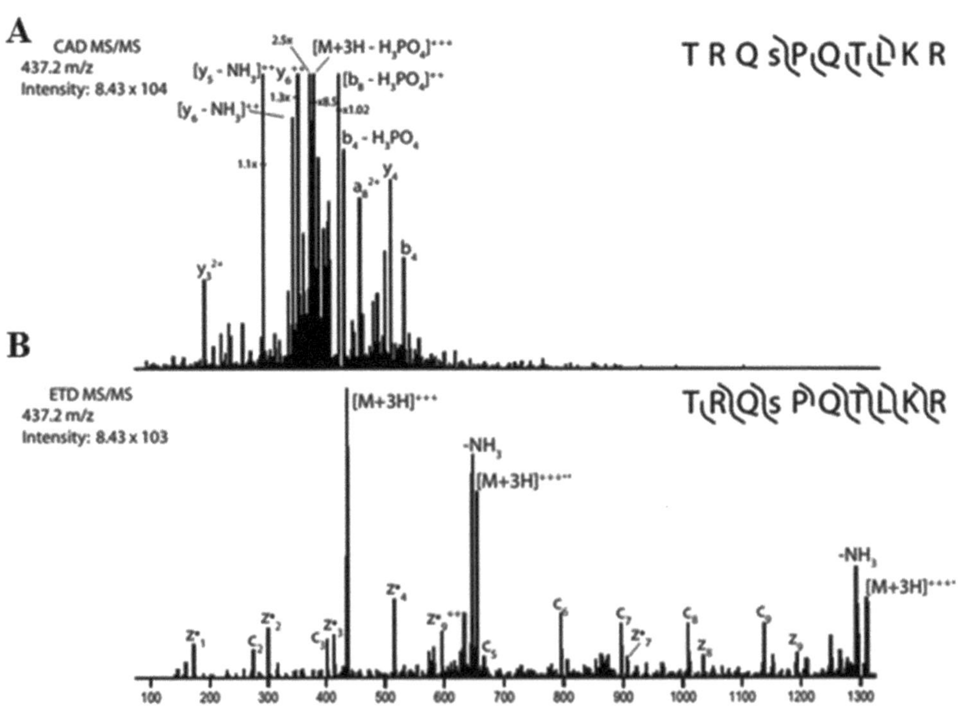

Fig. 17.12 Analysis of phosphopeptide TRQsPQTLKR ("s" indicates phosphoserine). (**a**) CAD mass spectra of the sequence. (**b**) ETD mass spectra of the same sequence (Adapted from [89])

implemented on various instruments, such as Q-TOF [114], and linear ion trap-Orbitrap hybrid instruments [115].

Recently, novel hybrid fragmentation techniques are developed based on the charge state and *m/z* value of the precursor ion. As reported by Heck et al. [121], a novel hybrid fragmentation technique, termed as EThcD (combining electron-transfer and higher-energy collision dissociation), is used for unambiguous phosphorylation site localization. ETD could not cleave the N–Cα bond N-terminal to proline, which hinders the phosphosite localization of the proline-rich peptides. However, EThcD can specifically address this issue by generating b/y- and c/z-type ions.

A data-dependent decision tree (DT) method was developed by Coon and co-workers. They designed and embedded a data-dependent decision tree algorithm (DT) in QLT-Orbitrap capable of both the CAD and ETD dissociation. Following the MS1 analysis in the Orbitrap, six data dependent MS/MS activation was performed either in ETD-only, CAD-only or DT-based selection mode with product ion analysis performed in the QLT. In the DT-based selection for every MS/MS event, CID or ETD is utilized to fragment precursors depending on its charge state and *m/z* in real time automatically. They compared the CAD-only or ETD-only analyses with DT-based selection in large-scale proteome analyses. Their data showed the DT approach identified more phosphopeptides (7422), compared with either CAD (2801) or ETD (5874) phosphopeptides alone [122]. Currently, Jyoti S. et al. have compared a data-dependent neutral loss-triggered-ETD (DDNL) strategy to DT. In a DDNL method performed on an LTQ Orbitrap Velos hybrid mass spectrometer, all peptides were fragmented by CID and if a prominent neutral loss peaks corresponding to the loss of a phosphoric acid were observed the precursor ion was isolated for fragmentation with ETD [123].

Detection of Other Types of PTMs

Apart from phosphorylation, considering different degrees of lability in other types of PTMs, variable polarities, charge states of precursor ions etc., it is important to choose and optimize fragmentation strategies and scan modes suitable for each PTM analysis. A number of strategies have been proposed to improve sequence and site diagnostic fragmentation, including the use of neutral loss-triggered MS3 and MSA in ion traps, HCD, ETD/ECD, or a combination of these approaches [121, 122]. For example, monitoring specific diagnostic ion of aDMA in a precursor ion scan can differentiate it from its isomer sDMA. That is because dimethylarginine is sufficiently stable under CID conditions, resulting in cleavage of the peptide backbone to support sequence information [104].

Compared to phosphorylation, glycosylation is structurally a much more complex PTM with highly heterogeneous glycan structures. Most of the major fragmentation methods described above have been used for characterizing intact glycopeptides. HCD with higher fragmentation energies and higher mass accuracy is also a favorite approach for glycosylation characterization. HCD, which generates diagnostic oxonium ions and Y1 ions (e.g. peptide + acetylglucosamine), has been proven highly effective for locating glycosylation sites [124]. ETD and ECD yield sequence fragments while maintaining the carbohydrate structure, thereby enabling site localization [110, 125].

Recently, Cooper et al. developed an HCD product ion-triggered ETD approach which effectively improves the accuracy and sensitivity in identifying both glycosylation site and peptide sequence simultaneously [126]. As shown in Fig. 17.13, after a full MS scan of a N-linked glycopeptide in Orbitrap, HCD MS/MS scan was triggered for a precursor ion at *m/z* 645.6194. If the diagnostic ions of HexNAc (N-acetylhexoseamine) oxonium ions (*m/z* 204.09) and HexHexNAc oxonium ions (*m/z* 366.14) were among the top 20 most abundant peaks, ETD MS/MS was then triggered in the linear ion trap to fragment the precursor ion (*m/z* 645.62) (Fig. 17.13c). The advantage of this approach is that the structure of the glycan and the sequence of the glycopeptide were determined simultaneously from HCD and ETD spectra

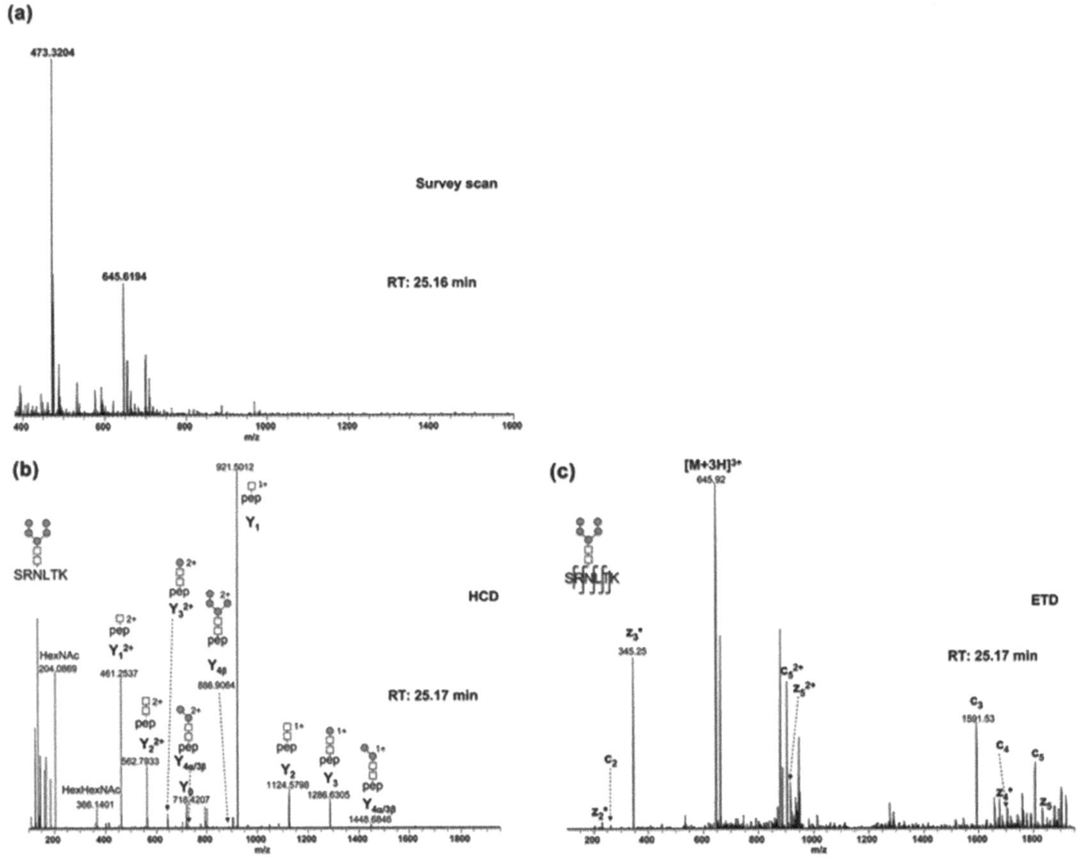

Fig. 17.13 HCD product ion-triggered ETD MS/MS of Lys-C digest of ribonuclease B. (**a**) A full MS scan (*m*/*z* 380–1600) recorded in the Orbitrap at retention time of 25.16 min. (**b**) HCD MS/MS spectrum of precursor ions at *m*/*z* 645.6194. (**c**) In linear ion trap supplemental activation ETD MS/MS of precursor ions with *m*/*z* 645.62 (Adapted from [126])

respectively. This approach is also routinely used in site mapping of O-linked glycosylated peptides. However, O-GalNAc and O-GlcNAc cannot be distinguished from each other by their signature ions only [127]. In a recent study reported by Hart et al. [128], O-GlcNAc-modified peptides were specifically labeled with AMT-GalNAc for both enrichment and better fragmentation in ETD scan. As a result, the enriched peptides appeared in charge states of +3 or higher, which increased the fragment efficiency in ETD, compared to untagged, native O-GlcNAc peptides. The AMT-GalNAc-GlcNAc modification brings in three major diagnostic oxonium fragment ions, which can be readily detected by HCD.

In summary, the advancement of both hardware and software in hybrid mass spectrometry, a combination of variable fragmentation modes (e.g. HCD plus ETD) and different scan modes is a powerful approach for mapping various PTMs. In conclusion, the features of different fragmentation methods are summarized in Table 17.2.

Recently ion mobility spectrometry (IMS) was applied in PTM analysis. This technology can potentially separate isomers and/or variants, such as phosphorylated variants [129], glycol-isoforms [130], variants of histone methylated and acetylated peptides [131]. Additionally, pulsed Q dissociation (PQD) combined with ETD can be applied to analyze the O-GlcNAc peptide [132].

Table 17.2 Features of different fragmentation methods for PTM detection

Fragment methods	Advantages	Limitations	Type of fragment ions
CAD/CID	1. High speed 2. High sensitivity [106]	1. Lower net charged peptides required 2. Labile PTMs are lost [89] 3. Selective cleavage [116]	Neutral losses, less b-, y- product ions compared with MSA [94]
HCD	1. No low mass cutoff and multiple cleavage events leading to richer fragments [106, 107] 2. Higher resolution, higher dynamic range and less noisy compared with CID spectra [107] 3. The neutral loss of phosphoric acid is unproductive and more sequence-specific fragment ions [106]	1. Slower scan cycle [107]	b- and y- ions, b/y- additional neutral losses of NH_3, H_2O, HPO_3, H_3PO_4 [117]; a2/b2 fragment ion pair; internal fragments, immonium ion [98, 117]
ETD/ ECD	1. Labile PTMs preserved [89] 2. Break randomly for longer peptides, such as peptides generated by the proteinase Lys C [113, 118]	1. Multiply charged (charge state > 2) peptides required [89, 119] 2. The activation time is longer [111] 3. Limited fragmentation efficiency of doubly charged species [111]	Mainly c- and z- fragment ions [89, 120]
MS3/ MSA	1. "Neutral loss" issue largely addressed 2. Especially useful in low-mass resolution instrumentation [91]	1. Need extra analysis time 2. Low-abundance, sequence informative product ions are lost after isolation of the major neutral loss product in MS3 [94]. 3. Most effective for singly and doubly charged peptides [94]	b- and y-ions, along with several new cleavages (e.g. b/y-H_3PO_4), devoid of the major neutral loss fragment ion [94]

Note: CID/CAD indicates Collision-induced dissociation/collision-activated decomposition; HCD indicates High-energy collisional dissociation or higher-energy C-trap dissociation; ETD indicates Electron transfer dissociation; ECD indicates Electron capture dissociation; MS3 indicates data-dependent MS3 method; MSA indicates Multistage activation, also name Pseudo-MS3 method

17.3.4 Bioinformatics Methods for Predicting and Identifying PTMs

MS-based methods provide tools for efficient localization of PTM sites in a global scale. Large amounts of data generated by modern mass spectrometers can lead to false identifications and should be interpreted carefully with detailed statistical analysis. In this section, several MS data interpretation software packages for confident localization of various PTM sites are discussed. Following that, a series of databases containing large datasets from global PTM analyses are introduced. These databases are useful resources for MS-based proteomic studies. Furthermore, we will sum up related bioinformatic tools for sophisticated analysis of biological pathways associated with PTMs. In this section we will cover the latest bioinformatics approaches that can be used to mine and analyze large datasets generated by MS-based workflows.

17.3.4.1 Localization of PTM Sites

Protein PTMs are important for understanding cell signaling and other important biological mechanisms. The raw MS data should be carefully processed to localize PTM sites with minimal errors. Traditional manual validation is hardly practical in large-scale PTM localization. In the past few years, a number of probability-based scoring systems have been developed for accurately localizing PTM sites on peptide sequences. In this part, several commonly used automated PTM localization algorithms, such as Ascore and PTMscore, will be discussed.

Ascore is an algorithm for localizing protein phosphorylation sites. When a peptide with multiple possible phosphosites (S/T/Y residues) is identified, Ascore measures the probability of each possible site being the phosphosite using "site-determining ions" extracted from MS/MS spectra. Site-determining ions are the critical b/y ions that can distinguish the accurate phosphosite. However, it is always possible for an irrelevant peak to be randomly annotated as a site-determining ion in the MS/MS spectra. Ascore utilizes a cumulative binomial model to calculate the probability of a peak randomly matched to one of the site-determining ions in the MS/MS spectra. A higher score implies a smaller probability for a random match and a higher confidence in phosphosite determination. Phosphopeptides with Ascore ≥ 19 indicates that there is a 99 % or more chance for a correct phosphorylation site localization. Ascore between 15 and 19 can ensure >90 % certainty for the localization and those with score of 3–15 has a success rate around 80 %. A Ascore < 3 means that peptide MSMS spectra contain little or no site-determining ions for proper phosphosite localization. Figure 17.14 illustrates a processing example of Ascore [133].

PTMscore is another localization tool that is widely used. *PTMscore* has a similar algorithm

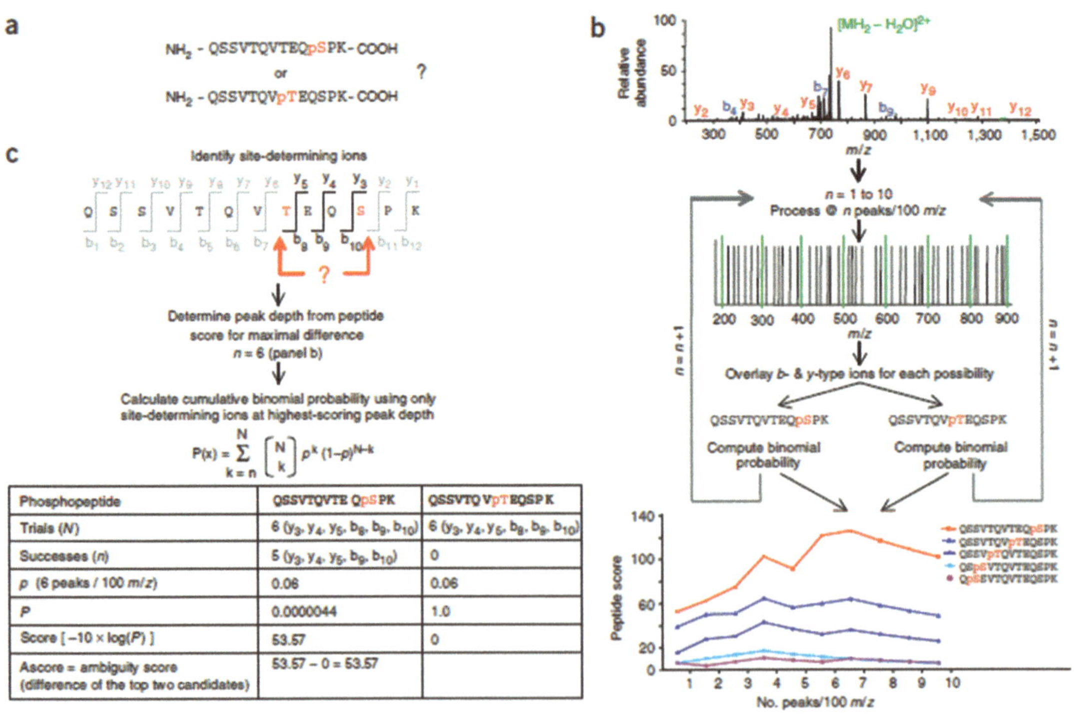

$$P(x) = \sum_{k=n}^{N} \binom{N}{k} p^k (1-p)^{N-k}$$

Phosphopeptide	QSSVTQVTE QpS PK	QSSVTQ VpT EQSP K
Trials (N)	6 ($y_3, y_4, y_5, b_8, b_9, b_{10}$)	6 ($y_3, y_4, y_5, b_8, b_9, b_{10}$)
Successes (n)	5 ($y_3, y_4, y_5, b_9, b_{10}$)	0
p (6 peaks / 100 m/z)	0.06	0.06
P	0.0000044	1.0
Score [$-10 \times \log(P)$]	53.57	0
Ascore = ambiguity score (difference of the top two candidates)	53.57 − 0 = 53.57	

Fig. 17.14 Localizing a PTM site with Ascore [133]. Different possible PTM sites within a single peptide can be differentiated with the site-determining ions (Fig. 16.14c). Ascore measures the probability of all detected site-determining ions to be random matches and the two candidate sites with the highest score (lowest chance of their site-determining ions being random matches) are picked to calculate the Ascore (the score difference between them). (Cited from *Nature Biotechnology, 24*(10), 1285–1292 with permission [133])

as Ascore which is also based on the random binomial distributions [33]. Putative site-determining b and y ions are generated to match with the actual MS spectra. The four most intense fragment ions in every 100 Da m/z intervals of MS^2 or MS^3 spectra are picked out. All possible combinations of the phosphorylation sites are tested (the putative ions and the actual spectra). *PTMscore* algorithm then generates a score (PTM score) for each combination. According to the PTM score and the motifs, all the testing peptides can be classified into four categories. Class I collects the phosphorylation sites with highest localization probability (>0.75). In class II & III, the sites have a localization probability which varies from 25 to 75 %. The sites in class II have to match at least one of 22 kinase motives whereas in class III this criterion is removed. If one site has a probability less than 25 %, it will be sorted into class IV with the lowest confidence level.

Mascot Delta Score is another phosphorylation site localization scoring tool that is similar to Ascore in terms of sensitivity and specificity. However, it is worth mentioning that Mascot outperforms the Ascore in tyrosine (Y) phosphosite localization. The Mascot Delta Score results from calculating the differences between the Mascot scores of the two top ions that are used to identify the peptide sequence but can hardly localize the PTM site. Mascot Delta Score can deal with ions from various fragmentation techniques (need to optimize respectively) [134].

PhosphoRS, is a newly developed PTM localization tool that is compatible with all common fragmentation techniques, such as ECD, ETD, HCD, and CID. Compared to Mascot Delta Score and Ascore, *PhosphoRS* can identify more phosphorylation sites at the same confidence level (>99 %) [135]. The fundamental algorithm of PhosphoRS applies a binomial probability. However, with different fragmentation techniques, the algorithm should be optimized respectively. The result comparison between PhosphoRS and other scoring systems is shown in Fig. 17.15a. This result show that various fragmentation modes lead to similar accuracy with PhosphoRS analysis (Fig. 17.15b).

sLoMo (Site Localization of Modification) is a localization tool developed from the Ascore algorithm capable of analyzing both CID and ECD/ETD generated ions. Furthermore, sLoMo can be used to perform site localization on a variety of modifications, such as oxidation and phosphorylation. The scoring algorithm uses a Poisson random distribution which is similar to the accumulated binomial distribution. sLoMo is

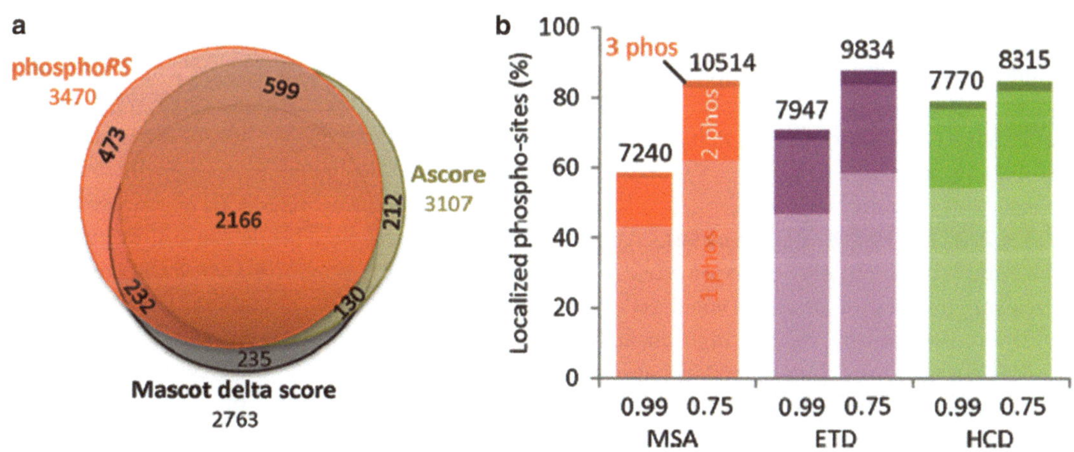

Fig. 17.15 (**a**) The diagram reveals the numbers of non-redundant phosphorylation sites in the same sample using various localization tools. (**b**) The comparison among MSA, ETD, and HCD generated data. The percentage and the absolute numbers of phospho-sites are visualized. (Cited from *Journal of Proteome Research, 10(12), 5354–5362* with permission [135])

also compatible with different data formats, such as Sequest and OMSSA [136].

Protein prospector is a search engine that reports all modifications present in an identified peptide [137]. The core localization tool in Protein prospector is called *SLIP (Site Localization in Peptide)*. SLIP scores are generated by comparing the tightness of the match between hypothetical MSMS spectra generated from *in silico* fragmentation of peptides modified at all possible sites and the acquired MSMS spectra.

Oscore is a tool that exclusively differentiates the O-GlcNAc peptides from the unmodified peptides. It utilizes the eight O-GlcNAcylation spectral features to calculate the sum of the normalized intensities divided by a rank value [132]. The score is tested by inputting more than 700 GlcNAc spectra from the O-GlcNAc peptide database and about 11,300 non-O-GlcNAc spectra. An Oscore lower than 2.0 indicates the existence of an O-GlcNAc peptide.

PTMap is a sequence alignment software designed for accurate identification of full-spectrum from posttranslationally modified proteins [138]. This software integrates two logical score systems, $S_{Unmatched}$ and PTMap score. A high $S_{Unmatched}$ score indicates that there are a large number of unmatched peaks with significant intensities in the MSMS spectra, whereas PTMap score estimates how well the sequence and the MSMS spectra explain each other. A PTMap score with a value over 1.0 indicates a confident match between the identified PTM sites and the hypothetical site. It worth mentioning that PTMap is the only software that can identify novel PTMs with high accuracy.

17.3.4.2 PTMs Related Databases

Proteomics is a rapidly evolving field with increasing number of datasets generated daily for different PTMs by various high-throughput LC-MS/MS platforms. Currently, MS-based proteomics approaches can map about 50,000 phosphorylation sites directly in a single cell line [81]. Table 17.3 summarizes various large-scale MS based PTM studies in the past few years. With increasing number of datasets, well curated databases are becoming an indispensable resource for PTM localization and biological analysis.

Table 17.3 Statistics of large-scale PTM mapping

PTM type	Sites	Proteins	Reference
Phosphorylation	50,000	7832	[139]
Ubiquitination	20,000	5000	[140]
Acetylation	3600	1750	[81]
Methylation	1160	N/A	[141]
N-Glycosylation	6367	2352	[142]
O-Glycosylation	177	602	[143]
Sumoylation	N/A	593	[144]

There are a variety of databases with specific features and emphasis that contain large amounts of MSMS data covering various PTMs. Comprehensive databases, such as PhosphoSitePlus or SysPTM2.0 aim at providing coverage for multiple PTMs. PhosphoSitePlus (http://www.phosphosite.org) is an updated version of PhosphoSite, which covers other common PTMs such as acetylation, methylation, ubiquitination and O-linked glycosylation in addition to phosphorylation [145]. Currently, 245,509 phosphorylation sites are stored in PhosphoSitePlus, which is higher than any other database. This database includes crucial information regarding various modified proteins' biological functions and structures. PhosphoSitePlus is one of the most dynamic and continuously updated databases, covering protein PTM information. In addition to PhosphoSitePlus, some newly developed databases also provide comprehensive information about various PTMs. dbPTM3.0 (http://dbptm.mbc.nctu.edu.tw) houses integrated data from 11 public resources along with manually curated data from MS/MS PTM extracted from research literature. This database stores information for more than 200,000 PTM sites with related information including PTM regulated protein-protein interactions and the topologies of the PTM carrying transmembrane proteins [146]. Compendium of Protein Lysine Modifications (CPLM) is a specific database for protein lysine modifications (PLMs) which occur at ε-amino groups of lysine residues. CPLM stores information for 200,000 sites from 12 different types of PLMs and the co-occurrences of various PLMs on the same modification site (http://cplm.biocuckoo.org) [147].

Other than general databases, specific databases have also been developed to store information about specific types of PTMs. Due to the significant role that protein phosphorylation plays in crucial cellular processes, such as cellular growth, intercellular signaling etc., several databases are dedicated to this modification entirely. PhosphoSitePlus is one of the largest resources of PTMs which have been mentioned above. Another commonly-used phosphorylation database is a eukaryotic protein database called Phospho.ELM (http://phospho.elm.eu.org/) that provides information such as phosphopeptide sequence, absolute position, Uniprot accession number and the upstream motif information [148]. In the latest update in 2012, Phospho. ELM had collected 42,914 non-redundant phospho-sites. PHOsphorylation SIte DAtabase (PHOSIDA) is another database build using MS data from screening datasets of phosphosites (http://www.phosida.com) [149]. In addition to the original 6600 phosphosites that are observed from *HeLa* cells, the database also has gathered information from other species. PHOSIDA has stored 70,095 phosphorylation sites and around 10,000 acetylation and N-Glycosylation sites.

Like phosphorylation, different types of glycosylation modification play significant roles in many biological processes. O-GLYCBASE is one of the earliest glycoprotein databases, which has in its collection 243 experimentally verified O-glycosylated proteins including 2413 different sites [150]. Unipep (http://www.unipep.org) is a N-linked glycosylation database that covers 9651 N-Glycosylated peptides including parent protein sequences and modification sites (predicted and identified) along with the relative motifs (if available) [151]. GlycoProtDB (http://jcggdb.jp/rcmg/gpdb) is another database that was constructed using data collected from a series of experiments in which nine mouse tissues and samples from other species such as *Homo sapiens* were systematically analyzed for detection of glycopeptides and their sites [152]. Another publicly available knowledgebase, UniCarbKB, is built based on data stored in a previous database called GlycoSuiteDB. It also integrates data from other protein glycosylation resources, such as EUROCarbDB (structural), UniCarb-DB (experimental LC-MS/MS data), etc. [153].

Several ubiquitination/ubiquitin-like conjugation databases have also been introduced in recent years including UUCD, SCUD and Ubiprot [154–156]. Ubiquitin and Ubiquitin-like Conjugation Database 2.0 (UUCD: http://uucd.biocuckoo.org/) stores 117,703 proteins from 144 eukaryotic species [154]. 1831 different ubiquitin-related enzymes and protein domains are collected from manually curated data and classified into various families respectively. *Saccharomyces Cerevisiae* Ubiquitination Database (SCUD: http://scud.kaist.ac.kr/) is another ubiquitination database that specifically records 940 ubiquitinated proteins and 73 related enzymes in Baker's yeast [155]. Another resource named UbiProt (http://ubiprot.org.ru/) is a database that summarizes various ubiquitination protein substrates [156]. Each protein contains information about ubiquitination sites, conjugation cascade (polyubiquitin topography), literature reference and links to related databases.

17.3.4.3 Pathway Analysis

PTM site localization tools and databases provide detailed PTM-related information and distinct understanding of specific PTM distributions in specific proteins or cells. However, in proteomics research, scientists often aim at solving specific biological problems that often require mapping a specific signaling pathway in the target organisms. As a result, signaling pathway analysis is a crucial step in hypothesis generation or testing that combines PTM sites information with system's interaction dynamics that may regulate a series of biological activities. Among various protein PTMs, phosphorylation signaling based on kinase and phosphotase activities is critical to nearly all cellular regulatory processes in archea, prokaryotic and eukaryotic organisms [157]. The pathway repository and analysis tools are essential in proteomics research and hence, some of the representative tools will be introduced in the following section.

Kyoto Encyclopedia of Genes and Genomes (KEGG) is a database storing genomic and relative functional information provided by bioinformatic analysis of genomics, proteomics and metabolomics data [158]. KEGG PATHWAY is one of KEGG's sub-databases that stores graphical representations of various cellular signaling pathways (http://www.kegg.jp/kegg/pathway.html). The database collects not only the metabolic pathways that are largely conserved in various species, but also the more complex regulatory pathways, such as signal transduction, cell cycle, etc. Furthermore, KEGG PATHWAY can automatically generate pathway diagrams differing from existing reference pathways using manually provided data. As of October 5th, 2014, KEGG pathway stored 465 manually drawn reference pathway maps and 318,245 computationally generated pathway maps in total.

KinomeXPlorer is another useful platform for modeling interactions between various kinase-substrates present in human and other major eukaryotic model organisms [159]. The platform includes an improved NetWorkIN, which is an algorithm that systematically predicts the motif-based network of kinases and their substrates, and an algorithm called NetPhorest for classifying phospho-sites according to the kinases and phospho-binding domains. The NetWorkIN algorithm first identifies the kinase motif in a phosphoprotein sequence. Then, a tool named STRING is used to construct a network of specific interactions for each substrate [160]. NetPhorest also includes a comprehensive online atlas of linear motifs from specific kinases and the phospho-binding domains. It also includes a series of probability-based classifiers for sorting out the phosphorylation sites in terms of their linear motifs [161]. The KinomeXPlorer is also able to calculate the likelihood of various kinase-substrate yielding desired information from NetWorkIN and NetPhorest algorithm analysis and gives a most possible kinase for a specific phosphorylation site with a largest calculated score.

To sum up, pathway analysis provides crucial information to guide downstream research by mapping large-scale PTM identifications especially phosphorylation sites for better understanding the biological functions of PTMs.

17.4 PTM Crosstalk

In the past decade various enrichment strategies have been developed for global analysis of various protein posttranslational modifications. Immobilized metal affinity chromatography (IMAC) and TiO_2 affinity enrichment methods are ubiquitously used for phosphopeptide enrichment. Antibodies have been raised to specifically recognize acetyllysine containing peptides to study protein acetylation. Ubiquitinated peptides are enriched with antibodies against a diGly moieties reminiscent of a ubiquitin chain. To reduce the complexity of PTM samples and increase the coverage, peptide fractionation methods such as strong cation exchange, HILIC or isoelectric focusing are usually used. Orthogonal combinations of different enrichment and fractionation approaches have been examined to study crosstalk between various PTMs.

Using *S. cerevisiae* as a model organism in a study focused on crosstalk between phosphorylation and ubiquitination, researchers identified about 2100 phosphorylation sites co-localizing with 2189 ubiquitination sites in about 466 proteins [162], using two different serial enrichment methodologies (Fig. 17.16). The first PTM purification step used cobalt-NTA (nitrilotriacetic acid) affinity media to purify His-tagged ubiquitinated proteins, followed by trypsin digestion of half the flow-through and enrichment of di-Gly peptides with a monoclonal antibody against lysine-diGly. The rest of the proteins after Ub-enrichment were then digested with another specific enzyme lysC, and exposed to subsequent phosphorylated peptides enrichment, with IMAC or TiO_2. In the second enrichment strategy SCX chromatography is used to separate tryptic peptides by their solution charge after trypsin digestion, followed by diGly peptides enrichment. Bioinformatic investigation

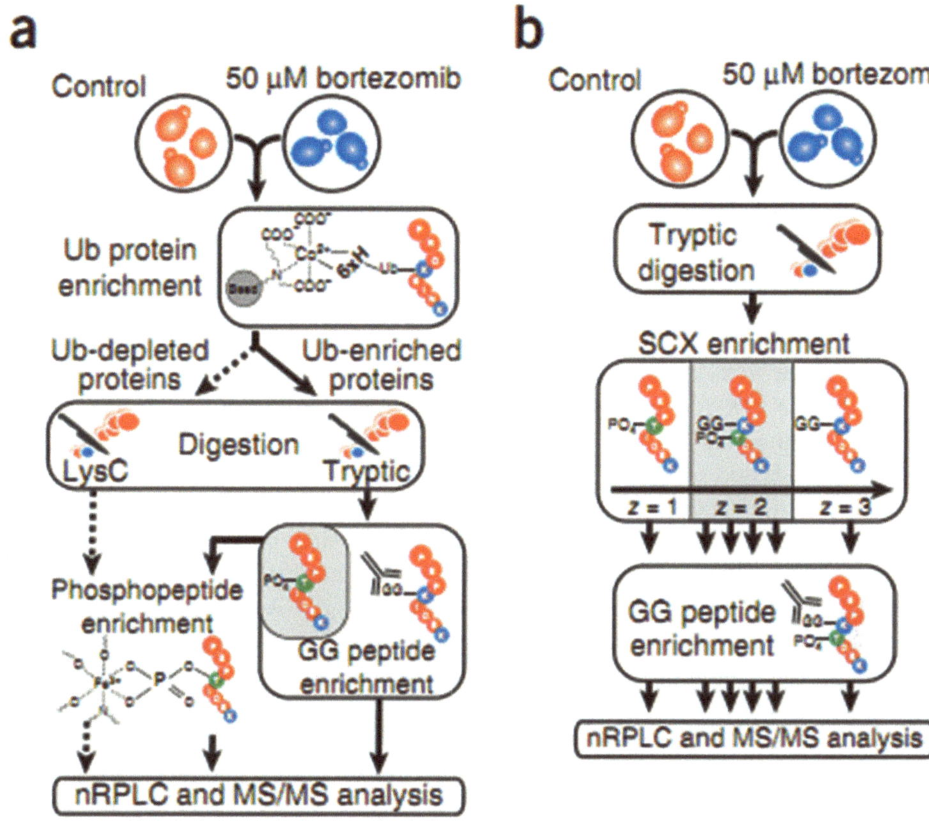

Fig. 17.16 Two enrichment strategies in the context of the proteasome inhibition experiment [162]

of the data suggests that phosphorylation sites co-localized with ubiquitination sites were more conserved than the rest, demonstrating the functional importance of PTM crosstalk.

The prevalence of co-occurring modifications and the role they might play in regulating protein function is not fully understood. The same study also showed that certain proteasome substrates require specific phosphorylation for degradation, denoted as phosphodegrons. SILAC experiment with Btz-mediated proteasome inhibition caused on average more than twofold increase in 12.9 % of ubiquitination sites and 3.4 % of phosphorylation sites on ubiquitinated proteins, suggesting that already ubiquitinated proteins may get further ubiquitinated to increase the stoichiometry for faster degradation or ubiquitination regulates the phosphorylation state of proteins [162]. In some kinases, enrichment of ubiquitination sites near the domain activation loop and in the glycine-rich region is a mechanism for kinase regulation.

In another study focused on studying crosstalk between phosphorylation, ubiquitination and acetylation, a fine-tuned method for serial enrichment of these PTMs from the same sample (SEPTM), has been described. Serial enrichment from high pH reverse phase chromatography fractions [163] greatly increases the quality and quantity of peptide coverage. A small percentage (5 %) of the fractionated peptides were analyzed by LC-MS/MS and the remaining (95 %) were subjected to subsequent finely designed serial PTM enrichments. The original 24 fractions were internally mixed into 12 fractions for phosphorylation enrichment (IMAC) and then into 6 fractions for ubiquitination enrichment (anti-K (GG) antibody) and the rest for acetylation peptides enrichment (anti-K (Ac) antibody). This serial enrichment combined with LC-MS/

MS allowed detection of more than 20,000 phosphorylation, 15,000 ubiquitination and 3000 acetylation sites in about 8000 proteins, uncovering the mysteries of PTM crosstalk.

Possible crosstalk between O-GlcNAcylation and phosphorylation-mediated signaling have been explored numerous times in the past with limited success. Protein phosphorylation is catalyzed by hundreds of distinct kinases but glycosylation is catalyzed mainly by two enzymes: polypeptide beta-N-acetylglucosaminyl transferase (OGT) and beta-D-N-acetylglucosaminidase (OGA), both of which gain specificity via transient associations with many other proteins. Based on this knowledge, a study in 2008 investigated the crosstalk between phosphorylation and glycosylation by detecting the changes in site-specific phosphorylation when GlcNAcylation is globally increased by inhibition of OGA [164]. As a result of GlcNAcylation up-regulation, more than 280 phosphorylation sites were found down-regulated and 148 sites found up-regulated, suggesting an elaborate interplay between these two posttranslational modifications. This study also yielded the hypothesis that there might be competition between these two PTMs for the occupancy of the same or proximal sites, by which regulating each other's the activity.

Though the above mentioned study helped to understand the principles behind crosstalk between O-GlcNAcylation and phosphorylation, study of crosstalk between glycosylation and other PTMs has been of limited success, as glycosylation site mapping is still limited by the state of technology. These limitations are in large part due to the following factors: (1) low stoichiometry of O-GlcNAcylation at each site on proteins; (2) low ionization efficiency of O-GlcNAcylated peptides; (3) the lability of β-linkage between O-GlcNAc moiety and Ser/Thr. These problems have been investigated for a long time which have resulted in sample enrichment method optimization and new generation of mass spectrometry fragmentation methods, such as electron capture dissociation (ECD) and electron transfer dissociation (ETD). A study reported in 2010 by Zihao Wang's group described an efficient enrichment protocol specifically for O-GlcNAc-modified proteins and peptides using a small amount of sample for comprehensive mapping of O-GlcNAc-modified amino acids [128]. This method made use of a new biotin reagent named PC-PEG-biotin-alkyne for O-GlcNAc-modified peptides enrichment, [165]. This reagent contains a photo-cleavable 1, 2-(nitro-phenyl) ethyl moiety that reacts with O-GlcNAc-modified peptides but can later be released by photoactive cleavage (UV 254 nm), leaving a basic amino-methyltriazole tag at the O-GlcNAc modification site. The biotinylated peptides were enriched by affinity chromatography, and finally released from the solid carrier, and analyzed by ETD-MS. A heavy isotope labeled version of the photo-cleavable biotin alkyne is currently under synthesis for site-specific O-GlcNAc quantification. It was later shown that the flow-through from the avidin chromatography can be further enriched for other posttranslational modifications. This idea was first applied to investigate crosstalk between glycosylation and phosphorylation, in which phosphatase was inhibited during the labeling process to prevent the loss of phospho sites [128]. The investigation of the interplay between phosphorylation and GlcNAcylation using a serial enrichment protocol, combined with SILAC, has mapped and quantified over 120 specific O-GlcNAc-modified residues and over 350 phosphorylated residues from only 15 μg of sample by MS/MS analysis.

References

1. Scott JD, Pawson T (2009) Cell signaling in space and time: where proteins come together and when they're apart. Science 326(5957):1220–1224
2. Bensimon A, Heck AJ, Aebersold R (2012) Mass spectrometry-based proteomics and network biology. Annu Rev Biochem 81:379–405
3. Christopher W (2006) Posttranslational modification of proteins: expanding nature's inventory. Colo.: Roberts and Co. Publishers, Englewood, p xxi
4. Bakri Y et al (2005) Balance of MafB and PU.1 specifies alternative macrophage or dendritic cell fate. Blood 105(7):2707–2716

5. Macek B, Mann M, Olsen JV (2009) Global and site-specific quantitative phosphoproteomics: principles and applications. Annu Rev Pharmacol Toxicol 49:199–221
6. Mann M et al (2002) Analysis of protein phosphorylation using mass spectrometry: deciphering the phosphoproteome. Trends Biotechnol 20 (6):261–268
7. Ihara Y, Nukina N, Miura R, Ogawara M (1986) Phosphorylated tau protein is integrated into paired helical filaments in Alzheimer's disease. J Biochem 99(6):1807–1810
8. Pedersen B, Holscher T, Sato Y, Pawlinski R, Mackman N (2005) A balance between tissue factor and tissue factor pathway inhibitor is required for embryonic development and hemostasis in adult mice. Blood 105(7):2777–2782
9. Spiro RG (2002) Protein glycosylation: nature, distribution, enzymatic formation, and disease implications of glycopeptide bonds. Glycobiology 12(4):43R–56R
10. Lechner J, Wieland F (1989) Structure and biosynthesis of prokaryotic glycoproteins. Annu Rev Biochem 58:173–194
11. Trombetta ES (2003) The contribution of N-glycans and their processing in the endoplasmic reticulum to glycoprotein biosynthesis. Glycobiology 13(9):77R–91R
12. Gemmill TR, Trimble RB (1999) Overview of N- and O-linked oligosaccharide structures found in various yeast species. Biochim Biophys Acta 1426 (2):227–237
13. Rudd PM, Elliott T, Cresswell P, Wilson IA, Dwek RA (2001) Glycosylation and the immune system. Science 291(5512):2370–2376
14. Kravtsova-Ivantsiv Y, Ciechanover A (2012) Non-canonical ubiquitin-based signals for proteasomal degradation. J Cell Sci 125 (Pt 3):539–548
15. Hicke L (1999) Gettin' down with ubiquitin: turning off cell-surface receptors, transporters and channels. Trends Cell Biol 9(3):107–112
16. Hicke L (2001) Protein regulation by monoubiquitin. Nat Rev Mol Cell Biol 2(3):195–201
17. Impens F, Radoshevich L, Cossart P, Ribet D (2014) Mapping of SUMO sites and analysis of SUMOylation changes induced by external stimuli. Proc Natl Acad Sci U S A 111(34):12432–12437
18. Kamitani T, Kito K, Nguyen HP, Yeh ET (1997) Characterization of NEDD8, a developmentally down-regulated ubiquitin-like protein. J Biol Chem 272(45):28557–28562
19. Ohsumi Y (2001) Molecular dissection of autophagy: two ubiquitin-like systems. Nat Rev Mol Cell Biol 2(3):211–216
20. Loeb KR, Haas AL (1992) The interferon-inducible 15-kDa ubiquitin homolog conjugates to intracellular proteins. J Biol Chem 267(11):7806–7813
21. Zhao S et al (2010) Regulation of cellular metabolism by protein lysine acetylation. Science 327 (5968):1000–1004
22. Chaurasia MK et al (2014) A prawn core histone 4: derivation of N- and C-terminal peptides and their antimicrobial properties, molecular characterization and mRNA transcription. Microbiol Res 170:78
23. Yang XJ, Seto E (2008) Lysine acetylation: codified crosstalk with other posttranslational modifications. Mol Cell 31(4):449–461
24. Huang DT, Walden H, Duda D, Schulman BA (2004) Ubiquitin-like protein activation. Oncogene 23(11):1958–1971
25. Black JC, Van Rechem C, Whetstine JR (2012) Histone lysine methylation dynamics: establishment, regulation, and biological impact. Mol Cell 48 (4):491–507
26. Jellinger KA (2010) The neuropathologic substrate of Parkinson disease dementia. Acta Neuropathol 119(1):151–153
27. Munshi A, Shafi G, Aliya N, Jyothy A (2009) Histone modifications dictate specific biological readouts. J Genet Genomics 36(2):75–88
28. Rabilloud T, Chevallet M, Luche S, Lelong C (2010) Two-dimensional gel electrophoresis in proteomics: past, present and future. J Proteomics 73 (11):2064–2077
29. Wang P, Giese RW (1998) Phosphate-specific fluorescence labeling with BO-IMI: reaction details. J Chromatogr A 809(1–2):211–218
30. Abu-Lawi KI, Sultzer BM (1995) Induction of serine and threonine protein phosphorylation by endotoxin-associated protein in murine resident peritoneal macrophages. Infect Immun 63(2):498–502
31. Arad-Dann H, Beller U, Haimovitch R, Gavrieli Y, Ben-Sasson SA (1993) Immunohistochemistry of phosphotyrosine residues: identification of distinct intracellular patterns in epithelial and steroidogenic tissues. J Histochem Cytochem 41(4):513–519
32. MacDonald JA, Mackey AJ, Pearson WR, Haystead TAJ (2002) A strategy for the rapid identification of phosphorylation sites in the phosphoproteome. Mol Cell Proteomics 1(4):314–322
33. Olsen JV et al (2006) Global, in vivo, and site-specific phosphorylation dynamics in signaling networks. Cell 127(3):635–648
34. Sugiyama N et al (2007) Phosphopeptide enrichment by aliphatic hydroxy acid-modified metal oxide chromatography for nano-LC-MS/MS in proteomics applications. Mol Cell Proteomics 6(6):1103–1109
35. Ficarro SB, Parikh JR, Blank NC, Marto JA (2008) Niobium (V) oxide (Nb2O5): application to phosphoproteomics. Anal Chem 80(12):4606–4613
36. Larsen MR, Thingholm TE, Jensen ON, Roepstorff P, Jørgensen TJD (2005) Highly selective enrichment of phosphorylated peptides from peptide mixtures using titanium dioxide microcolumns. Mol Cell Proteomics 4(7):873–886

37. Bodenmiller B, Mueller LN, Mueller M, Domon B, Aebersold R (2007) Reproducible isolation of distinct, overlapping segments of the phosphoproteome. Nat Methods 4(3):231–237
38. Wu J, Shakey Q, Liu W, Schuller A, Follettie MT (2007) Global profiling of phosphopeptides by titania affinity enrichment. J Proteome Res 6 (12):4684–4689
39. Villen J, Gygi SP (2008) The SCX/IMAC enrichment approach for global phosphorylation analysis by mass spectrometry. Nat Protoc 3(10):1630–1638
40. Zhou H et al (2013) Robust phosphoproteome enrichment using monodisperse microsphere-based immobilized titanium (IV) ion affinity chromatography. Nat Protoc 8(3):461–480
41. Feng S et al (2007) Immobilized zirconium ion affinity chromatography for specific enrichment of phosphopeptides in phosphoproteome analysis. Mol Cell Proteomics 6(9):1656–1665
42. Posewitz MC, Tempst P (1999) Immobilized gallium (III) affinity chromatography of phosphopeptides. Anal Chem 71(14):2883–2892
43. Andersson L, Porath J (1986) Isolation of phosphoproteins by immobilized metal (Fe3+) affinity chromatography. Anal Biochem 154(1):250–254
44. Ficarro SB et al (2002) Phosphoproteome analysis by mass spectrometry and its application to Saccharomyces cerevisiae. Nat Biotechnol 20(3):301–305
45. Engholm-Keller K et al (2012) TiSH–a robust and sensitive global phosphoproteomics strategy employing a combination of TiO2, SIMAC, and HILIC. J Proteome 75(18):5749–5761
46. Thingholm TE, Jensen ON, Robinson PJ, Larsen MR (2008) SIMAC (sequential elution from IMAC), a phosphoproteomics strategy for the rapid separation of monophosphorylated from multiply phosphorylated peptides. Mol Cell Proteomics 7 (4):661–671
47. Zhou H et al (2008) Specific phosphopeptide enrichment with immobilized titanium Ion affinity chromatography adsorbent for phosphoproteome analysis. J Proteome Res 7(9):3957–3967
48. Beltran L, Casado P, Rodriguez-Prados JC, Cutillas PR (2012) Global profiling of protein kinase activities in cancer cells by mass spectrometry. J Proteome 77:492–503
49. Hunter T, Sefton BM (1980) Transforming gene-product of Rous-sarcoma virus phosphorylates tyrosine. Proc Natl Acad Sci U S A-Biol Sci 77 (3):1311–1315
50. Matsuoka S et al (2007) ATM and ATR substrate analysis reveals extensive protein networks responsive to DNA damage. Science 316(5828):1160–1166
51. Gronborg M et al (2002) A mass spectrometry-based proteomic approach for identification of serine/threonine-phosphorylated proteins by enrichment with phospho-specific antibodies – Identification of a novel protein, Frigg, as a protein kinase A substrate. Mol Cell Proteomics 1(7):517–527
52. Beausoleil SA et al (2004) Large-scale characterization of HeLa cell nuclear phosphoproteins. Proc Natl Acad Sci U S A 101(33):12130–12135
53. Han G et al (2008) Large-scale phosphoproteome analysis of human liver tissue by enrichment and fractionation of phosphopeptides with strong anion exchange chromatography. Proteomics 8 (7):1346–1361
54. Gilar M, Olivova P, Daly AE, Gebler JC (2005) Orthogonality of separation in two-dimensional liquid chromatography. Anal Chem 77(19):6426–6434
55. Reinders J, Sickmann A (2005) State-of-the-art in phosphoproteomics. Proteomics 5(16):4052–4061
56. Alpert AJ (2008) Electrostatic repulsion hydrophilic interaction chromatography for isocratic separation of charged solutes and selective isolation of phosphopeptides. Anal Chem 80(1):62–76
57. Villén J, Beausoleil SA, Gerber SA, Gygi SP (2007) Large-scale phosphorylation analysis of mouse liver. Proc Natl Acad Sci 104(5):1488–1493
58. Zhai B, Villen J, Beausoleil SA, Mintseris J, Gygi SP (2008) Phosphoproteome analysis of drosophila metanogaster embryos. J Proteome Res 7 (4):1675–1682
59. McNulty DE, Annan RS (2008) Hydrophilic interaction chromatography reduces the complexity of the phosphoproteome and improves global phosphopeptide isolation and detection. Mol Cell Proteomics 7(5):971–980
60. Song CX et al (2010) Reversed-phase-reversed-phase liquid chromatography approach with high orthogonality for multidimensional separation of phosphopeptides. Anal Chem 82(1):53–56
61. Sano A, Nakamura H (2004) Chemo-affinity of titania for the column-switching HPLC analysis of phosphopeptides. Anal Sci 20(3):565–566
62. Kaji H et al (2003) Lectin affinity capture, isotope-coded tagging and mass spectrometry to identify N-linked glycoproteins. Nat Biotechnol 21 (6):667–672
63. Wang L et al (2006) OK—Concanavalin A-captured glycoproteins in healthy human urine. Mol Cell Proteomics 5(3):560–562
64. Wisniewski JR, Nagaraj N, Zougman A, Gnad F, Mann M (2010) Brain phosphoproteome obtained by a FASP-based method reveals plasma membrane protein topology. J Proteome Res 9(6):3280–3289
65. Yang Z, Hancock WS (2005) Monitoring glycosylation pattern changes of glycoproteins using multi-lectin affinity chromatography. J Chromatogr A 1070(1–2):57–64
66. Madera M, Mechref Y, Novotny MV (2005) Combining lectin microcolumns with high-resolution separation techniques for enrichment of glycoproteins and glycopeptides. Anal Chem 77 (13):4081–4090
67. Kaji H, Yamauchi Y, Takahashi N, Isobe T (2007) Mass spectrometric identification of N-linked glycopeptides using lectin-mediated affinity capture

and glycosylation site-specific stable isotope tagging. Nat Protoc 1(6):3019–3027
68. Zhang H, X-j L, Martin DB, Aebersold R (2003) Identification and quantification of N-linked glycoproteins using hydrazide chemistry, stable isotope labeling and mass spectrometry. Nat Biotechnol 21(6):660–666
69. Sun B et al (2007) Shotgun glycopeptide capture approach coupled with mass spectrometry for comprehensive glycoproteomics. Mol Cell Proteomics 6 (1):141–149
70. Alley WR Jr, Mann BF, Novotny MV (2013) High-sensitivity analytical approaches for the structural characterization of glycoproteins. Chem Rev 113 (4):2668–2732
71. Sun B, Hood L (2014) Protein-centric N-glycoproteomics analysis of membrane and plasma membrane proteins. J Proteome Res 13 (6):2705–2714
72. Wollscheid B et al (2009) Mass-spectrometric identification and relative quantification of N-linked cell surface glycoproteins. Nat Biotechnol 27 (4):378–386
73. Teo CF et al (2010) Glycopeptide-specific monoclonal antibodies suggest new roles for O-GlcNAc. Nat Chem Biol 6(5):338–343
74. Alfaro JF et al (2012) Tandem mass spectrometry identifies many mouse brain O-GlcNAcylated proteins including EGF domain-specific O-GlcNAc transferase targets. Proc Natl Acad Sci 109 (19):7280–7285
75. Anonsen JH, Vik A, Egge-Jacobsen W, Koomey M (2012) An extended spectrum of target proteins and modification sites in the general O-linked protein glycosylation system in Neisseria gonorrhoeae. J Proteome Res 11(12):5781–5793
76. Peng J et al (2003) A proteomics approach to understanding protein ubiquitination. Nat Biotechnol 21 (8):921–926
77. Tagwerker C et al (2006) A tandem affinity tag for two-step purification under fully denaturing conditions – Application in ubiquitin profiling and protein complex identification combined with in vivo cross-linking. Mol Cell Proteomics 5 (4):737–748
78. Xu G, Paige JS, Jaffrey SR (2010) Global analysis of lysine ubiquitination by ubiquitin remnant immunoaffinity profiling. Nat Biotechnol 28 (8):868–873
79. Kim W et al (2011) Systematic and quantitative assessment of the ubiquitin-modified proteome. Mol Cell 44(2):325–340
80. Kim SC et al (2006) Substrate and functional diversity of lysine acetylation revealed by a proteomics survey. Mol Cell 23(4):607–618
81. Choudhary C et al (2009) Lysine acetylation targets protein complexes and co-regulates major cellular functions. Science 325(5942):834–840
82. Mertins P et al (2013) Integrated proteomic analysis of post-translational modifications by serial enrichment. Nat Methods 10(7):634–637
83. Mann M, Jensen ON (2003) Proteomic analysis of post-translational modifications. Nat Biotechnol 21 (3):255–261
84. Tian R (2014) Exploring intercellular signaling by proteomic approaches. Proteomics 14(4–5):498–512
85. Gropengiesser J, Varadarajan BT, Stephanowitz H, Krause E (2009) The relative influence of phosphorylation and methylation on responsiveness of peptides to MALDI and ESI mass spectrometry. J Mass Spectrom 44(5):821–831
86. Gao Y, Wang Y (2007) A method to determine the ionization efficiency change of peptides caused by phosphorylation. J Am Soc Mass Spectrom 18 (11):1973–1976
87. Witze ES, Old WM, Resing KA, Ahn NG (2007) Mapping protein post-translational modifications with mass spectrometry. Nat Methods 4 (10):798–806
88. Tuytten R et al (2006) Stainless steel electrospray probe: a dead end for phosphorylated organic compounds? J Chromatogr A 1104(1–2):209–221
89. Swaney DL, Wenger CD, Thomson JA, Coon JJ (2009) Human embryonic stem cell phosphoproteome revealed by electron transfer dissociation tandem mass spectrometry. Proc Natl Acad Sci 106(4):995–1000
90. Syka JEP, Coon JJ, Schroeder MJ, Shabanowitz J, Hunt DF (2004) Peptide and protein sequence analysis by electron transfer dissociation mass spectrometry. Proc Natl Acad Sci U S A 101(26):9528–9533
91. Villen J, Beausoleil SA, Gygi SP (2008) Evaluation of the utility of neutral-loss-dependent MS3 strategies in large-scale phosphorylation analysis. Proteomics 8(21):4444–4452
92. Palumbo AM, Tepe JJ, Reid GE (2008) Mechanistic insights into the multistage gas-phase fragmentation behavior of phosphoserine- and phosphothreonine-containing peptides. J Proteome Res 7(2):771–779
93. Boersema PJ, Mohammed S, Heck AJR (2009) Phosphopeptide fragmentation and analysis by mass spectrometry. J Mass Spectrom 44(6):861–878
94. Schroeder MJ, Shabanowitz J, Schwartz JC, Hunt DF, Coon JJ (2004) A neutral loss activation method for improved phosphopeptide sequence analysis by quadrupole ion trap mass spectrometry. Anal Chem 76(13):3590–3598
95. Palumbo AM, Reid GE (2008) Evaluation of Gas-phase rearrangement and competing fragmentation reactions on protein phosphorylation site assignment using collision induced dissociation-MS/MS and MS3. Anal Chem 80(24):9735–9747
96. Cain JA, Solis N, Cordwell SJ (2014) Beyond gene expression: the impact of protein post-translational modifications in bacteria. J Proteome 97:265–286
97. Hung C-W, Schlosser A, Wei J, Lehmann WD (2007) Collision-induced reporter fragmentations for identification of covalently modified peptides. Anal Bioanal Chem 389(4):1003–1016
98. Olsen JV et al (2007) Higher-energy C-trap dissociation for peptide modification analysis. Nat Methods 4(9):709–712

99. Li X et al (2007) Large-scale phosphorylation analysis of alpha-factor-arrested Saccharomyces cerevisiae. J Proteome Res 6(3):1190–1197
100. Myung S et al (2011) High-capacity ion trap coupled to a time-of-flight mass spectrometer for comprehensive linked scans with no scanning losses. Int J Mass Spectrom 301(1–3):211–219
101. Chaze T et al (2014) O-Glycosylation of the N-terminal region of the serine-rich adhesin Srr1 of streptococcus agalactiae explored by mass spectrometry. Mol Cell Proteomics 13(9):2168–2182
102. Larsen MR, Trelle MB, Thingholm TE, Jensen ON (2006) Analysis of posttranslational modifications of proteins by tandem mass spectrometry. Biotechniques 40(6):790–798
103. Melo-Braga MN et al (2012) Modulation of protein phosphorylation, N-Glycosylation and Lys-Acetylation in grape (Vitis vinifera) mesocarp and exocarp owing to lobesia botrana infection. Mol Cell Proteomics 11(10):945–956
104. Rappsilber J, Friesen WJ, Paushkin S, Dreyfuss G, Mann M (2003) Detection of arginine dimethylated peptides by parallel precursor ion scanning mass spectrometry in positive ion mode. Anal Chem 75 (13):3107–3114
105. Na CH, Peng J (2012) Analysis of ubiquitinated proteome by quantitative mass spectrometry. Methods Mol Biol 893:417–429
106. Jedrychowski MP et al (2011) Evaluation of HCD- and CID-type fragmentation within their respective detection platforms for murine phosphoproteomics. Mol Cell Proteomics 10(12):M111 009910
107. Nagaraj N, D'Souza RCJ, Cox J, Olsen JV, Mann M (2010) Feasibility of large-scale phosphoproteomics with higher energy collisional dissociation fragmentation. J Proteome Res 9(12):6786–6794
108. Syrstad EA, Turecek F (2005) Toward a general mechanism of electron capture dissociation. J Am Soc Mass Spectrom 16(2):208–224
109. Chi A et al (2007) Analysis of phosphorylation sites on proteins from Saccharomyces cerevisiae by electron transfer dissociation (ETD) mass spectrometry. Proc Natl Acad Sci U S A 104(7):2193–2198
110. Mikesh LM et al (2006) The utility of ETD mass spectrometry in proteomic analysis. Biochim Biophys Acta 1764(12):1811–1822
111. Frese CK et al (2011) Improved peptide identification by targeted fragmentation using CID, HCD and ETD on an LTQ-Orbitrap Velos. J Proteome Res 10 (5):2377–2388
112. Good DM, Wirtala M, McAlister GC, Coon JJ (2007) Performance characteristics of electron transfer dissociation mass spectrometry. Mol Cell Proteomics 6(11):1942–1951
113. Molina H, Horn DM, Tang N, Mathivanan S, Pandey A (2007) Global proteomic profiling of phosphopeptides using electron transfer dissociation tandem mass spectrometry. Proc Natl Acad Sci U S A 104(7):2199–2204
114. Xia Y et al (2006) Implementation of ion/ion reactions in a quadrupole/time-of-flight tandem mass spectrometer. Anal Chem 78(12):4146–4154
115. McAlister GC et al (2008) A proteomics grade electron transfer dissociation-enabled hybrid linear ion trap-Orbitrap mass spectrometer. J Proteome Res 7 (8):3127–3136
116. Wysocki VH, Tsaprailis G, Smith LL, Breci LA (2000) Special feature: commentary – mobile and localized protons: a framework for understanding peptide dissociation. J Mass Spectrom 35 (12):1399–1406
117. Michalski A, Neuhauser N, Cox J, Mann M (2012) A systematic investigation into the nature of tryptic HCD spectra. J Proteome Res 11(11):5479–5491
118. Zubarev RA, Kelleher NL, McLafferty FW (1998) Electron capture dissociation of multiply charged protein cations. A nonergodic process. J Am Chem Soc 120(13):3265–3266
119. Cooper HJ, Hakansson K, Marshall AG (2005) The role of electron capture dissociation in biomolecular analysis. Mass Spectrom Rev 24(2):201–222
120. Bakhtiar R, Guan ZQ (2005) Electron capture dissociation mass spectrometry in characterization of post-translational modifications. Biochem Biophys Res Commun 334(1):1–8
121. Frese CK et al (2013) Unambiguous phosphosite localization using Electron-Transfer/Higher-Energy collision Dissociation (EThcD). J Proteome Res 12 (3):1520–1525
122. Swaney DL, McAlister GC, Coon JJ (2008) Decision tree-driven tandem mass spectrometry for shotgun proteomics. Nat Methods 5(11):959–964
123. Collins MO, Wright JC, Jones M, Rayner JC, Choudhary JS (2014) Confident and sensitive phosphoproteomics using combinations of collision induced dissociation and electron transfer dissociation. J Proteome 103:1–14
124. Hart-Smith G, Raftery MJ (2012) Detection and characterization of low abundance glycopeptides via higher-energy C-Trap dissociation and orbitrap mass analysis. J Am Soc Mass Spectrom 23 (1):124–140
125. Hakansson K et al (2001) Electron capture dissociation and infrared multiphoton dissociation MS/MS of an N-glycosylated tryptic peptide to yield complementary sequence information. Anal Chem 73 (18):4530–4536
126. Singh C, Zampronio CG, Creese AJ, Cooper HJ (2012) Higher Energy Collision Dissociation (HCD) product ion-triggered Electron Transfer Dissociation (ETD) mass spectrometry for the analysis of N-linked glycoproteins. J Proteome Res 11 (9):4517–4525
127. Zhao P et al (2011) Combining high-energy C-trap dissociation and electron transfer dissociation for protein O-GlcNAc modification site assignment. J Proteome Res 10(9):4088–4104

128. Wang Z et al (2010) Enrichment and site mapping of O-linked N-acetylglucosamine by a combination of chemical/enzymatic tagging, photochemical cleavage, and electron transfer dissociation mass spectrometry. Mol Cell Proteomics 9(1):153–160
129. Shvartsburg AA, Singer D, Smith RD, Hoffmann R (2011) Ion mobility separation of isomeric phosphopeptides from a protein with variant modification of adjacent residues. Anal Chem 83 (13):5078–5085
130. Creese AJ, Cooper HJ (2012) Separation and identification of isomeric glycopeptides by high field asymmetric waveform Ion mobility spectrometry. Anal Chem 84(5):2597–2601
131. Shvartsburg AA, Zheng Y, Smith RD, Kelleher NL (2012) Ion mobility separation of variant histone tails extending to the "middle-down" range. Anal Chem 84(10):4271–4276
132. Hahne H, Kuster B (2011) A novel two-stage tandem mass spectrometry approach and scoring scheme for the identification of O-GlcNAc modified peptides. J Am Soc Mass Spectrom 22(5):931–942
133. Beausoleil SA, Villen J, Gerber SA, Rush J, Gygi SP (2006) A probability-based approach for high-throughput protein phosphorylation analysis and site localization. Nat Biotechnol 24(10):1285–1292
134. Savitski MM et al (2011) Confident phosphorylation site localization using the mascot delta score. Mol Cell Proteomics 10(2):M110.003830
135. Taus T et al (2011) Universal and confident phosphorylation site localization using phosphoRS. J Proteome Res 10(12):5354–5362
136. Bailey CM et al (2009) SLoMo: automated site localization of modifications from ETD/ECD mass spectra. J Proteome Res 8(4):1965–1971
137. Baker PR, Trinidad JC, Chalkley RJ (2011) Modification site localization scoring integrated into a search engine. Mol Cell Proteomics 10(7): M111.008078
138. Chen Y, Chen W, Cobb MH, Zhao YM (2009) PTMap-A sequence alignment software for unrestricted, accurate, and full-spectrum identification of post-translational modification sites. Proc Natl Acad Sci U S A 106(3):761–766
139. Sharma K et al (2014) Ultradeep human phosphoproteome reveals a distinct regulatory nature of Tyr and Ser/Thr-based signaling. Cell Rep 8:1583
140. Udeshi ND et al (2013) Refined preparation and use of anti-diglycine remnant (K-epsilon-GG) antibody enables routine quantification of 10,000 s of ubiquitination sites in single proteomics experiments. Mol Cell Proteomics 12(3):825–831
141. Guo AL et al (2014) Immunoaffinity enrichment and mass spectrometry analysis of protein methylation. Mol Cell Proteomics 13(1):372–387
142. Zielinska DF, Gnad F, Wisniewski JR, Mann M (2010) Precision mapping of an in vivo N-glycoproteome reveals rigid topological and sequence constraints. Cell 141(5):897–907
143. Mommsen TP, Plisetskaya EM (1991) Insulin in fishes and agnathans – history, structure, and metabolic-regulation. Rev Aquat Sci 4 (2–3):225–259
144. Owens DR (2002) New horizons – alternative routes for insulin therapy. Nat Rev Drug Discov 1 (7):529–540
145. Hornbeck PV et al (2012) PhosphoSitePlus: a comprehensive resource for investigating the structure and function of experimentally determined post-translational modifications in man and mouse. Nucleic Acids Res 40(D1):D261–D270
146. Lu CT et al (2013) dbPTM 3.0: an informative resource for investigating substrate site specificity and functional association of protein post-translational modifications. Nucleic Acids Res 41 (D1):D295–D305
147. Liu ZX et al (2014) CPLM: a database of protein lysine modifications. Nucleic Acids Res 42(D1): D531–D536
148. Dinkel H et al (2011) Phospho.ELM: a database of phosphorylation sites-update 2011. Nucleic Acids Res 39:D261–D267
149. Gnad F, Gunawardena J, Mann M (2011) PHOSIDA 2011: the posttranslational modification database. Nucleic Acids Res 39:D253–D260
150. Gupta R, Birch H, Rapacki K, Brunak S, Hansen JE (1999) O-GLYCBASE version 4.0: a revised database of O-glycosylated proteins. Nucleic Acids Res 27(1):370–372
151. Zhang H et al (2006) UniPep – a database for human N-linked glycosites: a resource for biomarker discovery. Genome Biol 7(8):R73
152. Kaji H et al (2012) Large-scale identification of N-glycosylated proteins of mouse tissues and construction of a glycoprotein database, GlycoProtDB. J Proteome Res 11(9):4553–4566
153. Campbell MP et al (2014) UniCarbKB: building a knowledge platform for glycoproteomics. Nucleic Acids Res 42(D1):D215–D221
154. Gao TS et al (2013) UUCD: a family-based database of ubiquitin and ubiquitin-like conjugation. Nucleic Acids Res 41(D1):D445–D451
155. Lee WC, Lee M, Jung JW, Kim KP, Kim D (2008) SCUD: Saccharomyces Cerevisiae Ubiquitination Database. BMC Genomics 9:7
156. Chernorudskiy AL et al (2007) UbiProt: a database of ubiquitinated proteins. Bmc Bioinf 8:126
157. Fiedler D et al (2009) Functional organization of the S-cerevisiae phosphorylation network. Cell 136 (5):952–963
158. Kanehisa M, Goto S (2000) KEGG: Kyoto Encyclopedia of Genes and Genomes. Nucleic Acids Res 28 (1):27–30
159. Horn H et al (2014) KinomeXplorer: an integrated platform for kinome biology studies. Nat Methods 11(6):603–604
160. Linding R et al (2008) NetworKIN: a resource for exploring cellular phosphorylation networks. Nucleic Acids Res 36:D695–D699
161. Miller ML et al (2008) Linear motif atlas for phosphorylation-dependent signaling. Sci Signal 1 (35):ra2

162. Swaney DL et al (2013) Global analysis of phosphorylation and ubiquitination cross-talk in protein degradation. Nat Methods 10(7):676–682
163. Wang Y et al (2011) Reversed-phase chromatography with multiple fraction concatenation strategy for proteome profiling of human MCF10A cells. PROTEOMICS 11(10):2019–2026
164. Wang Z, Gucek M, Hart GW (2008) Cross-talk between GlcNAcylation and phosphorylation: site-specific phosphorylation dynamics in response to globally elevated O-GlcNAc. Proc Natl Acad Sci U S A 105(37):13793–13798
165. Olejnik J, Sonar S, Krzymanska-Olejnik E, Rothschild KJ (1995) Photocleavable biotin derivatives: a versatile approach for the isolation of biomolecules. Proc Natl Acad Sci U S A 92 (16):7590–7594

Protein-Protein Interaction Detection Via Mass Spectrometry-Based Proteomics

18

Benedetta Turriziani, Alexander von Kriegsheim, and Stephen R. Pennington

Abstract

Analysis of protein-protein interactions is one of the mainstays of mass spectrometry-based proteomics and recent developments, which have simplified the methodology, have permitted non-specialised laboratories to adopt the approach. We introduce and review three complimentary methods which allow for the targeted, global and site-specific analysis of protein complexes. Co-precipitation of endogenous or ectopically expressed proteins and their complexes followed by proteomic analysis allows for the discovery and accurate quantification of specific protein interactions. Whereas complimentary methods, such as co-purification of entire complexes based on physico-chemical attributes, can give a snapshot of the composition and dynamics of protein complexes on a global scale. Cross-linking on the other hand can pinpoint the amino acids involved in protein-protein interactions to such a resolution that the likely complex can be reconstructed computationally.

Keywords

Protein complex • Co-purification • Cross-linking • Co-elution • Interaction proteomics

18.1 Introduction

High-throughput DNA sequencing allowed for the first time, the correlation of pathologies to specific genes aberrations. Unfortunately, the genomic information itself is not enough for a comprehensive understanding of the mechanisms that bring about the pathological transformations. The number of proteins is larger than the number of codifying genes, due to the presence of

B. Turriziani • A. von Kriegsheim
Systems Biology Ireland, Conway Institute, University College Dublin, Belfield, Dublin 4, Ireland

S.R. Pennington (✉)
School of Medicine and Medical Sciences, UCD Conway Institute of Biomolecular and Biomedical Research, University College Dublin, Dublin 4, Ireland
e-mail: stephen.pennington@ucd.ie

H. Mirzaei and M. Carrasco (eds.), *Modern Proteomics – Sample Preparation, Analysis and Practical Applications*, Advances in Experimental Medicine and Biology 919, DOI 10.1007/978-3-319-41448-5_18

additional levels of regulation during and after protein translation. Events such as splicing and post-translational modifications, which regulate protein activation, and localization and degradation, all add to proteome complexity. All these factors are regulated by the interaction of proteins with other proteins, working cooperatively in intricate signalling networks. For these reasons, dynamic and static protein-protein interactions are essential cornerstones of all signalling networks. Thus, identifying and quantifying protein interactions is of great interest within the biological and medical sciences as well as the emerging field of systems biology.

Interaction proteomics aims to study protein interactions in an unbiased manner and the information obtained by this approach can be used to determine how proteins assemble in complexes and form networks. These methods must not only aim to identify novel interactions, but should ideally reveal precise protein collocation in the specific biological system investigated. In order to maximize the information gained from interaction screens, interaction studies are ideally designed not only to identify specific binders but to accurately quantify dynamic changes in protein-protein interactions in response to perturbations.

Since the focus on research shifted from genome to proteome, a wide number of methods have been developed to detect protein-protein interactions [1]. One of the most widely used methods is the yeast two-hybrid screening (Y2H). The system was firstly described by Fields and Song in 1989 [2] using a *Saccharomyces cerevisiae* model. The yeast two-hybrid consists of a genetic screening system in which protein interactions are detected by the signalling of a reporter gene. Two plasmids are required to perform the screening. In the first plasmid the protein of interest, called bait, is fused with a DNA-binding domain (BD) of a yeast transcriptional factor, generally Gal4. The BD binds a region upstream to the promoter of the reporter gene. In the other plasmid a second protein, called the prey, is fused with the activation domain (AD) for the transcriptional factor Gal4. Both the BD and AD of Gal4 are required for the activation of the reporter gene. Thus, the transcription of the reporter gene only occurs when bait and prey interact with each other and form a functional transcription factor complex. Therefore, the interaction between bait and prey can be detected by the signal resultant of the reporter gene expression [3].

Although genome-wide interaction screens using Y2H have been undertaken, the method has several drawbacks. One of the major limitations of Y2H is the reliability of the data generated. The rate of false positives and false negatives among the interactions identified can be higher than 60 % in some cases [4]. The main reasons for this unreliability are connected to both biological and technical factors. The biochemical differences in protein translation and regulation between the host, which is generally yeast, and the subject of study is the first parameter to keep in mind. Expressing two proteins in yeast may lead to a physiologically irrelevant interaction as in physiological conditions these two proteins may be expressed in different compartments, in different cells, or even at different times in cell cycle. Crucially, protein interactions are frequently regulated by post-translational modifications (PTMs), which may not occur in yeast. For instance, if the interaction between two proteins requires the phosphorylation of one of them, in yeast this interaction may not occur. Since such regulation by PTMs is frequent, this is an aspect that dramatically affects the efficiency of Y2H as a tool for interaction proteomics studies. A further limitation concerns the fact that Y2H fails to provide information about the dynamics of an interaction. The last aspect that makes Y2H–generated data unreliable is that it only detects direct binary interactions. Taken together, since there is interest in studying dynamic protein complexes in their physiological context, all these limitations highlight the need for complimentary methods.

The possibility to work in near physiological conditions and, more importantly, to provide information about protein complexes stoichiometry and dynamics, have made affinity purification coupled to mass spectrometry (AP-MS) [5] a suitable complimentary method to Y2H. The AP requires purification of proteins and their complexes by enriching the bait protein by affinity purification of endogenous or tagged proteins with specific antibodies or other affinity tags. In essence the tagged or endogenous bait-protein and its complexes are captured by an immobilized matrix, processed and subsequently analysed by mass spectrometry. Commonly used tags are Flag, GFP, Myc, His, and HA; each of which vary in terms of efficiency of purification [6]. In order to reduce the complexity of the sample, eluted proteins are subsequently separated by PAGE, the bands of interest excised, in-gel digested and then identified by mass spectrometry [7].

Even though this analysis workflow allows us to overcome some of the limitation inherent to Y2H, due to the gel-based fractionation the method requires significant machine time, numerous handling steps [8] and loss of sensitivity due to contamination and sample loss. Overall these limitations have made gel-based fractionation impractical for large-scale interactome analysis in the average biological laboratory. Thus, new gel-free methods which reduced handling and analysis time have been developed.

18.2 Co-immunoprecipitation (Co-IP)

Recent advances in mass spectrometry technology have resulted in increasingly more sensitive instruments, perfectly suitable for protein complex identification. These advancements have given a big boost to the application of AP-MS in interaction proteomics studies, but also have highlighted the necessity for novel purification protocols, which appear to be the main constraining step towards new improvements. Several purification strategies have been improved since then, especially protocols involving immunoprecipitation. One of the most efficient technologies, which is still in use, was developed in 1999 by Rigaut et al. using a tandem affinity purification tag (TAP) [9]. This method requires expression of a fusion protein consisting of a bait protein and two epitope tags used for the affinity purification. The TAP plasmid contains three modules, the tags, the *Staphylococcus aureus* proteinA (ProtA) IgG binding domain and a calmodulin binding domain (CBD), separated by a TEV (tobacco etch virus) protease cleavage site. Completing this procedure requires a two-step affinity purification. In the first step the lysate is incubated with IgG Sepharose beads. The ProtA of the TAP tag interacts with the IgG domain in a way that only the bait and its interactors remain bound to the beads. The rest is removed by several washes. The complexes are then treated with TEV protease. Treatment with TEV has two roles, first to disrupt the link between IgG beads and bait, and the second is to expose the CBD for the next purification steps. The eluate is then incubated a second time with calmodulin beads. After a second elution, samples are further fractionated and separated by SDS polyacrylamide gel or directly analysed by mass spectrometry. A general protocol for tandem affinity purification was described by Puig et al. [10]. Double step purification improves the purity of the target complex compared to a single-step affinity purification protocol. Despite improved sensitivity in the identification of protein-protein interactions in TAP purification methods, major limitations still remain. Primarily, due to low yields of overall recovery and the large number of cells or lysate that is required. In addition, as a consequence of the lengthy protocol, only stable complexes remain intact and dynamic interactions are all but lost. Novel, more rapid methods have been developed using alternative affinity tags. Although these improvements have reduced the time required for performing the experiments, none of the methods have been able to improve the recovery of weak and dynamic interactions. Thus, classical TAP purification methods have been largely abandoned.

A good alternative strategy to TAP was established by Rees et al. [11], which still retains the specificity of the double step purification. In the method they developed, a parallel affinity capture (iPAC) method was coupled to mass spectrometry using *D. melanogaster* as a model organism. Similar to the previous protocol, the method is based on a double affinity tag system. A construct is generated as previously described [12], containing tags in an exon flanked by splicing sites. This way, only plasmids in which the target gene was inserted correctly will be translated into a functional protein. The tags used in this type of approach are a marker of expression to check if the bait is expressed in the cells, in this case the yellow fluorescent protein (YFP), and two affinity tags for the immunoprecipitation and purification protocol, StrepII and the FLAG. Unlike the TAP methods, in which the two purifications are done sequentially, they are performed in parallel in the iPAC approach. In one sample the bait is isolated by an immobilized anti-FLAG antibody. In parallel, a second, identical sample is incubated with StrepII beads. The principle of this method is that since the bait expresses both tags, the two independent immunoprecipitations should give similar results or at least contain a subset of common proteins. A workflow of the method is illustrated in (Fig. 18.1a).

The interactomes isolated from the two samples after mass spectrometry analysis are compared with tagless controls purified with both the FLAG and the StrepII beads. Contaminants are identified by a cross analysis

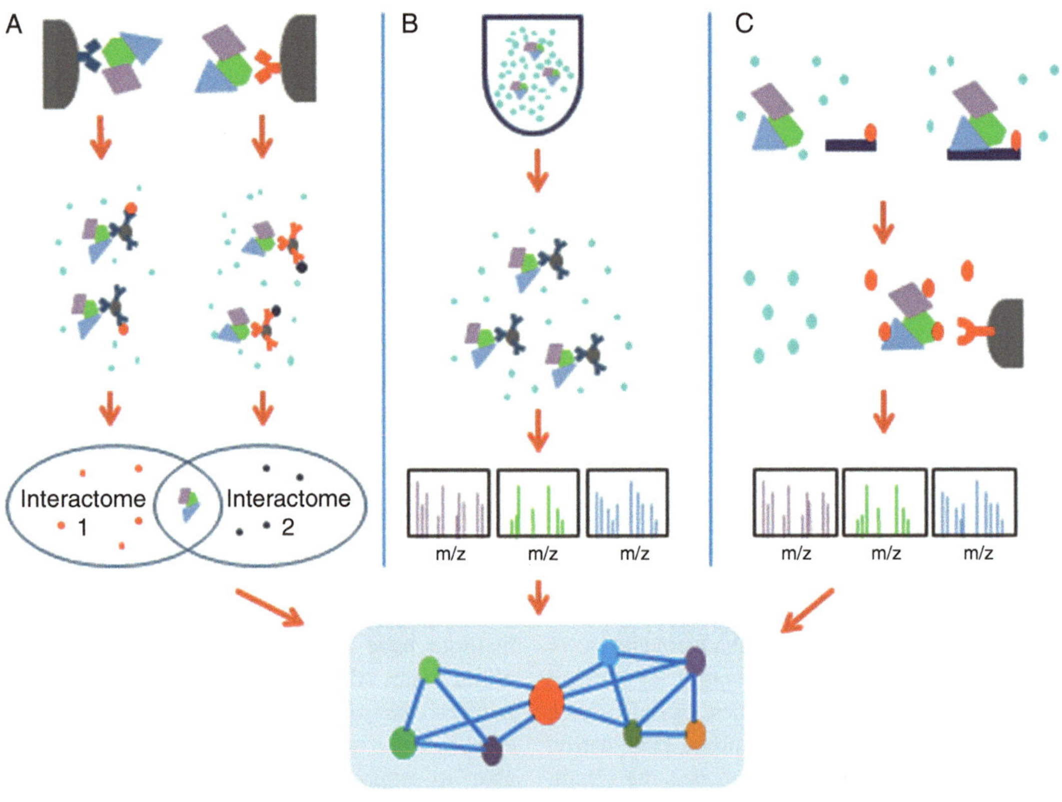

Fig. 18.1 Schematic representation of co-affinity-precipitation workflows. (**a**) In the IPAC method the bait is double tagged with Flag and streptavidin and is expressed in cells. The cell lysate is then incubated with beads recognising either Flag or streptavidin in two parallel incubations. After binding of the complexes to the beads, unspecific interactors are washed away, whereas the protein complexes are retained and analysed by mass spectrometry. Proteins identified in both reactions are considered specific interactors. (**b**) On-bead digestion workflow, the lysate is incubated with a bait-specific antibody followed by washes and digestion. After digestion the interacting proteins are identified by mass spectrometry. (**c**) In BioID the bait is tagged with a promiscuous biotin–ligase leading to the selective biotinylation of proteins proximate to the bait-ligase fusion protein. Proteins forming complexes with the bait are likely to be biotinylated, while non-interacting proteins will remain untagged. The biotinylated proteins are purified by affinity purification. The proteins are subsequently identified via mass spectrometry

among the different controls. The data sets from FLAG, StrepII and controls are compared and proteins common between the three sets of data are classified as contaminants and as a group are called the BEADome. Secondly, the FLAG and StrepII pull-downs are compared for proteins specifically binding to the affinity matrices independently of the bait. Thus, all the proteins that are identified in one pull-down and not in the other are classified as contaminants. The remaining identified proteins that overlap between the FLAG and StrepII set of data are then classified as specific interactions. The protocol can be performed using different tags according to specific experimental exigencies. To prove the effectiveness of the protocol, both methods have been performed and their efficiency evaluated in terms of stability of the bait, interactome recovery after all of the procedural steps and in terms of identified interactome. Overall, iPAC combines the benefits of a double purification with different tags in terms of efficiency and quality of the results, with the ability to detect weak and transient interactions.

More recently, another approach has been developed that is based on a single purification step using biotin ligase as a tag. The work by Couzens et al. was focused on the mammalian Hippo pathway and how the phosphorylation state affects its interaction network [13]. Two complimentary methods were performed in parallel, FLAG affinity purification (FLAG-AP) and a new method of biotinylation followed by streptavidin affinity purification (BioID). Nineteen proteins were tagged with a promiscuous biotin ligase (BirA) that promotes the covalent linkage of a biotin moiety not only to the proteins directly interacting with the bait, but also to proteins present in its proximity (Fig. 18.1c). In this way, when the bait forms a protein complex, all the proteins of this complex are labelled with biotin. Incubation with streptavidin beads allows purification of biotinylated proteins and their identification by mass spectrometry. A comparative analysis of the data sets generated with FLAG-AP and BioID revealed only a partial overlap. The BioID data set contains a larger number of potential interactors compared to the FLAG-AP in native conditions. More interestingly, the major part of those interactors identified using the BioID, but not the FLAG-AP, were proteins of specific cellular compartments which are generally more difficult to isolate due to their poor solubility, including proteins associated with membranes, the centrosome, chromatin and cell junctions. This last observation makes the BioID a suitable method to study proteins from specific localizations, which would otherwise be difficult to analyse. Another advantage of the BioID seems to be the more gentle conditions in which the experiment is performed that allow the recovery of weak interactions usually lost during FLAG-AP. One limitation of the BioID protocol is the incubation time necessary to label the cell with biotin, which in the cited work was about 24 h. From this point of view it may be difficult to reconcile this method with dynamic interactions.

In recent years there has been a trend to abandon gel-based fractionation for unfractionated in-solution or on-bead digestions. This trend has been triggered by improvements in the acquisition speed of mass spectrometers and the resolution of uHPLC chromatographic systems. These new approaches have several advantages. Chiefly, the reduced need for fractionation, which drastically reduces the acquisition time and sample handling [14, 15]. Indeed, just a few steps are required to go from a lysate to the final analysis on the mass spectrometer. This not only shortens the analysis time, but also reduces the overall level of contamination. In addition, sample recovery is improved as in-gel digestion results in loss of material due to inefficient elution and recovery from gel slices. Finally, streamlined protocols are compatible with high throughput, automated robotic handling stations, which will permit large-scale interaction screens required for systems biology. Excitingly, using a "double barrel" column LC method coupled to an ultra-fast Q-Exactive HF, the Mann group has recently published a method which uses 96-well plates for all handling steps and, by using fast gradients, has broken through the 24 h per-plate machine time [16]. This unparalleled speed and, therefore, reduced cost per-sample, finally puts mass spectrometry based interaction proteomics in the same ball park as targeted antibody approaches, such as western blotting, in terms of cost, sensitivity and sample throughput.

Quantitative proteomics has proven essential for well-controlled proteomics experiments and interactome proteomics is no exception. But unfortunately, Stable isotope labeling by amino acids in cell culture (SILAC), which is one of the most widely utilized methodologies for quantitative proteomics, is not applicable in IP-MS experiments. SILAC, which allows the mixing of the samples prior to any processing, cannot be used because in IP-MS, samples can only be combined after the IP has been performed. This limitation is due to the dynamic exchanges that could occur between heavy and light complexes during the incubation step with the affinity matrix. Thus, alternative methods, such as label free quantification (LFQ) [17] are increasingly used. The LFQ method works with any source material, as it does not require any form of labelling. An example of a strategy using LFQ in conjunction with on-beads digestion was elaborated by Turriziani et al. [18]. The protocol is based on classical IP protocols implemented in most biological laboratories and is therefore easily implemented without specialist knowledge or equipment.

All of the strategies described in this section have advantages and weaknesses, but in general they offer good alternatives to the original IP-MS protocols and each contribute their strengths to overcome the main limitations of IP-MS methods.

Regardless of all these improvements, all of these strategies use exogenous, tagged proteins as baits, which are often overexpressed. As highlighted before, there could be many issues related to the process of tagging a protein or simply overexpressing it.

On the other hand, the immunoprecipitation of endogenous proteins has several pitfalls as well. Even if an antibody is found which efficiently enriches the protein, the antibody may have cross-reactivity to other unrelated proteins. Thus, endogenous IPs have to be well controlled in order to avoid false-positives. An efficient method to overcome this shortcoming is represented by an approach which combines quantitative immunoprecipitation with a knock-down strategy (QUICK) [19]. In this work the authors use a SILAC-based quantification [20] and a control in which the bait protein is transiently knocked down by silencing RNA (siRNA). As a result, proteins interacting with the bait are reduced in the control, whereas the concentrations of unspecific and cross-reacting proteins are not affected by the knockdown. Alternatively, a negative control can be generated by using knock-out cell line generated by CRISPR-Cas9. The system is based on the nuclease Cas9 that introduces a double-strands break in the target gene and can lead to silencing of the gene by introducing frame-shift mutations [21].

18.3 Co-elution

All previously mentioned approaches isolate a protein of interest and its interactors by (immuno)-precipitation. Alternatively, protein correlation profiling (PCP) aims not to detect how a single protein relates with its interactome, but rather determines the composition of all intact complexes (Fig. 18.2). The underlying assumption is that protein complexes co-purify when separated based on their biochemical characteristic, such as size, density or hydrophobicity [22]. The conditions in which the separation is performed are tailored to preserve the interactions and integrity of the protein complex. Complexes are frequently separated via chromatography using disparate gradients and solid phases. As mentioned, the principle of the method is based on the fact that interacting proteins and protein complexes will co-elute with the same profile. After fractionating individual complexes, the proteins are digested, analysed and quantified by mass spectrometry. Several protocols have been developed to demonstrate the efficiency of this approach. An example is the work of Andersen et al., which focusses on the human centrosome [23]. The isolated centrosomes were separated by a sucrose

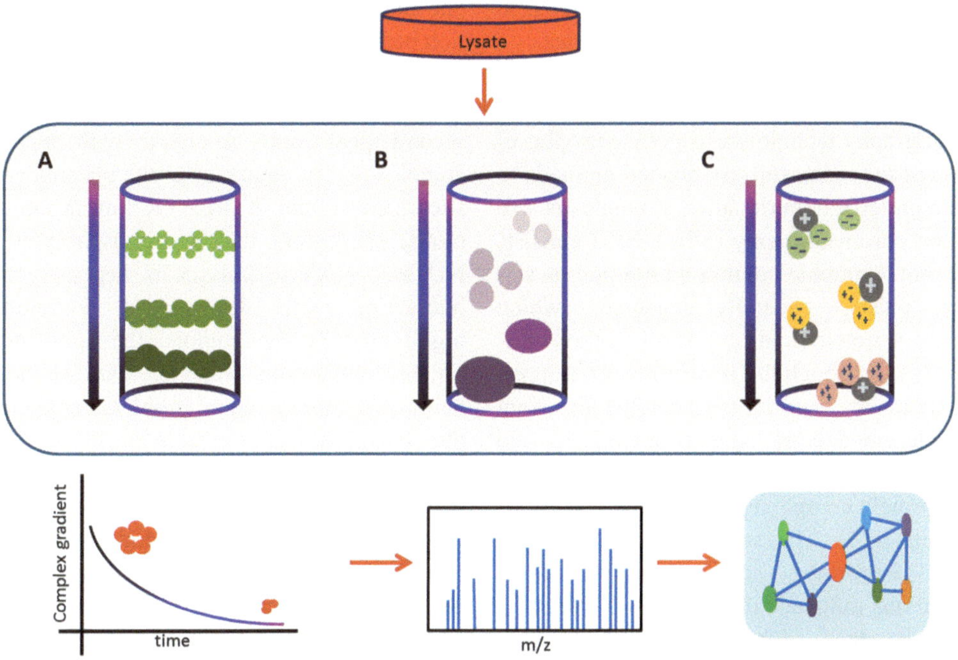

Fig. 18.2 General Summary of co-elution principles. After lysis, protein complexes are separated based on their various biochemical properties. The complexes are collected into different fractions, digested and analyzed by the mass spectrometry. Proteins identified in the same fraction are likely to in the same complex. (**a**) The complexes are separated according to their density in a sucrose gradient. The heavier complexes will be located towards the bottom of the tube, whereas lighter ones will remain at the top. (**b**) Schematization of a size-exclusion chromatography. The column is filled with a porous gel. Bigger complexes travel faster through the gel and are the first ones to be collected. Smaller complexes are trapped in the porous gel and are collected towards the end. (**c**) Representation of the ion exchange chromatography. The complexes pass through the column and interact with the charged matrix. The retention time depends on the overall charge of the complexes. Positively charged complexes travel faster through the column, whereas negatively charged complexes are retained longer due to their interaction with the positively charged matrix

gradient (Fig. 18.2a). Different fractions were collected, analysed and individually quantified. The results were validated using two orthogonal approaches. They initially compared the protein content of different fractions with the elution profiles of centrosome markers. Additionally, they also selected a few candidates and individually validated them by Co-IP. Overall, the combination of these approaches allowed the authors to identify most of the known centrosomal complexes, and groups of likely novel complexes were further partially validated by co-immunofluorescence. This study showed that PCP is a reliable method to characterise multiprotein complexes by co-elution profiling. In particular, PCP showed good efficiency in the characterization of cellular structures, which are difficult to isolate, especially those associated with organelles. As highlighted by the authors, this technique is compatible with isotope-labelling, which can provide quantitative information about the protein complex dynamics over time. Nevertheless, this strategy failed to detect a conspicuous part of proteins associated with the centrosome. Possible causes include low abundance of missed proteins or their participation in complexes that are not stably associated with the centrosome, raising questions about the effectiveness of this method for detecting weak and dynamic interactions.

Although the PCP approaches have some advantages over AP-MS, they also present a set of technical challenges that need to be overcome in order to improve the depth and efficiency of this method. The first complication is related to the effectiveness of the separation strategies.

New methods of separation used in combination with mass spectrometry range from various chromatography techniques to gel electrophoresis. One of the most well-established methods to separate proteins in purification strategies is size exclusion chromatography (SEC). SEC consists of a chromatographic column with a porous stationary phase, generally agarose, which fractionates proteins and complexes by their ability to migrate through the pores or being trapped in the stationary material (Fig. 18.2b). In recent work, Kirkwood et al. have coupled SEC and mass spectrometry to isolate and characterize soluble protein complexes from human osteosarcoma cells [24]. The study focussed on how protein isoforms and post-translational modifications influence the association with distinct complexes. The experiments were performed in the absence of detergent in order to preserve the interactions. In addition, a lysate collected in denaturing conditions was used as negative control. The complexes were separated in a SEC column and 40 fractions were collected. After separation, the different fractions were digested separately and analysed by mass spectrometry. Overall, 8000 proteins were identified and clustered according to the elution profile. Subsequently, protein complexes were assigned to clusters by correlating known interaction to the elution profiles. Interestingly, it was possible to define differential behaviour of various protein isoforms. Authors reported the case of the heterochromatin protein-1 binding protein 3 (HP1BP3), which is essential for the modulation of chromatin functions. This protein has four known isoforms, three of which eluted with similar elution profiles, while isoform 3 migrated differently, indicating it might be part of a distinct complex. Isoform 3 lacks a particular protein interaction domain that is present in the other three isoforms, which might explain why it behaves differently. Similarly, the authors of this study identified posttranslational modifications that alter complex formation. As an example, NUDT5 was identified with 13 peptides, one of them showing a serine phosphorylation. The protein presented two elution peaks, in fractions 21 and 28. Unlike the rest of the peptides, which eluted in both peaks, the phosphorylated peptide was only present in the second peak and co-eluted with a specific complexes. The sequence of the phosphorylation site matched the ATM/ATR kinase consensus motif, suggesting that the phosphorylation of NUTD5 by these kinases is necessary for its interaction with the second complex. As highlighted by the authors, the application of efficient separation strategies, like SEC to PCP, helps to overcome some problems connected to the sensitivity and effectiveness of gradient separations. The improved separation of complexes by SEC, in comparison to sucrose gradients, improves the resolution and specificity of detected complexes. Additionally, clustering of co-eluting proteins with components of known complexes facilitates the detection of new interactors of known complexes.

Recently, Kristensen et al. have described a strategy to study the dynamic of interactome in HeLa cells as a result of epidermal growth factor (EGF) treatment. In this paper they combined the SILAC labelling method for protein quantification with SEC separation and mass spectrometry [25]. The aim of the study was to identify and quantify the changes in protein-protein interactions following stimulation with EGF. The light medium was used to label an internal standard for the identification of proteins in each fraction, while medium and heavy SILAC media were used to label control and treated cells, respectively. The ratio of heavy proteins over medium was used to quantify the dynamic changes in protein interactions following EGF treatment. As expected, the co-elution profile of some proteins represented by their corresponding chromatography peaks in SEC was changed after the treatment. Some of these proteins were bound to different complexes, while others were associated with the same complex but their stoichiometry was altered upon EGF stimulation. About 350 proteins showed a different behaviour compared to the control, among which were a number of well-known components of the EGF signalling pathway.

Beyond the advantages already listed for the general PCP protocols, the combination of PCP with a quantitative method like SILAC is an effective method to track changes in dynamic protein interactions as a result of various perturbations.

In a recent study Havugimana et al. [26] have performed an extensive analysis of the soluble protein interactome in mammalian cell lines (HeLa S3 and HEK293) using a combination of different fractionation methods. Initially, protein complexes were separated via ion exchange chromatography (IEX-HPLC) in non-denaturing conditions (Fig. 18.2c). To study protein interactions that could be disrupted by salt, a second method of fractionation was used in parallel, involving sucrose gradient centrifugation coupled with isoelectric focusing. By combining these methods, 364 previously unannotated complexes were identified and linked to the pathologies they were studying. The quality of the method was assessed by comparing the co-elution profiles of 20 well-known complexes as references. The issue of overlapping profiles and consequential false positives was solved by the development of a computational algorithm that correlated the data with previous functional genomic and evolutionary correlations [27, 28]. This robust computational method improved the reliability of the data by identifying and filtering false positives, which resulted in a high confidence physical interaction network. Using this strategy, an accurate characterization of protein complexes can be achieved. The main concern about the described method is the large amount of fractions collected (over 1000), which was necessary to achieve the desired resolution. In addition, the bioinformatics analysis is challenging. These issues limit the use of this approach for most groups. More recently, a computational method has been developed to facilitate the analysis of protein interaction data generated with PCP. The study combined a SEC approach with SILAC and was aimed at detecting protein complexes altered after infection with *Salmonella enterica*. The authors developed analytical tools which allowed the generation of interaction maps with a single script. Overall, this reduced the time for the analysis from weeks to days and the details of the computational method is described in the paper published by Scott et al. [29]. This method is not specific for SEC and can be applied to other fractionation techniques.

18.4 Cross-Linking

A remarkable number of new methods have been developed to improve the quality of interactome data in terms of sensitivity, reliability and high throughput. While co-immunoprecipitation and co-elution strategies have proven effective for the characterization of protein interactions, both suffer from a loss of weak and transitory interactions. One way to stabilize this type of interaction is to use chemical crosslinking to covalently link weakly interacting proteins. Chemical cross-linkers are molecules capable of generating a covalent bond between two polymers or macromolecules and their use is well established in chemistry. The use of chemical cross-linkers to characterize protein interactions was first reported in the 1970s [30], but what gave a bigger incentive to use this technology more widely were the developments of proteomic approaches based on mass spectrometry. The cross-linking strategy relies on converting protein interactions into strong covalent bonds that become resistant even in denaturing conditions. As such, cross-linking has been an attractive method to investigate weak and transient interactions. As a general principle, a cross-linker is constituted by two reactive groups separated by a spacer. The two groups can react with lateral chains of amino acids that are close to each other, especially thiols, carboxylic acids and amino residues that are more reactive (Fig. 18.3).

There is a large variety of cross-linkers available, each of them with specific features in terms of reactivity and mechanisms of action. One of the most commonly used crosslinking agents is formaldehyde (FA), which can be used not only to fix protein-protein interactions, but also the interaction between proteins and nucleic acids

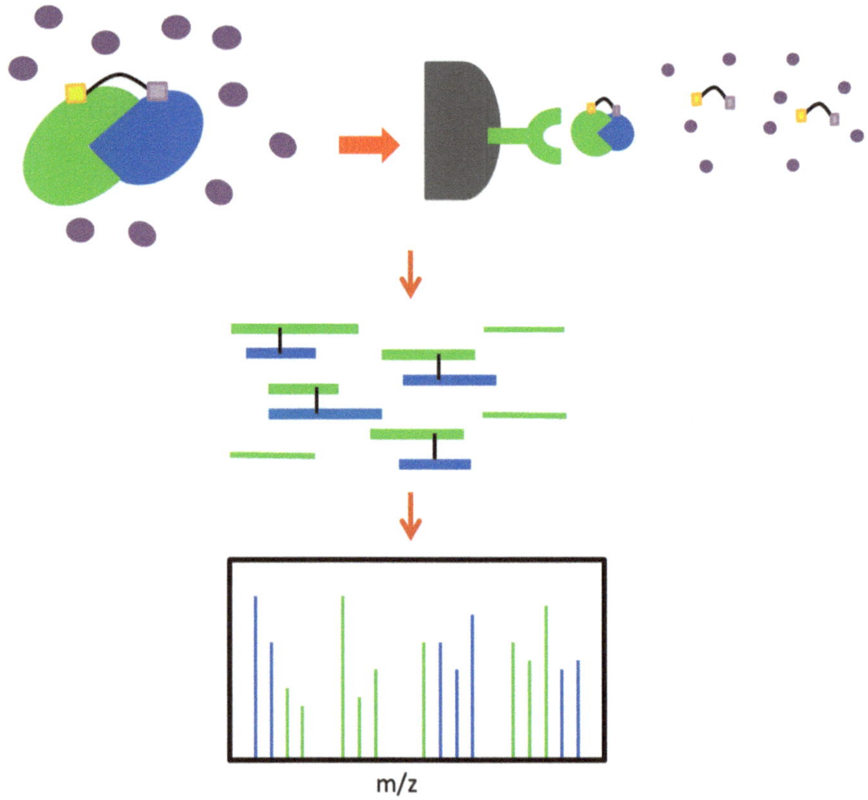

Fig. 18.3 Cross-linking principle Intact cells or cell lysate is incubated with a cross-linker composed of one or two reactive groups separated by a spacer of various lengths. The reactive groups bind the proteins in a complex and generate a covalent bond which stabilises the complexes. Complexes are then isolated by affinity purification, digested and identified by mass spectrometry. Individual, cross-linked peptides can indicate proximity of the peptides in the intact protein complexes, which can suggest a likely complex assembly

[31, 32]. FA has been used in several protocols in order to stabilize weaker and transient interactions prior to affinity purification [33], such as the interactome of a constitutively active form of M-Ras [34]. Myc-tagged M-Ras was expressed in cells and interactions between the bait and its interactors were fixed in cells by FA. After the cross-linking reaction, the proteins were purified by an anti-Myc IP. The analysis showed that the method was able to efficiently purify the tagged bait and its interacting complexes and the authors were able to identify several new interactors. Nevertheless, since the protocol is based on the tagging of the bait, it presents the same limitations previously described. More recently, a rapid immunoprecipitation mass spectrometry of endogenous proteins (RIME) procedure has been developed by Mohammed et al. [35]. The method couples cross-linking using FA with an in-solution digest. In contrast to the previous method, the bait is not tagged or exogenously expressed. Instead, the endogenous bait is enriched with a mixture of antibodies specifically raised against the protein. In this specific study, the interactome of the oestrogen receptor (ER) was analysed in breast cancer cell lines. In addition, SILAC labelling was used to quantify the interaction differences induced by either oestrogen or tamoxifen. The observation that the interaction between GREB1 and ER is predominant in ER^+ specimen and that its expression is decreased in tamoxifen resistant cells lead to speculation on a role of this protein in the hormonal response.

The presented methods are only an example of systematic analysis of protein-protein interactions by cross-linking and it is evident that these strategies are valuable tools to study low abundance proteins and transient interactions. More importantly, the sensitivity of these protocols allows working with a small amount of biological starting material, which makes the analysis feasible for primary cultures. Additionally, cross-linking with FA facilitates the elimination of major contaminant and non-specific interactors due to the attained stability that allows stringent lysis and washes.

Aside from FA, a large number of other cross-linker has been developed over the years, each with different characteristics to take under consideration when planning an experiment. The length of the cross-linker has a significant influence on its capability to perform an efficient reaction. For example, long cross-linkers are often associated with an increased rate of non-specific interactions [36] as longer cross-linker can not only link non-interacting proteins simply proximal to the bait but, due to their size, alter the structure of the linked protein by internal crosslinking. Another factor that influences the cross-linking efficiency is the hydrophobicity and the size of the cross linker itself. These factors have to be considered to find cross-linkers that can permeate the membrane or access the inner surface of protein complexes efficiently. In addition, one main feature to consider is the specificity of the reactive groups on the cross-linker that targets specific amino acid residues [37]. Chemical cross-linking methods also require a lot of optimization, cross-linker concentration and incubation times must all be specifically tested for each experimental set-up and target.

Chemical cross-linking is a good strategy to stabilize weak interactions to help detect low abundance proteins in interaction proteomics studies, but as discussed above, the process of adding a reactive chemical cross-linker to the sample can generate artefacts if the cross-linker is not carefully selected. Photo-cross-linking is an alternative strategy which can overcome some of the limitations and has been recently applied to the study of protein interactions. In contrast to chemical cross-linking strategies, a natural amino acid is replaced by a modified, reactive analogue. In essence, specific amino acids contain a photo-reactive diazirine group, which can be activated upon exposure to ultraviolet light [38] to become a reactive intermediate that covalently bind to an acceptor group within a neighbouring protein. The amino acids most frequently modified are Leucine and Methionine. A good example of photo-cross-linking applied to interaction proteomics is the work of Suchanek et al. [39]. Photo reactive isoleucines, leucines and methionines were introduced into COS7 cells by using engineered tRNA. The expression of the modified amino acids did not interfere with protein biosynthesis and the exposure to UV light did not affect cell viability as assessed by the authors. The new method was used to study protein-protein interactions of a regulatory complex involved in lipid homeostasis. A HA-tagged PGRMC1 and a Myc-tagged Insig1 were expressed in COS7 with or without photo-methionine (photo-Met) and their interaction was validated via SDS-PAGE after both HA or Myc immunoprecipitation. As highlighted by the authors, photo-cross-linking showed the same specificity as the chemical approach, with an advantage over conventional chemical cross-linking in the characterization of large complexes. Since modified amino acids are already part of the protein sequence, photo-crosslinking doesn't have the two major limitations of chemical cross-linking, namely risk of altering the protein structure and accessibility to the core of the protein complex. On the other hand, the specificity may be altered due to the irradiation and duration of photo-activation which could generate unspecific interactions due to reactive intermediates.

In addition to facilitating the detection of interactors, cross-linking can generate further information. Since photo cross-linkers have the ability to connect amino acids in their proximity, they can link residues in separate domains of the same protein. The requirement of physical proximity for cross-linking can help provide structural information [40]. One example of where structural information is extracted using photo cross-linking experiments is the characterization of the RNA polymerase II/TFIIF transcriptional complex in *Saccharomyces cerevisiae* [41]. In this study the authors used bis(sulfosuccinimidyl)suberate (BS3), a cross-linker that reacts with the amine groups of lysines, and were able to identify different linkage sites between the subunits of the large RNA polymerase complex and the transcriptional factors TFIIF. The method provided information about the direct interactions between the two complexes, and identified the regions within TFIIF which directly binds the RNA polymerase surface. This approach is especially suited for studying the native configuration of transcriptional factors complexes that are generally challenging to study. Nevertheless, this method still has some limitation [42]. Although the rate of false positive is low, there is still the risk of generating artefacts. The cross-linking has to be well calibrated in order to decrease the amount of unspecific links. Furthermore, the identification of cross-linked peptides is challenging, despite the emergence of new analytical tools, due to the increased complexity of the linked peptides [42]. In our opinion it would be advantageous to develop cross-linkers which can be reliably split in an ion trap. The linked peptides could then be fragmented individually by MS^3b, which would reduce the complexity while still retaining the information about the linkage. A fragmentable cross-linker, in conjunction with targeted enrichment methods for linked peptides, would allow researchers to determine protein-protein interactions and more importantly, the exact site of interaction.

18.5 Conclusions

While it is clear that proteomics, especially the interactome studies, can play a critical role in determining how pathological events are initiated by molecular events, initial attempts to develop such methodologies have suffered from a number of technical limitations. These limitations have been addressed to a large extent in recent years to make proteomic network mapping reliable, fast and able to cope with dynamic changes in the variety of networks. In this sense, large-scale studies have been carried out to develop new protocols to attempt to define the interactome of several organisms.

Novel protocols to process samples in an easier, faster way in conjunction with a new generation of reliable, sensitive LC-MS/MS platforms has democratized the mass spectrometry based detection of protein complexes. Systematic analysis, which a few years ago were only accessible to large, specialized groups are now within the reach of more applied biological researchers. The possibility to study protein complexes in every laboratory without the need to work with specialized facilities has opened up new opportunities. Initial proteomics studies have already given us a glimpse of how protein interactions link up to build intricate and complex dynamic networks which are difficult, if not impossible, to decipher without these advances. Moreover, due to rapid developments of new methods, per-sample costs have been reduced and can now easily compete with other high and medium throughput proteomics methods. In contrast with other techniques, mass spectrometry-based proteomics still retains the unique ability to assess and quantify what is known, but also to discover new interactions and links in a systematic and unbiased fashion.

References

1. Phizicky EM, Fields S (1995) Protein-protein interactions: methods for detection and analysis. Microbiol Rev 59(1):94–123
2. Fields S, Song O (1989) A novel genetic system to detect protein-protein interactions. Nature 340 (6230):245–246
3. Parrish JR, Gulyas KD, Finley RL Jr (2006) Yeast two-hybrid contributions to interactome mapping. Curr Opin Biotechnol 17(4):387–393
4. Vidalain PO et al (2004) Increasing specificity in high-throughput yeast two-hybrid experiments. Methods 32(4):363–370
5. Gingras AC et al (2007) Analysis of protein complexes using mass spectrometry. Nat Rev Mol Cell Biol 8(8):645–654

6. Chang IF (2006) Mass spectrometry-based proteomic analysis of the epitope-tag affinity purified protein complexes in eukaryotes. Proteomics 6 (23):6158–6166
7. Aebersold R, Mann M (2003) Mass spectrometry-based proteomics. Nature 422(6928):198–207
8. Alvarado R et al (2010) A comparative study of in-gel digestions using microwave and pressure-accelerated technologies. J Biomol Tech 21(3):148–155
9. Rigaut G et al (1999) A generic protein purification method for protein complex characterization and proteome exploration. Nat Biotechnol 17(10):1030–1032
10. Puig O et al (2001) The tandem affinity purification (TAP) method: a general procedure of protein complex purification. Methods 24(3):218–229
11. Rees JS et al (2011) In vivo analysis of proteomes and interactomes using Parallel Affinity Capture (iPAC) coupled to mass spectrometry. Mol Cell Proteomics 10(6):M110.002386
12. Morin X et al (2001) A protein trap strategy to detect GFP-tagged proteins expressed from their endogenous loci in Drosophila. Proc Natl Acad Sci U S A 98(26):15050–15055
13. Couzens AL et al (2013) Protein interaction network of the mammalian Hippo pathway reveals mechanisms of kinase-phosphatase interactions. Sci Signal 6(302):rs15
14. Hubner NC et al (2010) Quantitative proteomics combined with BAC TransgeneOmics reveals in vivo protein interactions. J Cell Biol 189(4):739–754
15. Domon B, Aebersold R (2006) Mass spectrometry and protein analysis. Science 312(5771):212–217
16. Hosp F et al (2015) A double-barrel liquid chromatography-tandem mass spectrometry (LC-MS/MS) system to quantify 96 interactomes per day. Mol Cell Proteomics 14(7):2030–2041
17. Zhu W, Smith JW, Huang CM (2010) Mass spectrometry-based label-free quantitative proteomics. J Biomed Biotechnol 2010:840518
18. Turriziani B et al (2014) On-beads digestion in conjunction with data-dependent mass spectrometry: a shortcut to quantitative and dynamic interaction proteomics. Biology (Basel) 3(2):320–332
19. Selbach M, Mann M (2006) Protein interaction screening by quantitative immunoprecipitation combined with knockdown (QUICK). Nat Methods 3 (12):981–983
20. Ong SE et al (2002) Stable isotope labeling by amino acids in cell culture, SILAC, as a simple and accurate approach to expression proteomics. Mol Cell Proteomics 1(5):376–386
21. Waldrip ZJ et al (2014) A CRISPR-based approach for proteomic analysis of a single genomic locus. Epigenetics 9(9):1207–1211
22. Larance M, Lamond AI (2015) Multidimensional proteomics for cell biology. Nat Rev Mol Cell Biol 16 (5):269–280
23. Andersen JS et al (2003) Proteomic characterization of the human centrosome by protein correlation profiling. Nature 426(6966):570–574
24. Kirkwood KJ et al (2013) Characterization of native protein complexes and protein isoform variation using size-fractionation-based quantitative proteomics. Mol Cell Proteomics 12(12):3851–3873
25. Kristensen AR, Gsponer J, Foster LJ (2012) A high-throughput approach for measuring temporal changes in the interactome. Nat Methods 9(9):907–909
26. Havugimana PC et al (2012) A census of human soluble protein complexes. Cell 150(5):1068–1081
27. Alberts B (1998) The cell as a collection of protein machines: preparing the next generation of molecular biologists. Cell 92(3):291–294
28. Hartwell LH et al (1999) From molecular to modular cell biology. Nature 402(6761 Suppl):C47–C52
29. Scott NE et al (2015) Development of a computational framework for the analysis of protein correlation profiling and spatial proteomics experiments. J Proteomics 118:112–129
30. Clegg C, Hayes D (1974) Identification of neighbouring proteins in the ribosomes of Escherichia coli. A topographical study with the cross-linking reagent dimethyl suberimidate. Eur J Biochem 42 (1):21–28
31. Sutherland BW, Toews J, Kast J (2008) Utility of formaldehyde cross-linking and mass spectrometry in the study of protein-protein interactions. J Mass Spectrom 43(6):699–715
32. Toth J, Biggin MD (2000) The specificity of protein-DNA crosslinking by formaldehyde: in vitro and in drosophila embryos. Nucleic Acids Res 28(2), e4
33. Bousquet-Dubouch MP et al (2009) Affinity purification strategy to capture human endogenous proteasome complexes diversity and to identify proteasome-interacting proteins. Mol Cell Proteomics 8(5):1150–1164
34. Vasilescu J, Guo X, Kast J (2004) Identification of protein-protein interactions using in vivo cross-linking and mass spectrometry. Proteomics 4 (12):3845–3854
35. Mohammed H et al (2013) Endogenous purification reveals GREB1 as a key estrogen receptor regulatory factor. Cell Rep 3(2):342–349
36. Hwang YJ, Granelli J, Lyubovitsky J (2012) Effects of zero-length and non-zero-length cross-linking reagents on the optical spectral properties and structures of collagen hydrogels. ACS Appl Mater Interfaces 4(1):261–267
37. Zybailov BL et al (2013) Large scale chemical cross-linking mass spectrometry perspectives. J Proteomics Bioinform 6(Suppl 2):001
38. Gomes AF, Gozzo FC (2010) Chemical cross-linking with a diazirine photoactivatable cross-linker investigated by MALDI- and ESI-MS/MS. J Mass Spectrom 45(8):892–899

39. Suchanek M, Radzikowska A, Thiele C (2005) Photo-leucine and photo-methionine allow identification of protein-protein interactions in living cells. Nat Methods 2(4):261–267
40. Rappsilber J et al (2000) A generic strategy to analyze the spatial organization of multi-protein complexes by cross-linking and mass spectrometry. Anal Chem 72 (2):267–275
41. Chen ZA et al (2010) Architecture of the RNA polymerase II-TFIIF complex revealed by cross-linking and mass spectrometry. EMBO J 29(4):717–726
42. Rappsilber J (2011) The beginning of a beautiful friendship: cross-linking/mass spectrometry and modelling of proteins and multi-protein complexes. J Struct Biol 173(3):530–540

Protein Structural Analysis via Mass Spectrometry-Based Proteomics

19

Antonio Artigues, Owen W. Nadeau, Mary Ashley Rimmer, Maria T. Villar, Xiuxia Du, Aron W. Fenton, and Gerald M. Carlson

Abstract

Modern mass spectrometry (MS) technologies have provided a versatile platform that can be combined with a large number of techniques to analyze protein structure and dynamics. These techniques include the three detailed in this chapter: (1) hydrogen/deuterium exchange (HDX), (2) limited proteolysis, and (3) chemical crosslinking (CX). HDX relies on the change in mass of a protein upon its dilution into deuterated buffer, which results in varied deuterium content within its backbone amides. Structural information on surface exposed, flexible or disordered linker regions of proteins can be achieved through limited proteolysis, using a variety of proteases and only small extents of digestion. CX refers to the covalent coupling of distinct chemical species and has been used to analyze the structure, function and interactions of proteins by identifying crosslinking sites that are formed by small multi-functional reagents, termed crosslinkers. Each of these MS applications is capable of revealing structural information for proteins when used either with or without other typical high resolution techniques, including NMR and X-ray crystallography.

Keywords

Protein structural analysis • Hydrogen/Deuterium Exchange (HDX) • Limited proteolysis • Chemical Crosslinking (CX)

A. Artigues (✉) • O.W. Nadeau • M.A. Rimmer
M.T. Villar • A.W. Fenton • G.M. Carlson
Department of Biochemistry and Molecular Biology, University of Kansas Medical Center, Kansas City, KS, USA
e-mail: aartigues@kumc.edu

X. Du
Department of Bioinformatics and Genomics, University of North Carolina at Charlotte, Charlotte, NC, USA

H. Mirzaei and M. Carrasco (eds.), *Modern Proteomics – Sample Preparation, Analysis and Practical Applications*, Advances in Experimental Medicine and Biology 919, DOI 10.1007/978-3-319-41448-5_19

19.1 Hydrogen/Deuterium Exchange

19.1.1 Introduction

Protein functions commonly rely on conformational changes within the protein. In some cases these conformational changes include large sections or entire domains of the protein. In other cases, protein conformational changes are restricted to small specific regions of the protein. Extensive conformational changes are associated with protein folding immediately during or after their synthesis *in vivo*, when they fold to acquire their native conformational structure. Knowledge of the location of functionally relevant conformational changes within the protein and the magnitude and rates of conformational interconversion among various protein conformations (i.e. dynamics) are of great importance to the understanding of protein function.

Direct or indirect evidence of protein conformational changes have been deduced through the use of several spectroscopic techniques, including circular dichroism, electron paramagnetic resonance, intrinsic protein fluorescence, UV–vis and IR spectroscopy, and it is not uncommon to use a combination of these techniques to obtain a general description of the structure and dynamics of the protein system under consideration. Measurements of protein dynamics traditionally have been done by determining ^{15}N NMR relaxation times and calculating S2, the average order parameter, a measure of the motion of the N-H vector at peptide amide linkages. Higher order parameters indicate less freedom of movement. Motions measured by these techniques are on the pico- to nano-second time scale but may also indicate if slower motions might be occurring. To fully understand the dynamics of a particular protein, it is desirable to span as wide a time range as possible.

Hydrogen exchange is a well-understood phenomenon, and in conjunction with mass spectrometry (MS) is a useful method for studying protein dynamics and structure. This exchange was first discovered in the early 1950s by Kaj Ulrik Linderstrom-Lang and Aase Hvidt, scientists at the Carlsberg Laboratory in Copenhagen. They discovered that both the polar side chain hydrogens and the peptide group hydrogens undergo continual exchange with the hydrogens from the solvent. Using density gradient tubes, Lang and Hvidt developed a novel method to measure this exchange of amide backbone hydrogens with a heavier isotope, deuterium [1, 2]. With this method, they were able to show that the newly discovered α-helices and β-sheets in native proteins do indeed have the proposed hydrogen-bonded backbone structures [1, 3]. Despite having extremely limited resolution and accuracy at this time, Lang and Hvidt were able to derive equations and propose mechanisms that are still being used today in hydrogen/deuterium exchange (HDX) methodologies [1].

During the following 40 years, many developments were made using hydrogen exchange in conjunction with different techniques, including NMR, tritium gel-filtration, and circular dichroism. Some of these advances include showing that the chemical nature of adjacent side chains has a major effect on the exchange rate [4], measuring the rates of both acid- and base-catalyzed exchange [5], developing protein fragmentation and HPLC separation methods [6], and site-resolved HDX [7]. Finally, in 1991 Katta and Chait showed that HDX could be used with electrospray ionization mass spectrometry, removing many of the limitations associated with applications of HDX, including the size of the protein that can be studied [8]. With the use of MS to analyze the exchange, the use of HDX to study protein structure continues to advance, with the development of faster and more automated software for both analyzing data and running samples [9], and cold boxes for HPLC to maintain low temperatures during injection to avoid back exchange [10]. As a result, the size and type of proteins being studied with HDX, as well as the number of people employing this method, continue to grow.

Recently, HDX in combination with MS has been used to characterize protein movements in solution over a time range from milliseconds to several hours. This technique has become increasingly popular to characterize conformational changes and the dynamic transitions between the conformations in proteins. The purpose of this chapter is to describe the procedures and methods for HDX MS to novice users. Thus, we will first describe a basic methodology and a simple experimental set up. Then, we will discuss alternative workflows, caveats and potential problems, and complementary techniques.

HDX MS is a method in which deuterium atoms present in buffer replace hydrogen atoms in the protein [11–16]. Of all the hydrogen atoms present in a protein polypeptide, only hydrogen atoms in O-H, N-H and S-H groups can be replaced with deuterium atoms present in the buffer. As a further limitation, only those present in the amide linkages can be measured by HDX MS (all other hydrogens exchange too rapidly during sample handling to be detected by mass spectrometry). The amino acid sequences of peptides (and thereby their locations in the protein) and their mass (and thereby the identification of which peptides undergo deuterium exchange) are detected by enzymatic (most often pepsin) digestion of the protein into peptides and peptide mass evaluation by MS. The total number of exchangeable protons and the rate of exchange events are both dependent on the equilibrated protein conformational average and the rate of dynamic transitions between conformations. Therefore, HDX MS is a sensitive technique for evaluating both changes in average conformation of the peptide backbone chain and changes in its dynamics.

A number of attributes of HDX MS make it ideal for evaluating macromolecular systems:

1. Mass spectrometry requires low concentrations of protein. This can remove some of the ambiguity at higher protein concentrations (such as those required for many NMR study).
2. Deuterium-labeling of a protein results in the introduction of multiple reporting groups (one reporter/protein residue) with minimum structural modification of the protein. This results in higher resolution information compared to many other techniques.
3. The number of exchanging protons can be determined. Each proton that is exchanged for a deuteron adds 1 atomic mass unit (amu) to the average molecular mass of a protein. Thus, the increase in the mass determines the number of deuterium incorporated.
4. By observing the isotopic pattern for a given protein or peptide fragment (discussed in detail below), HDX MS can distinguish between localized unfolding events (referred to as EX2 kinetics and seen as a binomial isotope pattern) and more global, or cooperative unfolding events (referred to as EX1 kinetics and seen as a bimodal isotopic pattern).
5. There is no upper limit to the size of the macromolecule that can be analyzed by HDX MS analysis. This is due to the fact that for detailed analysis of deuterium content in specific regions (peptides) of the protein, the protein is proteolyzed before mass analysis.
6. Measurements are for proteins in solution with no dependency on crystal growth, as is required for X-ray crystallography.
7. As mentioned above, protein dynamics can be probed on a much longer time scale than is accessible with many other techniques (e.g. NMR relaxation). HDX MS can probe dynamics ranging from milliseconds to several hours, and perhaps longer. As a result, HDX MS can increase significantly the overall description of dynamic motions within a protein.

19.1.2 Theoretical Basis and Experimental Design for HDX MS

The theory and methodology used to study protein conformation and dynamics using HDX MS have been described in several reviews [12, 16–20]. In the absence of secondary structure restraints, HDX for a specific polypeptide is dependent on the temperature and pH of the reaction. The most common experimental procedure for HDX is *continuous labeling*. In this method, the exchange is initiated by making a large dilution of a concentrated stock of the protein into deuterated buffer. The progress of the exchange reaction is sampled at different times. Under these conditions, the chemistry and the thermodynamic parameters of HDX are well established [21–23]. The rate of HDX at the protein amide linkages is acid or base catalyzed, and can be expressed as follows:

$$k_{hdx} = k_H[H] + k_{OH}[OH] \qquad (19.1)$$

Thus, the rate of HDX for a specific polypeptide is dependent on the pH and temperature of the reaction. This rate, as determined experimentally, has a minimum in the pH range 2.3–2.5. Figure 19.1 illustrates the theoretical rates of HDX for rat liver mitochondrial aspartate aminotransferase, a 49,000 Da globular protein, in the absence of secondary structure restraints, calculated at 0 °C and at both pH 7.5 and 2.3. At pH 7.5 HDX is very fast ($t_{1/2} = 0.014$ min) and the exchange is completed almost instantly. However, there is a minimum exchange rate at pH 2.4. At this pH, minutes are required before complete exchange occurs. This sensitivity of exchange rates to pH requires careful control of pH during exchange. However, the same pH sensitivity provides a tool to quench exchange by quickly lowering temperature and pH, a step necessary during mass analysis.

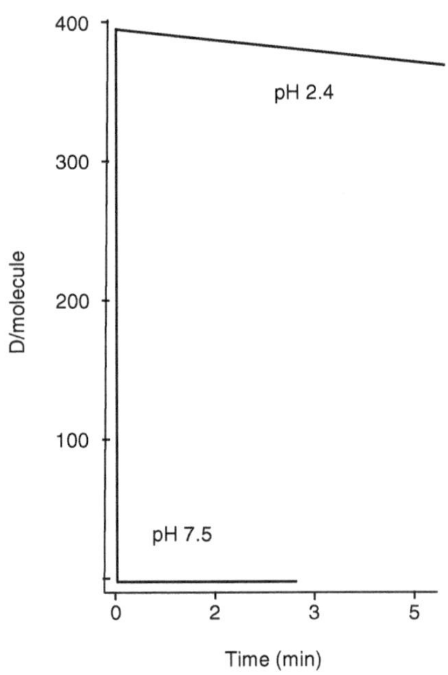

Fig. 19.1 *Theoretical rates of hydrogen/deuterium exchange of mitochondrial aspartate aminotransferase.* The theoretical rate of HDX at 0 °C and pH 7.5 or 2.4 was calculated for mitochondrial aspartate aminotransferase (MW 44,597 Da) according to a previously published algorithm [22] using HXPEP, written and kindly provided by Zhongqi Zhang (Amgen, Thousand Oaks, CA)

In the absence of any structural constraints, the hydrogen atoms of solvent exposed amide linkages exchange at their free, unmodified rates. However, if the amide hydrogen atoms are involved in stable internal hydrogen-bonding, or are not exposed to solvent, they will exchange more slowly. In native proteins, the local differences in these rates are due to the fact that the structure of these molecules is not rigid, but has a certain degree of mobility. This mobility has been called "breathing", and can be visualized as shown in Fig. 19.2. The kinetics of HDX can be described according to the following kinetic equation:

$$k_{hdx} = \frac{k_{cl}k_{op}}{k_{cl} + k_{op} + k_e} \qquad (19.2)$$

where k_{cl}, k_{op} and k_e are the constants of closing, opening and chemical hydrogen/deuterium exchange, respectively. For proteins in their native state, a common assumption is that $k_{cl} >> k_{op}$ and $k_e >> k_{op}$.

Depending on the relative values of the kinetic constants, two extreme kinetic behaviors can be found. When $k_{cl} << k_e$, the exchange rate is determined by the first order rate constant k_{op}. Thus, k_{hdx} is dependent exclusively on the conformation of the protein. This first extreme behavior is defined as EX1. EX1 kinetics are rarely observed. However, EX1 exchange can be observed under experimental conditions that favor the unfolded state [24, 25] of proteins (high temperature or in the presence of chaotropic or

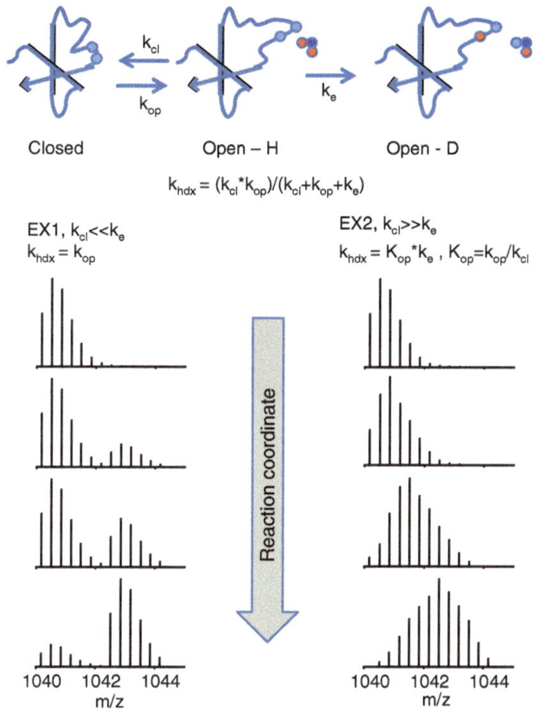

Fig. 19.2 *Schematic representation of the mechanism of hydrogen/deuterium exchange.* Hydrogen atoms in the peptide backbone (*top panel*) can exchange with hydrogen (*blue*) or deuterium (*red*) atoms in water in a process dependent on accessibility and breathing (*opening and closing*) of the protein. In the EX1 regime (*left panel*), opening is faster than closing ($k_{op}>> k_{cl}$), and the rate of exchange is determined by the rate constant of opening. In the EX2 regime (*right panel*), closing is faster than opening ($k_{cl}>> k_{op}$), and the reaction is dependent on the rate of opening and chemical exchange. The isotopic patterns shows the theoretical exchange pattern of a triply charged peptide with m/z = 1040.08 under EX1 or EX2 exchange regimes

unfolding reagents). On a mass spectrometer, EX1 is characterized by a binomial transition from one mass (i.e., undeuterated) to the final (deuterated) species (Fig. 19.2). In other words, two isotopic envelopes are detected, one for the undeuterated peptide-ion and a second one for the fully deuterated ion. The relative intensity of these two isotopic envelopes changes over time as the exchange reaction proceeds.

In contrast to the conditions that define EX1, when $k_{cl} >> k_e$, the k_{hdx} is second order and depends exclusively on the factors determining the chemical hydrogen/deuterium exchange. In this case the rate of exchange is measured by $k_{hdx} = \mathrm{K}_{op}k_e$; and $\mathrm{K_{op}} = k_{op}/k_{cl}$. This second extreme behavior is defined as EX2. The EX2 mechanism is most commonly observed for proteins in the folded state. EX2 behavior is characterized by a monotonic change of the isotopic envelope with the progress of the exchange reaction (Fig. 19.2). The EX1 kinetic mechanism reflects the activation energy for segmental opening and the EX2 represents the sum of all energies of opening and proton transfer. In EX2, the free energy difference (ΔG^0) of the opening event can be described according to the following equation:

$$\Delta G^0 = -RT\ln K_{op} = -RT\ln(k_{hdx}/k_e) \quad (19.3)$$

where K_{op} is the equilibrium constant of the opening/closing reaction (Fig. 19.2).

Based on these concepts, most HDX MS experimental designs rely on two different stages [14]: exchange and quenching. In the first stage, reaction conditions (i.e., pH and temperature) are designed to allow HDX while the protein undergoes normal folding/function. In the second stage, the HDX is quenched by rapidly decreasing the temperature (to 0 °C or below) and pH (to pH 2.3–2.5). Deuterium content in the protein is then analyzed by mass spectrometry.

19.1.3 Equipment

- Cooling HPLC interface. To reduce back exchange during mass analysis of the intact protein or its peptides, all experimental steps after HDX are performed at low pH and temperature. The simplest instrumental set-up consists of immersing the solvents, columns and all parts of an HPLC in an ice bath, or enclosing the entire HPLC set-up in a refrigerated chamber. For better control of temperature, we designed a Semi-Automatic Interface for Deuterium Exchange (SAIDE, Fig. 19.3) [10]. This interface consists of a TVC –S2 box (Mecour) equipped with a 6-port valve (Cheminert, N60 SS) and a 4-port valve (Cheminert, C2). The 6-port valve is equipped with a through-the-handle

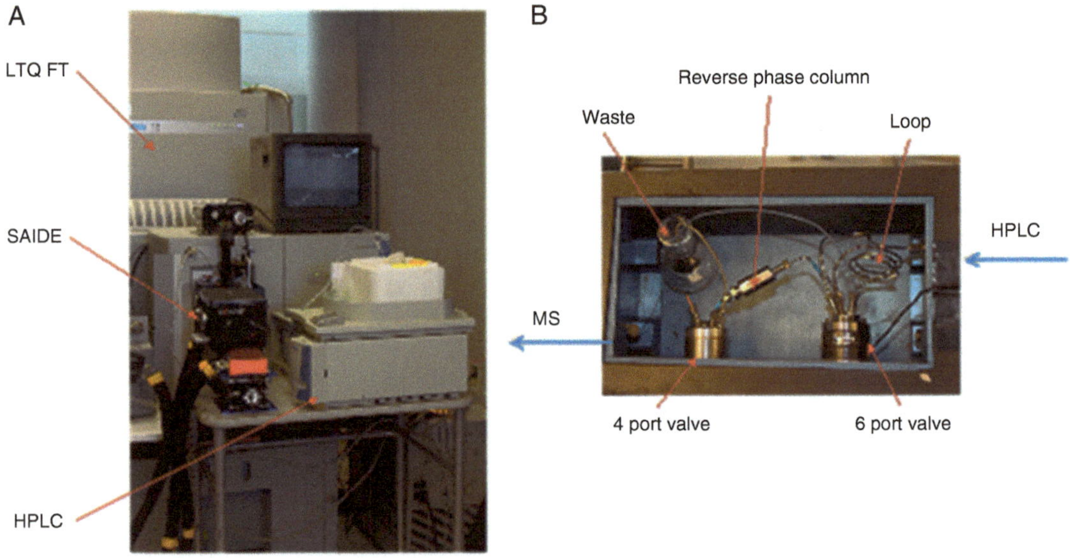

Fig. 19.3 *Mass spectrometer rigged for HDX MS*. (**A**) The cooling box (SAIDE) is located right before the ESI source of a high resolving mass spectrometer (LTQ FT) and after the HPLC pumps (HPLC). (**B**) Detail of the SAIDE box showing the internal components of the unit: two valves, one reverse phase column, loop and fluid lines. The box is used for temperature control during all stages of protein digestion, peptide desalting and chromatographic elution of peptides

external loop injector and holds the sample loop (10 μL). The sample loop acts as the reaction vessel during protease digestion. The reversed phase column bridges the two valves, and the 4-port valve directs flow to either waste or to the mass spectrometer. Other specialized equipment is available that performs automatic sample pick up, mixing, injection and data acquisition, although at a considerable expense [18].

- High performance liquid chromatograph (HPLC). The system should be able to deliver flows between 20 and 50 μL/min. We use a quaternary HPLC MS pump (ThermoFisher Scientific).
- Chromatographic columns. A reverse phase C8 (MicroTech Scientific, Zorbax C8 SB Wide Pore Guard Column 2.5 cm × 0.2 cm) is needed to desalt the protein when measuring global rate of the exchange in the intact protein. As an alternative, a reverse phase C4 may be required to desalt highly hydrophobic proteins (MicroTech Scientific, Zorbax C4 SB Wide Pore Guard Column 2.5 cm × 0.2 cm). A reverse phase C18 column (MicroTech Scientific, Zorbax C18 SB Wide Pore Guard Column 2.5 cm × 0.2 cm) is needed to resolve peptic peptides and identify regions with deuterium incorporation.
- Mass spectrometer. The mass spectrometers useful for HDX MS characterization of macromolecular complexes are Tandem Mass Spectrometers. That is, those that allow for at least two different stages of mass analysis: one to scan for the peptide-ions (parent ions) present in the sample, and the second to scan for the fragment ions produced after a specific parent ion has undergone a stage of fragmentation (see Section II: Mass Spectrometry). A high resolving power mass spectrometer, such as an FT-ICR or Orbitrap, is recommended. However, other mass spectrometers with lower resolving power have been used. Because of the high flows used for peptide elution, the ESI tip must be chosen carefully. A 100 μm ID tip with an opening of 30 μm has proved to be ideal for our experimental set-up.

19.1.4 Materials

- Protein or protein system of interest.
- Pepsin. Make a pepsin (Worthington) stock solution by diluting an appropriate weighed amount of pepsin and dilute it in 200 mM ammonium formate, pH 2.3 at a final protein concentration of 1.6 mg/ml. Pepsin concentration can be estimated from its absorbance at 280 nm using a 1 % absorptivity coefficient of 1.4.
- Protein stock buffer. A buffer appropriate for your particular protein system.
- Deuteration buffers. Buffers adequate for your protein system made in D_2O (99.9 % D_2O, ACROS Organics). Note that a correction factor must be introduced when measuring pH of deuterated buffers to account for the differences in activity of protium *vs.* deuterium: pH = pD + 0.4.
- Quench buffer. Quench buffer is 200 mM NH_4CH_3COOH, pH 2.3, ice cold. Other buffer composition can be used (ammonium phosphate). Note in some cases it might be required to supplement the quench buffer with a low amount of a denaturing or chaotropic agent (i.e., 0.6 M guanidine hydrochloride) to achieve full unfolding of the protein and efficient pepsin digestion.
- HPLC solvents. Two solvents are needed to create a gradient. Solvent A is 0.05 % trifluoroacetic acid in H_2O (TFA, MS grade). Solvent B is 0.05 % TFA acid in acetonitrile.

19.1.5 Experimental Procedure

Figure 19.4A outlines the procedures involved in a *continuous labeling experiment*. Usually, a stock solution of the protonated protein is diluted into a deuterated buffer and the direction of the exchange is H→D (*on-exchange*). Figure 19.4B outlines the reverse procedure (*off-exchange*), when a protein is first fully exchanged with deuterium, and the exchange reaction proceeds in the opposite direction. This method has been used to study the reversible unfolding of a protein. The procedure outlined below describes the steps involved in a *continuous labeling, on-exchange* procedure. Other experimental procedures are possible, however, and the particular design will depend on the question of interest.

19.1.5.1 Initiate Exchange Reaction

(A) The exchange reaction is initiated by making a 1:10 (or higher) dilution of a concentrated stock solution of the protein or protein system of interest into a buffer made in D_2O

(B) At different time points, the exchange reaction is sampled by taking an aliquot. Two mass measurements can be made: the global rate of exchange in the intact protein (see Sect. 19.1.5.2. Global rate of exchange) or rate of exchange in pepsin generated peptides (see Sect. 19.1.5.3 Location of deuterium exchange along the peptide backbone).

19.1.5.2 Global Rate of Exchange

To obtain a global rate of exchange, the change in mass of the protein at different times following the initiation of the exchange reaction is measured. For the measurement of the mass of the intact protein, mass analysis is performed by direct injection of an aliquot of the labeling reaction mixture on a C4 or C8 nano-column. Following desalting at 0–15 % B at high flow, the protein is eluted using a step gradient of acetonitrile (0–60 % B in B + A in 15 min) and analyzed on-line by mass spectrometry.

19.1.5.3 Location of Deuterium Exchange Along the Peptide Backbone

To identify the residues involved in the hydrogen/deuterium exchange reaction, it is first required to identify the peptides resulting from the proteolysis of the protein. This first stage is performed under control conditions; that is, in the absence of deuterium in the buffers but under identical conditions to be used to measure the exchange. This results in a list of peptides of

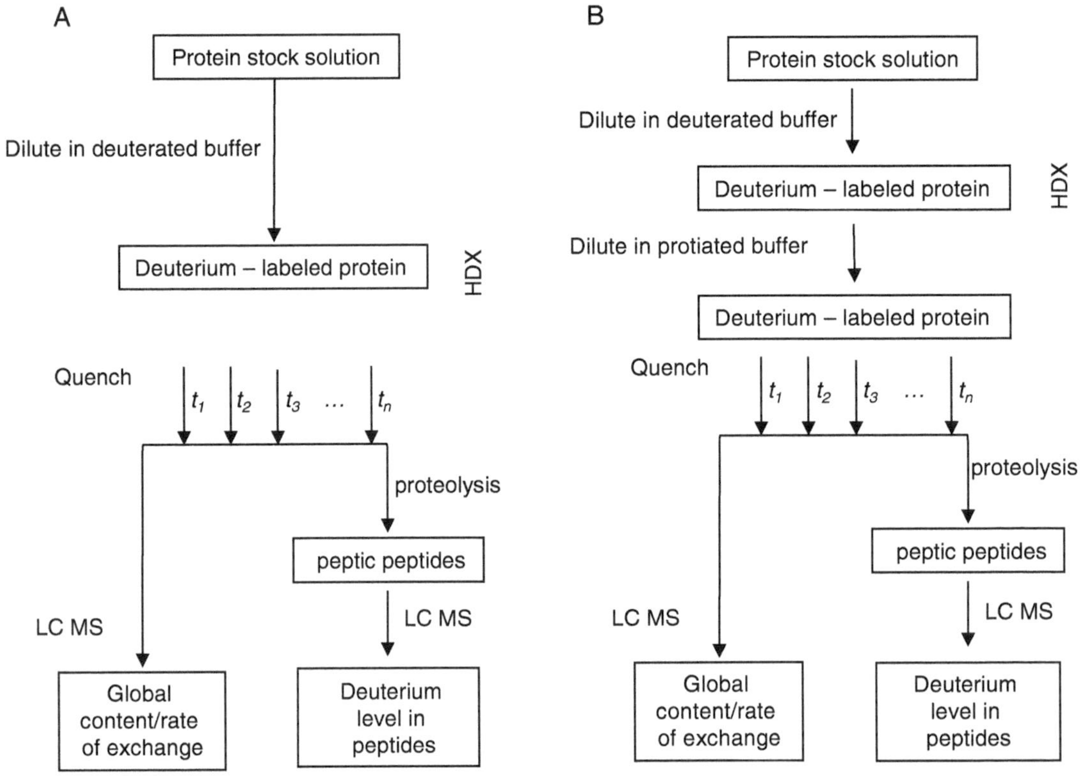

Fig. 19.4 *HDX MS general experimental procedure.* The scheme shows the steps to perform *continuous labeling* HDX *on-exchange* (**A**) or *off-exchange* (**B**) experimental procedures. HDX is initiated by making a dilution of a concentrated stock of the protein into a deuterated buffer. At different time points the reaction is sampled by taking an aliquot and measuring the mass of the intact protein (global exchange) or of the proteolytic fragments (deuterium level in peptides) with the aid of a mass spectrometer

interest. Then, the experiment is repeated under the exchange conditions using deuterated buffers.

Peptic mass maps

(a) The first step is to make a dilution (1:10–1:20) of the protein stock in the protonated buffer. This dilution is equivalent to the dilution that will be made later in deuterated buffer to initiate the exchange reaction.
(b) Peptic digestions of the protein are performed by making a second 1:10 dilution of an aliquot of the protein in ice cold 200 mM ammonium formate (pH 2.3) containing pepsin at a final protein:protease ratio of 1:1 (w:w). Note that the ratio protein:protease must be optimized experimentally.
(c) Inject the reaction sample immediately into the loop of the 6-port valve on the SAIDE interface.
(d) Allow pepsin digestion to proceed for 2–5 min (time of digestion must be optimized for each protein).
(e) Switch the 6-port valve, start HPLC gradient. The resulting enzyme digest is desalted on a C18 nano-column at 75–100 μL/min for 2 min while the flow on the 4-port valve is diverted to waste.
(f) Following desalting, switch the 4-port valve to direct the flow to the mass spectrometer ESI source for peptide detection.

(g) Elute peptides using a 2–40 % gradient of 0.05 % TFA acetonitrile in 0.05 % TFA in 15 min. The peptides are detected on-line using a high resolving power mass spectrometer. Figure 19.5 shows a representative elution profile of a peptide digest using our chromatographic system. The MS settings should be been optimized for detection of peptides using high flow mobile phase. Data are acquired under automatic control to perform MS followed by tandem mass scans of the four to six most intense ions, using an exclusion list of 2–4 min, depending on the capabilities of your mass spectrometer and chromatographic system.

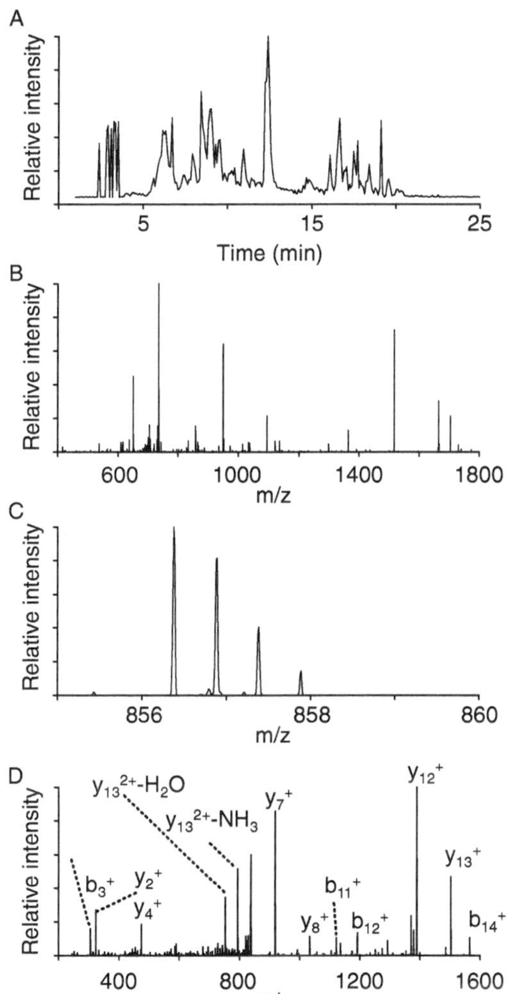

Fig. 19.5 *Representative chromatographic profile and data analysis* (**A**) Base line chromatographic profile of a peptide digest of mitochondrial aspartate aminotransferase. (**B**) Mass spectrum at 8.5 min of elution. (**C**) A magnification of the mass spectrum of panel B, showing the doubly charged ion with m/z of 856.5 corresponding to the peptic peptide AHNPTGTDPTEEEWK. (**D**) Tandem mass spectrum of the same peptide; to simplify the figure, only the most prominent *b*- and *y*- ions are indicated

Measurement of deuterium content in peptides

(a) The exchange reaction is initiated as indicated above, using deuterated buffer instead of the protonated buffer.

(b) At different times during the exchange reaction, remove an aliquot and dilute it in the quench buffer in the presence of pepsin, as before (see Sect. 19.1.5.3.A. b–g).

(c) MS analysis is performed as above with the exception that the mass spectrometer is operated to perform mass analysis only (no MS/MS).

19.1.6 Data Analysis

19.1.6.1 Peptide Identification

When working with pure proteins, as is the case in HDX MS, statistical tools for False Discovery Rate (FDR) and peptide/protein probabilities calculation are, as a general rule, not useful. Instead, peptide identification is based on parameters that rely on the quality of the tandem mass spectra. When data are acquired on a high resolving power mass spectrometer and Proteome Discoverer is used to analyze them, peptide identifications are made using an in-house protein database. This database includes the protein of interest, pepsin and common contaminant protein sequences. The database is made assuming that pepsin has no specificity, using a fragment ion mass tolerance of 20 ppm, and a parent ion tolerance of 0.30 Da. Peptide identifications are accepted if they can be established at Xcorr score of at least 1.5, 2.0 or 2.5 for peptides with 1, 2 or 3 charges, respectively, with a ΔCorrelation score larger than 0.08. Note, manual inspection and validation of some tandem mass spectra may be required. See Chap. 14 for more information on tandem mass spectrometry peptide/protein sequencing and identification.

19.1.6.2 Deuterium Content

The change in deuterium content is measured as the change in mass of the deuterated and undeuterated averaged masses of the protein. Many software packages can be used, and usually the instrument manufacturer will provide a program to obtain this measurement. Specialized software is recommended. HDExaminer (Sierra Analytics) is a commercial software that performs automatic isotopic envelope isolation, measurement of the average mass and deuterium content of the peptides, and can plot the results in a variety of formats, including the comparison of multiple states of a protein. There are, however, several free tools for the same purpose: HDXFinder [26], HD desktop [27] and its successor HDX Workbench [9], HX Express [28], Hexicon [29, 30] and MagTran [31], among others.

19.1.6.3 Mathematical Analysis

A. Curve fitting – Eq. 19.2 describes the exchange reaction for a single amide linkage. In theory, one could expect one phase per amide linkage. However, in practice, multiple protons in the peptide might exchange and individual rate constants of exchange cannot be measured. In practice, the exchange reaction is fitted to an exponential rise (*on-exchange*) or decay (*off-exchange*):

$$D_t = \sum_{i=1}^{n} A_i\left(1 - e^{-k_i t}\right) \qquad (19.4)$$

Where D_t is the deuterium content at time t, A_i and k_i are the amplitude and the rate constant for the ith phase. In practice, multiple HDX reactions are grouped into fast, medium and slow phases (n = 3).

B. Use of overlapping peptides – Because pepsin has low selectivity for cleavage site, pepsin digestion results in the production of multiple overlapping peptides. Statistical and logical analysis of the deuterium content of these overlapping peptides can provide higher spatial resolution than that obtained at the peptide level. Some of the programs mentioned above will apply logical restrictions and will provide a value for the amount of deuterium incorporated/retained in smaller units than obtained at the peptide level.

C. Additional considerations – When calculating the total number of exchanged H/D, one must keep several things in mind. (1) Any HDX at the N-terminal end of the peptide is lost during proteolysis. (2) Previous studies have demonstrated that any HDX at the second amide linkage is also lost very quickly during the chromatographic step [22, 32]. (3) Proline in peptic bonds does not have an exchangeable proton at its amide linkage.

19.1.7 Alternative Workflows

As mentioned above, the generic experimental protocol outlined in Fig. 19.4 can be modified to fit specific questions. In most cases these require additional equipment. For example, manual mixing, as indicated in the protocol outlined above, allows the measurement of deuterium content after the first few seconds of exchange (10 s), but exchange reactions that occur below that threshold cannot be measured. For rapid mixing and quenching of the reaction in the time range below seconds, a quench flow instrument is required. In this situation, *quench flow* in combination with HDX MS has been used to access these very fast rates of exchange of enzymes during catalysis [33]. In *pulse labeling* experiments, an additional pump is used to expose briefly the protein sample to a pulse of deuterium and quench it quickly. This method has been used to study intermediates of folding pathways of proteins [34–36].

Most HDX MS studies make use of *in-solution* pepsin digestion. However, immobilized pepsin columns have been shown to improve digestion efficiency [37, 38]. In some cases there is too much back exchange, rendering the data unusable. Care must be taken on the choice of support used to conjugate the protein [39].

19.1.8 Complementary Methodology

It has been observed that ESI of proteins in an unfolded state will produce higher charged envelopes than those produced by ESI of proteins in native conditions. This indicates that the protein ions in gas phase retain some of the structure that the protein had in solution, thus the charge distribution of the protein ions is an indication of the global structure of the protein. This is thought to be a consequence of the higher exposure of potentially charged residues that are otherwise protected in the core of the protein in the native state.

To obtain higher spatial resolution it would be necessary to interpret the tandem mass spectrum. However, due to the low energy of fragmentation of CID, this fragmentation method results in scrambling of deuterium among the resulting fragment ions [40–46]. Thus, the CID mass spectrum of these peptides cannot be used to determine the position of the deuterium in amide linkages. The development of ETD, a more energetic method of fragmentation, results in the efficient fragmentation of peptides with little or no scrambling and interpretation of the tandem mass spectra of these peptides results in amino acid resolution.

In *hydroxyl radical labeling* [47], a protein solution is exposed briefly to oxidative conditions. This results in oxidative modifications of solvent exposed amino acid side chains. This can be achieved by either chemical reaction using Fenton chemistry [48] or by UV cleavage of hydrogen peroxide in fast photo oxidation of proteins (FPOP) [49]. The appearance of covalently modified amino acid residues with oxygen can be identified by tandem mass spectrometry following trypsin digestion. When interpreting these data, it is important to keep in mind that reactivity of individual amino acid residues is determined not only by their accessibility to solvent but also by their individual reactivity. The reactivity of amino acid side chains is as follows: Cys > Met > Trp > Tyr > Phe > Cystine > His > Leu ~ Ile > Arg ~ Lys ~ Val > Ser ~ Thr ~ Pro > Gln ~ Glu > Asp ~ Asn > Ala > Gly [47]. For detailed discussion of this methodology see the reviews by Chance [50] and Konermann [51].

19.1.9 Problems and Caveats

19.1.9.1 Back Exchange

A primary concern in mass analysis is the loss of deuterium during sample handling for mass analysis. Reducing pH to quench exchange requires the addition of acid. This quenching results in a reduced exchange rate, not a complete absence of exchange. Furthermore, reduced pH also exposes the now deuterated protein to additional protons. Also, the deuterated protein is further exposed to protonated buffer during the HPLC stage of desalting/peptide separation. Therefore, deuterons can be replaced with buffer protons during data acquisition steps in a process known as back-exchange. In order to minimize loss of deuterium, mass measurement must be taken quickly, usually within the first few minutes following quenching. Despite efforts to work quickly, the back exchange of side chains is too rapid to be assessed with normal mass spectrometry methodologies and is the reason that HDX MS is limited to detecting information about the peptide backbone.

In most cases, two states of the protein are compared (control and experimental condition). Thus, assuming that the experimental conditions are maintained constant for each state, the differences in both total deuterium content and/or rate constants in identical peptides are used to describe different states of the protein. However, if a fully deuterated form of the protein is available, the following equation can be used to correct for the loss of deuterium during the analytical stages [13]:

$$D_t = (m_t - m_H)/(m_D - m_H) \qquad (19.5)$$

where D_t is the content of deuterium at time t, m_t is the average mass at time t, m_H the mass of the undeuterated peptide and m_D the mass of the fully deuterated peptide.

19.1.9.2 Overlapping Peptides

To reduce back exchange, peptides are eluted using sharp gradients. In most cases there are only 30 min for data collection after quenching of the exchange reaction, which includes protease digestion, desalting and peptide separation. Moreover, the use of an enzyme with low selectivity results in the co-elution of multiple peptides. The isotopic envelopes of these peptides are changing in shape and average mass as the exchange reaction proceeds. This often results in the overlapping of peptide isotopic envelopes. Most software applications resolve this problem by either extracting ion chromatograms (HDFinder) or by curve fitting a theoretical envelope to the experimental data (HDExaminer, HD Desktop). The use of high resolving power mass spectrometer alleviates this problem. However, each peptide assignment must be validated individually.

19.1.9.3 Spatial Resolution

The spatial resolution of HDX MS detected with simple mass measurements is at the peptide level. Most HDX MS studies published to-date have been made using this mode of operation. As a result, such experimental designs provide medium resolution, i.e. deuterium content is measured at the peptide level. To gain more spatial resolution using this experimental design, multiple overlapping peptides are required and deuterium assignment content is provided by logical analysis of these multiple overlapping peptides. However, this is not always possible, since certain regions of the protein may not produce the necessary number of overlapping peptides to obtain the degree of resolution desired.

19.2 Limited Proteolysis

19.2.1 Introduction

The development of the concept of "limited proteolysis" is widely attributed to work from the Linderstrom-Lang laboratory in the 1940s [3]. Among other studies, his laboratory demonstrated that proteins could be "enzymatically modified without serious degradation" by restricting proteolysis [52]. Subsequently, the Neurath laboratory also made extensive use of this technique to study the structure of proteins [53, 54]. Unlike the complete proteolysis that is normally used for mass spectrometry, limited proteolysis refers to proteolysis that is halted by some means, so that complete degradation of the protein does not occur (see Sect. 19.2.3.2 for details on quenching proteolysis). Limited, controlled, *in vitro* proteolysis is a simple, but powerful, tool to study the conformation of proteins.

Proteases have a variety of specificities, i.e., residues at which they preferentially cleave. This specificity controls the sites of cleavage based on the primary structure of proteins not showing higher order structure. With the added dimension of folding, however, the normal specificities of proteases are no longer the only factor dictating cleavage location. Secondary structure will obscure sites from proteases, regardless of exposure, as will any additional structure that hides regions within folds or causes stereochemical constraints [55]. Accessibility to the protease active site by the protein target becomes more restricted upon folding, thus the structure of the substrate contributes to the selectivity of the protease.

As an experimental technique, limited proteolysis was initially used to cleave larger proteins or complexes into separate domains to study them individually. It was first used to probe protein structure by Neurath in 1980, when he observed that most globular proteins were relatively resistant to proteolysis until they were denatured [53]. He proposed that, as with other enzymatic reactions, optimal proteolytic activity occurred when there was complete complementarity between the substrate structure and the active site of the protease. The ability of the protease to cleave the substrate also depends on the location of a potential cleavage site within the structure, as only solvent exposed regions will be accessible in a tightly folded protein. Neurath proposed a model in which functional domains of proteins are tightly packed, and therefore relatively protease resistant, whereas linker regions or loops are surface exposed and more susceptible to proteolysis [53, 54]. Using crystal structures and limited proteolysis to confirm correlations between flexibility and cleavage, this model became the basis of limited proteolysis theory: that is, limited proteolysis occurs only at sites on a protein's structure that are solvent accessible and flexible enough to conform and fit within the active site of the protease [56–59]. Generally, this solvent accessibility and flexibility occurs at specific region(s) of a protein; so that even when multiple proteases with different specificities are used, the cleavage sites are clustered together, although not necessarily with cleavage at the same residue [55].

Because protease specificity still plays a role in determining cleavage sites, it is important to use proteases with broad specificities, along with multiple proteases with differing specificities. This will ensure that the regions being targeted reflect their exposure in the tertiary structure, rather than their primary structure. Therefore, it is also imperative to maintain the protein's higher order structure. When planning and executing an experiment, it is essential to keep in mind the basic premise of limited proteolysis: brief proteolysis of surface exposed regions while maintaining the protein core. Because proteolysis of a protein can cause conformational changes, it should not be allowed to proceed for too long, as regions that were not originally surface exposed may become so, causing results to be skewed. If the protein core becomes compromised, information about the structure is no longer reliable.

Limited proteolysis was initially analyzed using SDS-PAGE and Edman degradation; however, with the development of MS to study proteins in the late 1980s [60, 61], it became the preferred method of analysis for limited proteolysis. MS has allowed the applications and capabilities of limited proteolysis to greatly increase. With the use of MS, it is now possible to easily identify the exact sites where proteolysis occurs, providing a map of the regions cleaved by the brief proteolysis, allowing for detailed identification of the flexible and surface exposed regions. Unlike NMR or crystallography, MS requires only a minimal amount of protein to obtain structural information, and the ratio of protein to protease is key, rather than the absolute amount of either. Limited proteolysis and MS can also be used on proteins of any size, as there are no minimum or maximum protein size restrictions. It can be used on single-domain proteins, multi-domain proteins, multi-subunit proteins, *etc.* Another advantage of limited proteolysis/MS is the ability of MS to analyze complex mixtures [62, 63].

19.2.2 Limited Proteolysis Applications

Limited proteolysis can be used to study different aspects of protein structure, several of which are described below. Because surface accessibility and flexibility are required for proteolysis to occur, the most obvious application of limited proteolysis is the identification of exposed loops and disordered regions. By employing proteases of different specificities and limiting proteolysis, while maintaining the protein core, it is possible to map exposed loops and identify regions of disorder. This can be used to complement NMR or crystallography data [64, 65], or even to replace these techniques if they cannot be used on the protein of interest. Crystallography can be especially difficult for disordered or dynamic regions, as it results in low resolution. Limited proteolysis can be used to confirm the disorder and dynamic properties of these regions [66, 67].

Likewise, as first proposed by Neurath, multidomain proteins often have flexible and disordered linker segments joining the domains, and these will be preferentially cleaved during partial proteolysis [57, 68]. Therefore, identification of domains and their exact boundaries is possible. This separation of domains was one of the first applications of limited proteolysis, as seen in several early papers [54, 69, 70]. More recently, this application has been used in conjunction with MS for the specific identification of linker regions. For example, applying these techniques to the *E. coli* transcriptional activator protein NtrC, a protein with three separate domains, Bantscheff et al. [57] developed a system combining limited proteolysis, MS, and SDS-PAGE to identify two flexible linker regions. Limited proteolysis can also be used to cleave flexible linker regions to produce separate domains, making feasible the study of single domains and potential folding intermediates [71].

Another application of limited proteolysis is the study of complexes formed between proteins and their targets. This is possible because the interface regions of the complex will initially be solvent accessible on the surface of the protein, but become protected when the complex forms. Therefore, by first performing limited proteolysis on an isolated protein and then on the protein in complex, it is possible to identify the interface regions, although regions affected by conformational changes prompted by the interaction may also show changes in the level of protection. Different peptide maps for the two protein states, free and in complex, will be observed by MS following the limited proteolysis. An example of this approach is the study of the calmodulin-melittin complex [72]. The authors performed limited proteolysis on free calmodulin, free melittin, and the calmodulin-melittin complex, observing different peptide maps for the free proteins *vs.* the proteins in complex. From the regions that changed, they designed a model showing the sites of interactions between melittin and calmodulin. A similar application of limited proteolysis to study protein complexes is to identify regions of protein-DNA, protein-RNA interactions, and antibody epitope identification [73–75].

Regardless of the experimental design – identifying domain linkers, mapping exposed loops, or interactions – another use of limited proteolysis is comparing changes in those regions upon protein activation, mutagenesis, or ligand binding. In these cases, the limited proteolysis of the protein in its basal state is compared to that of the altered protein. If there are conformational changes occurring on the surface of the protein, the resultant peptide maps can show regions of differential proteolysis, indicating they are more or less flexible or exposed.

19.2.3 Methodology

19.2.3.1 Optimization

The most basic rule to keep in mind when designing and executing a limited proteolysis experiment is that the protein core must remain intact, or it is no longer "limited" proteolysis, and information about the protein structure may no longer be valid. Because this is so essential, experiments must be performed under conditions that maintain the stability and structure of the protein being studied, regardless of the optimal conditions for the proteases being used.

Because it is important to ensure that the higher order structure, and not the protease specificity, dictates the sites of cleavage, it is advisable to use multiple proteases with varying specificities and some with broad specificities. This means, however, that the individual proteases will most likely not be cleaving under their optimal conditions (pH, temperature, *etc.*). Given that maintaining target protein stability is the most important factor, one must first identify conditions that are optimal for protein stability. This will include conditions such as buffer, pH, temperature, and duration of proteolysis. Once these conditions are determined, the concentration of proteases required for sufficient, yet limited proteolysis, can be optimized. Because sub-optimal conditions will undoubtedly be used for some of the proteases, it will likely be necessary to use different ratios of protein to protease for each protease in order to ensure similar levels of proteolysis with minimal cleavage. Examples of this are shown in Table 19.1.

Another important experimental variable to optimize is the quenching step, because different proteases may be typically inhibited differently. The ideal quenching step, however, is one that can

Table 19.1 Protease specificities and final concentrations[a]

Protease	Specificity	Kinase: protease ratio
Thermolysin	Hydrophobic	15:1
Chymotrypsin	Aromatic	2000:1
Protease V8 (Glu C)	Asp and Glu	150:1
Trypsin	Arg and Lys	5000:1
Ficin	Nonspecific	10,000:1
Arg C (Clostripain)	Arg	10:1
Lys C	Lys	50:1
Papain	Nonspecific	10,000:1
Proteinase A	Nonspecific	100:1
Subtilisin	Nonspecific	200,000:1
Pepsin	Aromatic, acidic, hydrophobic	10:1

[a]Different proteases can be, and should be, used in limited proteolysis experiments. Listed above are examples of proteases and the protein:protease ratios that were used in limited proteolysis experiments at pH 6.8 on the glycogenolytic enzyme phosphorylase kinase [76]. While these ratios will likely differ for other proteins, these are reasonable starting points for optimization. Other proteases that are commonly used include Proteinase K, elastase, and Asp-N

be used for all proteases in the study. If more than one quenching method is used, it should be shown that neither the results nor the protein are affected. Finally, as discussed further in Sect. 19.2.3.2, quenching must be both rapid and complete.

19.2.3.2 Quenching of Proteolysis

For reproducibility and to avoid too much proteolysis, it is important to ensure hydrolysis is quenched effectively. In ideal quenching conditions, proteases should be stopped instantly and completely. The requirement for instant protease deactivation excludes many protease inhibitors that inhibit chemically, e.g., active-site-directed affinity labels, as they may act relatively slowly. Quenching by changing conditions, such as pH, can be useful; however, if the quenching pH must be altered prior to analysis, the possibility for renewed proteolysis must be considered. Often quenching is achieved by adding trifluoroacetic acid or acetonitrile, although protein precipitation may occur. Denaturants can also be used to quench; however, some proteases still show residual activity in the presence of denaturants. The denatured protein that is being studied will likely be an even better substrate for proteolysis than its native counterpart. When analysis is performed by MALDI, proteolysis has sometimes been quenched by addition of the matrix solution or by pipetting an aliquot of the hydrolysis mixture directly onto the plate [66, 73]. The bottom line is that whatever quenching condition one chooses to employ, it is imperative to experimentally test it to establish with certainty that quenching does, in fact, occur.

19.2.3.3 Mass Spectrometry

MALDI and ESI MS are both capable of analyzing limited proteolysis data. MALDI-MS is tolerant of buffers and does not require desalting the samples, both desirable features. ESI-MS does require desalting, but chromatographic separation of complex mixtures allows for sequencing of more peptides, particularly desirable in complex mixtures.

19.2.3.4 Peptide Identification

Given that limited proteolysis is typically used on purified, known proteins, the use of standard protocols, which employ probabilities and false discovery rates, is not essential. Peptide identification in limited proteolysis is similar to that used in HDX-MS (Sect. 19.1.6.1) and general peptide identification is discussed in more detail in Chap. 14. Typically, a region will be targeted, rather than a specific residue, and if different proteases with different specificities are used, it is likely there will be overlapping peptides covering the same region. This indicates consistency of the data and the flexibility and exposure of that region. Proteolysis will likely result in sub-digestions, i.e., after a region has been initially cleaved, the protease may continue to act on that peptide, resulting in multiple smaller peptides from the same region. These sub-digestions can be ignored in favor of the longer peptides that cover the same region. In fact, by considering sub-digestions cautiously, one can avoid over-interpreting the putative importance of specific residues within the larger region that encompasses them.

19.2.3.5 General Protocol

(A) Proteolysis – Incubate protein with protease at the optimized ratio determined previously (Sect. 19.2.3.1) under conditions (buffer, pH, and temperature) best suited for protein stability
(B) Quenching – After incubation for appropriate time(s), remove aliquot and quench reaction (Sect. 19.2.3.2)
(C) MS – Prepare samples following protocol established for the MS to be used. Be aware of maintaining quenched conditions, so as not to resume proteolysis. Keep all peptides for analysis. See Sect. 19.2.5 for discussion on peptide release.

19.2.4 Representative Results and Data Presentation

Organization and presentation of data are largely dependent on the main point of the experiment, the type of experiment performed, and the protein(s) involved. Table 19.2 and Fig. 19.6 show several possible ways to present results.

19.2.5 Caveats Concerning Limited Proteolysis

A possibility that is not often considered is whether all peptides formed during limited proteolysis are actually released from the parent protein following quenching. This is not an important concern when product analysis is carried out by MALDI, as all peptides should be observed; however, the binding of proteolyzed peptides may be a concern with other analytical methods, as some peptides could be missing in the final product analysis. The non-covalent binding of otherwise free peptides by a proteolyzed parent has been observed with the protein phosphorylase kinase, a 1.3 MDa complex of multiple subunits. Following selective chymotryptic hydrolysis of its largest subunit (to the extent that no remaining trace of it was observed on SDS-PAGE), there were only small changes in the structure of the proteolyzed parent as observed by electron microscopy [77], despite the fact that the degraded subunit accounts for 43 % of that parent complex's total mass. Consequently, evaluating a variety of conditions for the quenching of proteolysis, or between proteolysis and the removal of remaining parent protein prior to analysis, could prove advantageous in assuring maximum release of generated peptides. Note also that if the parent protein is precipitated prior to analysis, peptides derived from it could be trapped within the precipitant.

A caveat that was discussed in Sect. 19.2.3.4 is the production of smaller peptides from the sub-digestion of initially released larger parent peptides, which may potentially produce peptides too small to detect. If a proteolysis time-course is run, these sub-digestion peptides are likely to be observed later than their parent peptides. A time course can also show the later secondary appearance of less readily cleaved peptides from different regions of the protein. A caveat concerning interpretation of the appearance of these unique secondary peptides is that, instead of representing regions less readily cleaved, they could also represent a new conformation of the protein induced by an initial proteolysis event. A new proteolytically induced conformational change is especially problematic for proteins whose function is controlled by so-called intrasteric regulation [78] (i.e., a region of primary structure in the protein is autoregulatory through its interaction with other regions of the protein, generally the active site)

Table 19.2 Representative data[a]

Trypsin		Pepsin		Arg C	
WT	Mutant	WT	Mutant	WT	Mutant
1–20		1–23			
83–90	83–90	83–89	83–89	83–90	83–90
150–161	150–161	150–158	150–158	150–161	150–161

[a]When comparing mutants, activated proteins, or complex formation, it is necessary that all conformers or states are included in the table in a format that makes comparison easy. Because using multiple proteases is advisable, shown here is a table in which the results from different proteases are compared side by side for the different conformers of protein. The titles (wild type and mutant) can be exchanged for non-activated *vs.* activated protein forms, complexed protein *vs.* free, *etc*

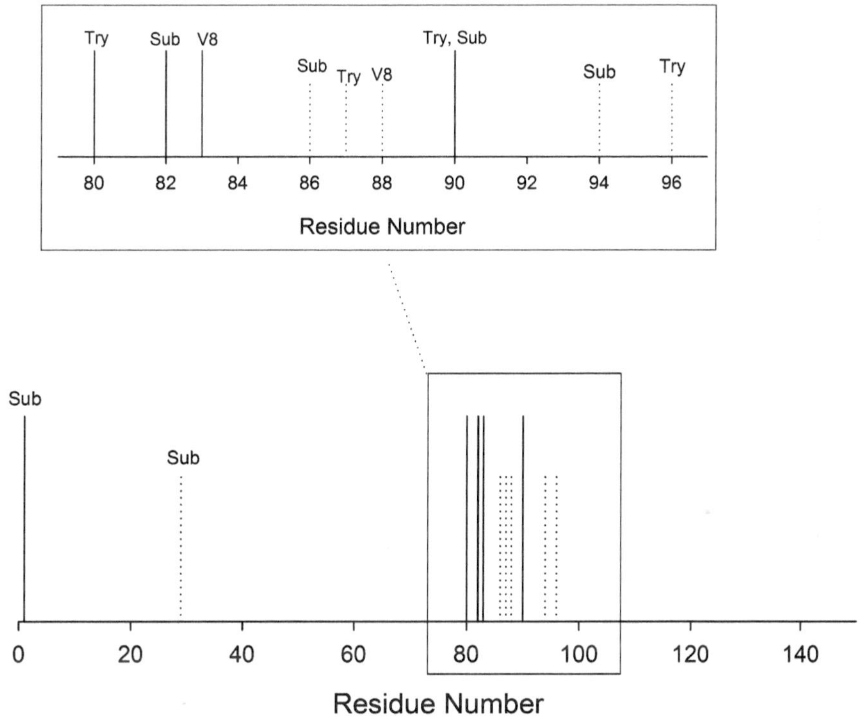

Fig. 19.6 *Representative results.* When mapping exposed loops and regions of disorder, it is helpful to visually demonstrate the protein structure and sites of cleavage. Demonstrated here is a way to conveniently show regions that are targeted by various proteases. This figure also demonstrates clearly that different proteases are targeting the same region, further implying flexibility and exposure. Depending on the size of the protein, the line representing residues could be substituted for the actual sequence. Alternatively, if the protein is too large, the representative residue lines used in this figure may more clearly portray the results, and regions that are cleaved can be magnified to highlight cleavage details

[79]. For many of these proteins, the auto-regulation can be overcome by limited proteolysis, resulting in a new conformation with a different activity. Thus, an important control to include in limited proteolysis experiments is the determination of functional changes following proteolysis. This concern also suggests that keeping the extent of proteolysis relatively limited is prudent.

19.2.6 Side Chain Modification as a Complementary Technique to Partial Proteolysis

Historically the goal of this method has been to identify relatively reactive nucleophilic amino acid side chains that are accessible to the electrophilic reagent used to covalently modify them. Thus, the residues modified are likely to be on the surface of the protein and could be within, or adjacent to, the exposed loops implicated by partial proteolysis. Identification of modified residues can, therefore, corroborate results from partial proteolysis. Over the years, more complex methods of side chain modification having a considerably wider range of amino acid targets, such as oxidation by hydroxyl radicals [80–82], have been developed, but the underlying idea of preferentially modifying surface residues remains the same. An increase in the variety of side chains that can be modified does, however, add greatly to the power of the technique, making it complementary to HDX. Unlike HDX, however, the covalent modifications are irreversible, potentially facilitating analysis.

The general method of side chain modification could reasonably be called chemical or protein "footprinting". Historically the term footprinting has connoted protection of DNA chains from cleavage by DNA-binding proteins. Similarly, the term "protein footprinting" has been used to denote cleavage of a protein at specific residues subsequent to its modification by a chemical reagent [83, 84]. The same term has also been used, however, to describe the analyses through side chain modification of nearly every characteristic of proteins (structure, dynamics, binding, etc.) with cleavage occurring after modification prior to MS analyses. Consequently, to avoid potential confusion in terminology, we call this approach side chain modification, rather than protein footprinting.

There are few variables in carrying out side chain modification experiments: choices of modifying reagent and of modifying conditions (time, pH and concentration of modifier with respect to protein). The conditions used will affect the rate, and perhaps the extent, of modification, and deciding on which conditions to use is an empirical process. One wants to obtain a reasonable amount of modification in a reasonable amount of time; however, what represents a reasonable amount of modification is not always obvious. Certainly, enough product should be formed to be able to argue that it truly represents the conformation of a large population of the native protein as opposed to the conformations of minor components produced by denaturation, oxidation, post-translational modification, or minor proteolysis during protein purification. On the other hand, one doesn't want so much modification that the conformation of the protein could be altered by the modifications or the conditions employed to modify it. Consequently control analyses should be carried out to characterize the properties of the protein following modification. Evaluating full retentions of biological function and the higher order structure of the protein after modification are two necessary controls. Many studies do not address the extent of modification, nor its reproducibility. The latter is necessary to assure that similar results are obtained with multiple independent protein preparations. If one is comparing two conformations, e.g. apo-protein *vs.* ligand-bound, misleading information is less likely if modifications of both are kept in the linear phase.

19.3 Crosslinking

19.3.1 Introduction

Chemical crosslinking refers to the covalent coupling of separate functional moieties. This technique has been used for over 50 years to analyze the structure, function and interactions of proteins by identifying crosslinking sites formed by small multi-functional reagents, termed crosslinkers. The coupling of protein crosslinking with modern MS techniques (CXMS) has led to resurgence in the field, with new instruments and crosslinking technologies being developed to facilitate identification of conjugates (crosslinked proteins and/or peptides) from ever smaller amounts (nmole to pmole) of sample. CXMS is a bottom-up approach, in that the protein is first crosslinked and then digested specifically with proteases to generate peptides for detection by MS. A limiting factor in the analysis of proteins by CXMS is the extensive array of products (including many side products) that are possible from such digests. These product arrays are too complex to be annotated manually and require the use of search engines that have been developed specifically to identify cross-linked peptides. Our intent in this chapter is to expose novice users to: (*A*) CXMS approaches that minimize the generation of side products and maximize structurally useful conjugates, (*B*) available conventional, mass and affinity tag crosslinking reagents, and (*C*) search engine technologies for identifying conjugates.

19.3.1.1 Advantages and Applications

Crosslinking provides low to medium structural information for proteins that are not amenable to high resolution techniques such as NMR and X-ray crystallography. It is a versatile technique that, in its simplest form, has been used to determine nearest neighbors and the minimal subunit stoichiometry in multi-oligomeric complexes [85]. And in its more complex from in combination with Western blotting, immuno-precipitation, various protein labeling methods, top-down MS and CXMS approaches, it has been used successfully to study protein-protein interactions (PPI) in transient and stable identification of crosslinked amino acid side chains. CX sites may be complexes [86], providing maximum distance information for these targets in both *in vitro* and *in vivo* studies (reviewed in [87, 88]). Recent advances in the detection of peptides from complex mixtures by modern MS and supporting search engine technologies have provided a robust platform for the development of CXMS and its primary use in the identification of crosslinked peptides from digests of crosslinked proteins. CXMS provides a range of structural information, and the resolution of this information is dependent on how specifically a crosslinking (CX) site can be localized. Identification of crosslinking sites which provids the highest structural resolution requires the used to determine the proximity of domains and amino acid side chains in protein monomers or complexes, to identify potential intra or intermolecular protein binding sites, and to provide structural constraints for theoretical protein models [89–91]. Many search algorithms and specialized reagents have been developed to enrich and enhance the detection of conjugates and more numerous side products from the digests of crosslinked proteins [90, 92–94], making this approach readily accessible to researcher with access to MS and proteomics facilities.

19.3.1.2 Chemistry of Crosslinking

Crosslinking Reagents

The range of structural information gained from CXMS is inherently dependent upon the type of cross-linking reagent (CXR) used. The largest and most commonly used classes of CXRs are bifunctional molecules containing two reactive groups that are connected by an intervening spacer group. Bifunctional CXRs are further divided into two subgroups, based on whether they contain identical (homobifunctional) or different (heterobifunctional) reactive groups. Many different reactive groups with varying chemistries have been incorporated into CXRs (Table 19.3). However, there are five functional groups that are commonly used, because they react with protein side chains in aqueous solutions at near physiological pH [85]. N-hydroxysuccinimide (NHS) and imidoester groups react preferentially with the N-termini of proteins, as well as the pyrrole and ε-amines of histidine and lysine, respectively. Sulfo-derivatives of the NHS group are also available to increase the solubility of CXRs with large hydrophobic spacer groups. Maleimide and alkyl halide groups are targeted primarily by the free thiols of cysteine. As opposed to the functional groups above, aryl azides are promiscuous, and upon exposure to UV, insert non-selectively as nitrenes at active hydrogen-carbon bonds or undergo ring expansion to form dehydroazepines [87], which react both with nucleophiles and active hydrogen-containing species.

Table 19.3 Selected reactive groups of typical crosslinking reagents

Reactive group chemical structure	Group name	Amino acid preferentially targeted
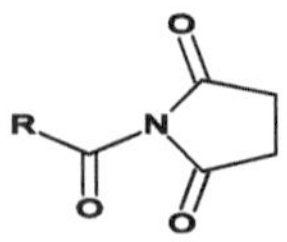	N-Hydroxysuccinimide ester (NHS)	Lysine
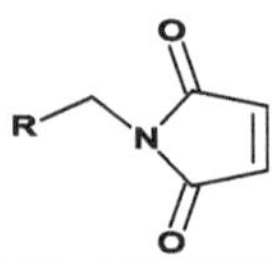	Maleimide	Cysteine
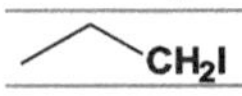	Alkylhalide	Cysteine
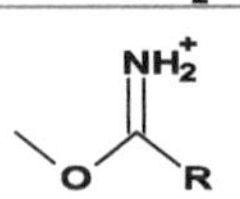	Imidoester	Lysine
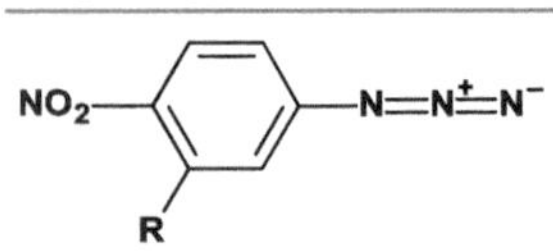	Phenylazide	Non-specific
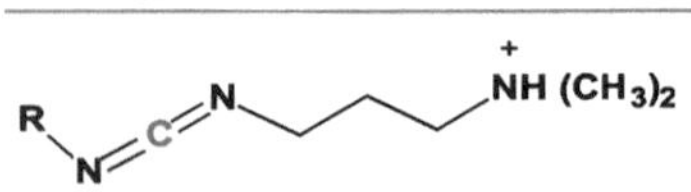	Carbodiimide[b]	Aspartic and Glutamic acid

[a]R denotes spacer and second reactive group, except for the carbodiimide
[b]Zero-length crosslinking reagent that activates carboxyl groups for subsequent attack by proximal amines

Spacers or linkers are chemical moieties that covalently couple the reactive functional groups of a crosslinker. Besides determining the distance between the reactive groups, spacers also influence the geometry of crosslinking and the solubility of the CXR. CXRs with long alkyl spacers are generally hydrophobic and cover a broad range of crosslinking distances between potential nucleophiles due to the flexibility of the linker. Spacers also contain functional groups that allow for their cleavage by specific reagents, such as periodate or DTT, which cleave intervening glycol or disulfide groups, respectively. Crosslinkers that contain these groups are members of a subclass of bifunctional reagents, termed cleavable CXRs. In addition to chemical cleavage sites, CXRs with spacers containing MS-cleavable functional groups have been developed to facilitate bond breaking by collision-induced dissociation (CID) and/or electron transfer dissociation (ETD) in mass spectrometers. Such reagents are used as reporter groups to aid in the identification of crosslinked peptides from complex mixtures [95, 96]. Spacers comprising affinity tags such as biotin and Click chemistry labels are employed to enrich low abundant conjugates [97, 98], and even more complex forms that contain both affinity and mass tags have been synthesized to simultaneously enhance enrichment and identification of crosslinked peptides [99]. CXRs containing functional spacers are often identified as trifunctional or multifunctional reagents; however, the term trifunctional also refers to CXRs that contain three reactive groups that emanate from a central spacer group or atom, each of which is capable of reacting with three distinct sites on protein targets.

Zero-length CXRs refer to molecules that directly couple amino acid side chains without an intervening spacer. These reagents generally modify and activate functional groups on specific side chains for subsequent attack by an adjacent protein nucleophile, such as the ε-amine of lysine. For example, N-substituted carbodiimides react with the carboxylates of Asp and Glu residues to form acylisourea intermediates that facilitate the formation of isoamide bonds with proximal lysine residues (Table 19.3). Free thiols may also target these reactive intermediates to form thioester linkages; however, these conjugates are relatively unstable by comparison with the corresponding amide linkage. For complete and thorough reviews of crosslinking reagents see the works of Wong and Hermanson [87, 88].

Proteins as Reactants

Proteins as reactants add to the complexity of products generated in cross-linking reactions, because they are polyvalent structures, containing multiple reactive amino acid side chains with varying reactivities that are dependent upon their microenvironments in the protein complex. The microenvironment of an amino acid depends on the dynamics of the region encompassing the location of that amino acid in the tertiary structure, its solvent accessibility, and its interactions with and chemical composition of its nearest neighbors. On the basis of hydrophobicity, amino acids may be divided into two major classes, nonpolar and polar. Polar residues can be separated into those containing side chains with nonionizable (asparagine, glutamine, serine and threonine) and ionizable (histidine, lysine, arginine, tyrosine, cysteine, aspartate and glutamate) functional groups. With the occasional exception of tryptophan, the latter group is primarily targeted by CXRs.

Products of Crosslinking

As previously mentioned, crosslinking and subsequent digestion of a protein and/or protein complexes generate a vast array of products that must be accounted for to detect crosslinked peptides. Figure 19.7 shows examples of the types of products that are typically observed when two proteins are treated with a bifunctional crosslinker. In addition to crosslinking between the two proteins (intermolecular) and within each protein (intramolecular), numerous mono-modifications occur as well. Moreover, crosslinking is a continuous process, and if not carefully controlled, results in the formation of multiple protein conjugates, progressing from crosslinked dimers to large polymeric arrays. Subsequent digestion of the crosslinked proteins significantly increases the number of possible products, particularly if the CXR targets side chains that are also substrates of the protease used which results in incomplete digestion of the crosslinked protein targets [100]. Estimates suggest that the number of potential peptide products from such digests increases exponentially with the size of the protein [101] necessitating the use of bioinformatics approaches to annotate all possible products.

Data Analysis

For two reasons, analysis of CXMS data is not trivial and requires dedicated software tools. The first is that the number of candidates that must be considered is enormous in comparison to regular proteome-wide peptide analyses. The second is that the abundance of crosslinked proteins is much lower than that of non-modified proteins, and the data analysis algorithm must be sufficiently sensitive to identify small signal peaks amongst dominating neighboring peaks.

A number of software tools have been developed in the past decade for CXMS data analysis. In the following sections, we will explain the basic data analysis principles, look into the computational algorithms behind these tools, examine their pros and cons, and finally provide our perspectives on future development of data analysis algorithms and software tools for CXMS analysis.

Fig. 19.7 Products of protein crosslinking

19.3.2 Methodology

Crosslinking is a specialized form of general protein chemical modification, both of which are empirical processes. It is simply impossible to predict under which conditions and with which CXR a given protein will undergo crosslinking. Variables such as time, reaction component concentrations, pH and CXRs must be screened to maximize the specificity and selectivity of crosslinking. Specificity refers to the preferential stable modification of a protein side chain functional group by a specific class of CXR reactive group. Selectivity on the other hand, denotes the potential for detecting observed protein interactions by crosslinking. Both of these parameters are inter-related and the extent to which one is controlled significantly affects the other. Ultimately, successful crosslinking of proteins to obtain maximum yields of a desired conjugate depends on these two factors. Crosslinking is the first step in the CXMS pathway to identifying CX sites in any protein or complex of interest. Optimization of subsequent proteolysis and detection steps is also critical and the corresponding protocols, instrumentations, and software will be discussed in the following sections.

19.3.2.1 pH

Most CXRs contain electrophilic reactive groups that are targeted by protein nucleophiles in reactions. These reactions generally involve either displacement of a leaving group or direct addition to a double bond with adjacent electron withdrawing groups on the CXR to form a covalent bond between it and the attacking amino acid side-chain. In terms of Lewis acid–base theory, the reactivity of an amino acid side chain is directly related to the nucleophilicity (or electron-pair donating capability) of its side chain functional group, which in turn can be expressed in terms of the ratio of its electron donor/base (A^-) and electron acceptor/acid (HA) forms in solution. This ratio can be estimated theoretically using the Henderson-Hasselback equation, which implies mathematically that for a nucleophile to exist equally in its conjugate base and acid forms, the pH value must equal its pK_a.

$$\mathrm{pH} = \mathrm{pK_a} + \log([\mathrm{A}^-]/[\mathrm{HA}])$$

For one and two unit increases in pH, the percentage of the basic form increases correspondingly from 50 to 95 and 99 %, respectively. Thus at alkaline pH values, the nucleophilicity for basic $R\text{-}S^-$ and $R\text{-}NH_2$ protein nucleophiles is greater than their corresponding acid forms ($R\text{-}SH$ and $R\text{-}NH_3^+$) at low pH values.

The relative order of nucleophilicity for protein functional groups involved in crosslinking reactions is: $R\text{-}S^- > R\text{-}NH_2 > R\text{-}COO^- \cong R\text{-}O^-$. With the exception of zero-length crosslinkers, most conventional, commercially available CXRs are designed to react preferentially with thiolate or amine-containing protein nucleophiles. Examining the range of theoretical pKa's for cysteine (8.8–9.1) and lysine (9.3–9.5) [87], one might conclude that they are poor nucleophiles at neutral pH. However, in the microenvironments of proteins, these side chains are often reactive and covalently modified by CXRs. Thus optimization of pH is critical in controlling the outcome of crosslinking. For example, crosslinking at high pH values might seem prudent to increase the reactivity of amino acid side chains; however, it also significantly diminishes the selectivity of a CXR for its intended target and may diminish the specificity of crosslinking by increasing unwanted side reactions and the formation of large conjugates, rendering the results uninterpretable. Moreover, hydrolysis of many CXR reactive groups increases significantly and competes with crosslinking at high pH values, generating excessive dead-end modifications. A general rule of thumb is that pH and all other variables in the crosslinking reaction should be adjusted through screening to maximize the formation of detectable desirable low mass conjugates.

19.3.2.2 Uses of Conventional and Mass/Affinity Tag CXRs

In the following section, conventional CXRs are defined as those not containing mass tag/reporter and/or affinity tags. Because crosslinking is an empirical process, a CXR is generally chosen for a protein target from screens of reagents with multiple chemistries under multiple conditions. That having been said, there are many commercially available CXRs with properties that are advantageous for specific types of analyses. Zero to short (2–3 Å) length CXRs are preferable for detecting protein interactions, in that their product conjugates are more likely to represent an actual interaction within or between protein targets, i.e. the specificity of crosslinking is maximized. For such analyses, conventional or specialized mass/affinity tag-containing reagents with large crosslinking spans (>2–3 Å) should be avoided. In low resolution crosslinking experiments in which the identification of one or more binding partners is sufficient, longer span CXRs with affinity or mass tags are more advantageous, simply because they generally increase the likelihood of isolating and/or detecting crosslinked products. Hydrophobic, water insoluble CXRs are typically used for screening protein interactions that are stabilized by hydrophobic binding surfaces, whereas hydrophilic, water soluble, reagents are often employed for labeling charged, solvent accessible residues on the surfaces of proteins. Homobifunctional CXRs are used primarily in one-step crosslinking reactions, in which all components are present in the reaction. Heterobifunctional reagents containing two chemically distinct functional groups are often exploited for use in two-step crosslinking experiments. In such experiments, a protein target is first modified under conditions that favor the reactivity of one functional group, followed by purification of the labeled complex to remove non-covalently bound reagent and to facilitate its exchange into conditions that favor reaction of the second CXR functional group. For example, CXRs containing both photo- and thermo-reactive functional groups (Table 19.1) are typically used in these reactions, with the protein first labeled with the thermo-reactive group and then purified in the dark, following which the modified complex is exposed to UV radiation to activate and promote crosslinking by the photoactive group.

Some specialized CXRs contain affinity or mass tags. Affinity tagged crosslinked proteins are enriched using affinity purification media. Mass tagged crosslinked proteins, on the other hand, generate peptides with unique isotopic signatures that aid in the detection and identification of crosslinks and dead-end side modifications. For the most part these reagents use the same chemistries as conventional crosslinkers, most of which incorporate NHS groups to target lysine ε-amines. Many strategies have been introduced to follow sequentially labeled precursor ions (ionized intact crosslinked peptide) and their collision products with mass/affinity tag combination CXRs created to reduce the complexity of the product pool and to facilitate cleavage of both peptide arms of the crosslink. Several notable examples include CLIP [98], which utilizes a *bis*-NHS CXR, with a terminal Click alkyne tag for enrichment (using biotin azide) and an NO_2 reporter group that both enhances water solubility and acts as a neutral loss reporter during MS-induced fragmentation. Several groups have developed CXRs that fragment during MS/MS to release small molecules that provide mass signatures for crosslinked peptides [95], termed protein interaction reporters [102]. Using a different ligation approach, Trnka and Burlingame synthesized a novel CXR, diformyl ethynlbenzene (DEB), which forms Schiff bases with lysine ε-amines that are subsequently reduced to secondary amines with cyanoborohydride [91]. The authors demonstrate that reduction to an amine, rather than an acetylation product formed by NHS groups, provides two additional protonation centers. Additionally, incorporation of the DEB an intervening rigid ring spacer, decreases the m/z ratio of the conjugate for more optimal fragmentation by high resolution ETD and electron capture dissociation (ECD), providing more complete fragmentation along the peptide backbone. Moreover, the reagent contains a clickable moiety for addition of affinity or mass tags for purification or generation of diagnostic reporter ions during MS/MS fragmentation. Digestion of DEB crosslinked proteins also generates high charge state gas phase precursor ions (4–6^+), which allows for their exclusion from native and dead-end modified peptide ions using charge dependent precursor selection [91]. More recently Ihling et al. have developed a CXR with a spacer that contains an N-oxy-tetramethylpiperidine linked to benzene (TEMPO), which contains a CID-labile NO-C bond [103]. This reagent facilitates free radical initiated peptide sequencing (FRIPS), generating open shell radicals that provide signatures for determining the sequence and location of the CX site on crosslinked peptides by successive MS2 and MS3 analyses. More solutions for reducing the complexity of crosslinking products are likely as the list of these reagents that exploit high resolution tandem MS continues to grow.

19.3.2.3 Equipment

The initial stages in the analysis of proteins by crosslinking require very basic equipment, commonly found in most biochemical laboratories. These include various SDS-PAGE apparati (large and mini gel versions) to analyze protein crosslinking products, circulating water baths and incubators to control for temperature and light boxes for viewing stained gels. In-gel proteolysis techniques require a laminar flow hood, bench-top centrifuge and vacuum centrifuge. After optimizing the yield and proteolysis for a crosslinked protein, MS technologies are employed to analyze the digest. There are many different configurations for mass spectrometers; however, high resolution instruments with fast duty cycles almost always produce the best data for analysis by search engines, because considerable mass accuracy is required to sort out the mass degeneracy resulting from the diversity of the peptide pool generated after the digestion of crosslinked proteins [100]. High resolution instruments also have faster acquisition time and shorter duty cycles (percentage of a time window required to make a measurement) which increase the potential for analyzing low intensity ions typically associated with crosslinked peptides during a given run. Orbitrap MS instruments best fulfill these requirements [104]. In addition to the parameters listed above, Orbitraps come in different tandem MS configurations, with the most advance being capable of carrying out CID, ETD and higher-energy C-trap dissociation (HCD) fragmentation of precursor ions. See Chap. 6 for more detailed descriptions of mass spectrometers.

19.3.2.4 Data Analysis Using Search Engines

The goal of data analysis is to identify crosslinked peptides. Crosslinked peptides include inter-crosslinked peptides, intra-crosslinked peptides, and dead-end crosslinked peptides. Identification of intra-crosslinked peptides and dead-end crosslinked peptides may be achieved by using software tools that were designed originally to identify regular (i.e. non-crosslinked) peptides from shotgun proteomics experiments; however, their identification is extremely difficult. This is because inter-crosslinked peptides include two peptides and the search algorithm must search each experimental spectrum (i.e. query spectrum) against all of the possible pairs of peptides. Figure 19.8 illustrates a general data analysis procedure that comprises several steps that are explained in detail below.

In the first step, sample proteins are digested in *silico* to generate all of the possible peptides using a digestion rule, which uses the known chemistry of the protease selected to determine where cleavage should take place along the amide backbone. For example, if trypsin is selected, then the algorithm generates all possible peptides arising from cleavage C-terminal to lysine and arginine, except when these residues are located N-terminal in the primary sequence to proline. Experimentally it is not uncommon for trypsin to miss one or more of its cleavage sites so peptides with 1–2 miscleavages are also considered. Peptides that are too short or extremely hydrophilic are often lost in wash steps prior to injection in the mass spectrometer and large peptides with masses greater than 4000 Da are often not efficiently cleaved and transmitted. Therefore algorithms must be flexible enough to

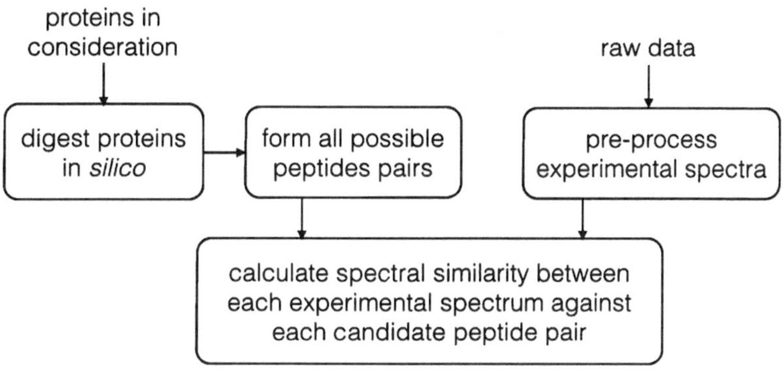

Fig. 19.8 Data analysis workflow

accommodate these and other results that deviate from ideal theoretical conditions. To accommodate these possibilities, search engines typically incorporate two user input parameters that may be adjusted to narrow the range of peptide chain length, depending upon the capabilities of the instrument being used, and the number of miscleavage sites (NMC).

In the second step, the peptides generated in the first step are combined in pairs, and their masses are calculated and annotated in extremely large databases, based on sequence and potential chemical modifications. Possible modifications are defined from rule sets that take into account the possible chemistries and residues that are potentially targeted by any given CXR. Depending on the flexibility of the search engine, a user may manually limit the number of potential crosslinked products from any given peptides during the run. More powerful programs may also include products that may result from crosslinking between three or more peptides using the parameters described above.

Experimental spectra are pre-processed in the third step to account for variations in noise in tandem MS signals and to normalize low and high abundance peaks, both of which are generally important in conjugate identification. Pre-processing of an experimental spectrum separates signal peaks from noise peaks, removes the latter, and normalizes the resulting signal peaks so that low and high intensity peaks are scaled differently. Normalization permits amplification of low intensity peaks, which are often characteristic of crosslinked peptides and thus allows them to be weighted to a greater extent in subsequent scoring rounds. Programs that are designed to carry out this procedure are generally capable of detecting more conjugates than those that simply analyze a given number of the most intense peaks in the spectrum.

Usually, the terminal step in processing is to score the spectral similarity between processed experimental spectra from step three and theoretically generated spectra for all potential candidates generated in step two. Existing programs calculate spectral similarity in different ways, either by cross-correlation or simply summarizing the number of matches detected between peaks from experimental MS/MS and theoretical fragmentation spectra. Candidates are first generated, and these consist of all of the crosslinks with calculated masses that fall within a defined range bracketing the precursor mass measured in the experimental MS/MS spectrum. For each of the candidate crosslinks, a theoretical fragmentation spectrum is generated. As opposed to the general processing of non-modified peptides, only *b*- and *y*-ions are generally considered for crosslinks fragmented by CID and only *c*- and *z*-ions are considered for crosslinks fragmented by ETD. This is because each crosslink contains two or more peptides and their theoretical fragmentation spectra become very complicated if other ion types, such as *a*-ions and those arising from loss of H_2O and NH_3 are also considered. Existing software tools are summarized in Table 19.4.

Table 19.4 Search engines

Name	Publication year	Reference
PeptideMap	1997	[105]
ASAP and MS2Asign	2000	[106]
GPMAW	2001	[107]
X-Link	2002	[108]
Popitam	2003	[109]
MS2PRO	2003	[110]
Links and MS2Link	2004	[111]
CLPM	2005	[112]
XLINK	2006	[113]
VIRTUAL-MSLAB	2006	[114]
SearchXLinks	2006	[115, 116]
Pro-Crosslink	2006	[117]
X!Link	2007	[118, 119]
X-Links	2007	[120]
CrossSearch	2008	[100]
MS-3D	2008	[121]
xComb	2010	[122]
xQuest	2010	[123, 124]
Mass-Matrix	2010	[125]
CRUX	2010	[126]
MS-Bridge	2010	[127]
Xlink-Identifier	2011	[128]
CrossWork	2011	[129]
StavroX	2012	[130]
pLink	2012	[131]
SQID-XLinK	2012	[132]
Hekate	2013	[133]
XLPM	2014	[134]
MXDB	2014	[135]
AnchorMS	2014	[136]
SIM-XL	2015	[137]

19.3.3 General Protocols

Because crosslinking is an entirely empirical process, the following sections will focus primarily on developing screens, rather than explicit protocols, to determine the best conditions and reagents for optimizing the yield and digestion of a desired conjugate from protein targets. Because CXMS is a bottom-up process, we will assume in these screens that protein reactants are purified to near homogeneity.

19.3.3.1 Crosslinking Screens

Ideally in any reagent screen, it is advisable to first analyze the target protein under conditions that allow its native state or that support a known function and/or interaction with a specific partner. The concentration of protein should be sufficient for visualization using general gel staining procedures. Under such conditions, either the time or concentration can be varied for the CXR. In one-step crosslinking screens, the concentration of CXR is generally varied in molar excess from 10 to 500 over the protein target, initially for a fixed time of 15 min. Conversely, greater than 500 M excesses of CXR are incubated with the protein for short time periods, ranging between 1 and 10 min. Using small gel formats with 15–20 wells, 3–4 reagents may be assessed per gel, and as many as 16 reagents may be tested in 1 day. If any conjugates are observed, then reaction conditions may be varied to optimize formation of the desired conjugate. During the screening process, care must be taken to assure that accessory components (e.g., buffers, salts, stabilizing reagents) are compatible with the crosslinker being used. For example, amine containing buffers such as TRIS should be avoided when using NHS-substituted CXRs or any other functional group that targets amines. To avoid large quantities of side-product formation, excessive amounts of crosslinker should be avoided, and only the amount required to generate sufficient amounts of the desired conjugate should be used. Additionally, extremely high pH values should be avoided, because most conventional CXR reactive groups are susceptible to hydrolysis and are either rapidly deactivated or preferentially mono-modify the protein target to form dead-end side products.

Screens using heterobifunctional CXRs to form conjugates in two-step crosslinking protocols are more complicated than one-step screens, because of the intermediate purification step required between successive modification steps with each of the two functional groups of the CXR (see Sect. 19.3.2.2). A rapid assessment of conditions required for two-step crosslinking can be achieved by using small one-mL spin columns loaded with desalting gel media to partially purify the complex after the first modification step and to exchange it into reaction media that are compatible with the second modification step. For example when screening conditions for optimizing crosslinking with a heterobifunctional CXR containing photo-reactive azido and NHS functional groups, time- and concentration-dependent modification of the protein target with the NHS group is carried out in the first step, as described above under one-step crosslinking. To avoid activating the azido group, reactions should be carried out using Eppendorf tubes that are not transparent to light and the total volume for each condition should not exceed 100 μl. Reactions are simply quenched by removing excess reagent with the desalting spin column using a benchtop centrifuge in dark. The desalting gel should be equilibrated in a buffer solution that is compatible with the second photolysis step. Multiple samples may be loaded onto crystallization trays that contain shallow wells, and exposed simultaneously to UV light using a simple hand-held lamp that is placed over the tray for 2–3 min. The reactions are then quenched using SDS-buffer and loaded onto gels to analyze product formation. Although the spin columns do not remove all excess CXR and do not permit a complete exchange of conditions, they provide an efficient method for narrowing the conditions required for optimal crosslinking of the target.

19.3.3.2 Digestion of Conjugates for MS Analyses

In-gel or hetero-phase and in-solution digestions are the two most common approaches for hydrolyzing crosslinked proteins, and MS facilities generally provide basic protocols to follow for sample submission or provide services to perform these procedures. However, the preparation of protein samples, and specifically crosslinked proteins, for MS analyses is a critical and often overlooked component of CXMS. The ultimate goal of this process is to maximize the coverage of the crosslinked protein, which requires optimal cleavage and recovery of the peptide components of the conjugate. Both in-gel and in-solution methods require similar components, which include a targeting protease or chemical to catalyze hydrolysis at specific sites along the amide chain, a denaturant to unfold the protein to enhance maximal cleavage along the backbone, a reducing agent (typically either DTT or 2-mercaptoethanol) to reduce cysteine disulfides and an alkylating agent (iodoacetic acid or iodoacetamide) to modify free thiols generated by reduction. The latter two steps are carried out to prevent refolding. Proteins have unique properties and are targeted to different extents by specific proteases. Covalent attachments introduced by crosslinking usually further complicate proteolysis by affecting the reproducibility and completeness of the digestion. With crosslinking, proteolysis is an empirical process and must be optimized by varying solution conditions and the general components discussed above [138]. Typically, the reaction steps are carried out in the following order: denaturation, reduction, alkylation and proteolysis. Historically, denaturants such as urea, guanidinium hydrochloride and SDS were used and subsequently diluted after reduction and alkylation steps to concentrations tolerated by the protease; however, they interfere and are poorly tolerated by MS. To address this problem, more MS-friendly denaturants such as Rapigest™ (Waters, Milford MA) [139], sodium deoxycholate (SDC) [139] or sodium 3-[(2-methyl-2-undecyl-1,3-dioxolan-4-yl)methoxyl]-1-propanesulfonate (ALS) [140] have been developed. Alternatively, spin concentrators and various filters have been developed to facilitate exchange of secondary chemicals and denaturants without significant loss of the conjugate prior to proteolysis [139, 141]. Additional methods for improving protein denaturation, including thermal (IR and microwave radiation), ultrasonic and solvent-based techniques, are summarized in an excellent review by Hustoft et al. [142]. After denaturation, engineered forms of trypsin are generally used to carry out proteolysis, because they specifically cleave amide backbones after lysine and arginine, function well in low concentrations of multiple denaturants, and are relatively resistant to autolysis. A recent report suggests that tandem application of Lys-C (lysine–specific protease) and trypsin promotes more efficient cleavage of protein substrates than trypsin alone [138]. Despite all the possible choices in such reactions, some of the following parameters are good starting points for in solution digestions. First, the conjugate may be reduced with DTT (10 M excess) and denatured concomitantly in either 0.1 % ALS or SDC at elevated temperatures (~50–85 °C) for 1 h. This is followed by alkylation with iodoacetic acid (40 M excess) in the dark for 30 min at 30 °C. After alkylation, DTT is added in excess of iodoacetic acid to prevent alkylation of trypsin. The denatured protein may then be exchanged into 25 mM ammonium bicarbonate using a 3000 MW cutoff spin concentrator (EMD Millipore) and digested overnight at 30 °C using a 25-fold excess (w/w) of sequencing grade trypsin (Promega). Peptides may be recovered by several rounds of centrifugation and washes with 10 % acetonitrile in 25 mM ammonium bicarbonate. Peptides are then concentrated to remove acetonitrile or lyophilized in a vacuum concentrator.

In-gel digestion uses the resolving power of SDS-PAGE to isolate the desired conjugate from complex mixtures of crosslinking products, significantly reducing the number of products to be analyzed. On the other hand, gels can hinder peptide recovery, depending to a great extent on the type of extraction procedure used. Several aspects of this technique are unique compared to in-solution methods, based on the polyacrylamide matrix, which limits diffusion of the reactants and protease necessary for generating peptides [143]. Thus, the ratios of protease to substrate are generally much higher than those typically used in solution. Additionally, the gel sections containing the conjugate must be treated with solvents (typically 50 % acetonitrile in 25 mM ammonium bicarbonate) to remove SDS and other gel solution components that inhibit the activity of the protease. Another important consideration is that there are many handling steps that can potentially introduce contaminants, particularly keratins. Thus all reagents must be prepared carefully and any instruments used must be cleaned scrupulously before carrying out the procedure. Gloves and sterile sleeve protectors should be worn at all times. Specific details for gel-phase proteolysis conditions are outlined in a published protocol [143] and can be accessed online at the UCSF mass spectrometry website.

19.3.3.3 Data Input

As discussed in Sect. 19.3.2.3, the use of search engines generally requires little from the user. Most are designed with interfaces that allow the user to upload the sequence(s) and reagents being tested. Additionally, some parameters such as the number of allowed side products and crosslinks per conjugate may also be adjustable. Typically, one should limit these parameters in the first round of an analysis; first to minimize computing space and time, and second to avoid extensive data output. Some programs ask the user to specify the reactive groups of the CXR and the mass of the intervening spacer (after modification), as well as the mass of dead-end products. Users with limited knowledge of cross-linking or chemistry should avoid the latter programs.

19.3.4 Caveats

Perhaps the greatest mistake made by even experienced users of the crosslinking technique is to over interpret results. First, there is a tendency in the literature for users to define a detected CX site on a protein as a binding site, no matter the span of the CRX. The specificity and, therefore, the probability that crosslinking represents an actual binding event is greatest when zero-length chemically coupled residues on opposing binding partners are identified. CX sites that are detected using CXR reagents with crosslinking spans greater than 2–3 Å should be discussed in terms of the proximity of the linked residues, defined by the range of distances that the spacer can occupy in solution [144]. Another common misconception is that the absence of crosslinking indicates absence of interaction [145]. In this case there are many more reasons why crosslinking does not occur, based on incompatibility of the CXR with the chemistry, geometry and/or solvent accessibility of the protein-protein interaction surface(s). There are many examples for which CX sites are purportedly identified by simply matching the experimental mass of a precursor ion with the theoretical mass of a crosslinked peptide. With large proteins it has been demonstrated that a large number of dead-end and crosslinked peptides may account for a single precursor ion within the resolving limits (~3 ppm) of a high resolution FT instrument [100]. Even when the masses of a precursor and its corresponding fragments are well matched to those for a theoretically generated candidate, there is a reasonable potential for misidentification, based on the limited resolving capabilities (>200 ppm) of typical collision cells, i.e. significant error in the identification fragment ions generated by tandem MS. High resolution measurements of fragment ions are now possible in new Orbitrap MS instruments using HCD, significantly increasing the potential for boosting confidence levels in matching assignments. Finally, corroborating evidence from alternative methods is always desirable for any interaction that is detected or suggested by crosslinking.

19.3.5 Representative Results and Complementary Techniques

Because of its versatility, CXMS has been used in combination with many complementary techniques developed to detect protein-protein interactions. These studies often are focused on determining the structure of proteins and their complexes and to theoretically model non-homologous proteins. Several examples of these combined approaches will be discussed in terms of how each complements the other. With the development of MS instruments that are capable of transmitting large macromolecular complexes [146], top-down MS has become a well-established method for analyzing the interaction of proteins and/or subunits in large protein complexes that are not amenable to NMR and X-ray crystallographic methods [147], providing a potent alternative and complementary approach to crosslinking [94]. The basic approach relies on the transmission of a partially hydrated protein complex in near-native conditions, in which the topological arrangement and interactions of its protein components are probed either by CID after injection [148] or the introduction of sub-stoichiometric amounts of small molecules that destabilize the complex prior to injection [149]. Maps defining the interactions of integral subunits in the complex are constructed based on the composition and number of subcomplexes detected [150]. Top-down MS also is capable of detecting differences in the stability of a complex in different conformational states [151]. For example Lane et al. showed that the native, non-activated form of the $(\alpha\beta\gamma\delta)_4$ phosphorylase kinase complex (PhK) is more stable than its active phosphorylated form by demonstrating that the percentage of intact phosphorylated complex decreased with respect to that of the native under identical conditions [151]. In that study, phosphorylation of the complex also perturbed interactions of the subunits in the complex, resulting in preferential interactions among the regulatory β subunits, also detected by crosslinking in a previous study [152]. In a parallel study, these investigators combined CXMS, immuno EM, cryoEM, modeling and biochemical data to determine the location of the regulatory β subunits in the PhK complex [89]. The topology and location of the subunits in the connecting bridge region of the bilobal complex was determined using top-down MS, and CXMS was used to constrain an atomic model of the β subunit (generated by I-TASSER [153]) to facilitate its docking in the bridges of the cryoEM 3D structure. Aebersold and coworkers have also used CXMS to provide distance constraints in combination with tandem affinity purification to model a protein phosphatase 2A network of interactions [154]. CXMS has become the method of choice for constraining theoretical models [155], and is widely used in integrative structural modeling (ISM), an approach in which theoretical models of variable resolution are scored, based on their agreement with constraints provided by different forms of experimental data, commonly referred to as input data [156]. ISM approaches using CXMS have been used to model complex macromolecular assemblies, including the yeast eIF3:eIF5 complex [90] and the photoreceptor phosphodiesterase hetero-oligomer [157].

References

1. Baldwin RL (2011) Early days of protein hydrogen exchange: 1954–1972. Proteins 79:2021–2026
2. Hvidt A, Linderstrom-Lang K (1954) Exchange of hydrogen atoms in insulin with deuterium atoms in aqueous solutions. Biochim Biophys Acta 14:574–575
3. Schellman JA, Schellman CG (1997) Kaj Ulrik Linderstrom-Lang (1896–1959). Protein Sci 6:1092–1100
4. Sheinblatt M (1970) Determination of an acidity scale for peptide hydrogens from nuclear magnetic resonance kinetic studies. J Am Chem Soc 92:2505–2509
5. Molday RS, Englander SW, Kallen RG (1972) Primary structure effects on peptide group hydrogen exchange. Biochemistry 11:150–158
6. Rosa JJ, Richards FM (1979) An experimental procedure for increasing the structural resolution of chemical hydrogen-exchange measurements on proteins: application to ribonuclease S peptide. J Mol Biol 133:399–416
7. Wagner G, Wuthrich K (1982) Amide protein exchange and surface conformation of the basic pancreatic trypsin inhibitor in solution. Studies with two-dimensional nuclear magnetic resonance. J Mol Biol 160:343–361
8. Katta V, Chait BT (1991) Conformational changes in proteins probed by hydrogen-exchange electrospray-ionization mass spectrometry. Rapid Commun Mass Spectrom 5:214–217
9. Pascal BD, Willis S, Lauer JL, Landgraf RR, West GM, Marciano D, Novick S, Goswami D, Chalmers MJ, Griffin PR (2012) HDX workbench: software for the analysis of H/D exchange MS data. J Am Soc Mass Spectrom 23:1512–1521
10. Villar MT, Miller DE, Fenton AW, Artigues A (2010) SAIDE: A Semi-Automated Interface for Hydrogen/Deuterium Exchange Mass Spectrometry. Proteomica 6:63–69
11. Englander JJ, Rogero JR, Englander SW (1985) Protein hydrogen exchange studied by the fragment separation method. Anal Biochem 147:234–244
12. Wales TE, Engen JR (2006) Hydrogen exchange mass spectrometry for the analysis of protein dynamics. Mass Spectrom Rev 25:158–170
13. Zhang Z, Smith DL (1993) Determination of amide hydrogen exchange by mass spectrometry: a new tool for protein structure elucidation. Protein Sci 2:522–531
14. Smith DL, Deng Y, Zhang Z (1997) Probing the non-covalent structure of proteins by amide hydrogen exchange and mass spectrometry. J Mass Spectrom 32:135–146
15. Hamuro Y, Coales SJ, Southern MR, Nemeth-Cawley JF, Stranz DD, Griffin PR (2003) Rapid analysis of protein structure and dynamics by hydrogen/deuterium exchange mass spectrometry. J Biomol Tech 14:171–182
16. Englander SW (2006) Hydrogen exchange and mass spectrometry: a historical perspective. J Am Soc Mass Spectrom 17:1481–1489
17. Busenlehner LS, Armstrong RN (2005) Insights into enzyme structure and dynamics elucidated by amide H/D exchange mass spectrometry. Arch Biochem Biophys 433:34–46
18. Chalmers MJ, Busby SA, Pascal BD, He Y, Hendrickson CL, Marshall AG, Griffin PR (2006) Probing protein ligand interactions by automated hydrogen/deuterium exchange mass spectrometry. Anal Chem 78:1005–1014
19. Chalmers MJ, Busby SA, Pascal BD, Southern MR, Griffin PR (2007) A two-stage differential hydrogen deuterium exchange method for the rapid characterization of protein/ligand interactions. J Biomol Tech 18:194–204
20. Hoofnagle AN, Resing KA, Ahn NG (2004) Practical methods for deuterium exchange/mass spectrometry. Methods Mol Biol 250:283–298
21. Englander SW, Downer NW, Teitelbaum H (1972) Hydrogen exchange. Annu Rev Biochem 41:903–924
22. Bai Y, Milne JS, Mayne L, Englander SW (1993) Primary structure effects on peptide group hydrogen exchange. Proteins 17:75–86
23. Weis DD, Wales TE, Engen JR, Hotchko M, Ten Eyck LF (2006) Identification and characterization of EX1 kinetics in H/D exchange mass spectrometry by peak width analysis. J Am Soc Mass Spectrom 17:1498–1509
24. Ferraro DM, Lazo N, Robertson AD (2004) EX1 hydrogen exchange and protein folding. Biochemistry 43:587–594
25. Krishna MM, Hoang L, Lin Y, Englander SW (2004) Hydrogen exchange methods to study protein folding. Methods 34:51–64
26. Miller DE, Prasannan CB, Villar MT, Fenton AW, Artigues A (2012) HDXFinder: automated analysis and data reporting of Deuterium/Hydrogen exchange mass spectrometry. J Am Soc Mass Spectrom 23:425–429
27. Pascal BD, Chalmers MJ, Busby SA, Griffin PR (2009) HD desktop: an integrated platform for the analysis and visualization of H/D exchange data. J Am Soc Mass Spectrom 20:601–610
28. Weis DD, Engen JR, Kass IJ (2006) Semi-automated data processing of hydrogen exchange mass spectra using HX-express. J Am Soc Mass Spectrom 17:1700–1703
29. Lou X, Kirchner M, Renard BY, Kothe U, Boppel S, Graf C, Lee CT, Steen JA, Steen H, Mayer MP,

Hamprecht FA (2010) Deuteration distribution estimation with improved sequence coverage for HX/MS experiments. Bioinformatics 26:1535–1541
30. Lindner R, Lou X, Reinstein J, Shoeman RL, Hamprecht FA, Winkler A (2014) Hexicon 2: automated processing of hydrogen-deuterium exchange mass spectrometry data with improved deuteration distribution estimation. J Am Soc Mass Spectrom 25:1018–1028
31. Zhang Z, Marshall AG (1998) A universal algorithm for fast and automated charge state deconvolution of electrospray mass-to-charge ratio spectra. J Am Soc Mass Spectrom 9:225–233
32. Connelly GP, Bai Y, Jeng MF, Englander SW (1993) Isotope effects in peptide group hydrogen exchange. Proteins 17:87–92
33. Liu YH, Konermann L (2006) Enzyme conformational dynamics during catalysis and in the 'resting state' monitored by hydrogen/deuterium exchange mass spectrometry. FEBS Lett 580:5137–5142
34. Hu W, Walters BT, Kan ZY, Mayne L, Rosen LE, Marqusee S, Englander SW (2013) Stepwise protein folding at near amino acid resolution by hydrogen exchange and mass spectrometry. Proc Natl Acad Sci U S A 110:7684–7689
35. Wintrode PL, Rojsajjakul T, Vadrevu R, Matthews CR, Smith DL (2005) An obligatory intermediate controls the folding of the alpha-subunit of tryptophan synthase, a TIM barrel protein. J Mol Biol 347:911–919
36. Yang H, Smith DL (1997) Kinetics of cytochrome c folding examined by hydrogen exchange and mass spectrometry. Biochemistry 36:14992–14999
37. Busby SA, Chalmers MJ, Griffin PR (2007) Improving digestion efficiency under H/D exchange conditions with activated pepsinogen coupled cloumns. Int J Mass Spectrom 259:130–139
38. Ahn J, Jung MC, Wyndham K, Yu YQ, Engen JR (2012) Pepsin immobilized on high-strength hybrid particles for continuous flow online digestion at 10,000 psi. Anal Chem 84:7256–7262
39. Wu Y, Kaveti S, Engen JR (2006) Extensive deuterium back-exchange in certain immobilized pepsin columns used for H/D exchange mass spectrometry. Anal Chem 78:1719–1723
40. Tsybin YO, Haselmann KF, Emmett MR, Hendrickson CL, Marshall AG (2006) Charge location directs electron capture dissociation of peptide dications. J Am Soc Mass Spectrom 17:1704–1711
41. Demmers JA, Rijkers DT, Haverkamp J, Killian JA, Heck AJ (2002) Factors affecting gas-phase deuterium scrambling in peptide ions and their implications for protein structure determination. J Am Chem Soc 124:11191–11198
42. Ferguson PL, Konermann L (2008) Nonuniform isotope patterns produced by collision-induced dissociation of homogeneously labeled ubiquitin: implications for spatially resolved hydrogen/deuterium exchange ESI-MS studies. Anal Chem 80:4078–4086
43. Ferguson PL, Pan J, Wilson DJ, Dempsey B, Lajoie G, Shilton B, Konermann L (2007) Hydrogen/deuterium scrambling during quadrupole time-of-flight MS/MS analysis of a zinc-binding protein domain. Anal Chem 79:153–160
44. Jorgensen TJ, Bache N, Roepstorff P, Gardsvoll H, Ploug M (2005) Collisional activation by MALDI tandem time-of-flight mass spectrometry induces intramolecular migration of amide hydrogens in protonated peptides. Mol Cell Proteomics 4:1910–1919
45. Jorgensen TJ, Gardsvoll H, Ploug M, Roepstorff P (2005) Intramolecular migration of amide hydrogens in protonated peptides upon collisional activation. J Am Chem Soc 127:2785–2793
46. Kim MY, Maier CS, Reed DJ, Deinzer ML (2001) Site-specific amide hydrogen/deuterium exchange in E. coli thioredoxins measured by electrospray ionization mass spectrometry. J Am Chem Soc 123:9860–9866
47. Xu G, Takamoto K, Chance MR (2003) Radiolytic modification of basic amino acid residues in peptides: probes for examining protein-protein interactions. Anal Chem 75:6995–7007
48. Sharp JS, Becker JM, Hettich RL (2003) Protein surface mapping by chemical oxidation: structural analysis by mass spectrometry. Anal Biochem 313:216–225
49. Hambly DM, Gross ML (2005) Laser flash photolysis of hydrogen peroxide to oxidize protein solvent-accessible residues on the microsecond timescale. J Am Soc Mass Spectrom 16:2057–2063
50. Takamoto K, Chance MR (2006) Radiolytic protein footprinting with mass spectrometry to probe the structure of macromolecular complexes. Annu Rev Biophys Biomol Struct 35:251–276
51. Konermann L, Stocks BB, Pan Y, Tong X (2010) Mass spectrometry combined with oxidative labeling for exploring protein structure and folding. Mass Spectrom Rev 29:651–667
52. Linderstrom-Land K, Ottesen M (1947) A new protein from ovalbumin. Nature 159:807
53. Neurath H (1979) Limited proteolysis, protein folding and physiological regulation. In: Jaenicke R (ed) Protein folding. Elsevier/North-Holland Biomedical Press, University of Regensburg, Regensburg
54. Bloxham DP, Ericsson LH, Titani K, Walsh KA, Neurath H (1980) Limited proteolysis of pig heart citrate synthase by subtilisin, chymotrypsin, and trypsin. Biochemistry (Mosc) 19:3979–3985
55. Fontana A, de Laureto PP, Spolaore B, Frare E (2012) Identifying disordered regions in proteins by limited proteolysis. Methods Mol Biol 896:297–318
56. Fontana A, Fassina G, Vita C, Dalzoppo D, Zamai M, Zambonin M (1986) Correlation between

sites of limited proteolysis and segmental mobility in thermolysin. Biochemistry (Mosc) 25:1847–1851
57. Bantscheff M, Weiss V, Glocker MO (1999) Identification of linker regions and domain borders of the transcription activator protein NtrC from Escherichia coli by limited proteolysis, in-gel digestion, and mass spectrometry. Biochemistry (Mosc) 38:11012–11020
58. Hubbard SJ (1998) The structural aspects of limited proteolysis of native proteins. Biochim Biophys Acta 1382:191–206
59. Hubbard SJ, Eisenmenger F, Thornton JM (1994) Modeling studies of the change in conformation required for cleavage of limited proteolytic sites. Protein Sci 3:757–768
60. Karas M, Hillenkamp F (1988) Laser desorption ionization of proteins with molecular masses exceeding 10,000 daltons. Anal Chem 60:2299–2301
61. Fenn JB, Mann M, Meng CK, Wong SF, Whitehouse CM (1989) Electrospray ionization for mass spectrometry of large biomolecules. Science 246:64–71
62. Suh MJ, Pourshahian S, Limbach PA (2007) Developing limited proteolysis and mass spectrometry for the characterization of ribosome topography. J Am Soc Mass Spectrom 18:1304–1317
63. Feng Y, De Franceschi G, Kahraman A, Soste M, Melnik A, Boersema PJ, de Laureto PP, Nikolaev Y, Oliveira AP, Picotti P (2014) Global analysis of protein structural changes in complex proteomes. Nat Biotechnol 32:1036–1044
64. Orru S, Dal Piaz F, Casbarra A, Biasiol G, De Francesco R, Steinkuhler C, Pucci P (1999) Conformational changes in the NS3 protease from hepatitis C virus strain Bk monitored by limited proteolysis and mass spectrometry. Protein Sci 8:1445–1454
65. Zappacosta F, Pessi A, Bianchi E, Venturini S, Sollazzo M, Tramontano A, Marino G, Pucci P (1996) Probing the tertiary structure of proteins by limited proteolysis and mass spectrometry: the case of Minibody. Protein Sci 5:802–813
66. Bothner B, Dong XF, Bibbs L, Johnson JE, Siuzdak G (1998) Evidence of viral capsid dynamics using limited proteolysis and mass spectrometry. J Biol Chem 273:673–676
67. Fontana A, Zambonin M, Polverino de Laureto P, De Filippis V, Clementi A, Scaramella E (1997) Probing the conformational state of apomyoglobin by limited proteolysis. J Mol Biol 266:223–230
68. Villa JA, Cabezas M, de la Cruz F, Moncalian G (2014) Use of limited proteolysis and mutagenesis to identify folding domains and sequence motifs critical for wax ester synthase/acyl coenzyme A: diacylglycerol acyltransferase activity. Appl Environ Microbiol 80:1132–1141
69. Graves DJ, Hayakawa T, Horvitz RA, Beckman E, Krebs EG (1973) Studies on the subunit structure of trypsin-activated phosphorylase kinase. Biochemistry (Mosc) 12:580–585
70. Potter RL, Taylor SS (1980) The structural domains of cAMP-dependent protein kinase I. Characterization of two sites of proteolytic cleavage and homologies to cAMP-dependent protein kinase II. J Biol Chem 255:9706–9712
71. Fontana A, de Laureto PP, Spolaore B, Frare E, Picotti P, Zambonin M (2004) Probing protein structure by limited proteolysis. Acta Biochim Pol 51:299–321
72. Scaloni A, Miraglia N, Orru S, Amodeo P, Motta A, Marino G, Pucci P (1998) Topology of the calmodulin-melittin complex. J Mol Biol 277:945–958
73. Cohen SL, Ferre-D'Amare AR, Burley SK, Chait BT (1995) Probing the solution structure of the DNA-binding protein Max by a combination of proteolysis and mass spectrometry. Protein Sci 4:1088–1099
74. Monti M, Pucci P (2006) Limited proteolysis mass spectrometry of protein complexes. In: Mass spectrometry of protein interactions. Wiley, Hoboken, pp 63–82
75. Suckau D, Kohl J, Karwath G, Schneider K, Casaretto M, Bitter-Suermann D, Przybylski M (1990) Molecular epitope identification by limited proteolysis of an immobilized antigen-antibody complex and mass spectrometric peptide mapping. Proc Natl Acad Sci U S A 87:9848–9852
76. Trempe MR, Carlson GM (1987) Phosphorylase kinase conformers. Detection by proteases. J Biol Chem 262:4333–4340
77. Trempe MR, Carlson GM, Hainfeld JF, Furcinitti PS, Wall JS (1986) Analyses of phosphorylase kinase by transmission and scanning transmission electron microscopy. J Biol Chem 261:2882–2889
78. Kemp BE, Pearson RB (1991) Intrasteric regulation of protein kinases and phosphatases. Biochim Biophys Acta 1094:67–76
79. Kobe B, Kemp BE (1999) Active site-directed protein regulation. Nature 402:373–376
80. Xu G, Chance MR (2005) Radiolytic modification and reactivity of amino acid residues serving as structural probes for protein footprinting. Anal Chem 77:4549–4555
81. Kiselar JG, Chance MR (2010) Future directions of structural mass spectrometry using hydroxyl radical footprinting. J Mass Spectrom 45:1373–1382
82. Zhang H, Gau BC, Jones LM, Vidavsky I, Gross ML (2011) Fast photochemical oxidation of proteins for comparing structures of protein-ligand complexes: the calmodulin-peptide model system. Anal Chem 83:311–318
83. Hanai R, Wang JC (1994) Protein footprinting by the combined use of reversible and irreversible lysine modifications. Proc Natl Acad Sci U S A 91:11904–11908
84. Tu BP, Wang JC (1999) Protein footprinting at cysteines: probing ATP-modulated contacts in cysteine-substitution mutants of yeast DNA

topoisomerase II. Proc Natl Acad Sci U S A 96:4862–4867
85. Nadeau OW, Carlson GM (2005) Protein interactions captured by chemical cross-linking. In: Golemis E, Adams PD (eds) Protein-protein interactions : a molecular cloning manual, 2nd edn. Cold Spring Harbor Laboratory Press, Cold Spring Harbor, pp 105–127
86. Nadeau OW (2006) Protein interaction analysis: chemical cross-linking. In: Ganten D, Ruckpaul K (eds) Encyclopedic reference of genomics and proteomics in molecular medicine. Springer, Berlin
87. Hermanson GT (2008) Bioconjugate techniques, 2nd edn. Elsevier Academic Press, Amsterdam/Boston
88. Wong SS (1993) Chemistry of protein conjugation and cross-linking. CRC Press, Boca Raton
89. Nadeau OW, Lane LA, Xu D, Sage J, Priddy TS, Artigues A, Villar MT, Yang Q, Robinson CV, Zhang Y, Carlson GM (2012) Structure and location of the regulatory beta subunits in the (alphabetagammadelta)4 phosphorylase kinase complex. J Biol Chem 287:36651–36661
90. Politis A, Schmidt C, Tjioe E, Sandercock AM, Lasker K, Gordiyenko Y, Russel D, Sali A, Robinson CV (2015) Topological models of heteromeric protein assemblies from mass spectrometry: application to the yeast eIF3:eIF5 complex. Chem Biol 22:117–128
91. Trnka MJ, Burlingame AL (2010) Topographic studies of the GroEL-GroES chaperonin complex by chemical cross-linking using diformyl ethynylbenzene: the power of high resolution electron transfer dissociation for determination of both peptide sequences and their attachment sites. Mol Cell Proteomics 9:2306–2317
92. Paramelle D, Miralles G, Subra G, Martinez J (2013) Chemical cross-linkers for protein structure studies by mass spectrometry. Proteomics 13:438–456
93. Singh P, Panchaud A, Goodlett DR (2010) Chemical cross-linking and mass spectrometry as a low-resolution protein structure determination technique. Anal Chem 82:2636–2642
94. Stengel F, Aebersold R, Robinson CV (2012) Joining forces: integrating proteomics and cross-linking with the mass spectrometry of intact complexes. Mol Cell Proteomics 11:R111.014027
95. Chowdhury SM, Munske GR, Tang X, Bruce JE (2006) Collisionally activated dissociation and electron capture dissociation of several mass spectrometry-identifiable chemical cross-linkers. Anal Chem 78:8183–8193
96. Tang X, Munske GR, Siems WF, Bruce JE (2005) Mass spectrometry identifiable cross-linking strategy for studying protein-protein interactions. Anal Chem 77:311–318
97. Rostovtsev VV, Green LG, Fokin VV, Sharpless KB (2002) A stepwise huisgen cycloaddition process: copper(I)-catalyzed regioselective "ligation" of azides and terminal alkynes. Angew Chem Int Ed Engl 41:2596–2599
98. Chowdhury SM, Du X, Tolic N, Wu S, Moore RJ, Mayer MU, Smith RD, Adkins JN (2009) Identification of cross-linked peptides after click-based enrichment using sequential collision-induced dissociation and electron transfer dissociation tandem mass spectrometry. Anal Chem 81:5524–5532
99. Vellucci D, Kao A, Kaake RM, Rychnovsky SD, Huang L (2010) Selective enrichment and identification of azide-tagged cross-linked peptides using chemical ligation and mass spectrometry. J Am Soc Mass Spectrom 21:1432–1445
100. Nadeau OW, Wyckoff GJ, Paschall JE, Artigues A, Sage J, Villar MT, Carlson GM (2008) CrossSearch, a user-friendly search engine for detecting chemically cross-linked peptides in conjugated proteins. Mol Cell Proteomics 7:739–749
101. Maiolica A, Cittaro D, Borsotti D, Sennels L, Ciferri C, Tarricone C, Musacchio A, Rappsilber J (2007) Structural analysis of multiprotein complexes by cross-linking, mass spectrometry, and database searching. Mol Cell Proteomics 6:2200–2211
102. Hoopmann MR, Weisbrod CR, Bruce JE (2010) Improved strategies for rapid identification of chemically cross-linked peptides using protein interaction reporter technology. J Proteome Res 9:6323–6333
103. Ihling C, Falvo F, Kratochvil I, Sinz A, Schafer M (2015) Dissociation behavior of a bifunctional tempoactive ester reagent for peptide structure analysis by free radical initiated peptide sequencing (FRIPS) mass spectrometry. J Mass Spectrom 50:396–406
104. Jedrychowski MP, Huttlin EL, Haas W, Sowa ME, Rad R, Gygi SP (2011) Evaluation of HCD- and CID-type fragmentation within their respective detection platforms for murine phosphoproteomics. Mol Cell Proteomics 10:M111.009910
105. Fenyo D (1997) A software tool for the analysis of mass spectrometric disulfide mapping experiments. Comput Appl Biosci 13:617–618
106. Young MM, Tang N, Hempel JC, Oshiro CM, Taylor EW, Kuntz ID, Gibson BW, Dollinger G (2000) High throughput protein fold identification by using experimental constraints derived from intramolecular cross-links and mass spectrometry. Proc Natl Acad Sci U S A 97:5802–5806
107. Peri S, Steen H, Pandey A (2001) GPMAW–a software tool for analyzing proteins and peptides. Trends Biochem Sci 26:687–689
108. Taverner T, Hall NE, O'Hair RA, Simpson RJ (2002) Characterization of an antagonist interleukin-6 dimer by stable isotope labeling, cross-linking, and mass spectrometry. J Biol Chem 277:46487–46492
109. Hernandez P, Gras R, Frey J, Appel RD (2003) Popitam: towards new heuristic strategies to improve protein identification from tandem mass spectrometry data. Proteomics 3:870–878

110. Kruppa GH, Schoeniger J, Young MM (2003) A top down approach to protein structural studies using chemical cross-linking and Fourier transform mass spectrometry. Rapid Commun Mass Spectrom 17:155–162
111. Kellersberger KA, Yu E, Kruppa GH, Young MM, Fabris D (2004) Top-down characterization of nucleic acids modified by structural probes using high-resolution tandem mass spectrometry and automated data interpretation. Anal Chem 76:2438–2445
112. Tang Y, Chen Y, Lichti CF, Hall RA, Raney KD, Jennings SF (2005) CLPM: a cross-linked peptide mapping algorithm for mass spectrometric analysis. BMC Bioinf 6 Suppl 2:S9
113. Seebacher J, Mallick P, Zhang N, Eddes JS, Aebersold R, Gelb MH (2006) Protein cross-linking analysis using mass spectrometry, isotope-coded cross-linkers, and integrated computational data processing. J Proteome Res 5:2270–2282
114. de Koning LJ, Kasper PT, Back JW, Nessen MA, Vanrobaeys F, Van Beeumen J, Gherardi E, de Koster CG, de Jong L (2006) Computer-assisted mass spectrometric analysis of naturally occurring and artificially introduced cross-links in proteins and protein complexes. FEBS J 273:281–291
115. Schnaible V, Wefing S, Resemann A, Suckau D, Bucker A, Wolf-Kummeth S, Hoffmann D (2002) Screening for disulfide bonds in proteins by MALDI in-source decay and LIFT-TOF/TOF-MS. Anal Chem 74:4980–4988
116. Wefing S, Schnaible V, Hoffmann D (2006) SearchXLinks. A program for the identification of disulfide bonds in proteins from mass spectra. Anal Chem 78:1235–1241
117. Gao Q, Xue S, Doneanu CE, Shaffer SA, Goodlett DR, Nelson SD (2006) Pro-CrossLink. Software tool for protein cross-linking and mass spectrometry. Anal Chem 78:2145–2149
118. Lee YJ, Lackner LL, Nunnari JM, Phinney BS (2007) Shotgun cross-linking analysis for studying quaternary and tertiary protein structures. J Proteome Res 6:3908–3917
119. Lee YJ (2009) Probability-based shotgun cross-linking sites analysis. J Am Soc Mass Spectrom 20:1896–1899
120. Anderson GA, Tolic N, Tang X, Zheng C, Bruce JE (2007) Informatics strategies for large-scale novel cross-linking analysis. J Proteome Res 6:3412–3421
121. Yu ET, Hawkins A, Kuntz ID, Rahn LA, Rothfuss A, Sale K, Young MM, Yang CL, Pancerella CM, Fabris D (2008) The collaboratory for MS3D: a new cyberinfrastructure for the structural elucidation of biological macromolecules and their assemblies using mass spectrometry-based approaches. J Proteome Res 7:4848–4857
122. Panchaud A, Singh P, Shaffer SA, Goodlett DR (2010) xComb: a cross-linked peptide database approach to protein-protein interaction analysis. J Proteome Res 9:2508–2515
123. Leitner A, Walzthoeni T, Aebersold R (2014) Lysine-specific chemical cross-linking of protein complexes and identification of cross-linking sites using LC-MS/MS and the xQuest/xProphet software pipeline. Nat Protoc 9:120–137
124. Leitner A, Walzthoeni T, Kahraman A, Herzog F, Rinner O, Beck M, Aebersold R (2010) Probing native protein structures by chemical cross-linking, mass spectrometry, and bioinformatics. Mol Cell Proteomics 9:1634–1649
125. Xu H, Hsu PH, Zhang L, Tsai MD, Freitas MA (2010) Database search algorithm for identification of intact cross-links in proteins and peptides using tandem mass spectrometry. J Proteome Res 9:3384–3393
126. McIlwain S, Draghicescu P, Singh P, Goodlett DR, Noble WS (2010) Detecting cross-linked peptides by searching against a database of cross-linked peptide pairs. J Proteome Res 9:2488–2495
127. Chu F, Baker PR, Burlingame AL, Chalkley RJ (2010) Finding chimeras: a bioinformatics strategy for identification of cross-linked peptides. Mol Cell Proteomics 9:25–31
128. Du X, Chowdhury SM, Manes NP, Wu S, Mayer MU, Adkins JN, Anderson GA, Smith RD (2011) Xlink-identifier: an automated data analysis platform for confident identifications of chemically cross-linked peptides using tandem mass spectrometry. J Proteome Res 10:923–931
129. Rasmussen MI, Refsgaard JC, Peng L, Houen G, Hojrup P (2011) CrossWork: software-assisted identification of cross-linked peptides. J Proteome 74:1871–1883
130. Gotze M, Pettelkau J, Schaks S, Bosse K, Ihling CH, Krauth F, Fritzsche R, Kuhn U, Sinz A (2012) StavroX–a software for analyzing crosslinked products in protein interaction studies. J Am Soc Mass Spectrom 23:76–87
131. Yang B, Wu YJ, Zhu M, Fan SB, Lin J, Zhang K, Li S, Chi H, Li YX, Chen HF, Luo SK, Ding YH, Wang LH, Hao Z, Xiu LY, Chen S, Ye K, He SM, Dong MQ (2012) Identification of cross-linked peptides from complex samples. Nat Methods 9:904–906
132. Li W, O'Neill HA, Wysocki VH (2012) SQID-XLink: implementation of an intensity-incorporated algorithm for cross-linked peptide identification. Bioinformatics 28:2548–2550
133. Holding AN, Lamers MH, Stephens E, Skehel JM (2013) Hekate: software suite for the mass spectrometric analysis and three-dimensional visualization of cross-linked protein samples. J Proteome Res 12:5923–5933
134. Jaiswal M, Crabtree N, Bauer MA, Hall R, Raney KD, Zybailov BL (2014) XLPM: efficient algorithm for the analysis of protein-protein contacts using

chemical cross-linking mass spectrometry. BMC Bioinf 15 Suppl 11:S16

135. Wang J, Anania VG, Knott J, Rush J, Lill JR, Bourne PE, Bandeira N (2014) Combinatorial approach for large-scale identification of linked peptides from tandem mass spectrometry spectra. Mol Cell Proteomics 13:1128–1136
136. Mayne SL, Patterton HG (2014) AnchorMS: a bioinformatics tool to derive structural information from the mass spectra of cross-linked protein complexes. Bioinformatics 30:125–126
137. Lima DB, de Lima TB, Balbuena TS, Neves-Ferreira AG, Barbosa VC, Gozzo FC, Carvalho PC (2015) SIM-XL: a powerful and user-friendly tool for peptide cross-linking analysis. J Proteomics 129:51
138. Glatter T, Ludwig C, Ahrne E, Aebersold R, Heck AJ, Schmidt A (2012) Large-scale quantitative assessment of different in-solution protein digestion protocols reveals superior cleavage efficiency of tandem Lys-C/trypsin proteolysis over trypsin digestion. J Proteome Res 11:5145–5156
139. Leon IR, Schwammle V, Jensen ON, Sprenger RR (2013) Quantitative assessment of in-solution digestion efficiency identifies optimal protocols for unbiased protein analysis. Mol Cell Proteomics 12:2992–3005
140. Nomura E, Katsuta K, Ueda T, Toriyama M, Mori T, Inagaki N (2004) Acid-labile surfactant improves in-sodium dodecyl sulfate polyacrylamide gel protein digestion for matrix-assisted laser desorption/ionization mass spectrometric peptide mapping. J Mass Spectrom 39:202–207
141. Wisniewski JR, Zougman A, Nagaraj N, Mann M (2009) Universal sample preparation method for proteome analysis. Nat Methods 6:359–362
142. Hustoft HK, Reubsaet L, Greibrokk T, Lundanes E, Malerod H (2011) Critical assessment of accelerating trypsination methods. J Pharm Biomed Anal 56:1069–1078
143. Shevchenko A, Tomas H, Havlis J, Olsen JV, Mann M (2006) In-gel digestion for mass spectrometric characterization of proteins and proteomes. Nat Protoc 1:2856–2860
144. Green NS, Reisler E, Houk KN (2001) Quantitative evaluation of the lengths of homobifunctional protein cross-linking reagents used as molecular rulers. Protein Sci 10:1293–1304
145. Nadeau OW, Carlson GM (2002) Chemical cross-linking in studying protein-protein interactions. In: Golemis E (ed) Protein-protein interactions : a molecular cloning manual. Cold Spring Harbor Laboratory Press, New York, pp 75–91
146. Sobott F, Hernandez H, McCammon MG, Tito MA, Robinson CV (2002) A tandem mass spectrometer for improved transmission and analysis of large macromolecular assemblies. Anal Chem 74:1402–1407
147. Benesch JL, Robinson CV (2006) Mass spectrometry of macromolecular assemblies: preservation and dissociation. Curr Opin Struct Biol 16:245–251
148. Benesch JL, Ruotolo BT, Simmons DA, Robinson CV (2007) Protein complexes in the gas phase: technology for structural genomics and proteomics. Chem Rev 107:3544–3567
149. Hernandez H, Robinson CV (2001) Dynamic protein complexes: insights from mass spectrometry. J Biol Chem 276:46685–46688
150. Hernandez H, Dziembowski A, Taverner T, Seraphin B, Robinson CV (2006) Subunit architecture of multimeric complexes isolated directly from cells. EMBO Rep 7:605–610
151. Lane LA, Nadeau OW, Carlson GM, Robinson CV (2012) Mass spectrometry reveals differences in stability and subunit interactions between activated and nonactivated conformers of the (alphabetagammadelta)4 phosphorylase kinase complex. Mol Cell Proteomics 11:1768–1776
152. Fitzgerald TJ, Carlson GM (1984) Activated states of phosphorylase kinase as detected by the chemical cross-linker 1,5-difluoro-2,4-dinitrobenzene. J Biol Chem 259:3266–3274
153. Zhang Y (2008) I-TASSER server for protein 3D structure prediction. BMC Bioinf 9:40
154. Herzog F, Kahraman A, Boehringer D, Mak R, Bracher A, Walzthoeni T, Leitner A, Beck M, Hartl FU, Ban N, Malmstrom L, Aebersold R (2012) Structural probing of a protein phosphatase 2A network by chemical cross-linking and mass spectrometry. Science 337:1348–1352
155. Rappsilber J (2011) The beginning of a beautiful friendship: cross-linking/mass spectrometry and modelling of proteins and multi-protein complexes. J Struct Biol 173:530–540
156. Schneidman-Duhovny D, Pellarin R, Sali A (2014) Uncertainty in integrative structural modeling. Curr Opin Struct Biol 28:96–104
157. Zeng-Elmore X, Gao XZ, Pellarin R, Schneidman-Duhovny D, Zhang XJ, Kozacka KA, Tang Y, Sali A, Chalkley RJ, Cote RH, Chu F (2014) Molecular architecture of photoreceptor phosphodiesterase elucidated by chemical cross-linking and integrative modeling. J Mol Biol 426:3713–3728

Part V

Clinical Proteomics

Introduction to Clinical Proteomics 20

John E. Wiktorowicz and Allan R. Brasier

Abstract

Within the context of this section, biomarkers are defined as a panel of proteins and peptides that are predictive of the risk for developing a pathological condition. It is important to note here that the use of the descriptor 'panel' is purposeful in that single "biomarkers" are rarely sufficient to permit accurate prediction of a pathological condition. More specifically, the primary application of a biomarker panel is that it serves as a molecular indicator of the severity of a disease or its early response to treatment. In this way, biomarkers enable the application of precision medicine, an approach that tailors specific interventions to those individuals that would most benefit. For a recent comprehensive review of the proteomic-based biomarker development process with a focus on bladder cancer, the reader is directed to Frantzi et al. [Clin Transl Med 3:7, 2014], or a special issue with multiple reviews [Stuhler and Poschmann, Biochim Biophys Acta Proteins Proteomics 1844:859–1058, Elsevier, B V, 2014].

Keyword

Clinical proteomics

20.1 Overview

Within the context of this section, biomarkers are defined as a panel of proteins and peptides that are predictive of the risk for developing a pathological condition. It is important to note here that the use of the descriptor 'panel' is purposeful in that single "biomarkers" are rarely sufficient to permit accurate prediction of a pathological condition. More specifically, the primary application of a biomarker panel is that it serves as a molecular indicator of the severity of a disease or its early response to treatment. In this way, biomarkers enable the application of precision medicine, an approach that tailors specific interventions to those individuals that would

J.E. Wiktorowicz (✉) • A.R. Brasier
The University of Texas Medical Branch,
Galveston, TX, USA
e-mail: jowiktor@utmb.edu

H. Mirzaei and M. Carrasco (eds.), *Modern Proteomics – Sample Preparation, Analysis and Practical Applications*, Advances in Experimental Medicine and Biology 919, DOI 10.1007/978-3-319-41448-5_20

most benefit. For a recent comprehensive review of the proteomic-based biomarker development process with a focus on bladder cancer, the reader is directed to Frantzi et al. [1], or a special issue with multiple reviews [2].

Despite the great interest in biomarkers and their potential impact in clinical practice, there have been surprisingly few biomarker panels that have been translated to clinical practice. A recent survey (2001–2014) of PubMed yielded 241 papers describing biomarker studies using a proteomic approach. Unfortunately, as has been noted extensively, very few have advanced to a Verification stage, much less Validation. The reasons for this barren landscape are manifold, and we will examine a few significant issues below.

The currently accepted process [3] that leads to biomarker approval for clinical use is summarized in the column labeled "Phase" in Fig. 19.1. The biomarker development process proceeds in distinct phases in which protein markers are initially identified, assayed, and selected for optimal performance. For the purposes of this work, Discovery is a phase that employs a broad survey of proteins using semi-high throughput assays. Qualification is a phase involving independent measurement of differentially expressed proteins, typically within the Discovery samples. Verification refers to confirmation of the differentially expressed proteins within an independent, second clinical cohort. As noted, there is an inverse relationship between the number of candidates and the number of samples as the candidates move through the confirmation process. Survival of a candidate marker is dependent upon the quality of the quantitative analysis and statistical tools used to narrow the field in this analysis, the authors emphasized a mass spectrometry approach for discovery through verification, as well as argued for analyses of proximal biofluids. Published in 2006, the conclusions drawn were optimistic in that if the suggested approaches were utilized by the proteomics biomarker community, greater numbers of biomarkers would survive the

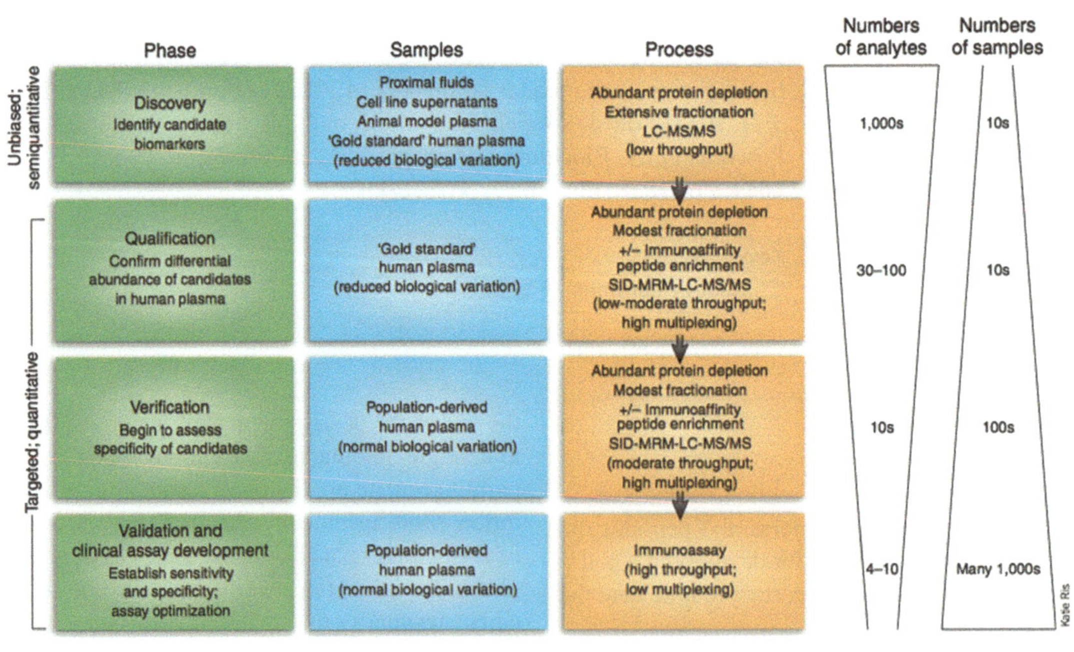

Fig 20.1 Process flow for the development of novel protein biomarker candidates [3]. 'Numbers of analytes' refers to the number of proteins expected to be evaluated as candidate biomarkers in each phase of development. 'Numbers of samples' refers to the sample requirements for each phase. *LC-MS/MS* liquid chromatography tandem mass spectrometry, *SID* stable isotope dilution, *MRM* multiple reaction monitoring (Reprinted by permission from Macmillan Publishers, Ltd: Rafai, N. et al., Nature Biotechnology, 24:971–983, 2006)

development process. Many of these suggestions were implemented in the proteomic biomarker publications since then, unfortunately, up until this year, the barren landscape of proteomically derived validated biomarkers remains essentially unchanged. Clearly, additional factors have confounded the attempts to bring candidate biomarkers to full biomarker status, and our purpose here is to provide an overview of the challenges that we have encountered in our own research.

Our experience has led to some modification of the development workflow described in Fig. 19.1. To summarize the salient points, success in biomarker identification critically relies on disease definition, consideration of the goals of the biomarker panel, the strategy for selection of candidate biomarkers that will constitute the panel, statistical modeling, and alternative quantitative proteomic tools to identify the most robust markers. These steps are then followed by feature evaluation, model evaluation, and if necessary, model refinement prior to Verification (Fig. 19.2).

20.2 Candidate Biomarker Selection

The initial assembly of candidate biomarkers will involve a combination of both prior knowledge of pathophysiology with and quantitative (or semi-quantitative) proteomics surveys of relevant animal models or patient derived biofluids from well-designed clinical studies. At the outset, it is critical to define the disease for which the biomarker is being developed. A "disease" must be identifiable using objective criteria that are reproducible across multiple sites and are independent of observer bias. It is not uncommon in clinical practice for diseases to be diagnosed

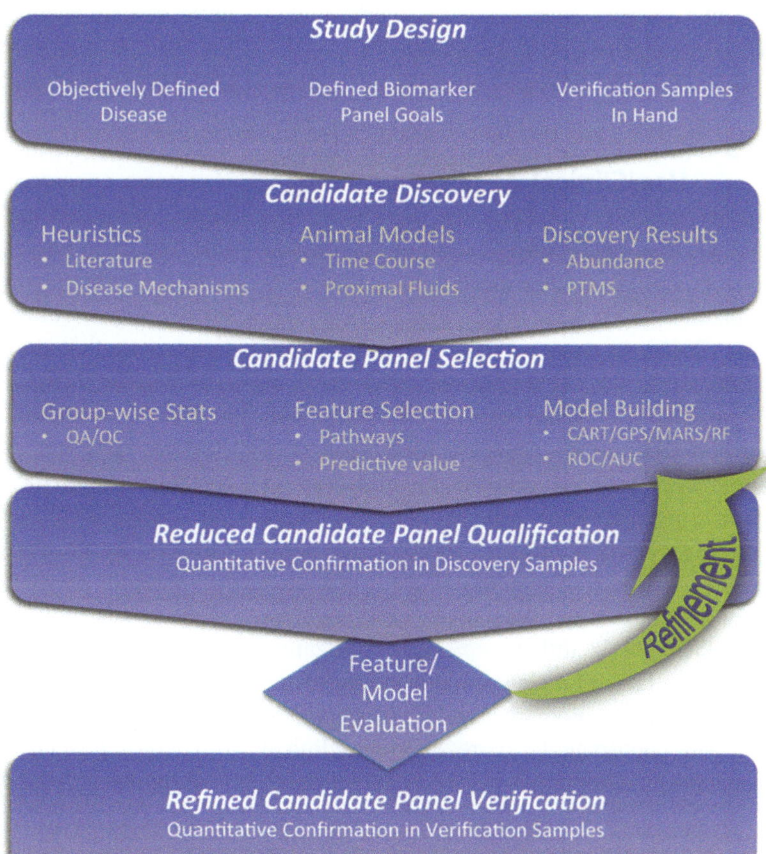

Fig. 20.2 Modified biomarker development process. The general terminology remains unchanged, however, greater attention is paid to study design, where investigated diseases should be well defined with clear diagnostic criteria, and with second sample set for a statistically powered Verification phase. Further modifications include the use of a heuristics and animal models to supplement candidate markers identified in the Discovery phase. Finally, the discovery marker set is reduced with appropriate statistical tools and used to create models correlating their pattern of abundance with the relevant goals of the study. This model is evaluated and refined upon confirmation after the Qualification phase

using a combination of criteria. For example, the diagnosis of dengue hemorrhagic fever is made on the basis of variable types of hemorrhage, plasma leak syndromes, and hemoconcentration [4]. The diagnosis of severe asthma is made on the basis of a constellation of symptoms, pharmacological responses, and symptomatic controls [5]. Rheumatoid arthritis is a syndrome of joint stiffness, cutaneous manifestations, and variable amounts of joint destruction. The important point here is that despite exhibiting similar constellations of signs and symptoms that satisfy a clinical definition of a disease, the patients may exhibit distinct pathophysiologic processes that will contribute to variability in the selection of biomarker panels. For example, petechiae in dengue hemorrhagic fever may be due to antibody-induced thrombocytopenia, whereas the plasma leak may be due to complement-mediated endothelial damage. Qualifying or verifying biomarkers in populations with these distinct pathophysiologies may result in markers that may generalize only to a subpopulation.

20.3 Considerations in Biomarker Use/Clinical Application

Another important consideration in biomarker development is whether the biomarker panel is actionable. In the case of dengue hemorrhagic fever (DHF), in endemic dengue regions, DHF is a relatively rare event, estimates of 5–10 % of all patients with acute dengue infections will manifest DHF. In this case, the identification of a biomarker panel is valuable for clinicians in resource poor areas to prioritize which patients should be closely monitored, and/or given intravenous hydration. Conversely if the application of the biomarker will not impact case management, there will be little acceptance or utilization of the test. Having a clear understanding of the application of the biomarker and how its application will contribute to more efficient clinical management is important in project selection.

Candidate biomarkers are selected from multiple sources of information. An important source of knowledge is prior pathophysiological studies, when this information is available. Information from relevant animal models, when these are available, can be valuable to select candidate markers for Qualification. Animal models can be useful for several reasons:

- Inbred animal strains have reduced genetic variations that contribute to distinct protein expression patterns
- The timing and onset of disease can be more precisely controlled than possible in human clinical studies
- Proximal biofluid sampling is possible

When these sources of information are limited, the final source of candidates comes from quantitative or semi-quantitative proteomics samples from observational human studies. This latter domain requires an objective definition of the disease and meticulous control of sample collection protocols. Sample collection protocols must be standardized and assiduously implemented, and patient information (day symptoms appeared, etc.) must be noted accurately. Many biomarker studies are on diseases that are relatively rare and therefore require multi-site clinical design. In this setting, the presence of disease must be identifiable using objective criteria that are reproducible across multiple sites and are independent of observer bias.

Finally, because the current publication environment requires confirmation of candidate markers through the Verification stage, it is critical to have identified a second, larger cohort of samples to be used for confirmation before initiating the biomarker Discovery stage.

As a side note, it is appropriate to point out that in the past, the published panels of Qualified markers have been used to inform targeted proteomic studies and these have led to Verified biomarker panels currently commercialized or undergoing clinical trials [6, 7]. However, this source of potential targets is now in jeopardy, due to new journal publication guidelines that require confirmation of candidate markers through the Verification phase before manuscripts are even accepted for review. Continuation of this policy will impose severe consequences to the field due to two factors:

- Few academic researchers have the resources to fund the large effort required for candidate Verification (second, larger sample cohort)
- The policy will hinder the acquisition of such funding by preventing the necessary publication(s) of supportive preliminary studies

A more considered approach to break the vicious circle would permit publication at the Qualification stage (same samples, alternative quantification/identification tools) provided that confirmation is robust and statistically valid; otherwise, this historically rich source of potential targets will disappear, greatly increasing the difficulty and cost of developing new effective biomarker panels.

20.4 Discovery (Chap. 20)

As a starting point, we define proteomics biomarker discovery as the global proteomic analysis of a sufficient number samples that can ensure a power of at least 0.8 that will result in a panel of candidate biomarkers. This usually results in the estimate of 30+ samples each for case and control.

The type and source of samples (biofluids, proximal fluids, tissues, cell, etc.) dictate the range of analytical options that can be applied. Since proteomics discovery encompasses a multi-step, multi-technique workflow, each sample type requires a customized strategy for separations, quantification, and identification. Since there are only 20,000+ genes and, by some estimates, more than 1,000,000 protein isoforms, the vast majority of proteomic complexity is encompassed by post-transcriptional mechanisms. Accordingly, comparisons of case and controls to extract only differences in abundance should not be the only goal pursued. Careful selection of analytical approaches must reflect the need to detect and quantify these post-transcriptional modifications. These considerations also drive separation, quantification, and identification approaches. Finally, a comprehensive search of the literature can provide additional inputs into the list of candidates.

As a statement of general principles for discovery, protein losses must be minimized so that quantification can be accurate and precise. We have used all liquid fractionation to limit the possibility of irreversible surface binding in the presence of denaturant (e.g., urea). In the case of biofluids, where high abundance proteins bind large numbers of peptides and proteins, urea also serves to dissociate any protein interactions. To track and permit normalization of protein losses, internal standards must be added as early as possible in the sample extraction/preparation phase.

20.5 Candidate Panel Selection/ Statistical Approaches (Chap. 21)

Typically upon quantification of protein/peptide signals, a "first level" of statistical analysis (e.g., non-parametric t-test or ANOVA) will establish a level of statistical significance to a narrowed list of candidates, decreasing the demands placed upon Qualification. Statistical methods involve not only difference testing, but need to inform candidate biomarker selection by incorporating additional information, including group-wise variance and identification of correlated markers. An important source of candidate biomarkers includes incorporation of heuristics to assemble candidate markers for Qualification and Verification. These and other factors will be discussed in detail in Chap. 20 – Discovery. Similarly, after each phase of candidate biomarker development, a combination of statistical approaches, including non-parametric hypothesis testing, feature reduction, hierarchical and non-hierarchical cluster analysis, and model building with receiver-operator analysis is used to confirm selection of candidate panels and appropriate predictive models.

20.6 Qualification (Chap. 22)

Qualification is defined as the workflow for confirming (or rejecting) the accuracy of the statistically selected list of proteins and peptides and/or the PTMs developed in the Discovery phase. By definition, the samples to be assayed are the identical samples used in Discovery, but analyzed by an alternate, quantitative, higher-throughput technique. While the expectation that the exact levels of abundance or PTM changes will be confirmed is not expected, a statistically derived *trend* consistent with the analytes' behaviors in the Discovery phase is required for confirmation. Finally to determine if biological mechanism(s) may be rationalized pathway analysis may be applied to examine networks and multivariate classification of patients and variably expressed proteins identified in the Discovery phase.

The process of feature selection, statistical modeling and Qualification may be an iterative process. It sometimes is the case that features initially identified in Discovery do not exhibit significant differences between cases and controls, or that the models do not perform well. In this case, the selection, statistical modeling and Qualification process may be repeated (indicated by yellow arrow in Fig. 19.2).

20.7 Verification (Chap. 22)

Verification is likewise defined as a confirmatory analysis of the qualified surviving candidates, but performed on an entirely different set of samples, whose numbers satisfy statistical power considerations for the analytical approach to be taken. Obviously, a critical consideration is the need for the samples to have been selected according to the identical clinical endpoints, objectively derived. Any deviation from the original selection criteria will lead to errors and non-confirmation of the qualified candidates. We will discuss the approach of Verification not of individual markers, but of the marker panel.

20.8 UTMB Clinical Proteomics Center (CPC)

The UTMB CPC was composed of 11 investigators organized into seven technology teams funded through a 6 year contract mechanism. The two major goals were to:

1. develop, standardize, and apply a protein biomarker discovery pipeline that incorporates quantitative pre-fractionation, 2-dimensional gel electrophoresis (2DE) and tandem liquid chromatography (LC)-mass spectrometry (MS), including MS-based identification
2. Develop predictive models of infectious outcomes that will drive further studies in Validation by collaborating investigators

The scope of our work was to serve as a proteomics resource for early stages of biomarker development (Discovery through Verification) for human cohort studies proposed by clinical investigators in the scientific community. During the conduct of the CPC contract, five projects were approved:

1. To identify a predictive panel of severity of dengue infection
2. To identify predictors associated with *Helicobacter pylori* induced peptic ulcers
3. To identify predictors of chagasic cardiomyopathy
4. To identify diagnostics of invasive aspergillosis in immunosuppressed patients
5. To identify predictors of acute rickettsial disease in acute spotted fever cases

Each project was unique in starting material and proteomic discovery platform and the development process followed the path described above. Our contract did not provide resources for Validation. During the conduct of the program, the biomarker development program evolved to better address the challenges in biomarker candidate selection and model development/refinement.

The following chapters will describe our refinement of the biomarker development strategy, and includes the separations, statistical, and mass spectrometric approaches we used to identify and confirm candidate biomarkers for the projects enumerated above. Our goals were to utilize a broad spectrum of proteomics tools to generate predictive candidate markers in concert with our NIAID Clinical Proteomics Center mandate to provide a panel of effective candidates that could be carried through to the Validation phase by a subsequent funding mechanism.

References

1. Frantzi M, Bhat A, Latosinska A (2014) Clinical proteomic biomarkers: relevant issues on study design & technical considerations in biomarker development. Clin Transl Med 3:7
2. Stuhler K, Poschmann G (2014) Biomarkers: a proteomic challenge. Biochim Biophys Acta Proteins Proteomics 1844:859–1058, Elsevier, B V
3. Rifai N, Gillette MA, Carr SA (2006) Protein biomarker discovery and validation: the long and uncertain path to clinical utility. Nat Biotechnol 24:971–983
4. World Health, O (2009) Dengue: guidelines for diagnosis, treatment, prevention and control. World Health Organization, Geneva
5. ad-hoc writing committee of the Assembly on Allergy, I. a. I (2000) Proceedings of the ATS workshop on refractory asthma. Current understanding, recommendations, and unanswered questions. Am J Respir Crit Care Med 162:2341–2351
6. Sun W, Hu G, Long G, Wang J, Liu D, Hu G (2014) Predictive value of a serum-based proteomic test in non-small-cell lung cancer patients treated with epidermal growth factor receptor tyrosine kinase inhibitors: a meta-analysis. Curr Med Res Opin 30:2033–2039
7. Li X-j, Hayward C, Fong P-Y, Dominguez M, Hunsucker SW, Lee LW, McLean M, Law S, Butler H, Schirm M, Gingras O, Lamontagne J, Allard R, Chelsky D, Price ND, Lam S, Massion PP, Pass H, Rom WN, Vachani A, Fang KC, Hood L, Kearney P (2013) A blood-based proteomic classifier for the molecular characterization of pulmonary nodules. Sci Transl Med 5:207ra142

Discovery of Candidate Biomarkers **21**

John E. Wiktorowicz and Kizhake V. Soman

Abstract

Properly performed, biomarker discovery can lead to effective candidates that can ultimately serve as predictors of disease, medical condition, define therapeutic parameters, and many other applications in medicine. Preferably, biomarkers comprise a panel of indicators, e.g. proteins and/or peptides that can be predictive or diagnostic of the medical condition of interest. Emphasis here is placed on "panel," as single candidates are rarely sufficient to provide the necessary sensitivity and specificity. To develop an effective panel that survives the development process described in Chap. 19, proper experimental design and attention to important statistical parameters are critical to ensure success. Errors in discovery can lead to an inefficient use of expensive resources, as these may not be uncovered until the latter stages in biomarker development. Hence, accuracy, precision, and an estimate of the power of the proposed analyses are critical in the discovery of the panel of candidate biomarkers by proteomic methods, as is the selection of statistical approaches to refine and appropriately reduce the dataset for subsequent confirmatory assays.

Keywords

Biomarker discovery • Plasma • Serum • Antibody depletion • Protein pool

21.1 Introduction

Properly performed, biomarker discovery can lead to effective candidates that can ultimately serve as predictors of disease, medical condition, define therapeutic parameters, and many other applications in medicine. Preferably, biomarkers comprise a panel of indicators, e.g. proteins and/or peptides that can be predictive or diagnostic of the medical condition of interest. Emphasis here is placed on "panel," as single candidates are rarely sufficient to provide the necessary

J.E. Wiktorowicz (✉) • K.V. Soman
The University of Texas Medical Branch,
Galveston, TX, USA
e-mail: jowiktor@utmb.edu

H. Mirzaei and M. Carrasco (eds.), *Modern Proteomics – Sample Preparation, Analysis and Practical Applications*, Advances in Experimental Medicine and Biology 919, DOI 10.1007/978-3-319-41448-5_21

sensitivity and specificity. To develop an effective panel that survives the development process described in Chap. 19, proper experimental design and attention to important statistical parameters are critical to ensure success. Errors in discovery can lead to an inefficient use of expensive resources, as these may not be uncovered until the latter stages in biomarker development. Hence, accuracy, precision, and an estimate of the power of the proposed analyses are critical in the discovery of the panel of candidate biomarkers by proteomic methods, as is the selection of statistical approaches to refine and appropriately reduce the dataset for subsequent confirmatory assays.

Simply put, the power of a study is an estimate of the chance of detecting a real difference of a given size [1]. In statistical terms and commonly used in proteomics, the power of a study is the number of samples necessary to achieve >80 % power that the null hypothesis is false and depends upon the desired level of significance, and the sample assay variance (mean and standard deviation). There are a number of software and web-based resources that can be used to estimate the number of samples necessary to achieve a certain power, given the parameters enumerated above. One very important caveat is that the power analysis must be performed *a priori*, or before, the actual experiment is performed.

In summary, care should be taken to understand the collection nuances for the tissues or biofluids to be used as the sample source, as well as selection of proper pre-separation treatments to maximize recovery and minimize artifacts. Along with these considerations, the proteomic strategy and quantification should reflect carefully chosen methods tailored to the nuances and number of samples and goals of the study.

In our discussion of these factors in this chapter, and consistent with the other chapters in this Section, we will focus on biofluid samples, in particular, plasma and serum, used to develop candidate biomarkers for our NIAID-funded Clinical Proteomics Center for Infectious Disease and Biodefense (CPC). Moreover, the goals of the following studies were to "discover" panels of proteins and peptides that could serve as a potential predictors for the risk of developing clinically severe sequelae of infectious disease (Dengue Fever and Chagas cardiomyopathy), or that could serve as diagnostic tool for the infectious agent (invasive aspergillosis). All investigations proceeded through Discovery, Qualification, and Verification as highlighted in Chap. 19. In this chapter, we will discuss Discovery in pursuit of candidates for these three diseases.

21.2 Sample Source: Plasma and Serum

In 2005, the Human Proteome Organization (HUPO) published a compendium of studies resulting from the years-long Plasma Proteome Initiative [2]. In it, the authors highlighted the tactical successes with the following recommendations:

- Selection of plasma over serum
- EDTA over heparin
- Minimum number of freeze-thaw cycles
- A number of other important procedural recommendations for the use of biofluids in biomarker discovery

They also notably highlighted significant tactical flaws, including the use of bottom-up, label-free mass spectrometric approaches, among others. The compendium was notable in its honest appraisal and wide-ranging recommendations for the improvement of biofluid-based candidate biomarker discovery, and the potential biomarker investigator would be well-served to examine this work carefully, despite its growing age.

Several important features are worth highlighting as they formed our discovery strategy for our Center projects. As had been widely noted, these include the enormous concentration range of plasma proteins (10–12 orders of dynamic range), the under-sampling that ensues upon depletion of the most highly abundant proteins that bind over 200 lower abundance proteins, and the endogenous proteases in plasma that may degrade important proteins to peptides. In addition, it is noteworthy that as much as 12 % of plasma peptides may constitute the natural peptidome [3]. As a result the peptide "degradome" [4] consists of naturally occurring peptides, which may serve as legitimate candidate markers, and those generated through artifactual influences, which may not.

The last issue becomes increasingly relevant in emphasizing the importance of standard operating procedures in plasma and sera collection. Unless specified, most plasma collection protocols do not include collection containers that have protease inhibitors or are specified for use for protein analysis (e.g., Becton-Dickinson BD™ P100 Kit). Therefore, variations in the collection and/or processing parameters may result in differences in plasma protein stability, leading to false signals in the differential analyses. To date, there is no widely accepted method for evaluating the quality of plasma for proteomic purposes. Our attempts to use capillary electrophoresis to gauge plasma quality by monitoring the four most abundant protein peaks did not lead to obvious correlation with plasma quality. Even the most highly abundant proteins show considerable variation from individual to individual, and so, at this point, the only way to assure quality is for meticulous adherence to the standard collection protocols, and limiting freeze-thaw cycles to at most two.

Typically, unbiased proteomic investigations of biofluids consist of the comparison of biofluids from normal (control) *versus* affected individuals (case). Since there is no opportunity for multiplexing *via* discriminating reagents at the pre-fractionation stage (presumably due to the expense of labeling the high abundance proteins that constitute 95–99 % of the largely uninformative total protein that will be removed in later steps), generating critically reproducible fractions for differential comparison from sequential or parallel fractionation is a challenge.

As described in Chap. 19, replicate analyses with readily available samples should be performed to estimate variances for power analysis before precious samples are processed. Because most projects involve dozens of biological samples, technical replicates at this stage are cost- and time-prohibitive, and not recommended; hence *a priori* power analysis is critical to provide an estimate of the number of samples statistically necessary.

21.2.1 Fractionation

In consideration of the above and other challenges, we have devised an efficient and reproducible platform for fractionation and quantification of both proteins and peptides from biofluids that achieves differential analysis from the same low-volume samples, called the Biofluids Analytical Platform (BAP) that we used routinely in our NIAID CPC studies (Fig. 21.1). The BAP utilizes a fluorescent internal size standard that defines a protein/peptide molecular size cutoff by which biofluid proteins are pooled separately from peptides after size-exclusion chromatography (SEC). Denatured samples (to disrupt molecular interactions) are mixed with a fluorescence-labeled protein standard and fractionated by SEC. Protein and peptide pools are created automatically according to the elution profile of the internal standard by a programmable UV monitor that controls a compatible fraction collector, thereby ensuring reproducible collection and allowing multiple columns to be used simultaneously (Fig. 21.2).

Since SEC is a non-adsorptive fractionation support, recoveries are routinely very high, and in our hands, reproducibly result in >95 % recovery of input proteins and peptides. Thus SEC permits fractionation before quantification with high recoveries, a necessary consideration in quantitative biomarker discovery. In addition, the ability to fractionate by size allows the use of urea as a denaturant prior to sample injection to ensure dissociation of peptides and proteins with the high abundance proteins. Other advantages of SEC include the ability to exchange buffers and remove small molecules (including urea) and plasma electrolytes, and the dilution of proteins as they pass through the column, minimizing the

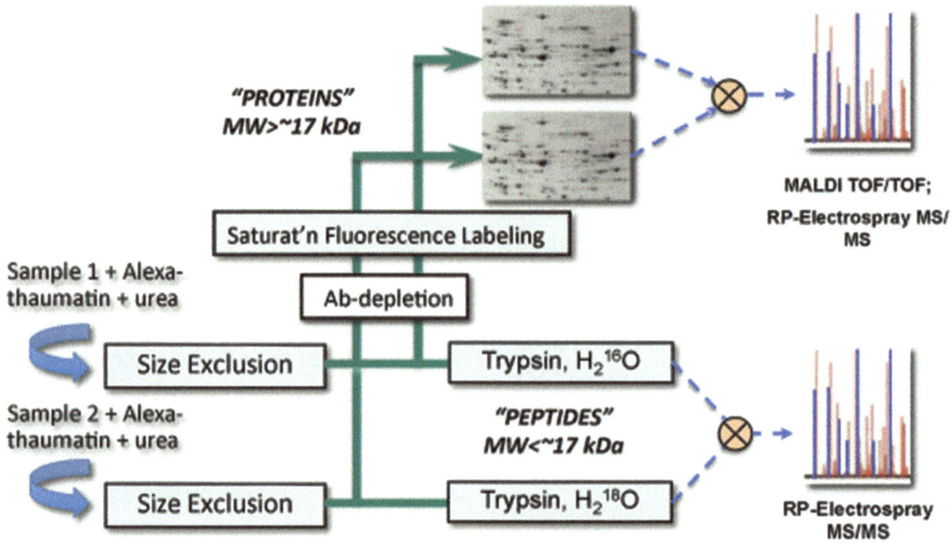

Fig. 21.1 Biofluids analytical platform. Samples are denatured with urea and thaumatin (a plant protein) labeled with Alexa-488, is added. The samples are separately loaded onto a size-exclusion column through an HPLC pump controlled by a computer. The effluent is monitored by UV spectrometer that controls a fraction collector. When the fluorescent dye is detected, the fraction collector is triggered to advance. Protein pools are defined by all fractions collected before the end of the fluorescent thaumatin peak, while peptide pools are defined by all fractions between the end of the thaumatin peak and the beginning of the free Alexa peak. The protein pools are antibody depleted of the 14 most abundant proteins in human plasma and saturation-labeled with BODIPY-Fl. Proteins are separated by 2DGE, analyzed, and identified by MS. Peptides are labeled via trypsin-mediated oxygen exchange, pools from sample 1 and 2 mixed, and separated and analyzed by on-line RP-LC-MS/MS

re-association of high and other abundant proteins.

Since potential biomarkers may be proteins, peptides, or both, comprehensive differential analysis must permit their recovery, ideally from the same sample. A further potential benefit of analyzing both pools from the same sample is that the concordant appearance of an observed peptide and its parent protein may signal that the peptide is an artifact of proteolysis in the biofluid, and therefore unlikely as an accurate candidate biomarker.

Finally, with only 20,300 genes in the human genome, the great complexity of proteins within any biological sample primarily reflects the plethora of post-transcriptional/translational modifications (PTM). There is little theoretical justification in the assumption that candidate biomarkers will not reflect one or more PTMs, including naturally occurring proteolytic clipping or enzymatically catalyzed cross-linking. This also justifies the top-down strategy of fractionating and separating intact proteins and peptides, where their PTM status might be lost in a peptide-centric, bottom-up approach. As will be seen below, we identified both PTM classes as engendering candidate biomarkers in several of our CPC projects.

21.2.2 Antibody Depletion

After BAP fractionation, proteins are largely disassociated and diluted, so depletion of the most abundant of them is performed without fear of excessive losses. Optimization of the depletion, however, is critical, and typically is monitored by 1D SDS-PAGE of the untreated protein pool, the depleted pool, and the proteins recovered from the depletion columns. To establish enrichment of non-abundant proteins, equal amounts of proteins are loaded in each lane of the gel. If depletion was efficient, comparison with the undepleted lane should reveal that the high

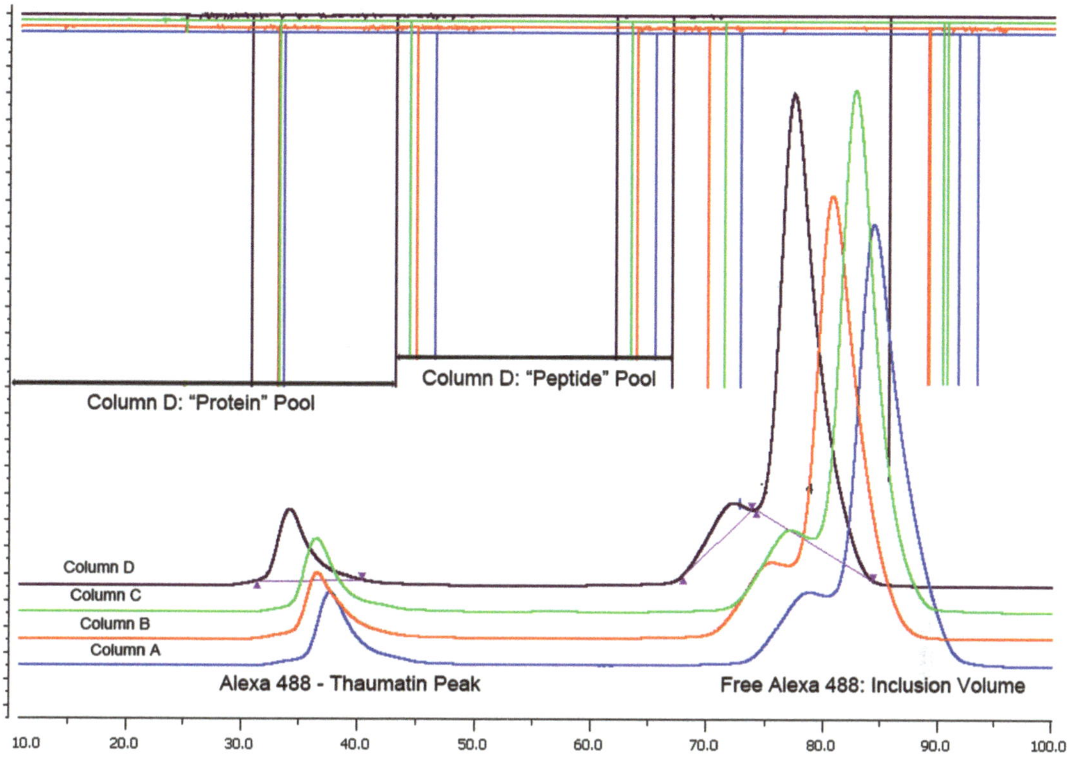

Fig. 21.2 Biofluids analytical platform separations output. In its current configuration, the BAP consists of four low-pressure columns (**A–D**) and generate four separate protein and peptide pools, as described in the text and Fig. 20.1 legend. When peaks are detected by the UV monitor, an event marker (vertical line) is placed on the chart. The event markers and the fractions used for the protein and peptide pools for "Column D" are highlighted in the Figure as examples. Note, no plasma-specific protein signals are detected, and therefore pool compositions are strictly governed by the thaumatin retention time internal standard, providing maximum fractionation reproducibility from sample to sample.
Note: UV tracings are purposely shifted to permit uncomplicated viewing

abundance proteins should be diminished in intensity, while the faint or undetectable proteins should be enriched in the depleted lane. The depleted proteins should appear in equal intensity in the recovered gel lane.

21.3 Analysis

We segment our discussion into analyses specific to the protein pool and those specific to the peptide pool. At the completion of the statistical reduction of the candidate protein and peptide biomarkers, we convolve the lists to determine overlaps and uniqueness of these panels. Where overlaps exists, we select the protein partner as the candidate biomarker, to be added to the list of other proteins and remaining uniquely appearing peptides to generate a comprehensive set of candidate biomarkers for subsequent confirmation.

21.3.1 Protein Pools

The protein pool, consisting of diluted proteins eluting before the end of the internal standard peak (23 kDa ≥ Alexa-thaumatin ≥ ~17 kDa; Fig. 21.1), is permitted to partially renature during its slow elution and subsequent storage at 4 °C overnight. After this period, the pool undergoes antibody depletion (IgY) by two successive passages through a column of antibodies specific for the 14 most abundant plasma proteins (Sigma-Aldrich).

Because technical replicates are not performed, we select covalent saturation labeling for protein pools that are specific for cysteine residues [5, 6]. Since all proteomic analyses alkylate cysteines before analysis, and >92 % of human proteins contain at least one cysteine residue [7], we simply alkylate with a fluorescent dye of high extinction coefficient at saturating ratios of dye: protein thiol (>50-fold) [5, 8] as determined by amino acid analysis. Saturation labeling with an *uncharged* (as opposed to neutral) dye, e.g. Bodipy-Fl, ensures reproducible quantification with no change in electrophoretic mobility. Separation followed by fluorescence quantification is accomplished by 2D gel electrophoresis (2DGE), and identification of differentially abundant proteins by MS/MS.

21.3.2 Peptide Pools

The peptides in the peptide pools obtained from the BAP to be compared are differentially labeled with $^{16/18}O$ by trypsin-mediated exchange under conditions previously established to ensure maximum incorporation of two oxygen isotopes at each carboxy-terminus [9]. Under these circumstances, matched controls and cases are ^{16}O and ^{18}O labeled, respectively. The incorporation of the first oxygen during tryptic exposure is catalytic and performed at pH 8.0. The second exchange is slow and non-catalytic, and is performed at pH 6.0 [10]. Maximum incorporation of ^{18}O is dependent upon peptide length and slow, so incubations are performed over 24 h (Fig. 21.3). Peptides thus labeled are mixed with their ^{16}O labeled controls, separated by RP-HPLC, and quantified and identified by a tandem electrospray MS/MS.

21.4 NIAID CPC Project 1-Dengue Fever

21.4.1 Introduction

Dengue Fever (DF) is a mosquito-borne disease caused by a single-stranded RNA virus of

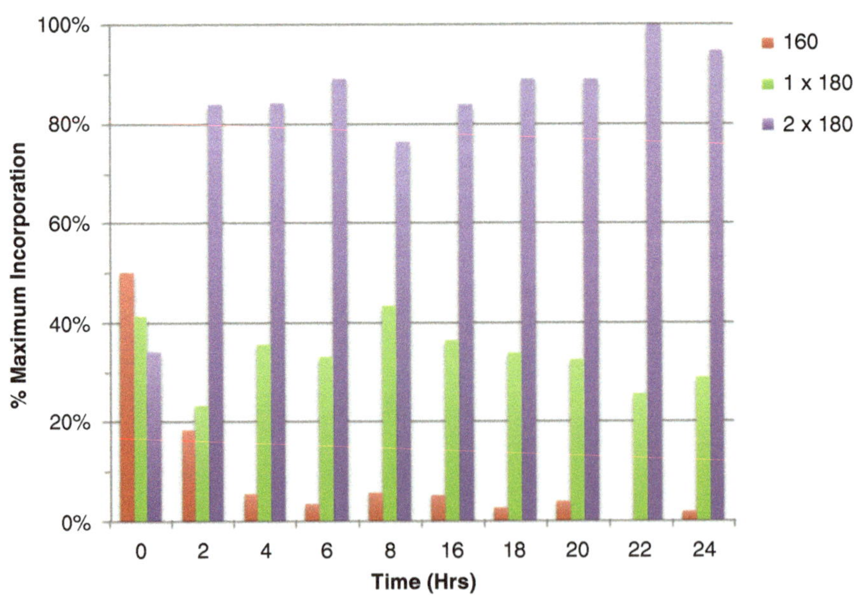

Fig. 21.3 Time course optimization of $H_2{}^{18}O$ peptide labeling. Seven peptides of varying length (945, 985, 1580, 1742, 1929, 2256, and 2759 Da) were exposed to trypsin-mediated isotope exchange for varying times as indicated in the figure. At the appropriate time, solutions were acidified with TFA and analyzed by RP-LC-MS/MS. Incorporation of the stable isotope was calculated and the averages of the seven peptides for each isotope substitution at each time point was normalized against the highest incorporation value. The maximum level of incorporation of the doubly labeled ^{18}O peptide can be seen after 22 h

4 serotypes. The mosquito thrives in tropical and sub-tropical environs in which 1/3 to 1/2 of the world's population lives. Initial infection confers life-long immunity against subsequent identical serotype infection, however, not against heterologous infection. Most infections are self-limiting, however, a small percentage of infected individuals develop a life-threatening syndrome characterized by vascular leakage and hemorrhage (Dengue Hemorrhagic Fever-DHF)—defined by classical WHO criteria—or severe complicated Dengue disease (DFC), defined as not satisfying WHO criteria for DHF, but who nevertheless exhibited hemorrhagic or thrombocytopenia within 7 days of the onset of symptoms. Early intervention of supportive therapy for these individuals significantly enhances survival. However, in challenging environments, the establishment of risk for developing DHF immediately upon clinical presentation would be of critical importance for the health of the patient. Because of this, the pursuit of biomarkers has resulted in literature suggestive of several candidate biomarker leads [11–14], although none have been confirmed by quantitative studies as required for proper candidate biomarker Qualification and Verification (see Chap. 22). Our challenge was therefore to identify a panel of candidate biomarkers that could accurately define the risk of developing the survivable hemorrhagic form of the infection and by extension the normal Qualification and Verification stages of biomarker development.

21.4.2 Study

Our Center was presented with two DF proposals, both representing Latin American cohorts. Only one (DF-Brazil), however, had sufficient numbers of patients for Discovery as well as a second cohort to take through qualification [15, 16]. As described in Chap. 22, the goal of our studies was to develop candidate biomarkers that might define the risk of developing DHF or DFC, so our discovery study design focused on plasma collected from 110 patients presenting in the acute stage with normally resolving DF ($n = 59$) compared to those who later developed DHF ($n = 22$), or DFC ($n = 29$).

21.4.3 Analysis

After 2DGE separations and analysis of the BAP protein and peptide pools, 1311 proteins were quantified, and 121 were judged significantly changed in DHF with respect to DF using non-parametric statistical analysis [16]. To reduce the candidate panel further, statistical tools were used (Chap. 19) and as a result, the panel was reduced to 15 proteins that accurately classified patients into DF, DHF, and DFC phenotypes. The significant proteins identified are listed in Table 21.2 and the feature-reduced set of 15 proteins in Table 21.3. The significant peptides are found in the analysis are listed in Table 21.4. Note that the Dengue NS1 protein and Complement factors in Table 21.3 were not obtained from Discovery, but resulted from heuristics methods.

The peptides from the BAP peptide pools were quantified by $^{16/18}O$ ratios, i.e. acute (^{18}O-labeled) and convalescent (^{16}O-labeled) samples from the same patient were mixed before MS, and the log2 normalized ratios for each peptide detected from DF and DHF samples were compared by t-test (Table 21.4). The three statistically significant peptides from DF and DHF all showed increased abundance in DHF.

Of the proteins and peptides identified, several are most useful as justification of our strategy to pursue post-translational modifications with analysis of intact proteins. One notable example is the high molecular size albumin (~200+ kDa). This protein is not depleted by the depletion antibodies, and is diminished in the patients suffering from the hemorrhagic sequel of DF compared to normal plasma or patients who resolved their uncomplicated DF [15, 16]. While the biochemistry is currently under investigation in our facility, the protein is likely covalently cross-linked by some cross-linking agent that appears depleted from the viral infection. It is clear that a "bottom-up"

Table 21.1 Summary of discovery results from the analysis of proteins and peptides in the NIAID-CPC projects discussed in the chapter

Project	Total number analyzed		Significant from discovery		Predictive features by *MARS* or other methods	
	Protein spots	Peptides[a]	Protein spots	Unique Peptides	Protein spots	Peptides
Dengue fever, Brazil	1311	8873	121	639	6	3
Invasive Aspergillosis	556	4402	66	360	9	3
Chagasic cardiomyopathy[b]	635	ND	36	ND	7	ND

[a]Probability ≥ 0.95; FDR < 0.01; Analysis by ProteoIQ™ (Premier Biosoft)
[b]Discovery was performed on purified PBMCs

Table 22.2 Proteins identified from 2DGE gel spots in the dengue fever project

Spot #	Spot pI[a]	Spot MW (kDa)[a]	Protein name	UniProt accession	Peptide count[b]	Seq. coverage[c]	Fold change	p (t-test)
Identified by MALDI TOF/TOF								
73	7.20	>250	Complement C3	P01024	21	18.3	−1.81	0.03335
80	6.18	>250	Serum albumin; Flags: Precursor	P02768	14	29.2	−1.33	0.06969
83	7.30	200	Complement C3	P01024	26	22.7	−1.72	0.01575
201	5.74	119	Alpha-2-macroglobulin	P01023	21	20.6	−1.43	0.01962
204	5.80	117	Alpha-2-macroglobulin	P01023	22	21.2	−1.46	0.01788
224	3.55	100	Keratin, type I cytoskeletal 10	P13645	16	32.4	−2.80	0.06689
306	7.30	80	Complement C3	P01024	31	27.2	−2.18	0.04624
330	9.19	71	Complement C4-B	POCOL5	18	13.8	−1.74	0.05309
335	9.09	70	Complement C4-B	POCOL5	20	15.9	−1.79	0.05494
434	9.20	49	Ig gamma-1 chain C region	P01857	6	23.6	−2.41	0.08858
444	9.09	48	Ig gamma-1 chain C region	P01857	6	23.6	−2.71	0.06011
486	5.27	44	Antithrombin-III	P01008	13	34.9	−1.64	0.04141
719	9.92	31	Keratin, type I cytoskeletal 10	P13645	16	33.9	1.38	0.01334
784	6.89	28	Complement C3	P01024	15	8.9	−1.52	0.06435
1434	4.11	13	Keratin, type II cytoskeletal 1	P04264	16	30.4	2.03	0.10884
1483	4.13	12	Keratin, type I cytoskeletal 9	P35527	19	49.4	1.91	0.01366
1516	7.96	11	Keratin, type I cytoskeletal 9	P35527	20	50.08	1.43	0.05599
Identified by LC-MS								
6	4.76	>250	Alpha-2-macroglobulin	P01023	5	3.7	−1.54	0.06468
81	6.68	>250	Complement C3	P01024	11	9.0	−1.69	0.05986
85	3.62	191	Desmoplakin	P15924	7	2.8	−2.00	0.02492
90	3.49	184	Alpha-2-macroglobulin	P01023	15	12.1	−2.01	0.01044
94	6.40	184	Alpha-2-macroglobulin	P01023	12	10.4	−1.15	0.07054
108	6.55	181	Alpha-2-macroglobulin	P01023	10	7.6	−1.56	0.04054
127	4.86	169	Alpha-2-macroglobulin	P01023	6	4.3	−1.52	0.01788
221	6.53	102	Complement factor B	P00751	3	2.2	−1.43	0.03297
303	7.69	81	Isoform 2 of Complement C4-A	POCOL4	7	4.1	−1.93	0.03246
325	9.39	73	Isoform 2 of Complement C4-A	POCOL4	10	6.2	−1.72	0.05882

(continued)

Table 22.2 (continued)

Spot #	Spot pI[a]	Spot MW (kDa)[a]	Protein name	UniProt accession	Peptide count[b]	Seq. coverage[c]	Fold change	p (t-test)
350	9.78	69	Isoform 2 of Complement C4-A	P0C0L4	7	4.2	1.38	0.00536
373	3.36	63	Alpha-1-antichymotrypsin	P01011	2	5.7	−2.04	0.00162
385	8.08	61	Complement C3	P01024	2	1.4	−1.52	0.03671
394	3.36	60	Alpha-1-antichymotrypsin	P01011	4	9.9	−2.48	0.02543
411	3.40	56	Ig gamma-1 chain C region	P01857	5	19.4	−3.23	0.00698
421	3.47	52	Ig gamma-1 chain C region	P01857	3	12.1	−2.69	0.01273
441	4.73	48	Alpha-1-antichymotrypsin	P01011	2	5.7	−1.23	0.03502
450	8.66	48	Ig gamma-1 chain C region	P01857	2	9.4	−1.62	0.04675
451	8.91	48	Ig gamma-1 chain C region	P01857	5	19.4	−2.11	0.04242
457	5.73	47	Fibrinogen gamma chain	P02679	3	6.5	−1.16	0.03002
458	8.06	47	Ig gamma-1 chain C region	P01857	3	13.0	−1.60	0.05222
465	4.49	46	Leucine-rich alpha-2-glycoprotein	P02750	2	5.5	−1.43	0.02089
493	5.07	44	Alpha-1-antitrypsin	P01009	2	4.8	−2.61	0.00261
506	9.47	43	Isoform 2 of Complement C4-A	P0C0L4	2	1.1	1.48	0.02166
539	6.51	39	Plasma serine protease inhibitor	P05154	4	10.6	−1.75	0.00800
546	4.79	39	Desmoplakin	P15924	24	10.9	−1.42	0.02830
556	5.38	38	Complement C3	P01024	5	3.7	−1.41	0.05469
563	5.07	38	Zinc-alpha-2-glycoprotein	P25311	3	10.4	−1.43	0.02424
564	5.24	38	Complement C3	P01024	3	2.8	−1.41	0.01681
565	5.29	38	Haptoglobin	P00738	2	6.4	−1.54	0.02545
566	5.52	38	Haptoglobin	P00738	1	3.0	−1.74	0.02893
567	5.67	38	Complement C3	P01024	5	4.4	−1.85	0.05336
584	5.68	36	Apolipoprotein E	P02649	3	12.0	−1.69	0.06733
604	7.78	36	Complement C3	P01024	11	9.3	−1.62	0.02817
721	4.83	31	Isoform 2 of Clusterin	P10909	55	2.0	1.60	0.01231
776	6.55	29	Complement C3	P01024	4	3.3	−1.44	0.03974
891	3.45	23	60 kDa heat shock protein, mitochondrial	P10809	2	9.1	−2.08	0.07175
964	5.77	20	Junction plakoglobin	F5GWP8	6	19.6	1.65	0.00919
1031	9.82	19	Ig kappa chain C region	P01834	2	35.8	1.98	0.00745
1138	6.38	18	Haptoglobin	P00738	2	6.2	−1.99	0.02162
1159	5.35	18	Haptoglobin	P00738	2	6.2	−1.93	0.05555
1232	8.03	17	60 kDa heat shock protein, mitochondrial	P10809	1	5.2	1.41	0.02449
1256	7.06	16	Desmoplakin	P15924	11	4.6	1.57	0.01826
1318	7.54	15	Isoform 2 of Dermcidin	P81605	3	21.5	2.42	0.00207
1416	9.06	13	Serum amyloid A-4 protein	P35542	2	14.6	1.47	0.01291
1459	9.32	12	Alpha-1-antitrypsin	P01009	2	4.8	1.42	0.00300
1490	8.44	12	Isoform 2 of Dermcidin	P81605	2	11.6	1.67	0.03968

[a]The pI and MW are from 2DGE spots by gel calibration
[b]"Peptide count" for MALDI and "Exclusive unique peptide count" for LC-MS identifications
[c]Percent of the protein sequence covered by the mapped peptides

Table 21.3 Predictive biomarkers for dengue fever[a]

Biological function	Biomarker candidate	Short name	Swiss protein accession
Dengue	Dengue NS1[b]	NS1	Q67431
Complement	Complement factor 4A[b]	CO4A	POCOL4
	Complement factor H[b]	CFH	P08603
	Complement factor D[b]	CFD	P00746
Acute phase reactant	A2-macroglobulin	A2M	P01023
	Alpha 1 anti-trypsin	A1AT	P00760
	Fibrinogen, alpha	FIBA	P02671
	Fibrinogen, beta	FIBB	P002675
	Ferritin, light chain	FRIL	P02792
	Haptoglobin	HPT	P00738
Plasma protein	Leucine-rich alpha2 glycoprotein	AG2L	P02750
	High MW albumin	HMWAIb	P02761
Immunoglobulin	Immunoglobulin J	IGJ	P01591
	Immunoglobulin kappa, C region	IGKC	P01834
	IgG-gamma-1, C region	IGHG1	P01857
Cytoskeletal	Keratin 1	KRT1	P04264
	Tropomyosin 4	TMPH	P67936
	Low MW desmoplakin	DESP	P15924
	Vimentin	VIME	P08670

[a]Adapted from Table 3 (Ref. [16]), with permission
[b]NS1 and complement factors were obtained by heuristics as outlined in the introduction

Table 21.4 Significant BAP peptides found in dengue fever discovery

Peptide	Protein ID	Gene name	Ratio[a]	p-value[b] (t-test)
YWGVASFLQK	Retinol-binding protein 4	RET-4	−1.53	0.028
YAASSYLSLTPEQWK	Ig Lambda-7 chain C-region	LAC-7	−1.59	0.031
DLATVYVDVLK	Serum albumin	ALBU	−2.54	0.013

[a]DHF/DF from acute vs convalescent peptide from the same patient
[b]Comparison of log2 normalized ratios from DF and DHF

approach to a candidate biomarker discovery effort would likely have missed this molecule—it would have appeared as simply albumin.

We determined from our analyses that several factors initially confounded the goals of our study, and were likely to have potentially confounded the other studies mentioned in the previous paragraph. We determined that gender plays a large role in differentiating DF and DHF phenotypes (Fig. 21.4). As can be seen from the principal component analysis (PCA), male and female DF vs DHF did not obviously cluster into separate groups, while the separate genders clearly clustered between DF and DHF.

To further investigate our findings, we took advantage of the samples that were collected at intermediate time points between the acute and convalescent times. In those studies, we observed considerable differences in the proteins identified as statistically significant, as well as sample data clustering by PCA, regardless of gender (Fig. 21.4d). However, as these times approached the onset of hemorrhage, we surmised that we were observing the development of the severe symptoms of DHF and were less predictive than diagnostic. Thus it could be argued that our analysis was somewhat underpowered, due to these additional factors.

These and other factors provide evidence for the importance of strict adherence to collection standard operating procedures, sensitivity to gender effects, collection times, and other

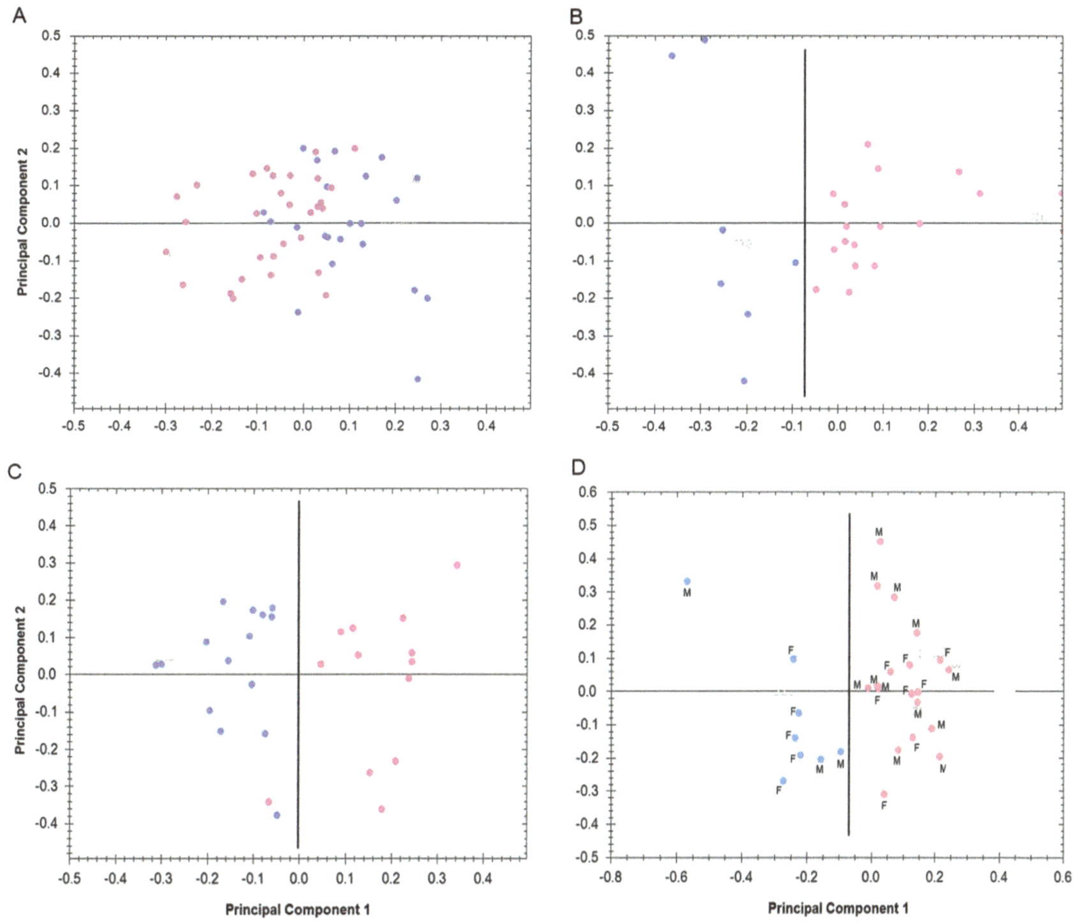

Fig. 21.4 Principal component analysis of DF and DHF, male and female. (**a**). Combined male and female DF (*pink*) and DHF (*blue*) analyses. (**b**). Male DF vs DHF. Clear separation of the two sample cohorts can be observed. (**c**). Female DF vs. DHF. Separation of the two disease states are clearly observable. (**d**). Analysis of intermediate time points for collection (days 3–6 after initial clinical presentation). Here, male (M) and female (F) samples are indistinguishable, but clear separation between DF and DHF are. Key: Each *dot* represents the behavior of each sample (gel data). DF gel data are *pink*, DHF gel data are *blue*

potentially confounding variables, as emphasized in Chap. 22.

21.5 NIAID CPC Project 2-Infectious *Aspergillosis* (IA)

21.5.1 Introduction

Aspergillus is one of the most common invasive fungal pathogens in hospitalized patients in the United States. IA accounts for the most deaths due to fungal pathogens, and is among the top three fungal killers in the world. The most common victims are patients who are immunocompromised after organ transplants, or due to AIDS, or neutropenic cancer. Infection most commonly occurs from the inhalation of airborne fungal spores. After the initial pulmonary disease, IA spreads via blood to other organs. Since blood cultures are rarely positive for the fungus, IA is difficult to diagnose and to control. The goals of this project were to:

1. Identify a panel of candidate biomarkers from plasma for rapid and accurate diagnosis

2. Confirm differential protein and peptide abundance by a targeted, qualitative and quantitative approach
3. Verify these candidate biomarkers by testing their ability to discriminate between uninfected and *Aspergillus*-infected samples, and samples infected by non-*Aspergillus* molds that produce similar clinical symptoms.

21.5.2 Study

We received and analyzed plasma samples from 34 patients clinically diagnosed with IA ("case"), from 17 patients of the cohort prior to their developing IA ("autocontrol"), and 34 subjects uninfected but matched to the infected by gender, disease, and immunosuppressed state ("matched control"). We proceeded with the sample processing and analysis protocols described in the above in the Dengue Fever.

From the analysis of the protein pools by 2D electrophoresis, we observed a total of 556 aligned spots that satisfied our criteria for quantitative analysis. An ANOVA comparison of the three sample groups—case, autocontrol, and matched control—yielded 66 differential spots ($p \leq 0.05$), which were identified by MALDI-TOF/TOF MS. These proteins are listed in Table 21.5. A feature reduction and biomarker panel development approach similar to the one used in the Dengue Fever project described above led to a predictive panel of six proteins listed in Table 21.6 (for a generalized discussion of these statistical tools, see Chap. 19).

The peptide pool was analyzed by the stable isotope method ($^{16/18}O$) as described in the "Peptide Pools" section above to compare case vs. autocontrol. Only the 15 peptides detected in at least 50 % of the samples were included in the comparison (Table 21.7). Three of these peptides (highlighted in the table) were found to be significantly different between case and control.

21.6 NIAID CPC Project 3-Chagasic Cardiomyopathy

21.6.1 Introduction

Chagas disease is a parasitic disease caused by *Trypanosoma cruzi* (*T. Cruzi*) infection, and is a serious health threat in Latin America. According to a WHO report [17], there are 16-–18 million people infected, and 25 % of the population of Latin America, i.e., ~120 million people, are at risk of infection. Due to migration and organ transplantation, it is estimated that about 300,000 infected patients live in the United States. The disease exhibits acute and chronic clinical forms. The acute phase starting several days post-infection is characterized by nonspecific symptoms, although skin reactions at the site of infection (chagoma) may be suggestive. Occasionally cardiac symptoms appear, but resolve normally within 6–8 weeks with the production of anti *T. cruzi* antibodies [18]. After recovery from the acute phase, patients enter an asymptomatic, chronic phase. About one-third of the chronic patients progress to develop cardiomyopathy in the form of an apical aneurysm as long as 30 years later, which can result in heart failure and death, or gastrointestinal abnormalities [18]. Because chronic infections are difficult to treat, early detection and treatment of those who are at high risk of developing chagasic cardiomyopathy is critical. The goal of this project was to identify a panel of candidate biomarkers that was capable of defining those at risk of developing chagasic cardiomyopathy.

21.6.2 Study

We obtained peripheral blood mononuclear cells (PBMCs) from four groups: healthy volunteers (Group 1), Chagas seropositive but cardio-asymptomatic (Group 2), Chagas seropositive cardio-symptomatic (Group 3), and non-Chagas, cardio-symptomatic patients. Our

Table 21.5 Proteins identified from 2D gel analysis in the invasive Aspergillosis project

Gel spot No.	pI (2D gel)	MW, kD (2D gel)	Protein name	Swiss prot accession	Protein score	Abund. ratio[a]	Abund ratio[b]	p-value (ANOVA)[c]
183	7.66	51	Fibrinogen beta chain	P02675	202	1.20	1.16	0.03015
187	5.99	49	Fibrinogen beta chain	P02675	177	1.16	1.14	0.04969
282	7.51	36	Alpha-mannosidase 2	Q16706	30	−2.56	−1.41	0.00311
300	8.19	35	Ig kappa chain V-III region SIE	P01620	153	−1.01	−1.54	0.00034
303	7.15	35	Ig kappa chain V-III region SIE	P01620	81	1.12	−1.82	0.00067
346	7.04	31	Ferritin light chain	P02792	167	2.29	2.12	0.00049
586	8.12	17	Leucine-rich alpha-2-glycoprotein	P02750	281	1.36	1.71	0.00178
103	7.31	73	Complement factor B	P00751	124	−1.37	−1.31	0.03761
356	7.38	31	Hemopexin HPX	P02790	65	−1.49	−1.36	0.00574
412	3.76	25	Serum amyloid A-4 protein	P35542	46	−2.04	−1.44	0.00169
508	5.37	19	Fibrinogen alpha chain	P02671	405	1.08	1.25	0.03188
588	8.21	17	Leucine-rich alpha-2-glycoprotein	P02750	302	1.28	1.43	0.00239
747	3.96	12	Fibrinogen alpha chain	P02671	180	−1.10	−1.37	0.00566
200	5.60	48	Complement C3	P01024	341	−1.12	−1.31	0.00530
245	7.25	42	Complement C4-A	POCOL4	40	1.33	−1.35	0.01312
348	7.55	31	Histidine protein methyltransferase 1 homolog METTL18	O95568	27	−1.33	−2.32	0.04394
359	5.22	30	Hemopexin HPX	P02790	79	−1.29	−1.28	0.01005
360	6.35	30	Hemopexin HPX	P02790	31	−1.21	−1.28	0.03737
364	9.26	29	Keratin, type II cytoskeletal 1 KRT1	P04264	68	−1.71	−1.12	0.04369
408	6.68	26	Serum amyloid A-4 protein SAA4	P35542	91	−1.57	−1.31	0.04750
458	8.10	20	MEF2-activating motif and SAP domain-containing transcriptional regulator MAMSTR	Q6ZN01	29	−2.21	−1.58	0.04684
468	7.03	20	Apolipoprotein A-II APOA2	P02652	85	−1.22	−1.36	0.02118
494	5.00	19	Alpha-1-antichymotrypsin SERPINA3	P01011	640	1.17	1.38	0.00531
496	7.74	19	Alpha-1-antichymotrypsin SERPINA3	P01011	633	1.17	1.33	0.02074
502	5.51	19	Alpha-1-antichymotrypsin SERPINA3	P01011	228	1.27	1.60	0.00440
568	4.87	17	Alpha-1-antitrypsin SERPINA1	P01009	43	1.28	1.33	0.00301
580	6.33	17	Leucine-rich alpha-2-glycoprotein LRG1	P02750	375	1.12	1.29	0.00869
581	9.18	17	Leucine-rich alpha-2-glycoprotein LRG1	P02750	399	1.17	1.37	0.00998
650	3.85	16	Alpha-1-antitrypsin SERPINA1	P01009	162	1.60	1.29	0.03102
653	4.81	16	Alpha-1-antitrypsin SERPINA1	P01009	272	1.37	1.37	0.04244
695	5.05	14	Alpha-1-acid glycoprotein 1 ORM1	P02763	91	1.44	1.54	0.00113

(continued)

Table 21.5 (continued)

Gel spot No.	pI (2D gel)	MW, kD (2D gel)	Protein name	Swiss prot accession	Protein score	Abund. ratio[a]	Abund ratio[b]	p-value (ANOVA)[c]
696	7.82	14	Alpha-1-acid glycoprotein 1 ORM1	P02763	51	1.45	1.51	0.00384
735	8.83	13	Apolipoprotein A-I APOA1	P02647	239	−1.07	−1.35	0.02470
737	7.59	13	Apolipoprotein A-I APOA1	P02647	97	−1.28	−1.44	0.00622
739	6.74	12	Apolipoprotein A-I APOA1	P02647	368	−1.02	−1.33	0.02157
764	4.81	12	Alpha-1-acid glycoprotein 1 ORM1	P02763	113	1.42	1.51	0.00112
828	6.70	35	Serum albumin ALB	P02768	97	1.63	1.26	0.02209
498	5.60	19	Alpha-1-antichymotrypsin SERPINA3	P01011	587	1.25	1.35	0.03639
499	5.75	19	Alpha-1-antichymotrypsin SERPINA3	P01011	521	1.45	1.42	0.04196
501	6.14	19	Alpha-1-antichymotrypsin SERPINA3	P01011	522	1.16	1.43	0.00648
503	5.25	19	Alpha-1-antichymotrypsin SERPINA3	P01011	503	1.17	1.50	0.00426
583	8.99	17	Leucine-rich alpha-2-glycoprotein LRG1	P02750	359	1.11	1.35	0.00294
585	9.09	17	Leucine-rich alpha-2-glycoprotein LRG1	P02750	337	1.14	1.33	0.00450
589	5.90	17	Leucine-rich alpha-2-glycoprotein LRG1	P02750	377	1.20	1.31	0.00479
639	3.52	16	Alpha-1-antitrypsin SERPINA1	P01009	433	1.38	1.33	0.04594
654	8.16	16	Alpha-1-antitrypsin SERPINA1	P01009	368	1.29	1.48	0.00094
691	4.57	14	Alpha-1-acid glycoprotein 1 ORM1	P02763	268	1.24	1.51	0.01454
697	3.55	14	Alpha-1-acid glycoprotein 1 ORM1	P02763	227	1.24	1.48	0.00586
766	3.99	12	Alpha-1-acid glycoprotein 1 ORM1	P02763	243	1.25	1.45	0.00207
767	4.64	12	Alpha-1-acid glycoprotein 1 ORM1	P02763	228	1.29	1.39	0.02195
49	9.47	109	Serum albumin ALB	P02768	455	1.09	−1.12	0.02791
115	8.14	73	Serum albumin ALB	P02768	207	−1.50	−1.62	0.00015
116	8.20	73	Serum albumin ALB	P02768	385	−1.40	−1.39	0.00398
117	6.22	73	Complement C4-A C4A	P0C0L4	157	−1.20	−1.56	0.03590
748	6.22	12	Fibrinogen alpha chain FGA	P02671	180	1.16	−1.42	0.04536
176	5.52	51	Fibrinogen beta chain FGB	P02675	303	1.34	1.29	0.00182
178	5.60	51	Fibrinogen beta chain FGB	P02675	403	1.28	1.27	0.00770
399	6.25	26	Putative uncharacterized protein C6orf50 C6orf50	Q9HD87	28	−1.88	−1.76	0.00120
401	6.95	26	Transthyretin TTR	P02766	40	−1.66	−1.48	0.00137
406	4.00	26	Transthyretin TTR	P02766	231	−1.68	−1.50	0.00034
407	5.35	26	Transthyretin TTR	P02766	239	−1.86	−1.62	0.00012
447	5.36	20	Apolipoprotein C-III APOC3	P02656	148	−1.56	−1.63	0.01367
560	7.65	17	Zinc-alpha-2-glycoprotein AZGP1	P25311	100	1.02	1.25	0.03762

(continued)

Table 21.5 (continued)

Gel spot No.	pI (2D gel)	MW, kD (2D gel)	Protein name	Swiss prot accession	Protein score	Abund. ratio[a]	Abund ratio[b]	p-value (ANOVA)[c]
850	5.56	12	Histidine protein methyltransferase 1 homolog METTL18	O95568	30	−1.48	−1.39	0.02281
857	6.84	15	Annexin A10 ANXA10	Q9UJ72	29	−2.17	−2.21	0.00010
860	6.76	14	Transthyretin TTR	P02766	42	−2.60	−1.97	0.00001

[a]Case vs. Auto control (Same individual, sampled before and after infection)
[b]Case vs. Matched control
[c]Case vs. Auto control vs. Matched control

Table 21.6 Predictive proteins identified from 2DGE analysis in the *Aspergillosis* project

Spot No.	Protein name	UniProt accession
115	Serum albumin ALB	P02768
200	Complement C3	P01024
494	Alpha-1-antichymotrypsin SERPINA3	P01011
654	Alpha-1-antitrypsin SERPINA1	P01009
359	Hemopexin HPX	P02790
399	Putative uncharacterized protein C6orf50	Q9HD87

Table 21.7 Statistical analysis of discovery BAP peptides in the invasive *Aspergillosis* project

Peptide	Mean Case	Mean Control	P	Valid N Case	Valid N Control
VPQVSTPTLVEVSR	**20.155**	**23.47225**	**0.000016**	**18**	**17**
DALSSVQESQVAQQAR	24.8204	24.31583	0.632213	16	14
DALSSVQESQVAQQAR	24.6258	24.55044	0.924253	24	23
TTPPVLDSDGSFFLYSK	21.05433	21.31827	0.632961	19	18
GWVTDGFSSLK	23.90974	24.27239	0.651187	18	17
AVMDDFAAFVEK	21.9167	22.31393	0.589658	17	15
STAAM STYTGIFTDQVLSVLKG EE	20.32244	20.67465	0.638155	20	18
SPELQAEAK	21.61976	22.37701	0.21187	22	20
GPSVFPLAPSSK	**21.49173**	**23.33829**	**0.001563**	**24**	**22**
MGPTELLIEMEDWK	20.35843	19.69993	0.206339	18	16
MGPTELLIEMEDWK	20.70003	20.52631	0.618689	19	19
YAASSYLSLTPEQWK	25.73156	24.65533	0.083302	16	15
TEGDGVYTLNDK	21.69819	22.15719	0.504701	20	19
TEGDGVYTLNNEK	21.85964	20.63309	0.482638	21	19
SVLGQLGITK	**20.69461**	**22.10105**	**0.000647**	**16**	**15**

Peptides in bold underline are statistically significant ($p \leq 0.05$)

goal was to characterize the proteomic differences between Groups 2 and 3 seeking candidate biomarkers that would allow classification of infected noncardio-symptomatic patients who are most likely to develop chagasic cardiomyopathy.

This project departs from the others in that PBMCs were analyzed, rather than plasma. The primary rationale is that blood has been shown to reflect the progress of infection [19], and that *T. cruzi* infection induces activation of inflammatory cells (macrophages, neutrophils) that release cytotoxic reactive oxygen species (ROS) and reactive nitrogen species (RNS) for the control of the parasite [20]. We, and our collaborator (Dr. N. Garg, UTMB) reasoned that an effective approach would be to globally investigate the oxidative status of PBMC proteins, namely, cysteinyl-S-nitrosylation (SNO), a widely recognized prototype of redox signaling.

In this 2DGE study, we employed the SNO by fluorescence approach (SNOFlo) to measure differential SNO [8, 21, 22]. Protein differential abundance was measured using the saturation fluorescence approach as in the other two projects described above [5]. Briefly, the SNOFlo analysis involves treating one half of each sample with ascorbate (Asc) to remove existing SNO modifications, labeling both the treated (Asc+) and untreated (Asc-) aliquots with the Bodipy-FL dye, and comparing spot intensities of the Asc + and Asc- 2-D gels from Group 2 samples with those of Group 3. The degree of differential Cys-S nitrosylation is obtained by the calculation of a p-value and a "Ratio of Ratios" (or RoR; see [8]). Differential abundance between groups 2 and 3 was calculated as in the other two projects. In the asymptomatic and symptomatic groups we had $n = 25$ and $n = 28$ samples, respectively, that we could take through the entire 2-DE analyses, leading to a total of 53 Asc + and 53 Asc- gel images for analysis. After gel alignment, spot filtering, and editing, there were 635 spots for quantitative sample group comparisons. Based on t-test p-values and fold-changes (RoR for SNOFlo), there were a total of 33 spots that were significantly different either in abundance, or nitrosylation, or both. These spots were picked and identified by MALDI TOF/TOF mass spectrometry. These 33 proteins are listed in Table 21.8. The identified proteins are marked on the reference gel in Fig. 21.5.

A predictive set of proteins was arrived at by an approach involving classification modeling with MARS, Ensemble methods, Treenet, Generalized pathseeker (GPS) and Random Forests (RF) which are described in detail in Chap. 19 in this volume. Using this approach, we reduced the protein set to the seven proteins listed in Table 21.9.

21.7 Conclusions

We have described the importance of analyzing both proteins and peptides from plasma, as well as the importance of recognizing that post-translational modifications can convert so called "nuisance" proteins (e.g., high-abundance, high molecular weight proteins) into potential biomarkers by virtue of ionic and/or size isomerization. This is not surprising, as many plasma proteins represent leakage of cellular proteins into the plasma due to the molecular pathology of the disease, and by their very nature, suggest structural modifications that facilitate their leakage. We have also seen that modifications resulting in either increased or decreased molecular size, including notably those of high-abundance proteins, by virtue of their unique size qualified them as candidates. This character would have been lost in a bottom-up strategy.

Table 21.8 Significant proteins identified by 2DGE analysis in the chagasic cardiomyopathy project

Gel Spot No.	Spot pI	Spot MW (kD)	Protein name	UniProt accession	MS ID protein score	Abundance ratio	p-Value	SNO[a]	p-Value
63	7.53	99	Vinculin GN = VCL PE = 1 SV = 4	P18206	236	1.32	0.182504	−1.32	0.033111
141	5.97	63	Serum albumin (Fragment) GN = ALB PE = 4 SV = 1	H0YA55	167	−1.25	0.204533	1.69	0.027258
165	4.19	55	Isoform 3 of Integrin alpha-IIb GN = ITGA2B	P08514-3	268	1.17	0.339162	−1.06	0.040421
170	9.41	55	Isoform H7 of Myeloperoxidase GN = MPO	P05164-3	79	1.75	0.125809	−1.71	0.034137
261	6.57	41	POTE ankyrin domain family member F OS = Homo sapiens GN = POTEF PE = 1 SV = 2	A5A3E0	121	1.39	0.011500	−1.35	0.037733
267	6.41	41	Actin, cytoplasmic 2, N-terminally processed GN = ACTG1 PE = 3 SV = 1	F5H0N0	106	1.30	0.034774	−1.34	0.024770
273	8.84	41	Isoform 2 of Fibrinogen alpha chain OS = Homo sapiens GN = FGA	P02671-2	54	−1.46	0.035493	1.04	0.044564
339	6.33	34	Tubulin beta chain OS = Homo sapiens GN = TUBB PE = 3 SV = 1	Q5JP53	121	−1.20	0.034097	1.19	0.012075
355	5.18	32	Talin 1 GN = TLN1 PE = 2 SV = 1	Q5TCU6	187	−1.07	0.350146	1.26	0.037653
382	4.68	30	Actin, cytoplasmic 1, N-terminally processed GN = ACTB PE = 2 SV = 1	B4E335	468	−1.10	0.291819	1.17	0.024110
385	5.72	29	Vimentin OS = Homo sapiens GN = VIM PE = 3 SV = 1	F5H288	171	1.43	0.006230	−1.41	0.006024
389	6.09	28	Annexin A3 OS = Homo sapiens GN = ANXA3 PE = 1 SV = 3	P12429	549	−1.41	0.027378	1.34	0.021102
404	7.51	26	Actin, cytoplasmic 1, N-terminally processed GN = ACTB PE = 2 SV = 1	B4DW52	133	1.13	0.116141	−1.27	0.020184
411	7.98	26	Unconventional myosin-IXa GN = MYO9A PE = 4 SV = 1	H3BMM1	46	−1.02	0.516115	−1.13	0.043125
425	8.63	25	Peptidyl-prolyl cis-trans isomerase A GN = PPIA PE = 1 SV = 2	P62937	110	−1.10	0.196105	1.14	0.045972
438	8.84	23	WD repeat-containing protein 49 GN = WDR49 PE = 4 SV = 1	F8WBC8	46	1.37	0.413441	1.09	0.047502
502	7.12	20	Keratin, type II cytoskeletal 1 GN = KRT1 PE = 1 SV = 6	P04264	153	−1.09	0.619820	−1.27	0.023729
506	8.92	20	Parathyroid hormone 2 receptor (Fragment) GN = PTH2R PE = 4 SV = 1	H7C0B0	42	−1.33	0.051258	1.50	0.042012
524	9.34	19	Keratin, type I cytoskeletal 10 OS = Homo sapiens GN = KRT10 PE = 1 SV = 6	P13645	286	1.55	0.046182	−1.35	0.033418
535	7.96	19	Proteasome subunit beta type-2 GN = PSMB2 PE = 1 SV = 1	P49721	82	−1.08	0.532056	−1.23	0.023404
563	5.99	18	Actin, cytoplasmic 2, N-terminally processed (Fragment) OS = Homo sapiens GN = ACTG1 PE = 4 SV = 1	I3L1U9	137	−1.83	0.030927	2.47	0.045743
572	5.80	18	Ferritin light chain GN = FTL PE = 1 SV = 2	P02792	74	1.23	0.067410	−1.15	0.032468

(continued)

Table 21.8 (continued)

Gel Spot No.	Spot pI	Spot MW (kD)	Protein name	UniProt accession	MS ID protein score	Abundance ratio	p-Value	SNO[a]	p-Value
592	4.38	17	Myosin regulatory light chain 12B OS = Homo sapiens GN = MYL12B PE = 1 SV = 2	O14950	320	−1.19	0.047581	1.07	0.040693
605	5.64	16	ATP synthase subunit alpha OS = Homo sapiens GN = ATP5A1 PE = 2 SV = 1	A8K092	95	1.27	0.038451	1.17	0.058813
627	5.47	15	Annexin GN = ANXA6 PE = 3 SV = 1	E5RIU8	103	1.44	0.150651	−1.38	1.040693
640	5.12	15	Actin, cytoplasmic 1, N-terminally processed GN = ACTB PE = 3 SV = 1	G5E9R0	193	1.06	0.915841	−1.19	0.402444
644	7.71	15	Keratin, type I cytoskeletal 10 GN = KRT10 PE = 1 SV = 6	P13645	47	−1.00	0.214532	−1.20	2.040693
650	7.26	15	Heterogeneous nuclear ribonucleoprotein A1 (Fragment) GN = HNRNPA1 PE = 4 SV = 1	F8W646	102	−1.47	0.016761	1.52	0.913503
735	4.32	10	SH3 domain-binding glutamic acid-rich-like protein 3 GN = SH3BGRL3 PE = 1 SV = 1	Q9H299	349	1.29	0.029492	−1.30	3.040693
744	4.40	0	Ras-related protein Rap-1b GN = RAP1B PE = 2 SV = 1	B4DQI8	58	1.23	0.180346	−1.22	0.800487
758	5.44	38	Actin, cytoplasmic 1, N-terminally processed GN = ACTB PE = 2 SV = 1	B4E335	624	−1.01	0.956162	1.15	4.040693
816	7.47	78	Actin, cytoplasmic 1, N-terminally processed GN = ACTB PE = 2 SV = 1	B4E335	454	1.50	0.158319	−1.44	0.576176
878	6.90	10	Protein S100-A11 OS = Homo sapiens GN = S100A11 PE = 1 SV = 2	P31949	246	−1.60	0.037919	1.19	5.040693

[a]Ratio of ratios (Change in SNO normalized against change in abundance) [8]

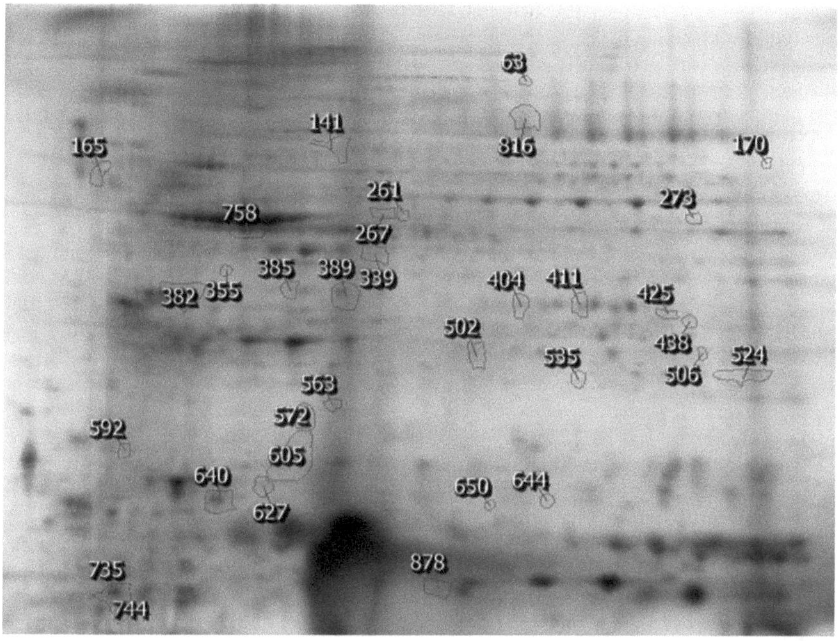

Fig. 21.5 The significant proteins found by 2DGE analysis and identified by MALDI TOF/TOF in the project are shown on the reference gel used in the experiment. The protein spot numbers are marked in the figure, and the corresponding protein names are in Table 20.8

Table 21.9 Predictive proteins identified from 2DGE analysis in the chagasic cardiomyopathy project

Spot No.	Protein name	Swiss prot accession
141	Serum albumin (Fragment)	H0YA55
273	Isoform 2 of Fibrinogen alpha chain	P02671-2
339	Tubulin beta chain	Q5JP53
385	Vimentin	F5H288
389	Annexin A3	P12429
650	Heterogeneous nuclear ribonucleoprotein A1 (Fragment)	F8W646
735	SH3 domain-binding glutamic acid-rich-like protein 3	Q9H299

References

1. Glanz SA (2005) Primer of biostatistics. McGraw-Hill, New York
2. (2005) Special issue: exploring the human plasma proteome. The HUPO Plasma Proteome Project (HPPP). Proteomics 5:3223–3549
3. Richter R, Schulz-Knappe P, Schrader M, Standker L, Jurgens M, Tammen H, Forssmann W-G (1999) Composition of the peptide fraction in human blood plasma: database of circulating human peptides. J Chromatogr B Biomed Sci Appl 726:25–35
4. Villanueva J, Martorella AJ, Lawlor K, Philip J, Fleisher M, Robbins RJ, Tempst P (2006) Serum peptidome patterns that distinguish metastatic thyroid carcinoma from cancer-free controls are unbiased by gender and age. Mol Cell Proteomics 5:1840–1852
5. Pretzer E, Wiktorowicz JE (2008) Saturation fluorescence labeling of proteins for proteomic analyses. Anal Biochem 374:250–262
6. Tyagarajan K, Pretzer EL, Wiktorowicz JE (2003) Thiol-reactive dyes for fluorescence labeling of proteomic samples. Electrophoresis 24:2348–2358
7. Miseta A, Csutora P (2000) Relationship between the occurrence of cysteine in proteins and the complexity of organisms. Mol Biol Evol 17:1232–1239
8. Wiktorowicz JE, Stafford S, Rea H, Urvil P, Soman K, Kurosky A, Perez-Polo JR, Savidge TC (2011) Quantification of cysteinyl S-nitrosylation by fluorescence in unbiased proteomic studies. Biochemistry 50:5601–5614

9. Yao X, Freas A, Ramirez J, Demirev PA, Fenselau C (2001) Proteolytic 18O labeling for comparative proteomics: model studies with two serotypes of adenovirus. Anal Chem 73:2836–2842
10. Miyagi M, Rao KCS (2007) Proteolytic 18O-labeling strategies for quantitative proteomics. Mass Spectrom Rev 26:121–136
11. Khedr A, Hegazy M, Kamal A, Shehata MA (2015) Profiling of esterified fatty acids as biomarkers in the blood of dengue fever patients using a microliter-scale extraction followed by gas chromatography and mass spectrometry. J Sep Sci 38:316–324
12. Poole-Smith BK, Gilbert A, Gonzalez AL, Beltran M, Tomashek KM, Ward BJ, Hunsperger EA, Ndao M (2014) Discovery and characterization of potential prognostic biomarkers for dengue hemorrhagic fever. Am J Trop Med Hyg 91:1218–1226
13. Thayan R, Huat TL, See LL, Tan CP, Khairullah NS, Yusof R, Devi S (2009) The use of two-dimension electrophoresis to identify serum biomarkers from patients with dengue haemorrhagic fever. Trans R Soc Trop Med Hyg 103:413–419
14. Lee CY, Seet RC, Huang SH, Long LH, Halliwell B (2009) Different patterns of oxidized lipid products in plasma and urine of dengue fever, stroke, and Parkinson's disease patients: cautions in the use of biomarkers of oxidative stress. Antioxid Redox Signal 11:407–420
15. Brasier AR, Garcia J, Wiktorowicz JE, Spratt HM, Comach G, Ju H, Recinos A 3rd, Soman K, Forshey BM, Halsey ES, Blair PJ, Rocha C, Bazan I, Victor SS, Wu Z, Stafford S, Watts D, Morrison AC, Scott TW, Kochel TJ (2012) Discovery proteomics and nonparametric modeling pipeline in the development of a candidate biomarker panel for dengue hemorrhagic fever. Clin Transl Sci 5:8–20
16. Brasier AR, Zhao Y, Wiktorowicz JE, Spratt HM, Nascimento EJM, Cordeiro MT, Soman KV, Ju H, Recinos A, Stafford S, Wu Z, Marques ETA, Vasilakis N (2015) Molecular classification of outcomes from dengue virus −3 infections. J Clin Virol 64:97–106
17. WHO (2002) Control of chagas' disease. Second report of the WHO Expert Committee. WHO Tech Rep Ser 905:1–109, Geneva
18. Duran-Rehbein GA, Vargas-Zambrano JC, Cuellar A, Puerta CJ, Gonzalez JM (2014) Mammalian cellular culture models of Trypanosoma cruzi infection: a review of the published literature. Parasite 21:38
19. Wen JJ, Dhiman M, Whorton EB, Garg NJ (2008) Tissue-specific oxidative imbalance and mitochondrial dysfunction during Trypanosoma cruzi infection in mice. Microbes Infect 10:1201–1209
20. Gupta S, Wen JJ, Garg NJ (2009) Oxidative stress in chagas disease. Interdiscip Perspect Infect Dis 190354
21. Savidge TC, Urvil P, Oezguen N, Ali K, Choudhury A, Acharya V, Pinchuk I, Torres AG, English RD, Wiktorowicz JE, Loeffelholz M, Kumar R, Shi L, Nie W, Braun W, Herman B, Hausladen A, Feng H, Stamler JS, Pothoulakis C (2011) Host S-nitrosylation inhibits clostridial small molecule-activated glucosylating toxins. Nat Med 17:1136–1141
22. Sheffield-Moore M, Wiktorowicz JE, Soman KV, Danesi CP, Kinsky MP, Dillon EL, Randolph KM, Casperson SL, Gore DC, Horstman AM, Lynch JP, Doucet BM, Mettler JA, Ryder JW, Ploutz-Snyder LL, Hsu JW, Jahoor F, Jennings K, White GR, McCammon SD, Durham WJ (2013) Sildenafil increases muscle protein synthesis and reduces muscle fatigue. Clin Transl Sci 6:463–468

Statistical Approaches to Candidate Biomarker Panel Selection

22

Heidi M. Spratt and Hyunsu Ju

Abstract

The statistical analysis of robust biomarker candidates is a complex process, and is involved in several key steps in the overall biomarker development pipeline (see Fig. 22.1, Chap. 19). Initially, data visualization (Sect. 22.1, below) is important to determine outliers and to get a feel for the nature of the data and whether there appear to be any differences among the groups being examined. From there, the data must be pre-processed (Sect. 22.2) so that outliers are handled, missing values are dealt with, and normality is assessed. Once the processed data has been cleaned and is ready for downstream analysis, hypothesis tests (Sect. 22.3) are performed, and proteins that are differentially expressed are identified. Since the number of differentially expressed proteins is usually larger than warrants further investigation (50+ proteins versus just a handful that will be considered for a biomarker panel), some sort of feature reduction (Sect. 22.4) should be performed to narrow the list of candidate biomarkers down to a more reasonable number. Once the list of proteins has been reduced to those that are likely most useful for downstream classification purposes, unsupervised or supervised learning is performed (Sects. 22.5 and 22.6, respectively).

Keywords

Candidate biomarker selection • Data inspection • Data consistency • Outlier detection • Data normalization • Data transformations • Data clustering • Machine learning

The statistical analysis of robust biomarker candidates is a complex process, and is involved in several key steps in the overall biomarker development pipeline (see Fig. 22.1, Chap. 19). Initially, data visualization (Sect. 22.1, below) is

H.M. Spratt (✉) • H. Ju, Ph.D
The University of Texas Medical Branch, 301 University Blvd, Galveston, TX 77555-1148, USA
e-mail: hespratt@utmb.edu; hsjuser@gmail.com

H. Mirzaei and M. Carrasco (eds.), *Modern Proteomics – Sample Preparation, Analysis and Practical Applications*, Advances in Experimental Medicine and Biology 919, DOI 10.1007/978-3-319-41448-5_22

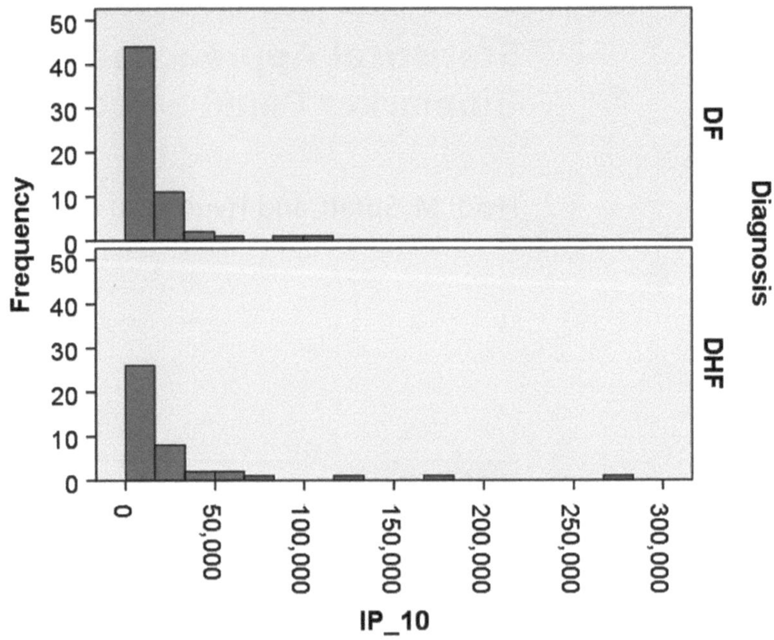

Fig. 22.1 Histograms for IP_10 Cytokine data. Dengue Fever is on the top; Dengue Hemorrhagic Fever is on the bottom

important to determine outliers and to get a feel for the nature of the data and whether there appear to be any differences among the groups being examined. From there, the data must be pre-processed (Sect. 22.2) so that outliers are handled, missing values are dealt with, and normality is assessed. Once the processed data has been cleaned and is ready for downstream analysis, hypothesis tests (Sect. 22.3) are performed, and proteins that are differentially expressed are identified. Since the number of differentially expressed proteins is usually larger than warrants further investigation (50+ proteins versus just a handful that will be considered for a biomarker panel), some sort of feature reduction (Sect. 22.4) should be performed to narrow the list of candidate biomarkers down to a more reasonable number. Once the list of proteins has been reduced to those that are likely most useful for downstream classification purposes, unsupervised or supervised learning is performed (Sects. 22.5 and 22.6, respectively).

The statistical analysis of proteomics data to identify candidate biomarkers and ultimately, the development of predictive models is a complex, multi-step and iterative process. Candidate biomarker selection requires involvement by dedicated statisticians and bioinformaticians with in-depth knowledge of experimental design, insight about how experimental data was generated, as well as a grasp of the types of data structures that the proteomics experiment generated. For these reasons, analysts should be involved in the biomarker study design from the very beginning. Doing so also allows them to obtain a better understanding of the resultant data and any nuances associated with them. Further, they can also assist with experimental details to ensure that the proper analyses can be performed at the end of the experiment. Such an appreciation of the data obtained helps drive strategies for handling outliers or missing data, the pre-processing approaches frequently necessary when working with omics data, and the appropriate selection of hypothesis tests for analyzing the data.

The goal of learning methods is to classify the samples into two or more groups based on a subset of proteins that are most useful for distinguishing between the groups. This subset of proteins is commonly referred to as candidate biomarkers for the classification. The result of supervised learning is a variable importance list that ranks those proteins which are most likely to separate one group of interest from another. This variable importance list is ordered by each protein's ability to discriminate one group from another. In order for a classification task to generalize samples outside the initial discovery samples, some sort of resampling needs to be employed (Sect. 22.7). Resampling techniques can be as simple as setting aside a separate sample set to validate the performance of classification algorithm, or cross-validation techniques where some of the discovery data are left out of the training and are instead used for the testing the trained model. Additionally, methods exist for assessing the ability of a supervised learning algorithm to correctly classify samples from each of the groups of interest. Examining the prediction success or receiver operating characteristic (ROC) curves gives the user a feel for how well the classification algorithm performs (Sect. 22.8). Ideally, the classification algorithm should be able to predict class identity just as well on the training dataset as the testing dataset for a biomarker panel that can be used to distinguish one group from another.

The end result of the biomarker discovery pipeline mentioned above is a list of candidate biomarkers that can be used to distinguish a future sample as belonging to a particular group. However, the experimentation/data analysis process does not end with the creation of a predictive model. This is just the initial discovery phase where a candidate biomarker panel has been identified, and subjected to qualification using independent quantitative proteomics measurements. The next phase is the verification phase, wherein the same biomarker panel has to prove a successful predictor in an independent dataset. This step is critical to the survival of a biomarker panel for further study, by demonstrating the ability of said biomarker panel to generalize to additional samples. Once the biomarker panel has been verified in an independent dataset, further downstream steps can be taken to Validate (Chap. 19: Introduction), produce, and market a diagnostic test based on the discovered biomarker panel.

22.1 Data Inspection/Visualization

Proteomics data typically have a high degree of variability, due both to biological variability from one sample to another and technical variability relating to the technology used, as well as to inherent differences between proteins (e.g., isoforms and post-translational modifications). In addition, proteomics experiments are frequently performed on small sample sizes (less than ten samples per group). The resultant data typically contains over 1000 variables, which results in a wide data set – one that has small n (sample size) and large p (number of variables).

The first step in working with any data set should be data inspection and/or data visualization. This process involves checking the data for consistency of type, examining the dataset for missing values or outliers, as well as graphically displaying the data to better understand the nature and behavior of the various observations.

22.1.1 Data Consistency

Checking the data for consistency involves examining the values present for each individual variable. If the data is supposed to be numeric, one should check that all the values are actually numbers, and that there are no textual strings present. Bioplex cytokine assay data frequently are returned from the instrument with values such as "OOR<". It is up to the data analyst to determine what this value represents (while it is an actual value, but it is below the limit of detection of the instrument), and what to replace this value with. We will discuss data replacement in following sections. An example of this is presented in Table 22.1. If the data is supposed to be positive values only, do any of the columns have negative values? This can be easily checked simply by calculating the minimum for all variables. Another way to check data consistency is to make sure that the data is matched correctly

Table 22.1 Sample of initial data file

Sample No.	IP-10	MIP-1a	TNF-a	VEGF	TRAIL
1	36800.84	718.23	28017.48	44634.68	21562.09
2	13247.18	2675.18		−10569.1	5360.15
3	2682.51	OOR >	5006.67	2790.2	1359.8
4	10.28	5.4	18.75	9.04	1.9
5	3.34	1.57	5.33	3.37	2.39
6	0	*5.80	*7.11	167.62	OOR <

by subject, if matching is mentioned in the study design. Matching is a statistical technique where members of one group are "matched" to members of another group with regards to possibly confounding variables in order to minimize the effect the confounder has on the treatment effect. If the study design suggests that individuals will be matched for gender and age, then the data analyst should verify that males are matched with males, females are matched with females, and individuals with a similar age are matched to other individuals within that same age range. If any data consistency issues are present, they should be corrected before any type of analysis is performed. Doing so often involves communication with the PI as well as with the technical staff that generated the dataset.

Table 22.1 demonstrates several examples of data inconsistencies. For instance, the last value in the IP-10 column is a 0. This was a value that was initially missing, but the researcher changed all missing values to 0 for that cytokine. MIP-1a has two issues. The first is a value of "OOR >" (out of range positive) when a numeric value is expected. The second is a value of *5.80 when a strict numeric is expected. The analyst needs to best determine how to handle such instances, often in consultation with the lab performing the experiment or the PI of the project. TNF-a has a true missing value as well as an "*" that needs to be dealt with. VEGF has a negative value when only positive values are possible. Thus, further investigation is needed as to why the negative value is present. Lastly, TRAIL has a value of "OOR <" (out of range negative) that needs to be properly handled.

22.1.2 Missing Values/Outlier Detection

Dealing with missing values and outliers presents many challenges for the data analyst. Frequently, basic science experimentalists will replace a missing value with a value of 0. This value of 0 can have many different definitions. For instance, a value of 0 might indicate a plausible, real value, but one that fell below the detection limit of the instrument. Instead of placing a value for that particular data point, a researcher might opt to call it a 0. How the analyst handles a 0 value depends on the true meaning of that 0 value. In other situations, the researcher might opt to replace a missing value with a 0 value. This is done because some software packages are unable to handle missing values, and the researcher thinks missing values make the dataset look ugly. Thus, it is important to determine if any such substitutions have been made within a given dataset. If multiple 0 values are observed, the PI or research technician should be consulted to determine the true meaning of these 0 values.

Common ways to deal with missing values include simply leaving those samples out of the data analysis, data imputation, or choosing analysis methods that ignores missing values (such as those mentioned in Sects. 22.6.2, 22.6.3, and 22.6.4). Several methods exist for data imputation, which will be discussed in the following section.

Another common issue in proteomics data sets, as well as other omics data sets, is the presence of outliers. Outliers are individual data points, large or small, that lie further from the majority of the data than would ordinarily be expected and can have an exaggerated influence on the fit of a given algorithm. An outlier may indicate a bad data point: one that results from improper coding, or possibly an experiment gone awry. Outliers heavily influence descriptive statistics such as the mean, as well as impacting the types of hypothesis testing that can validly/reliably be performed on the data. Thus, their detection is an important step in the analysis pipeline. A simple method for detecting outliers is the creation of a boxplot, discussed in the next section, which will graphically display the absence/presence of any outliers. Specifically, a data value is often said to be an outlier if it lies further away from the mean or median of the dataset than ± 1.5*IQR (interquartile range). If the outlier is the result of improper data entry, its value should be easily corrected. If the outlier is the result of an experiment gone awry, then the value should be removed from the dataset and treated as a missing value. If, however, the cause of the outlier cannot be attributed to either of those two instances, the value must remain in the dataset and appropriate procedures should be utilized downstream, i.e. ones that are robust to the presence of outliers.

22.1.3 Graphical Methods

Many types of graphical methods exist to display proteomics data. These include histograms, boxplots, scatterplots, and quantile-quantile plots (also known as q-q plots), among others. Plots can tell one about the presence of outliers within the data, about possible relationships amongst variables within the dataset, about the validity of certain hypothesis test assumptions (such as whether the data is normally distributed), or about possible differences between groups.

Histograms arrange the data points into bins of equal width, where the height of each bar represents the number or proportion of data points that lie within each bin. Histograms are useful for determining whether there are outliers within the data (are there single bins which are separated by many empty bins from the rest of the data?), as well as giving a feel for the shape/distribution of the data. One can visually determine if the data is symmetric (possibly normally distributed) or skewed (not normally distributed). A skewed distribution is one that is not symmetrical, but rather has a long tail in one direction. Example histograms are presented in Fig. 22.1. These histograms represent Bioplex cytokine assay values for IP-10 for patients with Dengue Fever (DF) and Dengue Hemorrhagic Fever (DHF), taken from our NIAID Clinical Proteomics Center Dengue Fever (CPC) project (see Chap. 20 for description). These histograms represent data that is highly skewed, as the shape of the histogram is not symmetric, but rather shifted towards the left. In addition, outliers are present within the DHF subjects as there are three bins that are separated from the rest of the data.

Boxplots show the shape of the distribution, the central value of a dataset, and the variability within the dataset, by displaying the median, the interquartile range (IQR), as well as any potential outliers. As the name implies, the graphs have a box shape. The middle 50 % of the data is displayed within the central rectangle, the median value is frequently displayed as a line within the central rectangle, and whiskers are displayed above and below the central rectangle, representing the range up to some multiple of the IQR away from the median. The upper hinge (edge) of the box indicates the 75th percentile of the data, and the lower hinge (edge) of the box indicates the 25th percentile of the data. In addition, individual outliers are displayed usually as stars or circles on the plot. If no outliers are present, the ends of the whiskers represent the largest and smallest value within a dataset. An example of a boxplot is presented in Fig. 22.2. Like histograms, boxplots can be used to assess the distribution of a given dataset. For data that is symmetric (and thus possibly normally distributed), the median line will lie roughly in the center of the rectangle. In addition, the whisker above the rectangle will be roughly the same length as the whisker below the rectangle. For skewed data (and thus data that is probably not normally distributed), the median line will lie much closer to the top or bottom of the rectangle than the middle, and the whiskers above and

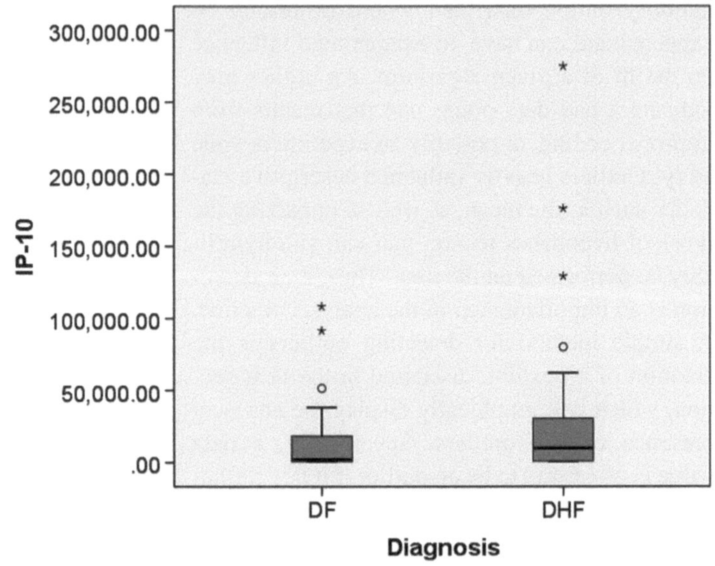

Fig. 22.2 Boxplots for IP_10 Cytokine data. Dengue Fever is on the left; Dengue Hemorrhagic Fever is on the right

below the rectangle will not be the same length. Boxplots can also be used to examine if there are differences between two or more groups by looking for overlapping rectangles when multiple boxplots are drawn on the same plot.

The example in Fig. 22.2 represents IP-10 Bioplex cytokine assay data for the Dengue Fever project (the same as is shown for the histogram in Fig. 22.1). The value of IP-10 for patients with Dengue Fever (DF) is shown in the left boxplot, and the value of IP-10 for patients with Dengue Hemorrhagic Fever (DHF) is shown in the right boxplot. Both of these groups have outliers, which are represented by the circles and the stars (figure created using SPSS v20). In addition, both of the boxplots represent data that is skewed as the median line is not in the middle of the rectangle. The whisker above both boxes is also much longer than the whisker below each box.

A **scatterplot** is a graphical representation of how two different variables relate to each other. The values for one variable are plotted along the x-axis, while the values for another variable are plotted along the y-axis. Scatterplots are useful for detecting a correlation between two variables, as well as if there appears to be some sort of functional relationship between the two variables. If the two variables are correlated, there will be an obvious trend in the location of the dots on the scatterplot. Scatterplots can be used to determine if the functional relationship is a linear one, a quadratic one, a logarithmic one, or one of many different types of functions. This is extremely useful if some sort of modeling is to be done later on.

Another mechanism for aiding with determining whether the dataset is normally distributed is the **q-q plot**. The q-q plot is a special scatterplot. As the name implies, it is one that uses the quantiles of the data to create the plot. Along the x axis are the quantiles of the experimental data. Along the y axis are the quantiles of a specified distribution that has the same mean and standard deviation as the experimental data. The specified distribution is a normal distribution if the assumption of normality is being assessed. If all the data points lie on the line $y = x$, then the experimental data perfectly matches data that is normally distributed. If however, there is deviation away from such a straight line, some amount of skewness is present.

Figure 22.3a, b represent q-q plots for the IP-10 Bioplex cytokine assay data discussed above. The shape of both of these plots is very non-linear, which indicates again that the data is highly skewed. Ideally, we would like all of the points to lie on the straight diagonal line which would mean the experimental data exactly follows a normal distribution.

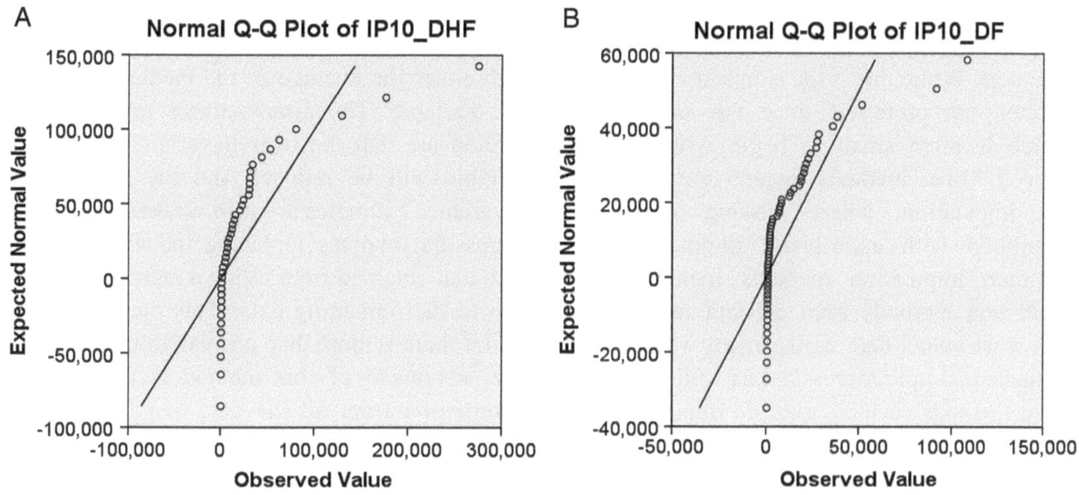

Fig. 22.3 (**a**) Q-Q plot of IP-10 cytokine data for Dengue Hemorrhagic Fever, (**b**) Q-Q plot of IP-10 cytokine data for Dengue Fever

22.2 Pre-processing

Data pre-processing methods refer to the addition, deletion, or transformation of the proteomics data in some fashion before downstream analysis is performed. Pre-processing of the data is a critical step to ensure that the results obtained from statistical testing are both valid as well as correct. How the data is pre-processed can sometimes have a dramatic effect on the output of the model creation process. Some procedures, such as classification and regression trees (Sect. 22.6.2) are fairly insensitive to data pre-processing, while other methods such as logistic regression are not [18]. Pre-processing includes dealing with outliers and missing values through techniques such as imputation as well as normalizing or transforming the dataset to meet hypothesis testing and/or modeling assumptions.

22.2.1 Missing Values/Imputation

Missing values within a dataset present an important, and often overlooked, challenge to downstream data analysis. The reasons for the missing data might bias the results, so the underlying mechanism needs to be considered when determining the most appropriate method for handling missing values. Rubin [25] defines three types of mechanisms that cause data to be missing: data missing completely at random, data missing at random, and data missing not at random. Data missing completely at random means the missing values are truly just randomly missing (and thus ignorable). What this means is that there is no relationship between the value that is missing and either the observed variables or the unobserved parameters of interest. This is the easiest mechanism for a data value to be missing and results in unbiased data analysis. Data missing at random (but not completely) happens when the probability of a missing value depends on some observed values but does not depend on any data that has not been observed (or the group assignment). Unfortunately, missing at random has a somewhat confusing name as it does not mean missing completely at random which is what the name implies. Data that is missing not at random means the probability of a missing value depends on the variable that is missing. This type of "missingness" is often a result in survey analysis where the respondent fails to answer a question because of the nature of the question (i.e. income level).

Some analytic techniques (such as those that deal with repeated measurements on the same sample) require that there are no missing values. Several methods exist that will allow the end user to overcome missing data. The first is to simply remove the sample which resulted in the incomplete data. While this is the simplest approach, it is often not preferred since the sample size (which is often small to begin with) will be reduced. Other methods include some form of data imputation, where missing values are substituted with appropriate imputed values. Common imputation methods include single imputation methods such as data replacement with a set value, data replacement with a mean or median, simple regression, as well as model-based methods such as multiple imputation and maximum likelihood [19].

List-wise deletion can occur in one of two ways: complete case elimination as mentioned above or pairwise deletion. List-wise deletion results in removing an entire observation from the dataset. The advantage of this method is that it is simple, and one can compare the results of one variable to that of another as the dataset is the same for all variables. The disadvantages include a reduction in the power of the analysis since the size of the dataset has been reduced, and that the loss of generality that downstream analysis is not based on all the information collected. Pairwise deletion involves just removing the data that is missing for a given variable, and leaving that subject in the analysis of other variables where information is present. The advantages of pairwise deletion are that this method uses all the data that has been collected for the analysis of a given variable as well as keeps as many cases/samples as possible. The disadvantage is that it becomes difficult to compare the results from one variable to another as the same samples were not used to generate all results.

Single imputation methods are fairly straightforward. For cytokine data, replacement of character data (such as OOR > or OOR<) is often done with ten times the largest value observed in the dataset or one tenths the lowest value observed within the dataset, respectively. This is because the definition of OOR > is "Out-of-Range High" which means that a numeric value should exist for that data point, but said value is above the detection range of the instrument. The approach is similar for OOR < values. Imputation using the mean or median value replaces any missing value for a given variable with either the average or the median value for that variable. The disadvantages to using this method are that the overall variability for that variable will be reduced and the correlation/covariance estimates are also weakened. Simple regression involves replacing the missing data with that obtained from fitting a regression equation to the remaining data. This method works well if there is more than one variable of interest. The advantage of this method is that it uses information from all the data that is obtained. The disadvantages are that the overall measure of variability within the dataset has been diminished as well as that the model fit and correlation estimates will likely be better than had a value been initially obtained.

Model-based imputation methods include multiple imputation and maximum likelihood. This method does not impute any data, but rather uses each cases available data to compute maximum likelihood estimates [9]. The maximum likelihood estimate of a parameter is the value of the parameter that is most likely to have resulted in the observed data. The likelihood is computed separately for those cases with complete data on some variables and those with complete data on all variables. These two likelihoods are then maximized together to find the estimates. Like multiple imputation, this method gives unbiased parameter estimates and standard errors. One advantage is that it does not require the careful selection of variables used to impute values that multiple imputation method requires. An additional imputation method is K-nearest neighbors (KNN) imputation [1]. With this technique, the k nearest neighbor algorithm is utilized where proteins behave as the neighbors and the distance between proteins is based on the correlation between two proteins.

22.2.2 Normalization

The most common form of data pre-processing is what is commonly known as normalizing (or standardizing) the data. This process centers and rescales the data. To center the data for a given variable, each value has the overall variable mean subtracted from the original value. To scale the data variable, the centered data is then divided by the standard deviation of each data variable. Centering results in the data variable having a mean of zero, and scaling results in the variable having a standard deviation of one. Normalizing the data is commonly done to improve the numeric stability of some classification techniques. Support Vector Machine modeling (Sect. 22.6.5) is one technique that requires the data to be normalized before analysis. Principal Component Analysis is another technique that benefits from centering and scaling the data. The downside to centering and scaling data is that the data are no longer in their original units, so it is sometimes challenging to interpret the normalized computer output. However, simple arithmetic manipulations can make the model fit into the original scale.

22.2.3 Transformations

Data transformations are frequently used when the data appear skewed. The most common forms of data transformations are the logarithmic transformation, the square root transformation, or the inverse transformation. Many statistical hypothesis tests have an underlying assumption that the data be normally distributed, the variances be equal (homoscedasticity), or both. Biological data, which are often skewed, frequently do not meet these assumptions. Data transformations often make the assumptions more valid. To perform a data transformation, simply take the logarithm (any base will do as long as the chosen base is consistent from one data value to another) of every entry for a given variable. Likewise, the square root of the data value or the inverse can be taken as well. The goal is to help transform the proteomic data into a dataset that is not skewed or one that has similar variance between two or more groups of interest. Both of these concepts are important assumptions for parametric testing mentioned in Sect. 22.3.1. Once the transformation has been applied to every data point, the transformed data should be used for downstream analysis. It is important to check whether the transformation helped make the data appear more normally distributed. Valid methods for examining normality of the data include checking boxplots as well as q-q plots, both of which were described in Sect. 22.1.3. If the data contain multiple groups (such as Dengue Fever vs Dengue Hemorrhagic Fever, or Chagas disease with cardiomyopathy versus Chagas disease without cardiomyopathy, Chap. 20), each group should be assessed for symmetry individually. The logarithmic transformation frequently works well for intensity data (such as that from 2D gel or cytokine experiments) while the square root transformation works well for count data.

22.3 Hypothesis Testing

While exploratory techniques are an important component to guide investigators to promising hypotheses about mechanisms and structure, the classical techniques for inference such as hypothesis testing and confidence interval construction provide a useful and generally accepted metric for validating or rejecting hypotheses of interest. Built on the classical adversarial construction of proof against a null hypothesis of no discovery, hypothesis testing provides researchers with a way to summarize and quantify evidence that is generally invariant across most fields of science. Statistical hypothesis tests falls into two categories: parametric and nonparametric. As mentioned above in Sect. 22.2.3, parametric tests require extra assumptions for their validity. These assumptions are that the data come from a simple random sample, are normally distributed, and also that the variances are homogeneous. If the normality or variance assumptions are violated, parametric tests are not appropriate for a dataset. Nonparametric techniques have no assumptions about the distribution of the data. However, they do require randomness of the data and independence of the samples.

22.3.1 Parametric Tests

Parametric tests include one sample t-tests, two-sample t-tests, paired t-tests, and multiple versions of analysis of variance (ANOVA). One-sample t-tests are the most simplistic form of a hypothesis test. They are used when researchers want to determine whether a parameter (such as the mean) of a variable matches that of one that was published. The null hypothesis is that the mean of the obtained variable is equal to the published or hypothesized value, and the alternative hypothesis is that the mean of the observed variable is not equal to the null value. This type of test is most often used when a researcher is new to a specific technique or instrument, and they want to check that they are performing the experiment properly or that they have used the correct settings for an instrument. As the name implies, the values of only one condition are being measured. In other words, only a control sample is examined.

The two-sample *t*-test is an extension of the one sample *t*-test. Instead of just one group being compared to some known value, the objective of a two-sample *t*-test is to look for differences between two groups: typically a control sample and an infected sample. Here, the two samples should be sampled independently from each other. This means that the samples are not matched or related in some fashion. An additional assumption for the two-sample *t*-test to be valid is that the variance of group 1 (i.e. controls) is similar to the variance of group 2 (e.g. Infected). If this assumption is violated, there is a version to control for unequal variances, called Welch's correction, which can be used instead. For either form of the *t*-test, each group needs to be checked for normality to meet that assumption. Thus, one should check to see if the controls samples are normally distributed and one should separately check to see if the infected/treated samples are also normally distributed. If either group violates the normality assumption, a parametric test may not be appropriate. However, a transformation should typically be attempted before reverting to a nonparametric test. If the transformation is applied to one group, that same transformation must be applied to all other groups. The transformed data should then be checked for normality. If the transformed data helps the samples look more normally distributed, then the transformed data should be analyzed via the student's *t*-test (not the raw data). For the Dengue Fever example data, the data was transformed using log base 2. This data was then analyzed via the Welch's correction for the two-sample *t*-test. The results indicate that 107 protein spots are differentially abundant between the DF & DHF samples at a significance level of 0.05. These 107 spots will be the input for the examples in Sects. 22.6.2, 22.6.3, 22.6.4, 22.6.5, 22.6.6, and 22.6.7.

Paired t-tests are similar to two-sample t-tests; however, instead of the two samples being independent, they are required to be dependent (matched). This means that they have either been matched to account for possible confounding factors such as on age, gender, race, etc. or that the sample is from the same patient over time, such as a pre- and post- measurement after the administration of some drug. Just as the previous t-tests had the assumption that the data be normally distributed, the paired *t*-test does as well. However, when checking for normality with paired data, the difference between groups is assessed instead of the normality of each group separately. This is because the formula for the test statistic is based on the difference instead of each group individually. Thus, to determine if the data is normally distributed for paired data, the pre measurement would be subtracted from the post measurement and that value would be plotted on a q-q plot.

Analysis of Variance (ANOVA) techniques are used when there are more than two groups being compared or there are multiple factors being investigated. There are many forms of ANOVA, including one-way ANOVA (three or more groups being compared at once), two-way ANOVA (at least two factors with at least two levels each), and repeated measures or mixed-model ANOVA (in which at least one factor has multiple measurements on the same individual over time). The basic premise for ANOVA is that instead of mean values being considered, the amount of variability both within a group and also between groups is being compared. For one-way ANOVA, a single null hypothesis is examined: whether there is a difference among multiple different group means. For two-way ANOVA, multiple null hypotheses are examined: whether there is a difference due to the first factor; whether (typically) there is a difference due to the second factor; and whether there is an interaction between the first and second factor.

Unlike with t-tests where you immediately can conclude if group 1 is significantly different from group 2 based on the observed p-value, with ANOVA all that is known based on the initial results of running the hypothesis test is that at least one group differs from the others within a given factor. Post-hoc tests, such as Tukey's or Dunnett's tests, can be used to determine exactly where the differences lie. Tukey's post-hoc test compares all levels of a factor to each other. Dunnett's test, on the other hand, compares each level to a control level only.

For example, if you are comparing a control strain of a disease to an attenuated strain to a virulent strain, the initial results of a one-way ANOVA will tell you that at least one of the strains is different from the others, but the exact differences will not be able to be determined. Running Tukey's test will compare control to attenuated, control to virulent, and attenuated to virulent to allow one to possibly conclude that control is different from virulent only. Dunnett's test, on the other hand, would only compare control to attenuated and also control to virulent, but will not compare attenuated to virulent (which may not be a hypothesis of interest for some studies).

22.3.2 Nonparametric Tests

Nonparametric tests include chi-square tests, the Mann–Whitney test, the Wilcoxon Signed Rank test, and the Kruskal-Wallis test. All of these tests do not require the data to have a specific shape. Because of the lack of assumptions, nonparametric tests should only be used if the data are highly skewed or the variances are not homogeneous between groups. Whereas the test statistic for a parametric test is based on the actual data value, a test statistic for a nonparametric test is based on the ranks of the data instead. As a result, if data meets the assumptions for using a parametric test, such a test should be preferred over a nonparametric equivalent.

22.3.3 Multiple Hypothesis Corrections

When dealing with proteomics experiments, and other "omics" experiments as well, instead of just testing one protein at a time, researchers typically examine many (often hundreds or thousands) of hypotheses at a time. Doing so increases the probability of false positives: that is, incorrectly rejecting a null hypothesis when no difference between groups exists. This is a serious problem in many basic science experiments, and needs to be dealt with accordingly. The method for correcting the number of false positives, and bringing number of false positives back to a more reasonable level, is known as multiple hypothesis corrections. There are two methods for controlling the false positive rate when one is testing multiple hypotheses simultaneously. They are known as the Family Wise Error Rate (FWER) and False Discovery Rate (FDR) corrections.

The FWER is the probability of wrongly rejecting any of the null hypotheses. The most common FWER correction is the Bonferroni correction [21], although Tukey's test corrects for FWER in the ANOVA setting. FWER corrections are considered to be conservative methods for controlling for multiple hypothesis tests, and frequently results in no proteins remaining significantly differentially abundant in a proteomics experiment.

FDR, on the other hand, seeks to control the proportion of false positives among the complete set of rejected null hypotheses (rather than the probability of any false positives). The most common FDR method is the Benjamini-Hochberg method [2]. FDR procedures allow for more potential false positives than FWER methods, but they have increased power when compared to FWER methods. As a result, FDR methods are less conservative than FWER methods, and usually result in more proteins being significantly differentially abundant between two groups.

22.4 Feature Reduction

A major problem in mining large datasets is the "curse of dimensionality": that is, model efficacy decreases as more variables are added. In many omics experiments, we not only want to learn about which genes/proteins are different from one group to another, but we would like to build a predictive model to determine possible biomarkers for things such as a disease progression or diagnosis. However, as more and more variables are added to the model, the computational time increases and the information gained becomes minimal. Feature reduction aims to decrease the number of input variables to the model; it moderates the effect of the curse of dimensionality by removing irrelevant or redundant variables or noisy data. Feature reduction has the following positive effects: speeding up processing time of the algorithm, enhancing the quality of the data, increasing the predictive power of the algorithm, and making the results more understandable.

22.4.1 Hypothesis Testing Results

One technique for reducing the dimension of the variables to be included in predictive analysis is to eliminate those variables which show no variable-wise significant difference between groups without adjustment for multiple testing. This means some form of hypothesis test has been run on the dataset, and the insignificant variables (p-value > 0.05) are removed. Frequently in omics data analysis, removing only those variables with a p-value > 0.05 still results in a large (greater than 100) variables of interest. In this case, a more restrictive p-value cut-off is used for the downstream analysis.

22.4.2 Significance Analysis of Microarrays (SAM)

Significance analysis of microarrays (SAM) is a widely used permutation-based approach to identify differentially expressed genes when assessing statistical significance using false discovery rate (FDR) adjustment in high dimensional datasets [23]. SAM can be applied to proteomics data since protein abundance microarrays are high-throughput technology capable of generating large quantities of proteomics. SAM algorithm is a great tool comparing t statistic with multiple hypothesis testing adjustments to determine which hypothesis to reject to minimize the number of false positives and negatives by permuting the columns of the protein abundance. Resampling method (permutation) can be used to estimate p values to avoid the joint distribution of the test statistics. Two sample *t*-test procedures require parametric Gaussian assumptions. There are attractive points to SAM using multiple testing procedures, that it does not rely on the parametric assumptions and it does not involve any complex estimation procedures. SAM uses the permutation methods (default 100 times) to estimate FDR and computes a modified t-statistics which measures the strength of the relationship between protein abundance and disease outcome. It also accounts for feature-specific fluctuations in signals and adjusts for increasing variation in features with low signal-to-noise ratios. Data are presented as a scatter plot of expected (x-axis) vs observed (y-axis) relative differences between group, where significant deviations that exceed a threshold from expected relative differences are identified and considered "significant". The solid line indicates the relative difference expression of group is identical, but the dotted line drawn at threshold delta value. The delta was chosen by minimal cross-validation errors. The high rank features of SAM results are marked red color (induced protein) and green color (suppressed protein). For our CPC aspergillosis study, the 110 spots among total 655 spots in 2D-gel data are selected for differentiating case vs. control by 100 permutations and FDR as 5 % for delta = 0.35. The Microsoft Excel add-in SAM package can be used with specific option filtering. There are several options, for example, multi-class, two-class paired, and two-class unpaired response types using the *t*-test, Wilcoxon test, and analysis of variance test. The limitation of SAM procedure is that this approach is a univariate version approach and not allowed to consider the correlated structure between features like a multivariate regression modeling. An example of a SAM result for aspergillosis data is shown in Fig. 22.4.

22.4.3 Fold-Change

Fold Change refers to the values for the control samples being divided by the obtained values for the treated samples. If this results in a value less than one, then the inverse value is taken and a negative sign is added. Thus, the value for fold changes range from -infinity to −1, and then also from 1 to + infinity. A fold-change cut-off value of ±2 is frequently used in proteomics experiments when looking for differential expression. Proteins with an absolute fold-change greater than 2 are thought to be differentially abundant between groups of interest. Thus, only proteins that exhibit such characteristics are considered for downstream analysis. The fold-change cut-off is sometimes increased (to 2.5 or 3) if the number of proteins that have such a fold-change is large.

22.4.4 PCA

Principal component analysis (PCA) is useful for the classification as well as compression of a dataset. The main goal of PCA is to decrease the dimensionality of the dataset by finding a new set of variables, called principal components that represent the majority of the information present within the original dataset. The information is related to the variation present within the original dataset and is calculated by the covariance among the original variables. The number of important principal components is typically smaller than the initial number of variables in the dataset. This new variable space will reduce the complexity and noise within the dataset and

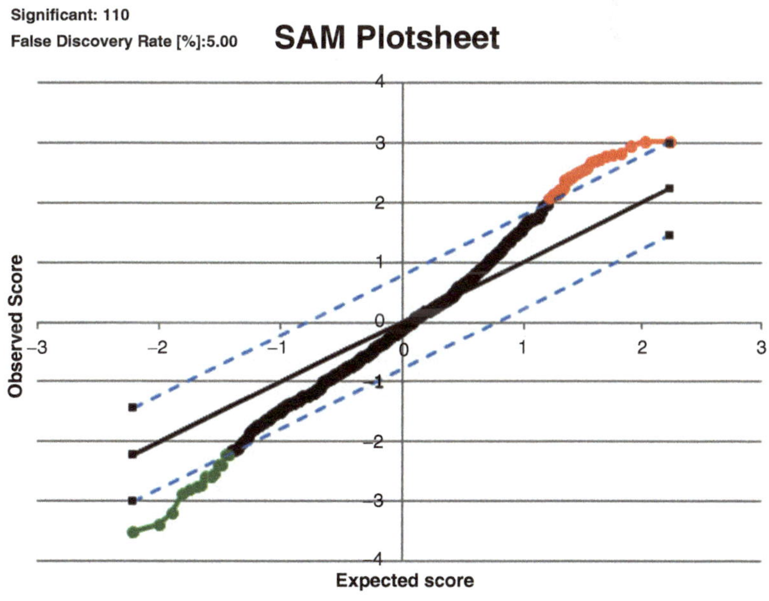

Fig. 22.4 SAM result for Aspergillosis dataset

reveal hidden characteristics within the data. The principal components are uncorrelated (orthogonal) with each other and are also ordered by the total fraction of information about the original dataset they contain. The first principal component accounts for as much of the variability in the original dataset as possible, and each subsequent component accounts for as much of the remaining variability as possible. The process for determining the principal components is one based on covariance eigenvalues and eigenvectors. The results are presented in the form of scores (projections of the eigenvectors) and loadings (eigenvalues).

22.5 Unsupervised Learning

Machine learning falls into two categories of methods: those that are considered to be unsupervised, and those that are considered to be supervised. The primary difference between the two methods is what is assumed to be known at the start of the process. For unsupervised learning, the "truth" is not assumed to be known, nor is it used in the process. "Truth", in our context, is knowing which group a sample belongs to. In an experiment distinguishing between Dengue Fever and Dengue Hemorrhagic Fever, the "truth" would be which patients have Dengue Fever, and which patients have Dengue Hemorrhagic Fever. For supervised methods, the "truth" is required for each algorithm. Hierarchical clustering, K-means clustering, and PCA are all examples of unsupervised learning methods.

22.5.1 Hierarchical Clustering

Hierarchical clustering seeks to group available data into clusters by the formation of a dendrogram. Hierarchical clustering is based on two key principles: (1) Members of each cluster are more closely related to other members of that cluster than they are to members of another cluster, and (2) Elements in different clusters are further apart from each other than they are from members of their own cluster. The process by which samples are grouped into clusters is determined by a measure of similarity between the objects. Various measures of similarity exist, including Euclidean distance, Manhattan (city-block) distance, and Pearson correlation. Euclidean distance is the most commonly used measure of similarity for proteomics experiments, but it is sensitive to outliers within the data. The Manhattan distance requires that the data be standardized before use. The Pearson correlation is a similarity measure that is scale invariant, but it is not as intuitive to use as the other measures of similarity.

Not only must one measure the similarity (distance) between two data points, but one must also determine how to measure the distance between two clusters. This distance can be calculated in at least three ways as: (1) the minimum distance between any two objects in the different clusters; (2) the maximum distance between any two objects in the different clusters; or (3) the average distance between all objects in one cluster and all objects in the other cluster. In addition to measures of similarity and distance, one can build the dendrogram either via top-down (divisive) methods or bottom-up (agglomerative) methods. For divisive methods, the process is reversed with each object first belonging to its own cluster. Figure 22.5 represents the results of hierarchical clustering on the Dengue Fever dataset. The input data is the log2 transformed 2D gel data using only the 107 spots that were significantly different based on the *t*-test analysis. As the reader can see, this dataset is challenging. Ideally, the DF subjects should cluster with the DHF subjects. Unfortunately, there is some amount of overlap between the diseases as the clusters are not solely one disease or the other.

22.5.2 K-means Clustering

K-means clustering is similar to hierarchical clustering; however, instead of obtaining n clusters at the end, the data samples are grouped into a pre-specified number, $k < n$, clusters. The goal of k-means clustering is to partition the data into k subsets which are significantly different from each other. K-means

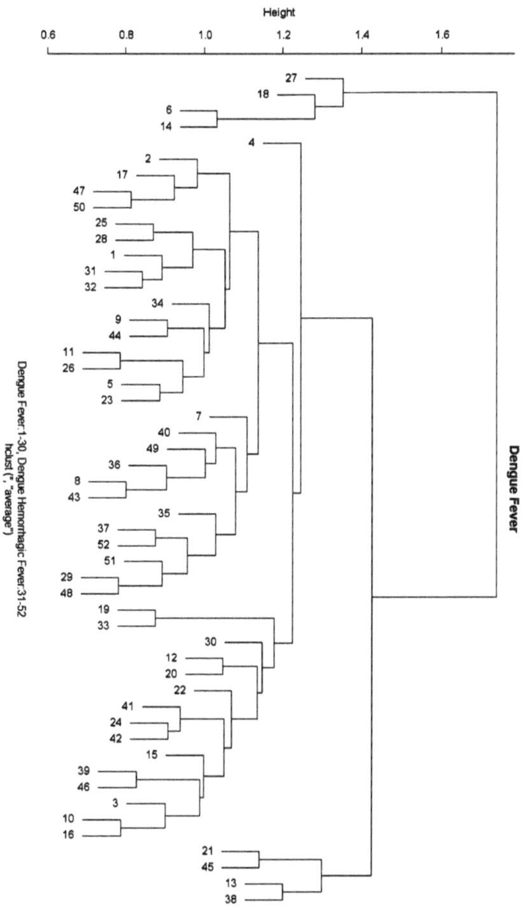

Fig. 22.5 Hierarchical clustering of Dengue Fever study. Subjects labeled 1–30 are subjects with Dengue Fever; Subjects labeled 31–52 are subjects with Dengue Hemorrhagic Fever

clustering is most useful when the user knows *a-priori* the number of clusters that the data should belong to, i.e. if the data samples come from control, attenuated, and virulent strains of a disease, one would expect three clusters to be created. Methods do exist, however, to aid the user in estimating the appropriate number of clusters. With both k-means clustering and hierarchical clustering, the user has the ability to examine in a graphical fashion how similar different groups of data are, and whether there are some proteins that will enable one to easily discriminate one group (i.e., control) from another group (i.e., infected).

22.5.3 PCA

As mentioned above in Sect. 21.4.4, PCA is used to identify patterns in the data. PCA expresses data in such a way that it highlights differences and similarities between both groups and samples within each group. In a data set with many correlations, an ordination technique is needed to look at overall structure of the available data. PCA is based on linear correlation between the data values, and transforms the original variables into new, uncorrelated variables. Consider m observations (e.g., protein abundance levels) on n variables (e.g., conditions/individuals). This results in an $m \times n$ data matrix. PCA reduces the dimensionality of the data matrix by identifying r new variables, where $r < n$. Each new variable, r, is a principal component (PC). Each PC is a linear combination of the original n variables.

To perform PCA, start with the $m \times n$ matrix of protein abundance data: m rows correspond to proteins (expression levels), n columns correspond to conditions/individuals. Apply data standardization, such as the logarithmic transformation or scaling and centering the data such that the mean value is 0 and the standard deviation is 1. Calculate the covariance matrix of the dataset, C. Find the eigenvectors and eigenvalues of the matrix C. Create n new variables, PC_n, that are linear functions of the original n observations:

$$PC_1 = a_{11}\times_1 + a_{12}\times_2 + \ldots + a_{1n}\times_n$$
$$PC_2 = a_{21}\times_1 + a_{22}\times_2 + \ldots + a_{2n}\times_n$$
$$PC_n = a_{n1}\times_1 + a_{n2}\times_2 + \ldots + a_{nn}\times_n$$

The coefficients above (referred to as "loadings"), a_{nn}, represent the linear correlation between the original variables, x_n, and the PC_n. The coefficients are chosen to satisfy three requirements: (1) the variance of PC_n is as large as possible; (2) all values of PC_n are uncorrelated; and (3) sum across rows = 1 ($a_{11}\times_1 + a_{12}\times_2 + \ldots + a_{11}\times_n = 1$). Thus, the end result of PCA is that the data has been transformed so it is expressed in terms of the patterns between samples and groups.

22.6 Supervised Learning/ Classification

Machine learning is the study of how to build systems that learn from experience. It is a subfield of artificial intelligence and utilizes theory from cognitive science, information theory, and probability theory. Machine learning usually involves a training set of data as well as a test set of data. These are both from the same dataset, and the system is "trained" using the training data, and then run on the test data to classify it, and test the model? There are two types of machine learning algorithms: supervised and unsupervised learning. In unsupervised learning, we simply have a set of data points. We do not know classes associated with these data points. In supervised learning, we also know which classes the training data belong to. Machine learning has recently been applied in the areas of medical diagnosis, bioinformatics, stock market analysis, classifying DNA sequences, speech recognition, and object recognition.

22.6.1 Logistic Regression

The main objective of logistic regression is to model the relationship between a set of continuous, categorical, or dichotomous variables and a dichotomous outcome that is modeled via the logit function. Whereas typical linear regression seeks to regress one variable onto another (typically continuous data), logistic regression seeks to model via a probability function of a binary outcome. Logistic regression is the method used when the outcome is a "yes/no" response versus a continuous one. Traditionally, such problems were solved by ordinary least squares regression or linear discriminant analysis. However, these approaches were found to be less than optimal due to their strict assumptions (normality, linearity, constant error variance, and continuity for ordinary least squares regression and multivariate normality with equal variances and covariances for discriminant analysis). A logistic regression equation takes the form:

$$\ln\left[\frac{p}{(1-p)}\right] = \alpha + \sum_{j=1}^{k} \beta j \mathrm{Xj} + \mathrm{e}$$

where p is the probability that event Y occurs P(Y = 1), p/(1 − p) is the odds ratio, and ln [p/(1 − p)] is the log odds ratio (the logit).

Thus, logistic regression is the method used for a binary, rather than a continuous outcome. The logistic regression model does not necessarily require the assumptions of some other regression models, such as the assumption that the variables are normally distributed in linear discriminant analysis. Maximum likelihood estimation is used to solve for the logistic regression equation estimates. Recent techniques such as penalized shrinkage and regularization estimation, and also lasso-type regularization logistic regression models have been developed to improve prediction accuracy in classification.

One of the advantages of using logistic regression is that there is assumed to be a linear association between the feature and response variables. However, one has the ability to add logarithmic transformations or squares of data to increase the performance of the model. One of the key disadvantages of logistic regression is that the method does not accommodate missing values. Additionally, logistic regression is unable to deal with variables that are highly correlated, except when using the lasso or ridge penalties. Lastly, including variables that are not important features can hinder (decrease) the performance of the model. For this reason, logistic regression cannot be used as an additional feature selection technique. It can, however, be used in combination with other feature selection techniques.

22.6.2 CART

Classification and regression trees (**CART**, [3]) are a nonparametric method for building decision trees to classify data. CART is highly useful for our applications because it does not require initial variable selection. The three main components of CART are creating a set of rules for splitting each node in a tree, deciding when a tree is fully grown, and assigning a classification to each terminal node of the tree [22]. Decision trees, such as CART, have a human readable split at each node which is a binary response of some feature in the data set. The basic algorithm for building the decision tree seeks some feature of the data which splits it (here into two groups) maximizing the difference between the classes contained in the parent node. CART is a recursive algorithm which means that once it has decided on an appropriate split resulting in two child nodes, the child nodes then become the new parent nodes, and the process is carried on down the branches of the tree. CART can use cross validation techniques to determine the accuracy of the decision trees.

To build a decision tree, the following need to be determined: (1) which variable should be tested at a node, (2) when should a node be declared a terminal node and further splitting stop, and (3) if a terminal node contains objects from different classes, how should the class of a terminal node be determined? The process for doing so is listed below.

1. Start with splitting a variable at all of its split points. Sample splits into two binary nodes at each split point.
2. Select the best split in the variable in terms of the reduction in impurity (heterogeneity).
3. Repeat steps 1 & 2 for all variables at the root node.
4. Assign classes to the nodes according to a rule that minimizes misclassification costs.
5. Repeat steps 1–5 for each non-terminal node.
6. Grow a very large tree $T_{\max}$ until all terminal nodes are either small or pure or contain identical measurement vectors.
7. Prune and choose final tree using cross validation.

Some of the advantages of CART are that it can easily handle data sets which are complex in structure, it is extremely robust and not very effected by outliers, and it can use a combination of both categorical and continuous data. Missing data values do not pose any obstacle to CART as it develops alternative split points for the data that can be used to classify the data when there are missing values. Additionally, variables used within the CART framework are not required to meet any distributional assumptions (such as being normally distributed or having equal variances within groups). CART can also handle correlated data.

CART also has several disadvantages. CART tends to overfit data, so one should plan to trim (prune) the model so that it can be most useful. Unfortunately, how much to prune the data/tree is one of personal choice. Many software implementations of CART have automatic pruning as an option. The tree structures within CART may be unstable. This means that even small changes in the sample data can result in a drastically different tree. Lastly, while the tree is optimal at each individual split, it might not be globally optimal.

CART was run on the Dengue Fever example data mentioned in prior sections of this chapter. Namely, the log2 transformed data from the 107 significant 2D gel spots was used as input to the CART algorithm. Tenfold cross-validation was selected since the sample size is fairly small (less than 30 subjects within each class). Figure 22.6a shows the representation of the Classification Tree that was produced that is best able to discriminate DF from DHF samples. Figure 22.6b shows the variable importance for the CART model. Table 22.2a shows the prediction success for the training data and Table 22.2b shows the prediction success for the testing data. Figure 22.7 shows the ROC Curves for both the training and testing datasets. The blue curve represents the training data, and the red curve represents the testing data. The AUC for the training data is 0.90 and the AUC for the testing data is 0.47.

22.6.3 RF

Random Forests (**RF**), developed by L. Breiman [4], offers several unique and extremely useful features which include built-in estimation of prediction accuracy, measures of feature importance, and a measure of similarity between sample inputs. RF improves upon classical

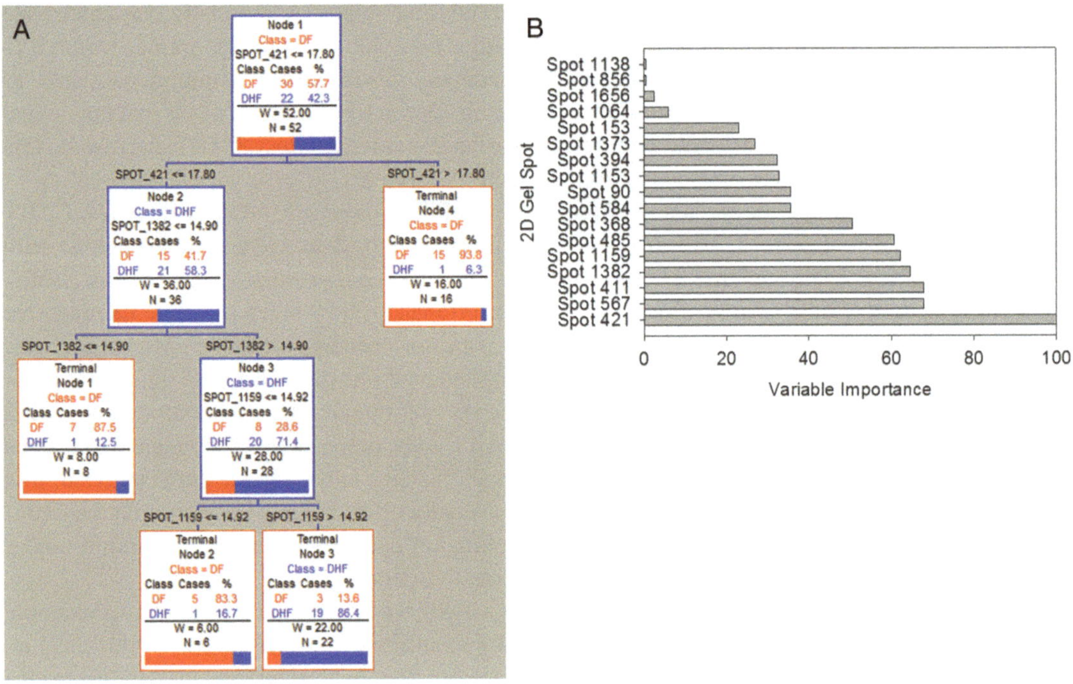

Fig. 22.6 (**a**) CART tree for DF vs DHF comparison, (**b**) Variable importance for the CART model

Table 22.2 (**a**) Prediction success for the training data, (**b**) Prediction success for the testing

A Class	Total	Prediction	
		DF (n = 30)	DHF (n = 22)
DF	30	27	3
DHF	22	3	19
Total	52	Correct = 90 %	Correct = 87 %
B Class	Total	Prediction	
		DF (n = 30)	DHF (n = 22)
DF	30	15	15
DHF	22	15	7
Total	52	Correct = 50 %	Correct = 32 %

decision trees such as CART while still keeping many of the appealing properties of tree methods. Decision trees are known for their ability to select the most informative descriptors among many and to ignore the irrelevant ones. By being an ensemble of trees, RF inherits this attractive property and exploits the statistical power of ensembles. The RF algorithm is very efficient, especially when the number of descriptors is very large. This efficiency over traditional CART methods arises from two general areas. The first is that CART requires some amount of pruning of the tree to reach optimal prediction strength; RF, however, does not do any pruning, which reduces performance time. Second, RF uses only a small number of descriptors to test the splitting performance at each node instead of doing an exhaustive search, as does CART.

RF thus builds many trees and determines the most likely splits based upon a comparison within the ensemble of trees. The procedure makes use of both a training dataset and a test dataset. It proceeds as follows: First, a sample is bootstrapped from the training dataset. Then, for each bootstrapped sample, a classification tree is grown. Here, RF modifies the CART algorithm by randomly selecting from a subset of the descriptors, instead of choosing the best split among all samples and variables. This means that at each node, a user defined number of variables are examined to determine the best split/variable amongst that list. This number is typically small, on the order of five to ten variables to choose from. At each node of the tree, a separate list of variables is considered.

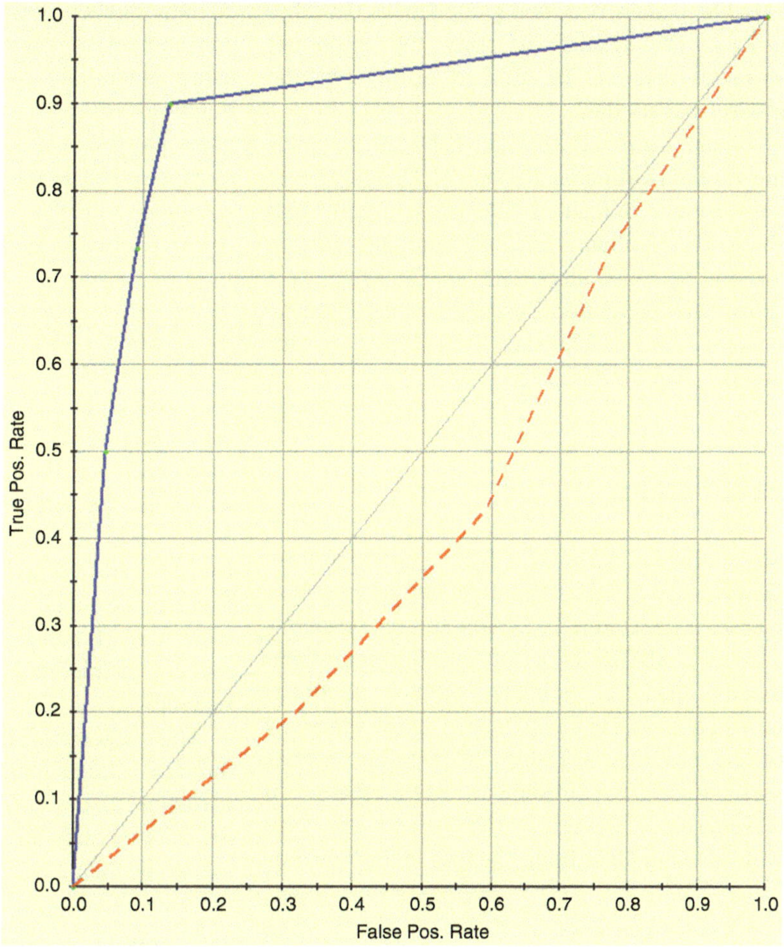

Fig. 22.7 ROC Curves for both the training and testing datasets. The blue curve represents the training data, and the red curve represents the testing data. The AUC for the training data is 0.90 and the AUC for the testing data is 0.47

This procedure of creating trees is repeated until a sufficiently large number of trees have been computed, usually 500 or more.

In practice, some form of cross-validation technique is used to test the prediction accuracy of any computational technique. RF performs a bootstrapping cross-validation procedure in parallel with the training step of its algorithm. This method allows some of the data to be left out at each step, and then used later to estimate the accuracy of the classifier after each instance (i.e. tree) has been completed.

The advantages of RF are that high levels of predictive accuracy are delivered automatically, and there are only a few control parameters to experiment with. Additionally, RF works equally well for classification situations as well as regression situations. RF is the most resistant to overfitting of the models discussed in this chapter. This means the algorithm typically generalizes well for new data. RF is a quick algorithm, which means it creates results rapidly even with thousands of potential predictors. This is because RF does not use all variables at each level of the tree building process. RF does not require prior feature reduction, as it can perform variable selection during the tree building process. RF also has the ability to handle missing values. There are only a couple disadvantages to RF. First, the algorithm can overfit some datasets that are extremely noisy. Additionally, the classifications created by RF can be difficult to interpret as the splits are not listed in the results file (i.e., the user does not know what value of a variable to classify as one group versus the other group). The results list only the important variables that can be used to distinguish one group from another.

Figures 22.8 and 22.9 and Table 22.3 depicts the results of running RF on the Dengue Fever data using a default of 500 trees. Figure 22.8 shows the resultant variable importance for the top twenty most important spots. Table 22.3 shows the prediction success for the models. Figures 22.8 and 22.9 shows the ROC curve for the data. The AUC for the ROC is 0.77.

22.6.4 MARS

Multivariate Adaptive Regression Splines (**MARS**) is a robust nonparametric modeling approach for feature reduction and model building [12]. MARS is a multivariate regression method that can estimate complex nonlinear relationships using a sequence of spline functions of the predictor variables. Regression splines seek to find thresholds and breaks in relationships between variables and are very well suited to identifying changes in the behavior of individuals or processes over time. The basic concept behind regression splines is to model using potentially discrete linear or nonlinear functions of given analytes over differing intervals. The resulting piecewise curve, referred to as a spline, is represented by basis functions within the model.

MARS builds models of the form

$$f(x) = \sum_{i=1}^{k} c_i B_i(x).$$

Each basis function $B_i(x)$ takes one of the following three forms: (1) a constant, there is just one such term, the intercept; (2) a *hinge* function, which has the form max(0,x − *const*) or max(0, *const* − x). MARS automatically selects variables and values of those variables for knots of the hinge functions; or (3) a product of two or more hinge functions. These basis functions can model interactions between two or more variables.

This algorithm has the ability to search through a large number of candidate predictor variables to determine those most relevant to the classification model. The specific variables to use and their exact parameters are identified by

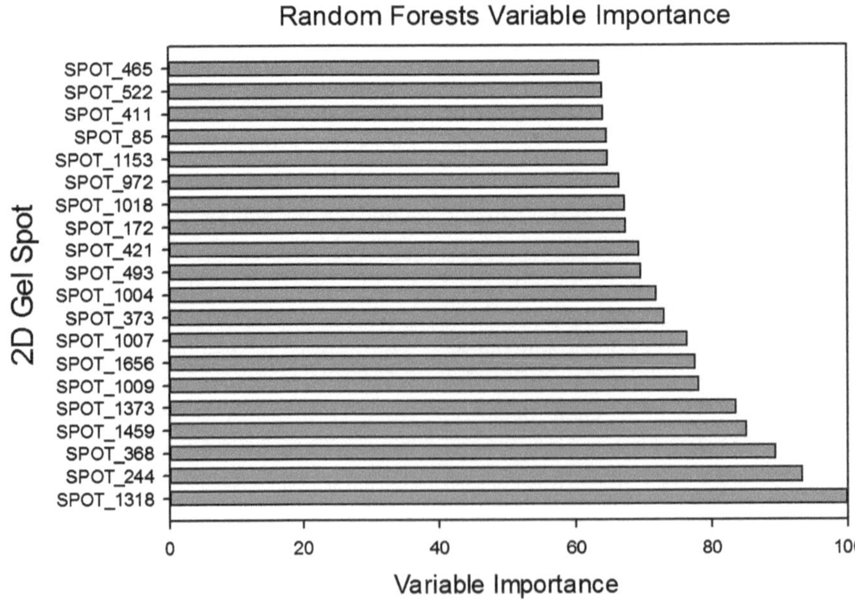

Fig. 22.8 Random forests variable importance for the top 20 most important spots

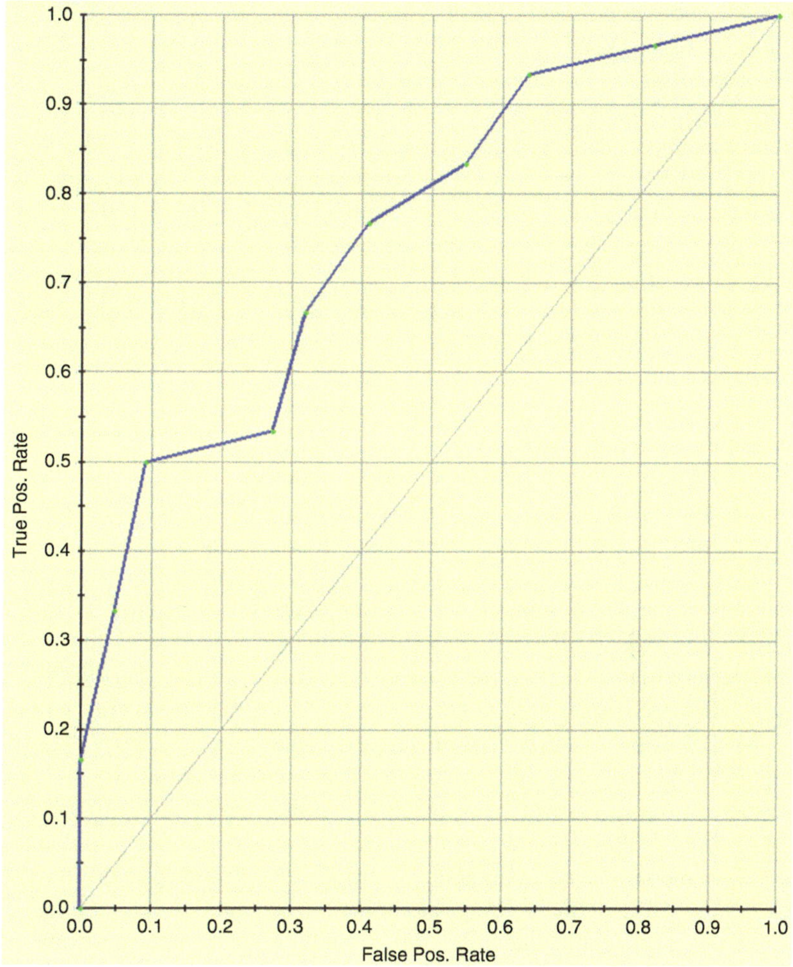

Fig. 22.9 ROC curve for the data. The AUC for the ROC is 0.77

an intensive search procedure that is fast in comparison to other methods. The optimal functional form for the variables in the model is based on regression splines called basis functions.

MARS uses a two-stage process for constructing the optimal classification model. The first half of the process involves creating an overly large model by adding basis functions that represent either single variable transformations or multivariate interaction terms. The model becomes more flexible and complex as additional basis functions are added. The process is complete when a user-specified number of basis functions have been added. In the second stage, MARS deletes basis functions in order of least contribution to the model until the optimum one is reached. By allowing for the model to take on many forms as well as interactions, MARS is able to reliably track the very complex data structures that are often present in high-dimensional data. By doing so, MARS effectively reveals important data patterns and relationships that other models often struggle to detect. Missing values are not a problem because they are dealt with via nested variable techniques. Cross-validation techniques are used within MARS to avoid over-fitting the

Table 22.3 Prediction success for the models

Class	Total	Prediction	
		DF (n = 11)	DHF (n = 41)
DF	30	10	20
DHF	22	1	21
Total	52	Correct = 33 %	Correct = 95 %

classification model, and randomly selected test data can also be used to avoid the issue as well. The end result is a classification model based on single variables and interaction terms which will determine class identity. Thus, MARS excels at finding thresholds and breaks in relationships between variables and as such is very well suited for identifying changes in the behavior of individuals or processes over time.

Some of the advantages of MARS are that it can model predictor variables of many forms, whether continuous or categorical, and that it can tolerate large numbers of input predictor variables. As a nonparametric approach, MARS does not make any underlying assumptions about the distribution of the predictor variables of interest. MARS is also a relatively fast algorithm, which means you can get results for large datasets in under a minute. In addition, like CART and RF, MARS also has the ability to handle missing values within a dataset so that imputation techniques are not necessary.

MARS also has several disadvantages. The algorithm performs in such a fashion that the results are easily overfit to a specific dataset. While MARS allows interactions terms to appear in the model, such interaction terms are extremely difficult to interpret biologically. In addition, confidence intervals for predictive variables cannot be calculated directly.

Table 22.4 and Figs. 22.10 and 22.11 show the results of running MARS on the log2 transformed Dengue Fever dataset. The model was created using tenfold cross-validation and allowed for 107 potential basis functions to be included. Table 22.4a shows the variable importance; Table 22.4b shows the prediction success rates for the training data; Figure X + 2: 6C shows the prediction success rates for the testing data; and Figure X + 2: 6D shows the ROC curves for the training and testing data. The blue curve represents the training data and the red curve represents the testing data. The AUC for the training data is 1.0 and the AUC for the testing data is 0.63.

Table 22.4 (**a**) MARS prediction success rates for the training data, (**b**) MARS prediction success rates for the testing data

A Class	Total	Prediction	
		DF (n = 29)	DHF (n = 23)
DF	30	29	1
DHF	22	0	22
Total	52	Correct = 97 %	Correct = 100 %
B Class	Total	Prediction	
		DF (n = 32)	DHF (n = 20)
DF	30	20	10
DHF	22	12	10
Total	52	Correct = 67 %	Correct = 45 %

22.6.5 SVM

Support vector machines (**SVMs**) are based on simple ideas that originated in the area of statistical learning theory [16]. SVMs apply a transformation to highly dimensional data to enable researchers to linearly separate the various features and classes. As it turns out, this transformation avoids calculations in high dimension space. The popularity of SVMs owes much to the simplicity of the transformation as well as their ability to handle complex classification and regression problems. SVMs are trained with a learning algorithm from optimization theory and tested on the remainder of the available data that were not part of the training dataset [6]. The main aim of support vector machines is to devise a computationally effective way of learning optimal separating parameters for two classes of data.

SVMs project the data into higher dimensional space where different classes or categories are linearly or orthogonally separable by locating a hyperplane (basically, a line or surface that linearly separates data) within the space of data points that can separate multiple classes of data. SVMs also maximize the width of a band separating the data from the hyperplane so that the linear separation is optimal. SVMs use an implicit mapping of the input data, commonly referred to as Φ, into a highly dimensional feature space defined by some kernel function. The learning then occurs in the feature space, and the data points appear in dot products with other data points [20]. One particularly nice property of

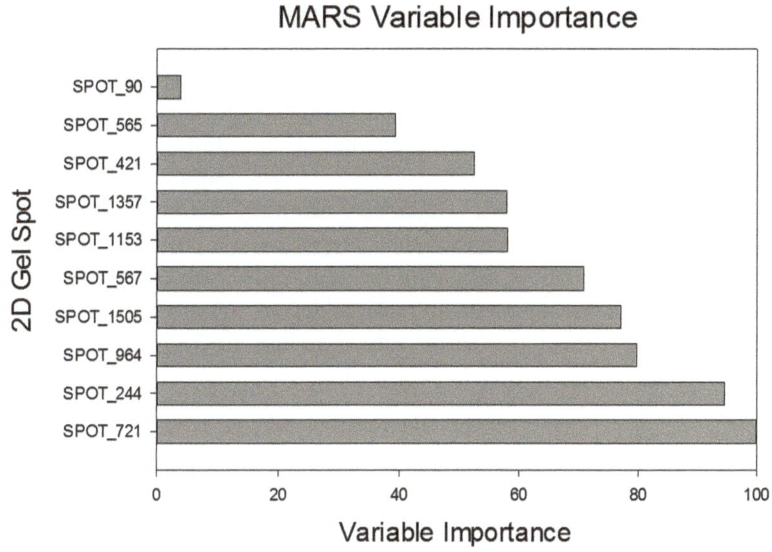

Fig. 22.10 MARS variable importance

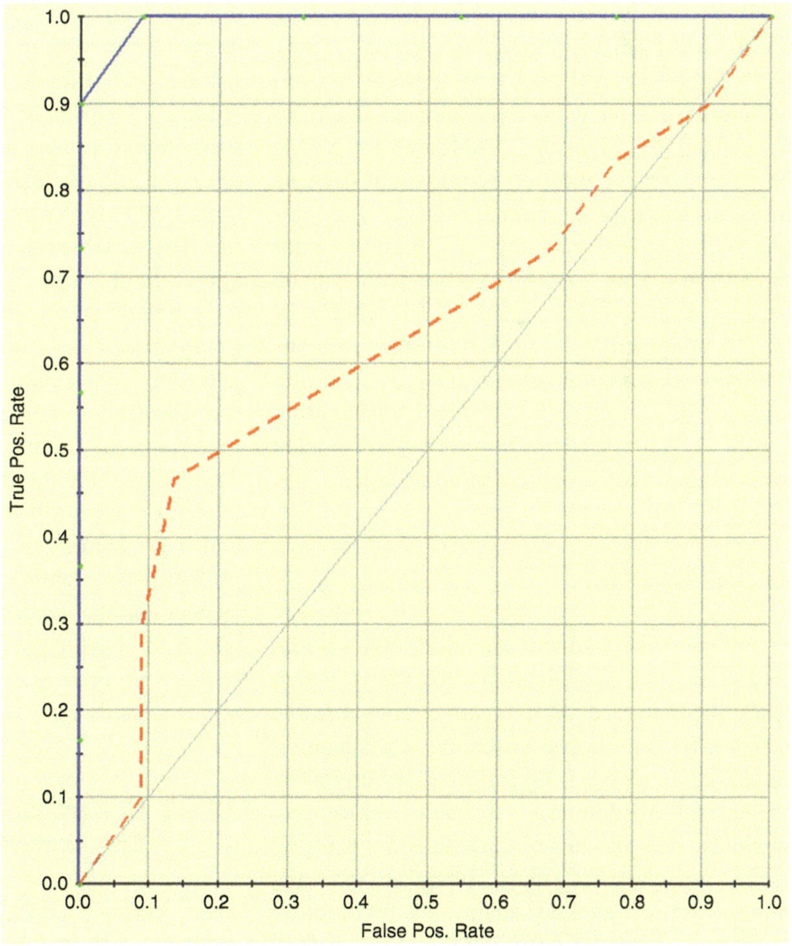

Fig. 22.11 ROC curves for the training and testing data. The blue curve represents the training data and the red curve represents the testing data. The AUC for the training data is 1.0 and theAUC for the testing data is 0.63

SVMs is that once a kernel function has been selected and validated, it is possible to work in spaces of any dimension. Thus, it is easy to add new data into the formulation since the complexity of the problem will not be increased by doing so.

The advantage of SVMs is that they are not data-type dependent. This means that categorical as well as quantitative data can be analyzed together. SVMs are also not dimension dependent. They have the ability to map the data into higher dimensions in order to find a dimension where the data appear to separate into different groups. Additionally, there are various kernel functions that can be used to map the input data into the feature space, the most popular being the radial basis function (or Gaussian) kernel. SVMs also can be used to classify more than just two groups at a time.

There are several disadvantages to using SVMs. Most notably, SVMs are seen as "black box" algorithms and thus fewer researchers are willing to use them because they do not fully understand the algorithm. Additionally, SVMs have an extensive memory requirement because of the quadratic programming necessary to complete the transformation to higher order dimensions. Thus, the SVM algorithm can be very slow in the testing phase. The choice of the kernel function is subjective, which means that choosing one kernel function over another will possibly result in a different classification. Lastly, the data needs to be normalized (scaled and centered) before using the algorithm.

22.6.6 TreeNet

Sometime CART algorithm build non smooth step function classification boundary which leads the variance of it is large and unstable results, so alternative ensemble classification modeling is needed to improve accuracy by increasing randomness through resampling methods. If in the binary classification, a fitting model misclassifies those observations, that model can be applied again, but with extra weight given to the observations misclassified. Then, after a large number of fitting attempts, each with difficult-to-classify observations given relatively more weight, overfitting can be reduced if the fitted values from the different fitting attempts are combined. Boosting is a weak learning algorithm which combines the outputs from many weak classifiers to produce a powerful classifier [15]. A stochastic gradient-boosted model (TreeNet) is a generalized tree boosting that produces an accurate and effective off-the-shelf procedure for data mining [10]. The algorithm generates thousands of small decision trees built in a sequential error-correcting process to converge to an actual model. At each iteration, a subsample of the training sample is drawn at random without replacement from the full training sample to improve robustness to outliers-contaminated data. The variance of the individual base learner is increased at each iteration, but the correlations between the estimates are decreased at different iterations, therefore the variance of the combined model would be reduced. TreeNet performs consistently well in predictive accuracy across many different kinds of data while maintaining the ability to train the model quickly comparing to one CART classifier. The variable importance measure in percentage scale provides how the variables contribute to predictions on the classification. TreeNet graph provides relative influence of each variable and root mean square (RMS) error, a measure of the differences between values predicted by a model and the values actually observed, to assess the power of the model. TreeNet model is also a black box approach how classifiers are complex and hard to interpret the results unlike general probabilistic framework to reach a particular answer and the weak classifiers are too complex, which can lead to over-fitting. Treenet algorithm requires no prior knowledge needed about weak learner, and is easy to run quickly.

22.6.7 Generalized Path Seeker (GPS) Based on AIC and BIC

The comparisons of penalized-regression methods in binary response and logistic regression such as the ridge penalty ($\alpha\sum\beta_i^2$.), lasso penalty ($\alpha\sum|\beta_i|$), and elastic net (combined $\alpha\sum|\beta_i| + (1 - \alpha) \sum\beta_i^2$) were conducted. The ridge regression can only shrink the coefficients, but the lasso regression can do both shrink and variable selection on the coefficients. The elastic net regression can identify the group effect where strongly correlated features tend to be in the model together. The corresponding log-likelihood function of β (L) is given by

$$L = \log L(\beta) = Y^T X\beta - \sum \log(1 + \exp(x_i\beta)).$$

The coefficient vector β that minimizes the penalized log-likelihood is $\beta = \text{argmin}_{\beta \in R^p} - \sum(y_i \log p_i + (1 - y_i) \log(1 - p_i)) + \text{Penalty}(\beta)$, where $p_i = P(y = 1|x)$. To estimate the coefficient, we perform generalized path seeker (GPS), a high speed lasso-style regression from Friedman [11] to regularize regression. GPS demonstrates the regularized regression based on the generalized elastic net family of penalties. The efficient least angle regression (LARS) algorithm of Efron et al. [8] finds the entire regularization paths in an iterative way with the computational effort. For a binary outcome variable and the logistic regression models, the lasso estimator is estimated by penalizing the negative log-likelihood with the L_1-norm through the absolute constraint of regression coefficients like $\alpha\|\beta\|_1 = \alpha\sum|\beta_i|$.

The Akaike information criterion (AIC) is given by

$$AIC = -2\ln(L) + 2(p + 1),$$

where L is the binomial log-likelihood for the model, and p is number of covariates estimated in the model.

The Bayesian information criterion (BIC) is given by

$$BIC = -2\ln(L) + \ln(n) \times (p + 1),$$

where n is the samples size, and p is defined as those variables in AIC.

Among the models having different number of covariates, the one yielding the smallest AIC and BIC values is selected as the optimal model. In AIC and BIC, the binomial log-likelihood may be viewed as a measure of the goodness-of-fit of a model with the number of parameters functioning as a penalty for model complexity. The complexity penalty α term is chosen by AIC or BIC criterion to evaluate the negative log-likelihood. The elastic net, $\alpha\|\beta\|_1 + (1 - \alpha)\|\beta\|^2_2$, combines the L_1 and L_2 penalizing terms and possesses a grouping effect, i.e., in a set of variables that have high pairwise correlations, the elastic net groups the correlated variables together. Lasso and elastic net are especially well suited to wide data, meaning data with more predictors than observations in linear regression model. The regularization model outputs provide piece-wise linear regression path plots along with cross validation to identify important predictors. This procedure is applicable for variable selection for the parametric linear components. If the parametric assumptions are not satisfied, we need nonparametric approach like MARS model beyond linearity of features related to disease outcomes.

Figures 22.12 and 22.13 and Tables 22.5a, b depict the results of running GPS on the Dengue Fever data using tenfold cross-validation. Figure 22.12 shows the resultant variable importance for the top twenty most important spots. Table 22.5a shows the prediction success for the training data; Table 22.5b shows the prediction success for the testing data. Figure 22.13 shows the ROC curve for the data. The blue curve represents the training data; the red curve represents the testing data. The AUC for the training ROC is 1.0; the AUC for the testing data is 0.92.

22.7 Resampling Techniques

22.7.1 Training/Testing Sets

A key concept in machine learning is the creation of a predictive model based on a training dataset, and then assessing the ability of the model to perform on an independent testing dataset (mentioned in Sect. 22.6). Ideally, the training data should be collected separately from the testing

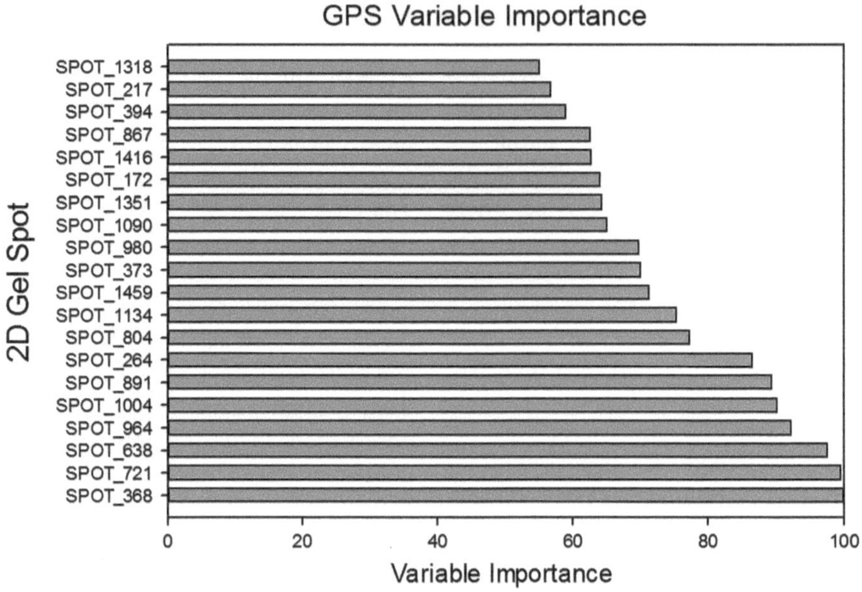

Fig. 22.12 GPS variable importance for the top 20 most important spots

data. This can mean that discovery samples are used for the training data and validation samples are used for the testing data. Another way to create training and testing datasets is to set aside some of the training data to be used instead for the testing data. If the study contains more than 60 samples in a given group, this is the preferred method for machine learning algorithms. How much of the training data to set aside for the testing data is up to the user. Frequently, 70–80 % of the dataset samples are retained for the training of the predictive model, with the additional 20–30 % being set aside to test the model performance. For the majority of the work performed by the Clinical Proteomics Center, the analysis was performed by using cross-validation techniques (mentioned below in G.3).

22.7.2 Bootstrapping

Bootstrap resampling [7] is a general method for inference that has been applied to a variety of statistical problems too difficult to solve analytically. The standard nonparametric bootstrap resampling treats the population data as a sample and samples with replacement repeatedly to produce an approximation to a statistic's sampling distribution. As a result, reliable confidence intervals and hypothesis tests are easily calculated, often with properties superior to standard parametric techniques. In predictive modeling, bootstrap resampling has been found to "smooth" out discontinuities in many fitting algorithms. The resulting model is typically less variable without a substantial increase in bias.

22.7.3 CV/k-fold CV

CV gives an accurate and robust indication of how well an algorithm can make new predictions [17]. CV is an important technique for avoiding testing hypotheses that may be inferred from the data, but don't actually exist. CV is appropriate for each of the classification methods we will discuss. One well-accepted method for cross validation is termed "k-fold" CV. Here the full dataset is divided into k subsets and the holdout method, where a set amount of data is withheld from the analysis, is repeated k times. Each time, one of the k subsets is used as the test set and the remaining subsets are used as the training sets.

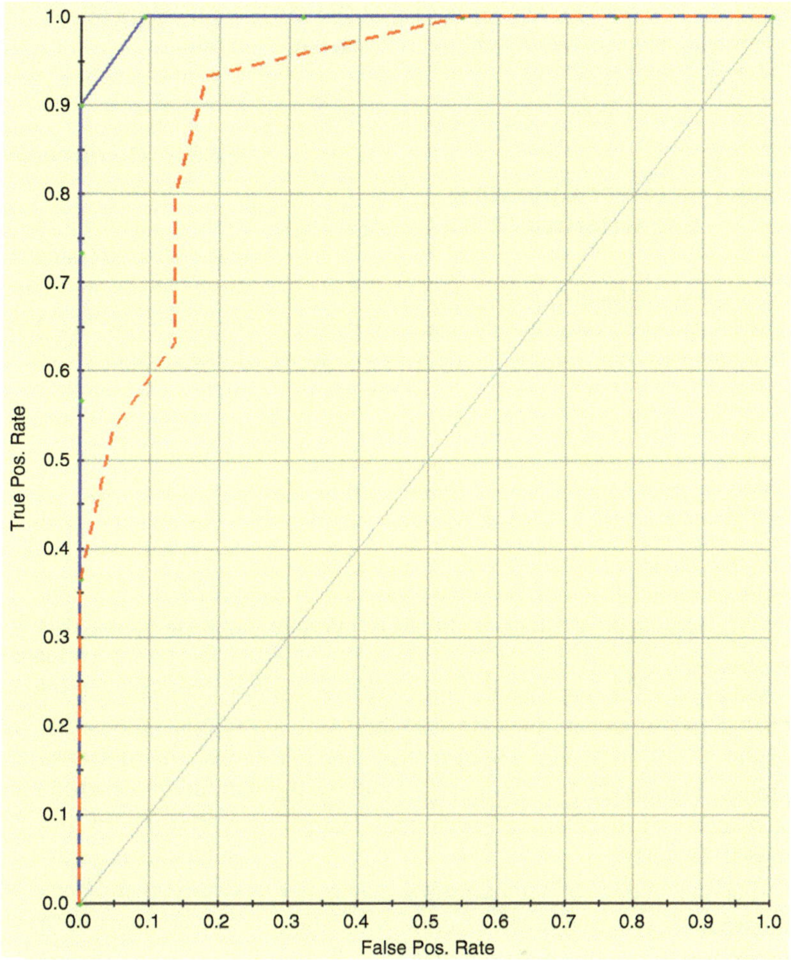

Fig. 22.13 ROC curve for the data. The blue curve represents the training data; the red curve represents the testing data. The AUC for the training ROC is 1.0; the AUC for the testing data is 0.92

Table 22.5 (**a**) Prediction success for the training data, (**b**) Prediction success for the testing data

A Class	Total	Prediction	
		DF (n = 31)	DHF (n = 21)
DF	30	30	0
DHF	22	1	21
Total	52	Correct = 100 %	Correct = 95 %
B Class	Total	Prediction	
		DF (n = 27)	DHF (n = 25)
DF	30	24	6
DHF	22	3	19
Total	52	Correct = 80 %	Correct = 86 %

The average error across all trials is then computed to assess the predictive power of the classification technique used. The advantage of the k-fold CV method is that many combinations of training set vs. test set trials are used to calculate an average predictive error; so this method provides an estimate of an algorithm's predictive power that is much less dependent upon the initial selection of members for the training set.

22.8 Model Diagnostics/Performance/Quality Assessment

In practice, it is often customary for a supervised classification to be conducted using several modeling approaches. The investigators then examine the model performance using a variety of criteria, as well as look for convergence of informative features. A widely accepted approach in model evaluation is to evaluate the area under the receiver operating characteristic (ROC) curve [14].

22.8.1 Receiver-Operating Characteristic (ROC) Curves/ AUC

For a given technique, multiple models are frequently created. One method used to evaluate and compare the various models is by ROC curves [24]. An ROC curve is a graphical plot of the sensitivity vs. 1-specificity of a binary classifier system as its discrimination threshold is varied. This is an equivalent representation of a plot of the fraction of true positives vs. the fraction of false positives. The assumption is that the samples on each side of the binary classifier are from a separate population, and the ROC curve is a graphical presentation of the validity of this assumption. The area under the ROC curve (AUC) measurements indicate the ability of a model to discriminate amongst the outcome groups. Figures 22.7, 22.9, and 22.11 show the ROC curve for the Dengue Fever study comparing DF vs. DHF.

For the choice of regularization parameter, information criterion such as cross validation, generalized cross validation (GCV), Akaike information criterion (AIC) and Bayesian information criterion (BIC) can be used. Generalized cross-validation can be viewed as an approximation to cross-validation,

$$GCV = \frac{1}{n}\sum_{i=1}^{n}\left[\frac{[y - f(x_i)]}{\left(1 - \frac{k}{n}\right)}\right]^2,$$

where n is the number of observations, y is dependent variable x is the independent variable (s), and k is the effective number of parameter or degree of freedom in the model. The effective degrees of freedom is the means by which the GCV error functions puts a penalty on adding variables to the model. The effective degrees of freedom is chosen by the modeler. The GCV can be used to rank the variables in importance. To rank the variables in importance, the GCV is computed with and without each variable in the model.

22.8.2 Deviance/Residual Plots

Model checking is an important procedure to check for assessing model adequacy in multiple linear or logistic regression. Often the interest is to assess the linear or non-linear association of binary responses on features. The logistic regression model assumes that the logit of the outcome is a linear combination of features. When model assumptions are not satisfied, we have problems, the confidence intervals of the coefficients are wide and the statistical tests are incorrect and inefficient. We examine whether our model has all of the relevant predictors and if the linear association of them is appropriate.

Next, we evaluate the partial residual plot as a diagnostic graphical tool for identifying the nonlinear relationship between the logit of the disease outcome and features for additive models. A partial residual plot (Fig. 22.14) is a scatterplot of the partial correlation of each independent with the dependent outcome after removing the linear effects of the other independent features in the model. The log-likelihoods ratio test statistic is twice the difference in log-likelihoods of linear and nonlinear of each feature. For each feature, we also examine the log-likelihood ratio-test p-values comparing the negative binomial log likelihood (i.e., deviance $d_i = 2\left[y_i \log\left(\frac{y_i}{\hat{p}_i}\right) + (n_i - y_i)\log\left(\frac{n_i - y_i}{n - \hat{p}_i}\right)\right]$) between the full model and the reduced model. After performing log-likelihoods ratio test on nonlinear models with smaller p-value less than 0.05, it is preferable to use a non-parametric fit like MARS model. An example of partial residual plot of lymphocytes clinical data for Dengue data is shown in Fig. 22.14. It shows the non-linearity of lymphocytes to logit of the Dengue Hemorrhagic Fever.

In proteomic studies, some proteins could not be accurately measured, so they lead measurement error problems. It is well known that ignoring measurement error in covariate leads to

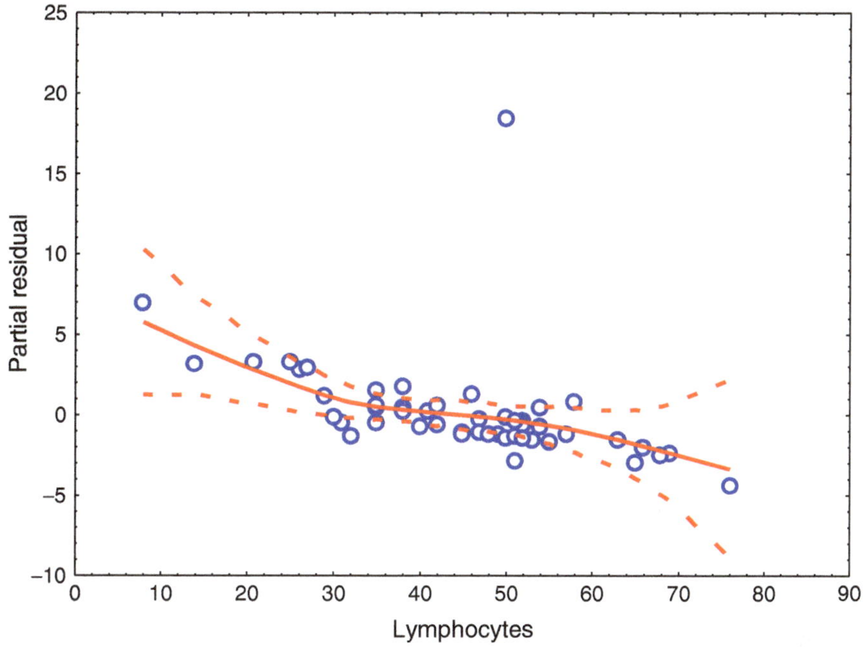

Fig. 22.14 Partial residual plot

biased estimate of the covariate effects. There are a number of measurement error models reported in the literature [5, 13].

Measurement error in the predictors, lack-of-fit error (under-fitting and over-fittings), and error due to omitting relevant important predictors can cause poor performance when building models, especially in terms of reproducibility of the training model into test data. Statistical methods include the random effects in linear mixed effect models could quantify between variation, within variation and unwanted noise variation. Therefore, the model performance estimators should be evaluated from a test set. We need to perform an examination process of similarity between training and test set samples for reproducibility of the model. We observed that verification sample variations in aspergillosis are much larger than in the qualification sample ones. We know that the final optimal classification model can be used to predict the probability of new data being in the disease group in the training samples. The final classification model could be optimized in terms of minimal noise in the predictors and response.

Bibliography

1. Batista G, Monard M (2002) A study of K-nearest neighbour as an imputation method. Hybrid Intelligent Systems, Santiago, Chile, pp 251–260
2. Benjamini Y, Hochberg Y (1995) Controlling the false discovery rate: a practical and powerful approach to multiple testing. J R Stat Soc Ser B 57:125–133
3. Breiman L, Friedman J, Olshen R, Stone C (1984) Classification and regression trees. Wadsworth, Belmont
4. Breiman L (2001) Random forests–random features. University of California, Berkeley
5. Carroll R, Ruppert A, Stefanski L, Crainiceanu C (2006) Measurement error in nonlinear models: a modern perspective, 2nd edn. CRC Press, London
6. Cristianini N, Shawe-Taylor J (2000) An introduction to support vector machines: and other kernel-based learning methods. Cambridge University Press, Cambridge
7. Efron B (1979) Bootstrap methods: another look at the jackknife. Ann Stat 7:1–26
8. Efron B, Hastie T, Johnstone I, Tibshirani R (2004) Least angle regression. Ann Stat 32:407–499
9. Enders C (2001) A primer on maximum likelihood algorithms available for use with missing data. Struct Equ Model Multidiscip J 8:128–141
10. Friedman J (1999) Greedy function approximation: a gradient boosting machine. Department of Statistics, Stanford University

11. Friedman J (2012) Fast sparse regression and classification. Int J Forecast 28:722–738
12. Friedman J (1991) Multivariate adaptive regression splines. Ann Stat 19:1–41
13. Fuller W (1987) Measurement error models. Wiley, New York
14. Hanley JA, McNeil BJ (1982) The meaning and use of the area under a receiver operating characteristic (ROC) curve. Radiology 143:29–36
15. Hastie T, Tibshirani R, Friedman J (2001) The elements of statistical learning; data mining, inference and prediction. Springer, New York
16. Karatzoglou A, Meyer D, Hornik K (2006) Support vector machines in R. J Stat Softw 15:1–28
17. Kohavi R (1995) A study of cross-validation and bootstrap for accuracy estimation and model selection. Fourteenth international joint conference on artificial intelligence, Montreal, Canada, pp 1137–1143
18. Kuhn M, Johnson K (2013) Applied predictive modeling. Springer, New York
19. Little R, Rubin D (2002) Statistical analysis with missing data, 2nd edn. Wiley & Sons, New York
20. Scholkopf B, Smola A (2002) Learning with kernels: support vector machines, regularization, optimization, and beyond. MIT Press, Cambridge, MA
21. Shaffer J (1995) Multiple hypothesis testing. Annu Rev Psychol 46:561–584
22. Steinberg D, Colla P (1995) CART: tree-structured nonparametric data analysis. Salford Systems, San Diego
23. Tusher V, Tibshirani R, Chu G (2001) Significance analysis of microarrays applied to the ionizing radiation response. Proc Natl Acad Sci U S A 98:5116–5121
24. Zweig MH, Campbell G (1993) Receiver-operating characteristic (ROC) plots: a fundamental evaluation tool in clinical medicine. Clin Chem 39:561–577
25. Rubin D (1976) Inference and missing data. Biometrika 63:581–592.

Qualification and Verification of Protein Biomarker Candidates

23

Yingxin Zhao and Allan R. Brasier

Abstract

The importance of biomarkers has long been recognized by the public, scientific community, and industry. Yet despite extensive efforts and funding investments in biomarker discovery, only 109 protein biomarkers in plasma or serum were approved by the US Food and Drug Administration throughout 2008 (Anderson NL. Clin Chem 56:177–185, 2010), and even fewer protein biomarkers are currently used routinely in the clinic. In recent years, the introduction of new protein biomarkers approved by the US Food and Drug Administration has fallen to an average of 1.5 per year (a median of only 1 per year) (Anderson NL. Clin Chem 56:177–185, 2010). The low efficiency of biomarker development is due to several reasons, including the poor quality of clinical samples, the gap between subjective clinical definition of a disease and objective protein measurements, and high false discovery rate of differentially expressed proteins identified in the initial discovery phase (Rifai N, Gillette MA, Carr SA. Nat Biotechnol 24:971–983, 2006). It has become clear that the vast majority of differentially expressed proteins identified in the discovery phase will ultimately fail as useful clinical biomarkers, and only few true positive candidates can move through the biomarker development pipeline. Isolation of true biomarkers from the large pool of differentially expressed proteins identified in the discovery phase becomes the greatest challenge and the bottleneck in most biomarker pipelines. To succeed, after the initial discovery study (see Chap. 20), the authenticity of biomarker candidates need to be tested in a pilot study with high throughput, high accuracy and reasonable cost. This essential process is addressed by qualification and verification phase of the biomarker development pipeline.

Y. Zhao (✉) • A.R. Brasier
The University of Texas Medical Branch, 301 University Blvd, 77555 Galveston, TX, USA
e-mail: yizhao@utmb.edu; arbrasie@utmb.edu

H. Mirzaei and M. Carrasco (eds.), *Modern Proteomics – Sample Preparation, Analysis and Practical Applications*, Advances in Experimental Medicine and Biology 919, DOI 10.1007/978-3-319-41448-5_23

Keywords
Biomarker verification • ELISA-based verification • Selected Reaction Monitoring (SRM) • Parallel reaction monitoring (PRM) • Accurate inclusion mass screening (AIMS) • Data-independent MS/MS acquisition (DIA-MS/MS) • Cohort selection

Abbreviations

AIF	All-ion fragmentation
AIMS	Accurate inclusion mass screening
AQUA	absolute quantification peptides standard
CART	classification and regression trees
CE	capillary electrophoresis
CFD	complement factor D
CV	coefficient of variation
DDA-MS/MS	data-dependent MS/MS acquisition
DIA-MS/MS	Data-independent MS/MS acquisition
ELISA	Enzyme-linked immunosorbent assay
FDR	false positive rate
FWHM	full width at half maximum
HCD	higher energy C-trap dissociation
HPLC	high performance liquid chromatography
HR/AM	high resolution and mass accuracy
IPed	immuno-precipitated
LC	liquid chromatography
LLOQ	lower limit of quantification
LOD	limit of detection
MARS	multivariate adaptive regression splines
MS	mass spectrometry
MS/MS	tandem mass spectrometry
PAcIFIC	Precursor acquisition independent from ion count
PRM	parallel reaction monitoring
QCAT	concatemer of standard peptides
QQQ-MS	triple quadrupole mass spectrometry
SAM	significance analysis of microarray
SEC	size-exclusive chromatography
SID	stable isotope dilution
SISCAPA	stable isotope-labeled standards with capture by anti-peptide antibodies
SOPs	standard operating protocols
SRM	selected reaction monitoring
SWATH	Sequential window acquisition of all theoretical mass spectra.

23.1 Overview

The importance of biomarkers has long been recognized by the public, scientific community, and industry. Yet despite extensive efforts and funding investments in biomarker discovery, only 109 protein biomarkers in plasma or serum were approved by the US Food and Drug Administration throughout 2008 [1], and even fewer protein biomarkers are currently used routinely in the clinic. In recent years, the introduction of new protein biomarkers approved by the US Food and Drug Administration has fallen to an average of 1.5 per year (a median of only one per year) [1]. The low efficiency of biomarker development is due to several reasons, including the poor quality of clinical samples, the gap between subjective clinical definition of a disease and objective protein measurements, and high false discovery rate of differentially expressed proteins identified in the initial discovery phase [2]. It has become clear that the vast majority of differentially expressed proteins identified in the discovery phase will ultimately fail as useful clinical biomarkers, and only few true positive candidates can move through the biomarker development pipeline. Isolation of true biomarkers from the large pool of differentially expressed proteins identified in the discovery phase becomes the greatest challenge and the bottleneck in most biomarker pipelines. To succeed, after the initial discovery study (see Chap. 20), the authenticity of biomarker candidates need to be tested in a pilot study with high throughput, high accuracy and reasonable cost. This essential process is addressed by qualification and verification phase of the biomarker development pipeline.

The aims of the qualification and verification phase in biomarker development pipeline are:

1. to confirm the differential expression of candidates observed in the discovery phase
2. To verify the correlation of the biomarker candidates to the disease over a relative large population of patients
3. To confirm the performance of the statistical model combining the biomarker panel.

The qualification and verification phase, therefore, is a critical phase in the transition from discovery to clinical applications. Three major factors influence the feasibility of a biomarker qualification and verification study:

1. The availability of biospecimens from a well-curated cohort
2. The availability of highly specific and quantitative assays for the biomarker candidates of interest
3. The expense for assay development and applying the assays to measure a large number of targeted analytes across many samples.

23.1.1 Biospecimens from Clinical Cohort

Success in the qualification and verification phase relies on a rigorous clinical study design and attention to detail in sample acquisition, archival and tracking. Biomarker studies typically seek to identify combinations of proteins whose measurement will serve as a molecular indicator of the severity of a disease or its early response to treatment. This application of biomarkers enables the application of precision medicine, an approach that tailors specific interventions to those individuals that would most benefit. Described in Chap. 20, the "discovery phase" entails the application of high throughput proteomics measurements to broadly sample proteins that distinguish between two disease states. The discovery phase typically is applied to a small number of representative cases and controls in a cohort. The qualification phase will measure the candidates in the samples used in the discovery phase. The verification phase involves measuring the candidates in an independent, larger sample of similar cases and controls, frequently from multiple collaborating clinical sites. In order for the verification phase to be meaningful, a reproducible, observer-independent criteria for case definition needs to be applied.

Moreover, significant attention to detail in uniform sample acquisition and storage is paramount. There is increasing recognition that "center effects", variations in sample acquisition, processing and storage may have profound impact on the discovery, qualification, and verification phases of the biomarker development pipeline. To overcome this issue, multi-site clinical studies should develop and rigorously adhere to standard operating protocols (SOPs) for sample acquisition/archival at the onset of the study. Although the techniques for quality assessment/quality control of proteomics samples are currently limited, sample quality should be monitored where possible prior to the application of qualification and verification assays.

The number of samples used in verification study needs to provide sufficient power to assess the sensitivity and specificity of a candidate biomarker panel. The sample size for verification stage depends on multiple factors including the analytical variation of the assays, the biological variation between patients, the concentration of biomarker candidates in clinical samples, and the effect size (the difference in the biomarker's abundance between cases and controls). The statistical design for biospecimen size in verification studies should take these factors into account [3].

23.1.2 Requirements for Qualification and Verification Assays

The transition from discovery study to qualification and verification usually requires the transition from the unbiased, quantitative or semi-quantitative approaches used in the discovery to a targeted and much more precise, reproducible, quantitative approach. If such assays for biomarker candidates are not readily available, they need to be established *de novo* within a short lead-time. The analytical performance of the biomarker qualification and verification assays including accuracy, precise, repeatability, reproducibility, sensitivity, specificity, and linear dynamic range should be validated to meet the predicted needs of the study. The assays need to have high selectivity and sufficient sensitivity to detect and quantify the analytes targeted in a highly complex matrix (such as human plasma). Because the goals of biomarker qualification and verification are to confirm and verify the relative changes observed in the discovery study and to evaluate the model performance in their combination, but not to measure the actual amount of analytes in biological samples, the true accuracy is usually not required in qualification/verification studies. However, the assays need to have high repeatability and reproducibility so that they can be used to precisely and consistently measure relative changes in a large numbers of targeted analytes across many samples. Ideally, the assay can be standardized across laboratories.

Because all biomarker candidates identified from a discovery phase need to be tested in hundreds of samples over a short period of time and with reasonable cost, confirmatory technologies should have a high throughput capability for analyzing hundreds of samples with good precision and accuracy, be capable of multiplexing to evaluate the significant number of biomarker candidates at a time, require minimal sample consumption (because samples amount may be limited), and have low assay cost.

23.2 Platforms for Qualification/Verification, Advantage and Disadvantage

1. Enzyme-linked immunosorbent assay (ELISA)

ELISA has been extensively used in verification of biomarkers. It is extraordinary sensitivity (low pg/mL) [4, 5]. This technique has high sample throughput, and is capable of analyzing hundreds of samples with good precision. For example, ELISA can reliably measure interleukin (IL)-6 at concentrations as low as 0.15 pg/mL with coefficient of variation (CV) of 5 % [2]. However, only a small number of potential biomarker candidates have immunoassay-grade antibody pairs available. Developing a new, clinical-grade ELISA assay is costly ($100,000–$4 million per biomarker candidate), time-consuming (1–1.5 years), and associated with a high failure rate [6]. And it is even more difficult to develop multiplex ELISA assays for a large number of protein targets because of the possible cross-reactivity between antibodies [7, 8]. Taken together, ELISA technology is not well-suited for quantifying a large number of protein candidates in the qualification and verification study.

2. Selected reaction monitoring (SRM)

A number of targeted mass spectrometry approaches have emerged recently, such as accurate inclusion mass screening (AIMS), parallel reaction monitoring (PRM), SRM, and data-independent acquisition (DIA-MS/MS) coupled with targeted data extraction. These approaches have tremendous promise for specific, reproducible, and quantitative measurements of changes of proteins of interest in clinical research. Among them, SRM is currently the most widely used approach for biomarker qualification and verification.

SRM-MS has emerged as a favorable alternative to immunoassays for qualification and verification of candidate biomarkers. In a SRM-MS assay, one or two signature proteotypic peptides are selected to stoichiometrically represent the protein candidate of interest. The SRM analysis of these signature peptides are performed on a triple quadrupole mass spectrometer (QQQ-MS). In SRM assays, the precursor ion of interest is preselected in the first mass filter (Q1), and stimulated to fragment by collision-induced dissociation in second quadrupole (Q2). Several preselected fragments are analyzed by the third mass filter (Q3). The signals of the fragment ions are then monitored over the chromatographic elution time. The SRM-MS offers several attractive features as a qualification/verification assay. First, because only preselected precursor-product ion transitions are monitored in SRM mode, the noise level is significantly reduced and thereby SRM assays decrease the lower detection limit for peptides by up to 100-fold in comparison to full scan MS/MS analysis. Second, if the precursor-product ion transition of one proteotypic peptide is unique to the protein of origin, it is not only distinguishable from other MS signals in one LC run, but it is a characteristic signature for the protein of interest. Therefore, the two filtering stages in SRM mode result in near-absolute structural specificity for the target protein, representing a significant advantage over immunoassays. Third, because no affinity reagent is typically needed, SRM assays can be rapidly and cost-efficiently developed in comparison to immunoassays. Finally, SRM assays have multiplexing capability. Hundreds of precursor-product ion transitions can be monitored in SRM mode over one LC run, allowing for the simultaneous quantification of tens-to one hundred protein biomarker candidates in parallel.

SRM-MS in combination with stable-isotope dilution (SID-SRM-MS) is a target-driven approach for direct quantification of target proteins in a complex mixture [9]. In stable isotope dilution experiments, ^{13}C-, or ^{15}N- labeled absolute quantification peptide standards (AQUA) [9], concatemer of standard peptides (QCAT) [10, 11], or isotope-labeled full-length target proteins [12, 13] are added to the sample as the internal standard. The sample is trypsin digested, and the resultant mix of unlabeled and labeled peptides are analyzed by SRM-MS. Absolute quantification of target protein can be done by comparing the abundance of the known internal standard peptide with its native peptide when well-qualified isotope-labeled full length protein standards are available.

The use of stable isotope-labeled peptides as internal standards has significantly increased the detection confidence and measurement precision in SRM experiments. In SRM, only 3–5 fragment ions from the preselected precursor ions are typically monitored. When it is used for analyzing the target analytes from a highly complex system such as plasma, this assay may be prone to matrix-related interference. Co-eluting matrix components can produce the same SRM transitions as the analytes of interest, resulting in false-positive identification and inaccurate quantification. Matrix components can also cause ion suppression by competing for available protons in the spray droplets. When matrix components co-elute with analytes of interest, they will cause variation in ion current response in different samples severely affecting the precision, accuracy, and sensitivity of quantification. The stable isotope-labeled peptides have identical structures as their endogenous peptide, and as a result, co-elute in LC fractionation. When ion suppression occurs, the suppression will affect both endogenous and stable isotope-labeled peptides at the same degree. Therefore, the ratio of analyte to its internal standard will not be affected by ion suppression. The LC retention time of stable isotope labeled peptides can also be used as the landmark to pinpoint the LC peak of endogenous peptide. Furthermore, stable isotope-labeled peptides generate identical sets of fragment ions as the endogenous counterparts. The relative abundances of the fragment ions of stable isotope labeled peptides can serve as reference to distinguish the true signal of targeted native peptides from other co-eluting isobaric peptides. It will be important to demonstrate that the LC retention time and the relative abundances of the fragment ions of the native peptide are near identical with the stable isotope labeled internal standards. This usually requires significant amount of time and effort to manually inspect the SRM data to ensure the accuracy of quantification [14–16]. Several bioinformatics tools mProphet [17] and AudIT [18] have been developed to overcome these problems. mProphet use criteria such as relative intensities from reference spectra, correlation with the reference spectra, retention time deviation, and co-elution to generate a single score and compute error rate of the measurement. AudIT identifies contaminated transitions. It relies on reference peptides and technical replicates.

SID-SRM is well-suited for highly reproducible quantification across many samples and, in fact, also across different mass spectrometers and laboratories. Recently, the Clinical Proteomic Tumor Analysis Consortium led a landmark multisite assessment study with a focus on the reproducibility of SID-SRM-MS assay between-run, between-laboratory, and between-mass spectrometer manufacturers [19]. In this study, the precision and reproducibility of SRM-based measurements of proteins spiked in a background of human plasma were assessed over nine different laboratories with mass spectrometers from two manufacturers. The results are very promising, with a 10–23 % inter-laboratory CV, a variance that includes variations in sample preparation and MS platforms.

Compared to ELISA, SRM-MS assay can be developed with a short lead-time (1–3 months). A critical step in SRM-MS assay development is the selection of suitable transitions for a target peptide [14]. The considerations are given to fragment ions that provide the highest signal intensity and lowest level of interfering signals. We previously reported a pathway for SRM assay development and optimization, an approach that requires both empirical and bioinformatics tools [14]. Several interfaces (for example, MRMaid [20], MRMer [21], and MaRiMba [22]) use fragment-spectra from shotgun experiments to help in designing favorable transitions for target peptides. For SRM assay design in analyses of complex samples it is also important to infer retention times. Software have been developed to realign and to predict elution times [3, 23]. Transitions extracted for an SRM assay need to be confirmed by addressing the likelihood that the chosen transitions and their intensity distributions are associated with target peptide. Several freely available software products (for example TIQAM, ATAQS [24], and Skyline [25]) integrate many of the above mentioned tasks and automate assay development for peptides (peptide and transition selection), data evaluation, and analyzing SRM traces.

A publicly available SRM assay database, SRMAtlas (www.srmatlas.org), features SRM assays for about 99 % of human proteins. This database was generated from high-quality measurements of natural and synthetic peptides conducted on a QQQ mass spectrometer and is intended as a resource for SRM-based proteomic workflows. Furthermore, to consider the detectability of the SRM assays, PASSEL [26] was created as a combined catalog of the best –available transitions selected from PeptideAtlas shotgun data and SRMAtlas, providing the validation information of all assays in the context of a specific sample. Huttenhain et al. [27] developed SRM assays for 1172 cancer-associated proteins. Using these SRM assays in the clinical samples, 182 proteins were detected in depleted plasma and 408 proteins were detected in urine. These databases of SRM assays are, therefore, valuable resources for designing and accelerating biomarker qualification/verification studies.

Some advancement in instrument design has helped to improve the sensitivity and specificity of SRM assays. For example, in most of SRM analysis, the first quadrupole (Q1) usually uses unit resolution (m/z window 0.7 full width at half maximum (FWHM)). This large m/z window allows other co-eluting sample constituents with similar m/z pass through Q1 and interfere with detection of the desired target. The frequency of these interferences increases as the complexity of the sample increases. Narrower mass windows Q1 will increase selectivity for precursor ions with the cost of a steep decline in signal as these windows are narrowed to <0.5 FWHM. The Thermo Scientific TSQ Quantum line of triple quadrupole mass spectrometers offers a new technique called highly selected reaction monitoring (H-SRM). With the advancement of the technology, the m/z window in Q1 can be narrowed to 0.1–0.2 FWHM to increase the specificity without sacrificing sensitivity. The practical advantage H-SRM is that it dramatically reduces isobaric chemical noise, thereby increasing the signal-to-noise (S/N) [15, 16], which translates to improved lower limit of quantification (LLOQs) and higher confidence in the quantification results. Improvements in the design of nano-electrospray source and interface and applications of the ion-funnel technology to triple-quadrupole mass spectrometers have been proven to increase the ionization efficiency and ion transmission, thus improving the LLOQ of SRM-MS [28, 29]. Application of further stages of ion filtering in QQQ MS increases the sensitivity and specificity of SRM in MRM3. This technique uses a hybrid quadrupole/linear ion trap instrument and monitors reconstructed ion chromatograms on secondary product ions derived from a trapped primary product ion [30, 31]. MRM3 can improve the limit of quantification by a factor of two to fourfold and enables protein biomarker quantification in the low ng/mL range in non-depleted human serum without using immunoaffinity enrichment. The drawback of this method is that it requires much longer acquisition times (350 ms) for each transition in comparison to regular SRM (6 ~ 20 ms), which reduces the number of data points that can be sampled over a given chromatographic peak and the number of peptides that can be monitored in one acquisition cycle.

3. Parallel reaction monitoring (PRM)

SRM is primarily performed on a triple quadrupole MS. With the newly introduced high resolution and mass accuracy (HR/AM) instruments (e.g., Q Exactive quadrupole-Orbitrap or quadrupole-TOF mass spectrometers), a new target proteomics approach referred as PRM has been developed [32, 33]. PRM has been used to measure amyloid-β, a biomarker for Alzheimer disease, in cerebrospinal fluid. The assay shows the similar performance as SRM, with a recovery of 100 % (15 %), intra-assay and inter-assay imprecision of 5 and 6.4 % [34]. The operation of PRM is similar to a SRM. The precursor ions of the target peptides are isolated in the quadrupole mass filter and transferred to higher energy C-trap dissociation (HCD) cells for fragmentation. The fragment ions are measured by HR/AM Orbitrap mass analyzer instead of a third quadrupole used in SRM. The use of an Orbitrap mass analyzer presents specific advantages. First, instead of only 3–5 transitions are monitored by the Q3 mass analyzer in SRM, PRM acquires a full MS/MS spectrum which contains all of potential fragment ions of one targeted peptide, which can significantly improve the confidence of identification of the LC peaks of target peptides. Second, the Orbitrap provides additional data on assay selectivity. In the case of complex samples, the interfering matrix ions co-isolated with the precursors of target peptides can sometimes generate fragment ions which have similar *m/z* values as those of the monitored transitions. These two signals sometimes cannot be separated by a quadrupole mass analyzer with isolation width of 0.7–1.0 *m/z* and may cause false positive identification and inaccurate quantification. The Orbitrap mass analyzer can separate fragment ions with *m/z* difference higher than 10 ppm; this mass accuracy and resolution is much greater than that of the quadrupole. This feature enables PRM technology to more effectively separate fragment ions of interest from interfering ions and improve the selectivity of quantification. The enhanced selectivity and specificity of the PRM method can result in better sensitivity of quantification [32, 35]. Performance comparison between PRM and SRM shows that the linearity and dynamic range of PRM can also rival the traditional SRM approach. However, it is clear that SRM has superior quantitative precision [33]. The imprecision of PRM is largely because the PRM relies on the Orbitrap mass analyzer, which is fundamentally less sensitive and has slower data acquisition rate than quadrupole mass analyzer. Quadrupole mass analyzers operate at a duty cycle nearing 100 % and have the ability to sample more points over a given chromatographic peak, thus provides a more accurate quantification of the LC peak and, in turn, greater precision and run-to-run repeatability. The Orbitrap requires much longer scan time and 40–120 ms Orbitrap injection time, which significantly decrease the duty cycle of acquisition. This reduces the data points sampled over a given chromatographic peak resulting in lower precision and repeatability of quantification. This feature limits the number of possible peptides that can be monitored in one PRM acquisition cycle. To increase multiplex capability, PRM requires time-scheduled acquisition, which relies on the availability of high-quality local spectral library with well-calibrated peptide chromatographic elution time. Unlike SRM, PRM does not require significant effort for assay development, but it requires the high-quality local spectral library to confirm the identity of the analytes and assess measurement quality, especially when the stable isotope labeled standards are not used.

4. Accurate inclusion mass screening (AIMS)

AIMS is another emerging targeted mass spectrometry-based proteomic technique [36]. In AIMS acquisition, a list of pre-selected precursor ions is used to generate an "inclusion list" for MS acquisition [37, 38]. Only precursors represented on the "inclusion list" will be selected for fragmentation if they are detectable in a survey scan. Compared to untargeted data-dependent LC-MS/MS acquisition (DDA-MS/MS) approach used in the discovery study, AIMS significantly improves the level of reproducibility, sensitivity, and dynamic range by restricting detection and fragmentation to only those peptides derived from proteins of interest. It is at least fourfold more efficient at detecting peptides of interest than DDA-MS/MS [36]. The analytical performance of AIMS is less satisfactory than SRM in terms of accuracy, sensitivity, specificity, and dynamic range. However, because AIMS has the ability for time-scheduled monitoring over 1000 peptides in a single LC-MS run, it can be used as a targeted approach for data-dependent triage and prioritization of hundreds of candidate biomarker in a time- and cost-effective manner [39]. In a newly developed targeted MS-based pipeline for biomarker verification, AIMS was implemented between discovery and SRM-based verification study to confirm the detectability of the candidates in plasma [39]. Only the candidates detected in the plasma by AIMS will be advanced to SRM-based assay development for more sophisticated quantitative comparison of the levels of the candidates in cases vs controls. This strategy allows one to test a much larger number of candidates than would have been possible over the traditional SID-SRM-MS based verification.

5. Data-independent MS/MS acquisition (DIA-MS/MS)

DIA-MS/MS is a new MS/MS acquisition technology [40, 41]. DIA-MS/MS carries the acronyms Precursor Acquisition Independent From Ion Count (PAcIFIC) [42], All-ion Fragmentation (AIF) [43], and Sequential window acquisition of all theoretical mass spectra (SWATH) [44, 45]. DIA-MS/MS is an approach where tandem mass spectra are acquired at every m/z value without regard for whether a precursor ion is observed or not. In DIA-MS/MS, the direct relationship between fragments and precursor from which they originate is lost, and assigning fragments to precursors can depend on the targeted data extraction and the availability of extensive spectral libraries such as PeptideAtlas [46, 47]. DIA-MS/MS demonstrates better sensitivity, reproducibility and dynamic range than DDA-MS/MS, and allows consistent quantification of proteins spanning a wide range of concentrations, e.g., 125–106 copies/cell [44], a range well within the needs for quantifying host cellular response profiles. Data-independent acquisition itself is not a targeted approach, but in combination with targeted data analysis, it can be used as an alternative approach of SRM assay in clinical research. In this approach, a quantitative, digitalized proteomic recording (SWATH maps) will be generated for each clinical sample as a personalized digital representation for each patient [48]. The profile of proteins of interest can then subsequently be extracted in a targeted fashion using assay information derived from mass spectrometric reference maps. In a recent study of N-linked glycoproteins in human plasma, N-linked glycoproteins in human plasma were enriched with solid phase extraction, then analyzed by both SWATH maps and SRM [45]. SWATH maps coupled with targeted data extraction shows less sensitivity than SRM, but achieved a higher analyte throughput, comparable dynamic range, reproducibility, and accuracy if stable isotope labeled peptides of analytes were used as internal standards. This finding indicates that SWATH maps can be used as targeted, reproducible quantitative approach for biomarker qualification/verification in less complicated samples [45]. Furthermore, SWATH maps are permanent digital maps and can be easily re-examined for qualification/verification of new sets of biomarker candidates without reanalyzing the sample physical samples [48]. Although SWATH maps require little assay development, it can be useful only when a high quality MS/MS spectra reference maps with well-calibrated elution times are available and can be replicated on the instrument used for SWATH MS analysis. SWATH generates highly complex and overlapping MS/MS spectra, and significant bioinformatic effort is required for analyzing SWATH data. Some special bioinformatic tools, such as openSWATH [49] and Spectronaut (www.biognosys.ch) have been developed for facilitating target data extraction from SWATH maps data and quantification.

We therefore summarize the benefits and tradeoffs inherent to each platform for biomarker verification with respect to the main factors characterizing measurements: accuracy, sensitivity, specificity, reproducibility, precision, dynamic range, sample throughput, analyte throughput, assay development easiness, and ease of data analysis (Fig. 23.1). Each method entails a compromise that maximizes the performance at some level, while reducing it at others. For example, PRM has higher specificity than SRM, but lower reproducibility and precision of quantification; SWATH can significantly improve analyte throughput but at the cost of specificity and accuracy. Given SRM has the best overall analytical performance, it is considered as the gold standard approach for biomarker qualification and verification.

Because the odds of discovery of a clinically useful biomarker or biomarker panel are extremely low, a large number of biomarker candidates must be tested in a qualification phase. Developing SID-SRM assays for every candidate identified by discovery study will become very costly and time consuming. A small number of candidates must be selected from the many hundreds of available candidates. Therefore, the qualification phase can be further divided into two steps: triage and quantification.

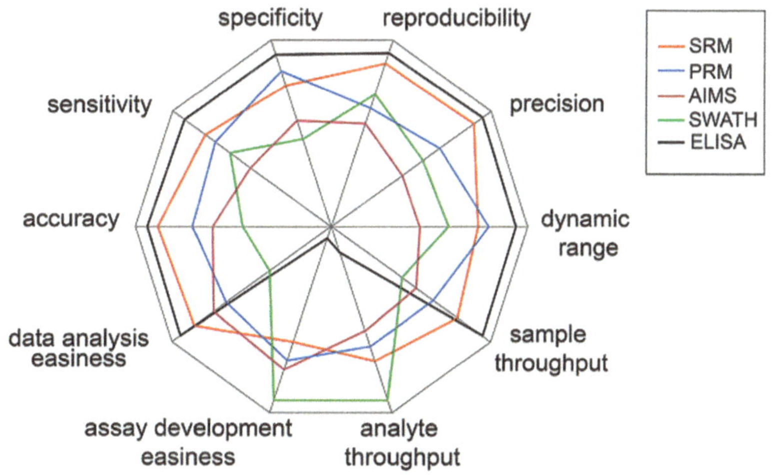

Fig. 23.1 Performance profiles comparing technical advantages and disadvantages of target MS platforms used in biomarker verification study

In the triage step, the biomarker candidates are measured by targeted, but less costly assays [39]. Among the platforms available for biomarker qualification/verification, PRM, AIMS, and SWATH have the capability to test and triage large number of candidates with lower expense and less lead-time for assay development. They can be easily developed if a local high quality MS/MS spectra reference maps with well-calibrated elution times are available and can be replicated on the instrument used for analyzing the clinical samples. Only the candidates that pass the triage step will be advanced to more expensive SID-SRM quantification. This staged qualification/verification strategy will enable one to test as many candidates as possible with reasonable cost and time to improve the chance of discovery of clinically useful biomarker panels.

23.3 Pre-fractionation and Enrichment Technologies

Ideally, SRM assays can be applied to verify biomarker candidates directly from plasma or serum without upfront sample fractionation. It is efficient, reproducible, high throughput, and less prone to errors and analytical variations. In recent studies, high and medium abundance human plasma proteins have been quantified by using multiplexed SRM approach without further sample preparation. Kuzyk et al. reported the simultaneous quantification of 45 major plasma proteins with a CV below 20 % for 94 % of the measured peptides [50]. Anderson et al. reported that 47 major plasma proteins were quantified with in-run CVs of 2–22 % [51]. The least abundant protein quantified, L-selectin, had a measured concentration of 0.67 μg/mL, a concentration 4–5 orders of magnitude lower than the concentration of albumin in plasma. Addnota et al. tested the LLOQ of SRM assays of target proteins in human plasma [18]. Eight of ten tested peptides had median LLOQ values between 0.66 and 2.0 fmol/μL when peptides were added into 1:60 diluted plasma (equivalent to a range of 0.70–3.34 μg/ml protein in plasma). These studies demonstrate SRM assay can reliably quantify the classic plasma protein biomarkers with concentration higher than 1 μg/mL directly in plasma. But this LLOQ of SRM assays is not sufficient for unambiguous detection and quantification of other types of protein biomarkers with lower concentration, such as tissue leakage products, interleukins, and cytokines, directly from plasma (Fig. 23.2). The lack of sensitivity by applying SRM assays directly to plasma is mainly caused by matrix-related interference and ion suppression. Plasma is an extremely complex mixture of proteins over a concentration range of 11 orders of magnitude in the presence of other endogenous salt, lipid, and metabolites. These matrix components have deleterious effect on the sensitivity of SRM assays. Competition for ionization between the analytes of interest and other endogenous (such as salt, lipid, and metabolite) or exogenous (such as polymers extracted from plastic tubes) species causes the ion suppression effect. When these interfering species elute at the same time as the analyte of interest, the signals of analytes will be suppressed [52]. Some matrix components can also produce the same product ions monitored for the analytes of interest, giving rise to chemical and biological noise, which reduce the S/N ratio necessary for detection and quantification. To overcome these sensitivity barriers, a variety of sample preparation strategies have been developed for target protein quantification aimed at reducing sample complexity while maintaining the requirements for high accuracy, reproducibility, and throughput.

23.3.1 Depletion of High-Abundance Proteins

Depletion of the highest abundance plasma proteins using affinity columns is the simplest way to reduce the sample complexity. In a study, Keshishian et al. reported that depletion of the 12 highest abundance plasma proteins improved the SRM assay LLOQ to 25 ng/mL [2]. The combination of depletion with strong cation exchange chromatography (SCX) further

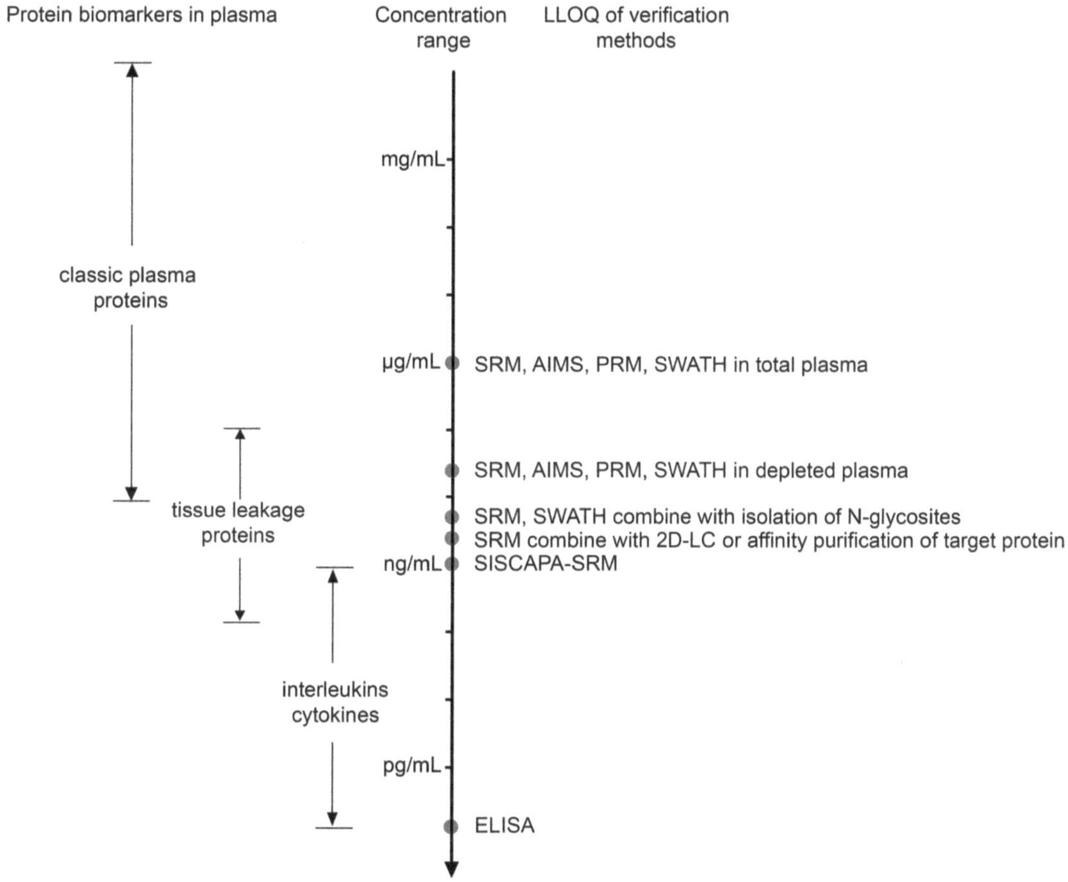

Fig. 23.2 Comparison of the LLOQ of different strategies for the quantification of protein biomarkers in plasma. A schematic diagram of the source and target concentration ranges of candidate plasma biomarkers. At right is LLOQ of current reported verification assay (Taken from Zhao, Current Proteomics, permission required)

improved the LLOQ of SRM assay to 1–10 ng/mL with CV below 15 % [53]. But this approach is impractical for biomarker qualification/verification because extensive prefractionation of samples into numbers of subfractions substantially reduces the throughput of the entire assay.

23.3.2 Enrichment of Target Proteins or Peptides Using Affinity Chromatography

Specifically isolating the target proteins or peptides from human plasma with affinity purification is the most efficient way to reduce the sample complexity. This approach is based on the highly specific interaction between the targeted proteins with affinity ligands, such as antibodies, aptamers, or lectins. Pre-fractionation is especially useful for quantification of low-abundance proteins in plasma. In our recent qualification and verification study of dengue fever biomarker panel, we found that the circulating level of one of the biomarker candidates, Complement Factor D (CFD), was below the LLOQ of the SID-SRM-MS assay and could not be detected in unfractionated plasma. To address this issue, we developed an assay in which the CFD was first immunoprecipitated (IPed) by anti-CFD antibody from plasma followed by quantification with SID-SRM-MS [54]. The CFD protein in each sample was IPed with biotin conjugated anti-CFD antibody. The complex of CFD and its antibody was captured by streptavidin magnetic beads. Stable isotope labeled CFD signature peptide was spiked into each sample, the proteins were trypsin-digested, and CFD abundance was quantified with SID-SRM-MS. By using this approach, we significantly improved the sensitivity of the assay.

IP-SRM can be multiplexed using a mixture of magnetic beads containing different antibodies to increase the throughput of the assay. Nicol et al. used this approach to quantify multiple proteins from human sera simultaneously [55]. The assays extend the LLOQ of SRM assay to low ng/ml range with good accuracy.

A newly emerging immuno-affinity-SRM approach termed stable isotope-labeled standards with capture by anti-peptide antibodies (SISCAPA) was developed by Anderson et al. [56], using immobilized anti-peptide antibodies to enrich the target peptides and the previously spiked synthetic stable isotope-labeled peptides. Using this method, more than 1000-fold enrichment for target peptides in a plasma digest can be achieved. In several studies, individual SISCAPA-SRM assays have been successfully configured for quantifying biomarkers in the ng/μL range in plasma with CV $< 20\ \%$ [56–58]. The protein concentration determined by this method with results obtained using a commercial immunoassay yield a high correlation of the two technologies [57, 59], demonstrating that the method can quantify low-abundance proteins with high accuracy. SISCAPA-SRM-MS has potential to multiplex the number of peptides measured in one assay by using a mixture of magnetic beads containing different antipeptide antibodies. Whiteaker et al. demonstrated that up to nine peptides have been enriched simultaneously with a LLOQ in the low ng/ml range (from 10 μl of plasma) and a median coefficient of variation of 12.6 % [58]. They also demonstrated that the LLOQ can be extended to low pg/ml range of protein concentration when larger volumes of plasma (1 ml) were used. This method holds great promise for verifying biomarker candidates. Interlaboratory evaluation of SISCAPA indicated that limits of detection of SISCAPA were at or below 1 ng/ml for the assayed proteins in 30 μl of plasma. Assay reproducibility was acceptable for verification studies, with median intra- and inter-laboratory CVs above the limit of quantification of 11 % and $<14\ \%$, respectively, for the entire immuno-MRM-MS assay process, including enzymatic digestion of plasma [60]. SISCAPA has several advantages over immunoaffinity capture of target proteins since; (1) it avoids potential interference from endogenous antibodies in the sample as they are digested to peptide by trypsin, and (2) anti-peptide antibodies are easier to generate in comparison to anti-protein antibodies. The limitation of this type of enrichment strategy is the requirement for specific antibody to be generated for each tryptic peptide used for a target protein. An alternative approach is the use of aptamers, oligonucleotide sequences with molecular recognition properties selected from combinatorial oligonucleotide libraries [61]. Aptamers bind protein ligands with high affinity and specificity [62]. They can be easily generated because they are chemically synthesized, enabling standardization of assays across multiple lots, a feature not possible with generation of polyclonal antibodies, for example.

23.3.3 Sample Fractionations for Protein Adduct or Fragments

Potential biomarkers may be proteins with post-translational modifications or peptide fragments derived from endogenous proteins. To unambiguously quantify these candidates, they have to be first separated from their canonical forms. In our recent biomarker discovery study of dengue fever, we identified a high molecular weight (>250 kDa) form of albumin is associated with dengue fever virus infection [63]. The nature of this protein is incompletely characterized, but is probably a covalently linked polymer [63]. To verify the high molecular weight albumin isoform, in our NIAID funded Clinical Proteomics Center, we developed a capillary electrophoresis

(CE) based fractionation approach. For CE fractionation, plasma samples were separated after spike-in with Beckman protein size standards. The 250 kDa fraction was collected into a receiving vial. The SDS in each collected CE fractions was removed by using SDS sample cleaning kit (Bio-Rad). The protein pellets were redissolved in 8 M urea. The proteins were digested with trypsin and quantified with SID-SRM-MS assay. Similarly, for the peptide fragments derived from endogenous proteins, size-based separation approaches such as size-exclusion chromatography (SEC) can be used. For example, in our recent biomarker discovery study of Aspergillosis (Discovery of Candidate Biomarkers, Chap. 20), we identified 26 small molecular sized peptides in plasma. These peptides are fragments of endogenous proteins such as albumin, apolipoprotein A-I, haptoglobin. To quantify these peptides, we first used size-exclusion chromatography to separate the denatured plasma into protein and peptide pools (MW $<$17 kDa). Then the concentration of these 26 peptide fragments in the peptide pool was quantified with SID-SRM-MS.

The qualification and verification strategies that were used for Dengue fever virus-3, infectious Aspergillosis, and Chagasic Cardiomyopathy are summarized in Table 23.1, 23.2, 23.3, and 23.4.

23.4 Feature Reduction/Candidate Selection

The qualification/verification phase seeks to reduce the number of candidate biomarkers to those most informative for general application in clinical setting. Another goal of qualification/verification is to test the statistical model that combines several of the informative features. Feature reduction aims to decrease the number of input variables to the model. Lower number of input variable enhances the quality of the data, increases the predictive power of the biomarker panel, and makes the results understandable and more robust for application to broader populations. This is a statistical approach that utilizes quantitative information derived from any of the qualification/verification assays described above. Approaches for feature reduction include pairwise statistical comparison, significance analysis of microarray (SAM), a technique that estimates false discovery rate (FDR) in high dimensional datasets, regression modeling, or machine learning techniques such as classification and regression trees (CART), multivariate adaptive regression splines (MARS) or ensemble methods. The application of these approaches is described more fully in the Chap. 20.

23.5 Consideration in Designing Quantification/Verification Study

1. Selection of sample cohorts for verification study

As described in the **Introduction to Proteomic-derived Biomarkers (Chap. 20)**, the samples in the qualification phase are the same samples used in the discovery phase. The verification phase involves measuring the candidates independently in a larger number of samples collected from patients with similar diagnosis and control patients from those that were assayed in the discovery phase of the biomarker pipeline. In order for the qualification/verification phase to be meaningful, a reproducible, observer-independent criteria for case definition needs to be applied. Samples should represent meaningful sampling of the patient cohort. Specifically the biospecimens should be derived from components of the cohort that meet the same objective criteria for cases and controls as those used for the discovery analysis.

2. Statistical design for verification study

The statistical design for the verification phase should be developed based on considerations of the effect size, outcomes (classes) in the experimental cohort, and experimental goal –e.g. is the focus to test the performance of a

Table 23.1 Qualification and verification strategies for candidate plasma proteins for Dengue fever virus-3

Biomarker candidates	Gene Name	Accession #	Qualification/ Verification strategy		SRM signature peptides
			Pre-fraction	Quantification	
Alpha-1-antitrypsin	SERPINA1	P01009	–	SID-SRM-MS	SVLGQLGITK
Leucine-rich alpha-2-glycoprotein	LRG1	P02750	–	SID-SRM-MS	GQTLLAVAK
Alpha-2-macroglobulin	A2M	P01023	–	SID-SRM-MS	QGIPFFGQVR
Serum albumin	ALB	P02768	–	SID-SRM-MS	LVNEVTEFAK
Apolipoprotein A-I	APOA1	P02647	–	SID-SRM-MS	DYVSQFEGSALGK
Apolipoprotein C-III	APOC3	P02656	–	SID-SRM-MS	DALSSVQESQVAQQAR
Complement factor D	CFD	P00746	–	SID-SRM-MS	VQVLLGAHSLSQPEPSK
Complement factor H	CFH	P08603	–	SID-SRM-MS	SPDVINGSPISQK
Complement C4-A	C4A	P0C0L4	–	SID-SRM-MS	VGDTLNLNLR
Desmoplakin	DSP	P15924	–	SID-SRM-MS	TLELQGLINDLQR
Fibrinogen alpha chain	FGA	P02671	–	SID-SRM-MS	GSESGIFTNTK
Fibrinogen beta chain	FGB	P02675	–	SID-SRM-MS	SILENLR
Ferritin light chain	FTL	P02792	–	SID-SRM-MS	LNQALLDLHALGSAR
Hemopexin	HPX	P02790	–	SID-SRM-MS	NFPSPVDAAFR
Haptoglobin	HP	P00738	–	SID-SRM-MS	VGYVSGWGR
Ig gamma-1 chain C region	IGHG1	P01857	–	SID-SRM-MS	GPSVFPLAPSSK
Immunoglobulin J chain	JCHAIN	P01591	–	SID-SRM-MS	ENISDPTSPLR
Ig kappa chain C region	IGKC	P01834	–	SID-SRM-MS	TVAAPSVFIFPPSDEQLK
Keratin	KRT1	P04264	–	SID-SRM-MS	SLDLDSIIAEVK
Dengue-2 virus NS1 nonstructural protein	NS1	Q67431	–	SID-SRM-MS	SCTLPPLR
Tropomyosin alpha-4 chain	TPM4	P67936	–	SID-SRM-MS	LVILEGELER
Vimentin	VIM	P08670	–	SID-SRM-MS	VELQELNDR
Complement Factor D	CFD	P00746	IP	SID-SRM-MS	VQVLLGAHSLSQPEPSK
Low MW Desmoplakin	DSP	P15924	CE	SID-SRM-MS	TLELQGLINDLQR
High MW albumin	ALB	P02761	CE	SID-SRM-MS	LVNEVTEFAK

For each of the candidate plasma proteins, SID-SRM-MS assays were developed. Shown is the protein accession number, common name, pre-fraction technology, and signature sequence

Table 23.2 Qualification and verification strategies for candidate plasma proteins for infectious Aspergillosis

			Qualification/Verification strategy		
Biomarker candidates	Gene Name	Accession #	Pre-fraction	Quantification	SRM signature peptides
Alpha-1-acid glycoprotein 1	ORM1	P02763	–	SID-SRM-MS	YVGGQEHFAHLLILR
Alpha-1-antitrypsin	SERPINA1	P01009	–	SID-SRM-MS	SVLGQLGITK
Alpha-1-antichymotrypsin	SERPINA3	P01011	–	SID-SRM-MS	EIGELYLPK
Serum albumin	ALB	P02768	–	SID-SRM-MS	LVNEVTEFAK
Apolipoprotein A-I	APOA1	P02647	–	SID-SRM-MS	DYVSQFEGSALGK
Apolipoprotein C-III	APOC3	P02656	–	SID-SRM-MS	DALSSVQESQVAQQAR
Fibrinogen alpha chain	FGA	P02671	–	SID-SRM-MS	GLIDEVNQDFTNR
Fibrinogen beta chain	FGB	P02675	–	SID-SRM-MS	SILENLR
Leucine-rich alpha-2-glycoprotein	LRG1	P02750	–	SID-SRM-MS	GQTLLAVAK

For each of the candidate plasma proteins, SID-SRM-MS assays were developed. Shown is the protein accession number, common name, pre-fraction technology, and signature sequence

biomarker to differentiate cases vs controls, or to evaluate the statistical model? The reader should refer to **Statistical Approaches (Chap. 22)** for more details.

3. Selection of assays – Fit-for-purpose concept

We propose to adopt staged, fit-for-purpose strategy for design a biomarker qualification/verification study [64, 65]. Depending on the number of biomarker candidates, the concentration of biomarker candidates in clinical samples, the feasibility of de novo assay development for the candidates, the analytical performance of the assays, and the cost of assay development and application for measuring a large numbers of targeted analytes across many samples, qualification/verification study can consist of three steps: triage, quantification, and verification (Fig. 23.3). The triage and quantification are performed in the qualification phase with the same samples used in the discovery phase. One important lesson learned from past 10-year's biomarker discovery studies is that the odds of identifying a clinically useful biomarker panel is extraordinarily low. To increase the chance of identifying a successful biomarker panel, researchers usually assemble a candidate pool for the qualification study from several sources including local proteomic and transcriptional profiling experiments, as well as data from the published literature. The candidate pool can become very large and these candidates may not directly associate with the disease of interest. In the case of that hundreds of candidates have to be tested in the qualification study, the study should start with a triage process to test these candidates while containing cost. The goal of this triage process is to reduce the initial list of candidates to a small subset that will be quantified with SID-SRM in the quantification stage. The technology used in this step should have higher capacity to triage large number of candidates with lower expense and shorter lead time for assay development. The assay should have enough specificity and precision to semi-quantitatively measure the relative changes in the level of large number of analytes across

Table 23.3 Qualification and verification strategies for candidate plasma peptides for infectious Aspergillosis

Biomarker candidates	Gene Name	Accession #	Qualification/Verification strategy: Pre-fraction	Qualification/Verification strategy: Quantification	SRM signature peptides
Serum albumin	ALBU_671	P02768	BAP	SID-SRM-MS	AVMDDFAAFVEK
Serum albumin	ALBU_734	P02768	BAP	SID-SRM-MS	RHPDYSVVLLLR
Serum albumin	ALBU_756	P02768	BAP	SID-SRM-MS	VPQVSTPTLVEVSR
Serum albumin	ALBU_820	P02768	BAP	SID-SRM-MS	KVPQVSTPTLVEVSR
Apolipoprotein A-I	APOA1_615	P02647	BAP	SID-SRM-MS	QGLLPVLESFK
Apolipoprotein A-I	APOA1_618	P02647	BAP	SID-SRM-MS	DLATVYVDVLK
Apolipoprotein A-II	APOA2_486	P02652	BAP	SID-SRM-MS	SPELQAEAK
Apolipoprotein A-II	APOA2_578	P02652	BAP	SID-SRM-MS	SKEQLTPLIK
Apolipoprotein A-II	APOA2_600	P02652	BAP	SID-SRM-MS	VKSPELQAEAK
Glutathione peroxidase 3	GPX3_657	P22352	BAP	SID-SRM-MS	FLVGPDGIPIMR
Glutathione peroxidase 3	GPX3_665	P22352	BAP	SID-SRM-MS	FLVGPDGIPIM[Oxid]R
Haptoglobin	HPT_720	P00738	BAP	SID-SRM-MS	TEGDGVYTLNNEK
Haptoglobin-related protein	HPTR_448	P00739	BAP	SID-SRM-MS	NPANPVQR
Haptoglobin-related protein	HPTR_656	P00739	BAP	SID-SRM-MS	TEGDGVYTLNDK
Ig kappa chain C region	IGKC_973	P01834	BAP	SID-SRM-MS	TVAAPSVFIFPPSDEQLK
Ig lambda-3 chain C regions	LAC3_495	P0CG06	BAP	SID-SRM-MS	AGVETTTPSK
Ig lambda-6 chain C region	LAC6_872	P0CF74	BAP	SID-SRM-MS	YAASSYLSLTPEQWK
Retinol binding protein 4	RBP4_599	Q5VY30	BAP	SID-SRM-MS	YWGVASFLQK

For each of the candidate plasma proteins, SID-SRM-MS assays were developed. Shown is the protein accession number, common name, pre-fraction technology, and signature sequence

large number of samples. The validation of the assays for triage will be minimal, including specificity, precision and run-to-run variation. The accuracy of quantification is not required. Although the use of stable-isotope labeled standards for each analytes are not required for triage process, a constant set of stable isotope labeled isotopic peptides corresponding to certain housekeeping proteins is recommended to be spiked into the samples in same amount. These standards can serve as benchmarks for normalization of run-to-run reproducibility and landmarks for calibration of LC retention time.

The targeted MS assays such as PRM, AIMS and SWATH with targeted data extraction are well-suited for this purpose. They can monitor the entire set of fragment ions for each analytes with high resolution and high mass accuracy.

Table 24.4 Qualification and verification strategies for candidate protein markers for Chagasic Cardiomyopathy

Biomarker candidates	Gene Name	Accession #	Qualification/ Verification strategy		SRM signature peptides
			Pre-fraction	Quantification	
Serum albumin	ALB	P02768	–	SID-SRM-MS	LVNEVTEFAK
Annexin A3	ANXA3	P12429	–	SID-SRM-MS	LTFDEYR
Fibrinogen alpha chain	FGA	P02871	–	SID-SRM-MS	GLIDEVNQDFTNR
Heterogeneous nuclear ribonucleoprotein A1	HNRNPA1	P09651	–	SID-SRM-MS	LFIGGLSFETTDESLR
SH3 domain-binding glutamic acid-rich-like protein 3	SH3BGRL3	Q9H299	–	SID-SRM-MS	VYSTSVTGSR
Tubulin-5	TUBB	P07437	–	SID-SRM-MS	YLTVAAVFR
Vimentin	VIM	P08670	–	SID-SRM-MS	VELQELNDR

For each of the candidate proteins, SID-SRM-MS assays were developed. Shown is the protein accession number, common name, pre-fraction technology, and signature sequence

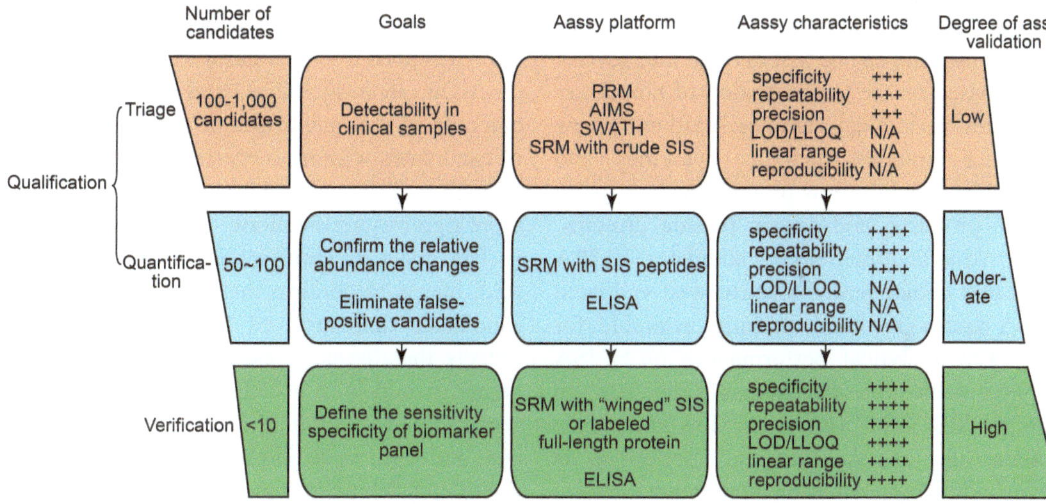

Fig. 23.3 Multistage, targeted proteomic workflow for biomarker qualification and verification

With the absence of stable isotope labeled peptides as internal standards for each target analyte, these approaches heavily rely on the reference database of standard spectra of each analyte to construct time-scheduled data acquisitions and confirm the identification of the analytes. The acquired MS/MS spectra will be compared with authentic standard spectra to examine the agreement of relative abundance of fragment ions and LC retention time. The identification confidence is determined by the number of fragment ion observed and the correlation of the observed LC retention of the analyte to its predicted retention time. It should be noted that SRM without stable isotope labeled peptide of each analyte is not a reliable tool for the triage process because SRM usually monitors only 3–5 transitions with moderate mass accuracy and unit resolution. This technique cannot provide sufficient confidence in detecting candidate biomarkers in the absence of stable isotope-labeled peptide standard. If SRM is the only platform available for the study, low cost, unpurified stable isotope labeled peptides for each targeted analyte should be used to provide the confidence needed for LC peak identification.

Measurements in triage step are semi-quantitative, only allowing rough estimations of relative abundance changes of targeted proteins. The small set of candidates derived from triage step require additional quantification with SID-SRM-MS to confirm the observed changes. In addition to prioritizing the candidates for more accurate quantification, triage step will also determine which protein candidates can be quantified directly from clinical samples, and which candidates need additional sample fractionation or enrichment to improve the limit of detection and quantification.

In the quantification step, the list of candidates for quantification can be first divided into several groups based on the concentration of biomarker candidates in clinical samples: extremely low abundance proteins such as cytokines and interleukins, medium-low proteins such as tissue leakage products, and classic plasma proteins. For cytokine and interleukin candidates, ELISA is the first choice assay because well-validated ELISA assays are commercially available for most. The analytical performances of ELISA are acceptable for the studies. The task to develop SID-SRM assays for low-abundance proteins such as cytokines and interleukins is very challenging, requiring significant amount time and effort to find suitable antibodies for the candidates. Even with antibody enrichment, the sensitivity of SRM will not be able to reach the required LLOQ of pg/ml in order to quantify cytokines and interleukins. As a result, a much larger biospecimen volume is required for their quantification by SRM. For tissue leakage products and classic plasma proteins, SID-SRM is the primary choice for quantification. SRM can be applied to verify classic plasma proteins directly from clinical samples without upfront sample fractionation. For tissue-leakage proteins, certain strategies for sample fractionation or enrichment are usually required in order to quantify the candidates with acceptable sensitivity and specificity (Fig. 23.2). If antibodies are not readily available, IP-SRM and SISCAPA-SRM are not recommended for less credentialed candidates because of tremendous effort required for developing suitable antibodies.

The use of stable internal standards in SRM assays are required to provide the highest level of detection confidence and measurement precision. Stable isotope labeled tryptic peptide standards are the most commonly used internal standards. They can provide sufficient precision and reproducibility to confirm the differential expression of candidates by the disease and eliminate the false positive candidates identified in the discovery phase. But in this approach the accuracy of quantification is only moderate because stable isotope-labeled peptide standards do not account for the differences in trypsin digestion efficiency. So assays using stable isotope-labeled peptide standards need to be validated to prove moderate precision, reproducibility, and specificity. The outcome of the quantification process is the list of candidates with high correlation with disease of interest. These candidates will then advance to more rigorous verifications.

The goals of verification process are three-fold; one is to confirm that the small subset of candidates that survived the triage step truly reflects the disease presence, severity, or outcome, second is to establish the specificity and sensitivity of the biomarker panel for its intended use; and third is to implement suitable sample fraction/enrichment approach for the targets, if applicable. It was found that trypsin digestion and its requisite sample handling usually contribute the most to assay variability. It has been shown that the use of stable isotope-labeled protein as an internal standard instead of stable isotope labeled peptides to account for losses in the digestion process nearly doubles assay accuracy [60]. Therefore, in verification phase to increase the accuracy of quantification, labeled, full-length proteins, or winged-peptides with 2–6 amino acids of native flanking sequence at the N-, and C- termini of tryptic peptide analyte, or concatemer of standard peptides should be added at the start of trypsin digestion to serve as more robust internal standards. The purity and quantity of internal standards must be established. For "winged" peptides, quantification is usually done by HPLC and amino acid analysis. If the concentration of targeted proteins are below the LLOQ of SID-SRM-MS assays and cannot be quantified directly from clinical samples, suitable strategies to enrich targeted proteins should be established. IP-SRM or SISCAPA are the first choice for this purpose because they are proven to be very efficient way to enrich the targeted proteins with high precision and repeatability compared to other approaches.

Similarly, the confidence in the accuracy of the qualification/verification assay should increases as the credential of the biomarker candidate increases. Although achieving total accuracy in mass spectrometry based protein quantification is not possible, the assays used for high credential candidates should have high specificity, reproducibility, precision (less than 35 % CV), and sensitivity for target quantification [65]. Analytical validation assays are evaluated based on their assay precision, linear dynamic range, and sensitivity (LOD and LLOQ). If a prefractionation/enrichment step is implemented prior to MS analysis, such steps also need to be validated as part of the overall assay validation for factors such as run-to-run variation, recovery, and carryover. Ideally, the assays for high credential candidates should be able to be standardized across laboratories and translated into clinical assays.

23.6 Summary

By far, the most challenging step in the biomarker development pipeline is isolating the true biomarkers from a large pool of differentially expressed proteins identified in discovery phase. The large size of the initial candidates pool is due to several factors including high false positive discovery rate, the poor quality of clinical samples, the high complexity of clinical samples, and the lack of highly specific and quantitative assays for quanitfying all protein in biofluids. Recent advances in targeted MS-based technologies such as AIMS, PRM, SWATH and SID-SRM-MS show the potential to alleviate the bottleneck in biomarker pipeline. Among them, SID-SRM-MS assays have been proven to be the most reliable approach for biomarker qualification/verification. With the progress that has been made in recent years, it is becoming more of a realistic possibility that SID-SRM-MS approach can also be developed into a FDA-approvable assay for clinical test. MS-based clinical assays can complement traditional immunoassays well especially for protein biomarkers that high quality ELISA assays cannot detect, or in cases where protein isoforms or posttranslational modifications constitute the biomarker. In this chapter, we proposed a fit-for-purpose, staged biomarker qualification/verification workflow to verify the hundreds of candidates generated from discovery phase with a cost-effective rapid manner. This workflow starts with a data-dependent biomarker candidate triage step by using semi-quantitative AIMS, PRM, or SWATH approaches followed by SID-SRM-MS based qualification and verification for candidates that survive the triage. The accuracy and precision of qualification/verification assays for final candidates need to be confirmed at every step. The rigor of biomarker assay validation should increase as the credential of biomarker candidate increases. This continuous and evolving fit-for-purpose strategy will conserve resources and efforts in the qualification/verification stages of biomarker development and increase the chance to identify a successful biomarker panel.

References

1. Anderson NL (2010) The clinical plasma proteome: a survey of clinical assays for proteins in plasma and serum. Clin Chem 56:177–185
2. Rifai N, Gillette MA, Carr SA (2006) Protein biomarker discovery and validation: the long and uncertain path to clinical utility. Nat Biotechnol 24:971–983
3. Spicer V, Grigoryan M, Gotfrid A, Standing KG, Krokhin OV (2010) Predicting retention time shifts

associated with variation of the gradient slope in peptide RP-HPLC. Anal Chem 82:9678–9685
4. Roobol MJ, Carlsson SV (2013) Risk stratification in prostate cancer screening. Nat Rev Urol 10:38–48
5. Del Mastro L, Lambertini M, Bighin C, Levaggi A, D'Alonzo A, Giraudi S, Pronzato P (2012) Trastuzumab as first-line therapy in HER2-positive metastatic breast cancer patients. Expert Rev Anticancer Ther 12:1391–1405
6. Carr SA, Anderson L (2008) Protein quantification through targeted mass spectrometry: the way out of biomarker purgatory? Clin Chem 54:1749–1752
7. Hoofnagle AN, Wener MH (2009) The fundamental flaws of immunoassays and potential solutions using tandem mass spectrometry. J Immunol Methods 347:3–11
8. Krastins B, Prakash A, Sarracino DA, Nedelkov D, Niederkofler EE, Kiernan UA, Nelson R, Vogelsang MS, Vadali G, Garces A, Sutton JN, Peterman S, Byram G, Darbouret B, Perusse JR, Seidah NG, Coulombe B, Gobom J, Portelius E, Pannee J, Blennow K, Kulasingam V, Couchman L, Moniz C, Lopez MF (2013) Rapid development of sensitive, high-throughput, quantitative and highly selective mass spectrometric targeted immunoassays for clinically important proteins in human plasma and serum. Clin Biochem 46:399–410
9. Gerber SA, Rush J, Stemman O, Kirschner MW, Gygi SP (2003) Absolute quantification of proteins and phosphoproteins from cell lysates by tandem MS. Proc Natl Acad Sci U S A 100:6940–6945
10. Beynon RJ, Doherty MK, Pratt JM, Gaskell SJ (2005) Multiplexed absolute quantification in proteomics using artificial QCAT proteins of concatenated signature peptides. Nat Methods 2:587–589
11. Pratt JM, Simpson DM, Doherty MK, Rivers J, Gaskell SJ, Beynon RJ (2006) Multiplexed absolute quantification for proteomics using concatenated signature peptides encoded by QconCAT genes. Nat Protoc 1:1029–1043
12. Dupuis A, Hennekinne JA, Garin J, Brun V (2008) Protein Standard Absolute Quantification (PSAQ) for improved investigation of staphylococcal food poisoning outbreaks. Proteomics 8:4633–4636
13. Brun V, Dupuis A, Adrait A, Marcellin M, Thomas D, Court M, Vandenesch F, Garin J (2007) Isotope-labeled protein standards: toward absolute quantitative proteomics. Mol Cell Proteomics 6:2139–2149
14. Zhao Y, Brasier AR (2013) Applications of selected reaction monitoring (SRM)-mass spectrometry (MS) for quantitative measurement of signaling pathways. Methods 61:313–322
15. Zhao Y, Tian B, Edeh CB, Brasier AR (2013) Quantification of the dynamic profiles of the innate immune response using multiplex selected reaction monitoring-mass spectrometry. Mol Cell Proteomics 12:1513–1529
16. Zhao Y, Widen SG, Jamaluddin M, Tian B, Wood TG, Edeh CB, Brasier AR (2011) Quantification of activated NF-kappaB/RelA complexes using ssDNA aptamer affinity-stable isotope dilution-selected reaction monitoring-mass spectrometry. Mol Cell Proteomics 10:M111 008771
17. Reiter L, Rinner O, Picotti P, Huttenhain R, Beck M, Brusniak MY, Hengartner MO, Aebersold R (2011) mProphet: automated data processing and statistical validation for large-scale SRM experiments. Nat Methods 8:430–435
18. Abbatiello SE, Mani DR, Keshishian H, Carr SA (2010) Automated detection of inaccurate and imprecise transitions in peptide quantification by multiple reaction monitoring mass spectrometry. Clin Chem 56:291–305
19. Addona TA, Abbatiello SE, Schilling B, Skates SJ, Mani DR, Bunk DM, Spiegelman CH, Zimmerman LJ, Ham AJ, Keshishian H, Hall SC, Allen S, Blackman RK, Borchers CH, Buck C, Cardasis HL, Cusack MP, Dodder NG, Gibson BW, Held JM, Hiltke T, Jackson A, Johansen EB, Kinsinger CR, Li J, Mesri M, Neubert TA, Niles RK, Pulsipher TC, Ransohoff D, Rodriguez H, Rudnick PA, Smith D, Tabb DL, Tegeler TJ, Variyath AM, Vega-Montoto LJ, Wahlander A, Waldemarson S, Wang M, Whiteaker JR, Zhao L, Anderson NL, Fisher SJ, Liebler DC, Paulovich AG, Regnier FE, Tempst P, Carr SA (2009) Multi-site assessment of the precision and reproducibility of multiple reaction monitoring-based measurements of proteins in plasma. Nat Biotechnol 27:633–641
20. Mead JA, Bianco L, Ottone V, Barton C, Kay RG, Lilley KS, Bond NJ, Bessant C (2009) MRMaid, the web-based tool for designing multiple reaction monitoring (MRM) transitions. Mol Cell Proteomics 8:696–705
21. Martin DB, Holzman T, May D, Peterson A, Eastham A, Eng J, McIntosh M (2008) MRMer, an interactive open source and cross-platform system for data extraction and visualization of multiple reaction monitoring experiments. Mol Cell Proteomics 7:2270–2278
22. Sherwood CA, Eastham A, Lee LW, Peterson A, Eng JK, Shteynberg D, Mendoza L, Deutsch EW, Risler J, Tasman N, Aebersold R, Lam H, Martin DB (2009) MaRiMba: a software application for spectral library-based MRM transition list assembly. J Proteome Res 8:4396–4405
23. Krokhin OV, Spicer V (2010) Predicting peptide retention times for proteomics. Curr Protoc Bioinformatics. Wiley
24. Brusniak MY, Kwok ST, Christiansen M, Campbell D, Reiter L, Picotti P, Kusebauch U, Ramos H, Deutsch EW, Chen J, Moritz RL, Aebersold R (2011) ATAQS: A computational software tool for high throughput transition optimization and validation for selected reaction monitoring mass spectrometry. BMC Bioinf 12:78
25. MacLean B, Tomazela DM, Shulman N, Chambers M, Finney GL, Frewen B, Kern R, Tabb

DL, Liebler DC, MacCoss MJ (2010) Skyline: an open source document editor for creating and analyzing targeted proteomics experiments. Bioinformatics 26:966–968
26. Farrah T, Deutsch EW, Kreisberg R, Sun Z, Campbell DS, Mendoza L, Kusebauch U, Brusniak MY, Huttenhain R, Schiess R, Selevsek N, Aebersold R, Moritz RL (2012) PASSEL: the PeptideAtlas SRMexperiment library. Proteomics 12:1170–1175
27. Huttenhain R, Soste M, Selevsek N, Rost H, Sethi A, Carapito C, Farrah T, Deutsch EW, Kusebauch U, Moritz RL, Nimeus-Malmstrom E, Rinner O, Aebersold R (2012) Reproducible quantification of cancer-associated proteins in body fluids using targeted proteomics. Sci Transl Med 4:142ra194
28. Kelly RT, Page JS, Zhao R, Qian WJ, Mottaz HM, Tang K, Smith RD (2008) Capillary-based multi nanoelectrospray emitters: improvements in ion transmission efficiency and implementation with capillary reversed-phase LC-ESI-MS. Anal Chem 80:143–149
29. Page JS, Tang K, Kelly RT, Smith RD (2008) Subambient pressure ionization with nanoelectrospray source and interface for improved sensitivity in mass spectrometry. Anal Chem 80:1800–1805
30. Fortin T, Salvador A, Charrier JP, Lenz C, Bettsworth F, Lacoux X, Choquet-Kastylevsky G, Lemoine J (2009) Multiple reaction monitoring cubed for protein quantification at the low nanogram/milliliter level in nondepleted human serum. Anal Chem 81:9343–9352
31. Jeudy J, Salvador A, Simon R, Jaffuel A, Fonbonne C, Leonard JF, Gautier JC, Pasquier O, Lemoine J (2014) Overcoming biofluid protein complexity during targeted mass spectrometry detection and quantification of protein biomarkers by MRM cubed (MRM3). Anal Bioanal Chem 406:1193–1200
32. Gallien S, Duriez E, Crone C, Kellmann M, Moehring T, Domon B (2012) Targeted proteomic quantification on quadrupole-Orbitrap mass spectrometer. Mol Cell Proteomics 11:1709–1723
33. Peterson AC, Russell JD, Bailey DJ, Westphall MS, Coon JJ (2012) Parallel reaction monitoring for high resolution and high mass accuracy quantitative, targeted proteomics. Mol Cell Proteomics 11:1475–1488
34. Leinenbach A, Pannee J, Dulffer T, Huber A, Bittner T, Andreasson U, Gobom J, Zetterberg H, Kobold U, Portelius E, Blennow K, proteins, I. S. D. W. G. o. C. (2014) Mass spectrometry-based candidate reference measurement procedure for quantification of amyloid-beta in cerebrospinal fluid. Clin Chem 60:987–994
35. Gallien S, Bourmaud A, Kim SY, Domon B (2014) Technical considerations for large-scale parallel reaction monitoring analysis. J Proteome 100: 147–159
36. Jaffe JD, Keshishian H, Chang B, Addona TA, Gillette MA, Carr SA (2008) Accurate inclusion mass screening: a bridge from unbiased discovery to targeted assay development for biomarker verification. Mol Cell Proteomics 7:1952–1962
37. Schmidt A, Gehlenborg N, Bodenmiller B, Mueller LN, Campbell D, Mueller M, Aebersold R, Domon B (2008) An integrated, directed mass spectrometric approach for in-depth characterization of complex peptide mixtures. Mol Cell Proteomics 7:2138–2150
38. Schmidt A, Claassen M, Aebersold R (2009) Directed mass spectrometry: towards hypothesis-driven proteomics. Curr Opin Chem Biol 13:510–517
39. Whiteaker JR, Lin C, Kennedy J, Hou L, Trute M, Sokal I, Yan P, Schoenherr RM, Zhao L, Voytovich UJ, Kelly-Spratt KS, Krasnoselsky A, Gafken PR, Hogan JM, Jones LA, Wang P, Amon L, Chodosh LA, Nelson PS, McIntosh MW, Kemp CJ, Paulovich AG (2011) A targeted proteomics-based pipeline for verification of biomarkers in plasma. Nat Biotechnol 29:625–634
40. Geromanos SJ, Vissers JP, Silva JC, Dorschel CA, Li GZ, Gorenstein MV, Bateman RH, Langridge JI (2009) The detection, correlation, and comparison of peptide precursor and product ions from data independent LC-MS with data dependant LC-MS/MS. Proteomics 9:1683–1695
41. Venable JD, Dong MQ, Wohlschlegel J, Dillin A, Yates JR (2004) Automated approach for quantitative analysis of complex peptide mixtures from tandem mass spectra. Nat Methods 1:39–45
42. Panchaud A, Scherl A, Shaffer SA, von Haller PD, Kulasekara HD, Miller SI, Goodlett DR (2009) Precursor acquisition independent from ion count: how to dive deeper into the proteomics ocean. Anal Chem 81:6481–6488
43. Geiger T, Cox J, Mann M (2010) Proteomics on an Orbitrap benchtop mass spectrometer using all-ion fragmentation. Mol Cell Proteomics 9:2252–2261
44. Gillet LC, Navarro P, Tate S, Rost H, Selevsek N, Reiter L, Bonner R, Aebersold R (2012) Targeted data extraction of the MS/MS spectra generated by data-independent acquisition: a new concept for consistent and accurate proteome analysis. Mol Cell Proteomics 11:O111
45. Liu Y, Huttenhain R, Surinova S, Gillet LC, Mouritsen J, Brunner R, Navarro P, Aebersold R (2013) Quantitative measurements of N-linked glycoproteins in human plasma by SWATH-MS. Proteomics 13:1247–1256
46. Deutsch EW, Lam H, Aebersold R (2008) PeptideAtlas: a resource for target selection for emerging targeted proteomics workflows. EMBO Rep 9:429–434
47. Deutsch EW (2010) The PeptideAtlas project. Methods Mol Biol 604:285–296
48. Liu Y, Huttenhain R, Collins B, Aebersold R (2013) Mass spectrometric protein maps for biomarker discovery and clinical research. Expert Rev Mol Diagn 13:811–825
49. Rost HL, Rosenberger G, Navarro P, Gillet L, Miladinovic SM, Schubert OT, Wolski W, Collins

BC, Malmstrom J, Malmstrom L, Aebersold R (2014) OpenSWATH enables automated, targeted analysis of data-independent acquisition MS data. Nat Biotechnol 32:219–223
50. Kuzyk MA, Smith D, Yang J, Cross TJ, Jackson AM, Hardie DB, Anderson NL, Borchers CH (2009) Multiple reaction monitoring-based, multiplexed, absolute quantification of 45 proteins in human plasma. Mol Cell Proteomics 8:1860–1877
51. Anderson L, Hunter CL (2006) Quantitative mass spectrometric multiple reaction monitoring assays for major plasma proteins. Mol Cell Proteomics 5:573–588
52. Furey A, Moriarty M, Bane V, Kinsella B, Lehane M (2013) Ion suppression; a critical review on causes, evaluation, prevention and applications. Talanta 115:104–122
53. Keshishian H, Addona T, Burgess M, Mani DR, Shi X, Kuhn E, Sabatine MS, Gerszten RE, Carr SA (2009) Quantification of cardiovascular biomarkers in patient plasma by targeted mass spectrometry and stable isotope dilution. Mol Cell Proteomics 8:2339–2349
54. Brasier AR, Zhao Y, Wiktorowicz JE, Spratt HM, Nascimento EJ, Cordeiro MT, Soman KV, Ju H, Recinos A 3rd, Stafford S, Wu Z, Marques ET Jr, Vasilakis N (2015) Molecular classification of outcomes from dengue virus −3 infections. J Clin Virol 64:97–106
55. Nicol GR, Han M, Kim J, Birse CE, Brand E, Nguyen A, Mesri M, FitzHugh W, Kaminker P, Moore PA, Ruben SM, He T (2008) Use of an immunoaffinity-mass spectrometry-based approach for the quantification of protein biomarkers from serum samples of lung cancer patients. Mol Cell Proteomics 7:1974–1982
56. Anderson NL, Anderson NG, Haines LR, Hardie DB, Olafson RW, Pearson TW (2004) Mass spectrometric quantification of peptides and proteins using Stable Isotope Standards and Capture by Anti-Peptide Antibodies (SISCAPA). J Proteome Res 3:235–244
57. Kuhn E, Addona T, Keshishian H, Burgess M, Mani DR, Lee RT, Sabatine MS, Gerszten RE, Carr SA (2009) Developing multiplexed assays for troponin I and interleukin-33 in plasma by peptide immunoaffinity enrichment and targeted mass spectrometry. Clin Chem 55:1108–1117
58. Whiteaker JR, Zhao L, Zhang HY, Feng LC, Piening BD, Anderson L, Paulovich AG (2007) Antibody-based enrichment of peptides on magnetic beads for mass-spectrometry-based quantification of serum biomarkers. Anal Biochem 362:44–54
59. Hoofnagle AN, Becker JO, Wener MH, Heinecke JW (2008) Quantification of thyroglobulin, a low-abundance serum protein, by immunoaffinity peptide enrichment and tandem mass spectrometry. Clin Chem 54:1796–1804
60. Kuhn E, Whiteaker JR, Mani DR, Jackson AM, Zhao L, Pope ME, Smith D, Rivera KD, Anderson NL, Skates SJ, Pearson TW, Paulovich AG, Carr SA (2012) Interlaboratory evaluation of automated, multiplexed peptide immunoaffinity enrichment coupled to multiple reaction monitoring mass spectrometry for quantifying proteins in plasma. Mol Cell Proteomics 11:M111 013854
61. Tuerk C, Gold L (1990) Systematic evolution of ligands by exponential enrichment: RNA ligands to bacteriophage T4 DNA polymerase. Science 249:505–510
62. Nery AA, Wrenger C, Ulrich H (2009) Recognition of biomarkers and cell-specific molecular signatures: aptamers as capture agents. J Sep Sci 32:1523–1530
63. Brasier AR, Garcia J, Wiktorowicz JE, Spratt HM, Comach G, Ju H, Recinos A 3rd, Soman K, Forshey BM, Halsey ES, Blair PJ, Rocha C, Bazan I, Victor SS, Wu Z, Stafford S, Watts D, Morrison AC, Scott TW, Kochel TJ, the Venezuelan Dengue Fever Working, G. (2012) Discovery proteomics and nonparametric modeling pipeline in the development of a candidate biomarker panel for dengue hemorrhagic fever. Clin Transl Sci 5:8–20
64. Lee JW, Devanarayan V, Barrett YC, Weiner R, Allinson J, Fountain S, Keller S, Weinryb I, Green M, Duan L, Rogers JA, Millham R, O'Brien PJ, Sailstad J, Khan M, Ray C, Wagner JA (2006) Fit-for-purpose method development and validation for successful biomarker measurement. Pharm Res 23:312–328
65. Carr SA, Abbatiello SE, Ackermann BL, Borchers C, Domon B, Deutsch EW, Grant RP, Hoofnagle AN, Huttenhain R, Koomen JM, Liebler DC, Liu T, MacLean B, Mani DR, Mansfield E, Neubert H, Paulovich AG, Reiter L, Vitek O, Aebersold R, Anderson L, Bethem R, Blonder J, Boja E, Botelho J, Boyne M, Bradshaw RA, Burlingame AL, Chan D, Keshishian H, Kuhn E, Kinsinger C, Lee JS, Lee SW, Moritz R, Oses-Prieto J, Rifai N, Ritchie J, Rodriguez H, Srinivas PR, Townsend RR, Van Eyk J, Whiteley G, Wiita A, Weintraub S (2014) Targeted peptide measurements in biology and medicine: best practices for mass spectrometry-based assay development using a fit-for-purpose approach. Mol Cell Proteomics 13:907–917

Protocol for Standardizing High-to-Moderate Abundance Protein Biomarker Assessments Through an MRM-with-Standard-Peptides Quantitative Approach

24

Andrew J. Percy, Juncong Yang, Andrew G. Chambers, Yassene Mohammed, Tasso Miliotis, and Christoph H. Borchers

Abstract

Quantitative mass spectrometry (MS)-based approaches are emerging as a core technology for addressing health-related queries in systems biology and in the biomedical and clinical fields. In several 'omics disciplines (proteomics included), an approach centered on selected or multiple reaction monitoring (SRM or MRM)-MS with stable isotope-labeled standards (SIS), at the protein or peptide level, has emerged as the most precise technique for quantifying and screening putative analytes in biological samples. To enable the widespread use of MRM-based protein quantitation for disease biomarker assessment studies and its ultimate acceptance for clinical analysis, the technique must be standardized to facilitate precise and accurate protein quantitation. To that end, we have developed a number of kits for assessing method/platform performance, as well as for screening proposed candidate protein biomarkers in various human biofluids. Collectively, these kits utilize a bottom-up LC-MS methodology with SIS peptides as internal standards and quantify proteins using regression analysis of standard curves. This chapter details the methodology used to quantify 192 plasma proteins of high-to-moderate abundance (covers a 6 order of magnitude range from 31 mg/mL for

A.J. Percy • J. Yang • A.G. Chambers
University of Victoria – Genome British Columbia Proteomics Centre, Vancouver Island Technology Park, #3101 – 4464 Markham St., Victoria, BC V8Z 7X8, Canada

Y. Mohammed
University of Victoria – Genome British Columbia Proteomics Centre, Vancouver Island Technology Park, #3101 – 4464 Markham St., Victoria, BC V8Z 7X8, Canada

Center for Proteomics and Metabolomics, Leiden University Medical Center, 2333 ZA, Leiden, Netherlands

T. Miliotis
AstraZeneca R&D, Innovative Medicines, S-431 83, Mölndal, Sweden

C.H. Borchers (✉)
University of Victoria – Genome British Columbia Proteomics Centre, Vancouver Island Technology Park, #3101 – 4464 Markham St., Victoria, BC V8Z 7X8, Canada

Department of Biochemistry and Microbiology, University of Victoria, Petch Building Room 207, 3800 Finnerty Rd., Victoria, BC V8P 5C2, Canada
e-mail: christoph@proteincentre.com

H. Mirzaei and M. Carrasco (eds.), *Modern Proteomics – Sample Preparation, Analysis and Practical Applications*, Advances in Experimental Medicine and Biology 919, DOI 10.1007/978-3-319-41448-5_24

albumin to 18 ng/mL for peroxidredoxin-2), and a 21-protein subset thereof. We also describe the application of this method to patient samples for biomarker discovery and verification studies. Additionally, we introduce our recently developed Qualis-SIS software, which is used to expedite the analysis and assessment of protein quantitation data in control and patient samples.

Keywords
Biomarker • Internal standards • MRM • Plasma • Proteomics • Quantitation • Standardization

24.1 Introduction

MS-based protein quantitation is increasingly utilized to determine differences between samples from healthy and diseased patients for biomarker (*i.e.*, biological indicators of disease or disorder) and systems biology studies. Although quantitation can be performed using a relative technique, such as iTRAQ (isobaric tags for relative and absolute quantitation [1]) or TMT (tandem mass tag [2]), techniques that provide exact endogenous concentrations (often reported in ng/mL units), as opposed to fold changes of abundance levels, are more informative and better suited for applications where the analysis of pre-clinical and clinical samples is the ultimate goal. Such quantitative techniques are commonly referred to as "absolute", and require the use of isotopically labeled standards (typically expressed in bacterial media, in the case of proteins [3], or chemically synthesized, for peptides) and a targeted form of MS detection (usually MRM-MS with electrospray ionization, ESI, for gas phase ionization of the chromatographic eluent) to be employed within a bottom-up analytical workflow [4–6]. In this generalized approach, proteotypic peptides serve as molecular surrogates for the target proteins. The isotopically labeled standards are typically labeled with ^{13}C and/or ^{15}N, as opposed to ^{18}O or ^{2}H, and these labels are incorporated into amino acids within a protein or the C-terminal residue of a tryptic peptide. Collectively, the standards are used for normalization of the peptide signal and LC-MS conditions. In MRM-MS, specific precursor-product ion pairs (referred to as transitions) are used for peptide detection. Generating peptide specific transitions requires *a priori* knowledge of the analyte and its dissociation upon collisional activation (also referred to as collision induced dissociation or CID). While the use of MRM is common and is classically performed on a triple quadrupole mass spectrometer, directed quantitation has also recently been accomplished by parallel reaction monitoring (PRM) on a hybrid quadrupole-Orbitrap (*i.e.*, Q Exactive) mass spectrometer [7–9] and by MS/MS^{ALL} with SWATH acquisition on a quadrupole time-of-flight (QTOF) [10] or a hybrid quadrupole linear ion trap (QTRAP) mass spectrometer [11]. Mechanistically in PRM, for instance, all product ions that lie within a specified mass range and emanate from a specifically fragmented precursor are detected in the high resolution, high mass accuracy Orbitrap analyzer. An attractive feature of this technique, as well as MS/MS^{ALL}, is that it allows the post-analysis mining of previously collected (or archived) MS/MS data, and therefore allows the selection of alternate quantitative transitions if interference with the target(s) is observed.

The most desirable sample sources for biomarker research and clinical measurement are ideally non-invasive, such as urine or saliva. Although blood plasma and serum are semi-invasive, they are still commonly used for monitoring and stratifying diseases. Plasma and serum are used because they are relatively inexpensive to collect and analyze, and carry a wide dynamic range of proteins (approximating or exceeding

10 orders of magnitude [12]) that are secreted, released, or leaked from neighboring cells, tissues, or organs into the systemic circulation. The fluid therefore paints a physiological picture of the health status of an individual, which is imperative for disease diagnosis and prognosis. It is important to note here that there is a distinction between plasma and serum since the two are often incorrectly used interchangeably by the proteomics field. Plasma and serum are both derived from whole blood, with serum collected from plasma after coagulation. It is through the coagulation process that an assembly of mid-abundance proteins (*e.g.*, fibrinogen, prothrombin, thrombin, and a host of coagulation factors – notably II, V, and VIII) are at least partially removed. Serum is, however, generally disfavored by the Human Proteome Organization (HUPO [13, 14]) since coagulation can cause additional proteins to be unintentionally removed through non-specific interactions and is also a highly variable process, with the results being dependent upon the coagulation conditions and the nature of the collection tube [15]. It is for these reasons that our blood-based assay developments and analyses are commonly conducted with plasma, with the exception being our dried fluid spot quantitative analyses where the spots originate from whole blood [16].

As inferred above, plasma is an inherently complex biofluid, carrying thousands of potentially measurable proteins spanning the low mg/mL (or millimolar; encompassing serum albumin and the immunoglobulins, among others) to low pg/mL (or attomolar; which includes the interleukins and cytokines) concentration range. An active area of biomedical research centers on developing sensitive methods to accurately and reproducibly quantify proteins at the lower end of the concentration range since these candidates are considered to have the greatest diagnostic potential. Targeted quantitative methods for detection of proteins with concentrations below the MRM detection limit often use anti-protein [17] or anti-peptide [12, 18, 19] antibodies for immunoaffinity enrichment or alternatively the implementation of multidimensional separations (increasingly with alkaline and acidic RPLC (reversed-phase liquid chromatography) [20, 21], less commonly with strong cation-exchange and RPLC configurations [22]) for peptide fractionation. Additional techniques developed for deeper protein quantification involves the upfront use of immunodepletion for high abundant protein removal via antibody-based, affinity interactions [22–26]. Depletion, however, is disfavored from a cost and throughput perspective, as well as for the potential of target protein loss through non-specific or non-covalent interactions with the depletion cartridge or depleted proteins. An added detraction of this technique is the potential underestimation of protein concentration, as was demonstrated recently by Percy et al. in the side-by-side comparison of a depletion-based and depletion-free, multiplexed quantitative proteomic assay of cerebrospinal fluid [27]. Nonetheless, despite the increasing emphasis on low-abundance proteins, antibody- and fractionation-free quantitative proteomic methods should also be developed for the screening of higher-abundance protein markers since these are also informative and correlate with multiple diseases such as cancer and cardiovascular disease (CVD) [28, 29]. This is why we have developed sets of highly-multiplexed (defined as enabling multi-analyte detection in a single analytical run) MRM assays for the precise quantitation of high-to-moderate abundance, candidate protein biomarkers in undepleted and non-enriched human plasma [20, 30–32].

The protein biomarker pipeline is essentially comprised of four stages – discovery, verification, pre-clinical validation, and clinical validation. Although quantitative MRM or PRM methods can be used to assess marker utility at all levels, their greatest value lies in the discovery and verification phases. Once the lengthy list of potential candidate markers has been screened and condensed according to statistical significance, resources can then be invested in the development of antibodies, which is a costly and developmentally intensive process [33]. At the validation stages of biomarker assessment, shorter lists of verified candidates (typically <10) are interrogated against a larger number of samples (on order of 1000s at the validation stage vs. 10s–100s in the preceding stages [33]). While ELISAs (enzyme linked immunosorbent assays) are often considered to be the "gold-standard" for clinical applications [34], emerging techniques, such as iMALDI (immuno matrix-assisted laser desorption/ionization; where peptide detection of captured peptides occurs by MALDI-TOF-MS without prior chromatographic separation [35]) and SISCAPA (stable isotope standards and capture with anti-peptide antibodies via LC-MS [18, 36] or MALDI-MS [37, 38] detection) could alternatively be employed.

To expedite biomarker verification, the targeted quantitative methods must be standardized. This should facilitate improved method reproducibility and transferability and lead to a more rapid and accurate evaluation of the candidate protein biomarkers in a given biological fluid [39, 40]. To this end, a variety of kits have been developed for the quantitative proteomics community. Stemming from work done in our laboratory, QC kits are developed to evaluate the performance of a LC-MS system and/or one type of sample preparation in a targeted quantitative proteomic workflow [41, 42]. Recently, we have also developed several biomarker assessment kits (BAKs) for screening various protein panels against patient plasma samples for biomarker discovery or verification studies. The methods collectively utilize an antibody-/fractionation-free approach, a rigorously optimized and evaluated bottom-up LC/MRM proteomic workflow, and our well characterized SIS peptides. The targeted proteins are either putative biomarkers for CVD and cancer or have unknown disease associations. Each BAK contains a collection of key starting materials (*i.e.*, reference plasma, trypsin, and SIS peptide mixture), a detailed protocol, a LC-MS acquisition method, data analysis software, and a troubleshooting guide. This chapter will detail the protocol and provide the rationale behind the development and application of two recent biomarker assessment kits – BAK-192 for discovery and a custom BAK-21 for verification – for MRM-based quantitative proteomic studies. Also provided is a description and implementation of our recently developed Qualis-SIS software [43] for quantitative proteomic applications.

24.2 Targeted Quantitation Method – Strategy, Description, and Rationale

The principle checkpoints we use in developing sensitive and specific MRM-based quantitative proteomic assays, such as the BAK-192 and BAK-21, involve protein/peptide target selection, SIS peptide production, solution/sample preparation, interference screening, and protein quantitation (see Fig. 24.1 for our generalized workflow). Additional important steps include balancing the concentrations of the mixture of SIS peptides to their corresponding natural (or NAT) peptide signals (balancing helps reduce analytical variation between analyses [44]), as well as optimizing the MRM transitions (includes their collision energies) and LC gradient. This section expands upon that basic framework developed to quantify multiplexed panels of plasma proteins for assessment as potential biomarkers via a bottom-up LC/MRM approach using SIS peptides. By outlining our strategy and rationale behind each development step, the user will obtain the necessary tools for extending the quantitative method to alternative panels and types of samples. Nonetheless, the applications that these BAKs are designed for is discussed in the section that follows.

24.2.1 Protein and Peptide Selection

The first step in our quantitative proteomic method development is generating a list of potential biomarkers in human plasma. These putative biomarkers are selected from prior discovery experiments or from literature reports, and typically exist in a wide range of concentrations. Tryptic peptides (ideally a minimum of 2) are then chosen to act as molecular surrogates for each biomarker. Selection is based on adherence

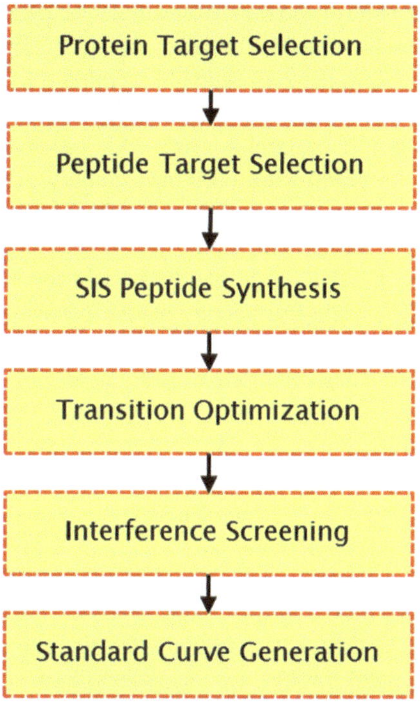

Fig. 24.1 General workflow for MRM assay development. Protein/peptide selection is a bioinformatics exercise aided by previously collected data or curated databases, as well as by software tools, such as PeptidePicker. The internal standards employed are SIS peptides, which are synthesized, purified, and characterized for more accurate protein quantitation. MRM transition optimization and screening for chemical interference in the sample matrix is performed empirically, while protein quantitation is performed on the interference-free peptides via standard curves

to a set of qualification criteria [45], with the most notable ones indicated below:

- Peptides must be unique to the target biomarker (human in this case; determined from a BLASTp search).
- Peptides must have been previously observed in tandem MS proteomic studies (revealed in the Global Proteome Machine and PeptideAtlas databases).
- Peptides must not contain a missed tryptic cleavage site (Kiel rules obeyed [46]).
- Peptides must be between 5 and 25 residues in length to ensure acceptable ionization and gas-phase fragmentation.

To reduce error and subjectivity, the rules have recently been assembled into a software tool we named PeptidePicker, which automates candidate identification and ranks the selected peptide(s) for a given protein within a specified proteome (human or mouse) [47]. This program, we note, is an advancement over the PeptideSieve tool (developed by the Seattle Proteome Centre), which predicts proteotypic propensity based solely on the physicochemical properties of the peptides expected to result from a digest of a given protein [48]. Due to the accuracy and enhanced speed of peptide selection in PeptidePicker (ca. 50 proteins per hour compared to 8 per day in peptideSieve [47]), the time devoted to bioinformatics is significantly reduced, allowing more time to be spent on the rest of assay development. Furthermore, PeptidePicker reduces human error and provides users with a standardized method for target peptide selection of any panel of biomarkers.

24.2.2 SIS Peptide Production

Once the proteotypic peptides have been selected, their heavy isotope labeled analogues are synthesized, purified, and characterized. These are essential steps for obtaining absolute and precise, but not necessarily accurate, endogenous protein concentrations. In our laboratory, synthesis is performed in-house on an Overture peptide synthesizer (Protein Technologies) using Fmoc chemistry. To enable chromatographic alignment of heavy isotope coded peptides with the regular NAT peptides (which greatly assists in the subsequent interference testing step), [^{13}C]/[^{15}N] isotopes (Cambridge Isotope Laboratories) are incorporated at the C-terminal residue of tryptic peptides, typically leading to +8 Da (from [$^{13}C_6$, $^{15}N_2$]-lysine) or + 10 Da ([$^{13}C_6$, $^{15}N_4$]-arginine) mass shifts. Purification is also performed in-house by RPLC, with the fractions of interest confirmed by MALDI-TOF-MS on an Ultraflex III TOF/TOF mass spectrometer (Bruker Daltonik). After lyophilization of the pooled target fractions, amino acid analysis (AAA) and capillary zone electrophoresis (CZE) are then performed for absolute concentration and purity determination, respectively. Of relevance here, the average purity of the 487 target peptides used in the discovery BAK-192 is 92 %.

24.2.3 Sample Preparation and LC-MS Processing

It is our general practice to prepare small sample sets (*i.e.*, <20) manually in polypropylene Maxymum recovery microtubes (Axygen), but automate the preparation of larger sets of samples with a robot (Freedom EVO 150 platform; Tecan) in 96-well microtiter plates. A generalized flow chart of our sample preparation and processing process is illustrated in Fig. 24.2. It should be noted that our robot is configured to automate *only* the liquid handling steps, with centrifugation and incubation occurring externally.

Toward the preparation of plasma proteolytic digests, a ten-fold diluted plasma sample (20 μL for the control and 6 μL of raw fluid per patient) is denatured, reduced, alkylated, and quenched with 1 % sodium deoxycholate (10 % initially), 5 mM tris(2-carboxyethyl) phosphine (50 mM initially), 10 mM iodoacetamide (100 mM initially), and 10 mM dithiothreitol (100 mM initially), respectively, all prepared in 25 mM ammonium bicarbonate. The protein denaturation and Cys-Cys reduction steps occur simultaneously for 30 min at 60 °C, while Cys alkylation and iodoacetamide quenching is performed subsequently for 30 min at 37 °C. Thereafter, proteolysis is achieved by the addition of 23.3 μL TPCK-treated trypsin (Worthington) (1.8 mg in 2 mL of 25 mM ammonium bicarbonate; prepared immediately before addition) at a 10:1 substrate:enzyme ratio. After overnight incubation at 37 °C, proteolysis is arrested by the step-wise addition of a chilled SIS peptide mixture (concentration balanced; 50 μL at 250 to 0.5 fmol/μL for the control or 50 μL at 25 fmol/μL for the patient plasma) and a chilled formic acid (FA) solution (277 μL at 1 %) to a digest aliquot (277 μL; pooled from 4 digests in the control prep). The SIS mixes used in the control will be used to prepare the calibration

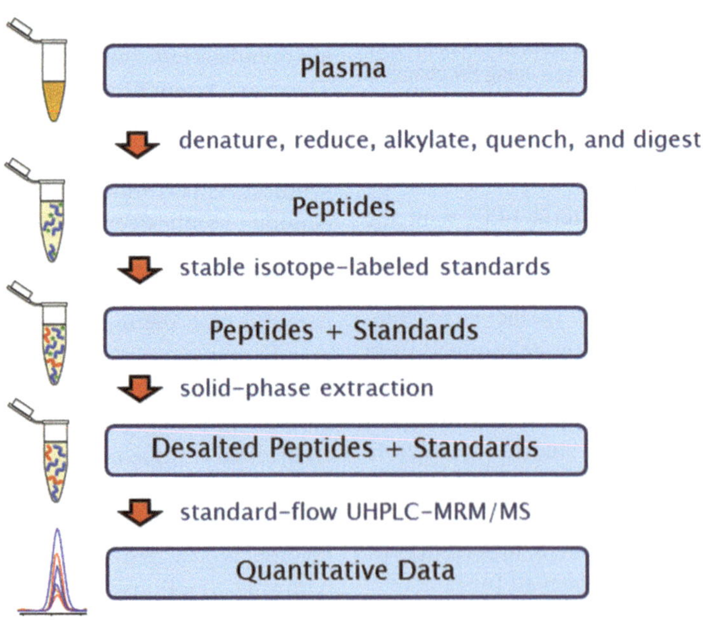

Fig. 24.2 Overview of our sample preparation and processing workflow. The plasma proteins are unfolded and the disulfide bridges are cleaved and capped by a series of denaturation, reduction, alkylation, and quenching steps prior to tryptic proteolysis. Labeled peptide standards are spiked post-digestion to prevent chemical modification which can occur during proteolysis. After the sample is concentrated by solid phase extraction, peptide mixture is separated by RPLC and detected by dynamic MRM on a QqQ mass spectrometer. Plasma protein quantitation is achieved through SPM or regression analysis of the standard control curve

curves. These mixtures each contain a fixed amount of endogenous peptide and an increasing concentration of synthetic peptide (over a 500-fold concentration range). The resulting dilution series prepared from each reference standard is as follows: 250 fmol/μL stock (standard F), 125 fmol/μL (standard E), 25 fmol/μL (standard D), 12.5 fmol/μL (standard C), 2.5 fmol/μL (standard B), and 0.5 fmol/μL (standard A; all prepared in 0.1 % FA). A merit of the deoxycholate surfactant is that is acid insoluble and therefore can be readily removed by simple centrifugation (10 min at 12,000 rpm). This is in contrast to sodium dodecyl sulfate which damages the LC column and causes signal suppression if not properly removed. Following centrifugation, the peptide supernatant is concentrated by solid phase extraction (SPE) using a polymeric RP sorbent (10 mg Oasis HLB; Waters). The extraction steps are as follows:

1. wash with 1 mL methanol,
2. condition with 1 mL water,
3. load with 556 μL of 0.1 % FA followed by 444 μL of digest supernatant,
4. wash with 1 mL water, and
5. elute with 300 μL of 50 % acetonitrile (ACN) in 0.1 % FA.

The eluate is then lyophilized and rehydrated in 100 μL of 0.1 % FA for LC-MRM/MS.

The LC-MS system we routinely use for the BAKs consists of a 1290 Infinity system that is interfaced to a 6490 triple quadrupole (QqQ) mass spectrometer (all from Agilent Technologies) via a standard-flow, ESI source (operated in the positive ionization mode). The LC column is a Zorbax Eclipse Plus RP-UHPLC column (2.1 × 150 mm, 1.8 μm particles). The separation occurs over a 43 min gradient (1.5–81 % mobile phase B; mobile phase compositions: 0.1 % FA in water for A and 0.1 % FA in ACN for B) at flow rates of 0.4 mL/min and a temperature of 50 °C. A 4 min post-acquisition step using mobile phase A is allotted for column equilibration. The specific gradient we employ is as follows (time in min, %B): 0, 1.5; 1.5, 6.3; 16, 13.5; 18, 13.77; 33, 22.5; 38, 40.5; 39, 81; 42.9, 81; and 43, 1.5. Note that standard flow rates are used instead of conventional nano-flow rates due to the superior analytical merits (in terms of reproducibility and sensitivity) found for the standard flow system when 10× material is loaded onto a wider-bore column [49]. The mass spectrometer is operated in the dynamic MRM mode (*i.e.*, scheduled retention times for enhanced analyte specificity and reduced duty cycle) with 1 min detection windows and cycle times approximating 850 ms (see [32] and its supplemental tables for the general and specific acquisition parameters).

24.2.4 Interference Reduction and Screening

Interference is commonly observed in the quantitative analysis of human plasma. These interferences exist despite the m/z and retention time filtering in scheduled MRM acquisitions, and is attributed largely to the inherent complexity of blood plasma, as well as to the low resolution QqQ mass spectrometer employed. Tryptic proteolysis further increases the complexity as it converts thousands of plasma proteins into millions of peptides. This increased complexity increases the possibility of non-target ion transmission in the quadrupole mass analyzers (Q1 and Q3) which necessitates utilizing interference reduction and screening techniques in quantitative proteomic studies.

Interferences can be reduced by minimizing concurrent MRM transitions, so our method development first involves optimizing the LC gradient, to produce an even distribution of peptides across the chromatographic space. To ensure the accuracy of quantitative results, the control and sample are first screened for interference. This is conducted empirically in our laboratory, as opposed to theoretically using a program such as SRM Collider [50]. In the analysis of the control (also referred to as the reference) sample digest, interferences are determined by monitoring the SIS and NAT responses (*i.e.*, peak areas) under matrix-free and matrix-containing conditions (both at n = 2). The variability in these calculated response ratios indicates the presence or absence of interferences in the MRM ion channels. For a given peptide to be interference-free, the average relative ratios between a SIS transition in buffer or plasma, and NAT transition in plasma, must have CVs below 20 %. Further, the NAT and SIS signals must be the same in both peak shape and retention time. Figure 24.3a shows a typical example of an interference-free and interference-containing peptide. In this example, the interference observed in the NAT transition of YWGVASFLQK and the high variability of two of its three average relative ratios precludes its use for protein quantitation.

The aforementioned approach is suitable for the inspection of control samples, but an alternative strategy must be adopted for interference screening in patient samples. Our recommended strategy requires a minimum of two peptides to be targeted for a given protein in order to construct peptide correlation plots (ratios of quantifier NAT/SIS relative responses), as first introduced by Agger et al. [51]. The linearity of each plot is then examined for outliers; with those that deviate requiring further inspection of their SIS and NAT peptide extracted ion chromatograms (XICs) to evaluate the level of interference. We recently demonstrated the implementation of this strategy in the quantitative analysis of 40 CVD-linked proteins (inferred from an average of three peptides per protein) across a small CVD patient cohort (n = 18; blood plasma supplied by Bioreclamation). As illustrated in Fig. 24.3b, the peptides SFNPNSPGK and IQNILTEEPK can effectively serve as surrogates for serum paraoxonase/arylesterase 1 (P27169) in all of the measured samples since they are interference-free, while peptide VVLSQGSK cannot be used to quantify sex hormone-binding globulin (P04278) in the CVD patient sample marked with the arrow due to interference. The advantage of this approach is that it requires the peptide responses of only the quantifier transitions, which enables BAK-192 to be processed with a single acquisition method. The use of multiple transitions (customarily with 1 quantifier and 2 qualifiers) for enhanced interference evaluation for BAK-192 discovery requires 3 LC-MS acquisition methods (2922 total transitions for the 487 peptides with 1461 transitions targeted for both peptide forms). In this case, multiple methods are required to reduce the duty cycle and obtain sufficient points across a chromatographic peak (defined as 10–15) for improved ion statistics.

24.2.5 Plasma Protein Quantitation

The MRM data is first examined with MassHunter Quantitative Analysis software (Agilent; Skyline can alternatively be used), for verification, peak selection and integration. Thereafter, the processed data is inputted into our in-house developed software tool – Qualis-SIS – for analysis. This tool requires two input files for each of the reference and sample data sets. These files carry peptide- and protein-related information, with SIS and NAT responses required for the former (retention time, peak width, symmetry but other metrics can additionally be included) and protein molecular weights and SIS peptide concentrations required for the latter. After defining a small number of criteria (*e.g.*, regression weighting, precision and accuracy requirements) for each concentration level of the standard curve, the tool automatically performs the following three functions: (1) generates and extracts assay information from standard control curves, (2) determines the endogenous protein concentrations in the patient samples, and (3) assesses the quality of the quantitative sample measurement with respect to the assay's linear dynamic range. The following information is provided by each control curve: endogenous protein concentration, dynamic range, lower and upper limits of quantitation (LLOQ and ULOQ), and regression equation (slope and y-intercept) with coefficient of determination (R^2). In the analysis of the samples, each measured concentration (derived from the relative response measurements also referred to as single point measurement –SPM- and linear regression analysis) is plotted on each peptide's standard curve. The quality assessment page

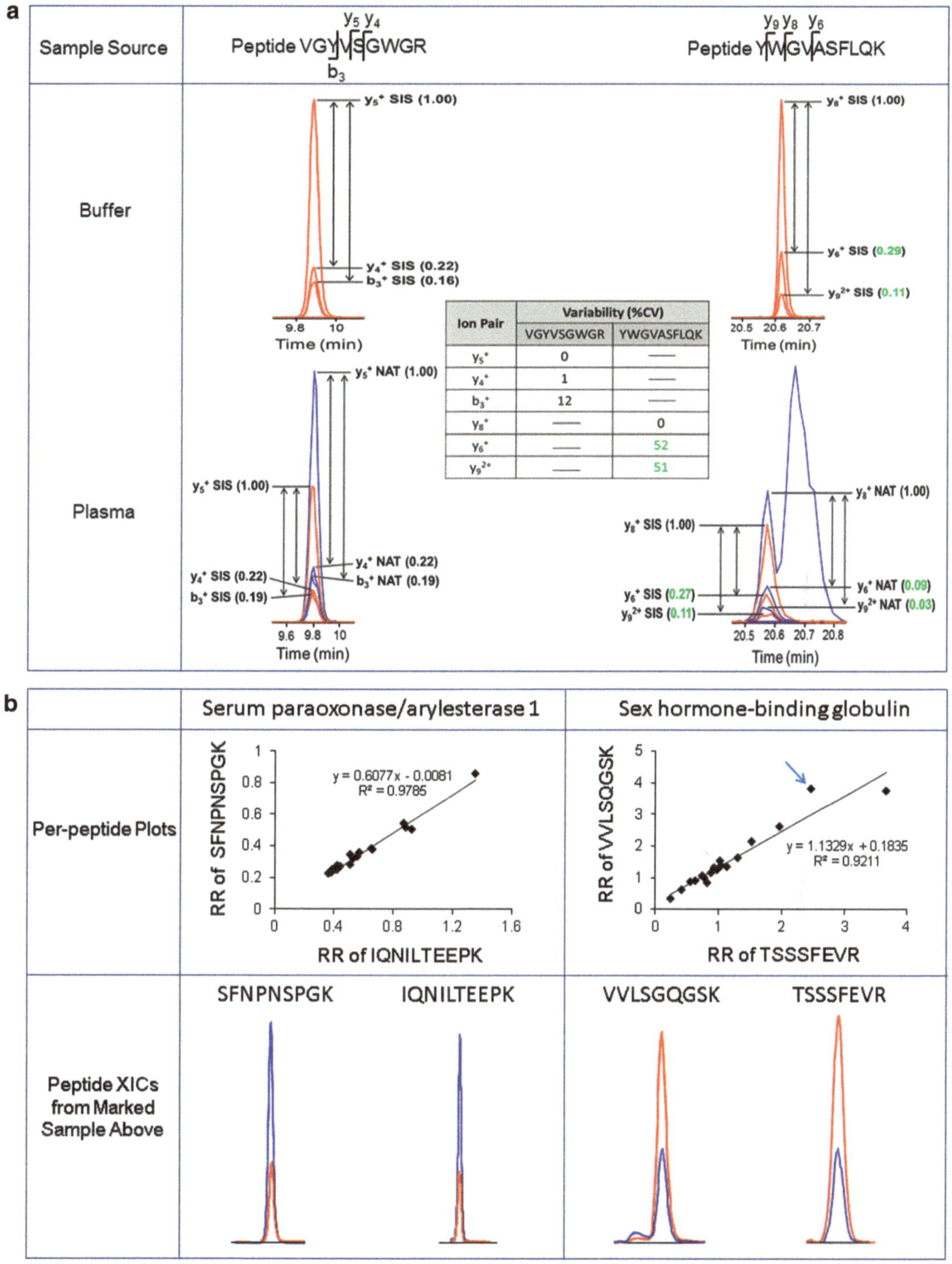

Fig. 24.3 Interference screening strategies for MRM transitions monitored in control and patient plasma digests. (**a**) Representative XICs of 3 SIS and NAT transitions measured in buffer and control plasma for the interference-free peptide VGYVSGWGR (from haptoglobin, P00738) and the interference-containing peptide YWGVASFLQK (from retinol-binding protein 4; P02753). (**b**) Relative response (RR) correlation plots

indicates whether or not the results should be trusted through a color-coded matrix. In the matrix, green denotes an acceptable quantitative value (due to its presence within the assay's range of linearity), yellow indicates that caution should be exercised, while red suggests that the value should be discarded (see Fig. 24.4 for an example of each classification type from the CVD-directed quantitative study indicated above). The assessment is based on the relationship of the concentration to the linear dynamic range as well as its deviation from the LOQ and the user-defined confidence threshold. The comprehensive and summarized results can then be exported for subsequent reporting and statistical treatment.

24.3 Method Implementation and Practical Biomarker Applications

Through rigorous evaluation and refinement, a well characterized set of MRM assays has been developed for quantifying a multiplexed panel of 192 candidate disease markers in unfractionated human plasma. The method centers on a bottom-up UHPLC/MRM workflow and uses concentration-balanced SIS peptides as internal standards. The quantified proteins are of high-to-moderate abundance, with concentrations spanning 6 orders of magnitude, from 31 mg/mL (for serum albumin, P02768) to 18 ng/mL (for peroxiredoxin-2, P32119) – see Fig. 24.5a for the quantitation range. These endogenous concentrations were derived from standard control curves (based on 144 proteins [52]) and/or individual XIC measurements (based on an additional 48 proteins [20]) using peptides as surrogates (487 interference-free in total). Regarding the curves, these were constructed from a strict set of qualification criteria, which our developed software – Qualis-SIS – accurately applies in an automated and rapid manner. The result of this analysis were a set of assays with average linear dynamic ranges of 10^2–10^3, protein LLOQs between 5 ng/mL and 260 ng/mL (based on quantifier peptides), and average R^2 values of 0.980. The assay reproducibility is high, with average relative responses of <6 % and average retention times of <0.1 % routinely obtained over replicate analyses [52]. These quantitative panels can now be applied in discovery- and verification-directed proteomic studies to help bridge the gap between biomarker discovery and validation.

In the classical sense, protein biomarker discovery is accomplished through bottom-up (or shotgun) LC-MS/MS using a multidimensional protein identification technology (MudPIT) in conjunction with data dependent acquisition (DDA). In DDA, a subset of peptide precursor ions, detected in the survey scan, are selected for CID based on abundance, yielding a collection of complete product ion spectra. Typical acquisition instruments for this include the quadrupole time-of-flight (QTOF) and hybrid ion trap-Orbitrap mass spectrometers. While technological advancements have enabled broad classes of putative protein biomarkers to be identified through DDA, their detection sensitivity and sample-to-sample reproducibility is limited due to the intensity-driven, stochastic nature of the precursor ion selection process [53, 54]. To overcome these inherent issues, data independent acquisition (DIA) strategies, such as MS/MS^{ALL} [55], have been proposed. This is based on the acquisition of complete product ion spectra generated from the dissociation of all precursors measured in given SWATH windows (typically 25 amu spanning from 400 to 1200 m/z) over the chromatographic run. While this may provide

Fig. 24.3 (continued) and peptide XICs for the interference-free peptides SFNPNSPGK and IQNILTEEPK from serum paraoxonase/arylesterase 1 (P27169) and the interference-containing peptide VVLSQGSK from sex hormone-binding globulin (P04278) in the CVD patient sample marked with the arrow. These figures were reprinted from [41] and [32], respectively, with permission

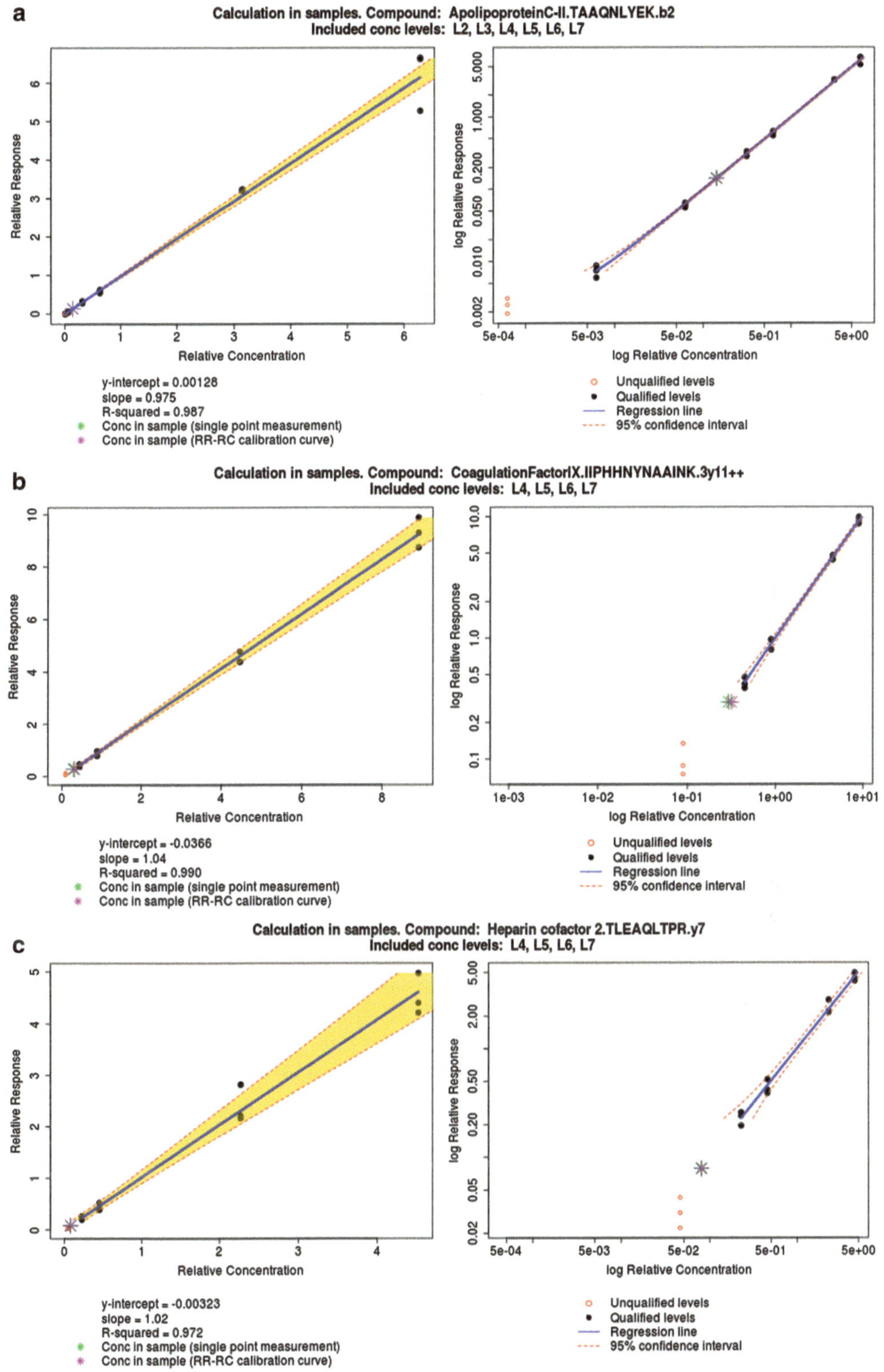

Fig. 24.4 Examples of patient sample results from the Qualis-SIS data analysis software tool. The examples show cases where the quantitative results are (**a**) acceptable (TAAQNLYEK from apolipoprotein

enhanced reproducibility and throughput over a DDA-based method, a MRM-based methodology, such as that described above, can instead be employed at the discovery stage for improved sensitivity, throughput, and reproducibility.

The discovery BAK-192 platform allows the interrogation of 192 proteins using 487 peptides as molecular surrogates. In this targeted application, the candidates will be assessed by quantitatively comparing the patient sample results with those from healthy controls. Ideally, a minimum of three process replicates (also referred to as "analytical replicates" that encompass the entire preparatory workflow) should be obtained. But only replicates that are quantitatively reproducible and interference-free should be used for comparison. To be statistically significant, a fold change ratio exceeding 1.5 and a p value <0.05 is desired [56]. While this biomarker panel is rather small, it covers a broad concentration range of proteins that can be consistently quantified without laborious pre-fractionation, which can in itself introduce variability. For more comprehensive biomarker discovery efforts, however, pre-fractionation is undoubtedly required. Using a scaled-up sample preparation method, we have recently developed a multidimensional LC-MRM workflow for quantifying a broader and deeper (by a 2 order of magnitude concentration range) panel of putative protein markers in human plasma [20]. In that method, the LCs are operated under alkaline and acidic mobile phase conditions for altered peptide selectivity, using an ACN gradient with constant 10 mM ammonium hydroxide (pH 10) in the former dimension and an ACN gradient with constant 0.1 % FA (pH 3) in the latter. Both dimensions additionally utilize RP stationary phases and standard-flow rates. Using SPM, and recently standard curves for a smaller protein panel (*e.g*., the low abundance targets osteopontin and matrix metalloproteinase 9 at 7 and 16 ng/mL, respectively), 253 proteins (inferred from 625 peptides) were quantified across an 8 order-of-magnitude concentration range. This panel can also represent a potentially useful starting point for assessing potential biomarker candidates at lower concentrations.

In a separate study focused on biomarker verification, a 21-plex protein assay was selected by a group of investigators based on their previous proteomic discovery results and our 1D LC-MRM/MS quantitative capabilities. The overall aim of their study was to determine whether these proteins play a role in the resolution and remission of type 2 diabetes after bariatric surgery. Bariatric surgery is of considerable research interest as it has rapid and dramatic effects on glycemic control. Recent studies by Mingrone G et al. [57] and Schauer P et al. [58] found bariatric surgery to be more effective than conventional medical therapy in controlling hyperglycemia in severely obese patients with type 2 diabetes, leading to long-term benefits on macro and microvascular disease [59]. Since some bariatric procedures, such as biliopancreatic diversion, improve glycemic control in people with diabetes, understanding this additional effect could provide insight into the pathogenesis of type 2 diabetes and assist in the development of new drug modalities. To address this unanswered question, we are currently engaged in a project involving a cohort of 20 morbidly obese, insulin-resistant patients whose plasma was collected over a 13-point time-course (from before surgery to 28 days post-surgery).

Sample preparation and analysis of the BAK-21 is as described above. This requires standard curves to be prepared for each of the 5 plates of 50 samples. Preliminary results for the concentration distribution from this study are shown in Fig. 24.5b. To aid in standardization, key starting materials (*i.e*., reference plasma, trypsin, and SIS peptide mixture) and

Fig. 24.4 (continued) C-II; P02655), (**b**) intermediate (IIPHHNYNAAINK from coagulation factor IX; P00740), or (**c**) unacceptable (TLEAQLTPR from heparin cofactor 2; P05546). The results were obtained from the same patient plasma sample used in the CVD study

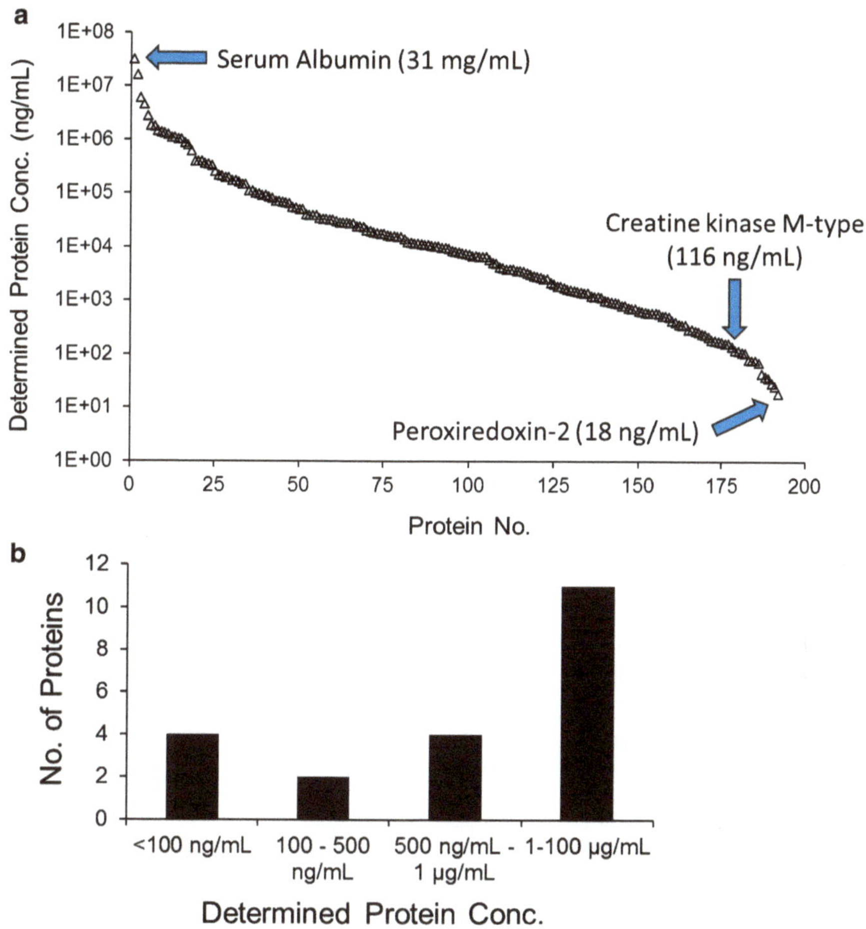

Fig. 24.5 Quantitation results from the multiplexed MRM analysis of control plasma. The range of protein concentrations shown in (**a**) was determined from the BAK-192 discovery analysis, while the concentration distribution in (**b**) is from the BAK-21 verification analysis

acquisition/analysis methods have been assembled. The final MRM acquisition method consists of a maximum of two proteotypic peptides per protein (39 total) and three transitions per peptide, which will be used for interference screening and protein quantitation of the patient samples, as outlined above. To ensure consistent performance of the LC-MS platform, daily/monthly QC kits will also be run before and after each plate. These kits require only simple rehydration of the lyophilized, SIS-spiked plasma digest(s) prior to LC-MRM/MS analysis, with evaluations achieved through value tracking and correlation to the reference values in the kits. These QC kits, we note, have already proven useful in diagnosing instrument errors and performance deficits in intra-/inter-lab studies in the past [41, 42], and should help again here to validate the experimental workflow and analytical system.

24.4 Summary

We have developed a set of highly specific and robust MRM-based assays for quantifying a large panel of 192 high-to-moderate abundance candidate protein markers in antibody- and fractionation-free human plasma. The 192 proteins (inferred from 487 peptides) are designed to be implemented in targeted, biomarker discovery-based studies, while a subset panel of 21 targets has been designed for biomarker verification in a diabetes-centric study. To help standardize the process, essential materials required to complete the entire protocol (from sample preparation and processing to quantitative analysis) have been assembled into kits, as described here for the BAK-192 and BAK-21. Additionally, our recently developed Qualis-SIS software offers an automated means of quantifying proteins in reference and patient samples through regression analysis of standard curves or through SPM. To aid in quality assessment, the results are illustrated in a color-coded matrix for rapid visualization and evaluation of the results. Continued developments are focused on extending these panels for more comprehensive discovery and verification of putative, or unknown, protein biomarkers. Nonetheless, the strategies, kits, and tools discussed here act as a useful starting point for biomarker evaluation of a panel of proteins of interest in patient samples.

Acknowledgements We wish to thank Genome Canada for STIC (Science and Technology Innovation Centre) funding and support. Carol Parker (UVic-Genome BC Proteomics Centre) is acknowledged for assisting in the manuscript editing process.

Competing Interests
CHB is the director of the Centre and the Chief Scientific Officer of MRM Proteomics, which has commercialized the performance kits (namely the PeptiQuant LC-MS Platform and PeptiQuant MRM/MS Workflow kits) and the assessment kits (PeptiQuant Human Discovery Assay kit, or BAK-192, and BAK-21) described here.

References

1. Ross PL, Huang YN, Marchese JN, Williamson B, Parker K, Hattan S et al (2004) Multiplexed protein quantitation in Saccharomyces cerevisiae using amine-reactive isobaric tagging reagents. Mol Cell Proteomics 3:1154–69
2. Dayon L, Sanchez JC (2012) Relative protein quantification by MS/MS using the tandem mass tag technology. Methods Mol Biol 893:115–27
3. Picard G, Lebert D, Louwagie M, Adrait A, Huillet C, Vandenesch F et al (2012) PSAQ™ standards for accurate MS-based quantification of proteins: from the concept to biomedical applications. J Mass Spectrom 47:1353–63
4. Villanueva J, Carrascal M, Abian J (2014) Isotope dilution mass spectrometry for absolute quantification in proteomics: concepts and strategies. J Proteomics 96:184–99
5. Gillette MA, Carr SA (2013) Quantitative analysis of peptides and proteins in biomedicine by targeted mass spectrometry. Nat Methods 10:28–34
6. Picotti P, Aebersold R (2012) Selected reaction monitoring-based proteomics: workflows, potential, pitfalls and future directions. Nat Methods 9:555–66
7. Gallien S, Duriez E, Crone C, Kellmann M, Moehring T, Domon B (2012) Targeted proteomic quantification on quadrupole-orbitrap mass spectrometer. Mol Cell Proteomics 11:1709–23
8. Gallien S, Bourmaud A, Kim SY, Domon B (2014) Technical considerations for large-scale parallel reaction monitoring analysis. J Proteomics 100:147–59
9. Peterson AC, Russell JD, Bailey DJ, Westphall MS, Coon JJ (2012) Parallel reaction monitoring for high resolution and high mass accuracy quantitative, targeted proteomics. Mol Cell Proteomics 11:1475–88
10. Gillet LC, Navarro P, Tate S, Röst H, Selevsek N, Reiter L et al (2012) Targeted data extraction of the MS/MS spectra generated by data-independent acquisition: a new concept for consistent and accurate proteome analysis. Mol Cell Proteomics 11:O111.016717
11. Picotti P, Clément-Ziza M, Lam H, Campbell DS, Schmidt A, Deutsch EW et al (2013) A complete mass-spectrometric map of the yeast proteome applied to quantitative trait analysis. Nature 494:266–70
12. Anderson NL, Anderson NG (2002) The human plasma proteome: history, character, and diagnostic prospects. Mol Cell Proteomics 1:845–67
13. Omenn GS (2007) The HUPO Human Plasma Proteome Project. Proteomics Clin Appl 1:769–79
14. Omenn GS (2004) The Human Proteome Organization Plasma Proteome Project pilot phase: reference specimens, technology platform comparisons, and standardized data submissions and analyses. Proteomics 4:1235–40
15. Caisey JD, King DJ (1980) Clinical chemical values for some common laboratory animals. Clin Chem 26:1877–9
16. Chambers AG, Percy AJ, Yang J, Camenzind AG, Borchers CH (2013) Multiplexed quantitation of endogenous proteins in dried blood spots by multiple reaction monitoring mass spectrometry. Mol Cell Proteomics 12:781–91
17. Berna M, Ott L, Engle S, Watson D, Solter P, Ackermann B (2008) Quantification of NTproBNP in rat serum using immunoprecipitation and LC/MS/MS: a biomarker of drug-induced cardiac hypertrophy. Anal Chem 80:561–6
18. Whiteaker JR, Zhao L, Lin C, Yan P, Wang P, Paulovich AG (2012) Sequential multiplexed analyte quantification using peptide immunoaffinity

enrichment coupled to mass spectrometry. Mol Cell Proteomics 11:M111.015347. doi:10.1074/mcp.M111
19. Whiteaker JR, Zhao L, Frisch C, Ylera F, Harth S, Knappik A et al (2014) High-affinity recombinant antibody fragments (Fabs) can be applied in peptide enrichment immuno-MRM assays. J Proteome Res 13:2187–96
20. Percy AJ, Simon R, Chambers AG, Borchers CH (2014) Enhanced sensitivity and multiplexing with 2D LC/MRM-MS and labeled standards for deeper and more comprehensive protein quantitation. J Proteomics 106:113–24
21. Shi T, Fillmore TL, Sun X, Zhao R, Schepmoes AA, Hossain M et al (2012) Antibody-free, targeted mass-spectrometric approach for quantification of proteins at low picogram per milliliter levels in human plasma/serum. Proc Natl Acad Sci U S A 109:15395–400
22. Keshishian H, Addona T, Burgess M, Kuhn E, Carr SA (2007) Quantitative, multiplexed assays for low abundance proteins in plasma by targeted mass spectrometry and stable isotope dilution. Mol Cell Proteomics 6:2212–29
23. Huttenhain R, Soste M, Selevsek N, Rost H, Sethi A, Carapito C et al (2012) Reproducible quantification of cancer-associated proteins in body fluids using targeted proteomics. Sci Transl Med 4:142ra94
24. Liu T, Hossain M, Schepmoes AA, Fillmore TL, Sokoll LJ, Kronewitter SR et al (2012) Analysis of serum total and free PSA using immunoaffinity depletion coupled to SRM: correlation with clinical immunoassay tests. J Proteomics 75:4747–57
25. Rezeli M, Végvári A, Ottervald J, Olsson T, Laurell T, Marko-Varga G (2011) MRM assay for quantitation of complement components in human blood plasma – a feasibility study on multiple sclerosis. J Proteomics 75:211–20
26. Keshishian H, Addona T, Burgess M, Mani DR, Shi X, Kuhn E et al (2009) Quantification of cardiovascular biomarkers in patient plasma by targeted mass spectrometry and stable isotope dilution. Mol Cell Proteomics 8:2339–49
27. Percy AJ, Yang J, Chambers AG, Simon R, Hardie DB, Borchers CH (2014) Multiplexed MRM with internal standards for cerebrospinal fluid candidate protein biomarker quantitation. J Proteome Res 13:3733–47
28. Chambers AG, Percy AJ, Simon R, Borchers CH (2014) MRM for the verification of cancer biomarker proteins: recent applications to human plasma and serum. Expert Rev Proteomics 11:137–48
29. Percy AJ, Byrns S, Chambers AG, Borchers CH (2013) Targeted quantitation of CVD-linked plasma proteins for biomarker verification and validation. Expert Rev Proteomics 10:567–78
30. Domanski D, Percy AJ, Yang J, Chambers AG, Hill JS, Cohen Freue GV et al (2012) MRM-based multiplexed quantitation of 67 putative cardiovascular disease biomarkers in human plasma. Proteomics 12:1222–43
31. Percy AJ, Chambers AG, Yang J, Borchers CH (2013) Multiplexed MRM-based quantitation of candidate cancer biomarker proteins in undepleted and non-enriched human plasma. Proteomics 13:2202–15
32. Percy AJ, Chambers AG, Yang J, Hardie D, Borchers CH (2014) Advances in multiplexed MRM-based protein biomarker quantitation toward clinical utility. Biochim Biophys Acta 1844:917–26
33. Paulovich AG, Whiteaker JR, Hoofnagle AN, Wang P (2008) The interface between biomarker discovery and clinical validation: the tar pit of the protein biomarker pipeline. Proteomics Clin Appl 2:1386–402
34. Wilson R (2013) Sensitivity and specificity: twin goals of proteomics assays. Can they be combined? Expert Rev Proteomics 10:135–49
35. Camenzind AG, van der Gugten JG, Popp R, Holmes DT, Borchers CH (2013) Development and evaluation of an immuno-MALDI (iMALDI) assay for angiotensin I and the diagnosis of secondary hypertension. Clin Proteomics 10:20
36. Anderson NL, Anderson NG, Haines LR, Hardie DB, Olafson RW, Pearson TW (2004) Mass spectrometric quantitation of peptides and proteins using stable isotope standards and capture by anti-peptide antibodies (SISCAPA). J Proteome Res 3:235–44
37. Anderson NL, Razavi M, Pearson TW, Kruppa G, Paape R, Suckau D (2012) Precision of heavy-light peptide ratios measured by MALDI-tof mass spectrometry. J Proteome Res 11:1868–78
38. Sparbier K, Wenzel T, Dihazi H, Blaschke S, Müller GA, Deelder AM et al (2009) Immuno-MALDI-TOF MS: new perspectives for clinical applications of mass spectrometry. Proteomics 9:1442–50
39. Kennedy JJ, Abbatiello SE, Kim K, Yan P, Whiteaker JR, Lin C et al (2014) Demonstrating the feasibility of large-scale development of standardized assays to quantify human proteins. Nat Methods 11:149–55
40. Addona TA, Abbatiello SE, Schilling B, Skates SJ, Mani DR, Bunk DM et al (2009) Multi-site assessment of the precision and reproducibility of multiple reaction monitoring-based measurements of proteins in plasma. Nat Biotechnol 27:633–41
41. Percy AJ, Chambers AG, Smith DS, Borchers CH (2013) Standardized protocols for quality control of MRM-based plasma proteomic workflow. J Proteome Res 12:222–33
42. Percy AJ, Chambers AG, Yang J, Jackson AM, Domanski D, Burkhart J et al (2013) Method and platform standardization in MRM-based quantitative plasma proteomics. J Proteomics 95:66–76
43. Mohammed Y, Percy AJ, Chambers AG, Borchers CH (2015) Qualis-SIS: automated standard curve generation and quality assessment for multiplexed targeted quantitative proteomic experiments with labeled standards. J Proteome Res 14:1137–46
44. Kuzyk MA, Smith D, Yang J, Cross TJ, Jackson AM, Hardie DB et al (2009) Multiple reaction monitoring-based, multiplexed, absolute quantitation of

45 proteins in human plasma. Mol Cell Proteomics 8:1860–77
45. Kuzyk MA, Parker CE, Domanski D, Borchers CH (2013) Development of MRM-based assays for the absolute quantitation of plasma proteins. Methods Mol Biol 1023:53–82
46. Rodriguez J, Gupta N, Smith RD, Pevzner PA (2008) Does trypsin cut before proline? J Proteome Res 7:300–5
47. Mohammed Y, Domanski D, Jackson AM, Smith DS, Deelder AM, Palmblad M et al (2014) PeptidePicker: a scientific workflow with web interface for selecting appropriate peptides for targeted proteomics experiments. J Proteomics 106:151–61
48. Mallick P, Schirle M, Chen SS, Flory MR, Lee H, Martin D et al (2007) Computational prediction of proteotypic peptides for quantitative proteomics. Nat Biotechnol 12:125–31
49. Percy AJ, Chambers AG, Yang J, Domanski D, Borchers CH (2012) Comparison of standard- and nano-flow liquid chromatography platforms for MRM-based quantitation of putative plasma biomarker proteins. Anal Bioanal Chem 404:1089–101
50. Röst H, Malmström L, Aebersold R (2012) A computational tool to detect and avoid redundancy in selected reaction monitoring. Mol Cell Proteomics 11:540–9
51. Agger SA, Marney LC, Hoofnagle AN (2010) Simultaneous quantification of apolipoprotein a-I and apolipoprotein B by liquid-chromatography-multiple-reaction-monitoring mass spectrometry. Clin Chem 56:1804–13
52. Percy AJ, Chambers AG, Yang J, Hardie DB, Borchers CH (1844) Advances in multiplexed MRM-based protein biomarker quantitation toward clinical utility. Biochim Biophys Acta 2014:917–26
53. Tabb DL, Vega-Montoto L, Rudnick PA, Variyath AM, Ham A-JL, Bunk DM et al (2010) Repeatability and reproducibility in proteomic identifications by liquid chromatography-tandem mass spectrometry. J Proteome Res 9:761–76
54. Domon B, Aebersold R (2010) Options and considerations when selecting a quantitative proteomics strategy. Nat Biotechnol 28:710–21
55. Röst HL, Rosenberger G, Navarro P, Gillet L, Miladinović SM, Schubert OT et al (2014) OpenSWATH enables automated, targeted analysis of data-independent acquisition MS data. Nat Biotechnol 32:219–23
56. Ni X, Li X, Guo Y, Zhou T, Guo X, Zhao C et al (2014) Quantitative proteomics analysis of altered protein expression in the placental villous tissue of early pregnancy loss using isobaric tandem mass tags. Biomed Res Int 2014:647143
57. Mingrone G, Iaconelli A, Leccesi L, Nanni G, Pomp A, Castagneto M et al (2012) Bariatric surgery versus conventional medical therapy for type 2 diabetes. N Engl J Med 366:1577–85
58. Schauer PR, Kashyap SR, Wolski K, Brethauer SA, Kirwan JP, Pothier CE et al (2012) Bariatric surgery versus intensive medical therapy in obese patients with diabetes. N Engl J Med 366:1567–76
59. Sjöström L, Peltonen M, Jacobson P, Ahlin S, Andersson-Assarsson J, Anveden Å et al (2014) Association of bariatric surgery with long-term remission of type 2 diabetes and with microvascular and macrovascular complications. JAMA 311:2297–304

Index

H. Mirzaei and M. Carrasco (eds.), *Modern Proteomics – Sample Preparation, Analysis and Practical Applications*, Advances in Experimental Medicine and Biology 919, DOI 10.1007/978-3-319-41448-5

FSC
www.fsc.org
MIX
Papier | Fördert gute Waldnutzung
FSC® C083411

Kolibri 360 GmbH
Ferdinand-Jühlke-Straße 7
99095 Erfurt, Deutschland
produktsicherheit@kolibri360.de